FOUNDATIONS OF

# SOLID MECHANICS

*Prentice-Hall International Series in Dynamics*

Y. C. Fung, *Editor*

PRENTICE-HALL, INC.
PRENTICE-HALL INTERNATIONAL, INC., UNITED KINGDOM AND EIRE
PRENTICE-HALL OF CANADA, LTD., CANADA

*Foundations of Solid Mechanics* by Y. C. Fung
*Principles of Dynamics* by Donald T. Greenwood

# FOUNDATIONS OF

# SOLID MECHANICS

Y. C. FUNG

Professor, California Institute of Technology

PRENTICE-HALL, INC.

*Englewood Cliffs, New Jersey*

PRENTICE-HALL INTERNATIONAL, INC., *London*
PRENTICE-HALL OF AUSTRALIA, PTY., LTD., *Sydney*
PRENTICE-HALL OF CANADA, LTD., *Toronto*
PRENTICE-HALL OF INDIA (PRIVATE) LTD., *New Delhi*
PRENTICE-HALL OF JAPAN, INC., *Tokyo*

 Library of Congress Catalog Card Number 65–18496.

Current printing (last digit):
11 10

Printed in the United States of America

32991—C

獻給
慈母張董老夫人
胡璉氏的慶七旬
上壽·一九六五·禎

*To*
*My Mother*

# PREFACE

Solid mechanics deals with the deformation and motion of "solids." The displacement that connects the instantaneous position of a particle to its position in an "original" state is of general interest. The preoccupation about particle displacements distinguishes solid mechanics from that of fluids.

This book is written for engineers and scientists who have had some exposure to the theory of strength of materials and elasticity. It deals with the mechanics of continuous media. The bulk of the text is concerned with the classical theory of elasticity, but the discussion also includes thermodynamics of solids, thermoelasticity, viscoelasticity, plasticity, and finite deformation theory. Fluid mechanics is excluded. Both dynamics and statics are treated; the concepts of wave propagations are introduced at an early stage. Variational calculus is emphasized, since it provides a unified point of view and is useful in formulating approximate theories.

Since the book is of an introductory nature, an introduction to the tensor analysis and the calculus of variations is included. The general tensor theory is presented here for those readers who are interested in advanced literature. However, to reach a wider circle of readers, I have used only Cartesian tensors in developing the theory (see footnote on p. 37). I believe that those who will take the time to study the general tensor analysis will find ample reward: the simplicity in conception and the efficiency of the notations gives the subject a great beauty.

The text was developed from my notes for a course offered to graduate students at the California Institute of Technology. Since the beginning of 1959, it was decided to modify the traditional elasticity course to one that gave greater emphasis to general methodology. This shift in emphasis was prompted by the broadening of engineering fields in recent years. It has been my experience in long association with the aerospace industry that young engineers are often asked to deal with subjects which were not taught in school. In view of the constantly changing problems in engineering, I believe that a broad course in solid mechanics is useful.

Of course, no one path can embrace the broad field of mechanics. As in mountain climbing, some routes are safe to travel, others more perilous; some may lead to the summit, others to different vistas of interest; some have popular claims, others are less traveled. In choosing a particular path for a tour through the field, one is influenced by the curricula, the trends in literature, the interest in engineering and science. Here, a particular way has been chosen to view some of the most beautiful vistas in classical mechanics. In making this choice I have aimed at straightforwardness and interest, and practical usefulness in the long run.

Holding the book to reasonable length did not permit inclusion of many numerical examples, which have to be supplemented through problems and references. Fortunately, there are many excellent references to meet this demand; a fairly comprehensive bibliography is included.

I am indebted to many authors whose writings are classics in this field. To Love, Lamb, Timoshenko, Southwell, von Karman, Prager, Synge, Biot, Green, Truesdell and many others, I am especially indebted. The study of classical mechanics is a profound experience. The deeper one delves into it, the more he appreciates the contributions of great masters.

It remains for me to record my gratitude to many of my teachers, colleagues, friends, and students. My appreciation of mechanics as a living subject was enhanced through many years of association with Prof. Ernest E. Sechler and Dr. Millard V. Barton. Their deep insight of many facets of engineering and broad knowledge about practical matters of design and construction inspires people around them always to seek a better or simpler solution. To the late Dr. Aristotle D. Michal, professor of mathematics at the California Institute of Technology, I am grateful for his gentle encouragement and guidance in earlier years. To Drs. Hans Krumhaar and Max L. Williams, who shared with me the teaching of the course from which this book was developed, I owe a special note of thanks. Sections 10.1–6 are based on Dr. Krumhaar's notes. Dr. John A. Morgan, professor of engineering at UCLA, read the entire manuscript and gave me many valuable suggestions for improvements. Dr. Benjamin Cummings checked part of the manuscript. Drs. Charles Babcock, jr., Wolfgang G. Knaus, Wei-Hsuin Yang, Gilbert A. Hegemier, Jerold L. Swedlow, and Messrs. Pin Tong, Jen-Shih Lee, and Jay-Chung Chen read the proofs and made many useful suggestions. I also wish to thank many of my students, not named here individually, without whose discussions the book would not have taken this form. The Prentice-Hall staff has been very cooperative. In particular, Mr. Nicholas Romanelli offered his competent knowledge and gave unstinted effort in editing this book. The preparation of the manuscript was helped by Helen Burrus, Joan Christensen, Jeanette Siefke, and Sandra Mann, who typed and retyped as the words were weighed and revised, without showing a hint of annoyance. To all of them I am truly thankful.

Y.C.F.

# CONTENTS

# 1

# PROTOTYPES OF THE THEORY OF ELASTICITY AND VISCOELASTICITY

Historically, the notion of elasticity was first announced in 1676 by Robert Hooke (1635–1703) in the form of an anagram, *ceiiinosssttuv*. He explained it in 1678 as

*Ut tensio sic vis*,

or "the power of any springy body is in the same proportion with the extension."† The mathematical theory developed and extended to other materials since that time is associated with the names of practically all great mathematical physicists of the last three centuries and forms one of the most important parts of classical physics. The line of inquiry has never been broken, and in recent times we witness the most vigorous developments.‡

Hooke's law is the constitutive law for a *Hookean*, or *linear elastic*, material. As stated in the original form, its meaning is not very clear. Our first task is, therefore, to give it a precise expression. This we shall do in two different ways. The first way is to make use of the common notion of "springs," and consider the load-deflection relationship. The second way is to state it as a tensor equation connecting the stress and strain. Although the second way is the proper way to start a general theory, the first, simpler and more restrictive, is not without interest. In this chapter we develop the first alternative.

## 1.1. HOOKE'S LAW AND ITS CONSEQUENCES

Let us consider the static equilibrium state of a solid body under the action of external forces (Fig. 1.1:1). Let the body be supported in some manner so that at least three points are fixed in a space which is described with respect to a rectangular Cartesian frame of reference. We shall make three basic hypotheses regarding the properties of the body under consideration.

† Edme Mariotte enunciated the same law independently in 1680.

‡ Some historical remarks are given in Notes 1.1, p. 473.

(H1) *The body is continuous and remains continuous under the action of external forces.*

Under this hypothesis the atomistic structure of the body is ignored and the body is idealized into a geometrical copy in Euclidean space whose points are identified with the material particles of the body. The continuity is defined in the mathematical sense with respect to this idealized continuum. Neighboring points remain as neighbors under any loading condition. No cracks or holes may open up in the interior of the body under the action of external load.

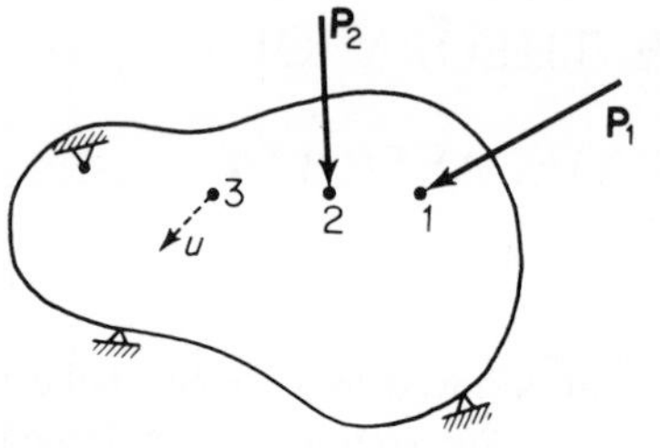


**Fig. 1.1:1.** Static equilibrium of a body under external forces.

A material satisfying this hypothesis is said to be a *continuum*. The study of the deformation or motion of a continuum is called the *continuum mechanics.*

To introduce the second hypothesis, let us consider the action of a set of forces on the body. Let every force be fixed in direction and in point of application, and let the magnitude of all the forces be increased or decreased together: always bearing the same ratio to each other. Let the forces be denoted by $\mathbf{P}_1, \mathbf{P}_2, \ldots, \mathbf{P}_n$ and their magnitude by $P_1, P_2, \ldots, P_n$. Then the ratios $P_1 : P_2 : \ldots : P_n$ remain fixed. When such a set of forces is applied on the body, the body deforms. Let the displacement at an arbitrary point in an arbitrary direction be measured with respect to a rectangular Cartesian frame of reference fixed with the supports. Let this displacement be denoted by $u$. Then our second hypothesis is

(H2) *Hooke's law,*

$$u = a_1P_1 + a_2P_2 + \ldots + a_nP_n, \tag{1}$$

where $a_1, a_2, \ldots, a_n$ are constants independent of the magnitude of $P_1, P_2, \ldots, P_n$. The constants $a_1, a_2, \ldots, a_n$ depend, of course, on the location of the point at which the displacement component is measured and on the directions and points of application of the individual forces of the loading.

Hooke's law in the form (H2) is one that can be subjected readily to direct experimental examination.

To complete the formulation of the theory of elasticity, we need a third hypothesis:

(H3) *There exists a unique unstressed state of the body, to which the body returns whenever all the external forces are removed.*

A body satisfying these three hypotheses is called a *linear elastic solid.*

A number of deductions can be drawn from these assumptions. We shall list a few important ones.

(A) *The principle of superposition for loads at the same point*

Let $\mathbf{P}_1$ and $\mathbf{P}_1'$ be forces in the same direction acting through the same point. Then the resultant displacement $u$ is equal to the sum of the displacements produced by $\mathbf{P}_1$ and $\mathbf{P}_1'$ acting individually, regardless of the order of application of $\mathbf{P}_1$ and $\mathbf{P}_1'$. This is evident from (H2) when applied to one load alone.

(B) *Principle of superposition*

By a combination of (H2) and (H3), we can show that Eq. (1) is valid not only for systems of loads for which the ratios $P_1:P_2:\ldots:P_n$ remain fixed as originally assumed, but also for an arbitrary set of loads $\mathbf{P}_1, \mathbf{P}_2, \ldots, \mathbf{P}_n$. In other words, Eq. (1) holds regardless of the order in which the loads are applied. The constant $a_1$ is independent of the loads $\mathbf{P}_2, \mathbf{P}_3, \ldots, \mathbf{P}_n$. The constant $a_2$ is independent of the loads $\mathbf{P}_1, \mathbf{P}_3, \ldots, \mathbf{P}_n$; etc. This is the principle of superposition of the load-and-deflection relationship.

*Proof.* If a proof of the statement above can be established for an arbitrary pair of loads, then the general theorem can be proved by mathematical induction.

Let $\mathbf{P}_1$ and $\mathbf{P}_2$ (with magnitudes $P_1$ and $P_2$) be a pair of arbitrary loads acting at points 1 and 2, respectively. Let the deflection in a specific direction be measured at a point 3 (see Fig. 1.1:1). According to (H2), if $\mathbf{P}_1$ is applied alone, then at the point 3 a deflection $u_3 = c_{31}P_1$ is produced. If $\mathbf{P}_2$ is applied alone, a deflection $u_3 = c_{32}P_2$ is produced. If $\mathbf{P}_1$ and $\mathbf{P}_2$ are applied together, with the ratio $P_1:P_2$ fixed, then according to (H2) the deflection can be written as

(a) $$u_3 = c_{31}'P_1 + c_{32}'P_2.$$

The question arises whether $c_{31}' = c_{31}$, $c_{32}' = c_{32}$. The answer is affirmative, as can be shown as follows. After $\mathbf{P}_1$ and $\mathbf{P}_2$ are applied, we take away $\mathbf{P}_1$. This produces a change in deflection, $-c_{31}''P_1$, and the total deflection becomes

(b) $$u_3 = c_{31}'P_1 + c_{32}'P_2 - c_{31}''P_1.$$

Now only $\mathbf{P}_2$ acts on the body. Hence, upon unloading $\mathbf{P}_2$ we shall have

(c) $$u_3 = c_{31}'P_1 + c_{32}'P_2 - c_{31}''P_1 - c_{32}P_2.$$

Now all the loads are removed, and $u_3$ must vanish according to (H3). Rearranging terms, we have

(d) $$(c_{31}' - c_{31}'')P_1 = (c_{32} - c_{32}')P_2.$$

Since the only possible difference of $c'_{31}$ and $c''_{31}$ must be caused by the action of $\mathbf{P}_2$, the difference $c'_{31} - c''_{31}$ can only be a function of $P_2$ (and not of $P_1$). Similarly, $c_{32} - c'_{32}$ can only be a function of $P_1$. If we write Eq. (d) as

$$\text{(e)} \qquad \frac{c'_{31} - c''_{31}}{P_2} = \frac{c_{32} - c'_{32}}{P_1},$$

then the left-hand side is a function of $P_2$ alone, and the right-hand side is a function of $P_1$ alone. Since $P_1$ and $P_2$ are arbitrary numbers, the only possibility for Eq. (e) to be valid is for both sides to be a constant $k$ which is independent of both $P_1$ and $P_2$. Hence,

$$\text{(f)} \qquad c'_{32} = c_{32} - kP_1.$$

But a substitution of (f) into (a) yields

$$\text{(g)} \qquad u_3 = c'_{31}P_1 + c_{32}P_2 - kP_1P_2.$$

The last term is nonlinear in $P_1$, $P_2$, and Eq. (g) will contradict (H2) unless $k$ vanishes. Hence, $k = 0$ and $c'_{32} = c_{32}$. An analogous procedure shows $c'_{31} = c''_{31} = c_{31}$.

Thus the principle of superposition is established for one and two forces. An entirely similar procedure will show that if it is valid for $m$ forces, it is also valid for $m + 1$ forces. Thus, the general theorem follows by mathematical induction. Q.E.D.

The constants $c_{31}$, $c_{32}$, etc., are seen to be of significance in defining the elastic property of the solid body. They are called *influence coefficients* or, more specifically, *flexibility influence coefficients.*

We have derived the principle of superposition, (B), from the hypotheses (H2) and (H3). Conversely, if we have assumed the general principle of superposition as part of Hooke's law, then by allowing Eq. (1) to be valid for an arbitrary set of loading, (H3) (that a unique unstressed state exists) follows as a consequence.

(C) *Corresponding forces and displacements and the unique meaning of the total work done by the forces*

Let us now consider a set of external forces $\mathbf{P}_1, \ldots, \mathbf{P}_n$ acting on the body and *define the set of displacements at the points of application and in the direction of the loads as the displacements "corresponding" to the forces at these points.* The reactions at the points of support are considered as external forces exerted on the body and included in the set of forces.

Under the loads $\mathbf{P}_1, \ldots, \mathbf{P}_n$, the corresponding displacements may be written as

$$\text{(2)} \qquad \begin{aligned} u_1 &= c_{11}P_1 + c_{12}P_2 + \ldots + c_{1n}P_n, \\ u_2 &= c_{21}P_1 + c_{22}P_2 + \ldots + c_{2n}P_n, \\ &\cdots\cdots\cdots\cdots\cdots\cdots \\ u_n &= c_{n1}P_1 + c_{n2}P_2 + \ldots + c_{nn}P_n. \end{aligned}$$

If we multiply the first equation by $P_1$, the second by $P_2$, etc., and add, we obtain

$$(3)\qquad P_1u_1 + P_2u_2 + \ldots + P_nu_n = c_{11}P_1^2 + c_{12}P_1P_2 + \ldots + c_{1n}P_1P_n + c_{21}P_1P_2 + c_{22}P_2^2 + \ldots + \ldots + c_{nn}P_n^2.$$

The quantity above is independent of the order in which the loads are applied. Hence, it has a definite meaning for each set of loads $\mathbf{P}_1, \ldots, \mathbf{P}_n$.

Now, in a special case, the meaning of the quantity on the left-hand side of Eq. (3) is clear. This is the case in which the ratios $P_1 : P_2 : P_3 : \ldots : P_n$ are kept constant and the loading increases very slowly from zero to the final value. In this case, the corresponding displacements also increase proportionally and slowly. It is then obvious that the work done by the force $\mathbf{P}_1$ is exactly $\frac{1}{2}P_1u_1$, that by $\mathbf{P}_2$ is $\frac{1}{2}P_2u_2$, etc. Hence, we conclude from (3) that the *total work done by the set of forces is independent of the order in which the forces are applied.*

(D) *Maxwell's reciprocal relation*

An important property of the influence coefficients of corresponding displacements follows immediately.

*The influence coefficients for corresponding forces and displacements are symmetric.*

$$(4)\qquad c_{ij} = c_{ji}.$$

In other words, *the displacement at a point i due to a unit load at another point j is equal to the displacement at j due to a unit load at i, provided that the displacements and forces "correspond," i.e., that they are measured in the same direction at each point.*

The proof is simple. Consider two forces $\mathbf{P}_1$ and $\mathbf{P}_2$ (Fig. 1.1:1). When the forces are applied in the order $\mathbf{P}_1, \mathbf{P}_2$, the work done by the forces is easily seen to be

$$W = \tfrac{1}{2}(c_{11}P_1^2 + c_{22}P_2^2) + c_{12}P_1P_2.$$

When the order of application of the forces is interchanged, the work done is

$$W' = \tfrac{1}{2}(c_{22}P_2^2 + c_{11}P_1^2) + c_{21}P_1P_2.$$

But according to (C) above, $W = W'$ for arbitrary $P_1, P_2$. Hence, $c_{12} = c_{21}$, and the theorem is proved.

The reciprocal relation may be put into a different form, sometimes more convenient in applications, as follows.

(E) *Betti-Rayleigh reciprocal theorem.*

Let a set of loads $\mathbf{P}_1, \mathbf{P}_2, \ldots, \mathbf{P}_n$ produce a set of corresponding displacements $u_1, u_2, \ldots, u_n$. Let a second set of loads $\mathbf{P}_1', \mathbf{P}_2', \ldots, \mathbf{P}_n'$, acting in the same directions and having the same points of application as those of the first, produce the corresponding displacements $u'_1, u'_2, \ldots, u'_n$. Then

$$P_1u'_1 + P_2u'_2 + \ldots + P_nu'_n = P'_1u_1 + \ldots + P'_nu_n. \tag{5}$$

In other words, *in a linear elastic solid, the work done by a set of forces acting through the corresponding displacements produced by a second set of forces is equal to the work done by the second set of forces acting through the corresponding displacements produced by the first set of forces.*

A straightforward proof is furnished by writing out the $u_i$ and $u'_i$ in terms of $P_i$ and $P'_i$, $(i = 1, 2, \ldots, n)$, with appropriate influence coefficients, comparing the results on both sides of the equation, and utilizing the symmetry of the influence coefficients.

In the form of Eq. (5), the reciprocal theorem can be generalized to include moments and rotations as the corresponding *generalized forces* and *generalized displacements.* An illustration is given in Fig. 1.1:2. These theorems are very useful in practical applications.

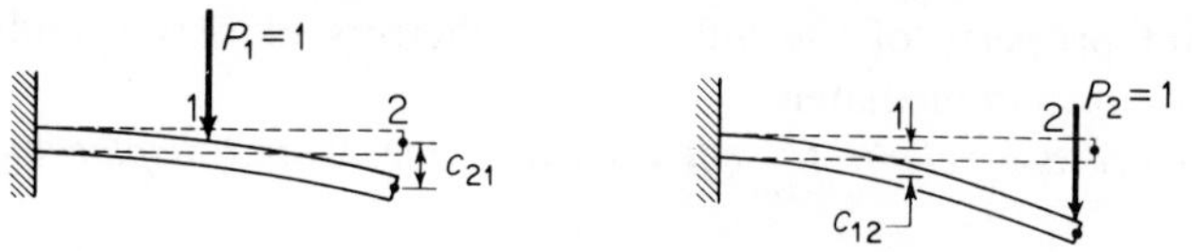


(a) Forces and corresponding displacements.

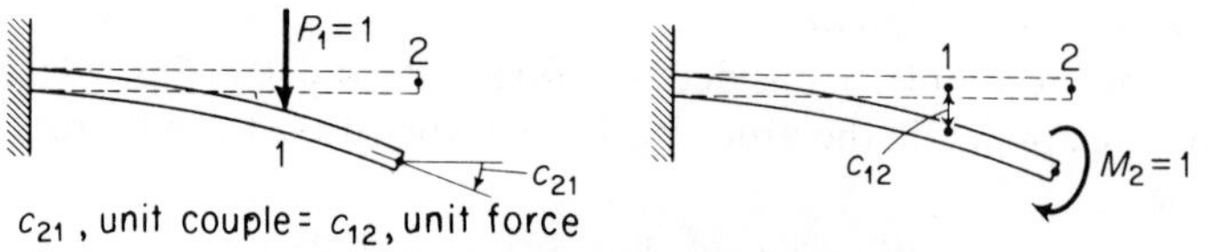


(b) Generalized force (moment) and the corresponding generalized displacement (moment ~ rotation of angle).

**Fig. 1.1:2.** Illustration of the reciprocal theorem.

### (F) *Strain energy*

Further insight can be gained from the first law of thermodynamics. When a body is thermally isolated and thermal expansions are neglected the first law states that the work done on the body by the external forces in a certain time interval is equal to the increase in the kinetic energy and internal energy in the same interval. If the process is so slow that the kinetic energy can be ignored, the work done is seen to be equal to the change in internal energy.

If the internal energy is reckoned as zero in the unstressed state, the stored internal energy shall be called strain energy. Writing $U$ for the strain energy, we have, from (3) and (4),

$$U = \tfrac{1}{2}\sum_{i=1}^{n}\sum_{j=1}^{n} c_{ij}P_iP_j = \tfrac{1}{2}\sum_{i=1}^{n} c_{ii}P_i^2 + \tfrac{1}{2}\sum_{i \neq j}\sum c_{ij}P_iP_j. \tag{6}$$

From what was said above in (C) about the expression on the right-hand side, we at once conclude that *the strain energy in a linear elastic solid is independent of the order in which the given forces are applied.*

If we differentiate Eq. (6) with respect to $P_i$, we obtain

$$\frac{\partial U}{\partial P_i} = c_{ii}P_i + \sum_{j \neq i} c_{ij}P_j, \qquad i = 1, 2, \ldots, n.$$

But, the right-hand side is precisely $u_i$; hence, we obtain

(G) *Castigliano's theorem*

$$\frac{\partial U}{\partial P_i} = u_i, \qquad i = 1, \ldots, n. \tag{7}$$

In other words, if a set of loads $P_1, \ldots, P_n$ is applied on a perfectly elastic body as described above and the strain energy is expressed as a function of the set $P_1, \ldots, P_n$, then the partial derivative of the strain energy, with respect to a particular load, gives the corresponding displacement at the point of application of that particular load in the direction of that load.

(H) *The principle of virtual work*

On the other hand, for a body in equilibrium under a set of external forces, the principle of virtual work can be applied to show that, *if the strain energy is expressed as a function of the corresponding displacements, then*

$$\frac{\partial U}{\partial u_i} = P_i, \qquad i = 1, \ldots, n. \tag{8}$$

The proof consists in allowing a virtual displacement $\delta u$ to take place in the body in such a manner that $\delta u$ is continuous everywhere but vanishes at all points of loading except under $P_i$. Due to $\delta u$, the strain energy changes by an amount $\delta U$, while the virtual work done by the external forces is the product of $P_i$ times the virtual displacement, i.e., $P_i\,\delta u_i$. According to the principle of virtual work, these two expressions are equal, $\delta U = P_i\,\delta u_i$. On rewriting it in the differential form, the theorem is established.

The important result (8) is established on the principle of virtual work as applied to a state of equilibrium under the additional assumption that a strain energy function that is a function of displacement exists. It is applicable also to elastic bodies that follow the nonlinear load-displacement relationship.

## 1.2. POSITIVE DEFINITENESS OF THE STRAIN ENERGY AND THE UNIQUENESS OF SOLUTION

The theorems established in the previous section [except (1.1:8)† which is more general] are based on the hypotheses (H1), (H2), and (H3). Of course, the theory can be modified by starting from some other hypothesis which implies (H2) or (H3) as consequences. For example, we have remarked that if we start from the assumption of principle of superposition, then (H2) and (H3) follow as consequences. But this is a rather trivial change. A much more significant alternative is to start with the assumption that a strain energy function $U$ exists which is expressible as a function of the displacements. If such a strain energy function exists, we could use Eq. (1.1:8), which is based on the principle of virtual work, and deduce the load-displacement relationship for the elastic body. Obviously, the load-deflection relation depends on the form of $U$. If $U$ is a *homogeneous quadratic function of* $u_1, u_2, \ldots u_n$ *(called a quadratic form), then* $P_i$ *will be a linear function of* $u_1, \ldots, u_n$; and the load-displacement relationship is linear, as assumed by Hooke's law. However, a nonlinear load-displacement relationship can be formulated if an appropriate $U$ other than a quadratic form in displacements is assumed. This second approach was first taken by the self-taught Nottingham genius, George Green (1793–1841), in 1837.

There is a theorem in thermodynamics which states that for a solid body to have a *stable* natural state (such as the unstressed state of a linear elastic solid) the strain energy function must be positive definite; i.e., it must be nonnegative, and it is zero only in the natural state. We shall discuss this further in Chap. 12. For the moment, let us assume the existence and the positive definiteness of the strain energy function, and see what are the consequences.

The positive definiteness of the strain energy function implies certain relationships between the influence coefficients. For example, if Eq. (1.1:6) which defines $U$ in terms of $P_1, \ldots, P_n$,

$$U = \tfrac{1}{2}\sum_{i=1}^{n} c_{ii}P_i^2 + \tfrac{1}{2}\sum_{i\neq j}\sum c_{ij}P_iP_j, \tag{1}$$

is positive definite, i.e., $U \geqslant 0$, and the equality sign holds when and only when all $P$'s vanish, then the sum of the coefficients $c_{ii}$ and the determinant $|c_{ij}|$ are both positive. The necessary and sufficient conditions for the positive definiteness of a quadratic form are well known. (See pp. 29, 30.) These conditions restrict the elastic constants of a material, as we shall see later.

Based on the assumption of the positive definiteness of the strain energy function $U$ as a function of the displacements, it is easy to establish the

† Equation (8) of Sec. 1.1. This notation will be used hereafter.

*uniqueness theorem*, (Kirchhoff, 1859). The theorem states that *for a linear elastic solid as defined in Sec. 1.1 and having a positive definite strain energy function, there exists a one-to-one correspondence between the elastic deformation and the forces acting on the body.* A proof runs as follows. Let Eqs. (1.1:2) define the displacements as functions of the corresponding forces so that the strain energy function is expressible as (1). Let us assume that two different sets of forces $P_1, P_2, \ldots, P_n$ and $P_1', \ldots, P_n'$ acting at points 1, 2, $\ldots, n$ of a body correspond to a single state of deformation of the body. Let us now first apply the loads $P_1, \ldots, P_n$ to deform the body; then let us apply the loads $-P_1', \ldots, -P_n'$ to annul the displacements. The final configuration is completely unstrained; hence, $U = 0$. However, there are the forces $P_1 - P_1', P_2 - P_2'$, etc., acting on the body. This is impossible, because of the assumed positive definiteness of $U$, unless $P_1 = P_1'$, $P_2 = P_2'$, etc. Thus, a contradiction is obtained. The assumption that two different sets of forces correspond to the same displacements is untenable, and the uniqueness of solution of the Eqs. (1.1:2) is proved.

Thus, Eqs. (1.1:2) can be solved uniquely. This implies, first, that the determinant of the coefficients does not vanish.

$$\begin{vmatrix} c_{11} & c_{12} & \cdots & c_{1n} \\ \cdot & \cdot & & \cdot \\ \cdot & \cdot & & \cdot \\ \cdot & \cdot & & \cdot \\ c_{n1} & c_{n2} & \cdots & c_{nn} \end{vmatrix} \neq 0, \tag{2}$$

and, second, that a unique solution

$$\begin{aligned} P_1 &= k_{11}u_1 + k_{12}u_2 + \ldots + k_{1n}u_n \\ &\cdots\cdots\cdots\cdots\cdots\cdots \\ P_n &= k_{n1}u_1 + k_{n2}u_2 + \ldots + k_{nn}u_n \end{aligned} \tag{3}$$

exists. Since $c_{ij} = c_{ji}$, we can show that

$$k_{ij} = k_{ji}, \qquad i, j = 1, \ldots, n. \tag{4}$$

A substitution of (3) into (1) yields the strain energy expression

$$U = \tfrac{1}{2}\sum_{i=1}^{n} k_{ii}u_i^2 + \tfrac{1}{2}\sum_{i \neq j}\sum k_{ij}u_iu_j, \tag{5}$$

which could, of course, be derived directly by multiplying the individual forces $P_1, \ldots, P_n$ given by Eqs. (3) by one-half of the corresponding displacements $u_1, u_2, \ldots, u_n$ and summing. From (5), we again obtain, by differentiation,

$$\frac{\partial U}{\partial u_i} = P_i, \qquad i = 1, \ldots, n. \tag{6}$$

The constants $k_{ij}$ are called *stiffness influence coefficients* or, simply, *spring constants.*

## 1.3. MINIMUM COMPLEMENTARY ENERGY THEOREM

Consider the following problem. A beam rests on five rigid posts as shown in Fig. 1.3:1. A set of lateral loads $P_6, \ldots, P_9$, parallel to each other, is applied. Find the reactions at the posts.

The problem can be solved as follows. We remove the posts at points 1, 2, and 3, and apply the reactions $P_1$, $P_2$, $P_3$ as external loads. This results in a statically determinate problem from which the reactions $P_4$ and $P_5$ can be expressed as a function of $P_1$, $P_2$, $P_3$, and $P_6$ to $P_9$. From the known influence function of the beam deflection, we can write down the deflection $u_1$, $u_2$, $u_3$ at the center posts in terms of the loads $P_1, \ldots, P_9$. But the conditions at the

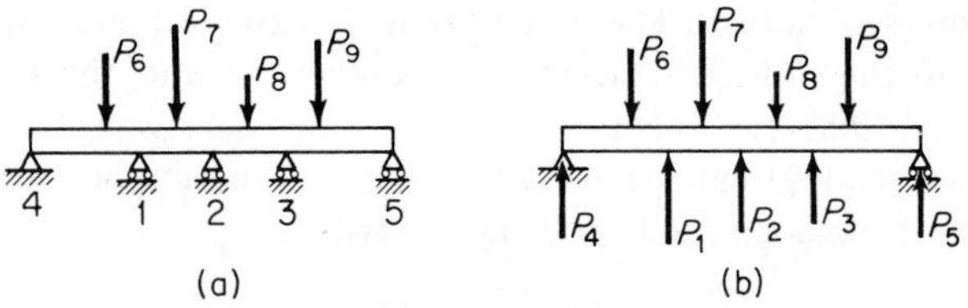


**Fig. 1.3:1.** A beam with redundant supports.

posts are $u_1 = u_2 = u_3 = 0$. Hence, three equations are obtained from which the three unknowns $P_1$, $P_2$, $P_3$ can be determined.

The method just described is called the *method of consistent deformation*. It is used very often. However, we shall now take a new point of view and formulate a different method of solution.

We begin by saying that the problem presented in Fig. 1.3:1 is to find the forces and deflections in the beam under the forces specified at points 6 to 9 and the deflections specified at points 1 to 5. Our attention shall be limited to the set of points 1, 2, . . . , 9 on the beam, the set of forces $P_1, \ldots, P_9$ acting at these points, and the "corresponding" displacements $u_1, \ldots, u_9$ in the direction of these forces. Because of the problem specification, we separate the set of points 1, 2, . . . , 9 into two subsets.

$S_u$: points 1, 2, 3, 4, 5, at which the deflections are specified;

$S_p$: points 6, 7, 8, 9, at which forces are specified.

We shall speak of the specified conditions as *boundary conditions*. Thus, the boundary conditions on forces are prescribed on $S_p$, those on displacements are prescribed on $S_u$.

Let us now consider the equations of equilibrium,

$$(1) \qquad \sum_{i=1}^{9} P_i = 0, \qquad \sum_{i=1}^{9} x_i P_i = 0,$$

where $x_i$ is the moment arm of $P_i$ about a point in the plane of the beam. Since $P_6$ to $P_9$ are known, we have two Eqs. (1) for the five unknowns $P_1$ to $P_5$. It is apparent that these equations can be satisfied by an arbitrary set of forces $P_1$, $P_2$, $P_3$. In other words, if we consider the equations of equilibrium alone, we have a triply infinite number of solutions. Now, among these solutions there is exactly one that will also satisfy the condition of consistent deformation (i.e., the specified support deflection). How can we characterize this one particular solution to separate it from the infinite number of possible solutions?

The answer is given by the *minimum complementary energy theorem.* We shall state this theorem for a set of $n$ points, which are separated into subsets $S_u$ and $S_p$ according to whether deflection or forces are specified at the point. The theorem states that *among all sets of forces* ($P_i$) *that satisfy the equations of equilibrium and the boundary conditions where forces are prescribed, the "actual" one that keeps a linear elastic solid continuous and compatible with all prescribed surface displacements is distinguished by the minimum value of the complementary energy:*

$$V^* = U - \sum_{S_u} u_i P_i, \tag{2}$$

where $U$ is the strain energy in the body expressed in terms of the loads $P_i$. When the flexibility influence coefficients $c_{ij}(= c_{ji})$ are used, we have

$$U = \tfrac{1}{2} \sum_i \sum_j c_{ij} P_i P_j. \tag{3}$$

The second term in Eq. (2) represents the potential energy of the loads. The sum $\sum_{S_u} u_i P_i$ needs to include only those points where the displacements $u_i$ are prescribed; a fact indicated by the notation $\sum_{S_u}$. In the example illustrated in Fig. 1.3:1, $u_i = 0$ on $S_u$ and the complementary energy is equal to the strain energy.

*Proof.* We first show that if the set of forces $P_i$ satisfies the equations of equilibrium and minimizes $V^*$, the solution will be compatible with the specified displacements over $S_u$. Let $P_i$ be the actual solution. All other solutions that satisfy the conditions of equilibrium can be written as $P_i + \alpha\, \delta P_i$, where $\alpha$ is a numerical factor, and $\delta P_i$ is a set of arbitrary loads satisfying the conditions of equilibrium and the boundary conditions. The set $\delta P_i$ must vanish on the set of points $S_p$, where the loads are prescribed. Such a set $\delta P_i$ is said to be admissible. The value of the complementary energy for $P_i + \alpha\, \delta P_i$ is

$$V^*(P_i + \alpha\delta P_i) = \tfrac{1}{2} \sum_{i=1}^{n} \sum_{j=1}^{n} c_{ij}(P_i + \alpha\, \delta P_i)(P_j + \alpha\, \delta P_j) - \sum_{S_u} u_i(P_i + \alpha\, \delta P_i). \tag{4}$$

Now, when $V^*$ is minimized with respect to arbitrary ($\alpha\, \delta P_i$), it is also minimized with respect to the numerical factor $\alpha$ for any specific set ($\delta P_i$).

But $V^*$ in (4) is a quadratic function in the variable $\alpha$; it can be exhibited as $V^*(\alpha)$. The condition for $V^*(\alpha)$ to be a minimum at $\alpha = 0$ is that its first derivative vanishes at $\alpha = 0$. Carrying out this differentiation, we have

$$\left(\frac{\partial V^*(\alpha)}{\partial \alpha}\right)_{\alpha=0} = \sum_{i=1}^{n}\left(\sum_{j=1}^{n} c_{ij}P_j - u_i\right)\delta P_i = 0. \tag{5}$$

If $\delta P_i$ were independently arbitrary, every coefficient of $\delta P_i$ in the equation above must vanish and we would obtain the desired result at once. However, $\delta P_i$ are subjected to the conditions of admissibility

$$\sum_{i=1}^{n}\delta P_i = 0, \qquad \sum_{i=1}^{n} x_i\,\delta P_i = 0, \quad \text{and} \quad \delta P_i = 0 \text{ over } S_P. \tag{6}$$

Since both (5) and (6) must be satisfied by $\delta P_i$, it is necessary that

$$\sum_{j=1}^{n} c_{ij}P_j - u_i + \lambda + \mu x_i = 0, \tag{7}$$

where $\lambda$ and $\mu$ are two constants (Lagrangian multipliers). It is clear that a displacement field represented by $\lambda + \mu x_i$ is a rigid-body motion. Since $P_i = 0$ implies $U_i = 0$ in our problem, we must have $\lambda = \mu = 0$. Equation (7) then becomes exactly the equation for consistent deformation at the points where displacements are specified. Hence, the minimal principle leads to the correct solution.

Conversely, we shall show that if $(P_i)$ satisfies all conditions of equilibrium and is compatible with the specified displacements over $S_u$, then $V^*$ is a minimum. We begin by considering the "complementary" virtual work $\sum_{i=1}^{n} u_i\,\delta P_i$ done by a set of small admissible variations $(\delta P_i)$. We obtain at once by Hooke's law that

$$\sum_{i=1}^{n} u_i\,\delta P_i = \sum_{i=1}^{n}\sum_{j=1}^{n} c_{ij}P_j\,\delta P_i. \tag{8}$$

Let us give the last term a different expression. Consider the strain energy $U(P_i)$ given by Eq. (3) and the difference

$$U(P_i + \delta P_i) - U(P_i) = \sum_i\sum_j c_{ij}P_j\,\delta P_i + \tfrac{1}{2}\sum_i\sum_j c_{ij}\,\delta P_i\,\delta P_j.$$

The last term is a small quantity of the second order if $\delta P_i$ are infinitesimal, and can be neglected in comparison with the first. When the last term is neglected, we shall write the difference above as $\delta U$. Similarly, since $\delta P_i = 0$ over $S_p$, we write

$$\sum_{i=1}^{n} u_i\,\delta P_i = \sum_{S_u} u_i(P_i + \delta P_i) - \sum_{S_u} u_iP_i = \delta\left(\sum_{S_u} u_iP_i\right).$$

Hence, Eq. (8) may be written as

$$\delta\left(\sum_{S_u} u_iP_i\right) = \delta U$$

or, transposing terms,

$$\delta V^* = 0, \tag{9}$$

with $V^*$ given by Eq. (2). This is a variational principle, the interpretation of which leads to the italicized theorem quoted above.

It remains to show that $V^*(P_i)$ actually is a minimum for the actual solution. This is easily done by computing the difference $V^*(P_i + \delta P_i) - V^*(P_i)$. From (2), (4), and (5) we obtain

$$V^*(P_i + \delta P_i) - V^*(P_i) = \tfrac{1}{2}\sum_{i=1}^{n}\sum_{j=1}^{n} c_{ij}\,\delta P_i\,\delta P_j \geqslant 0.$$

The last quantity is positive definite if the body is stable (Sec. 1.2). Hence, $V^*(P_i)$ is truly a minimum.

## 1.4. THE MINIMUM POTENTIAL ENERGY THEOREM

When a continuum deforms, it is subjected to the basic requirement that it remains a continuum. This requirement is fulfilled by any displacement function that is continuous and single-valued throughout the body.

Let us call $u_i$ a set of "admissible" displacements if $u_i$ leaves the body continuous and if $u_i$ satisfies the boundary conditions where displacements are prescribed. When a linear elastic solid is subjected to a set of loads $P_1, P_2, \ldots, P_n$, there corresponds a unique set of displacements $u_1, u_2, \ldots, u_n$ at the points of loading (see Sec. 1.2). Thus the conditions of equilibrium apparently single out one particular set of "corresponding" displacements from among all admissible sets of displacements. We shall now prove the following theorem.

*Of all admissible sets of displacements satisfying the boundary conditions where displacements are prescribed, the "actual" one which also satisfies the equations of equilibrium and Hooke's law is distinguished by the minimum value of the potential energy*

$$V = U - \sum_{S_p} u_i P_i, \tag{1}$$

where

$$U = \tfrac{1}{2}\sum_{i=1}^{n}\sum_{j=1}^{n} k_{ij} u_i u_j. \tag{2}$$

The constants $k_{ij}$ are the stiffness influence coefficients (Sec. 1.2); $k_{ij} = k_{ji}$, $i, j = 1, \ldots, n$. The first term on the right-hand side of (1) is the strain energy function; the second term is the potential energy of the external loads. The summation $\sum_{S} u_i P_i$ needs to include only those points where external loads are prescribed (a fact indicated by the notation $\sum_{S_p}$). The set of points $S_p$ represents the points where external loads are prescribed.

*Proof.* To prove this theorem, we denote by $(u_i)$ the actual displacements corresponding to the set of forces $(P_i)$. We shall compare $(u_i)$ with other admissible displacements, which may be written as $(u_i + \delta u_i)$. Since $(u_i + \delta u_i)$ satisfy the boundary conditions on displacements, $(\delta u_i)$ must vanish wherever $(u_i)$ are prescribed. Now consider the "virtual" work $\sum_{i=1}^{n} P_i \, \delta u_i$. From Eq. (1.2:3) we have

$$\text{(3)} \qquad \sum_{i=1}^{n} P_i \, \delta u_i = \sum_{i=1}^{n} \sum_{j=1}^{n} k_{ij} u_j \, \delta u_i = \delta\left(\tfrac{1}{2} \sum_{i=1}^{n} \sum_{j=1}^{n} k_{ij} u_i u_j\right) = \delta U.$$

After rearranging terms, this can be written as

$$\text{(4)} \qquad \delta V = 0.$$

Thus, the actual solution satisfies this variational equation, and $V$ is an extremum for $u_i$.

That $V$ is indeed a minimum is easily verified by showing that, on account of (4),

$$V(u_i + \delta u_i) - V(u_i) = \tfrac{1}{2} \sum_{i=1}^{n} \sum_{j=1}^{n} k_{ij} \, \delta u_i \, \delta u_j \geqslant 0.$$

The last inequality follows from the positive definiteness of the strain energy function.

The minimum principles are very important in the theory of elasticity, because they lead to practical approximate methods of solution of complicated problems.

## 1.5. SOME IMPORTANT REMARKS

Three important remarks should be added concerning the use of the theorems derived above.

First, the "force" in Hooke's law may be generalized to mean torque, couple, etc. The "corresponding" deformation can be so defined that the work done upon the body when the deformation changes by a small amount is equal to the force *times* the increment of the deformation. Thus, if $P$ is the generalized force and $\Delta u$ is the change of the corresponding generalized displacement, then

$$\Delta \text{ work} = P \cdot \Delta u.$$

Thus, for a force, torque, or couple, $u$ refers to displacement, angle of twist, and angle of bending, respectively.

When generalized forces are considered, it is very important to notice the word "corresponding" in the statement of the reciprocal relations. A generalized force *corresponds* to a generalized displacement if their product gives *exactly* the work done. Thus, for a beam, the change of slope corresponds to a moment, the deflection corresponds to a force, and the twisting angle corresponds to a torque. Although Maxwell's reciprocal relation asserts that the *deflection* at point 1 due to a unit couple at point 2 is equal to the rotation at point 2 due to a unit force at point 1, it is completely wrong to assert that the rotation at point 1 due to a unit force at point 2 must be equal to the rotation at point 2 due to a unit force at point 1. In the latter case the force and rotation do not correspond to each other.

The second remark is a truism: one may prove a theorem one way, but use it another way. We proved the reciprocal theorem, Castigliano's theorem, the minimum potential energy theorem, and the minimum complementary energy theorem with the influence coefficients $c_{ij}$ or $k_{ij}$, but we do not have to know these influence coefficients to apply these theorems. In fact, as long as we can write an expression for the strain energy, we can use these theorems. Indeed, this is why these energy theorems can form a basis for approximate theories in elasticity. For example, in the theory of simple beams we have the time-honored approximation that plane cross sections remain plane in bending and the Bernoulli-Euler approximation that the local curvature of the beam axis is proportional to the local moment in the beam. We can use these assumptions to derive an appropriate expression for strain energy. When we wish to apply the complementary energy theorem, the strain energy must be expressed in terms of forces and moments. When we wish to apply the minimum potential energy theorem, the strain energy must be expressed in terms of displacements. In particular, when the strain energy expressions are known in terms of forces, we can use Castigliano's theorem to derive deflections. (In practice, that is the usual way the flexibility influence coefficients are computed.)

In other words, the $c_{ij}$'s are the vehicle through which the general principles are established, but usually we do not know them. In applications, we try to obtain an expression for the strain energy with trustworthy accuracy and deduce the $c_{ij}$'s as well as other solutions from them. Establishing effective, simplifying assumptions which yield accurate strain energy expressions is indeed one of the most important objectives of engineering research.

Finally, we must comment on the sets of points $S_p$ and $S_u$ in Secs. 1.3 and 1.4. Although in the example in Sec. 1.3 the sum $S_p + S_u$ is equal to the total set of points engrossing our attention, such a limitation need not be imposed. Thus, the total set may be greater than $S_p + S_u$; the difference consists of *internal points*, whereas $S_p + S_u$ constitutes *boundary points*.

These remarks will be useful in solving the problems at the end of this section.

The theory presented in the foregoing sections forms the basis of the engineering theory of framed structures. Many ingenious applications of this theory have been made to the analysis of machines, trusses, rigid frames, reinforced concrete structures, aircraft structures, etc. In particular, the Maxwell or Betti-Rayleigh reciprocal relations and the first theorem of Castigliano are very useful in finding displacements in a structure when the stresses in it can be determined from statics. Limitation of space prevents us from going into any discussions of these applications. The reader is urged to study some of the standard references listed in Bibliography 1·3 at the end of the book in order to appreciate the many uses that can be made of this rudimentary theory.

But the elementary formulation has neither reduced the theory to its simplest form, nor given it sufficient generality to deal with more difficult problems. For example, to describe the elastic properties of a body by the method described above, we will need a complete set of influence coefficients (or influence functions) instead of a few elastic moduli of the material. The elementary theory cannot give a detailed picture of stress distribution in a body in response to external loads. Most problems in stress concentration, contact pressure, propagation and refraction of elastic waves, transient or steady oscillations, lie beyond the scope of the elementary theory.

Furthermore, Hooke's law expressed in terms of the linearity in load-displacement relationship is unnecessarily restrictive. There exist many cases in which the material remains elastic everywhere but the load-displacement relationship is nonlinear. Examples are a beam under simultaneous action of lateral and end loads, a large deflection of an elastic arch, and a large deflection of a thin plate or a thin shell.

Finally, practical engineering materials do not always obey Hooke's law. They may be elastic, but the stress-strain relationship may be nonlinear; or they may be viscoelastic or plastic. How shall we formulate the physical laws of nonlinear elasticity, of flow and motion? How can such laws be tested in the laboratory? What guidance can we obtain from such theoretical considerations as tensor analysis?

To answer these questions, a more general theory of continuum mechanics is developed, the presentation of which will occupy later chapters of this book.

## PROBLEMS

**1.1.** Prove that it is possible to generalize Hooke's law to deal with moments and angles of rotations by considering a concentrated couple as the limiting case of two equal and opposite forces approaching each other but maintaining a constant moment.

**1.2.** A pin-jointed truss is shown in Fig. P1.2. Every member of the truss is made of the same steel and has the same cross-sectional area. Find the tension or compression in every member when a load $P$ is applied at the point shown in the figure. *Note*: A joint is said to be a *pin joint* if no moment can be transmitted across it. This problem is statically determinate.

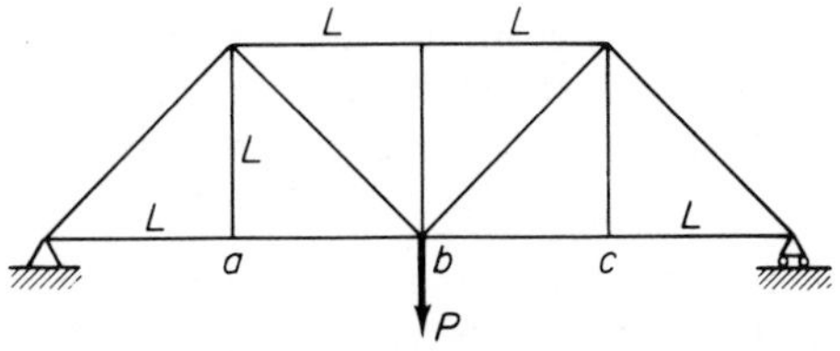


**P1.2**

**1.3.** Find the vertical deflections at $a$, $b$, $c$ of the truss in Fig. P1.2 due to a load $P$ at point $b$. Assume that $P$ is sufficiently small so that the truss remains linear elastic. Use the fact that for a single uniform bar in tension the total change in length of this bar is given by $PL/AE$, where $P$ is the load in the bar, $L$ is the bar length, $A$ is the cross-sectional area, and $E$ is the Young's modulus of the material. For structural steels $E$ is about $3 \times 10^7$ lb/sq in. *Hint*: Use Castigliano's theorem.

**1.4.** A "rigid" frame of structural steel shown in Fig. P1.4 is acted on by a horizontal force $P$. Find the reactions at the points of support $a$, and $d$. The cross-sectional area $A$ and the bending rigidity constant $EI$ of the various members are shown in the figure.

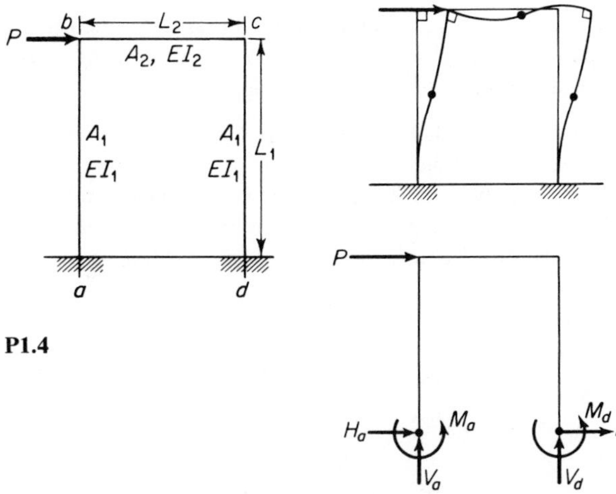


**P1.4**

By a "rigid" frame is meant a frame whose joints are rigid; e.g., welded. Thus, when we say that corner $b$ is a rigid joint, we mean that the two members meeting at $b$ will retain the same angle at that corner under any loading. The frame under loading will deform. It is useful to sketch the deflection line of the frame. Try to locate the points of inflection on the members. The points of *inflection* are points where the bending moments vanish.

The reactions to be found are the vertical force $V$, the horizontal force $H$, and the bending moment $M$, at each support.

Use engineering beam theory for this problem. In this theory, the change of curvature of the beam is $M/EI$, where $M$ is the local bending moment and $EI$ is the local bending rigidity. The strain energy per unit length due to bending is, therefore, $M^2/2EI$. The strain energy per unit length due to a tensile force $P$ is $P^2/2EA$. The strain energy due to transverse shear is negligible in comparison with that due to bending.

**1.5.** A circular ring of uniform linear elastic material and uniform cross section is loaded by a pair of equal and opposite forces at the ends of a diameter (Fig. P1.5). Find the change of diameters *aa* and *bb*. *Note*: This is a statically indeterminate problem. You have to determine the bending moment distribution in the ring.

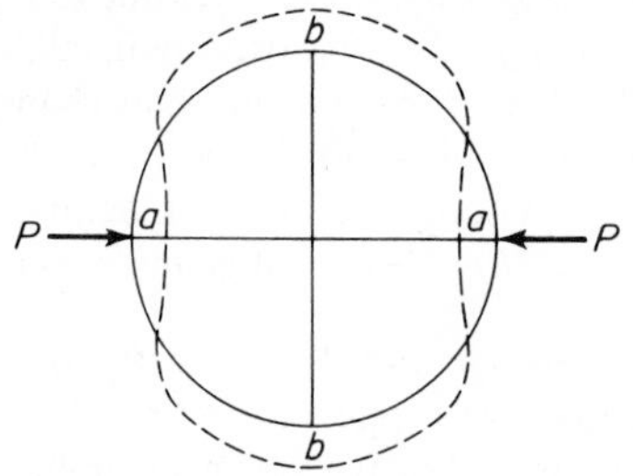


**P1.5**

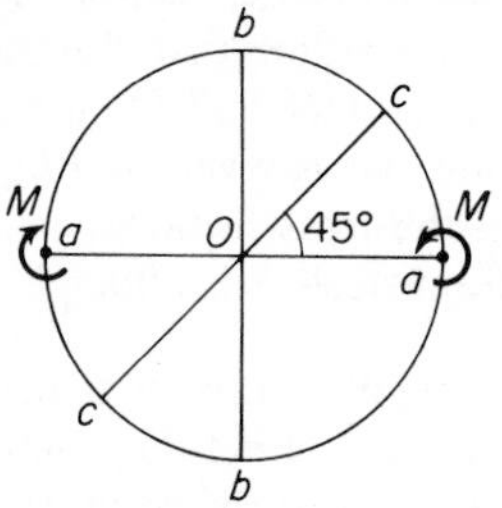


**P1.6**

**1.6.** Compare the changes in diameters *aa*, *bb*, and *cc* when a circular ring of uniform linear elastic material and uniform cross section is subjected to a pair of bending moments at the ends of a diameter (Fig. P1.6).

**1.7.** *Begg's Deformeter:* G. E. Beggs used an experimental model to determine the reactions of a statically indeterminate structure. For example, to determine the horizontal reaction $H$ at the right-hand support $b$ of an elastic arch under the load $P$, he imposes at $b$ a small displacement $\delta$ in the horizontal direction and measures at $P$ the deflection $\delta'$ in the direction corresponding with $P$, while preventing the vertical

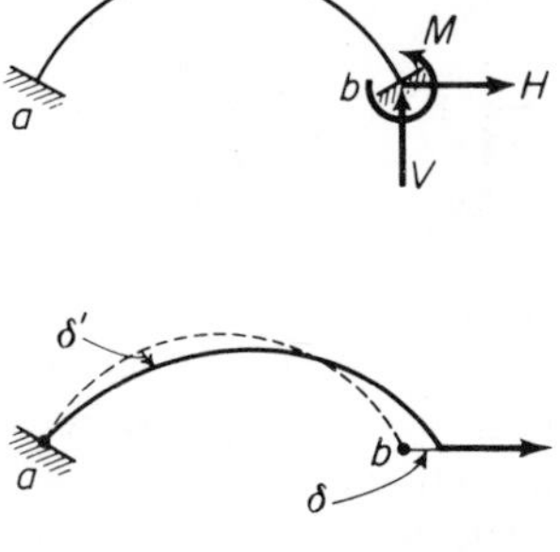


**P1.7**

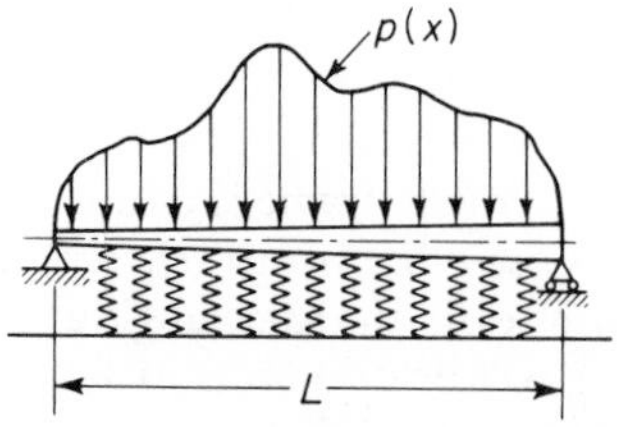


**P1.8**

displacement and rotation of the end $b$, Fig. P1.7. Show that $H = -P\delta'/\delta$. (Use the reciprocal theorem.) [G. E. Beggs, *J. Franklin Institute*, *203* (1927), pp. 375–386.]

**1.8.** A simply supported, thin elastic beam of variable cross section with bending rigidity $EI(x)$ rests on an elastic foundation with spring constant $k$ and is loaded by a distributed lateral load of intensity $p(x)$ per unit length (Fig. P1.8). Find an approximate expression for the deflection curve by assuming that it can be represented with sufficient accuracy by the expression

$$u(x) = \sum_{n=1}^{N} a_n \sin \frac{n\pi x}{L}.$$

*Note:* This expression satisfies the end conditions for arbitrary coefficients $a_n$. Use the minimum potential energy theorem.

**1.9.** Consider a uniform cantilever beam clamped at $x = 0$ (Fig. P1.9). According to Bernoulli-Euler theory of beams, the differential equation governing the deflection of the beam is

$$\frac{d^2w}{dx^2} = \frac{M}{EI},$$

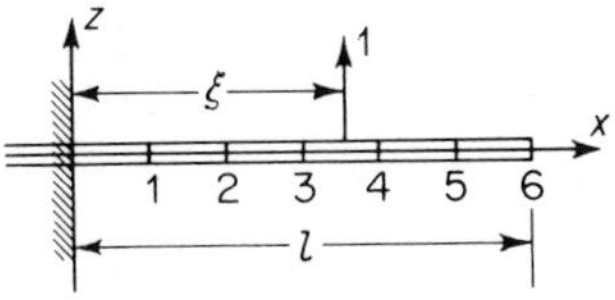


**P1.9**

where $w$ is the deflection parallel to the $z$-axis and $M$ is the bending moment at station $x$. This equation is valid if the beam is straight and slender and if the loading acts in a plane containing a principal axis of all cross sections of the beam, in which case the deflection $w$ occurs also in that plane.

Use this differential equation and the boundary conditions:

$$\textit{At clamped end:} \quad w = \frac{dw}{dx} = 0.$$

$$\textit{At free end:} \quad EI\frac{d^2w}{dx^2} = EI\frac{d^3w}{dx^3} = 0.$$

Derive the deflection curve of the beam when it is loaded by a unit force located at $x = \xi$.

*Ans.* For $0 \leqslant x \leqslant \xi \leqslant l$: $\quad w = \dfrac{x^2}{6EI}(3\xi - x).$

For $0 \leqslant \xi \leqslant x \leqslant l$: $\quad w = \dfrac{\xi^2}{6EI}(3x - \xi).$

The solution fits the definition of influence coefficients and is known as an influence function.

**1.10.** Let the beam of Problem 1.9 be divided into six equidistant sections as marked in Fig. P1.9. Compute the influence coefficients $c_{ij}$, where $i, j = 1, 2, \ldots, 6$.

**1.11.** Invert the $c_{ij}$ matrix of Prob. 1.10 to obtain the stiffness influence coefficients $k_{ij}$. *Note:* If you work this problem out, you will find that the inversion of the flexibility influence coefficients matrix $(c_{ij})$ is very difficult. The difficulty increases rapidly as the range of the indices $i, j$ increases. It arises from the fact that the $(c_{ij})$ matrix is nearly singular; i.e., neighboring columns of the matrix $(c_{ij})$ are nearly in linear proportion to each other. On the other hand, inversion of the stiffness influence coefficients matrix $(k_{ij})$, in general, will have no difficulty.

**1.12.** A cantilever wing of an airplane is subjected to a concentrated load $P$ at the wing tip (Fig. P1.12). By moving $P$ along the tip chord, a point $C$, called the *center of flexure*, can be found which has the property that if $P$ acts at $C$ the wing tip

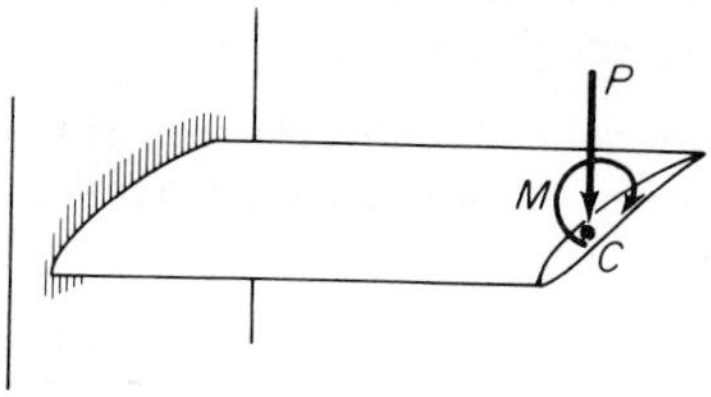


**P1.12**

section has no rotation (assuming the wing ribs to be rigid). On the other hand, if a couple $M$ is applied at the wing tip, the wing twists; but there will be one point in the tip section, say $C'$, which remains undisturbed and is called the *center of twist.* Show that *the center of flexure and the center of twist coincide*; i.e., $C$ and $C'$ are the same point.

## 1.6. LINEAR SOLIDS WITH MEMORY: MODELS OF VISCOELASTICITY

Most structural metals are nearly linear elastic under small strain, as measurements of load-displacement relationship reveal. The existence of normal modes of free vibrations which are simple harmonic in time, is often quoted as an indication (although not as a proof) of the linear elastic character of the material. However, when one realizes that the vibration of metal instruments does not last forever, even in a vacuum, it becomes clear that metals deviate somewhat from Hooke's law. Thus, other constitutive laws must be considered. The need for such an extension becomes particularly evident when organic polymers are considered.

In this section we shall consider a simple class of materials which retains linearity between load and deflection, but the linear relationship depends on a third parameter, time. For this class of material, the present state of deformation cannot be determined completely unless the entire history of loading is known.

A linear elastic solid may be said to have a simple memory: it remembers only one configuration; namely, the unstrained natural state of the body. Many materials do not behave this way: they remember the past. Among such materials with memory there is one class that is relatively simple in behavior. This is the class of materials named above, for which the cause and effect are linearly related.

To discuss the foregoing in concrete terms, let us consider a simple bar fixed at one end and subjected to a force in the direction of the axis at the other end. Let the force at time $t$ be $F(t)$ and the total elongation of the bar be $u(t)$. The elongation $u(t)$ is caused by the total history of the loading up to the time $t$. If the function $F(t)$ is continuous and differentiable, then in a small time interval $d\tau$ at time $\tau$ the increment of loading is $(dF/dt)\,d\tau$. This increment remains acting on the bar and contributes an element $du(t)$ to the elongation at the time $t$, with a proportionality constant $c$ depending on the time interval $t - \tau$. Hence, we may write†

$$du(t) = c(t - \tau)\frac{dF}{dt}(\tau)\,d\tau.$$

Let the origin of time be taken at the beginning of motion and loading. Then, on summing over the entire history, we have‡

$$u(t) = \int_0^t c(t - \tau)\frac{dF}{dt}(\tau)\,d\tau. \tag{1}$$

A similar argument, with the role of $F$ and $u$ interchanged, gives

$$F(t) = \int_0^t k(t - \tau)\frac{du}{dt}(\tau)\,d\tau. \tag{2}$$

These laws are linear, since a doubling of the load doubles the elongation, and vice versa.

Equations (1) and (2) are Boltzmann's formulation (Ludwig Boltzmann, 1844–1906) of the constitutive equation, in the case of a simple bar, for a material which has a linear load-deflection relationship. We may call such a material a *Boltzmann solid.* Vito Volterra (1860–1940), however, has introduced the more dramatic term *hereditary law* for any functional relation of the type exemplified in Eq. (1). To emphasize the linearity assumption, we shall call a material that follows such a law a *linear hereditary material.* There are, however, a host of other terms in common use, e.g, *viscoelasticity*, *creep*, *internal friction* or *damping*, *anelasticity*, *elastic after-effect*, *stress relaxation*, etc., each emphasizing a particular aspect or a particular model.

† The notation $(dF/dt)(\tau)$ means the value of $dF/dt$, which is a function of time, evaluated at the instant of time $t = \tau$. It is, of course, equal to the derivative $dF(\tau)/d\tau$, if the argument $t$ of $F(t)$ is replaced first by $\tau$.

‡ A simple modification of the formula is necessary if a finite load is applied suddenly at time $t = 0$, or if $F(t)$ is discontinuous in some other way. See Sec. 15.1, p. 413.

The function $k(t)$ is called the *relaxation function*. The function $c(t)$ is called the *creep function*. They are characteristic functions of the material.

Physically, $c(t)$ is the elongation produced by a sudden application at $t = 0$ of a constant force of magnitude unity; i.e., a unit-step forcing function which is zero when $t < 0$ and unity when $t > 0$. Similarly, $k(t)$ is the force that must be applied in order to produce an elongation which changes at $t = 0$ from zero to unity and remains unity thereafter.

It is generally accepted that the deformation at the *present time t* is due to forces that act in the *past*, and *not* in the future. This concept is expressed by the requirement that

$$c(t) = 0, \qquad \text{for } t < 0. \tag{3}$$

It is often referred to as the *axiom of nonretroactivity*. Similarly, by the same axiom,

$$k(t) = 0, \qquad \text{for } t < 0. \tag{4}$$

Of course, this nonretroactivity is implied already when we write $t$ for the upper limits of the integrals in Eqs. (1) and (2).

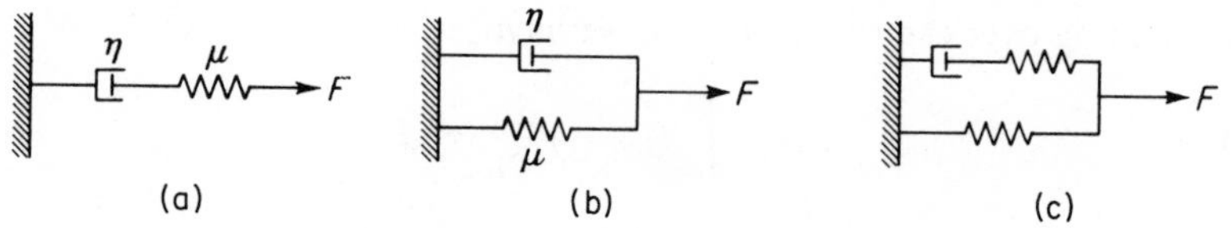


**Fig. 1.6:1.** Models of linear viscoelasticity: (a) Maxwell, (b) Voigt, (c) standard linear solid.

Before further discussions let us consider some simple examples. In Fig. 1.6:1 are shown three mechanical models of material behavior, namely, the Maxwell model, the Voigt model, and the "standard linear" model, all of which are composed of combinations of linear springs with spring constant $\mu$ and dashpots with coefficient of viscosity $\eta$. A *linear spring* is supposed to produce instantaneously a deformation proportional to the load. A *dashpot* is supposed to produce a velocity proportional to the load at any instant. The load-deflection relationship for these models are

(5) Maxwell model: $$\dot{u} = \frac{\dot{F}}{\mu} + \frac{F}{\eta}, \qquad u(0) = \frac{F(0)}{\mu},$$

(6) Voigt model: $$F = \mu u + \eta \dot{u}, \qquad u(0) = 0,$$

(7) Standard linear model:

$$F + \tau_\epsilon \dot{F} = E_R(u + \tau_\sigma \dot{u}), \qquad \tau_\epsilon F(0) = E_R \tau_\sigma u(0),$$

where $\tau_\epsilon$, $\tau_\sigma$ are two constants. When these equations are to be integrated, the initial conditions at $t = 0$ must be prescribed as indicated above.

The *creep functions* can be easily derived by solving Eqs. (5)–(7) for $u(t)$ when $F(t)$ is a unit-step function $\mathbf{1}(t)$. They are

(8) Maxwell solid:

$$c(t) = \left(\frac{1}{\mu} + \frac{1}{\eta} t\right)\mathbf{1}(t),$$

(9) Voigt solid: $c(t) = \frac{1}{\mu}(1 - e^{-(\mu/\eta)t})\mathbf{1}(t)$,

(10) Standard linear solid:

$$c(t) = \frac{1}{E_R}\left[1 - \left(1 - \frac{\tau_\epsilon}{\tau_\sigma}\right)e^{-t/\tau_\sigma}\right]\mathbf{1}(t),$$

where the *unit-step function* $\mathbf{1}(t)$ is defined as

$$\mathbf{1}(t) = \begin{cases} 1 & \text{when} \quad t > 0, \\ \frac{1}{2} & \text{when} \quad t = 0, \\ 0 & \text{when} \quad t < 0. \end{cases} \tag{11}$$

A body which obeys a load-deflection relation like that given by Maxwell's model is said to be a Maxwell solid, etc. Since a dashpot behaves as a piston moving in a viscous fluid, the above-named models are called models of *viscoelasticity*.

Interchanging the roles of $F$ and $u$, we obtain the *relaxation function* as a response $F(t) = k(t)$ corresponding to an elongation $u(t) = \mathbf{1}(t)$.

(12) Maxwell solid:

$$k(t) = \mu e^{-(\mu/\eta)t}\mathbf{1}(t),$$

(13) Voigt solid:

$$k(t) = \eta\, \boldsymbol{\delta}(t) + \mu\mathbf{1}(t),$$

(14) Standard linear solid:

$$k(t) = E_R\left[1 - \left(1 - \frac{\tau_\sigma}{\tau_\epsilon}\right)e^{-t/\tau_\epsilon}\right]\mathbf{1}(t).$$

Here we have used the symbol $\boldsymbol{\delta}(t)$ to indicate the *unit-impulse function*, or *Dirac-delta function*, which is defined as a function with a singularity at the origin:

$$\begin{aligned} \boldsymbol{\delta}(t) &= 0 && \text{for } t < 0, \text{ and } t > 0, \\ \int_{-\epsilon}^{\epsilon} f(t)\, \boldsymbol{\delta}(t)\, dt &= f(0), && \epsilon > 0, \end{aligned} \tag{15}$$

where $f(t)$ is an arbitrary function continuous at $t = 0$. These functions, $c(t)$ and $k(t)$, are illustrated in Figs. 1.6:2 and 1.6:3, respectively, for which we add the following comments.

For the Maxwell solid, a sudden application of a load induces an immediate deflection by the elastic spring, which is followed by "creep" of the dashpot. On the other hand, a sudden deformation produces an immediate reaction by the spring, which is followed by stress relaxation according to an

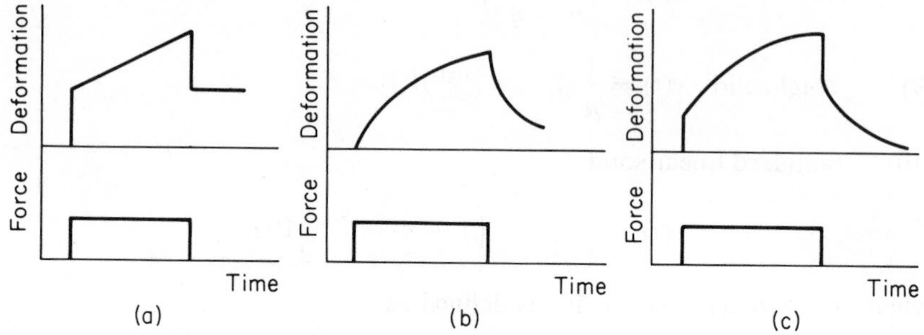


**Fig. 1.6:2.** Creep function of (a) Maxwell, (b) Voigt, (c) standard linear solid. A negative phase is superposed at the time of unloading.

exponential law (12). The factor $\eta/\mu$, with dimensions of time, may be called a *relaxation time*: it characterizes the rate of decay of the force.

For a Voigt solid, a sudden application of force will produce no immediate deflection because the dashpot, arranged in parallel with the spring, will not move instantaneously. Instead, as shown by (9) and Fig. 1.6:2(b), a deformation will be gradually built up, while the spring takes a greater and

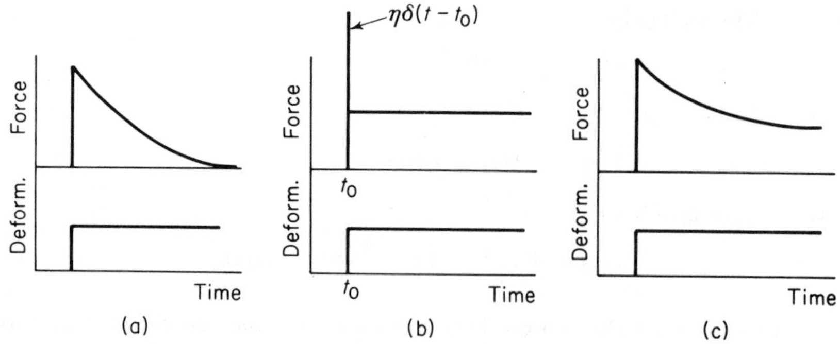


**Fig. 1.6:3.** Relaxation function, of (a) Maxwell, (b) Voigt, (c) standard linear solid.

greater share of the load. The dashpot displacement relaxes exponentially. Here the ratio $\eta/\mu$ is again a relaxation time: it characterizes the rate of decay of the deflection.

For the standard linear solid, a similar interpretation is applicable. The constant $\tau_\epsilon$ is the time of relaxation of load under the condition of constant deflection [see Eq. (14)], whereas the constant $\tau_\sigma$ is the time of relaxation of deflection under the condition of constant load [see Eq. (10)]. As $t \to \infty$, the dashpot is completely relaxed, and the load-deflection relation becomes that of the springs, as is characterized by the constant $E_R$ in Eqs. (10) and (14). Therefore, $E_R$ is called the *relaxed elastic modulus.*

Load-deflection relations such as (5)–(7) were proposed to extend the classical theory of elasticity to include anelastic phenomena. Lord Kelvin (Sir William Thomson, 1824–1907), on measuring the variation of the rate of dissipation of energy with frequency of oscillation in various materials, showed the inadequacy of the Maxwell and Voigt equations. A more successful generalization using mechanical models was first made by John H. Poynting (1852–1914) and Joseph John Thomson ("J.J.," 1856–1940) in their book *Properties of Matter* (London: C. Griffin and Co., 1902). The model shown in Fig. 1.6:1(c) identified with Eq. (7) is called the standard linear model because it is the most general relationship to include the load, the deflection, and their first derivatives (often known as linear derivatives!). It is, of course, a special case of Boltzmann's general linear relation, (1) or (2).

**Problem 1.13.** Derive Eqs. (8), (9), and (10), first by elementary methods of integration and then by Laplace transformation.

**Problem 1.14.** Consider Eqs. (1) and (2) of Sec. 1.6. Let $c(t)$ be a continuous function for $t \geqslant 0$, and $c(t) = 0$ for $t < 0$. Show that the condition for the existence of a continuous kernel $k(t)$, so that (2) is the inverse of (1), is that $c(0) \neq 0$. *Note:* In a Voigt solid $c(0) = 0$ and $k(t)$ contains a delta function, which is discontinuous.

## 1.7. SINUSOIDAL OSCILLATIONS IN A VISCOELASTIC MATERIAL

It is interesting to obtain the relationship between load and deflection when the body is forced to perform simple harmonic oscillations. We are interested in the steady-state response. Hence, we assume that a sinusoidal forcing function has been acting on the body for an indefinitely long time and that all initial transient disturbances have died out. Under this circumstance, it is convenient to put the beginning of motion at time $-\infty$. Hence, we shall replace the lower limits of Eqs. (1.6:1) and (1.6:2) by $-\infty$. Using complex representation for sinusoidal oscillations, we put $F(t) = F_0 e^{i\omega t}$ into these equations. It will be more convenient to make a change of variable $t - \tau = \xi$ first. This gives

$$u(t) = \int_0^\infty c(\xi) \frac{dF}{dt}(t - \xi)\, d\xi. \tag{1}$$

When $F(t) = F_0 e^{i\omega t}$, we obtain

$$u(t) = \int_0^\infty c(\xi) i\omega F_0 e^{i\omega(t-\xi)}\, d\xi$$
$$= i\omega F_0 e^{i\omega t} \int_0^\infty c(\xi) e^{-i\omega\xi}\, d\xi.$$

Since $c(t) = 0$ when $t < 0$, we can replace the lower limit of the integral above by $-\infty$ and write the integral in the conventional form of Fourier transformation

(2) $$\bar{c}(\omega) = \int_{-\infty}^{\infty} c(\tau) e^{-i\omega\tau}\, d\tau.$$

Assuming that the Fourier integral exists, we have

(3) $$u(t) = i\omega F_0 \bar{c}(\omega) e^{i\omega t}.$$

Hence, under a periodic forcing function, the displacement $u(t)$ is also periodic. On writing

$$u(t) = u_0 e^{i\omega t},$$

we obtain

(4) $$\frac{u_0}{F_0} = i\omega\bar{c}(\omega).$$

Thus the ratio $u_0/F_0$ is a complex number, which may be written as

(5) $$\frac{u_0}{F_0} = \frac{1}{\mathscr{M}} = i\omega\bar{c}(\omega) = |\omega\bar{c}(\omega)|\, e^{-i\delta},$$

where $\mathscr{M}$ is called the *complex modulus* of a viscoelastic material. The angle $\delta$ is of particular interest. It represents the phase angle by which the strain lags the stress. The tangent of $\delta$ is often used as a measure of *internal friction* of a linear viscoelastic material. Since

(6) $$\tan\delta = \frac{\text{imaginary part of } \mathscr{M}}{\text{real part of } \mathscr{M}},$$

we see that the internal friction can be easily computed when the Fourier transform of the creep function is known.

An entirely similar argument starting with Eq. (1.6:2) leads to

(7) $$\frac{F_0}{u_0} = i\omega\bar{k}(\omega) = \mathscr{M},$$

where $\bar{k}(\omega)$ is the Fourier transform of the relaxation function

(8) $$\bar{k}(\omega) = \int_{-\infty}^{\infty} k(\tau) e^{-i\omega\tau}\, d\tau.$$

From (4) and (7) we obtain a relation connecting $\bar{c}(\omega)$ and $\bar{k}(\omega)$,

(9) $$-\omega^2\bar{c}(\omega)\bar{k}(\omega) = 1.$$

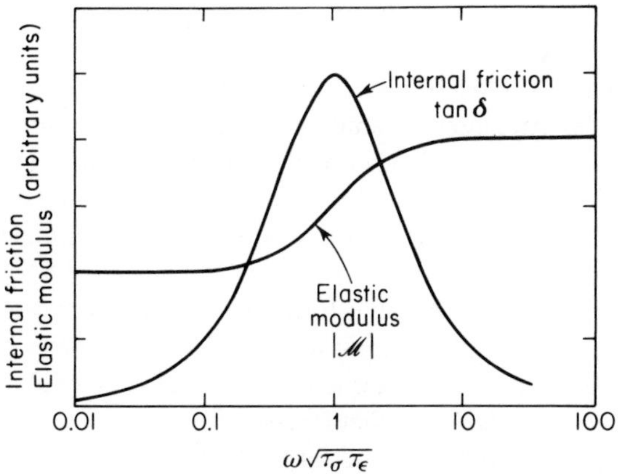


**Fig. 1.7:1.** Frequency dependence of internal friction and elastic modulus.

As an example, let us consider the standard linear solid. On putting $u = u_0 e^{i\omega t}$, $F = F_0 e^{i\omega t}$ into Eq. (1.6:7), we obtain

$$\mathscr{M} = \frac{1 + i\omega\tau_\sigma}{1 + i\omega\tau_\epsilon} E_R, \qquad |\mathscr{M}| = \left(\frac{1 + \omega^2\tau_\sigma^2}{1 + \omega^2\tau_\epsilon^2}\right)^{1/2} E_R, \tag{10}$$

$$\tan\delta = \frac{\omega(\tau_\sigma - \tau_\epsilon)}{1 + \omega^2(\tau_\sigma\tau_\epsilon)} = \frac{(\tau_\sigma - \tau_\epsilon)}{(\tau_\sigma\tau_\epsilon)^{1/2}} \frac{\omega(\tau_\sigma\tau_\epsilon)^{1/2}}{1 + \omega^2(\tau_\sigma\tau_\epsilon)}. \tag{11}$$

When $|\mathscr{M}|$ and $\tan\delta$ in (10) and (11) are plotted against the logarithm of $\omega$, curves as shown in Fig. 1.7:1 are obtained. Experiments with torsional oscillations of metal wires at various temperatures, reduced to room temperature according to certain thermodynamic formula, yield a typical "relaxation spectrum" as shown in Fig. 1.7:2. Many peaks are seen in the

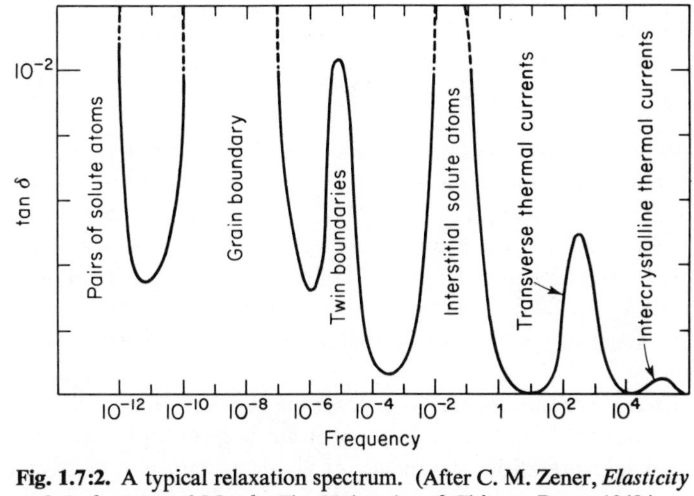


**Fig. 1.7:2.** A typical relaxation spectrum. (After C. M. Zener, *Elasticity and Anelasticity of Metals*, The University of Chicago Press, 1948.)

internal-friction-versus-frequency curve. It has been suggested that each peak should be regarded as representing an elementary process as described above, with a particular set of relaxation times $\tau_\sigma$, $\tau_\epsilon$. Each set of relaxation times $\tau_\sigma$, $\tau_\epsilon$ can be attributed to some process in the atomic or microscopic level. A detailed study of such a relaxation spectrum tells a great deal about the structures of metals; and the internal friction has been a very effective key to metal physics. Zener's fascinating book[1.4] is recommended to those interested in this subject.

A number of materials follow a linear viscoelastic stress-strain law very well when the strain is small. Among such materials are many high polymers. The damping characteristics of many metals can also be explained by such a model. The creep of metals at higher stresses, however, usually follows a non-linear law. See *Bibliography* 6.4 and 14.8.

## 1.8. STRUCTURAL PROBLEMS OF VISCOELASTIC MATERIALS

Several steps of generalization are necessary before the simple models discussed in Secs. 1.6 and 1.7 can be used in structural analysis. The simplest point of view is to idealize a solid body into a set of mass particles inter-connected by simple bars, for each of which a hereditary stress-strain law is specified. (For many structures and machines, this is actually not a very bad description!) The equations of motion of these mass particles and massless linkages can be written down and analyzed.

A second point of view is to generalize the Hooke's load-displacement law, Eq. (1.1:1), by a hereditary law. Thus, if a set of external forces $F_1, F_2, \ldots, F_n$ act on a body at points $1, 2, \ldots, n$ and produce the "corresponding" displacements $u_1, u_2, \ldots, u_n$ at these points, then, for a properly supported viscoelastic body, we may assume

$$u_i = \sum_{j=1}^{n} \int_0^t C_{ij}(t-\tau)\frac{d}{d\tau}F_j(\tau)\,d\tau, \qquad i = 1, \ldots, n.$$

In this specification of the constitutive property of the body we shall need the full set of creep functions $C_{ij}(t)$. These creep functions and the convolution integrals assume the role of the influence coefficients in linear elasticity. Physically, $C_{ij}(t)$ means the deflection $u_i(t)$ produced by a unit-step function $F_j = \mathbf{1}(t)$ acting at the point $j$. In reality, we cannot demand that all these functions $C_{ij}$ be determined experimentally. To do that would be too tedious. In general, we have to compute them from the basic equations of viscoelasticity.

Therefore, it is fundamental to express the concept of viscoelasticity in a stress-strain-history law. Such a generalization is given in Chap. 13 and 15. With proper interpretation, the theory of viscoelasticity parallels the theory of elasticity in every respect. However, the basic equations of viscoelasticity are more difficult to solve, and behavior of a body of viscoelastic material can be very complex.

## PROBLEMS†

**1.15.** Consider the quadratic form of two real variables,

$$f(x, y) = ax^2 + bxy + cy^2.$$

Show that $f(x, y)$ is positive definite if and only if $a > 0$, $4ac - b^2 > 0$.

**1.16.** Show that the necessary and sufficient condition for the quadratic form

$$(\mathbf{X}, \mathbf{KX}) \equiv \sum_{i,j=1}^{n} k_{ij} x_i x_j,$$

where $\mathbf{K}$ is real symmetric and $\mathbf{X}$ is real, to be positive definite is that all the eigenvalues of the eigenvalue problem $\mathbf{KX} = \lambda \mathbf{X}$ are real and positive. Here $\mathbf{K}$ is a $n \times n$ square matrix and $\mathbf{X}$ is a $1 \times n$ column matrix,

$$\mathbf{K} = \begin{pmatrix} k_{11} & k_{12} & \cdots & k_{1n} \\ \cdot & \cdot & & \cdot \\ \cdot & \cdot & & \cdot \\ \cdot & \cdot & & \cdot \\ k_{n1} & k_{n2} & \cdots & k_{nn} \end{pmatrix}, \qquad \mathbf{X} = \begin{pmatrix} x_1 \\ \cdot \\ \cdot \\ \cdot \\ x_n \end{pmatrix}.$$

$(\mathbf{X}, \mathbf{KX})$ is the scalar product of the vectors $\mathbf{X}$ and $\mathbf{KX}$. Similarly, $(\mathbf{X}, \mathbf{KX})$ is negative definite if all the eigenvalues are negative. (*Note:* Since

$$\sum_{i,j=1}^{n} k_{ij} x_i x_j = \tfrac{1}{2} \sum_{i,j=1}^{n} (k_{ij} + k_{ji}) x_i x_j,$$

no generality is lost by considering $\mathbf{K}$ as a symmetric matrix.)

**1.17.** Consider the determinantal equation

$$\begin{vmatrix} a_{11} - \lambda & a_{12} & \cdots & a_{1n} \\ a_{21} & a_{22} - \lambda & \cdots & a_{2n} \\ \cdot & \cdot & & \cdot \\ \cdot & \cdot & & \cdot \\ \cdot & \cdot & & \cdot \\ a_{n1} & a_{n2} & \cdots & a_{nn} - \lambda \end{vmatrix} = 0,$$

where $a_{ij} = a_{ji}$ are real numbers. Let the $n$ roots of this equation be denoted by $\lambda_1, \lambda_2, \ldots, \lambda_n$. These roots can be identified with the eigenvalues of the eigenvalue problem $\mathbf{AX} = \lambda \mathbf{X}$, where

$$\mathbf{A} = \begin{pmatrix} a_{11} & a_{12} & \cdots & a_{1n} \\ \cdot & \cdot & & \\ \cdot & \cdot & & \\ \cdot & \cdot & & \\ a_{n1} & a_{n2} & \cdots & a_{nn} \end{pmatrix}, \qquad \mathbf{X} = \begin{pmatrix} x_1 \\ \cdot \\ \cdot \\ \cdot \\ x_n \end{pmatrix}.$$

† These problems are not directly related to the previous sections. They form an appendix concerning the positive definiteness of quadratic forms, which are very important in continuum mechanics.

Let this determinantal equation be written as

$$F(\lambda) = (\lambda_1 - \lambda)(\lambda_2 - \lambda)\ldots(\lambda_n - \lambda)$$
$$= \alpha_0 - \alpha_1\lambda + \alpha_2\lambda^2 + \ldots + (-1)^\nu\alpha_\nu\lambda^\nu + \ldots + (-1)^n\lambda^n = 0.$$

The coefficients are symmetric functions of the roots $\lambda_\nu$:

$$\alpha_0 = \lambda_1\lambda_2\ldots\lambda_n = F(0) = \det|a_{\mu\nu}|$$
$$\alpha_\nu = \underbrace{\lambda_1\lambda_2\ldots\lambda_{n-\nu}} + \ldots + \underbrace{\lambda_{\nu+1}\lambda_{\nu+2}\ldots\lambda_n}, \qquad \nu = 0, 1, \ldots, n-1.$$

Prove the following statements:

1. All eigenvalues $\lambda_\nu$ are positive if and only if all $\alpha_\nu$ are positive.
2. All eigenvalues $\lambda_\nu$ are negative if and only if all $(-1)^{n-\nu}\alpha_\nu$ are positive.
3. There exist exactly $r$ eigenvalues equal to zero while the other $n - r$ eigenvalues are positive if and only if

$$\alpha_0 = \alpha_1 = \ldots = \alpha_{r-1} = 0, \qquad \alpha_\nu > 0, \qquad \nu = r, \ldots, n.$$

   *Hint:* Apply Descartes' rule of signs.

4. There exist exactly $r$ eigenvalues equal to zero while the other $n - r$ eigenvalues are negative if and only if

$$\alpha_0 = \alpha_1 = \ldots = \alpha_{r-1} = 0, \qquad (-1)^{n-\nu}\alpha_\nu < 0, \qquad \nu = r, \ldots, n.$$

5. If we strike out $\nu$ rows and $\nu$ columns of the same indices from the determinant $|a_{\mu\nu}|$, the determinant of the remaining elements is called a *principal minor* of order $n - \nu$. Show that $\alpha_\nu$ = sum of all principal minors of order $n - \nu$ of $|a_{\mu\nu}|$.
6. State the conditions for the positive or negative definiteness of a quadratic form in terms of the principal minors of the symmetric matrix of coefficients.

# 2

# TENSOR ANALYSIS

In attempting to make a generalization of the theory outlined in the previous chapter, we shall first learn to use the powerful tool of tensor calculus. At first sight, the mathematical analysis may seem involved, but a little study will soon reveal its simplicity.

## 2.1. NOTATION AND SUMMATION CONVENTION

Let us begin with the matter of notation. In tensor analysis one makes extensive use of indices. A set of $n$ variables $x_1, x_2, \ldots, x_n$ is usually denoted as $x_i, i = 1, \ldots, n$. A set of $n$ variables $y^1, y^2, \ldots, y^n$ is denoted by $y^i, i = 1, \ldots, n$. We emphasize that $y^1, y^2, \ldots, y^n$ are $n$ independent variables and *not* the first $n$ powers of the variable $y$.

Consider an equation describing a plane in a three-dimensional space $x^1, x^2, x^3$,

$$a_1x^1 + a_2x^2 + a_3x^3 = p, \tag{1}$$

where $a_i$ and $p$ are constants. This equation can be written as

$$\sum_{i=1}^{3} a_i x^i = p. \tag{2}$$

However, we shall introduce the *summation convention* and write the equation above in the simple form

$$a_i x^i = p. \tag{3}$$

The convention is as follows: *The repetition of an index* (*whether superscript or subscript*) *in a term will denote a summation with respect to that index over its range*. The *range* of an index $i$ is the set of $n$ integer values 1 to $n$. A lower index $i$, as in $a_i$, is called a *subscript*, and upper index $i$, as in $x^i$, is called a *superscript*. An index that is summed over is called a *dummy index*; one that is not summed out is called a *free index*.

Since a dummy index just indicates summation, it is immaterial which symbol is used. Thus, $a_i x^i$ may be replaced by $a_j x^j$, etc. This is analogous to the dummy variable in an integral

$$\int_a^b f(x)\, dx = \int_a^b f(y)\, dy.$$

The use of index and summation convention may be illustrated by other examples. Consider a unit vector $\boldsymbol{\nu}$ in a three-dimensional Euclidean space with rectangular Cartesian coordinates $x$, $y$, and $z$. Let the direction cosines $\alpha_i$ be defined as

$$\alpha_1 = \cos(\boldsymbol{\nu}, \mathbf{x}), \qquad \alpha_2 = \cos(\boldsymbol{\nu}, \mathbf{y}), \qquad \alpha_3 = \cos(\boldsymbol{\nu}, \mathbf{z}),$$

where $(\boldsymbol{\nu}, \mathbf{x})$ denotes the angle between $\boldsymbol{\nu}$ and the $x$-axis, etc. The set of numbers $\alpha_i$ $(i = 1, 2, 3)$ represents the projections of the unit vector on the coordinate axes. The fact that the length of the vector is unity is expressed by the equation

$$(\alpha_1)^2 + (\alpha_2)^2 + (\alpha_3)^2 = 1,$$

or simply

(4) $$\alpha_i \alpha_i = 1.$$

As a further illustration, consider a line element $(dx, dy, dz)$ in a three-dimensional Euclidean space with rectangular Cartesian coordinates $x, y, z$. The square of the length of the line element is

(5) $$ds^2 = dx^2 + dy^2 + dz^2.$$

If we define

(6) $$dx^1 = dx, \qquad dx^2 = dy, \qquad dx^3 = dz$$

and

(7) ▲ $$\begin{aligned} &\delta_{11} = \delta_{22} = \delta_{33} = 1, \\ &\delta_{12} = \delta_{21} = \delta_{13} = \delta_{31} = \delta_{23} = \delta_{32} = 0. \end{aligned}$$

Then (5) may be written as

(8) $$ds^2 = \delta_{ij}\, dx^i\, dx^j,$$

with the understanding that the range of the indices $i$ and $j$ is 1 to 3. Note that there are two summations in the expression above, one over $i$ and one over $j$. The symbol $\delta_{ij}$ as defined in (7) is called the *Kronecker delta*.

The following determinant illustrates another application

$$\begin{vmatrix} a_{11} & a_{12} & a_{13} \\ a_{21} & a_{22} & a_{23} \\ a_{31} & a_{32} & a_{33} \end{vmatrix} = \begin{aligned} &a_{11}a_{22}a_{33} + a_{21}a_{32}a_{13} + a_{31}a_{12}a_{23} \\ &- a_{11}a_{32}a_{23} - a_{12}a_{21}a_{33} - a_{13}a_{22}a_{31}. \end{aligned}$$

If we denote the general term in the determinant by $a_{ij}$ and write the determinant as $|a_{ij}|$, then the equation above can be written as

(9) $$|a_{ij}| = e_{rst} a_{r1} a_{s2} a_{t3},$$

where $e_{rst}$, the *permutation symbol*, is defined by the equations

$$e_{111} = e_{222} = e_{333} = e_{112} = e_{113} = e_{221} = e_{223} = e_{331} = e_{332} = 0,$$

(10) ▲ $$e_{123} = e_{231} = e_{312} = 1,$$

$$e_{213} = e_{321} = e_{132} = -1.$$

In other words, $e_{ijk}$ vanishes whenever the values of any two indices coincide; $e_{ijk} = 1$ when the subscripts permute like 1, 2, 3; and $e_{ijk} = -1$ otherwise.

The Kronecker delta and the permutation symbol are very important quantities which will appear again and again in this book. They are connected by the identity

(11) ▲ $$e_{ijk}e_{ist} = \delta_{js}\delta_{kt} - \delta_{jt}\delta_{ks}.$$

This $e$-$\delta$ identity is used sufficiently frequently to warrant special attention here. It can be verified by actual trial.

Finally, we shall extend the summation convention to differentiation formulas. Let $f(x^1, x^2, \ldots, x^n)$ be a function of $n$ numerical variables $x^1, x^2, \ldots, x^n$. Then its differential shall be written

(12) $$df = \frac{\partial f}{\partial x^i} dx^i.$$

## PROBLEMS

**2.1.** Show that, when $i, j, k$ range over 1, 2, 3,

(a) $\delta_{ij}\,\delta_{ij} = 3$

(b) $e_{ijk}e_{jki} = 6$

(c) $e_{ijk}A_jA_k = 0$

(d) $\delta_{ij}\,\delta_{jk} = \delta_{ik}$

**2.2.** The vector product of two vectors $\mathbf{x} = (x_1, x_2, x_3)$ and $\mathbf{y} = (y_1, y_2, y_3)$ is the vector $\mathbf{z} = \mathbf{x} \times \mathbf{y}$ whose three components are

$$z_1 = x_2y_3 - x_3y_2, \qquad z_2 = x_3y_1 - x_1y_3, \qquad z_3 = x_1y_2 - x_2y_1.$$

This can be shortened by writing

$$z_i = e_{ijk}x_jy_k.$$

Verify the following identity connecting three arbitrary vectors by means of the $e$-$\delta$ identity.

$$\mathbf{A} \times (\mathbf{B} \times \mathbf{C}) = (\mathbf{A} \cdot \mathbf{C})\mathbf{B} - (\mathbf{A} \cdot \mathbf{B})\mathbf{C}.$$

*Note:* The last equation is well known in vector analysis. After identifying the quantities involved as Cartesian tensors, this verification may be construed as a proof of the $e$-$\delta$ identity.

## 2.2. COORDINATE TRANSFORMATION

The central point of view of tensor analysis is to study the change of the components of a quantity such as a vector with respect to coordinate transformations.

A set of independent variables $x_1, x_2, x_3$ may be thought of as specifying the coordinates of a point in a frame of reference. A transformation from $x_1, x_2, x_3$ to a set of new variables $\bar{x}_1, \bar{x}_2, \bar{x}_3$ through the equations

$$\bar{x}_i = f_i(x_1, x_2, x_3), \qquad i = 1, 2, 3, \tag{1}$$

specifies a transformation of coordinates. The inverse transformation

$$x_i = g_i(\bar{x}_1, \bar{x}_2, \bar{x}_3) \qquad i = 1, 2, 3, \tag{2}$$

proceeds in the reverse direction. In order to insure that such a transformation is reversible and in one-to-one correspondence in a certain region $R$ of the variables $(x_1, x_2, x_3)$, i.e., in order that each set of numbers $(x_1, x_2, x_3)$ defines a unique set of numbers $(\bar{x}_1, \bar{x}_2, \bar{x}_3)$, for $(x_1, x_2, x_3)$ in the region $R$, and vice versa, it is sufficient that

(a) The functions $f_i$ are single-valued, continuous, and possess continuous first partial derivatives in the region $R$, and

(b) The *Jacobian determinant* $J = |\partial \bar{x}_i/\partial x_j|$ does not vanish at any point of the region $R$.

$$J = \left|\frac{\partial \bar{x}_i}{\partial x_j}\right| = \begin{vmatrix} \dfrac{\partial \bar{x}_1}{\partial x_1}, & \dfrac{\partial \bar{x}_1}{\partial x_2}, & \dfrac{\partial \bar{x}_1}{\partial x_3} \\ \dfrac{\partial \bar{x}_2}{\partial x_1}, & \dfrac{\partial \bar{x}_2}{\partial x_2}, & \dfrac{\partial \bar{x}_2}{\partial x_3} \\ \dfrac{\partial \bar{x}_3}{\partial x_1}, & \dfrac{\partial \bar{x}_3}{\partial x_2}, & \dfrac{\partial \bar{x}_3}{\partial x_3} \end{vmatrix} \neq 0 \qquad \text{in } R. \tag{3}$$

Coordinate transformations with the properties (a), (b) named above are called *admissible transformations*. If the Jacobian is positive everywhere, then a right-hand set of coordinates is transformed into another right-hand set, and the transformation is said to be *proper*. If the Jacobian is negative everywhere, a right-hand set of coordinates is transformed into a left-hand one, and the transformation is said to be *improper*. *In this book, we shall tacitly assume that our transformations are admissible and proper.*

## 2.3. EUCLIDEAN METRIC TENSOR

The first thing we must know about any coordinate system is how to measure length in that reference system. This information is given by the metric tensor.

Consider a three-dimensional Euclidean space, with the range of all indices 1, 2, 3. Let

$$\theta_i = \theta_i(x_1, x_2, x_3) \tag{1}$$

be an admissible transformation of coordinates from the rectangular Cartesian (in honor of Cartesius, i.e., Descartes) coordinates $x_1, x_2, x_3$ to some general coordinates $\theta_1, \theta_2, \theta_3$. The inverse transformation

$$x_i = x_i(\theta_1, \theta_2, \theta_3) \tag{2}$$

is assumed to exist, and the points $(x_1, x_2, x_3)$ and $(\theta_1, \theta_2, \theta_3)$ are in one-to-one correspondence.

Consider a line element with three components given by the differentials $dx^1, dx^2, dx^3$. Since the coordinates $x_1, x_2, x_3$ are assumed to be rectangular Cartesian, the length of the element $ds$ is determined by Pythagoras' rule:

(3) ▲ $$ds^2 = dx^i\, dx^i = \delta_{ij}\, dx^i\, dx^j.$$

When a coordinate transformation (1) is effected, we obtain from Eq. (2), according to ordinary rules of differentiation,

$$dx^i = \frac{\partial x_i}{\partial \theta_k}\, d\theta^k. \tag{4}$$

Substituting into (3), we have

$$ds^2 = \sum_{i=1}^{3} \frac{\partial x_i}{\partial \theta_k} \frac{\partial x_i}{\partial \theta_m}\, d\theta^k\, d\theta^m.$$

If we define the functions $g_{km}(\theta_1, \theta_2, \theta_3)$ by

(5) ▲ $$g_{km}(\theta_1, \theta_2, \theta_3) = \sum_{i=1}^{3} \frac{\partial x_i}{\partial \theta_k} \frac{\partial x_i}{\partial \theta_m},$$

then the square of the line element in the general $\theta_1, \theta_2, \theta_3$ coordinates takes the form

(6) ▲ $$ds^2 = g_{km}\, d\theta^k\, d\theta^m.$$

This, of course, stands for

$$\begin{aligned} ds^2 = g_{11}(d\theta_1)^2 + g_{12}\, d\theta_1\, d\theta_2 + g_{13}\, d\theta_1\, d\theta_3 & \\ + g_{21}\, d\theta_1\, d\theta_2 + g_{22}(d\theta_2)^2 + g_{23}\, d\theta_2\, d\theta_3 & \\ + g_{31}\, d\theta_1\, d\theta_3 + g_{32}\, d\theta_2\, d\theta_3 + g_{33}(d\theta_3)^2. & \end{aligned} \tag{7}$$

It is apparent from (5) that

(8) ▲ $$g_{km} = g_{mk} \qquad \text{for each } k \text{ and } m,$$

so the functions $g_{km}$ are symmetric in $k$ and $m$. The functions $g_{km}$ are called the components of the *Euclidean metric tensor* in the coordinate system $\theta_1, \theta_2, \theta_3$.

Let $\bar{\theta}_1, \bar{\theta}_2, \bar{\theta}_3$ be another general coordinate system. Let

$$\theta_i = \theta_i(\bar{\theta}_1, \bar{\theta}_2, \bar{\theta}_3) \tag{9}$$

be the transformation of coordinates from $\bar{\theta}_1, \bar{\theta}_2, \bar{\theta}_3$ to $\theta_1, \theta_2, \theta_3$. Now

$$d\theta^k = \frac{\partial \theta_k}{\partial \bar{\theta}_l}\, d\bar{\theta}^l. \tag{10}$$

Hence, from (6),

$$ds^2 = g_{km} \frac{\partial \theta_k}{\partial \bar{\theta}_l} \frac{\partial \theta_m}{\partial \bar{\theta}_n} d\bar{\theta}^l \, d\bar{\theta}^n. \tag{11}$$

If we define

$$\bar{g}_{ln}(\bar{\theta}_1, \bar{\theta}_2, \bar{\theta}_3) = g_{km}(\theta_1, \theta_2, \theta_3) \frac{\partial \theta_k}{\partial \bar{\theta}_l} \frac{\partial \theta_m}{\partial \bar{\theta}_n}. \tag{12}$$

Then (11) assumes a form which is the same as (3) or (6):

$$ds^2 = \bar{g}_{ln} \, d\bar{\theta}^l \, d\bar{\theta}^n. \tag{13}$$

Accordingly, we call $\bar{g}_{ln}$ the components of the Euclidean metric tensor in the coordinate system $\bar{\theta}_1, \bar{\theta}_2, \bar{\theta}_3$.

The quadratic differential forms (3), (6), or (13) are of fundamental importance since they define the length of any line element in general coordinate systems. We conclude that if $\theta_1, \theta_2, \theta_3$ and $\bar{\theta}_1, \bar{\theta}_2, \bar{\theta}_3$ are two sets of general coordinates, then Euclidean metric tensors $g_{km}(\theta_1, \theta_2, \theta_3)$ and $\bar{g}_{km}(\bar{\theta}_1, \bar{\theta}_2, \bar{\theta}_3)$ are related by means of the characteristic law of transformation (12).

The law of transformation of the components of a quantity with respect to coordinate transformation is an important property of that quantity. In the following section, we shall see that a quantity shall be called a tensor if and only if it follows certain specific laws of transformation.

All the results above apply as well to the plane (a two-dimensional Euclidean space), as can be easily verified by changing the range of all indices to 1, 2.

## PROBLEMS

**2.3.** Find the components of the Euclidean metric tensor in plane polar coordinates ($\theta_1 = r$, $\theta_2 = \theta$; see Fig. P2.3) and the corresponding expression for the length of a line element.

*Ans.* Let $x_1, x_2$ be a set of rectangular Cartesian coordinates. Then

$$x_1 = \theta_1 \cos \theta_2, \qquad \theta_1 = \sqrt{(x_1)^2 + (x_2)^2},$$

$$x_2 = \theta_1 \sin \theta_2, \qquad \theta_2 = \sin^{-1}\left(\frac{x_2}{\sqrt{(x_1)^2 + (x_2)^2}}\right),$$

$$g_{11} = \cos^2 \theta_2 + \sin^2 \theta_2 = 1,$$

$$g_{12} = (\cos \theta_2)(-\theta_1 \sin \theta_2) + (\sin \theta_2)(\theta_1 \cos \theta_2) = 0 = g_{21},$$

$$g_{22} = (-\theta_1 \sin \theta_2)^2 + (\theta_1 \cos \theta_2)^2 = (\theta_1)^2.$$

The line element is

$$ds^2 = (d\theta_1)^2 + (\theta_1)^2(d\theta_2)^2.$$

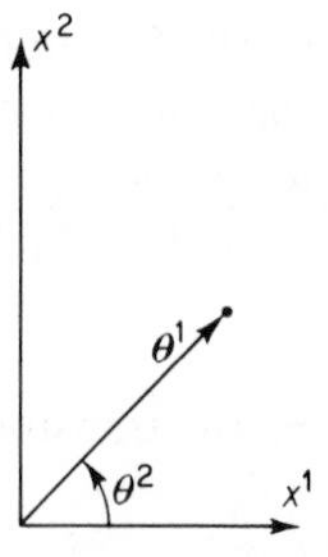


**P2.3**

**2.4.** Let $x_1, x_2, x_3$ be rectangular Cartesian coordinates and $\theta_1, \theta_2, \theta_3$ be spherical polar coordinates. (See Fig. P2.4.) Then

$$\theta_1 = \sqrt{(x_1)^2 + (x_2)^2 + (x_3)^2},$$

$$\theta_2 = \cos^{-1} \frac{x_3}{\sqrt{(x_1)^2 + (x_2)^2 + (x_3)^2}},$$

$$\theta_3 = \tan^{-1} \frac{x_2}{x_1},$$

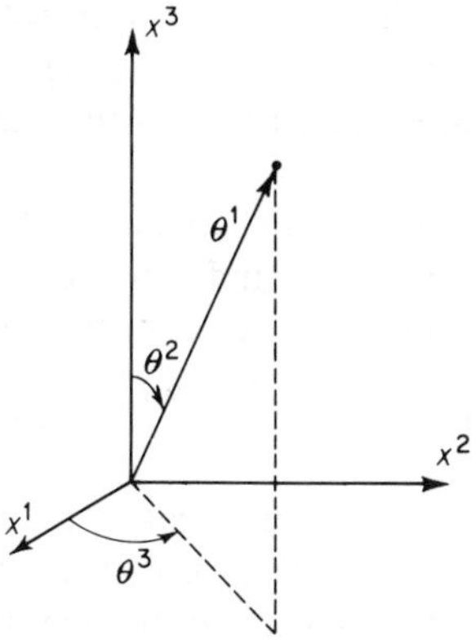

**P2.4**

and the inverse transformation is

$$x_1 = \theta_1 \sin \theta_2 \cos \theta_3,$$

$$x_2 = \theta_1 \sin \theta_2 \sin \theta_3,$$

$$x_3 = \theta_1 \cos \theta_2.$$

Show that the components of the Euclidean metric tensor in the spherical polar coordinates are

$$g_{11} = 1, \qquad g_{22} = (\theta_1)^2, \qquad g_{33} = (\theta_1)^2(\sin \theta_2)^2$$

and all other $g_{ij} = 0$. The square of the line element is, therefore,

$$ds^2 = (d\theta_1)^2 + (\theta_1)^2(d\theta_2)^2 + (\theta_1)^2(\sin \theta_2)^2(d\theta_3)^2.$$

**2.5.** Show that the length of the line element $(d\theta^1, 0, 0)$ is $\sqrt{g_{11}}\,|d\theta^1|$; that of the line element $(0, d\theta^2, 0)$ is $\sqrt{g_{22}}\,|d\theta^2|$.

**2.6.** Let the angle between the line elements $(d\theta^1, 0, 0)$ and $(0, d\theta^2, 0)$ be denoted by $\alpha_{12}$. Show that

$$\cos \alpha_{12} = \frac{g_{12}}{\sqrt{g_{11}}\sqrt{g_{22}}}.$$

*Hint:* Find the components of the line elements with respect to a system of rectangular Cartesian coordinates in which we know how to compute the angle between two vectors.

## 2.4. SCALARS, CONTRAVARIANT VECTORS, COVARIANT VECTORS†

In nonrelativistic physics there are quantities like mass and length which are independent of reference coordinates, and there are quantities like displacement and velocity whose components do depend on reference coordinates. The former are the scalars; the latter, vectors. Mathematically, we define them according to the way their components change under admissible transformations. In the following discussions we consider a system whose components are defined in the general set of variables $\theta^i$ and are

† One may pass over the rest of this chapter and proceed directly to Chap. 3 on first reading.

functions of $\theta^1, \theta^2, \theta^3$. If the variables $\theta^i$ can be changed to $\bar{\theta}^i$ by an admissible and proper transformation, then we can define new components of the system in the new variables $\bar{\theta}^i$. The system will be given various names according to the way in which the new and the old components are related.

A system is called a *scalar* if it has only a single component $\phi$ in the variables $\theta^i$ and a single component $\bar{\phi}$ in the variables $\bar{\theta}^i$, and if $\phi$ and $\bar{\phi}$ are numerically equal at the corresponding points,

$$\phi(\theta^1, \theta^2, \theta^3) = \bar{\phi}(\bar{\theta}^1, \bar{\theta}^2, \bar{\theta}^3). \tag{1}$$

A system is called a *contravariant vector field* or a *contravariant tensor field of rank one*, if it has three components $\xi^i$ in the variables $\theta^i$ and three components $\bar{\xi}^i$ in the variables $\bar{\theta}^i$, and if the components are related by the characteristic law

(2) ▲
$$\bar{\xi}^i(\bar{\theta}^1, \bar{\theta}^2, \bar{\theta}^3) = \xi^k(\theta^1, \theta^2, \theta^3)\frac{\partial \bar{\theta}^i}{\partial \theta^k}.$$

A contravariant vector is indicated by a superscript, called a *contravariant index*.

A differential $d\theta^i$ is a prototype of a contravariant vector:

$$d\bar{\theta}^i = \frac{\partial \bar{\theta}^i}{\partial \theta^k}\, d\theta^k. \tag{3}$$

Hence, a vector is contravariant if it transforms like a differential.†

A system is called a *covariant vector field*, or a *covariant tensor field of rank one*, if it has three components $\eta_i$ in the variables $\theta^1, \theta^2, \theta^3$ and three components $\bar{\eta}_i$ in the variables $\bar{\theta}^1, \bar{\theta}^2, \bar{\theta}^3$, and if the components in these two coordinate systems are related by the characteristic law

(4) ▲
$$\bar{\eta}_i(\bar{\theta}^1, \bar{\theta}^2, \bar{\theta}^3) = \eta_k(\theta^1, \theta^2, \theta^3)\frac{\partial \theta^k}{\partial \bar{\theta}^i}.$$

A covariant vector is indicated by a subscript, called a *covariant index*.

A gradient of a scalar potential $\phi$ transforms like (4),

$$\frac{\partial \phi}{\partial \bar{\theta}^i} = \frac{\partial \phi}{\partial \theta^k}\frac{\partial \theta^k}{\partial \bar{\theta}^i}. \tag{5}$$

Hence, a vector is covariant if it transforms like a gradient of a scalar potential.

† The variables $\theta^i$ and $\bar{\theta}^i$, in general, are not so related [see Eq. (2.3:9)]. Thus, although the differential $d\theta^i$ is a contravariant vector, the set of variables $\theta^i$ itself does not transform like a vector. Hence, in this instance, the position of the index of $\theta^i$ must be regarded as without significance.

## 2.5. TENSOR FIELDS OF HIGHER RANK

A system such as the three-dimensional Euclidean metric tensor $g_{ij}$ has nine components when $i$ and $j$ range over 1, 2, 3. Such quantities are given special names when their components in any two coordinate systems are related by specific transformation laws.

*Covariant tensor field of rank two, $t_{ij}$:*

$$\bar{t}_{ij}(\bar{\theta}^1, \bar{\theta}^2, \bar{\theta}^3) = t_{mn}(\theta^1, \theta^2, \theta^3)\,\frac{\partial \theta^m}{\partial \bar{\theta}^i}\,\frac{\partial \theta^n}{\partial \bar{\theta}^j}. \tag{1}$$

*Contravariant tensor field of rank two, $t^{ij}$:*

$$\bar{t}^{ij}(\bar{\theta}^1, \bar{\theta}^2, \bar{\theta}^3) = t^{mn}(\theta^1, \theta^2, \theta^3)\,\frac{\partial \bar{\theta}^i}{\partial \theta^m}\,\frac{\partial \bar{\theta}^j}{\partial \theta^n}. \tag{2}$$

*Mixed tensor field of rank two, $t^i_j$:*

$$\bar{t}^i_j(\bar{\theta}^1, \bar{\theta}^2, \bar{\theta}^3) = t^m_n(\theta^1, \theta^2, \theta^3)\,\frac{\partial \bar{\theta}^i}{\partial \theta^m}\,\frac{\partial \theta^n}{\partial \bar{\theta}^j}. \tag{3}$$

Generalization to tensor fields of higher ranks is immediate. Thus, we call a quantity $t^{\alpha_1 \cdots \alpha_p}_{\beta_1 \cdots \beta_q}$ a *tensor field of rank $r = p + q$, contravariant of rank $p$ and covariant of rank $q$*, if the components in any two coordinate systems are related by

$$\bar{t}^{\alpha_1 \cdots \alpha_p}_{\beta_1 \cdots \beta_q} = \frac{\partial \bar{\theta}^{\alpha_1}}{\partial \theta^{k_1}} \cdots \frac{\partial \bar{\theta}^{\alpha_p}}{\partial \theta^{k_p}} \cdot \frac{\partial \theta^{m_1}}{\partial \bar{\theta}^{\beta_1}} \cdots \frac{\partial \theta^{m_q}}{\partial \bar{\theta}^{\beta_q}}\, t^{k_1 \cdots k_p}_{m_1 \cdots m_q}. \tag{4}$$

Thus, the location of an index is important in telling whether it is contravariant or covariant. Again, if only *rectangular Cartesian coordinates* are considered, the distinction disappears.

These definitions can be generalized in an obvious manner if the range of the indices are 1, 2, . . . , $n$.

## PROBLEMS

**2.7.** Show that, *if all components of a tensor vanish in one coordinate system, then they vanish in all other coordinate systems which are in one-to-one correspondence with the given system.*

This is perhaps the most important property of tensor fields.

**2.8.** Prove the theorem: *The sum or difference of two tensors of the same type and rank (with the same number of covariant and the same number of contravariant indices) is again a tensor of the same type and rank.*

Thus, any linear combination of tensors of the same type and rank is again a tensor of the same type and rank.

**2.9.** *Theorem. Let* $A^{\beta_1\ldots\beta_s}_{\alpha_1\ldots\alpha_r}$, $B^{\beta_1\ldots\beta_s}_{\alpha_1\ldots\alpha_r}$ *be tensors. The equation*

$$A^{\beta_1\ldots\beta_s}_{\alpha_1\ldots\alpha_r}(\theta^1, \theta^2, \ldots, \theta^n) = B^{\beta_1\ldots\beta_s}_{\alpha_1\ldots\alpha_r}(\theta^1, \theta^2, \ldots, \theta^n)$$

*is a tensor equation;* i.e., *if this equation is true in some coordinate system, then it is true in all coordinate systems which are in one-to-one correspondence with each other.*

*Hint:* Use the results of the previous problems.

## 2.6. SOME IMPORTANT SPECIAL TENSORS

If we define the Kronecker delta and the permutation symbols introduced in Sec. 2.1 as components of covariant, contravariant, and mixed tensors of rank 2 and 3 in rectangular Cartesian coordinates $x_1, x_2, x_3$,

(1)
$$\delta_{ij} = \delta^{ij} = \delta^i_j = \delta^j_i \begin{cases} = 0 & \text{when } i \neq j, \\ = 1 & \text{when } i = j, j \text{ not summed,} \end{cases}$$

(2)
$$e_{ijk} = e^{ijk} \begin{cases} = 0 & \text{when any two indices are equal,} \\ = 1 & \text{when } i, j, k \text{ permute like 1, 2, 3,} \\ = -1 & \text{when } i, j, k \text{ permute like 1, 3, 2,} \end{cases}$$

what will be their components in general coordinates $\theta^i$? The answer is provided immediately by the tensor transformation laws. Thus,

(3)
$$g_{ij} = \frac{\partial x^m}{\partial \theta^i}\frac{\partial x^n}{\partial \theta^j}\delta_{mn} = \frac{\partial x^m}{\partial \theta^i}\frac{\partial x^m}{\partial \theta^j},$$

(4)
$$g^{ij} = \frac{\partial \theta^i}{\partial x^m}\frac{\partial \theta^j}{\partial x^n}\delta^{mn} = \frac{\partial \theta^i}{\partial x^m}\frac{\partial \theta^j}{\partial x^m},$$

(5)
$$g^i_j = \frac{\partial \theta^i}{\partial x^m}\frac{\partial x^n}{\partial \theta^j}\delta^m_n = \frac{\partial \theta^i}{\partial x^m}\frac{\partial x^m}{\partial \theta^j} = \delta^i_j,$$

(6)
$$\epsilon_{ijk} = \frac{\partial x^r}{\partial \theta^i}\frac{\partial x^s}{\partial \theta^j}\frac{\partial x^t}{\partial \theta^k} e_{rst} = e_{ijk}\left|\frac{\partial x^m}{\partial \theta^n}\right| = e_{ijk}\sqrt{g},$$

(7)
$$\epsilon^{ijk} = \frac{\partial \theta^i}{\partial x^r}\frac{\partial \theta^j}{\partial x^s}\frac{\partial \theta^k}{\partial x^t} e^{rst} = e^{ijk}\left|\frac{\partial \theta^m}{\partial x^n}\right| = \frac{e^{ijk}}{\sqrt{g}},$$

where $g$ is the value of the determinant $|g_{ij}|$ and is positive for any proper coordinate system.

(8)
$$g = |g_{ij}| > 0.$$

We see that *the proper generalizations of the Kronecker delta are the Euclidean metric tensors, and those of the permutation symbol are* $\epsilon_{ijk}$ *and* $\epsilon^{ijk}$, *which are called permutation tensors* or *alternators*. (See, however, Notes 2.2, p. 475.)

Note that the mixed tensor $g^i_j$ is identical with $\delta^i_j$ and is constant in all coordinate systems. From (3) and (4) we see that

$$g_{im}g^{mj} = \delta^j_{i}. \tag{9}$$

Hence, the determinant

$$|g_{im}g^{mj}| = |g_{ij}| \cdot |g^{ij}| = |\delta^i_j| = 1.$$

Using (3) and (4), we can write

$$\begin{aligned} g = |g_{ij}| &= \left|\frac{\partial x^i}{\partial \theta^j}\right|^2, \\ \frac{1}{g} = |g^{ij}| &= \left|\frac{\partial \theta^i}{\partial x^j}\right|^2. \end{aligned} \tag{10}$$

Since $g$ is positive, Eq. (9) may be solved to give

$$g^{ij} = \frac{D^{ij}}{g}, \tag{11}$$

where $D^{ij}$ is the cofactor of the term $g_{ij}$ in the determinant $g$. *The tensor $g^{ij}$ is called the associated metric tensor. It is as important as the metric tensor itself in the further development of tensor analysis.*

**Problem 2.10.** Prove Eqs. (6), (7), and (10). Write down explicitly $D^{ij}$ in Eq. (11) in terms of $g_{11}, g_{12}, \ldots, g_{33}$.

## 2.7. THE SIGNIFICANCE OF TENSOR CHARACTERISTICS

The importance of tensor analysis may be summarized in the following statement. The form of an equation can have general validity with respect to any frame of reference only if every term in the equation has the same tensor characteristics. If this condition is not satisfied, a simple change of the system of reference will destroy the form of the relationship. This form is, therefore, only fortuitous and accidental.

Thus, tensor analysis is as important as dimensional analysis in any formulation of physical relations. Dimensional analysis considers how a physical quantity changes with the particular choice of fundamental units. Two physical quantities cannot be equal unless they have the same dimensions; any physical equation cannot be correct unless it is invariant with respect to change of fundamental units.

Whether a physical quantity should be a tensor or not is a decision for the physicist to make. Why is a force a vector, a stress tensor a tensor? Because we say so! It is our judgement that endowing tensorial character to these quantities is in harmony with the world.

Once we decided upon the tensorial character of a physical quantity, we may take as the components of a tensor field in a given frame of reference any set of functions of the requisite number. A tensor field thus assigned in a given frame of reference then transforms according to the tensor transformation law when admissible transformations are considered. In other words, once the values of the components of a tensor are assigned in one particular coordinate system, the values of the components in any general coordinate system are fixed.

Why are the tensor transformation laws in harmony with physics? Because tensor analysis is designed so. For example, a tensor of rank one is defined in accordance with the physical idea of a vector. The only point new to the student is perhaps the distinction between contravariance and covariance. In elementary physics, natural laws are studied usually only in rectangular Cartesian coordinates of reference, in which the distinction between the contravariance and covariance disappears. When curvilinear coordinates are used in elementary physics, the vectorial components must be defined specifically in each particular case, and mathematical expressions of physical laws must be derived anew for each particular coordinate system. These derivations are usually quite tedious. Now, what is achieved by the definition of a tensor is a unified treatment, good for any curvilinear coordinates, orthogonal or nonorthogonal. This simplicity is obtained, however, at the expense of recognizing the distinction between contravariance and covariance.

In Sec. 2.14 we discuss the geometric interpretation of the tensor components of a vector in curvilinear coordinates. It will become clear that each physical vector has two tensor images: one contravariant and one covariant, depending on how the components are resolved.

## 2.8. CARTESIAN TENSORS

We have seen that it is necessary to distinguish the contravariant and covariant tensor transformation laws. However, *if only transformations between rectangular Cartesian coordinate systems are considered, the distinction between contravariance and covariance disappears*. To show this, let $x_1, x_2, x_3$ and $\bar{x}_1, \bar{x}_2, \bar{x}_3$ be two sets of rectangular Cartesian coordinates of reference. The transformation law must be simply

$$\bar{x}_i = \beta_{ij}x_j + a_i, \tag{1}$$

where $a_1, a_2, a_3$ are constants and $\beta_{ij}$ are the direction cosines of the angles between unit vectors along the coordinate axes $\bar{x}_i$ and $x_j$. Thus,

$$\beta_{21} = \cos(\bar{x}_2, x_1), \tag{2}$$

etc. The inverse transform is

$$x_i = \beta_{ji}\bar{x}_j + b_i. \tag{3}$$

Hence,

$$\frac{\partial x_k}{\partial \bar{x}_i} = \beta_{ik} = \frac{\partial \bar{x}_i}{\partial x_k}, \tag{4}$$

and the distinction between the characteristic transformation laws (2.4:2) and (2.4:4) disappears.

*When only rectangular Cartesian coordinates are considered, we shall write all indices as subscripts. This convenient practice will be followed throughout this book.*

## 2.9. CONTRACTION

We shall now consider some operations on tensors that generate new tensors.

Let $A^i_{jkl}$ be a mixed tensor so that, in a transformation from the coordinates $x^\alpha$ to $\bar{x}^\alpha$ ($\alpha = 1, 2, \ldots, n$), we obtain

$$\bar{A}^i_{jkl}(\bar{x}) = \frac{\partial \bar{x}^i}{\partial x^\alpha}\frac{\partial x^\beta}{\partial \bar{x}^j}\frac{\partial x^\gamma}{\partial \bar{x}^k}\frac{\partial x^\delta}{\partial \bar{x}^l} A^\alpha_{\beta\gamma\delta}(x).$$

If we equate the indices $i$ and $k$ and sum, we obtain the set of quantities

$$\begin{aligned}\bar{A}^i_{jil}(\bar{x}) &= \frac{\partial \bar{x}^i}{\partial x^\alpha}\frac{\partial x^\beta}{\partial \bar{x}^j}\frac{\partial x^\gamma}{\partial \bar{x}^i}\frac{\partial x^\delta}{\partial \bar{x}^l} A^\alpha_{\beta\gamma\delta}(x) \\ &= \frac{\partial x^\beta}{\partial \bar{x}^j}\frac{\partial x^\delta}{\partial \bar{x}^l}\delta^\gamma_\alpha A^\alpha_{\beta\gamma\delta}(x) \\ &= \frac{\partial x^\beta}{\partial \bar{x}^j}\frac{\partial x^\delta}{\partial \bar{x}^l} A^\alpha_{\beta\alpha\delta}(x).\end{aligned}$$

Let us write $A^\alpha_{\beta\alpha\delta}$ as $B_{\beta\delta}$. Then the equation above shows that

$$\bar{B}_{jl} = \frac{\partial x^\beta}{\partial \bar{x}^j}\frac{\partial x^\delta}{\partial \bar{x}^l} B_{\beta\delta}.$$

Hence, $B_{\beta\delta}$ satisfies the tensor transformation law and is therefore a tensor.

The process of equating and summing a covariant and a contravariant index of a mixed tensor is called a *contraction*. It is easy to see that the example above can be generalized to mixed tensors of other ranks. *The result of a contraction is another tensor.* If, as a result of contraction, there is no free index left, the resulting quantity is a scalar.

The following problem shows that, in general, equating and summing two covariant indices or two contravariant indices does not yield a tensor of lower order and is not a proper contraction. However, when only Cartesian coordinates are considered, we write all indices as subscripts and we contract by equating two subscripts and summing.

**Problem 2.11.** If $A^i_j$ is a mixed tensor of rank two, show that $A^i_i$ is a scalar. If $A^{ij}$ is a contravariant tensor of rank two show that in general $A^{ii}$ is not an invariant. Similarly, if $A_{ij}$ is a covariant tensor of rank two, $A_{ii}$ is not, in general, an invariant.

## 2.10. QUOTIENT RULE

Consider a set of $n^3$ functions $A(111)$, $A(112)$, $A(123)$, etc., or $A(i, j, k)$ for short, with the indices $i, j, k$ each ranging over $1, 2, \ldots, n$. Although the set of functions $A(i, j, k)$ has the right number of components, we do not know whether it is a tensor or not. Now, suppose that we know something about the nature of the product of $A(i, j, k)$ with an arbitrary tensor. Then there is a theorem which enables us to establish whether $A(i, j, k)$ is a tensor without going to the trouble of determining the law of transformation directly.

For example, let $\xi^\alpha(x)$ be an arbitrary tensor of rank 1 (a vector). Let us suppose that the product $A(\alpha, j, k)\xi^\alpha$ (summation convention used over $\alpha$) is known to yield a tensor of the type $A^j_k(x)$,

$$A(\alpha, j, k)\xi^\alpha = A^j_k.$$

Then we can prove that $A(i, j, k)$ is a tensor of the type $A^j_{ik}(x)$.

The proof is very simple. Since $A(\alpha, j, k)\xi^\alpha$ is of type $A^j_k$, it is transformed into $\bar{x}$-coordinates as

$$\bar{A}(\alpha, j, k)\bar{\xi}^\alpha = \bar{A}^j_k = \frac{\partial \bar{x}^j}{\partial x^r}\frac{\partial x^s}{\partial \bar{x}^k} A^r_s$$

$$= \frac{\partial \bar{x}^j}{\partial x^r}\frac{\partial x^s}{\partial \bar{x}^k}[A(\beta, r, s)\xi^\beta].$$

But $\xi^\beta = (\partial x^\beta/\partial \bar{x}^\alpha)\bar{\xi}^\alpha$. Inserting this in the right-hand side of the equation above and transposing all terms to one side of the equation, we obtain

$$\left[\bar{A}(\alpha, j, k) - \frac{\partial \bar{x}^j}{\partial x^r}\frac{\partial x^s}{\partial x^k}\frac{\partial x^\beta}{\partial \bar{x}^\alpha} A(\beta, r, s)\right]\bar{\xi}^\alpha = 0.$$

Now $\bar{\xi}^\alpha$ is an arbitrary vector. Hence, the bracket must vanish and we have

$$\bar{A}(\alpha, j, k) = \frac{\partial \bar{x}^j}{\partial x^r}\frac{\partial x^s}{\partial x^k}\frac{\partial x^\beta}{\partial x^\alpha} A(\beta, r, s),$$

which is precisely the law of transformation of the tensor of the type $A^j_{ik}$.

The pattern of the example above can be generalized to prove the theorem that, if $[A(i_1, i_2, \ldots, i_r)]$ *is a set of functions of the variables* $x^i$, *and if the product* $A(\alpha, i_2, \ldots, i_r)\xi^\alpha$ *with an arbitrary vector* $\xi^\alpha$ *be a tensor of the type* $A^{j_1 \ldots j_q}_{k_1 \ldots k_p}(x)$, *then the set* $A(i_1, i_2, \ldots i_r)$ *represents a tensor of the type* $A^{j_1 \ldots j_q}_{\alpha k_1 \ldots k_p}(x)$.

*Similarly, if the product of a set of $n^2$ functions $A(\alpha, j)$ with an arbitrary tensor $B_{\alpha k}$ (and summed over $\alpha$) is a covariant tensor of rank 2, then $A(i, j)$ represents a tensor of the type $A^i_j$.*

These and similar theorems that can be derived are called *quotient rules.* Numerous applications of these rules follow. See, for example, pp. 64, 85, 91.

## 2.11. PARTIAL DERIVATIVES IN CARTESIAN COORDINATES

The generation of a tensor of rank one from a tensor of rank zero by differentiation, such as the gradient of a scalar potential, indicates a way of generating tensors of higher rank. But, in general, the set of partial derivatives of a tensor does not behave like a tensor field. However, *if only Cartesian coordinates are considered, then the partial derivatives of any tensor field behave like the components of a tensor field under a transformation from Cartesian coordinates to Cartesian coordinates.* To show this, let us consider two Cartesian coordinates $(x_1, x_2, x_3)$ and $(\bar{x}_1, \bar{x}_2, \bar{x}_3)$ related by

$$\bar{x}_i = a_{ij}x_j + b_i, \tag{1}$$

where $a_{ij}$ and $b_i$ are constants. From (1), we have

$$\frac{\partial \bar{x}_i}{\partial x_j} = a_{ij}, \tag{2}$$

$$\frac{\partial^2 \bar{x}_i}{\partial x_j \, \partial x_k} = 0. \tag{3}$$

Now, if $\xi^i(x_1, x_2, x_3)$ is a contravariant tensor, so that

$$\bar{\xi}^i(\bar{x}_1, \bar{x}_2, \bar{x}_3) = \xi^\alpha(x_1, x_2, x_3) \frac{\partial \bar{x}_i}{\partial x_\alpha}. \tag{4}$$

Then, on differentiating both sides of the equation, one obtains

$$\frac{\partial \bar{\xi}^i}{\partial \bar{x}^j} = \left(\frac{\partial \xi^\alpha}{\partial x_\beta} \frac{\partial x_\beta}{\partial \bar{x}_j}\right) \frac{\partial \bar{x}_i}{\partial x_\alpha} + \xi^\alpha \left(\frac{\partial^2 \bar{x}_i}{\partial x_\alpha \, \partial x_\beta}\right) \frac{\partial x_\beta}{\partial \bar{x}_j}. \tag{5}$$

When $x_i$ and $\bar{x}_i$ are Cartesian coordinates, the last term vanishes according to (3). Hence,

$$\frac{\partial \bar{\xi}^i}{\partial \bar{x}_j} = \frac{\partial \xi^\alpha}{\partial x_\beta} \frac{\partial x_\beta}{\partial \bar{x}_j} \frac{\partial \bar{x}_i}{\partial x_\alpha}. \tag{6}$$

Thus, the set of partial derivatives $\partial \xi^\alpha / \partial x_\beta$ follows the transformation law for a mixed tensor of rank two under a transformation *from Cartesian coordinates to Cartesian coordinates.* However, the presence of the second derivative terms in (5) which does not vanish in curvilinear coordinates shows that $\partial\xi^\alpha/\partial x_\beta$ are not really the components of a tensor field in general coordinates. A similar situation holds obviously also for tensor fields of higher ranks.

In the following text, we shall often employ Cartesian coordinates to derive general results. *We shall use a comma to denote partial differentiation.* Thus,

$$\xi_{i,j} = \frac{\partial \xi_i}{\partial x_j}, \qquad \phi_{,i} = \frac{\partial \phi}{\partial x_i}, \qquad \sigma_{ij,k} = \frac{\partial \sigma_{ij}}{\partial x_k}.$$

When we restrict ourselves to Cartesian coordinates, $\phi_{,i}$, $\xi_{i,j}$, $\sigma_{ij,k}$ are tensors of rank one, two, three, respectively, provided that $\phi$, $\xi_i$, $\sigma_{ij}$ are tensors.

## 2.12. COVARIANT DIFFERENTIATION OF VECTOR FIELDS

The generalization of the concept of partial derivatives to the concept of *covariant derivative*, so that the covariant derivative of a tensor field is another tensor field, is the most important milestone in the development of tensor calculus. It is natural to search for such an extension in the form of a correction term that depends on the vector itself. Thus, if $\xi^i$ is a vector, we might expect the combination

$$\frac{\partial \xi^i}{\partial x^j} + \Gamma(i, j, \alpha)\xi^\alpha$$

to be a tensor. Here the suggested correction is linear function of $\xi^i$, and $\Gamma(i, j, \alpha)$ is some function with three indices. The success of this scheme hinges on the Euclidean Christoffel symbols, which are certain linear combinations of the derivatives of the metric tensor $g_{ij}$. This subject is beautiful, and the results are powerful in handling curvilinear coordinates. However, since the topic is not absolutely necessary for the development of solid mechanics, we shall not discuss it in detail, but merely outline below some of the salient results.

We discussed in Sec. 2.3 the metric tensor $g_{ij}$ in a set of general coordinates $(x^1, x^2, x^3)$, and in Sec. 2.6 the associated metric tensor $g^{ij}$. By means of these metric tensors, the *Euclidean Christoffel symbols* $\Gamma^i_{\alpha\beta}(x^1, x^2, x^3)$ *are defined as follows:*

(1) $$\Gamma^i_{\alpha\beta}(x^1, x^2, x^3) = \tfrac{1}{2} g^{i\sigma}\left(\frac{\partial g_{\sigma\beta}}{\partial x^\alpha} + \frac{\partial g_{\alpha\sigma}}{\partial x^\beta} - \frac{\partial g_{\alpha\beta}}{\partial x^\sigma}\right).$$

The $\Gamma^i_{\alpha\beta}$ is not a tensor. It transforms under a coordinate transformation $\bar{x}^i = f^i(x^1, x^2, x^3)$ as follows (see Prob. 2.24, p. 55)

(2) $$\bar{\Gamma}^i_{\alpha\beta}(\bar{x}^1, \bar{x}^2, \bar{x}^3) = \Gamma^\lambda_{\mu\nu}(x^1, x^2, x^3)\frac{\partial x^\mu}{\partial \bar{x}^\alpha}\frac{\partial x^\nu}{\partial \bar{x}^\beta}\frac{\partial \bar{x}^i}{\partial x^\lambda} + \frac{\partial^2 x^\lambda}{\partial \bar{x}^\alpha\,\partial \bar{x}^\beta}\frac{\partial \bar{x}^i}{\partial x^\lambda}.$$

This equation can be solved for $\partial^2 x^\lambda/\partial\bar{x}^\alpha\,\partial\bar{x}^\beta$ by multiplying (2) with $\partial x^m/\partial\bar{x}^i$ and sum over $i$ to obtain

$$(3)\qquad \frac{\partial^2 x^\lambda}{\partial\bar{x}^\alpha\,\partial\bar{x}^\beta} = \bar{\Gamma}^i_{\alpha\beta}(\bar{x})\,\frac{\partial x^\lambda}{\partial\bar{x}^i} - \Gamma^\lambda_{\mu\nu}(x)\,\frac{\partial x^\mu}{\partial\bar{x}^\alpha}\frac{\partial x^\nu}{\partial\bar{x}^\beta}\,.$$

Interchanging the roles of $x_i$ and $\bar{x}_i$ and with suitable changes in indices, we can substitute (3) into Eq. (2.11:5) to obtain

$$\frac{\partial\bar{\xi}^i}{\partial\bar{x}^\alpha} = \frac{\partial\xi^\lambda}{\partial x^\mu}\frac{\partial x^\mu}{\partial\bar{x}^\alpha}\frac{\partial\bar{x}^i}{\partial x^\lambda} + \xi^\lambda\frac{\partial x^\mu}{\partial\bar{x}^\alpha}\left[\Gamma^s_{\lambda\mu}(x)\,\frac{\partial\bar{x}^i}{\partial x^s} - \bar{\Gamma}^i_{mn}(\bar{x})\,\frac{\partial\bar{x}^m}{\partial x^\lambda}\frac{\partial\bar{x}^n}{\partial x^\mu}\right],$$

which can be reduced to

$$(4)\qquad \frac{\partial\bar{\xi}^i}{\partial\bar{x}^\alpha} + \bar{\Gamma}^i_{m\alpha}\bar{\xi}^m = \left(\frac{\partial\xi^\lambda}{\partial x^\mu} + \Gamma^\lambda_{s\mu}\xi^s\right)\frac{\partial x^\mu}{\partial\bar{x}^\alpha}\frac{\partial\bar{x}^i}{\partial x^\lambda}\,.$$

But this states that the functions $\partial\xi^\lambda/\partial x^\mu + \Gamma^\lambda_{s\mu}\xi^s$ are the components of a mixed tensor of rank two. Hence, the *functions*

$$(5)\ \blacktriangle\qquad \xi^i\big|_\alpha \equiv \frac{\partial\xi^i(x^1, x^2, x^3)}{\partial x^\alpha} + \Gamma^i_{\sigma\alpha}(x^1, {}_{\bullet}x^2, x^3)\xi^\sigma(x^1, x^2, x^3)$$

*are the components of a mixed tensor field of rank two, called the covariant derivative of the contravariant vector* $\xi^i$. We shall use *the notation* $\xi^i\big|_\alpha$ *for the covariant derivative of* $\xi^i$.

By a slight variation in the derivation, it can be shown that the *functions*

$$(6)\ \blacktriangle\qquad \xi_i\big|_\alpha \equiv \frac{\partial\xi_i}{\partial x^\alpha} - \Gamma^\sigma_{i\alpha}\xi_\sigma$$

*are the components of a covariant tensor field of rank two whenever* $\xi_i$ *are the components of a covariant vector field.* This is called the *covariant derivative of* $\xi_i$, and is denoted by $\xi_i\big|_\alpha$.

More generally, a long but quite straightforward calculation analogous to the above can be made to establish the *covariant derivative of a tensor* $T^{\alpha_1\cdots\alpha_p}_{\beta_1\cdots\beta_q}$ of rank $p+q$, contravariant of rank $p$, covariant of rank $q$:

$$T^{\alpha_1\ldots\alpha_p}_{\beta_1\ldots\beta_q}\Big|_\gamma = \frac{\partial T^{\alpha_1\ldots\alpha_p}_{\beta_1\ldots\beta_q}}{\partial x^\gamma} + \Gamma^{\alpha_1}_{\sigma\gamma}T^{\sigma\alpha_2\ldots\alpha_p}_{\beta_1\beta_2\ldots\beta_q} + \ldots$$
$$+ \Gamma^{\alpha_p}_{\sigma\gamma}Y^{\alpha_1\ldots\alpha_{p-1}\sigma}_{\beta_1\ldots\beta_{q-1}\beta_q} - \Gamma^\sigma_{\beta_1\gamma}T^{\alpha_1\alpha_2\ldots\alpha_p}_{\sigma\beta_2\ldots\beta_q} - \ldots - \Gamma^\sigma_{\beta_q\gamma}T^{\alpha_1\ldots\alpha_{p-1}\alpha_p}_{\beta_1\ldots\beta_{q-1}\sigma}$$

This derivative is contravariant of rank $p$, and covariant of rank $q+1$.

Since the components of the metric tensor $g_{ij}$ are constant in Cartesian coordinates, we see from Eq. (1) that the Euclidean Christoffel symbols are zero in Cartesian coordinates. The covariant derivative of a tensor field reduces to partial derivatives of the tensor field when the tensor field and the operations are evaluated in Cartesian coordinates.

## 2.13. TENSOR EQUATIONS

The theorems included in the problems at the end of Section 2.5 contain perhaps the most important property of tensor fields: *if all the components of a tensor field vanish in one coordinate system, they vanish likewise in all coordinate systems which can be obtained by admissible transformations.* Since the sum and difference of tensor fields of a given type are tensors of the same type, we deduce that *if a tensor equation can be established in one coordinate system, then it must hold for all coordinate systems obtained by admissible transformations.*

The last statement affords a powerful method for establishing equations in mathematical physics. For example, if a certain tensor relationship can be shown to be true in rectangular Cartesian coordinates, then it is also true in general curvilinear coordinates in Euclidean space. Thus, once an equation is established in rectangular Cartesian coordinates, the corresponding equation (stating a physical fact, such as a condition of equilibrium, of conservation of energy, etc.) in any specific curvilinear coordinates in Euclidean space can be obtained by a straightforward "translation" in the language of tensors.

As an example of the application of these remarks, let us consider the *successive covariant derivatives* of a tensor field. Since the covariant derivative of a tensor field is a tensor field, we can form the covariant derivative of the latter, which is called the *second covariant derivative of the original tensor field.* If $T^{ij}_{kl}$ denotes the original tensor field, we can consider the second covariant derivative $T^{ij}_{kl}|_{\gamma\delta}$. Now, if the space is Euclidean, then it can be described by a rectangular Cartesian coordinate system. In Cartesian coordinates, the Euclidean Christoffel symbols are all zero and the covariant derivatives of a tensor field reduce to partial derivatives of the tensor field. But partial derivatives are *commutative* if they are continuous. Hence, we see that the following equation is true in Cartesian coordinates if every component is continuous,

$$T^{ij}_{kl}|_{\gamma\delta} = T^{ij}_{kl}|_{\delta\gamma}, \tag{1}$$

and, therefore, it is true in all coordinates that can be obtained by admissible transformations from a Cartesian system.

We must remark that the commutativeness of the covariant differentiation operation is established above only in Euclidean space. A coordinate system in a more general Riemannian space may not be transformed into Cartesian coordinates and the method of proof used above cannot be applied. In fact, the theorem expressed in Eq. (1) is, in general, untrue in Riemannian space: successive covariant derivatives are, in general, not commutative in the Riemannian space. (See Prob. 2.31 below.)

As a second application we can prove the following theorem: *the covariant derivatives of the Euclidean metric tensor* $g_{ij}$ *and the associated contravariant tensor* $g^{ij}$ *are zero:*

(2) $$g_{ij}|_k = 0, \qquad g^{ij}|_k = 0.$$

Since $g_{ij}|_k$ and $g^{ij}|_k$ are tensors, the truth of the theorem can be established if we can demonstrate Eq. (2) in one particular coordinate system. But this is exactly the case in Cartesian coordinates, in which $g_{ij}$ and $g^{ij}$ are constants and, hence, their derivatives vanish. Thus, the proof is completed. In contrast to Eq. (1), however, it can be proved that Eqs. (2) remain true in Riemannian space.

Further examples are furnished in Probs. 2.17 to 2.23. See also Notes 2.3, p. 475.

## 2.14. GEOMETRIC INTERPRETATION OF TENSOR COMPONENTS

Before we conclude this chapter we shall consider briefly the geometric interpretation of tensor components. For this purpose we must use the concept of *base vectors*. We know that in a three-dimensional Euclidean space any three linearly independent vectors form a *basis* with which any other vectors can be expanded as a linear combination of these three vectors. When a rectangular Cartesian frame of reference is chosen, we can choose as base vectors the unit vectors $\mathbf{i}_1, \mathbf{i}_2, \mathbf{i}_3$ parallel to the coordinate axes: Thus, if a vector $\mathbf{A}$ has three components $(A_1, A_2, A_3)$, we can write

$$\mathbf{A} = A_1\mathbf{i}_1 + A_2\mathbf{i}_2 + A_3\mathbf{i}_3.$$

In a curvilinear coordinate system in a Euclidean space, we shall introduce the base vectors from the following consideration.

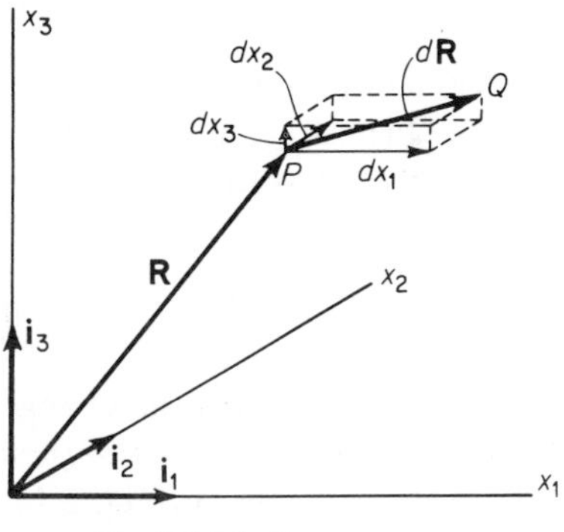


**Fig. 2.14:1.** Base vectors.

Let $d\mathbf{R}$ denote an infinitesimal vector $PQ$ joining a point $P = (x_1, x_2, x_3)$ to a point $Q = (x_1 + dx_1, x_2 + dx_2, x_3 + dx_3)$ where $x_1, x_2, x_3$ are referred to a rectangular Cartesian frame of reference. (See Figure 2.14:1.) Then, obviously,

(1) $$d\mathbf{R} = dx^r\,\mathbf{i}_r = dx_r\,\mathbf{i}^r,$$

where $\mathbf{i}_1, \mathbf{i}_2, \mathbf{i}_3$, or $\mathbf{i}^1, \mathbf{i}^2, \mathbf{i}^3$ denote the base vectors along coordinate axes. Here, since a rectangular Cartesian coordinate system is used, we can assign arbitrarily an index as contravariant or covariant. In the assignment chosen above, $dx^r$ and $dx_r$ are, respectively, contravariant and covariant differentials.

Now let us consider a transformation from the rectangular Cartesian coordinates $x^i$ to general coordinates $\theta^i$.

$$\theta^i = \theta^i(x_1, x_2, x_3). \tag{2}$$

According to the tensor transformation law, when $dx^r$ and $dx_r$ are regarded as tensors of order one, their components in the general coordinates become

$$d\theta^j = \frac{\partial \theta^j}{\partial x^i} dx^i, \qquad dx^i = \frac{\partial x^i}{\partial \theta^j} d\theta^j, \tag{3}$$

$$d\theta_j = \frac{\partial x^i}{\partial \theta^j} dx_i, \qquad dx_i = \frac{\partial \theta^j}{\partial x^i} d\theta_j. \tag{4}$$

Now in (3), $d\theta^j$ can be identified as the usual differential of the variable $\theta^j$, as specified in Eq. (2); hence, the superscript is justified. But in Eq. (4), $d\theta_j$ is not to be identified with the usual differential; $d\theta_j$ is a *covariant* differential that will have a different geometric meaning as will be seen later.

By (3) and (4), we may write Eq. (1) as

$$d\mathbf{R} = \mathbf{g}_r \, d\theta^r = \mathbf{g}^r \, d\theta_r, \tag{5 ▲}$$

where

$$\mathbf{g}_r = \frac{\partial x^s}{\partial \theta^r} \mathbf{i}_s, \qquad \mathbf{g}^r = \frac{\partial \theta^r}{\partial x^s} \mathbf{i}^s. \tag{6}$$

Since $\mathbf{g}_r$ and $\mathbf{g}^r$ are linear combinations of unit vectors, they are themselves vectors; they are known as the *covariant* and *contravariant base vectors*, respectively, or as the *base vectors* and *reciprocal base vectors*. Equation (5) shows that

$$\mathbf{g}_i = \frac{\partial \mathbf{R}}{\partial \theta^i}. \tag{7 ▲}$$

Hence $\mathbf{g}_i$ characterizes the change of the position vector $\mathbf{R}$ as $\theta^i$ varies. In other words, $\mathbf{g}_i$ *is directed tangentially along the coordinate curve* $\theta^i$. These vectors are illustrated in Fig. 2.14:2.

It is easily verified that

$$\mathbf{g}_r \cdot \mathbf{g}_s = g_{rs}, \qquad \mathbf{g}^r \cdot \mathbf{g}^s = g^{rs}, \tag{8 ▲}$$

$$\mathbf{g}^r \cdot \mathbf{g}_s = g^r_s = \delta^r_s, \tag{9 ▲}$$

$$\mathbf{g}^r = g^{rs}\mathbf{g}_s, \qquad \mathbf{g}_r = g_{rs}\mathbf{g}^s, \tag{10 ▲}$$

where $g_{rs}$ is the Euclidean metric tensor of the coordinates system, and $g^{rs}$ is the *associated*, or *conjugate metric tensor*. From (9) it is clear that *the contravariant base vectors* $\mathbf{g}^1, \mathbf{g}^2, \mathbf{g}^3$ *are, respectively, perpendicular to the planes of* $\mathbf{g}_2\,\mathbf{g}_3$, $\mathbf{g}_3\,\mathbf{g}_1$, $\mathbf{g}_1\,\mathbf{g}_2$. (See also Prob. 2.28.)

We can now deal with the contravariant and covariant components of a vector. Consider the expression $v^r\mathbf{g}_r$, where $v^r$ is a contravariant tensor of rank one, and $\mathbf{g}_r$ are the covariant base vectors at a generic point. This expression remains *invariant* under coordinate transformations, and, since it

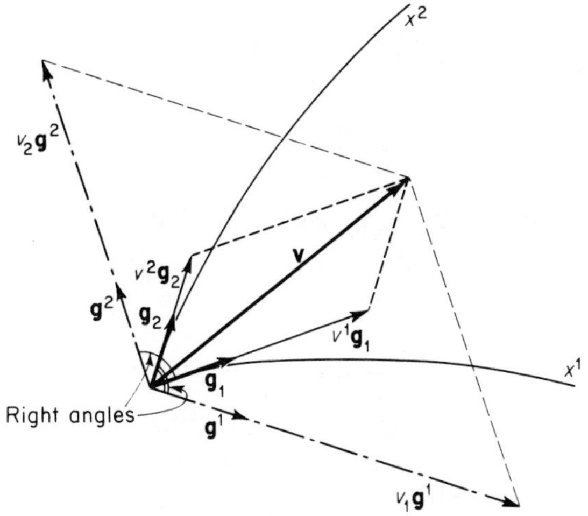


**Fig. 2.14:2.** Contravariant and covariant components of a vector $\mathbf{v}$ in two dimensions.

is a linear combination of the base vectors $\mathbf{g}_r$, it is a vector, which may be designated $\mathbf{v}$:

(11) $$\mathbf{v} = v^r\mathbf{g}_r.$$

By Eq. (10), replacing $\mathbf{g}_r$ by $g_{rs}\mathbf{g}^s$, we also have

(12) ▲ $$\mathbf{v} = v^r\mathbf{g}_r = v_s\mathbf{g}^s,$$

where

(13) ▲ $$v_s = g_{rs}v^r, \qquad v^r = g^{rs}v_s.$$

According to Eq. (12), if we represent $\mathbf{v}$ by a directed line, then the contravariant components $v^r$ are the components of $\mathbf{v}$ in the direction of the covariant base vectors, while the covariant components $v_r$ are the components of $\mathbf{v}$ in the direction of the contravariant base vectors. A two-dimensional illustration is shown in Fig. 2.14:2.

Equation (12) justifies naming $v^r$ the *contravariant components* of the vector $\mathbf{v}$, and $v_r$ the *covariant components* of $\mathbf{v}$. Thus, in our Euclidean space, the tensors $v^r$ and $v_r$ are two different representations of the same vector $\mathbf{v}$. Equations (13) establish the process of raising and lowering of indices.

## 2.15. GEOMETRIC INTERPRETATION OF COVARIANT DERIVATIVES

Consider now a vector field $\mathbf{v}$ defined at every point of space in a region $R$. Let the vector at the point $P(\theta^1, \theta^2, \theta^3)$ be

$$\mathbf{v}(P) = v^i(P)\mathbf{g}_i(P). \tag{1}$$

At a neighboring point $P'(\theta^1 + d\theta^1, \theta^2 + d\theta^2, \theta^3 + d\theta^3)$, the vector becomes

$$\begin{aligned}\mathbf{v}(P') &= \mathbf{v}(P) + d\mathbf{v}(P)\\ &= [v^i(P) + dv^i(P)][\mathbf{g}_i(P) + d\mathbf{g}_i(P)].\end{aligned}$$

On passing to the limit $d\theta^i \to 0$, we obtain the principal part of the difference

$$d\mathbf{v} = (v^i + dv^i)(\mathbf{g}_i + d\mathbf{g}_i) - v^i\mathbf{g}_i = \mathbf{g}_i dv^i + v^i d\mathbf{g}_i \tag{2}$$

and the derivative

(3) ▲
$$\frac{\partial \mathbf{v}}{\partial \theta^j} = \mathbf{g}_i \frac{\partial v^i}{\partial \theta^j} + \frac{\partial \mathbf{g}_i}{\partial \theta^j} v^i.$$

Thus the derivative of the vector $\mathbf{v}$ is resolved into two parts: one arising from the variation of the components $v^i$ as the coordinates $\theta^1, \theta^2, \theta^3$ are changed, the other arising from the change of the base vector $\mathbf{g}_i$ as the position of the point $\theta^i$ is changed. It can be shown (see Notes 2.1, p. 474) that, in a Euclidean space,

(4) ▲
$$\frac{\partial \mathbf{g}_i}{\partial \theta^j} = \Gamma^\alpha_{ij}\mathbf{g}_\alpha.$$

Hence

(5) ▲
$$\begin{aligned}\frac{\partial \mathbf{v}}{\partial \theta^j} &= \mathbf{g}_\alpha \frac{\partial v^\alpha}{\partial \theta^j} + v^i\Gamma^\alpha_{ij}\mathbf{g}_\alpha\\ &= \left(\frac{\partial v^\alpha}{\partial \theta^j} + v^i\Gamma^\alpha_{ij}\right)\mathbf{g}_\alpha = v^\alpha|_j \mathbf{g}_\alpha.\end{aligned}$$

*Thus, the covariant derivative $v^\alpha|_j$ represents the components of $\partial\mathbf{v}/\partial\theta^j$ referred to the base vectors $\mathbf{g}_\alpha$.*

## 2.16. PHYSICAL COMPONENTS OF A VECTOR

The base vectors $\mathbf{g}_r$ and $\mathbf{g}^r$ are in general not unit vectors. In fact, their lengths are

$$|\mathbf{g}_r| = \sqrt{g_{rr}}, \qquad |\mathbf{g}^r| = \sqrt{g^{rr}}, \qquad r \text{ not summed.}$$

Let us write Eq. (2.14:12) as

$$\mathbf{v} = \sum_{r=1}^{3} v^r\sqrt{g_{rr}}\,\frac{\mathbf{g}_r}{\sqrt{g_{rr}}} = \sum_{r=1}^{3} v_r\sqrt{g^{rr}}\,\frac{\mathbf{g}^r}{\sqrt{g^{rr}}}. \tag{1}$$

Then, since $\mathbf{g}_r/\sqrt{g_{rr}}$ and $\mathbf{g}^r/\sqrt{g^{rr}}$ are unit vectors, all components $v^r\sqrt{g_{rr}}$ and $v_r\sqrt{g^{rr}}$ ($r$ not summed) will have the same physical dimensions. It is seen that $v^r\sqrt{g_{rr}}$ are the components of $\mathbf{v}$ resolved in the direction of unit vectors $\mathbf{g}_r/\sqrt{g_{rr}}$ which are tangent to the coordinate lines; and that $v_r\sqrt{g^{rr}}$ are the components of $\mathbf{v}$ resolved in the direction of unit vectors $\mathbf{g}^r/\sqrt{g^{rr}}$ which are perpendicular to the coordinate planes. The components

$$v^r\sqrt{g_{rr}}, \qquad v_r\sqrt{g^{rr}}, \qquad\qquad r \text{ not summed},$$

are called the *physical components of the vector* $\mathbf{v}$. They do not transform according to the tensor transformation law and are not components of tensors.

We should remember that the tensor components of a physical quantity which is referred to a particular curvilinear coordinate system may or may not have the same physical dimensions. This difficulty (and it is also a great convenience!) arises because we would like to keep our freedom in choosing arbitrary curvilinear coordinates. Thus, in spherical polar coordinates for a Euclidean space, the position of a point is expressed by a length and two angles. In a four-dimensional space a point may be expressed in three lengths and a time. For this reason, we must distinguish the tensor components from the "physical components," which must have uniform physical dimensions.

## PROBLEMS

**2.12.** Show that in an orthogonal $n$-dimensional coordinates system, we have for each component $i = j$, $g^{ij} = 1/g_{ij}$.

**2.13.** If $a_{ij}$ is a tensor, and the components $a_{ij} = a_{ji}$, then the tensor $a_{ij}$ is called a *symmetric tensor*. If the components $a_{ij} = -a_{ji}$, then the tensor $a_{ij}$ is said to be *skew-symmetric*, or *antisymmetric*. *Show that the symmetry of a tensor with respect to two indices at the same level is conserved under coordinate transformations. Since* $a_{ij} = \frac{1}{2}(a_{ij} + a_{ji}) + \frac{1}{2}(a_{ij} - a_{ji})$, *any covariant (or contravariant) second-order tensor can be written as the sum of a symmetric and a skew-symmetric tensor.*

**2.14.** Show that the scalar product of a symmetric tensor $S^{ij}$ and a skew-symmetric tensor $W_{ij}$ vanishes indentically.

**2.15.** Show that the tensor $\omega_{ik} = e_{ijk}u_j$ is skew-symmetric, where $u_j$ is a vector.

**2.16.** Show that if $A_{jk}$ is a skew-symmetric Cartesian tensor, then the unique solution of the equation $\omega_i = \frac{1}{2}e_{ijk}A_{jk}$ is $A_{mn} = e_{mni}\omega_i$.

**2.17.** Let $\nabla$ be the operator

$$\nabla = \mathbf{g}^r \frac{\partial}{\partial \theta^r}.$$

Show that

$$\text{grad}\,\phi = \nabla\phi = \mathbf{g}^r \frac{\partial\phi}{\partial\theta^r},$$

$$\text{div}\,\mathbf{F} = \nabla\cdot\mathbf{F} = F^r|_r,$$

$$\text{curl}\,\mathbf{A} = \nabla\times\mathbf{A} = \epsilon^{rst}A_s|_r\mathbf{g}_t.$$

Show that these functions are invariant under coordinate transformations.

**2.18.** Let $g^{\alpha\beta}$ be the associated contravariant Euclidean metric tensor and $\psi(x^1, x^2, x^3)$ be a scalar field. Show that

(a) $g^{\alpha\beta}\psi|_{\alpha\beta}$ is a scalar field.

(b) In rectangular Cartesian coordinates $g_{\alpha\beta} = \delta_{\alpha\beta}$, the scalar $g^{\alpha\beta}\psi|_{\alpha\beta}$ reduces to the form (writing $x^1 = x, x^2 = y, x^3 = z$)

$$\frac{\partial^2\psi}{\partial x^2} + \frac{\partial^2\psi}{\partial y^2} + \frac{\partial^2\psi}{\partial z^2}.$$

(c) Hence, the Laplace equation in curvilinear coordinates with the scalar field $\psi(x^1, x^2, x^3)$ as unknown is given by

$$g^{\alpha\beta}(x^1, x^2, x^3)\,|\psi_{\alpha\beta}(x^1, x^2, x^3) = 0.$$

**2.19.** Let $y^1, y^2, y^3$ (or $x, y, z$) be rectangular Cartesian coordinates and $x^1, x^2, x^3$ (or $r, \phi, \theta$) be the spherical polar coordinates. Show that the Laplace equation in spherical polar coordinates, when the unknown function is a scalar field $\psi(r, \phi, \theta)$, is

$$\frac{\partial^2\psi}{\partial r^2} + \frac{1}{r^2}\frac{\partial^2\psi}{\partial\phi^2} + \frac{1}{r^2\sin^2\phi}\frac{\partial^2\psi}{\partial\theta^2} + \frac{2}{r}\frac{\partial\psi}{\partial r} + \frac{\cot\phi}{r^2}\frac{\partial\psi}{\partial\phi} = 0.$$

**2.20.** Let $\xi^i(y^1, y^2, y^3)$ be the components of an unknown vector in rectangular Cartesian coordinates. Let each component satisfy Laplace's equation in rectangular coordinates,

$$\frac{\partial\xi^i}{(\partial y^1)^2} + \frac{\partial^2\xi^i}{(\partial y^2)^2} + \frac{\partial^2\xi^i}{(\partial y^3)^2} = 0.$$

Show that a generalization of this equation is that $\xi^i(x^1, x^2, x^3)$ in curvilinear coordinates $x^1, x^2, x^3$ will satisfy the system of three differential equations

$$g^{\alpha\beta}\xi^i|_{\alpha\beta} = 0,$$

where $\xi^i|_{\alpha\beta}$ is the second covariant derivative of $\xi^i(x^1, x^2, x^3)$.

**2.21.** Prove that

$$F^r|_r = \frac{1}{\sqrt{g}}\frac{\partial(\sqrt{g}F^r)}{\partial x^r}.$$

*Hint:* Since $g = g_{i\alpha}G^{i\alpha}$ ($i$ not summed), $G^{ij}$ is the cofactor of the element $g_{ij}$ in $g = |g_{ij}|$, show that

$$\frac{\partial g}{\partial x^i} = g g^{\alpha\beta} \frac{\partial g_{\alpha\beta}}{\partial x^i}$$

and that

$$\frac{\partial g}{\partial x^i} = 2g\Gamma^{\alpha}_{\alpha i}.$$

Hence,

$$\Gamma^{\alpha}_{\alpha i} = \frac{\partial}{\partial x^i} \log \sqrt{g}.$$

**2.22.** Prove that the Laplacian of Prob. 2.18 can be written

$$g^{ij}\psi|_{ij} = \frac{1}{\sqrt{g}} \frac{\partial[\sqrt{g}g^{ij}(\partial\psi/\partial x^j)]}{\partial x^i}.$$

*Hint:* Use the results of Prob. 2.21.

**2.23.** Show that the covariant differentiation of sums and products follows the usual rules for partial differentiation. Thus,

$$\begin{aligned}
(\phi v^r)|_i &= \phi,_i v^r + \phi v^r|_i \\
(v^r v_r)|_i &= (v^r v_r),_i = v^r|_i v_r + v^r v_r|_i \\
(A_{ij}B^{mn})|_r &= A_{ij}|_r\, B^{mn} + A_{ij}B^{mn}|_r
\end{aligned}$$

where a comma indicates partial differentiation. Remember that the covariant derivatives of a scalar are the same as partial derivatives.

**2.24.** Derive the transformation law for the Euclidean Christoffel symbols $\Gamma^i_{\alpha\beta}(x^1, x^2, x^3)$. *Ans.* Under a transformation of coordinates from $x^i$ to $\bar{x}^i$, the Euclidean metric tensor transforms as

$$\bar{g}_{\sigma\beta}(\bar{x}^1, \bar{x}^2, \bar{x}^3) = g_{\mu\nu}(x^1, x^2, x^3) \frac{\partial x^\mu}{\partial \bar{x}^\sigma} \frac{\partial x^\nu}{\partial \bar{x}^\beta}.$$

Differentiating both sides of the equation with respect to $\bar{x}^\alpha$, considering the $\bar{x}^\alpha$ as the independent variable, we obtain

$$\frac{\partial \bar{g}_{\sigma\beta}}{\partial \bar{x}^\alpha} = \left(\frac{\partial g_{\mu\nu}}{\partial x^\rho} \frac{\partial x^\rho}{\partial \bar{x}^\alpha}\right) \frac{\partial x^\mu}{\partial \bar{x}^\sigma} \frac{\partial x^\nu}{\partial \bar{x}^\beta} + g_{\mu\nu} \frac{\partial^2 x^\mu}{\partial \bar{x}^\sigma\, \partial \bar{x}^\alpha} \frac{\partial x^\nu}{\partial \bar{x}^\beta} + g_{\mu\nu} \frac{\partial x^\mu}{\partial \bar{x}^\sigma} \frac{\partial^2 x^\nu}{\partial \bar{x}^\beta\, \partial \bar{x}^\alpha}.$$

On first interchanging $\alpha$ and $\sigma$ in this formula and then interchanging $\beta$ and $\alpha$, two similar expressions for $\partial \bar{g}_{\alpha\beta}/\partial \bar{x}^\sigma$ and $\partial \bar{g}_{\alpha\sigma}/\partial \bar{x}^\beta$ are obtained. Adding and subtracting, with several further interchange of dummy indices, and recalling $g_{\nu\mu} = g_{\mu\nu}$, we obtain

$$\frac{1}{2}\left(\frac{\partial \bar{g}_{\sigma\beta}}{\partial \bar{x}^\alpha} + \frac{\partial \bar{g}_{\alpha\sigma}}{\partial \bar{x}^\beta} - \frac{\partial \bar{g}_{\alpha\beta}}{\partial \bar{x}^\sigma}\right) = \frac{1}{2}\left(\frac{\partial g_{\rho\nu}}{\partial x^\mu} + \frac{\partial g_{\mu\rho}}{\partial x^\nu} - \frac{\partial g_{\mu\nu}}{\partial x^\rho}\right) \frac{\partial x^\mu}{\partial \bar{x}^\alpha} \frac{\partial x^\nu}{\partial \bar{x}^\beta} \frac{\partial x^\rho}{\partial \bar{x}^\sigma} + g_{\mu\nu} \frac{\partial x^\mu}{\partial \bar{x}^\sigma} \frac{\partial^2 x^\nu}{\partial \bar{x}^\beta\, \partial \bar{x}^\alpha}.$$

Now,

$$\bar{g}^{i\sigma} = g^{\lambda\omega} \frac{\partial \bar{x}^i}{\partial x^\lambda} \frac{\partial \bar{x}^\sigma}{\partial x^\omega}.$$

A multiplication of the corresponding sides of the last two equations leads to the desired result.

**2.25.** All the results obtained above can be applied to two-dimensional spaces by assigning the range of indices to be 1 to 2 instead of 1 to 3. Compute the Euclidean Christoffel symbols for the plane polar coordinates ($x^1 = r, x^2 = \theta$)
*Ans.* $\Gamma^1_{22} = -x^1$, $\Gamma^2_{12} = \Gamma^2_{21} = 1/x^1$, all other components $= 0$.

**2.26.** Show that $\Gamma^r_{mn}$ is symmetric in $m$ and $n$; i.e., $\Gamma^r_{mn} = \Gamma^r_{nm}$.

**2.27.** Show that the necessary and sufficient condition that a given curvilinear coordinate system be orthogonal is that $g_{ij} = 0$, if $i \neq j$, throughout the domain.

**2.28.** Prove that $\mathbf{g}_r \times \mathbf{g}_s = \epsilon_{rst}\mathbf{g}^t$, $\mathbf{g}^r \times \mathbf{g}^s = \epsilon^{rst}\mathbf{g}_t$, where $\epsilon_{rst}$, $\epsilon^{rst}$ are the permutation tensor of Sec. 2.6. Hence, if we denote the scalar product of the vectors $\mathbf{g}_1$ and $\mathbf{g}_2 \times \mathbf{g}_3$ by $[\mathbf{g}_1\mathbf{g}_2\mathbf{g}_3]$ or $(\mathbf{g}_1, \mathbf{g}_2 \times \mathbf{g}_3)$, we have

$$[\mathbf{g}_1\mathbf{g}_2\mathbf{g}_3] = (\mathbf{g}_1, \mathbf{g}_2 \times \mathbf{g}_3) = \sqrt{g}, \qquad [\mathbf{g}^1\mathbf{g}^2\mathbf{g}^3] = (\mathbf{g}^1, \mathbf{g}^2 \times \mathbf{g}^3) = 1/\sqrt{g}.$$

**2.29.** The element of area of a parallelogram with two adjacent edges $d\mathbf{s}_2 = \mathbf{g}_2\, d\theta^2$ and $d\mathbf{s}_3 = \mathbf{g}_3\, d\theta^3$ is

$$dS_1 = |d\mathbf{s}_2 \times d\mathbf{s}_3| = |\mathbf{g}_2 \times \mathbf{g}_3|\, d\theta^2\, d\theta^3.$$

Show that $dS_1 = \sqrt{(gg^{11})}\, d\theta^2\, d\theta^3$. In general, the element of area $dS_i$ of a parallelogram formed by the elements $\mathbf{g}_j\, d\theta^j$ and $\mathbf{g}_k\, d\theta^k$ on the $\theta_i$-surface is $dS_i = \sqrt{(gg^{ii})}\, d\theta^j\, d\theta^k$ ($i$ not summed, $i \neq j \neq k$).

**2.30.** With reference to Prob. 2.28, show that the volume element

$$dV = d\mathbf{s}_1 \cdot d\mathbf{s}_2 \times d\mathbf{s}_3 = [\mathbf{g}_1\mathbf{g}_2\mathbf{g}_3]\, d\theta^1\, d\theta^2\, d\theta^3 = \sqrt{g}\, d\theta^1\, d\theta^2\, d\theta^3.$$

**2.31.** If the space is non-Euclidean, we cannot find a coordinate system in which the metric tensor $g_{ij}$ has constant components everywhere throughout the space. (In a Euclidean space there do exist just such coordinate systems, namely, Cartesian coordinate systems.) In this case, we must compute the successive derivatives $\xi_r|_{st}$ and $\xi_r|_{ts}$ according to the covariant differentiation rules. Show that

$$\xi^i|_{st} - \xi^i|_{ts} = R^i_{pst}\xi^p,$$

where

$$R^i_{pst} = \frac{\partial \Gamma^i_{ps}}{\partial x^t} - \frac{\partial \Gamma^i_{pt}}{\partial x^s} + \Gamma^r_{ps}\Gamma^i_{rt} - \Gamma^r_{pt}\Gamma^i_{rs}.$$

Show that $R^i_{pst}$ is a tensor of rank 4, which is the famous *Riemann-Christoffel curvature tensor*. It is not a zero tensor in a general Riemannian space. Hence, in general,

$$\xi^i|_{st} \neq \xi^i|_{ts}$$

in a Riemannian space.

*Note:* The results obtained above hold true also for two-dimensional spaces, provided all indices range over 1 and 2 only. A curved shell in a three-dimensional

Euclidean space appears, in general, to be a two-dimensional non-Euclidean space to a "two-dimensional" animal who has to measure distances right on the shell surface and is never allowed to leave the shell surface to view the third dimension. For a spherical surface, the nonvanishing components of the two-dimensional Riemann-Christoffel curvature tensor are all equal to a constant, which may be written as 1. For a flat plate, they are all zero. For certain hyperboloidal surface all the nonvanishing components of curvature may take on the value $-1$. Since the spirit of the theory of thin elastic shells is to reduce all the properties of the shells into differential equations describing the middle surface of the shell, an engineer deals with non-Euclidean geometry rather frequently.

# 3

# STRESS TENSOR

The definitions of stress vector and stress components will be given and the equations of equilibrium will be derived. We shall then show how the stress components change when the frames of reference are changed from one rectangular Cartesian frame of reference to another, and in this way we will prove from physical standpoint that the stress components transform according to the tensor transformation rules. The symmetry of the stress tensor will then be discussed, and the consequences of the symmetry property will be derived. The principal stresses, the Mohr, Lamé, and Cauchy geometric representations, the stress deviations, the octahedral stresses, and finally, the stress tensor in general curvilinear coordinates, form the material for the remainder of the chapter.

Except for Sec. 3.14 et seq., we shall use only rectangular Cartesian frames of reference, whose coordinate axes will be denoted by $x_1$, $x_2$, $x_3$ and are rectilinear and orthogonal to each other. We shall use subscripts for all components unless stated otherwise.

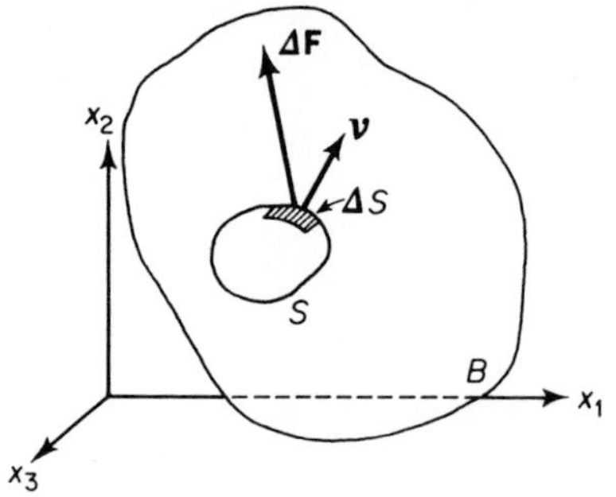


**Fig. 3.1:1.** Stress principle.

## 3.1. STRESSES

Consider a configuration occupied by a body $B$ at some time (Fig. 3.1:1). Imagine a closed surface $S$ within $B$. We would like to know the interaction between the material exterior to this surface and that in the interior. In this consideration, there arises the basic defining concept of continuum mechanics —the *stress principle* of Euler and Cauchy.

Consider a small surface element of area $\Delta S$ on our imagined surface $S$. Let us draw a unit vector $\boldsymbol{\nu}$ normal to $\Delta S$, with its direction outward from the interior of $S$. Then we can distinguish the two sides of $\Delta S$ according to the direction of $\boldsymbol{\nu}$. Consider the part of material lying on the positive side of the normal. This part exerts a force $\Delta \mathbf{F}$ on the other part, which is situated on the negative side of the normal. The force $\Delta \mathbf{F}$ is a function of the area and the orientation of the surface. We introduce *the assumption that as* $\Delta S$

*tends to zero, the ratio $\Delta\mathbf{F}/\Delta S$ tends to a definite limit $d\mathbf{F}/dS$ and that the moment of the forces acting on the surface $\Delta S$ about any point within the area vanishes in the limit.* The limiting vector will be written as

$$\overset{\nu}{\mathbf{T}} = \frac{d\mathbf{F}}{dS}, \tag{1}$$

where a superscript $\nu$ is introduced to denote the direction of the normal $\boldsymbol{\nu}$ of the surface $\Delta S$. The limiting vector $\overset{\nu}{\mathbf{T}}$ is called the *stress vector*, or *traction*, and represents the force per unit area acting on the surface.

The assertion that there is defined upon any imagined closed surface $S$ in the interior of a continuum a stress vector field whose action on the material occupying the space interior to $S$ is equipollent to the action of the exterior material upon it, is the stress principle of Euler and Cauchy.

Consider now a special case in which the surface $\Delta S_k$ is parallel to one of the coordinate planes. Let the normal of $\Delta S_k$ be in the positive direction of the $x_k$-axis. Let the stress vector acting on $\Delta S_k$ be denoted by $\overset{k}{\mathbf{T}}$, with three components $\overset{k}{T}_1, \overset{k}{T}_2, \overset{k}{T}_3$ along the direction of the coordinate axes $x_1, x_2, x_3$, respectively, the index $i$ of $\overset{k}{T}_i$ denoting the components of the force, and the symbol $k$ indicating the surface on which the force acts. In this special case, we introduce a new set of symbols for the stress components,

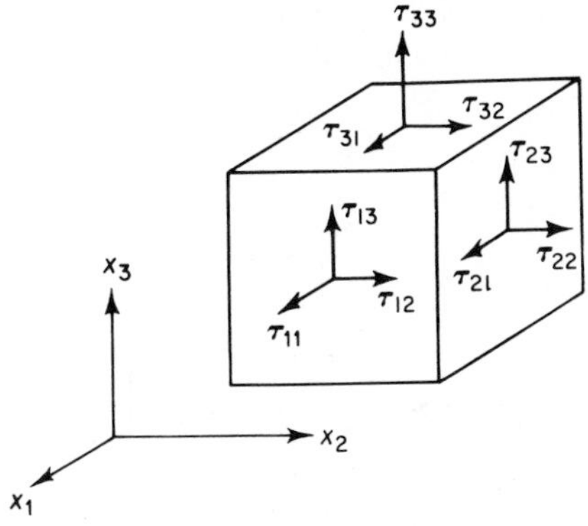


**Fig. 3.1:2.** Notations of stress components.

$$\overset{k}{T}_1 = \tau_{k1}, \qquad \overset{k}{T}_2 = \tau_{k2}, \qquad \overset{k}{T}_3 = \tau_{k3}. \tag{2}$$

If we arrange the components of tractions acting on the surfaces $k = 1$, $k = 2$, $k = 3$ in a square matrix, we obtain

| | Components of Stresses | | |
|---|---|---|---|
| | 1 | 2 | 3 |
| Surface normal to $x_1$ | $\tau_{11}$ | $\tau_{12}$ | $\tau_{13}$ |
| Surface normal to $x_2$ | $\tau_{21}$ | $\tau_{22}$ | $\tau_{23}$ |
| Surface normal to $x_3$ | $\tau_{31}$ | $\tau_{32}$ | $\tau_{33}$ |

This is illustrated in Fig. 3.1:2. The components $\tau_{11}, \tau_{22}, \tau_{33}$ are called *normal stresses*, and the remaining components $\tau_{12}, \tau_{13}$, etc., are called *shearing stresses.* Each of these components has the dimension of force per unit area, or $M/LT^2$.

A great diversity in notations for stress components exist in the literature. The most widely used notations in American literature are, in reference to a system of rectangular Cartesian coordinates $x, y, z$,

$$
\begin{matrix}
\sigma_x & \tau_{xy} & \tau_{xz} \\
\tau_{yx} & \sigma_y & \tau_{yz} \\
\tau_{zy} & \tau_{zy} & \sigma_z
\end{matrix}
\tag{3}
$$

Love writes $X_x$, $Y_x$ for $\sigma_x$ and $\tau_{xy}$, and Todhunter and Pearson use $\widehat{xx}$, $\widehat{xy}$. *In this book we shall use both* $\tau_{ij}$ *and* $\sigma_{ij}$. *We use* $\tau_{ij}$ *to denote stress tensors in general, and we use* $\sigma_{ij}$ *to denote the physical components of stress tensors in curvilinear coordinates* (see p. 86). *In rectangular Cartesian coordinates the tensor components and the physical components coincide. Hence, we use* $\sigma_{ij}$ *for Cartesian stress tensors.* Although the lack of uniformity may seem awkward, little confusion will arise, and in many instances different notations actually result in clarity.

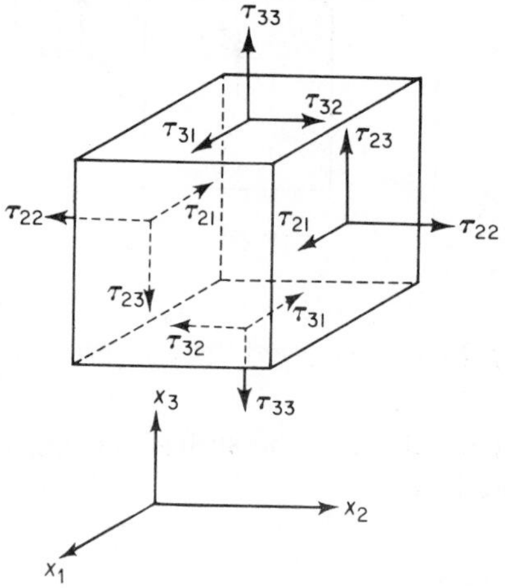


**Fig. 3.1:3.** Senses of positive stresses.

It is important to emphasize again that a stress will always be understood to be the force (per unit area) which the part lying on the positive side of a surface element (the side on the positive side of the outer normal) exerts on the part lying on the negative side. Thus, if the outer normal of a surface element points in the positive direction of the $x_1$-axis and $\tau_{11}$ is positive, the vector representing the component of normal stress acting on the surface element will point in the positive $x_1$-direction. But if $\tau_{11}$ is positive while the outer normal points in the negative $x_1$-axis direction, then the stress vector acting on the element also points to the negative $x_1$-axis direction (see Fig. 3.1:3).

Similarly, positive values of $\tau_{12}$, $\tau_{13}$ will imply shearing stress vectors pointing to positive $x_2$, $x_3$-axes if the outer normal agrees in sense with $x_1$-axis, whereas they point to the negative $x_2$, $x_3$-direction if the outer normal disagrees in sense with the $x_1$-axis, as illustrated in Fig. 3.1:3. A careful study of the figure is essential. Naturally, these rules agree with the usual notions of tension, compression, and shear.

## 3.2. LAWS OF MOTION

The fundamental laws of mechanics for bodies of all kinds are Euler's equations, which extend Newton's laws of motion for particles. Let the coordinate system $x_1, x_2, x_3$ be an inertial frame of reference. Let the space occupied by a material body at any time $t$ be denoted by $B(t)$. Let $\mathbf{r}$ be the position vector of a particle with respect to the origin of the coordinate system. Let $\mathbf{V}$ be the velocity vector of a particle at the point $x_1, x_2, x_3$. Let

$$\mathscr{P} = \int_{B(t)} \mathbf{V}\rho \, dv \tag{1}$$

be called the *linear momentum* of the body in the configuration $B(t)$, and let

$$\mathscr{H} = \int_{B(t)} \mathbf{r} \times \mathbf{V}\rho \, dv \tag{2}$$

be called the *moment of momentum*. In these formulas $\rho$ is the density of the material and the integration is over the volume $B(t)$. *Newton's laws, as stated by Euler for a continuum, assert that the rate of change of linear momentum is equal to the total applied force $\mathscr{F}$ acting on the body,*

$$\dot{\mathscr{P}} = \mathscr{F}, \tag{3}$$

*and that the rate of change of moment of momentum is equal to the total applied torque $\mathscr{L}$,*

$$\dot{\mathscr{H}} = \mathscr{L}. \tag{4}$$

The torque $\mathscr{L}$ is taken with respect to the same point as the origin of the position vector $\mathbf{r}$. It is easy to verify that if (3) holds, then if (4) holds for one choice of origin, it holds for all choices of origin.†

It is assumed that force and torque are quantities about which we have *a priori* information in certain frames of reference.

On material bodies considered in mechanics of continuous media, two types of external forces act:

1. Body forces, acting on elements of volume of the body.
2. Surface forces, or stresses, acting on surface elements.

Examples of body forces are gravitational forces and electromagnetic forces. Examples of surface forces are aerodynamic pressure acting on a body and pressure due to mechanical contact of two bodies.

To specify a body force, we consider a volume bounded by an arbitrary surface $S$ (Fig. 3.2:1). The resultant force vector contributed by the body

† The derivatives $\dot{\mathscr{P}}$ and $\dot{\mathscr{H}}$ are *material derivatives*; i.e., the time rate of change of $\mathscr{P}$ and $\mathscr{H}$ of a fixed set of particles (Cf. Secs. 5.2 and 5.3).

force is assumed to be representable in the form of a volume integral

$$\int_B \mathbf{X}\, dv.$$

The three components of **X**, namely, $X_1$, $X_2$, $X_3$, all of dimensions force per unit volume $M(LT)^{-2}$, are called the body force per unit volume. For example, in a gravitational field,

$$X_i = \rho g_i,$$

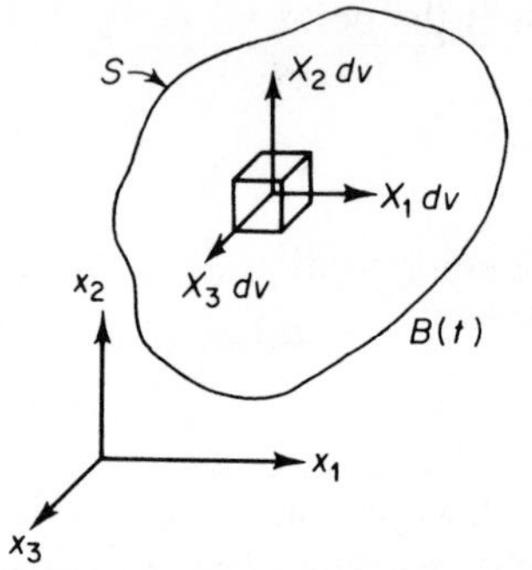


**Fig. 3.2:1.** Body forces.

where $g_i$ are components of a gravitational acceleration field and $\rho$ is the density (mass per unit volume) at a given point of the body.

The surface force acting on an imagined surface in the interior of a body is the stress vector conceived in Euler and Cauchy's stress principle. According to this concept, the total force acting upon the material occupying the region $B$ interior to a closed surface $S$ is

$$\mathscr{F} = \oint_S \overset{\nu}{\mathbf{T}}\, dS + \int_B \mathbf{X}\, dv, \tag{5}$$

where $\overset{\nu}{\mathbf{T}}$ is the stress vector acting on $dS$ whose outer normal vector is $\boldsymbol{\nu}$. Similarly, the torque about the origin is given by the expression

$$\mathscr{L} = \oint_S \mathbf{r} \times \overset{\nu}{\mathbf{T}}\, dS + \int_B \mathbf{r} \times \mathbf{X}\, dv. \tag{6}$$

In the following section we shall make some elementary applications of these equations to obtain the fundamental properties of the stress tensor.

## 3.3. CAUCHY'S FORMULA

With the equations of motion, we shall first derive a simple result which states that *the stress vector* $\mathbf{T}^{(+)}$ *representing the action of material exterior to a surface element on the interior is equal in magnitude and opposite in direction to the stress vector* $\mathbf{T}^{(-)}$ *which represents the action of the interior material on the exterior across the same surface element:*

(1) ▲
$$\mathbf{T}^{(-)} = -\mathbf{T}^{(+)}.$$

To prove this, we consider a small "pill box" with two parallel surfaces of area $\Delta S$ and thickness $\delta$, as shown in Fig. 3.3:1. When $\delta$ shrinks to zero, while $\Delta S$ remains small but finite, the volume forces and the linear momentum and its rate of change with time vanish, as well as the contribution of surface forces on the sides of the pill box. The equation of motion (3.2:3) implies, therefore, for small $\Delta S$,

$$\mathbf{T}^{(+)}\,\Delta S + \mathbf{T}^{(-)}\,\Delta S = 0.$$

Equation (1) then follows.

Another way of stating this result is that the stress vector is a function of the normal vector to a surface. When the sense of direction of the normal vector reverses, the stress vector reverses also.

Now we shall show that *knowing the components* $\tau_{ij}$, *we can write down at once the stress vector acting on any surface with unit outer normal vector* $\boldsymbol{\nu}$ *whose components are* $\nu_1, \nu_2, \nu_3$. *This stress vector is denoted by* $\overset{\nu}{\mathbf{T}}$, *with components* $\overset{\nu}{T}_1, \overset{\nu}{T}_2, \overset{\nu}{T}_3$ *given by Cauchy's formula*

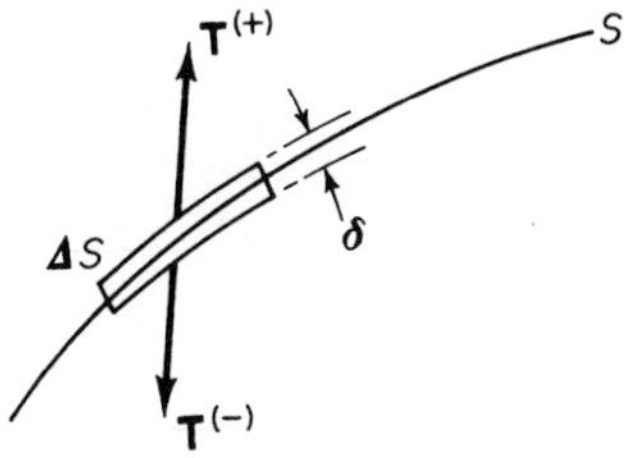


**Fig. 3.3:1.** Equilibrium of a "pill box" across a surface $S$.

(2) ▲ $$\overset{\nu}{T}_i = \nu_j \tau_{ji},$$

which can be derived in several ways. We shall give first an elementary derivation.

Let us consider an infinitesimal tetrahedron formed by three surfaces parallel to the coordinate planes and one normal to the unit vector $\boldsymbol{\nu}$ (see Fig. 3.3:2). Let the area of the surface normal to $\boldsymbol{\nu}$ be $dS$. Then the area of the other three surfaces are

$$\begin{aligned} dS_1 &= dS \cos(\boldsymbol{\nu}, \mathbf{x}_1) \\ &= \nu_1\, dS = \text{area of surface} \parallel \text{to the } x_2x_3\text{-plane}, \\ dS_2 &= \nu_2\, dS = \text{area of surface} \parallel \text{to the } x_3x_1\text{-plane}, \\ dS_3 &= \nu_3\, dS = \text{area of surface} \parallel \text{to the } x_1x_2\text{-plane}, \end{aligned}$$

and the volume of the tetrahedron is

$$dv = \tfrac{1}{3} h\, dS,$$

where $h$ is the height of the vertex $P$ from the base $dS$. The forces in the positive direction of $\mathbf{x}_1$ acting on the three coordinate surfaces can be written as

$$(-\tau_{11} + \epsilon_1)\, dS_1, \qquad (-\tau_{21} + \epsilon_2)\, dS_2, \qquad (-\tau_{31} + \epsilon_3)\, dS_3,$$

where $\tau_{11}, \tau_{21}, \tau_{31}$ are the stresses at the point $P$. The negative sign is obtained because the outer normals to the three surfaces are opposite in sense with respect to the coordinate axes, and the $\epsilon$'s are inserted because the tractions act at points slightly different from $P$. If we assume that the stress field is continuous, then $\epsilon_1, \epsilon_2, \epsilon_3$ are infinitesimal quantities. On the other hand, the force acting on the triangle normal to $\boldsymbol{\nu}$ has a component $(\overset{\nu}{T}_1 + \epsilon)\, dS$ in the $x_1$-axis direction, the body force has an $x_1$-component equal to $(X_1 + \epsilon')\, dv$,

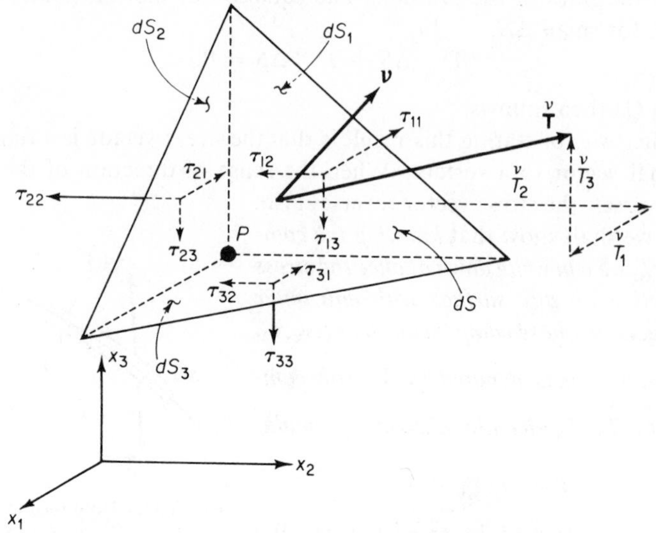


**Fig. 3.3:2.** Surface tractions on a tetrahedron.

and the rate of change of linear momentum has a component $\rho\dot{V}_1\,dv$. Here $\overset{\nu}{T}_1$ and $X_1$ refer to the point $P$ and $\epsilon$, $\epsilon'$ are again infinitesimal. The first equation of motion is thus

$$(-\tau_{11} + \epsilon_1)\nu_1\,dS + (-\tau_{21} + \epsilon_2)\nu_2\,dS + (-\tau_{31} + \epsilon_3)\nu_3\,dS$$

$$+ (\overset{\nu}{T}_1 + \epsilon)\,dS + (X_1 + \epsilon')\tfrac{1}{3}h\,dS = \rho\dot{V}_1\tfrac{1}{3}h\,dS.$$

Dividing through by $dS$ and taking the limit $h \to 0$, one obtains

$$\overset{\nu}{T}_1 = \tau_{11}\nu_1 + \tau_{21}\nu_2 + \tau_{31}\nu_3, \tag{3}$$

which is the first component of Eq. (2). Other components follow similarly.

*Cauchy's formula assures us that the nine components of stresses $\tau_{ij}$ are necessary and sufficient to define the traction across any surface element in a body. Hence the stress state in a body is characterized completely by the set of quantities $\tau_{ij}$. Since $\overset{\nu}{T}_i$ is a vector and Eq. (2) is valid for an arbitrary vector $\nu_j$, it follows from the quotient rule (Sec. 2.10) that $\tau_{ij}$ is a tensor.* Henceforth $\tau_{ij}$ will be called a *stress tensor.*

We note again that in the theoretical development up to this point we have assumed, first, that stress can be defined everywhere in a body, and, second, that the stress field is continuous. The same assumption will be made later with respect to strain. These are characteristic assumptions of continuum mechanics. Without these assumptions we can do very little indeed. However, in the further development of the theory, certain mathematical discontinuities will be permitted—often they are very useful tools—but one should always view these discontinuities with great care against the general basic assumptions of continuity of the stress and strain fields.

## 3.4. EQUATIONS OF EQUILIBRIUM

We shall now transform the equations of motion (3.2:3), (3.2:4) into differential equations. This can be done elegantly by means of Gauss'

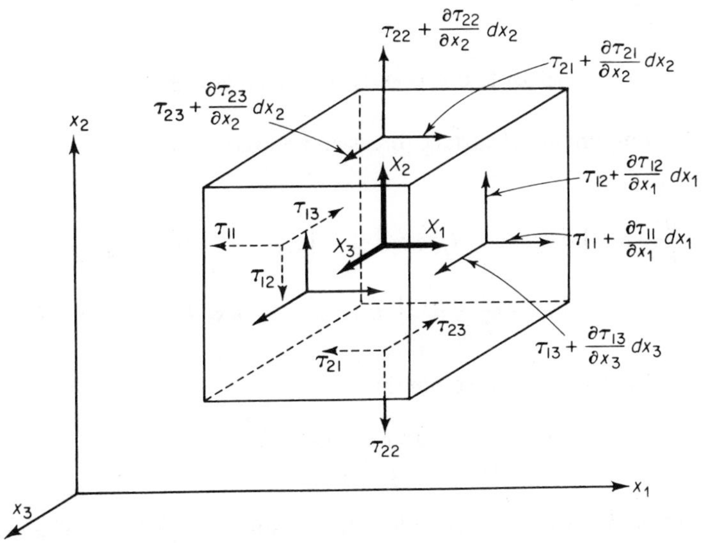


**Fig. 3.4:1.** Equilibrating stress components on an infinitesimal parallelepiped.

theorem and Cauchy's formula. But we shall pursue here an elementary course to assure physical clarity.

Consider the static equilibrium state of an infinitesimal parallelepiped with surfaces parallel to the coordinate planes. The stresses acting on the various surfaces are shown in Fig. 3.4:1. The force $\tau_{11}\, dx_2\, dx_3$ acts on the left-hand side, the force $\left(\tau_{11} + \frac{\partial \tau_{11}}{\partial x_1} dx_1\right) dx_2\, dx_3$ acts on the right-hand side, etc. These expressions are based on the assumption of continuity of the stresses. The body force is $X_i\, dx_1\, dx_2\, dx_3$.

Now, the equilibrium of the body demands that the resultant forces vanish. Consider the forces in the $x_1$-direction. As shown in Fig. 3.4:2, we

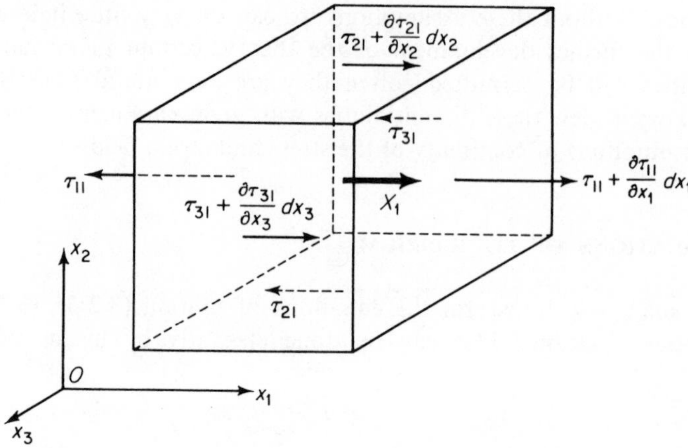


**Fig. 3.4:2.** Components of tractions in $x_1$ direction.

have six components of surface forces and one component of body force. The sum is

$$\left(\tau_{11} + \frac{\partial \tau_{11}}{\partial x_1} dx_1\right) dx_2\,dx_3 - \tau_{11}\,dx_2\,dx_3 + \left(\tau_{21} + \frac{\partial \tau_{21}}{\partial x_2} dx_2\right) dx_3\,dx_1 - \tau_{21}\,dx_3\,dx_1$$
$$+ \left(\tau_{31} + \frac{\partial \tau_{31}}{\partial x_3} dx_3\right) dx_1\,dx_2 - \tau_{31}\,dx_1\,dx_2 + X_1\,dx_1\,dx_2\,dx_3 = 0.$$

Dividing by $dx_1\,dx_2\,dx_3$, we obtain

(1) $$\frac{\partial \tau_{11}}{\partial x_1} + \frac{\partial \tau_{21}}{\partial x_2} + \frac{\partial \tau_{31}}{\partial x_3} + X_1 = 0.$$

A cyclic permulation of symbols leads to similar equations of equilibrium of forces in $x_2$, $x_3$-directions. The whole set, written concisely, is

(2) ▲ $$\frac{\partial \tau_{ji}}{\partial x_j} + X_i = 0.$$

This is an important result. A shorter derivation will be given later in Sec. 5.5.

The equilibrium of an element requires also that the resultant moment vanish. If there do not exist external moments proportional to a volume, the consideration of moments will lead to the important conclusion that *the stress tensor is symmetric*,

(3) ▲ $$\tau_{ij} = \tau_{ji}.$$

This is demonstrated as follows. Referring to Fig. 3.4:3 and considering the moment of all the forces about the axis $Ox_3$, we see that those components of forces parallel to $Ox_3$ or lying in planes containing $Ox_3$ do not contribute any

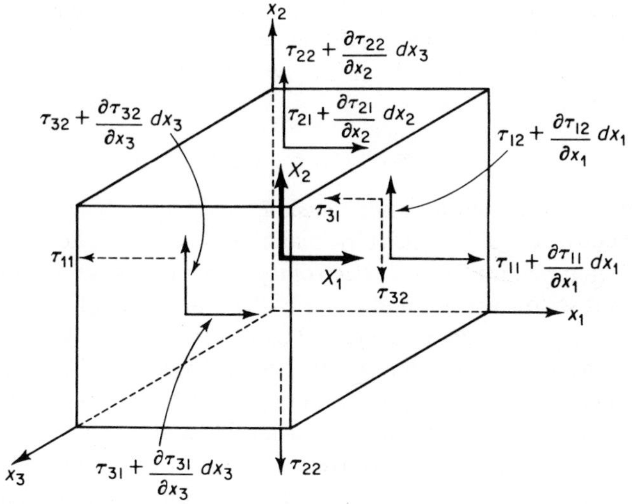


**Fig. 3.4:3.** Components of tractions that contribute moment about $Ox_3$-axis.

moment. The components that do contribute a moment about the $Ox_3$-axis are shown in Fig. 3.4:3. Therefore, properly taking care of the moment arm, we have

$$
\begin{aligned}
&-\left(\tau_{11} + \frac{\partial \tau_{11}}{\partial x_1} dx_1\right) dx_2\, dx_3 \cdot \frac{dx_2}{2} + \tau_{11}\, dx_2\, dx_3 \cdot \frac{dx_2}{2} \\
&+\left(\tau_{12} + \frac{\partial \tau_{12}}{\partial x_1} dx_1\right) dx_2\, dx_3 \cdot dx_1 - \left(\tau_{21} + \frac{\partial \tau_{21}}{\partial x_2} dx_2\right) dx_1\, dx_3 \cdot dx_2 \\
&+\left(\tau_{22} + \frac{\partial \tau_{22}}{\partial x_2} dx_2\right) dx_1\, dx_3 \cdot \frac{dx_1}{2} - \tau_{22}\, dx_1\, dx_3 \cdot \frac{dx_1}{2} \\
&+\left(\tau_{32} + \frac{\partial \tau_{32}}{\partial x_3} dx_3\right) dx_1\, dx_2 \cdot \frac{dx_1}{2} - \tau_{32}\, dx_1\, dx_2 \cdot \frac{dx_1}{2} \\
&-\left(\tau_{31} + \frac{\partial \tau_{31}}{\partial x_3} dx_3\right) dx_1\, dx_2 \cdot \frac{dx_2}{2} + \tau_{31}\, dx_1\, dx_2 \cdot \frac{dx_2}{2} \\
&- X_1\, dx_1\, dx_2\, dx_3 \cdot \frac{dx_2}{2} + X_2\, dx_1\, dx_2\, dx_3 \cdot \frac{dx_1}{2} = 0.
\end{aligned}
$$

On dividing through by $dx_1\, dx_2\, dx_3$ and passing to the limit $dx_1 \to 0$, $dx_2 \to 0$, $dx_3 \to 0$, we obtain

(4) $$\tau_{12} = \tau_{21}.$$

Similar considerations of resultant moments about $Ox_2$, $Ox_3$ lead to the general result given by Eq. (3). Again a shorter derivation will be given later in Sec. 5.5.

It should be noted that if an external moment proportional to the volume does exist, then the symmetry condition does not hold. For example, if there is a moment $c_3\, dx_1\, dx_2\, dx_3$ about the axis $Ox_3$, then we obtain in place of Eq. (4) the result

$$\tau_{12} - \tau_{21} + c_3 = 0. \tag{5}$$

Maxwell pointed out that nonvanishing body moments exist in a magnet in a magnetic field and in a dielectric material in an electric field with different planes of polarization. If the electromagnetic field is so intense and the stress level is so low that $\tau_{12}$ and $c_3$ are of the same order of magnitude, then, according to (5), $\tau_{12}$ cannot be equated to $\tau_{21}$. In this case, we have to admit the stress tensor $\tau_{ij}$ as asymmetric. If $c_3$ is very much smaller in comparison with $\tau_{12}$, then we can omit $c_3$ in (5) and consider (4) as valid approximately.

In developing a physical theory, particularly for the purpose of engineering, one of the most important objectives is to obtain the simplest formulation consistent with the desired degree of accuracy. A decision of whether or not we shall treat the stress tensor as symmetric must be based on the purpose of the theory. Since electromagnetic fields pervade the universe, the stress tensor is in general unsymmetric. But, if the theory is formulated for the purpose of a structural or mechanical engineer who studies the stress distribution in a structure or a machine with a view towards assessing its strength, stability, or rigidity, then a stress is important when it is of the order of the yielding stress of the material. Even for a structure which is designed primarily on the basis of stability, such as a column, an arch, or a thin-walled shell, a good design should produce a critical stress of the order of the yielding stress under the critical conditions, for otherwise the material is not economically used. When $\tau_{ij}$ is comparable to the yielding stress in magnitude (of order 10,000 to 100,000 lb/sq in. for a steel, or 50 to 5,000 lb/sq in. for a concrete), there are few circumstances in which the assumption of symmetry in stress tensor should cause concern.

However, if one wants to study the influence of a strong electromagnetic field on the propagation of elastic waves, or such influence on some high-frequency phenomenon in the material, then the stress level may be very low and the body moment may be significant. In such problems the stress tensor may not be assumed symmetric.

In the rest of this book the stress tensor will be assumed to be symmetric unless stated otherwise.

### *Notes on Couple-stresses*

If, following Voigt, we assume that across any infinitesimal surface element in a solid the action of the exterior material upon the interior is equipollent to a force *and a couple*, (in contrast to the assumption made in Sec. 3.1) then in addition to the traction $\overset{\nu}{\mathbf{T}}$ that act on the surface we must have also a *couple-stress* vector $\overset{\nu}{\mathbf{M}}$.

These two vectors $\overset{\nu}{\mathbf{T}}$ and $\overset{\nu}{\mathbf{M}}$, together, are now equipollent to the action of the exterior upon the interior. Similarly, one might have body couples as pointed out by Maxwell, i.e., couple per unit mass, $\mathbf{c}$, with components $c_i$, $(i = 1, 2, 3)$. If we accept these possibilities, then we must define a couple-stress tensor, $\mathcal{M}_{ij}$, in addition to the stress tensor $\tau_{ij}$. The tensor $\mathcal{M}_{ij}$ is related to the couple-stress vector by a linear transformation like Eq. (3.3:2):

$$\overset{\nu}{M}_i = \mathcal{M}_{ji}\nu_j$$

An analysis of the angular momentum then leads to the equation

$$\frac{\partial \mathcal{M}_{ji}}{\partial x_j} + \rho c_i = e_{ijk}\tau_{jk},$$

i.e.,

$$\frac{\partial \mathcal{M}_{xx}}{\partial x} + \frac{\partial \mathcal{M}_{xy}}{\partial y} + \frac{\partial \mathcal{M}_{xz}}{\partial z} + \rho c_x = \tau_{yz} - \tau_{zy}, \quad \text{etc.}$$

Thus, the antisymmetric part of the stress tensor is determined by the body couples and the divergence of the couple-stress tensor. When couples of both kinds are absent, the stress tensor must be symmetric.

Couple-stresses and body couples are useful concepts in dealing with materials whose molecules have internal structure and in the dislocation theory of metals.

## 3.5. TRANSFORMATION OF COORDINATES

In the previous section, the components of stress $\tau_{ij}$ are defined with respect to a rectangular Cartesian system $\mathbf{x}_1$, $\mathbf{x}_2$, $\mathbf{x}_3$. Let us now take a second set of rectangular Cartesian coordinates $\mathbf{x}_1'$, $\mathbf{x}_2'$, $\mathbf{x}_3'$ with the same origin but oriented differently, and consider the stress components in the new reference system (Fig. 3.5:1). Let these coordinates be connected by the linear relations

$$x_k' = \beta_{ki}x_i, \qquad k = 1, 2, 3. \tag{1}$$

where $\beta_{ki}$ are the direction cosines of the $\mathbf{x}_k'$-axis with respect to the $\mathbf{x}_i$-axis.

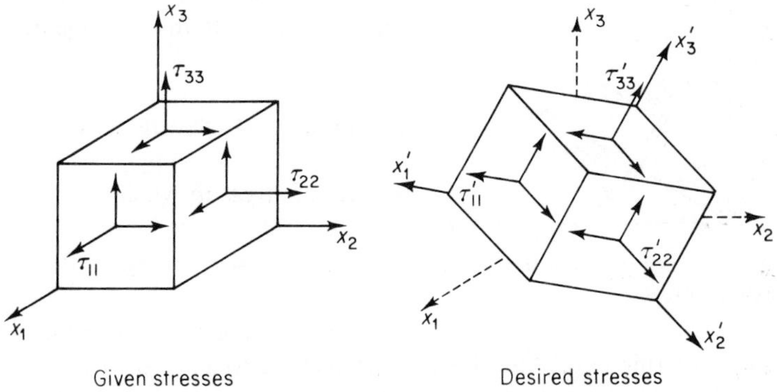


**Fig. 3.5:1.** Transformation of stress components under rotation of coordinates system.

Since $\tau_{ij}$ is a tensor (Sec. 3.3) we can write down the transformation law at once. However, in order to emphasize the importance of the result we shall insert an elementary derivation based on Cauchy's formula derived in Sec. 3.3, which states that if $dS$ is a surface element whose unit outer normal vector $\boldsymbol{\nu}$ has components $\nu_i$, then the force per unit area acting on $dS$ is a vector $\overset{\nu}{\mathbf{T}}$ with components

$$\overset{\nu}{T_i} = \tau_{ji}\nu_j.$$

If the normal $\boldsymbol{\nu}$ is chosen to be parallel to the axis $\mathbf{x}'_k$, so that

$$\nu_1 = \beta_{k1}, \qquad \nu_2 = \beta_{k2}, \qquad \nu_3 = \beta_{k3},$$

then

$$\overset{k}{T'_i} = \tau_{ji}\beta_{kj}.$$

The component of the vector $\overset{k}{T'_i}$ in the direction of the axis $x'_m$ is given by the product of $\overset{k}{T'_i}$ and $\beta_{mi}$. Hence, the stress component

$$\begin{aligned}\tau'_{km} &= \sum_i \text{proj. of } \overset{k}{T'_i} \text{ on } x'_m \\ &= \overset{k}{T'_1}\beta_{m1} + \overset{k}{T'_2}\beta_{m2} + \overset{k}{T'_3}\beta_{m3} \\ &= \tau_{11}\beta_{m1}\beta_{k1} + \tau_{21}\beta_{m1}\beta_{k2} + \dots,\end{aligned}$$

i.e.,

(3) ▲ $$\tau'_{km} = \tau_{ji}\beta_{kj}\beta_{mi}.$$

If we compare Eq. (3) with Eq. (2.5:2) we see that the stress components transform like a Cartesian tensor of rank two. Thus, the physical concept of stress which is described by $\tau_{ij}$ agrees with the mathematical definition of a tensor of rank two in a Euclidean space.

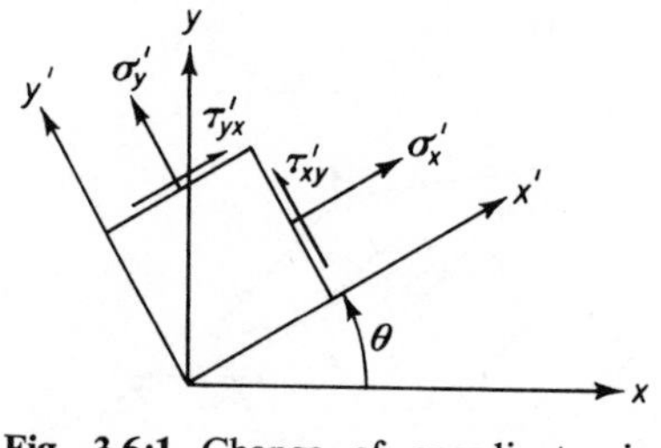


Fig. 3.6:1. Change of coordinates in plane state of stress.

### 3.6. PLANE STATE OF STRESS

A state of stress in which

(1) $$\tau_{33} = \tau_{31} = \tau_{32} = 0$$

is called a *plane state of stress* in the $x_1x_2$-plane. In this case, the direction cosines between two systems of rectangular Cartesian coordinates can be expressed in terms of a single angle $\theta$, as shown in Fig. 3.6:1. Writing $x, y$ and $x', y'$ in place of $x_1, x_2$ and $x'_1, x'_2$; $\sigma_x$

for $\tau_{11}$; $\tau_{xy}$ for $\tau_{12}$, etc. we have

$$\sigma'_x = \sigma_x \cos^2\theta + \sigma_y \sin^2\theta + 2\tau_{xy}\sin\theta\cos\theta,$$

$$\sigma'_y = \sigma_x \sin^2\theta + \sigma_y \cos^2\theta - 2\tau_{xy}\sin\theta\cos\theta,$$

$$\tau'_{xy} = (-\sigma_x + \sigma_y)\sin\theta\cos\theta + \tau_{xy}(\cos^2\theta - \sin^2\theta).$$

Since

$$\sin^2\theta = \tfrac{1}{2}(1 - \cos 2\theta), \qquad \cos^2\theta = \tfrac{1}{2}(1 + \cos 2\theta),$$

we may also write

$$\sigma'_x = \frac{\sigma_x + \sigma_y}{2} + \frac{\sigma_x - \sigma_y}{2}\cos 2\theta + \tau_{xy}\sin 2\theta,$$

(3)
$$\sigma'_y = \frac{\sigma_x + \sigma_y}{2} - \frac{\sigma_x - \sigma_y}{2}\cos 2\theta - \tau_{xy}\sin 2\theta,$$

$$\tau'_{xy} = -\frac{\sigma_x - \sigma_y}{2}\sin 2\theta + \tau_{xy}\cos 2\theta.$$

Note that

(4)
$$\sigma'_x + \sigma'_y = \sigma_x + \sigma_y,$$

(5)
$$\frac{\partial \sigma'_x}{\partial \theta} = 2\tau'_{xy},$$

6)
$$\tau'_{xy} = 0 \quad \text{when} \quad \tan 2\theta = \frac{2\tau_{xy}}{\sigma_x - \sigma_y}.$$

The directions given by the particular values of $\theta$ given by (6) are called ıe *principal directions;* the corresponding normal stresses are called the ·*incipal stresses* (see Sec. 3.7). Following (5) and (6), the principal stresses ·e extreme values of the normal stresses,

$$\left.\begin{matrix}\sigma_{\max}\\ \sigma_{\min}\end{matrix}\right\} = \frac{\sigma_x + \sigma_y}{2} \pm \sqrt{\left(\frac{\sigma_x - \sigma_y}{2}\right)^2 + \tau_{xy}^2}.$$

fferentiating $\tau'_{xy}$ with respect to $\theta$ and setting the derivative to zero, we can ɪ the angle $\theta$ at which $\tau'_{xy}$ attends its extreme value. This angle is easily n to be $\pm 45°$ from the principal directions given by (6), and the maximum ue of $\tau'_{xy}$ is

$$\tau_{\max} = \frac{\sigma_{\max} - \sigma_{\min}}{2} = \sqrt{\left(\frac{\sigma_x - \sigma_y}{2}\right)^2 + \tau_{xy}^2}.$$

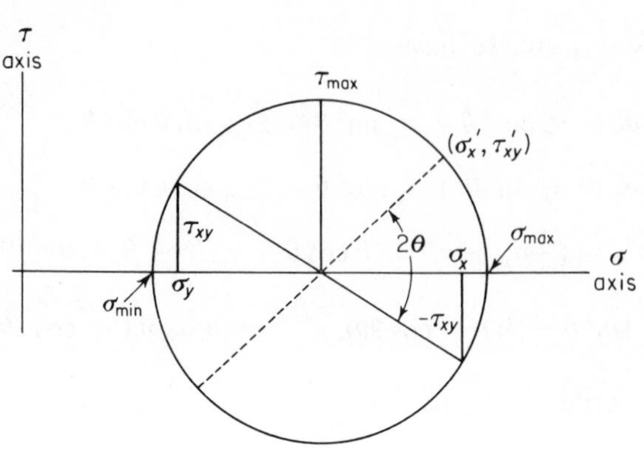


**Fig. 3.6:2.** Mohr's circle in plane state of stress.

Figure 3.6:2 is a geometric representation of the relations above. It is the well-known *Mohr's circle.*

**Problem 3.1.** Find the transformation law for the moments of inertia and products of inertia of an area about a set of rectangular Cartesian coordinates in a plane,

$$I_{xx} = \int y^2 \, dA, \qquad I_{xy} = \int xy \, dA, \qquad I_{yy} = \int x^2 \, dA,$$

with respect to rotation of the coordinate axes about the origin. Mohr's circle was invented in 1887 for the transformation of the inertia tensor.

**Problem 3.2.** Let $v_i$, $i = 1, 2, 3$, be the velocity vector field of a continuum, and let $\overline{v_i v_j}$ be the average value of the product $v_i v_j$ over a period of time. Show that the correlation function $\overline{v_i v_j}$, with components $\overline{u^2}$, $\overline{uv}$, $\overline{uw}$, $\overline{v^2}$, etc. in unabridged notations, is a symmetric tensor of the second order.

**Problem 3.3.** Show that the mass moment of inertia of a set of particles,

$$I_{ij} = e_{ipq} e_{jkq} \int x_p x_k \, dm, \qquad i, j = 1, 2, 3,$$

is a tensor, where $dm$ is an element of mass and the integration is extended over the entire set of particles. Write out the matrix of the inertia tensor $I_{ij}$. Show that $I_{ii}$ ($i$ not summed) is the moment of inertia about the $x_i$ axis, whereas $I_{ij}$ ($i \neq j$) is equal to the *negative* of the product of inertia about the axes $x_i$ and $x_j$. Show that for a rigid body rotating at an angular velocity $\omega_j$, the angular momentum vector of the body is $I_{ij}\,\omega_j$.

## 3.7. PRINCIPAL STRESSES

Let us now generalize some of the results obtained in the previous section, which was restricted to the plane state of stress.

In a general state of stress, the stress vector acting on a surface with outer normal $\boldsymbol{\nu}$ depends on the direction of $\boldsymbol{\nu}$. Let us ask in what direction $\boldsymbol{\nu}$ the stress vector becomes normal to the surface, on which the shearing stress vanishes. Such a surface shall be called a *principal plane*, its normal a *principal axis*, and the value of normal stress acting on the principal plane shall be called a *principal stress*.

Let $\boldsymbol{\nu}$ define a principal axis and let $\sigma$ be the corresponding principal stress. Then the stress vector acting on the surface normal to $\boldsymbol{\nu}$ has components $\sigma\nu_i$. On the other hand, this same vector is given by the expression $\tau_{ji}\nu_j$. Hence, writing $\nu_i = \delta_{ji}\nu_j$, we have, on equating these two expressions and transposing to the same side,

$$(1) \qquad (\tau_{ji} - \sigma\delta_{ji})\nu_j = 0.$$

The three equations, $i = 1, 2, 3$, are to be solved for $\nu_1, \nu_2, \nu_3$. Since $\boldsymbol{\nu}$ is a unit vector, we must find a set of nontrivial solutions for which $\nu_1^2 + \nu_2^2 + \nu_3^2 = 1$. Thus, Eq. (1) poses an eigenvalue problem. Since $\tau_{ij}$ as a matrix is real and symmetric, we need only to recall a result in the theory of matrices to assert that *there exist three real-valued principal stresses and a set of orthonormal principal axes. Whether the principal stresses are all positive, all negative, or mixed depends on whether the quadratic form* $\sum_{i,j=1}^{3} \tau_{ij}x_i x_j$ *is positive definite, negative definite, or uncertain, respectively*. Because of the importance of these results, we shall give the details of the reasoning below.

Equation (1) has a set of nonvanishing solutions $\nu_1, \nu_2, \nu_3$ if and only if the determinant of the coefficients vanishes, i.e.,

$$(2) \qquad |\tau_{ij} - \sigma\delta_{ij}| = 0.$$

Equation (2) is a cubic equation in $\sigma$; its roots are the principal stresses. For each value of the principal stress, a unit normal vector $\boldsymbol{\nu}$ can be determined.

On expanding Eq. (2), we have

(3)

$$|\tau_{ij} - \sigma\delta_{ij}| = \begin{vmatrix} \tau_{11} - \sigma & \tau_{12} & \tau_{13} \\ \tau_{21} & \tau_{22} - \sigma & \tau_{23} \\ \tau_{31} & \tau_{32} & \tau_{33} - \sigma \end{vmatrix} = -\sigma^3 + I_1\sigma^2 - I_2\sigma + I_3 = 0,$$

where

$$I_1 = \tau_{11} + \tau_{22} + \tau_{33},$$

$$(4) \qquad I_2 = \begin{vmatrix} \tau_{22} & \tau_{23} \\ \tau_{32} & \tau_{33} \end{vmatrix} + \begin{vmatrix} \tau_{11} & \tau_{13} \\ \tau_{31} & \tau_{33} \end{vmatrix} + \begin{vmatrix} \tau_{11} & \tau_{12} \\ \tau_{21} & \tau_{22} \end{vmatrix},$$

$$I_3 = \begin{vmatrix} \tau_{11} & \tau_{12} & \tau_{13} \\ \tau_{21} & \tau_{22} & \tau_{23} \\ \tau_{31} & \tau_{32} & \tau_{33} \end{vmatrix}.$$

On the other hand, if $\sigma_1, \sigma_2, \sigma_3$ are the roots of Eq. (3), which can be written as

$$(\sigma - \sigma_1)(\sigma - \sigma_2)(\sigma - \sigma_3) = 0,$$

it can be seen that the following relations between the roots and the coefficients must hold:

$$\begin{aligned} I_1 &= \sigma_1 + \sigma_2 + \sigma_3, \\ I_2 &= \sigma_1\sigma_2 + \sigma_2\sigma_3 + \sigma_3\sigma_1, \\ I_3 &= \sigma_1\sigma_2\sigma_3. \end{aligned} \tag{5}$$

Since the principal stresses characterize the physical state of stress at a point, they are independent of any coordinates of reference. Hence, Eq. (3) and the coefficients $I_1, I_2, I_3$ are invariant with respect to the coordinate transformation; $I_1, I_2, I_3$ are the *invariants* of the stress tensor. The importance of invariants will become evident when physical laws are formulated (see, for example, Chap. 6).

We shall show now that for a symmetric stress tensor the three principal stresses are all real and that the three principal planes are mutually orthogonal. These important properties can be established when the stress tensor is symmetric,

$$\tau_{ij} = \tau_{ji}. \tag{6}$$

The proof is as follows. Let $\overset{1}{\mathbf{v}}, \overset{2}{\mathbf{v}}, \overset{3}{\mathbf{v}}$ be unit vectors in the direction of the principal axes, with components $\overset{1}{\nu}_i, \overset{2}{\nu}_i, \overset{3}{\nu}_i$ $(i = 1, 2, 3)$ which are the solutions of Eq. (1) corresponding to the roots $\sigma_1, \sigma_2, \sigma_3$, respectively;

$$\begin{aligned} (\tau_{ij} - \sigma_1\,\delta_{ij})\overset{1}{\nu}_j &= 0, \\ (\tau_{ij} - \sigma_2\,\delta_{ij})\overset{2}{\nu}_j &= 0, \\ (\tau_{ij} - \sigma_3\,\delta_{ij})\overset{3}{\nu}_j &= 0. \end{aligned} \tag{7}$$

Multiplying the first equation by $\overset{2}{\nu}_i$, the second by $\overset{1}{\nu}_i$, summing over $i$ and subtracting the resulting equations, we obtain

$$(\sigma_2 - \sigma_1)\overset{1}{\nu}_i\overset{2}{\nu}_i = 0, \tag{8}$$

on account of the symmetry condition (6), which implies that

$$\tau_{ij}\overset{1}{\nu}_j\overset{2}{\nu}_i = \tau_{ji}\overset{1}{\nu}_j\overset{2}{\nu}_i = \tau_{ij}\overset{2}{\nu}_j\overset{1}{\nu}_i, \tag{9}$$

the last equality being obtained by interchanging the dummy indices $i$ and $j$.

Now, if we assume tentatively that Eq. (3) has a complex root, then, since the coefficients in Eq. (3) are real-valued, a complex conjugate root must also exist and the set of roots may be written as

$$\sigma_1 = \alpha + i\beta, \qquad \sigma_2 = \alpha - i\beta, \qquad \sigma_3,$$

where $\alpha$, $\beta$, $\sigma_3$ are real numbers and $i$ stands for the imaginary number $\sqrt{-1}$. In this case, Eqs. (7) show that $\overset{1}{\nu}_j$ and $\overset{2}{\nu}_j$ are complex conjugate to each other and can be written as

$$\overset{1}{\nu}_j \equiv a_j + ib_j, \qquad \overset{2}{\nu}_j \equiv a_j - ib_j.$$

Therefore,

$$\begin{aligned}\overset{1}{\nu}_j\overset{2}{\nu}_j &= (a_j + ib_j)(a_j - ib_j) \\ &= a_1^2 + a_2^2 + a_3^2 + b_1^2 + b_2^2 + b_3^2 \neq 0.\end{aligned}$$

It follows from (8) that $\sigma_1 - \sigma_2 = 2i\beta = 0$ or $\beta = 0$. This contradicts the original assumption that the roots are complex. Thus, the assumption of the existence of complex roots is untenable, and the roots $\sigma_1$, $\sigma_2$, $\sigma_3$ are all real.

When $\sigma_1 \neq \sigma_2 \neq \sigma_3$, Eq. (8) implies

$$\overset{1}{\nu}_i\overset{2}{\nu}_i = 0, \qquad \overset{2}{\nu}_i\overset{3}{\nu}_i = 0, \qquad \overset{3}{\nu}_i\overset{1}{\nu}_i = 0; \tag{10}$$

i.e., the principal vectors are mutually orthogonal to each other. If $\sigma_1 = \sigma_2 \neq \sigma_3$, we can determine an infinite number of pairs of vectors $\overset{1}{\nu}_i$ and $\overset{2}{\nu}_i$ and define $\overset{3}{\nu}_i$ as a vector orthogonal to $\overset{1}{\nu}_i$ and $\overset{2}{\nu}_i$. If $\sigma_1 = \sigma_2 = \sigma_3$, then any set of orthogonal axes may be taken as the principal axes.

If the reference axes $x_1$, $x_2$, $x_3$ are chosen to coincide with the principal axes, then the matrix of stress components becomes

$$(\tau_{ij}) = \begin{pmatrix} \sigma_1 & 0 & 0 \\ 0 & \sigma_2 & 0 \\ 0 & 0 & \sigma_3 \end{pmatrix}. \tag{11}$$

## 3.8. LAMÉ STRESS ELLIPSOID

On any surface element with a unit outer normal vector $\boldsymbol{\nu}$, $(\nu_i)$, there acts a traction vector $\overset{\nu}{\mathbf{T}}$, $(\overset{\nu}{T}_i)$, with components given by

$$\overset{\nu}{T}_i = \tau_{ji}\nu_j.$$

Let the coordinate axes $x_1, x_2, x_3$ be chosen as the principal axes of the stress tensor, and let the principal stresses be written as $\sigma_1$, $\sigma_2$, $\sigma_3$. Then,

$$\tau_{ij} = 0 \quad \text{if} \quad i \neq j,$$

and

$$\overset{\nu}{T}_1 = \sigma_1\nu_1, \qquad \overset{\nu}{T}_2 = \sigma_2\nu_2, \qquad \overset{\nu}{T}_3 = \sigma_3\nu_3.$$

Since $\nu_i$ is a unit vector, we have

$$(\nu_1)^2 + (\nu_2)^2 + (\nu_3)^2 = 1.$$

Hence, the components of $\overset{\nu}{T}_i$ satisfy the equation

(1) $$\frac{(\overset{\nu}{T}_1)^2}{(\sigma_1)^2} + \frac{(\overset{\nu}{T}_2)^2}{(\sigma_2)^2} + \frac{(\overset{\nu}{T}_3)^2}{(\sigma_3)^2} = 1,$$

which is the equation of an ellipsoid with reference to a system of rectangular Cartesian coordinates with axes labeled $\overset{\nu}{T}_1, \overset{\nu}{T}_2, \overset{\nu}{T}_3$. This ellipsoid is the locus of the end points of vectors $\overset{\nu}{T}_i$ issuing from a common center (Fig. 3.8:1). Its existence was noted by Cauchy and by Lamé during the period of founding of the theory of elasticity (1820–1830).

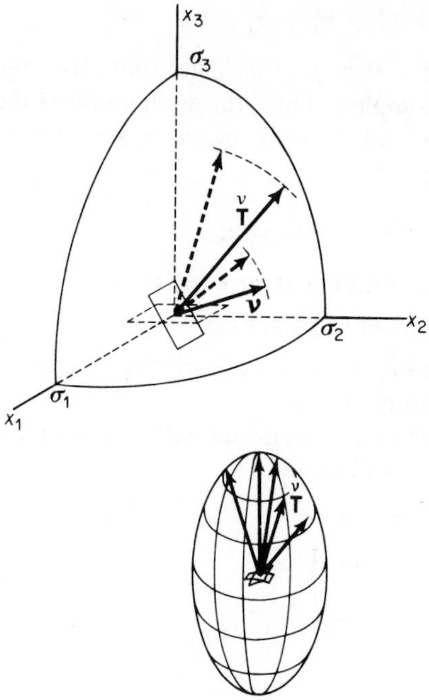


**Fig. 3.8:1.** Stress ellipsoid as the locus of the end of the vector $\overset{\nu}{\mathbf{T}}$ as $\boldsymbol{\nu}$ varies.

When negative values of the principal radii $\sigma_1, \sigma_2, \sigma_3$ of the Lamé ellipsoid of stress are admitted as a possibility, the three stress invariants $I_1, I_2, I_3$ in Eqs. (3.7:4) and (3.7:5) have the following geometric meaning: $I_1$ is the sum of the three principal radii of the ellipsoid, $I_2$ is proportional to the sum of the three principal diametral areas, and $I_3$ is proportional to the volume of the ellipsoid.

## 3.9. CAUCHY'S STRESS QUADRIC

We have seen that on an element of surface with a unit outer normal $\boldsymbol{\nu}$, $(\nu_i)$, there acts a traction $\overset{\nu}{\mathbf{T}}$, $(\overset{\nu}{T}_i = \tau_{ji}\nu_j)$. The component of $\overset{\nu}{\mathbf{T}}$ in the direction of $\boldsymbol{\nu}$ is the normal stress acting on the surface element. Let this normal stress be denoted by $\sigma_{(n)}$. Since the component of a vector in the direction of another vector is given by the scalar product of the two vectors, we obtain

(1) $$\sigma_{(n)} = \tau_{ij}\nu_i\nu_j.$$

Let a point $P$ be chosen along the outer normal (Fig. 3.9:1). If we multiply $\nu_i$ by the length $OP$, a vector

(2) $$x_i = |OP|\, \nu_i$$

is obtained. Multiplying Eq. (1) by $|OP|^2$, we obtain

(3) $$\sigma_{(n)}\, |OP|^2 = \tau_{ij}x_ix_j.$$

Cauchy considers the quadratic surface

(4) $$f(x_1, x_2, x_3) = \tau_{ij}x_ix_j \mp k_0^2 = 0, \qquad k_0 = \text{constant},$$

where the ambiguous sign is to be chosen so that (4) represents a real surface. A radius vector drawn from the origin to the quadratic surface will have a length $|OP|$,

(5) $$|OP|^2 = \frac{\pm k_0^2}{\sigma_{(n)}}.$$

If a radius vector drawn from the origin to the quadratic surface in the direction of $\nu_i$ intersects the surface at a point $P$, then the normal stress acting

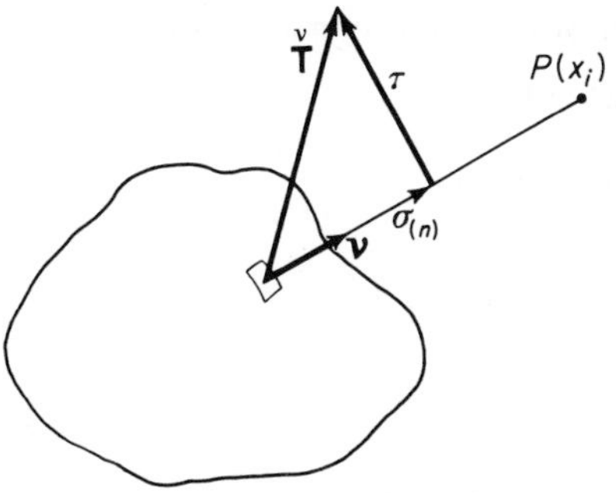


Fig. 3.9:1. Notations.

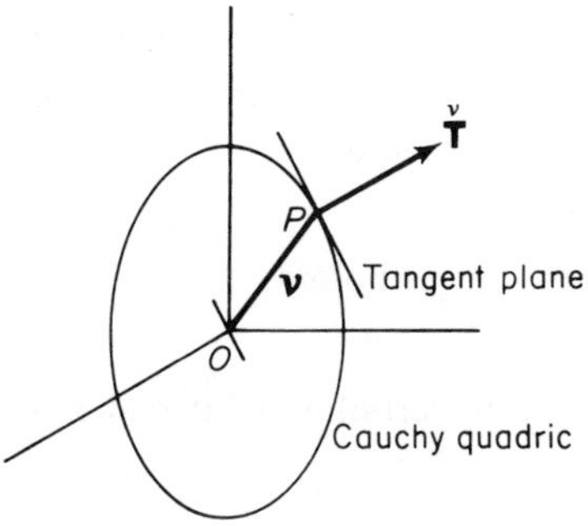


Fig. 3.9:2. Cauchy's stress quadric.

on the surface perpendicular to $\nu_i$ is given by

(6) $$\sigma_{(n)} = \frac{\pm k_0^2}{|OP|^2}.$$

Furthermore, since the direction cosines of a normal to a surface

(7) $$f(x_1, x_2, x_3) = 0$$

are proportional to the partial derivatives $\partial f/\partial x_1$, $\partial f/\partial x_2$, $\partial f/\partial x_3$, we can see that the components of the normal to the stress quadric (4) are

(8) $$\frac{\partial}{\partial x_k}(\tau_{ij}x_ix_j) = \tau_{kj}x_j + \tau_{ik}x_i.$$

This is equal to $2\overset{\nu}{T}_k\,|OP|$, if $\tau_{ij} = \tau_{ji}$. Hence, for a symmetric stress tensor, the outer normal to the stress quadric agrees in direction with the traction $\overset{\nu}{\mathbf{T}}$ acting on a surface with normal $\boldsymbol{\nu}$ (Fig. 3.9:2).

From the geometry of quadratic surfaces, it is known that a set of coordinate axes can be determined so that the cross-product terms in (4) vanish. The proof for this statement is formally identical with what we did in Sec. 3.7, although we now speak in the language of geometry. The axes so determined are the principal axes of the quadratic surface. With respect to the principal axes, the Cauchy quadric assumes the form

(9) $$\sigma_1(x_1)^2 + \sigma_2(x_2)^2 + \sigma_3(x_3)^2 = \pm k_0^2.$$

The coefficients $\sigma_1, \sigma_2, \sigma_3$ are the principal stresses. Let the axes be so numbered that

(10) $$\sigma_1 \geqslant \sigma_2 \geqslant \sigma_3.$$

Then (9) represents

(a) an ellipsoid if

$$\sigma_1 > \sigma_2 > \sigma_3 > 0, \quad (+k_0^2); \qquad 0 > \sigma_1 > \sigma_2 > \sigma_3, \quad (-k_0^2);$$

(b) an ellipsoid of revolution if $\sigma$'s are all of the same sign and either

$$\sigma_1 = \sigma_2, \quad \text{or} \quad \sigma_2 = \sigma_3, \quad \text{or} \quad \sigma_3 = \sigma_1;$$

(c) a sphere if

$$\sigma_1 = \sigma_2 = \sigma_3;$$

(d) a hyperboloid of one sheet if

$$\sigma_1 \geqslant \sigma_2 > 0, \qquad \sigma_3 < 0, \qquad +k_0^2;$$

(e) a hyperboloid of two sheets if

$$\sigma_1 \geqslant \sigma_2 > 0, \qquad \sigma_3 < 0, \qquad -k_0^2.$$

where $\pm k_0^2$ indicates the right-hand side term in Eq. (9).

The last two cases show that the ambiguous sign in Eq. (9) must be chosen as positive when the normal stress is in tension, negative when the normal stress is in compression. As shown in Fig. 3.9:3, it takes both sets of the hyperboloids to describe the stress state at a point $P$ in the last cases. The case $\sigma_1 > 0, 0 > \sigma_2 \geqslant \sigma_3$ is similar to the last named case.

If the stress tensor is asymmetric because body moments must be considered, then the normal to quadric (9) does not coincide with the surface traction $\overset{\nu}{\mathbf{T}}$, in which case the principal axes of the Cauchy quadric are not the principal axes for the stress.

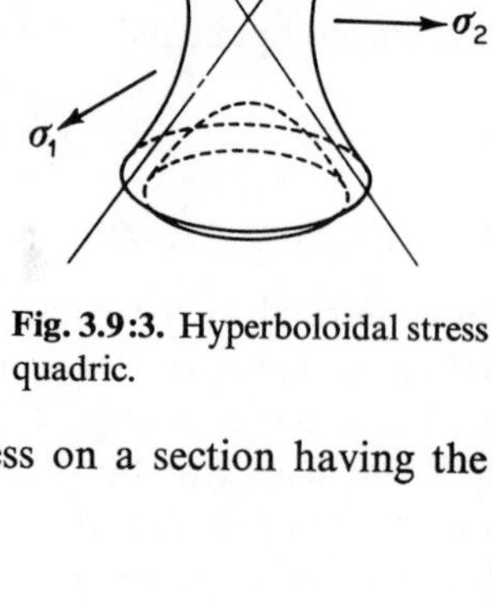

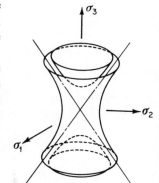

Fig. 3.9:3. Hyperboloidal stress quadric.

## 3.10. SHEARING STRESSES

The magnitude of the resultant shearing stress on a section having the normal $\nu_i$ is given by the equation

(1) $$\tau^2 = |\overset{\nu}{T}_i|^2 - \sigma_{(n)}^2$$

(see Fig. 3.9:1). Let the principal axes be chosen as the coordinate axes, and let $\sigma_1$, $\sigma_2$, $\sigma_3$ be the principal stresses. Then

$$|\overset{\nu}{T}_i|^2 = (\sigma_1\nu_1)^2 + (\sigma_2\nu_2)^2 + (\sigma_3\nu_3)^2,$$

and, from Eq. (3.9:1),

$$\sigma_{(n)}^2 = [\sigma_1(\nu_1)^2 + \sigma_2(\nu_2)^2 + \sigma_3(\nu_3)^2]^2.$$

On substituting into (1) and noting that

$$(\nu_1)^2 - (\nu_1)^4 = (\nu_1)^2[1 - (\nu_1)^2] = (\nu_1)^2[(\nu_2)^2 + (\nu_3)^2],$$

we see that

$$(2) \qquad \tau^2 = (\nu_1)^2(\nu_2)^2(\sigma_1 - \sigma_2)^2 + (\nu_2)^2(\nu_3)^2(\sigma_2 - \sigma_3)^2 + (\nu_3)^2(\nu_1)^2(\sigma_3 - \sigma_1)^2.$$

For example, if $\nu_1 = \nu_2 = 1/\sqrt{2}$ and $\nu_3 = 0$, then $\tau = \pm\frac{1}{2}(\sigma_1 - \sigma_2)$.

**Problem 3.4.** Show that $\tau_{\max} = \frac{1}{2}(\sigma_{\max} - \sigma_{\min})$ and that the plane on which $\tau_{\max}$ acts makes an angle of 45° with the direction of the largest and the smallest principal stresses.

## 3.11. MOHR'S CIRCLES

Let $\sigma_1$, $\sigma_2$, $\sigma_3$ be the principal stresses at a point. The stress components acting on any other sections can be obtained by the tensor transformation

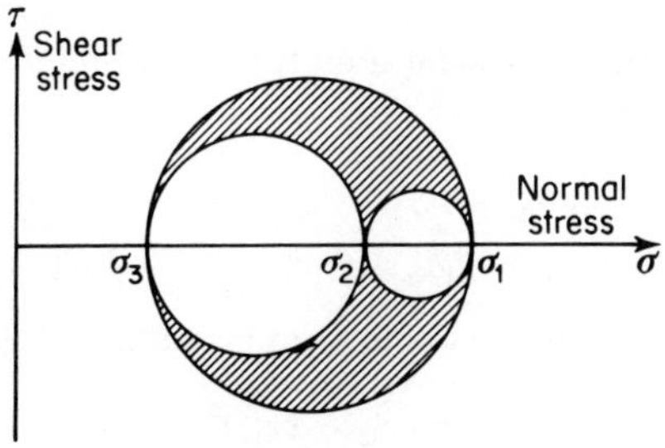


**Fig. 3.11:1.** Mohr's circles.

law, Eq. (3.5:3). Otto Mohr, in papers published in 1882 and 1900, has shown the interesting result that if the normal stress $\sigma_{(n)}$ and the shearing stress $\tau$ acting on any section be plotted on a plane, with $\sigma$ and $\tau$ as coordinates as shown in Fig. 3.11:1, the locus necessarily falls in a closed domain represented by the shaded area bounded by the three semicircles with centers on the $\sigma$-axis. A detailed proof can be found in Westergaard,[1.2] *Elasticity and Plasticity*, pp. 61–64; or Sokolnikoff,[1.2] *Elasticity*, p. 52. The practical problem of graphical construction of Mohr's circle from strain-gage data is discussed in Biezeno and Grammel,[1.2] *Engineering Dynamics*, Vol. 1, p. 245; Pearson,[1.4] *Theoretical Elasticity*, p. 64. See Bibliography on pp. 481–485.

### 3.12. STRESS DEVIATIONS

The tensor

(1) ▲ $$\tau'_{ij} = \tau_{ij} - \sigma_0\, \delta_{ij}$$

is called the *stress deviation tensor*, where $\delta_{ij}$ is the Kronecker delta and $\sigma_0$ is the mean stress

(2) $$\sigma_0 = \tfrac{1}{3}(\sigma_1 + \sigma_2 + \sigma_3) = \tfrac{1}{3}(\tau_{11} + \tau_{22} + \tau_{33}) = \tfrac{1}{3}I_1,$$

where $I_1$ is the first invariant of Sec. 3.7 and $\tau'_{ij}$ specifies the deviation of the state of stress from the mean stress.

The first invariant of the stress deviation tensor always vanishes:

(3) $$I'_1 = \tau'_{11} + \tau'_{22} + \tau'_{33} = 0.$$

To determine the principal stress deviations, the procedure of Sec. 3.7 may be followed. The determinental equation

(4) $$|\tau'_{ij} - \sigma'\, \delta_{ij}| = 0$$

may be expanded in the form

(5) $$\sigma'^3 - J_2\breve{\sigma}' - J_3 = 0.$$

It is easy to verify the following equations relating $J_2$, $J_3$ to the invariants $I_2$, $I_3$ defined in Sec. 3.7,

(6) $$J_2 = 3\sigma_0^2 - I_2,$$

(7) $$J_3 = I_3 - I_2\sigma_0 + 2\sigma_0^3 = \mathrm{I}_3 + J_2\sigma_0 - \sigma_0^3,$$

and the alternative expressions below on account of Eq. (3),

(8) $$\begin{aligned} J_2 &= -\tau'_{11}\tau'_{22} - \tau'_{22}\tau'_{33} - \tau'_{33}\tau'_{11} + (\tau_{12})^2 + (\tau_{23})^2 + (\tau_{31})^2 \\ &= \tfrac{1}{2}[(\tau'_{11})^2 + (\tau'_{22})^2 + (\tau'_{33})^2] + (\tau_{12})^2 + (\tau_{23})^2 + (\tau_{31})^2 \\ &= \tfrac{1}{6}[(\tau_{11} - \tau_{22})^2 + (\tau_{22} - \tau_{33})^2 + (\tau_{33} - \tau_{11})^2] + (\tau_{12})^2 + (\tau_{23})^2 + (\tau_{31})^2 \\ &= \tfrac{3}{2}\tau_0^2. \end{aligned}$$

The $\tau_0$ in the last equation is the octahedral stress, which will be defined in the next section.

We note also the simple expressions

(9) ▲ $$J_2 = \tfrac{1}{2}\tau'_{ij}\tau'_{ij},$$

(10) ▲ $$J_3 = \tfrac{1}{3}\tau'_{ij}\tau'_{jk}\tau'_{ki}.$$

**Problem 3.5.** Show that the principal stresses as given by the three roots of Eq. (5) can be written as

$$\sigma_1' = \tau_0\sqrt{2}\cos\alpha, \qquad \sigma_2' = \tau_0\sqrt{2}\cos\left(\alpha + \frac{2\pi}{3}\right), \qquad \sigma_3' = \tau_0\sqrt{2}\cos\left(\alpha - \frac{2\pi}{3}\right),$$

where $\cos 3\alpha = J_3\sqrt{2}/\tau_0^3$, and $J_2 = 3\tau_0^2/2$.

## 3.13. OCTAHEDRAL SHEARING STRESS

The octahedral shearing stress $\tau_0$ is the resultant shearing stress on a plane that makes the same angle with the three principal directions. Such a plane is called an *octahedral plane*; eight such planes can form an octahedron. See Fig. 3.13:1. The direction cosines $\nu_i$ of a normal to the octahedral plane relative to the principal axes are such that

$$(\nu_1)^2 = (\nu_2)^2 = (\nu_3)^2 = \tfrac{1}{3}.$$

Hence, Eq. (3.10:2) gives

$$9\tau_0^2 = (\sigma_1 - \sigma_2)^2 + (\sigma_2 - \sigma_3)^2 + (\sigma_3 - \sigma_1)^2,$$

which is proportional to the sum of the areas of Mohr's three semicircles. It can be easily verified that the octahedral stress can be expressed in terms of the two invariants $I_1$ and $I_2$ of Sec. 3.7,

$$9\tau_0^2 = 2I_1^2 - 6I_2.$$

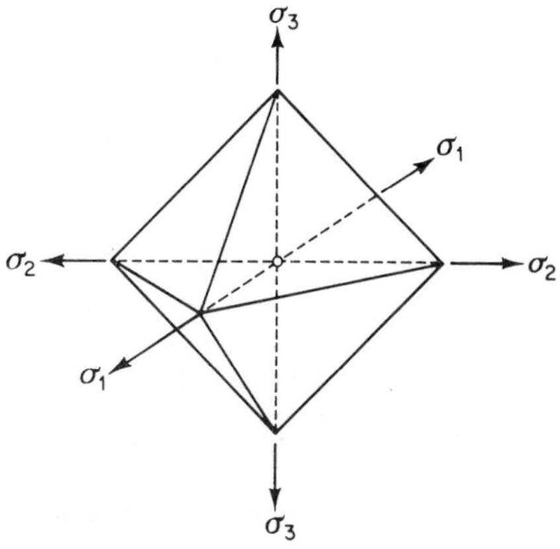


**Fig. 3.13:1.** Octahedral planes.

The square of the octahedral stress happens to be proportional to the second invariant $J_2$ of the stress deviation, Eqs. (3.12:6) and (3.12:8). In 1913, Richard von Mises proposed the hypothesis that yielding of some of the most important materials occurs at a constant value of the quantity $J_2$. Nadai then introduced the interpretation of $J_2$ as proportional to the octahedral shearing stress. In this way, $J_2$ or $\tau_0$ enters into the basic equations of plasticity.

**Problem 3.6.** If $\sigma_1 > \sigma_2 > \sigma_3$ and $\sigma_1$, $\sigma_3$ are given, at what values of $\sigma_2$ does $\tau_0$ attain its extreme values?

**Problem 3.7.** Let $\sigma_x = -5c$, $\sigma_y = c$, $\sigma_z = c$, $\tau_{xy} = -c$, $\tau_{yz} = \tau_{zx} = 0$, where $c = 1{,}000$ lb/sq in. Determine the principal stresses, the principal stress deviations, the direction cosines of the principal directions, the greatest shearing stress, and the octahedral stress.

**Problem 3.8.** Consider a horizontal beam as shown in Fig. P3.8. According to the usual elementary theory of bending, the "fiber stress" is $\sigma_{xx} = -12My/bh^3$, where $M$ is the bending moment which is a function of $x$. Assume this value of $\sigma_{xx}$, and assume further that $\sigma_{zz} = \sigma_{zx} = \sigma_{zy} = 0$, that the body force is absent,

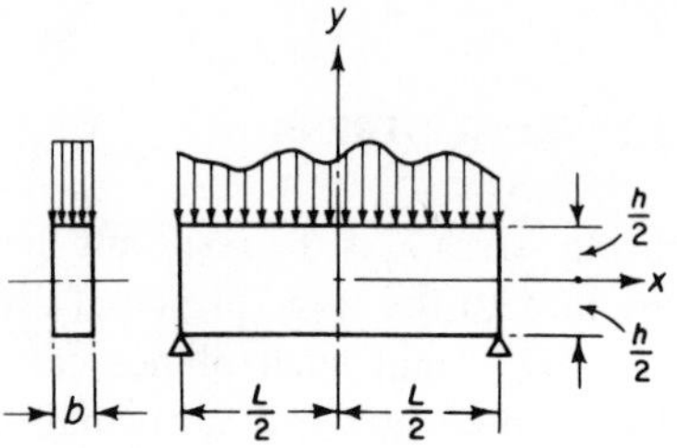


**P3.8**

that $\sigma_{xy} = 0$ at the top and bottom of the beam ($y = +h/2$), and that $\sigma_{yy} = 0$ at the bottom. Derive $\sigma_{xy}$ and $\sigma_{yy}$ from the equations of equilibrium. Compare the results with those derived in elementary mechanics of materials.

## 3.14. STRESS TENSOR IN GENERAL COORDINATES

So far we have discussed the stress tensor in rectangular Cartesian coordinates, in which there is no necessity to distinguish the contravariant and covariant transformations. The necessary distinction arises in curvilinear coordinates.

Just as a vector in three-dimensional Euclidean space may assume either a contravariant or a covariant form, a stress tensor can be either contravariant, $\tau^{ij}$, or mixed, $\tau^i_j$, or covariant, $\tau_{ij}$. The tensors $\tau^{ij}$, $\tau^i_j$, and $\tau_{ij}$ are related to each other by raising or lowering of the indices by forming inner products with the metric tensors $g_{ij}$ and $g^{ij}$:

$$
\begin{aligned}
\tau^i_j &= g_{\alpha j}\tau^{i\alpha} = g_{\alpha j}\tau^{\alpha i}, \\
\tau_{ij} &= g_{i\alpha}\tau^\alpha_j, \\
\tau^{ij} &= g^{i\alpha}\tau^j_\alpha.
\end{aligned}
\tag{1}
$$

The correctness of these tensor relations are again seen by specializing them into rectangular Cartesian coordinates. We have seen that $\tau_{ij}$ is symmetric in rectangular Cartesian coordinates. It is easy to see from the tensor transformation law that *the symmetry property remains invariant in coordinate transformation. Hence,* $\tau_{ij} = \tau_{ji}$ *in all admissible coordinates*†, and we are

† Similarly, *the contravariant stress tensor* $\tau^{ij}$ *is symmetric*, $\tau^{ij} = \tau^{ji}$. It does not make sense, however, to say that the mixed tensor $\tau^i_j$ is symmetric, since an equation like $\tau^i_j = \tau^j_i$, with the indices switching roles on the two sides of the equation, is not a tensor equation.

allowed to write the mixed tensor as $\tau^i_j$ and not $\tau^i{}_j$ or $\tau_j{}^i$. But what is the physical meaning of each of these components?

To clarify the meaning of the stress components in an arbitrary curvilinear coordinates system, let us first consider a geometric relationship. Let us form an infinitesimal tetrahedron whose edges are formed by the coordinate

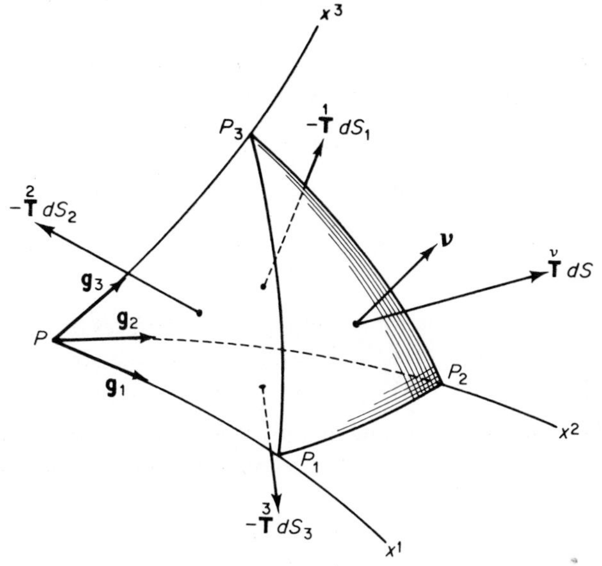


**Fig. 3.14:1.** Geometric relationship.

curves $PP_1$, $PP_2$, $PP_3$ and the curves $P_1P_2$, $P_2P_3$, $P_3P_1$, as shown in Fig. 3.14:1. Let us write, vectorially,

$$\overline{PP_1} = \mathbf{r}_1, \qquad \overline{PP_2} = \mathbf{r}_2, \qquad \overline{PP_3} = \mathbf{r}_3.$$

Then

$$\overline{P_1P_2} = \mathbf{r}_2 - \mathbf{r}_1, \qquad \overline{P_1P_3} = \mathbf{r}_3 - \mathbf{r}_1, \qquad \overline{P_2P_3} = \mathbf{r}_3 - \mathbf{r}_2,$$

and we have

$$\text{(2)} \qquad \overline{P_1P_2} \times \overline{P_1P_3} = (\mathbf{r}_2 - \mathbf{r}_1) \times (\mathbf{r}_3 - \mathbf{r}_1) = -\mathbf{r}_1 \times \mathbf{r}_3 - \mathbf{r}_2 \times \mathbf{r}_1 + \mathbf{r}_2 \times \mathbf{r}_3$$
$$= \mathbf{r}_2 \times \mathbf{r}_3 + \mathbf{r}_3 \times \mathbf{r}_1 + \mathbf{r}_1 \times \mathbf{r}_2.$$

Now the vector product $\mathbf{A} \times \mathbf{B}$ of any two vectors $\mathbf{A}$ and $\mathbf{B}$ is a vector perpendicular to $\mathbf{A}$ and $\mathbf{B}$, whose positive sense is determined by the *right-hand screw rule* from $\mathbf{A}$ to $\mathbf{B}$, and whose length is equal to the area of a parallelogram formed by $\mathbf{A}$, $\mathbf{B}$ as two sides. Hence, if we denote by $\boldsymbol{\nu}$, $\boldsymbol{\nu}_1$, $\boldsymbol{\nu}_2$, $\boldsymbol{\nu}_3$ the unit vectors normal to the surfaces $P_1P_2P_3$, $PP_2P_3$, $PP_3P_1$, $PP_1P_2$, respectively, and by $dS$, $dS_1$, $dS_2$, $dS_3$ their respective areas, Eq. (2) may be written as

$$\text{(3)} \qquad \boldsymbol{\nu}\, dS = \boldsymbol{\nu}_1\, dS_1 + \boldsymbol{\nu}_2\, dS_2 + \boldsymbol{\nu}_3\, dS_3.$$

Now let us recall that in Sec. 2.14 we defined the reciprocal base vectors $\mathbf{g}^1, \mathbf{g}^2, \mathbf{g}^3$ which are perpendicular to the coordinate planes and are of length $\sqrt{g^{11}}, \sqrt{g^{22}}, \sqrt{g^{33}}$, respectively. We see that the unit vectors $\boldsymbol{\nu}_1, \boldsymbol{\nu}_2, \boldsymbol{\nu}_3$ are exactly $\mathbf{g}^1/\sqrt{g^{11}}, \mathbf{g}^2/\sqrt{g^{22}}, \mathbf{g}^3/\sqrt{g^{33}}$, respectively. Hence,

$$(4) \qquad \boldsymbol{\nu}\, dS = \sum_{i=1}^{3} \frac{dS_i}{\sqrt{g^{ii}}} \mathbf{g}^i.$$

If the unit normal vector $\boldsymbol{\nu}$ is resolved into its covariant components with respect to the reciprocal base vectors, then

$$(5) \qquad \boldsymbol{\nu} = \nu_i \mathbf{g}^i.$$

We see from the last two equations that

$$(6) ▲ \qquad \nu_i \sqrt{g^{ii}}\, dS = dS_i, \qquad i \text{ not summed.}$$

This is the desired result.

Let us now consider the forces acting on the surfaces of the infinitesimal tetrahedron. We assume that the limiting value of the stress acting on any surface exists as the area of the surface tends to zero. Let the *stress vector* acting on $dS$ be denoted by $\overset{\nu}{\mathbf{T}}$, and let those acting on $dS_i$ be denoted by $-\overset{i}{\mathbf{T}}$. The equation of motion of this infinitesimal tetrahedron is, in the limit,

$$(7) \qquad \overset{\nu}{\mathbf{T}}\, dS = \overset{i}{\mathbf{T}}\, dS_i.$$

Volume forces and inertia (mass·acceleration) forces acting on the tetrahedron do not enter into this equation, because they are of higher order of smallness than the surface forces. On substituting (6) into (7) and canceling the non-vanishing factor $dS$, we have

$$(8) \qquad \overset{\nu}{\mathbf{T}} = \sum_{i=1}^{3} \overset{i}{\mathbf{T}} \nu_i \sqrt{g^{ii}}.$$

If the coordinates $x^i$ are changed to a new set $\bar{x}^i$ while the surface $P_1P_2P_3$ and the unit outer normal $\boldsymbol{\nu}$ remain unchanged, the stress vector $\overset{\nu}{\mathbf{T}}$ is invariant, but the vectors $\overset{i}{\mathbf{T}}$ will change because they will be associated with the new coordinate surfaces. (See Fig. 3.14.2, in which only one pair of vectors $\overset{2}{\mathbf{T}}, \overset{2}{\mathbf{T}}'$ are shown.) The covariant components $\nu_i$ will change also. Inasmuch as $\overset{\nu}{\mathbf{T}}$ is invariant and $\nu_i$ is a covariant tensor, Eq. (8) shows that $\overset{i}{\mathbf{T}}\sqrt{g^{ii}}$ transforms according to a contravariant type of transformation. Resolving the vectors $\overset{i}{\mathbf{T}}\sqrt{g^{ii}}$ into their components with respect to the base vectors $\mathbf{g}_i$ and the reciprocal base vectors $\mathbf{g}^i$, we have

$$(9) ▲ \qquad \sqrt{g^{ii}}\,\overset{i}{\mathbf{T}} = \tau^{ij}\mathbf{g}_j = \tau^i_{\,j}\mathbf{g}^j, \qquad i \text{ not summed.}$$

On the other hand, the components of $\overset{\nu}{\mathbf{T}}$ may be written as

(10) $$\overset{\nu}{\mathbf{T}} = \overset{\nu}{T}{}^j \mathbf{g}_j = \overset{\nu}{T}_j \mathbf{g}^j.$$

Substitution of (9) and (10) into (8) shows that

(11) ▲ $$\overset{\nu}{T}{}^j = \tau^{ij}\nu_i, \qquad \overset{\nu}{T}_j = \tau^i_{\ j}\nu_i.$$

The tensorial character of $\tau^{ij}$ and $\tau^i_j$ are demonstrated both by (9) and by (11) according to the quotient rule.

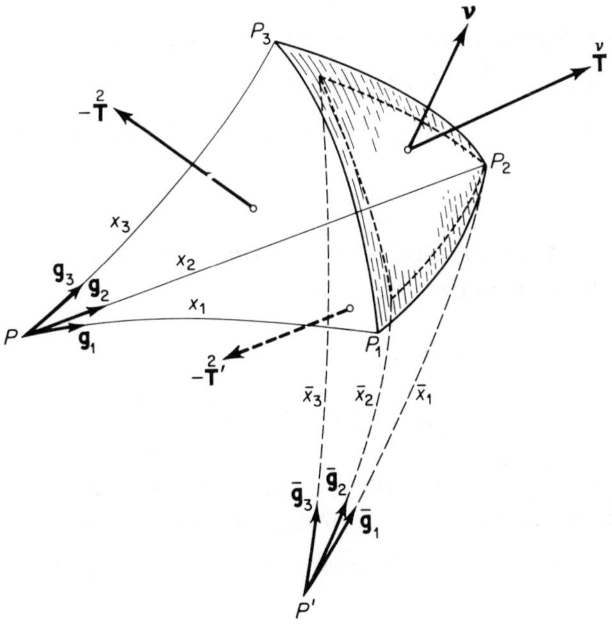


**Fig. 3.14:2.** Tractions referred to two different elementary tetrahedron with a common surface.

Let us recapitulate the important points. We consider an infinitesimal tetrahedron $P$—$P_1P_2P_3$ as shown in Fig. 3.14:1, where the surfaces $PP_2P_3$, $PP_3P_1$, $PP_1P_2$ are coordinate surfaces. The outer normal to the triangle $P_1P_2P_3$ is $\boldsymbol{\nu}$. Let the vector $\overset{\nu}{\mathbf{T}}$ represent the stress (force per unit area) acting on the face $P_1P_2P_3$ of the tetrahedron. Then the vectors $-\overset{1}{\mathbf{T}}$, $-\overset{2}{\mathbf{T}}$, $-\overset{3}{\mathbf{T}}$ represent the stress vectors acting on the coordinate surfaces $PP_2P_3$, $PP_3P_1$, $PP_1P_2$, respectively. We cannot say much about the vectors $-\overset{i}{\mathbf{T}}$, but we learn that when a scale factor is added the vectors $\overset{i}{\mathbf{T}}\sqrt{g^{ii}}$ ($i = 1, 2, 3$, not summed) transform according to the contravariant rule. When the vectors $\overset{i}{\mathbf{T}}\sqrt{g^{ii}}$ are resolved into components along the base vectors $\mathbf{g}_j$, the tensor components $\tau^{ij}$ are obtained. When the resolution is made with respect to the reciprocal base vectors $\mathbf{g}^j$, then the mixed tensor $\tau^i_j$ is obtained. On the other hand, if the stress vector $\overset{\nu}{\mathbf{T}}$ (acting on the triangle $P_1P_2P_3$ with outer normal $\boldsymbol{\nu}$) is resolved into the contravariant components $\overset{\nu}{T}{}^j$ along the base vectors $\mathbf{g}_j$ and the covariant components $\overset{\nu}{T}_j$ along the reciprocal base vectors $\mathbf{g}^j$, then

$$\overset{\nu}{T}{}^j = \tau^{ij}\nu_i, \qquad \overset{\nu}{T}_j = \tau^i_j\nu_i.$$

The contravariant stress tensor $\tau^{ij}$ and the mixed stress tensor $\tau^i_j$ are related to the stress vectors $\overset{i}{\mathbf{T}}$ by the Eq. (9). The covariant stress tensor $\tau_{ij}$ defined in Eq. (1) cannot be so simply related to the vectors $\overset{i}{\mathbf{T}}$ and are therefore of less importance.

## 3.15. PHYSICAL COMPONENTS OF A STRESS TENSOR IN GENERAL COORDINATES

If we write Eq. (3.14:9) as

(1)
$$\overset{i}{\mathbf{T}} = \sum_{j=1}^{3} \sqrt{\frac{g_{jj}}{g^{ii}}}\,\tau^{ij}\frac{\mathbf{g}_j}{\sqrt{g_{jj}}} = \sum_{j=1}^{3}\sigma^{ij}\frac{\mathbf{g}_j}{\sqrt{g_{jj}}}, \qquad i \text{ not summed.}$$

Then, since $\mathbf{g}_j/\sqrt{g_{jj}}$ ($j$ not summed) are unit vectors along the coordinate curves, the components $\sigma^{ij}$ are uniform in physical dimensions and represent the *physical components* of the stress vector $\overset{i}{\mathbf{T}}$ in the direction of the unit vectors $\mathbf{g}_j/\sqrt{g_{jj}}$,

(2)
$$\sigma^{ij} = \sqrt{\frac{g_{jj}}{g^{ii}}}\,\tau^{ij}, \qquad i, j \text{ not summed.}$$

But $\sigma^{ij}$ is not a tensor.

On the other hand, if we use the mixed tensor $\tau^i_j$ in (3.14:1), we have

(3)
$$\overset{i}{\mathbf{T}} = \tau^i_j\frac{\mathbf{g}^j}{\sqrt{g^{ii}}} = \tau^i_j\sqrt{\frac{g^{jj}}{g^{ii}}}\cdot\frac{\mathbf{g}^j}{\sqrt{g^{jj}}}, \qquad i \text{ not summed.}$$

Thus

(4)
$$\sigma^i_j = \sqrt{\frac{g^{jj}}{g^{ii}}}\,\tau^i_j, \qquad i, j \text{ not summed}$$

are the physical components of the tensor $\tau^i_j$, of uniform physical dimensions, representing the components of the stress vector $\overset{i}{\mathbf{T}}$ resolved in the directions of the reciprocal base vectors.

When curvilinear coordinates are used, we like to retain the liberty of choosing coordinates without regard to dimensions. Thus, in cylindrical polar coordinates $(r, \theta, z)$, $r$ and $z$ have the dimensions of length and $\theta$ is an angle. The corresponding tensor components of a vector referred to polar coordinates will have different dimensions. For physical understanding it is desirable to employ the physical components, but for the convenience of analysis it is far more expedient to use the tensor components.

## 3.16. EQUATIONS OF EQUILIBRIUM IN CURVILINEAR COORDINATES

In Sec. 3.4 we discussed the equations of equilibrium in terms of Cartesian tensors in rectangular Cartesian coordinates. To obtain these equations in any curvilinear coordinates, it is only necessary to observe that the equilibrium conditions must be expressed in a tensor equation. Thus the equations of equilibrium must be

$$\tau^{ij}|_j + X^i = 0, \quad \text{in volume,} \tag{1}$$

$$\tau^{ji}\nu_j = \overset{\nu}{T}{}^i, \quad \text{on surface.} \tag{2}$$

The truth is at once proved by observing that these are truly tensor equations and that they hold in the special case of rectangular Cartesian coordinates. Hence, they hold in any coordinates that can be derived from the Cartesian coordinates through admissible transformations.

The practical application of tensor analysis in the derivation of the equations of equilibrium in particular curvilinear coordinates will be illustrated in Secs. 4.11 and 4.12. It will be seen that these lengthy equations can be obtained in a routine manner without too much effort. Because the manipulation is routine, chances of error are minimized. This practical application may be regarded as the first dividend to be paid for the long process of learning the tensor analysis.

**Problem 3.9.** Let us recast the principal results obtained above into tensor equations in general coordinates of reference. In rectangular Cartesian coordinates, there is no difference in contravariant and covariant transformations. Hence, the Cartesian stress tensor may be written as $\tau_{ij}$, or $\tau^{ij}$, or $\tau^i_{j}$. In general frames of reference, $\tau_{ij}$, $\tau^{ij}$, $\tau^i_j$ are different. Their components may have different values. They are different versions of the same physical entity. Now prove the following results in general coordinates.

(a) The tensors $\tau^{ij}$, $\tau_{ij}$ are symmetric if there is no body moment acting on the medium; i.e.,

$$\tau^{ij} = \tau^{ji}, \qquad \tau_{ij} = \tau_{ji}.$$

(b) Principal planes are planes on which the stress vector $\overset{\nu}{\mathbf{T}}$ is parallel to the normal vector $\boldsymbol{\nu}$. If we use contravariant components, we have, on a principal plane, $\overset{\nu}{T}{}^j = \sigma\nu^j$, where $\sigma$ is a constant. If we use covariant components, we have, correspondingly $\overset{\nu}{T}_j = \sigma\nu_j$. Show that $\sigma$ must satisfy the characteristic determinantal equation

$$|\tau^i_j - \sigma\delta^i_j| = 0 \quad \text{or its equivalent} \quad |\tau^{ij} - \sigma g^{ij}| = 0.$$

(c) The first invariant of the stress tensor is $\tau^i_i$, or $\tau^{ij}g_{ij}$. However, $\tau_{ii}$ and $\tau^{ii}$ are in general not invariants.

(d) The stress deviation tensor $s^i_j$ is defined as

$$s^i_j = \tau^i_j - \tfrac{1}{3}\tau^\alpha_\alpha g^i_j.$$

The first invariant of $s^i_j$ is zero. The second invariant has the convenient form $J_2 = \frac{1}{2}s^i_k s^k_i$.

(e) The octahedral shearing stress has the same value in any coordinates system.

# 4

# ANALYSIS OF STRAIN

In this chapter we shall consider the deformation of a body as a "mapping" of the body from the original state to the deformed state. Strain tensors which are useful for finite strains as well as infinitesimal strains are then defined.

## 4.1. DEFORMATION

In the formulation of continuum mechanics the configuration of a solid body is described by a continuous mathematical model whose geometrical points are identified with the place of the material particles of the body. When such a continuous body changes its configuration under some physical action, we impose the assumption that the change is continuous; i.e., neighborhoods are changed into neighborhoods. Any introduction of new boundary surfaces, such as caused by tearing of a membrane or fracture of a test specimen, must be regarded as an extraordinary circumstance requiring special attention and explanation.

Let a system of coordinates $a_1, a_2, a_3$ be chosen so that a point $P$ of a body at a certain instant of time is described by the coordinates $a_i (i = 1, 2, 3)$. At a later instant of time, the body is moved (deformed) to a new configuration; the point $P$ is moved to $Q$ with coordinates $x_i (i = 1, 2, 3)$ with respect to a new coordinate system $x_1, x_2, x_3$. The coordinate systems $a_1, a_2, a_3$ and $x_1, x_2, x_3$ may be curvilinear and need not be the same (Fig. 4.1:1), but, of course, they both describe a Euclidean space.

The change of configuration of the body will be assumed to be continuous, and the *point transformation* (*mapping*) from $P$ to $Q$ is assumed to be one-to-one. The equation of transformation can be written as

(1) $$x_i = \hat{x}_i(a_1, a_2, a_3),$$

which has a unique inverse

(2) $$a_i = \hat{a}_i(x_1, x_2, x_3)$$

for every point in the body. The functions $\hat{x}_i(a_1, a_2, a_3)$ and $\hat{a}_i(x_1, x_2, x_3)$ are assumed to be continuous and differentiable.

We shall be concerned with the description of the strain of the body, i.e., with the stretching and distortion of the body. If $P$, $P'$, $P''$ are three neighboring points forming a triangle in the original configuration, and if they are transformed to points $Q$, $Q'$, $Q''$ in the deformed configuration, the change in area and angles of the triangle is completely determined if we know

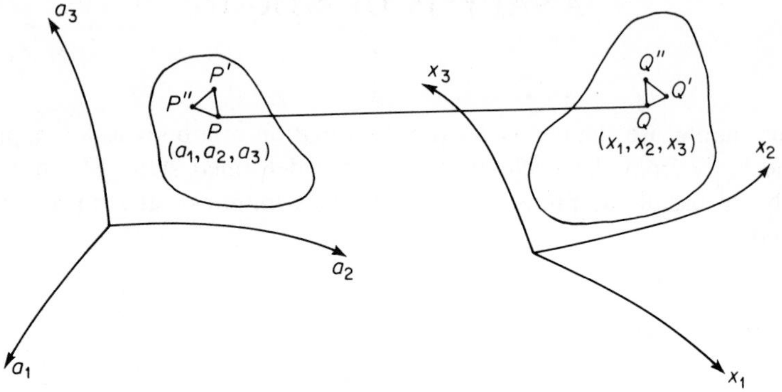


**Fig. 4.1:1.** Deformation of a body.

the change in length of the sides. But the "location" of the triangle is undetermined by the change of the sides. Similarly, if the change of length between any two arbitrary points of the body is known, the new configuration of the body will be completely defined except for the location of the body in space. In the following discussions our attention will be focused on the strain of the body, because it is the strain that is related to the stress. The description of the change in distance between any two points of the body is the key to the analysis of deformation.

Consider an infinitesimal line element connecting the point $P(a_1, a_2, a_3)$ to a neighboring point $P'(a_1 + da^1, a_2 + da^2, a_3 + da^3)$.† The square of the length $ds_0$ of $PP'$ in the original configuration is given by

$$ds_0^2 = a_{ij}\, da^i\, da^j, \tag{3}$$

where $a_{ij}$, evaluated at the point $P$, is the Euclidean metric tensor for the coordinate system $a_i$. When $P$, $P'$ is deformed to the points $Q(x_1, x_2, x_3)$ and $Q'(x_1 + dx^1, x_2 + dx^2, x_3 + dx^3)$, respectively, the square of the length $ds$ of the new element $QQ'$ is

$$ds^2 = g_{ij}\, dx^i\, dx^j, \tag{4}$$

where $g_{ij}$ is the Euclidean metric tensor for the coordinate system $x_i$.

† As remarked before in a footnote on p. 38, the point transformation between $a_i$ and $x_i$ does not follow tensor transformation law and the position of the indices has no significance. But the *differentials* $da^i$ and $dx^i$ do transform according to the contravariant tensor law and are contravariant vectors.

By Eqs. (1) and (2), we may also write

(5) $$ds_0^2 = a_{ij} \frac{\partial a_i}{\partial x_l} \frac{\partial a_j}{\partial x_m} dx^l dx^m,$$

(6) $$ds^2 = g_{ij} \frac{\partial x_i}{\partial a_l} \frac{\partial x_j}{\partial a_m} da^l da^m.$$

The difference between the squares of the length elements may be written, after several changes in the symbols for dummy indices, either as

(7) $$ds^2 - ds_0^2 = \left( g_{\alpha\beta} \frac{\partial x_\alpha}{\partial a_i} \frac{\partial x_\beta}{\partial a_j} - a_{ij} \right) da^i da^j,$$

or as

(8) $$ds^2 - ds_0^2 = \left( g_{ij} - a_{\alpha\beta} \frac{\partial a_\alpha}{\partial x_i} \frac{\partial a_\beta}{\partial x_j} \right) dx^i dx^j.$$

We define the *strain tensors*

(9) ▲ $$E_{ij} = \frac{1}{2}\left( g_{\alpha\beta} \frac{\partial x_\alpha}{\partial a_i} \frac{\partial x_\beta}{\partial a_j} - a_{ij} \right),$$

(10) ▲ $$e_{ij} = \frac{1}{2}\left( g_{ij} - a_{\alpha\beta} \frac{\partial a_\alpha}{\partial x_i} \frac{\partial a_\beta}{\partial x_j} \right),$$

so that

(11) ▲ $$ds^2 - ds_0^2 = 2E_{ij}\, da^i da^j,$$

(12) ▲ $$ds^2 - ds_0^2 = 2e_{ij}\, dx^i dx^j.$$

*The strain tensor $E_{ij}$ was introduced by Green and St. Venant, and is called Green's strain tensor. The strain tensor $e_{ij}$ was introduced by Cauchy for infinitesimal strains and by Almansi and Hamel for finite strains, and is known as Almansi's strain tensor.* In analogy with a terminology in hydrodynamics, $E_{ij}$ is often referred to as a strain tensor in Lagrangian coordinates and $e_{ij}$ as a strain tensor in Eulerian coordinates.

That $E_{ij}$ and $e_{ij}$ thus defined are tensors in the coordinate systems $a_i$ and $x_i$, respectively, follows from the quotient rule when it is applied to Eqs. (11) and (12). The tensorial character of $E_{ij}$ and $e_{ij}$ can also be verified directly from their definitions (9), (10) by considering further coordinates transformations in either the original configuration (from $a_1, a_2, a_3$ to $\bar{a}_1, \bar{a}_2, \bar{a}_3$), or the deformed configuration (from $x_1, x_2, x_3$ to $\bar{x}_1, \bar{x}_2, \bar{x}_3$). The details are left to the reader. The tensors $E_{ij}$ and $e_{ij}$ are obviously *symmetric*, i.e.,

(13) ▲ $$E_{ij} = E_{ji}, \qquad e_{ij} = e_{ji}.$$

An immediate consequence of Eqs. (11) and (12) is the fundamental result that a *necessary and sufficient condition that a deformation of a body be a rigid-body motion (consists merely of translation and rotation without changing distances between particles) is that all the components of the strain tensor $E_{ij}$ or $e_{ij}$ be zero throughout the body.*

In the discussion above we have used two sets of curvilinear coordinates to describe the position of each particle. One, $a_1, a_2, a_3$, is used in the original configuration, the other, $x_1, x_2, x_3$, is used in the deformed configuration. Now *there are two particularly favored choices of coordinates:*

**I.** *We use the same rectangular Cartesian coordinates for both the original and the deformed configurations of the body.* In this case, the metric tensors are extremely simple:

$$g_{ij} = a_{ij} = \delta_{ij}. \tag{14}$$

**II.** *We distort the frame of reference in the deformed configuration in such a way that the coordinates $x_1, x_2, x_3$, of a particle have the same numerical values $a_1, a_2, a_3$ as in the original configuration.* In this case, $x_i = a_i$, $\partial x_\alpha/\partial a_i = \delta_{\alpha i}$, $\partial a_\alpha/\partial x_i = \delta_{\alpha i}$, and Eqs. (9) and (10) are reduced to

$$E_{ij} = e_{ij} = \tfrac{1}{2}(g_{ij} - a_{ij}). \tag{15}$$

*Thus all the information about strain is contained in the change of the metric tensor as the frame of reference is distorted from the original configuration to the deformed configuration.* In many ways this is the most convenient choice in the study of large deformations. The coordinates so chosen are called *convected* or *intrinsic* coordinates.

In the following sections we wish to discuss the meaning of the individual components of the strain tensor. For this purpose the choice (I) above is the most appropriate.

## 4.2. STRAIN TENSORS IN RECTANGULAR CARTESIAN COORDINATES

If we use the *same rectangular Cartesian* (rectilinear and orthogonal) coordinate system to describe both the original and the deformed configurations of the body, then

$$g_{ij} = a_{ij} = \delta_{ij} = \begin{cases} 1 & \text{if } i = j, \\ 0 & \text{if } i \neq j. \end{cases} \tag{1}$$

Furthermore, if we introduce the *displacement vector* **u** with components

$$u_i = x_i - a_i, \qquad i = 1, 2, 3, \tag{2}$$

(see Fig. 4.2:1) then

$$\frac{\partial x_\alpha}{\partial a_i} = \frac{\partial u_\alpha}{\partial a_i} + \delta_{\alpha i}, \qquad \frac{\partial a_\alpha}{\partial x_i} = \delta_{\alpha i} - \frac{\partial u_\alpha}{\partial x_i},$$

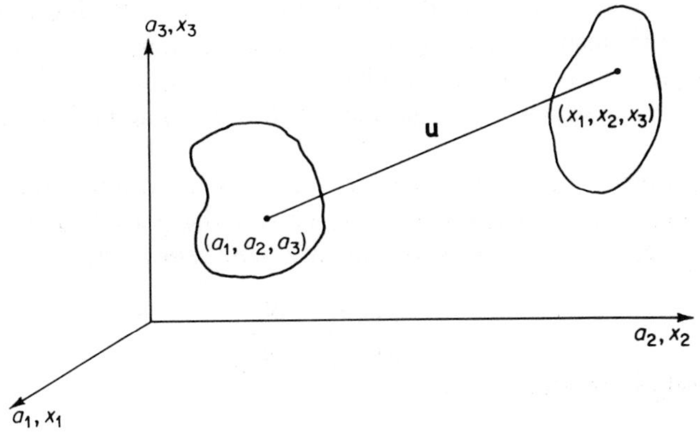


**Fig. 4.2:1.** Displacement vector.

and the strain tensors reduce to the simple form

(3) ▲
$$\begin{aligned} E_{ij} &= \frac{1}{2}\left[\delta_{\alpha\beta}\frac{\partial x_\alpha}{\partial a_i}\frac{\partial x_\beta}{\partial a_j} - \delta_{ij}\right] \\ &= \frac{1}{2}\left[\delta_{\alpha\beta}\left(\frac{\partial u_\alpha}{\partial a_i} + \delta_{\alpha i}\right)\left(\frac{\partial u_\beta}{\partial a_j} + \delta_{\beta j}\right) - \delta_{ij}\right] \\ &= \frac{1}{2}\left[\frac{\partial u_j}{\partial a_i} + \frac{\partial u_i}{\partial a_j} + \frac{\partial u_\alpha}{\partial a_i}\frac{\partial u_\alpha}{\partial a_j}\right] \end{aligned}$$

and

(4) ▲
$$\begin{aligned} e_{ij} &= \frac{1}{2}\left[\delta_{ij} - \delta_{\alpha\beta}\frac{\partial a_\alpha}{\partial x_i}\frac{\partial a_\beta}{\partial x_j}\right] \\ &= \frac{1}{2}\left[\delta_{ij} - \delta_{\alpha\beta}\left(-\frac{\partial u_\alpha}{\partial x_i} + \delta_{\alpha i}\right)\left(-\frac{\partial u_\beta}{\partial x_j} + \delta_{\beta j}\right)\right] \\ &= \frac{1}{2}\left[\frac{\partial u_j}{\partial x_i} + \frac{\partial u_i}{\partial x_j} - \frac{\partial u_\alpha}{\partial x_i}\frac{\partial u_\alpha}{\partial x_j}\right]. \end{aligned}$$

In unabridged notations ($x, y, z$ for $x_1, x_2, x_3$; $a, b, c$ for $a_1, a_2, a_3$; and $u, v, w$ for $u_1, u_2, u_3$), we have the typical terms

(5)
$$\begin{aligned} E_{aa} &= \frac{\partial u}{\partial a} + \frac{1}{2}\left[\left(\frac{\partial u}{\partial a}\right)^2 + \left(\frac{\partial v}{\partial a}\right)^2 + \left(\frac{\partial w}{\partial a}\right)^2\right], \\ e_{xx} &= \frac{\partial u}{\partial x} - \frac{1}{2}\left[\left(\frac{\partial u}{\partial x}\right)^2 + \left(\frac{\partial v}{\partial x}\right)^2 + \left(\frac{\partial w}{\partial x}\right)^2\right], \\ E_{ab} &= \frac{1}{2}\left[\frac{\partial u}{\partial b} + \frac{\partial v}{\partial a} + \left(\frac{\partial u}{\partial a}\frac{\partial u}{\partial b} + \frac{\partial v}{\partial a}\frac{\partial v}{\partial b} + \frac{\partial w}{\partial a}\frac{\partial w}{\partial b}\right)\right], \\ e_{xy} &= \frac{1}{2}\left[\frac{\partial u}{\partial y} + \frac{\partial v}{\partial x} - \left(\frac{\partial u}{\partial x}\frac{\partial u}{\partial y} + \frac{\partial v}{\partial x}\frac{\partial v}{\partial y} + \frac{\partial w}{\partial x}\frac{\partial w}{\partial y}\right)\right]. \end{aligned}$$

Note that $u, v, w$ are considered as functions of $a, b, c$, the position of points in the body in unstrained configuration, when the Lagrangian strain tensor is evaluated; thereas they are considered as functions of $x, y, z$, the position of points in the strained configuration, when the Eulerian strain tensor is evaluated.

If the components of displacement $u_i$ are such that their first derivatives are so small that the squares and products of the partial derivatives of $u_i$ are negligible, then $e_{ij}$ reduces to Cauchy's *infinitesimal strain tensor*,

$$(6) \quad \blacktriangle \qquad e_{ij} = \frac{1}{2}\left[\frac{\partial u_j}{\partial x_i} + \frac{\partial u_i}{\partial x_j}\right].$$

In unabridged notation,

$$(7) \qquad \begin{aligned} e_{xx} &= \frac{\partial u}{\partial x}, & e_{xy} &= \frac{1}{2}\left(\frac{\partial u}{\partial y} + \frac{\partial v}{\partial x}\right) = e_{yx}, \\ e_{yy} &= \frac{\partial v}{\partial y}, & e_{xz} &= \frac{1}{2}\left(\frac{\partial u}{\partial z} + \frac{\partial w}{\partial x}\right) = e_{zx}, \\ e_{zz} &= \frac{\partial w}{\partial z}, & e_{yz} &= \frac{1}{2}\left(\frac{\partial v}{\partial z} + \frac{\partial w}{\partial y}\right) = e_{zy}. \end{aligned}$$

*In the infinitesimal displacement case, the distinction between the Lagrangian and Eulerian strain tensor disappears*, since then it is immaterial whether the derivatives of the displacements are calculated at the position of a point before or after deformation.

## 4.3. GEOMETRIC INTERPRETATION OF INFINITESIMAL STRAIN COMPONENTS

Let $x, y, z$ be a set of rectangular Cartesian coordinates. Consider a line element of length $dx$ parallel to the $x$-axis ($dy = dz = 0$). The change of the square of the length of this element due to deformation is

$$ds^2 - ds_0^2 = 2e_{xx}(dx)^2.$$

Hence,

$$ds - ds_0 = \frac{2e_{xx}(dx)^2}{ds + ds_0}.$$

But $ds = dx$ in this case, and $ds_0$ differs from $ds$ only by a small quantity of the second order, if we assume the displacements $u, v, w$ and the strain components $e_{ij}$ to be infinitesimal. Hence,

$$\frac{ds - ds_0}{ds} = e_{xx},$$

and it is seen that $e_{xx}$ represents the *extension*, or change of length per unit length of a vector parallel to the $x$-axis. An application of the above discussion to a volume element is illustrated in Fig. 4.3:1, case 1.

To see the meaning of the component $e_{xy}$, let us consider a small rectangle in the body with edges $dx$, $dy$. It is evident from Fig. 4.3:1, cases 2, 3, and 4

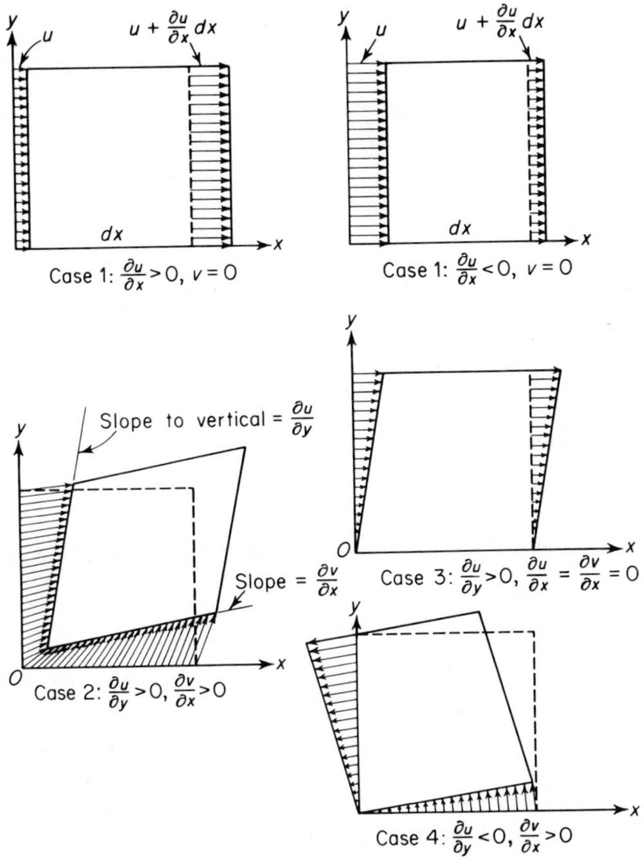


**Fig. 4.3:1.** Deformation gradients and interpretation of infinitesimal strain components.

that the sum $\partial u/\partial y + \partial v/\partial x$ represents the change of angle $xOy$ which was originally a right angle. Thus,

$$e_{xy} = \frac{1}{2}\left(\frac{\partial u}{\partial y} + \frac{\partial v}{\partial x}\right) = \tfrac{1}{2}(\text{change of angle } xOy).$$

In engineering usage, the strain components $e_{ij}$ $(i \neq j)$ doubled, i.e., $2e_{ij}$, are called the *shearing strains*, or *detrusions*. The name is perhaps particularly suggestive in Case 3 of Fig. 4.3:1, which is called the case of *simple shear*.

The quantity

$$\omega_z = \frac{1}{2}\left(\frac{\partial v}{\partial x} - \frac{\partial u}{\partial y}\right)$$

is called the *infinitesimal rotation* of the element $dx\,dy$. This terminology is suggested by Case 4 in Fig. 4.3:1. If

$$\frac{\partial v}{\partial x} = -\frac{\partial u}{\partial y},$$

then $e_{xy} = 0$ and $\omega_z$ is indeed the angle through which the rectangular element is rotated as a rigid body.

## 4.4. ROTATION

Consider an infinitesimal displacement field $u_i(x_1, x_2, x_3)$. From $u_i$, form the Cartesian tensor

(1) $$\omega_{ij} = \frac{1}{2}\left(\frac{\partial u_j}{\partial x_i} - \frac{\partial u_i}{\partial x_j}\right),$$

which is antisymmetric, i.e.,

(2) $$\omega_{ij} = -\omega_{ji}.$$

In three dimensions, from an antisymmetric tensor we can always build a *dual vector*,

(3) $$\omega_k = \tfrac{1}{2}e_{kij}\omega_{ij}, \quad \text{i.e.,} \quad \boldsymbol{\omega} = \tfrac{1}{2}\,\text{curl}\,\mathbf{u},$$

where $e_{kij}$ is the permutation tensor (Secs. 2.1, 2.6). On the other hand, from (3) and the $e$-$\delta$ identity [Eq. 2.1:11] it follows that $e_{ijk}\omega_k = \frac{1}{2}(\omega_{ij} - \omega_{ji})$. Hence, if $\omega_{ij}$ is antisymmetric, the relation (3) has a unique inverse,

(4) $$\omega_{ij} = e_{ijk}\omega_k.$$

Thus, $\omega_{ij}$ may be called the *dual* (antisymmetric) *tensor* of a vector $\omega_k$. We shall call $\omega_k$ and $\omega_{ij}$, repsectively, the *rotation vector* and *rotation tensor* of the displacement field $u_i$.

We shall consider the physical meaning of these quantities below.

At the end of sec. 4.3 we saw that $\omega_z$ represents the (infinitesimal) rotation of the element as a rigid body if $e_{xy}$ vanishes. Now we have the general theorem that *the vanishing of the symmetric strain tensor $E_{ij}$ or $e_{ij}$ is the necessary and sufficient condition for a neighborhood of a particle to be moved like a rigid body*. This follows at once from the definitions of strain tensors, Eqs. (4.1:11) and (4.1:12). For, if a neighborhood of a particle $P$ moves like a rigid body, the length of any element in the neighborhood will not change, so that $ds = ds_0$; it follows that $E_{ij} = e_{ij} = 0$. Conversely, if $E_{ij}$ or

$e_{ij}$ vanishes at $P$, the length of line elements joining any two points in a neighbourhood of $P$ will not change and the neighborhood moves like a rigid body.

*When the strain tensor vanishes at P, it can be shown that for an infinitesimal displacement field the infinitesimal rotation (angular displacement) of the rigid-body motion of a neighborhood of P is given by the vector* $\omega_i$. To show this, consider a point $P'$ in the neighborhood of $P$. Let the coordinates of $P$ and $P'$ be $x_i$ and $x_i + dx_i$, respectively. The relative displacement of $P'$ with respect to $P$ is

$$du_i = \frac{\partial u_i}{\partial x_j} dx_j. \tag{5}$$

This can be written as

$$du_i = \frac{1}{2}\left(\frac{\partial u_i}{\partial x_j} + \frac{\partial u_j}{\partial x_i}\right) dx_j + \frac{1}{2}\left(\frac{\partial u_i}{\partial x_j} - \frac{\partial u_j}{\partial x_i}\right) dx_j.$$

The first quantity in parentheses is the infinitesimal strain tensor, which is zero by assumption. The second quantity in parentheses may be identified with Eq. (1). Hence,

$$\begin{aligned} du_i &= -\omega_{ij}\, dx_j = \omega_{ji}\, dx_j \\ &= -e_{ijk}\omega_k\, dx_j, && \text{by Eq. (4),} \\ &= \boldsymbol{\omega} \times d\mathbf{x}, && \text{by definition.} \end{aligned} \tag{6}$$

Thus the relative displacement is the vector product of $\boldsymbol{\omega}$ and $d\mathbf{x}$. This is exactly what would have been produced by an infinitesimal rotation $|\boldsymbol{\omega}|$ about an axis through $P$ in the direction of $\boldsymbol{\omega}$.

It should be noted that we have restricted ourselves to infinitesimal angular displacements. Angular measures for finite displacements are related to $\omega_{ij}$ in a more complicated way.

## 4.5. FINITE STRAIN COMPONENTS

When the strain components are not small, it is no longer possible to give simple geometric interpretations of the components of the strain tensors.

Consider a set of rectangular Cartesian coordinates with respect to which the strain components are defined as in Sec. 4.1. Let a line element before deformation be $da^1 = ds_0$, $da^2 = 0$, $da^3 = 0$. Let the extension $E_1$ of this element be defined by

$$E_1 = \frac{ds - ds_0}{ds_0},$$

or

$$ds = (1 + E_1)\, ds_0. \tag{1}$$

From Eq. (4.1:11), we have

(2) $$ds^2 - ds_0^2 = 2E_{ij}\, da^i\, da^j = 2E_{11}(da^1)^2.$$

Combining (1) and (2), we obtain

$$(1 + E_1)^2 - 1 = 2E_{11},$$

or

(3) $$E_1 = \sqrt{1 + 2E_{11}} - 1.$$

This reduces to

(4) $$E_1 \doteq E_{11},$$

when $E_{11}$ is small.

To get the physical significance of the component $E_{12}$, let us consider two line elements $\underset{\rightarrow}{ds_0}$ and $\underset{\rightarrow}{d\bar{s}_0}$ which are at a right angle in the original state:

(5) $$\begin{aligned} &\underset{\rightarrow}{ds_0}: \quad da^1 = ds_0, \quad da^2 = 0, \quad da^3 = 0; \\ &\underset{\rightarrow}{d\bar{s}_0}: \quad da^1 = 0, \quad da^2 = d\bar{s}_0, \quad da^3 = 0. \end{aligned}$$

After deformation these line elements become $\underset{\rightarrow}{ds}$, $(dx^i)$ and $\underset{\rightarrow}{d\bar{s}}$, $(d\bar{x}^i)$. Forming the scalar product of the deformed elements, we obtain

$$\begin{aligned} ds\, d\bar{s} \cos\theta = dx^k\, d\bar{x}^k &= \frac{\partial x_k}{\partial a_i}\, da^i\, \frac{\partial x_k}{\partial a_j}\, d\bar{a}^j \\ &= \frac{\partial x_k}{\partial a_1}\frac{\partial x_k}{\partial a_2}\, ds_0\, d\bar{s}_0. \end{aligned}$$

Therefore, from the definition (4.1:9), on specializing to rectangular Cartesian coordinates we have

(6) $$ds\, d\bar{s}\cos\theta = 2E_{12}\, ds_0\, d\bar{s}_0.$$

But, from (1) and (3), we have

$$ds = \sqrt{1 + 2E_{11}}\, ds_0, \qquad d\bar{s} = \sqrt{1 + 2E_{22}}\, d\bar{s}_0.$$

Hence, (6) yields

(7) $$\cos\theta = \frac{2E_{12}}{\sqrt{1 + 2E_{11}}\sqrt{1 + 2E_{22}}}.$$

The angle $\theta$ is the angle between the line elements $\underset{\rightarrow}{ds}$ and $\underset{\rightarrow}{d\bar{s}}$ after deformation. The change of angle between the two line elements, which in the original state $\underset{\rightarrow}{ds_0}$ and $\underset{\rightarrow}{d\bar{s}_0}$ are orthogonal, is $\alpha_{12} = \pi/2 - \theta$. From (7) we therefore obtain

(8) $$\sin\alpha_{12} = \frac{2E_{12}}{\sqrt{1 + 2E_{11}}\sqrt{1 + 2E_{22}}}.$$

This reduces, in the case of infinitesimal strain, to

(9) $$\alpha_{12} \doteq 2E_{12}.$$

A completely analogous interpretation can be made for the Eulerian strain components. Defining the extension $e_1$ per unit *deformed* length as

(10) $$e_1 = \frac{ds - ds_0}{ds},$$

we find

(11) $$e_1 = 1 - \sqrt{1 - 2e_{11}}.$$

Furthermore, if the deviation from a right angle between two elements in the original state which after deformation become orthogonal be denoted by $\beta_{12}$, we have

(12) $$\sin \beta_{12} = \frac{2e_{12}}{\sqrt{1 - 2e_{11}}\sqrt{1 - 2e_{22}}}.$$

These again reduce to the familiar results

$$e_1 \doteq e_{11}, \qquad \beta_{12} \doteq 2e_{12},$$

in the case of infinitesimal strain.

## 4.6 COMPATIBILITY OF STRAIN COMPONENTS

The question of how to determine the displacements $u_i$ when the components of strain tensor are given naturally arises. In other words, how do we integrate the differential equations (in rectangular Cartesian coordinates)

(1) $$e_{ij} = \frac{1}{2}\left[\frac{\partial u_j}{\partial x_i} + \frac{\partial u_i}{\partial x_j} - \frac{\partial u_\alpha}{\partial x_i}\frac{\partial u_\alpha}{\partial x_j}\right]$$

to determine $u_i$?

Inasmuch as there are six equations for three unknown functions $u_i$, the system of Eqs. (1) will not have a single-valued solution in general, if the functions $e_{ij}$ were arbitrarily assigned. One must expect that a solution may exist only if the functions $e_{ij}$ satisfy certain conditions.

Since strain components only determine the relative positions of points in the body, and since any rigid-body motion corresponds to zero strain, we expect that the solution $u_i$ can be determined only up to an arbitrary rigid-body motion. But, if $e_{ij}$ were specified arbitrarily, we may expect that something like the cases shown in Fig. 4.6:1 may happen. Here a continuous triangle (portion of material in a body) is given. If we deform it by following an arbitrarily specified strain field starting from the point $A$, we might end at the point $C$ and $D$ either with a gap between them or with overlapping of material. For a single-valued continuous solution to exist (up to a rigid-body motion), the ends $C$ and $D$ must meet perfectly in the strained configuration. This cannot be guaranteed unless the specified strain field along the edges of the triangle obeys certain conditions.

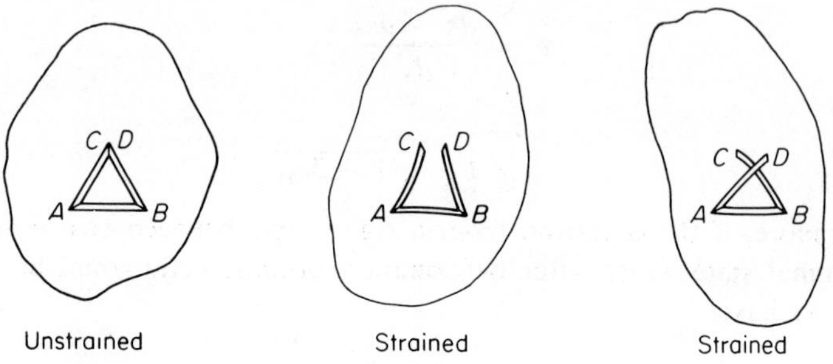


**Fig. 4.6:1.** Illustrations for the requirement of compatibility.

The conditions of integrability of Eqs. (1) are called the *compatibility conditions.* They are conditions to be satisfied by the strain components $e_{ij}$ and can be obtained by eliminating $u_i$ from Eqs. (1).

The nonlinear differential Eqs. (1) are difficult to handle, so first *let us consider the linear infinitesimal strain case*

(2) $$e_{ij} = \frac{1}{2}\left(\frac{\partial u_i}{\partial x_j} + \frac{\partial u_j}{\partial x_i}\right), \qquad \text{i.e.,} \quad e_{ij} = \tfrac{1}{2}(u_{i,j} + u_{j,i}).$$

By differentiation of (2), we have

$$e_{ij,kl} = \tfrac{1}{2}(u_{i,jkl} + u_{j,ikl}),$$

where an index $i$ following a comma indicates partial differentiation with respect to $x_i$. Interchanging subscripts, we have

$$e_{kl,ij} = \tfrac{1}{2}(u_{k,lij} + u_{l,kij}),$$
$$e_{jl,ik} = \tfrac{1}{2}(u_{j,lik} + u_{l,jik}),$$
$$e_{ik,jl} = \tfrac{1}{2}(u_{i,kjl} + u_{k,ijl}).$$

From these we verify at once that

(3) ▲ $$e_{ij,kl} + e_{kl,ij} - e_{ik,jl} - e_{jl,ik} = 0.$$

This is the *equation of compatibility* of St. Venant, first obtained by him in 1860.

Of the 81 equations represented by (3), only six are essential. The rest are either identities or repetitions on account of the symmetry of $e_{ij}$

and of $e_{kl}$. The six equations written in unabridged notation are

$$\frac{\partial^2 e_{xx}}{\partial y\,\partial z} = \frac{\partial}{\partial x}\left(-\frac{\partial e_{yz}}{\partial x} + \frac{\partial e_{zx}}{\partial y} + \frac{\partial e_{xy}}{\partial z}\right),$$

$$\frac{\partial^2 e_{yy}}{\partial z\,\partial x} = \frac{\partial}{\partial y}\left(-\frac{\partial e_{zx}}{\partial y} + \frac{\partial e_{xy}}{\partial z} + \frac{\partial e_{yz}}{\partial x}\right),$$

$$\frac{\partial^2 e_{zz}}{\partial x\,\partial y} = \frac{\partial}{\partial z}\left(-\frac{\partial e_{xy}}{\partial z} + \frac{\partial e_{yz}}{\partial x} + \frac{\partial e_{zx}}{\partial y}\right),$$

(4) ▲

$$2\frac{\partial^2 e_{xy}}{\partial x\,\partial y} = \frac{\partial^2 e_{xx}}{\partial y^2} + \frac{\partial^2 e_{yy}}{\partial x^2},$$

$$2\frac{\partial^2 e_{yz}}{\partial y\,\partial z} = \frac{\partial^2 e_{yy}}{\partial z^2} + \frac{\partial^2 e_{zz}}{\partial y^2},$$

$$2\frac{\partial^2 e_{zx}}{\partial z\,\partial x} = \frac{\partial^2 e_{zz}}{\partial x^2} + \frac{\partial^2 e_{xx}}{\partial z^2}.$$

These conditions are derived for infinitesimal strains referred to rectangular Cartesian coordinates. If the general curvilinear coordinates are used, we define the infinitesimal strain by

(5) $$e_{ij} = \tfrac{1}{2}(u_i|_j + u_j|_i),$$

where $u_i|_j$ is the covariant derivative of $u_i$, etc. The corresponding compatibility condition is then

(6) $$e_{ij}|_{kl} + e_{kl}|_{ij} - e_{ik}|_{jl} - e_{jl}|_{ik} = 0.$$

If, however, the strain is finite so that $e_{ij}$ is given by Eq. (1) in rectangular Cartesian coordinates, or by

(7) $$e_{ij} = \tfrac{1}{2}[u_i|_j + u_j|_i - u_\alpha|_i\, u^\alpha|_j]$$

in general coordinates, then it is necessary to use a new method of derivation. A successful method uses the basic concept that the compatibility conditions say that our body initially situated in a three-dimensional Euclidean space, must remain in the Euclidean space after deformation. A mathematical statement to this effect, expressed in terms of the strain components, gives the compatibility conditions. (See Probs. 2.31 and 4.9. Full expressions can be found in Green and Zerna,[1.2] *Theoretical Elasticity*, p. 62.)

Let us now return to the question posed at the beginning of this section and inquire whether conditions (3) or (4) are sufficient to assure the existence of a single-valued continuous solution of the differential Eqs. (2) up to a rigid-body motion. The answer is affirmative. Various proofs are available, the simplest having been given by E. Cesaro[4.1] in 1906. (See notes in Bib. 4.1.) The proof may proceed as follows.

Let $P_0(x_1^0, x_2^0, x_3^0)$ be a point at which the displacements $u_i^0$ and the components of rotation $\omega_{ij}^0$ are specified. The displacement $u_i$ at an arbitrary point $\bar{P}$ is obtained by a line integral along a continuous rectifiable curve $C$ joining $P_0$ and $\bar{P}$,

$$(8) \qquad u_i(\bar{x}_1, \bar{x}_2, \bar{x}_3) = u_i^0 + \int_{P_0}^{\bar{P}} du_i = u_i^0 + \int_{P_0}^{\bar{P}} \frac{\partial u_i}{\partial x_k} dx^k.$$

But

$$(9) \qquad \frac{\partial u_i}{\partial x_k} = \frac{1}{2}\left[\left(\frac{\partial u_i}{\partial x_k} + \frac{\partial u_k}{\partial x_i}\right) + \left(\frac{\partial u_i}{\partial x_k} - \frac{\partial u_k}{\partial x_i}\right)\right] = e_{ik} - \omega_{ik},$$

according to the definitions of the infinitesimal strain and rotation tensors, $e_{ij}$ and $\omega_{ij}$, respectively, as given by Eqs. (4.2:6) and (4.4:1). Hence, (8) becomes

$$(10) \qquad u_i(\bar{x}_1, \bar{x}_2, \bar{x}_3) = u_i^0 + \int_{P_0}^{\bar{P}} e_{ik}\, dx^k - \int_{P_0}^{\bar{P}} \omega_{ik}\, dx^k.$$

We must eliminate $\omega_{ik}$ in terms of $e_{ik}$. To achieve this, the last integral is integrated by parts to yield

$$(11) \qquad \begin{aligned} \int_{P_0}^{\bar{P}} \omega_{ik}\, dx^k &= \int_{P_0}^{\bar{P}} \omega_{ik}(x)\,(dx^k - d\bar{x}^k) \\ &= (\bar{x}_k - x_k^0)\omega_{ik}^0 + \int_{P_0}^{\bar{P}} (\bar{x}_k - x_k) \frac{\partial \omega_{ik}}{\partial x_l}(x)\, dx^l, \end{aligned}$$

where $\omega_{ik}^0$ is the value of $\omega_{ik}$ at $P_0$. In (8) we have replaced $dx^k$ by $dx^k - d\bar{x}^k$, to have $\omega_{ik}^0$ appear outside the integral instead of $\omega_{ik}$ at $\bar{P}$. This is permissible, since $\bar{P}$ is fixed with respect to the integration so that $d\bar{x}^k = 0$. Now,

$$(12) \qquad \begin{aligned} -\frac{\partial \omega_{ik}}{\partial x_l} &= \frac{1}{2}\left(\frac{\partial^2 u_i}{\partial x_l\, \partial x_k} - \frac{\partial^2 u_k}{\partial x_l\, \partial x_i}\right) \\ &= \frac{1}{2}\left(\frac{\partial^2 u_i}{\partial x_k\, \partial x_l} + \frac{\partial^2 u_l}{\partial x_k\, \partial x_i}\right) - \frac{1}{2}\left(\frac{\partial^2 u_k}{\partial x_i\, \partial x_l} + \frac{\partial^2 u_l}{\partial x_i\, \partial x_k}\right) \\ &= \frac{\partial}{\partial x_k}\frac{1}{2}\left(\frac{\partial u_i}{\partial x_l} + \frac{\partial u_l}{\partial x_i}\right) - \frac{\partial}{\partial x_i}\frac{1}{2}\left(\frac{\partial u_k}{\partial x_l} + \frac{\partial u_l}{\partial x_k}\right) \\ &= \frac{\partial e_{il}}{\partial x_k} - \frac{\partial e_{kl}}{\partial x_i}. \end{aligned}$$

Substitution of (11) and (12) into (10) yields

$$(13) \qquad u_i(\bar{x}_1, \bar{x}_2, \bar{x}_3) = u_i^0 - (\bar{x}_k - x_k^0)\omega_{ik}^0 + \int_{P_0}^{\bar{P}} U_{il}\, dx^l,$$

where

$$(14) \qquad U_{il} = e_{il} + (\bar{x}_k - x_k)\left(\frac{\partial e_{il}}{\partial x_k} - \frac{\partial e_{kl}}{\partial x_i}\right).$$

Now, if $u_i(\bar{x}_1, \bar{x}_2, \bar{x}_3)$ is to be single-valued and continuous, the last integral in (13) must depend only on the end points $P_0$, $\bar{P}$ and must not be dependent on the path of integration $C$. Therefore, the integrand must be an exact differential. (See Sec. 4.7 below.) The necessary condition that $U_{il}dx^l$ be an exact differential is

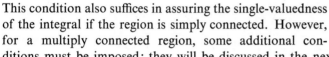

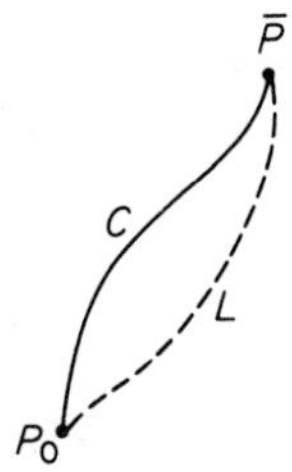


**Fig. 4.6:2.** Paths of integration.

$$\frac{\partial U_{il}}{\partial x_m} - \frac{\partial U_{im}}{\partial x_l} = 0. \tag{15}$$

This condition also suffices in assuring the single-valuedness of the integral if the region is simply connected. However, for a multiply connected region, some additional conditions must be imposed; they will be discussed in the next section.

Now we obtain, from (14) and (15),

$$\frac{\partial e_{il}}{\partial x_m} - \delta_{km}\left(\frac{\partial e_{il}}{\partial x_k} - \frac{\partial e_{kl}}{\partial x_i}\right) + (\bar{x}_k - x_k)\frac{\partial}{\partial x_m}\left(\frac{\partial e_{il}}{\partial x_k} - \frac{\partial e_{kl}}{\partial x_i}\right)$$
$$- \frac{\partial e_{im}}{\partial x_l} + \delta_{kl}\left(\frac{\partial e_{im}}{\partial x_k} - \frac{\partial e_{km}}{\partial x_i}\right) - (\bar{x}_k - x_k)\frac{\partial}{\partial x_l}\left(\frac{\partial e_{im}}{\partial x_k} - \frac{\partial e_{km}}{\partial x_i}\right) = 0.$$

The factors not multiplied by $\bar{x}_k - x_k$ cancel each other, and the factors multiplied by $\bar{x}_k - x_k$ form exactly the compatibility condition given by (3). Hence, Eq. (15) is satisfied if the compatibility conditions (3) are satisfied. Thus, we have proved that the satisfaction of the compatibility conditions (3) is necessary and sufficient for the displacement to be single-valued in a simply connected region. For a multiply connected region, Eq. (3) is necessary, but no longer sufficient. To guarantee single-valuedness of displacement for an assumed strain field, some additional conditions as described in Sec. 4.7 must be imposed.

**Problem 4.1.** Consider the two-dimensional case in unabridged notations,

$$\frac{\partial u}{\partial x} = e_{xx}, \qquad \frac{\partial v}{\partial y} = e_{yy}, \qquad \frac{\partial u}{\partial y} + \frac{\partial v}{\partial x} = 2e_{xy}.$$

Prove that in order to guarantee the solutions $u(x, y)$, $v(x, y)$ to be single-valued in a simply connected domain $D(x, y)$, the functions $e_{xx}(x, y)$, $e_{yy}(x, y)$, $e_{xy}(x, y)$ must satisfy the compatibility condition

$$\frac{\partial^2 e_{xx}}{\partial y^2} + \frac{\partial^2 e_{yy}}{\partial x^2} = 2\frac{\partial^2 e_{xy}}{\partial x\, \partial y}.$$

Prove the sufficiency of this condition in unabridged notations.

## 4.7. MULTIPLY CONNECTED REGIONS

As necessary conditions, the compatibility equations derived in Sec. 4.6 apply to any continuum. But the sufficiency proof at the end of the preceding section requires that the region be simply connected. For a multiply connected continuum, additional conditions must be imposed.

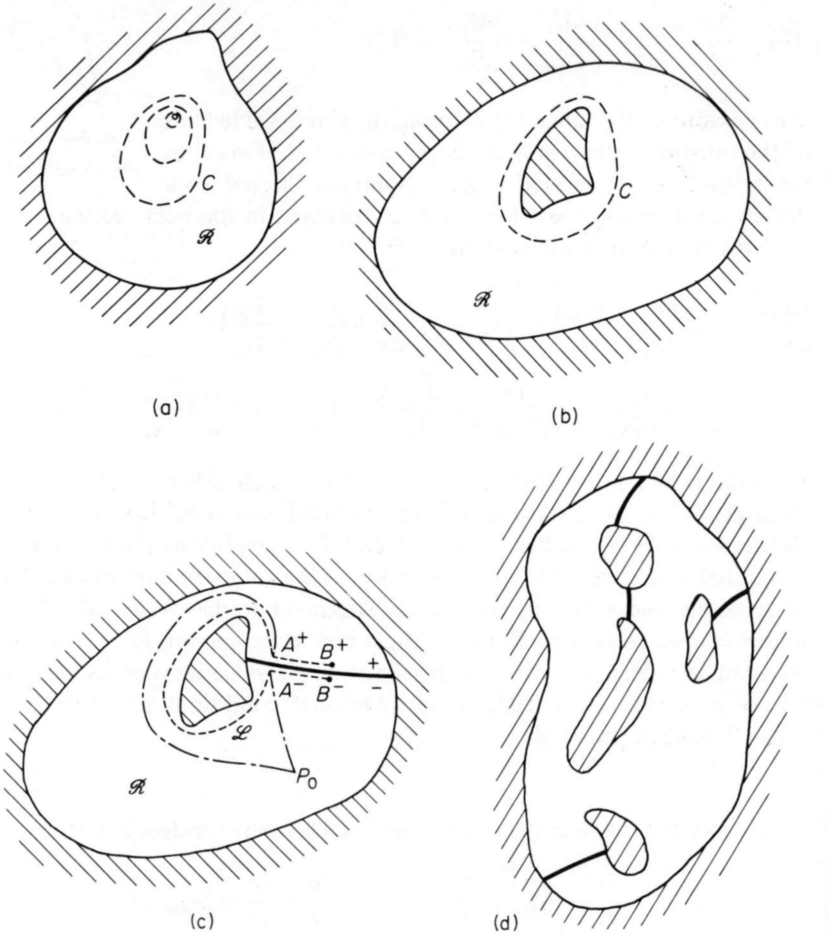


**Fig. 4.7:1.** Two-dimensional examples of simply and multiply connected regions. In (a), region $\mathscr{R}$ is simply connected; in (b), $\mathscr{R}$ is doubly connected

A region is *simply connected* if any simple closed contour drawn in the region can be shrunk continuously to a point without leaving the region; otherwise the region is said to be *multiply connected.*

Figure 4.7:1(a) illustrates a simply connected region $\mathscr{R}$ in which an arbitrary closed curve $C$ can be shrunk continuously to a point without leaving $\mathscr{R}$. Figures 4.7:1(b) and 4.7:2 illustrate doubly connected regions in two and three dimensions, respectively. Figures 4.7:1(c) and (d) show how multiply connected regions can be made simply connected by introducing cuts—imaginary boundaries.

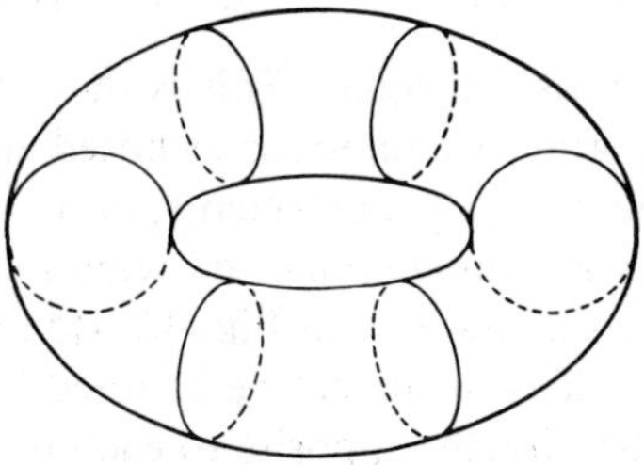

**Fig. 4.7:2.** A torus doubly connected in three dimensions.

The condition (4.6:15) is based on Green's theorem, and we shall review the reasoning in preparation for discussing multi-connected regions. In two dimensions, Green's theorem may be written as

(1) $$\int_C (P\,dx + Q\,dy) = \iint_{\mathscr{R}} \left(\frac{\partial Q}{\partial x} - \frac{\partial P}{\partial y}\right) dx\,dy$$

for a simply connected region $\mathscr{R}$ on a plane $x$, $y$ bounded by a closed contour $C$. In three dimensions, Green's theorem states

(2) $$\int_C (P\,dx + Q\,dy + R\,dz) = \iint_{\mathscr{R}} \left\{\left(\frac{\partial R}{\partial y} - \frac{\partial Q}{\partial z}\right) \cos(\nu, x) + \left(\frac{\partial P}{\partial z} - \frac{\partial R}{\partial x}\right) \cos(\nu, y) + \left(\frac{\partial Q}{\partial x} - \frac{\partial P}{\partial y}\right) \cos(\nu, z)\right\} dS$$

for a simply connected region $\mathscr{R}$ on a surface $S$ bounded by a closed contour $C$. The factor $\cos(\nu, x)$ is the cosine of the angle between the $x$-axis and the normal vector $\boldsymbol{\nu}$ normal to the surface $S$. When the conditions

(3) $$\frac{\partial R}{\partial y} - \frac{\partial Q}{\partial z} = 0, \qquad \frac{\partial P}{\partial z} - \frac{\partial R}{\partial x} = 0, \qquad \frac{\partial Q}{\partial x} - \frac{\partial P}{\partial y} = 0$$

are imposed, the line integral on the left-hand side of (2) vanishes for every possible closed contour $C$,

$$\int_C (P\,dx + Q\,dy + R\,dz) = 0.$$

When this result is applied to any two possible paths of integration between a fixed point $P_0$ and a variable point $\bar{P}$, as in Fig. 4.6:2, we see that the integral

(4) $$F(\bar{P}) = \int_{P_0}^{\bar{P}} (P\,dx + Q\,dy + R\,dz)$$

must be a single-valued function of the point $\bar{P}$, independent of the path of integration; for in Fig. 4.6:2 the curve $P_0CPLP_0$ forms a simple contour. Hence,

$$\int_{P_0C\bar{P}} + \int_{\bar{P}LP_0} = 0 = \int_{P_0C\bar{P}} - \int_{P_0L\bar{P}}$$

Thus, the last two integrals are equal. This is the result used in Sec. 4.6 to conclude that the compatibility conditions as posed are sufficient for uniqueness of displacements in a simply connected region.

Let us now consider a doubly connected region on a plane as shown in Fig. 4.7:1(b). After a cut is made as in Fig. 4.7:1(c), the region $\mathscr{R}$ is simply connected. Let the two sides of the cut be denoted by (+) and (−) and let the points $A^+$ and $A^-$ be directly opposite to each other on the two sides of the cut. The line integrals

$$F(A^-) = \int_{P_0}^{A^-} (P\,dx + Q\,dy), \qquad F(A^+) = \int_{P_0}^{A^+} (P\,dx + Q\,dy),$$

integrated from a fixed point $P_0$ along paths in the cut region, are both single-valued. But, since $A^+$ and $A^-$ are on the opposite sides of the cut, the values of the integrals need not be the same. Let the chain-dot path $P_0A^+$ in Fig. 4.7:1(c) be deformed into $P_0A^-$ plus a line $\mathscr{L}$ connecting $A^-$ to $A^+$, encircling the inner boundary. Then

$$\int_{P_0}^{A^+} = \int_{P_0}^{A^-} + \int_{\mathscr{L}}.$$

Hence,

$$F(A^+) - F(A^-) = \int_{\mathscr{L}} (P\,dx + Q\,dy). \tag{5}$$

Similar consideration for another pair of arbitrary points $B^+$ and $B^-$ across the boundary leads to the same result,

$$F(B^+) - F(B^-) = \int_{\mathscr{L}} (P\,dx + Q\,dy). \tag{6}$$

The right-hand sides of (5) and (6) are the same if $P$ and $Q$ are single-valued in the region $\mathscr{R}$, because a line $\mathscr{L}'$ connecting $B^-$ to $B^+$, encircling the inner boundary, may be deformed into one that goes from $B^-$ to $A^-$ along the cut, then follows $\mathscr{L}$ from $A^-$ to $A^+$, and finally moves from $A^+$ to $B^+$ along the + side of the cut. If $P$ and $Q$ were single-valued, the integral from $B^-$ to $A^-$ cancels exactly the one from $A^+$ to $B^+$.

From these considerations, it is clear that function $F(\bar{P})$ defined by the line integral (4) will be single-valued in a doubly connected region if the supplementary condition

$$\int_{\mathscr{L}} (P\,dx + Q\,dy) = 0 \tag{7}$$

is imposed, where $\mathscr{L}$ is a closed contour that goes from one side of a cut to the other, without leaving the region.

The same consideration can be applied generally to multiply connected regions in two or three dimensions. An $(m+1)$-ply connected region can be made simply connected by $m$ cuts. In the cut, simply-connected region, $m$ independent simple contours $\mathscr{L}_1, \mathscr{L}_2, \ldots, \mathscr{L}_m$ can be drawn. Each $\mathscr{L}_i$ starts from one side of a cut, and ends on the other side of the same cut. All cuts are thus embraced by the $\mathscr{L}$'s. Then the single-valuedness of the function $F(\bar{P})$ defined by the line integral (4) can be assured by imposing, in addition to the conditions (3), $m$ supplementary conditions

$$(8) \qquad \int_{\mathscr{L}_1} (P\,dx + Q\,dy + R\,dz) = \int_{\mathscr{L}_2} (\qquad) = \cdots = \int_{\mathscr{L}_m} (\qquad) = 0.$$

When these results are applied to the compatibility problem of Sec. 4.6 we see that if the body is $(m+1)$-ply connected, the single-valuedness of the displacement function $u_i(x_1, x_2, x_3)$ given by Eq. (4.6:13) requires $m \times 3$ supplementary conditions:

$$(9) \ ▲ \qquad \int_{\mathscr{L}_1} U_{il}\,dx^l = \cdots = \int_{\mathscr{L}_m} U_{il}\,dx^l = 0, \qquad i = 1, 2, 3,$$

where

$$(10) \ ▲ \qquad U_{il} = e_{il} + (\bar{x}_k - x_k)\left(\frac{\partial e_{il}}{\partial x_k} + \frac{\partial e_{kl}}{\partial x_i}\right)$$

and $\mathscr{L}_1, \ldots, \mathscr{L}_m$ are $m$ contours in $\mathscr{R}$ as described above.

## 4.8. MULTIVALUED DISPLACEMENTS

Some problems of thermal stress, initial strain, rigid inclusions, etc., can be formulated in terms of multivalued displacements. For example, if the horseshoe in Fig. 4.8:1(a) is strained so that the faces $S_1$ and $S_2$ come into contact and are then welded together, the result is the doubly connected body shown in Fig. 4.8:1(b). If the strain of the body (b) is known, we can use the procedure described in Sec. 4.6 to compute displacements and generally we will obtain the configuration of Fig. 4.8:1(a). In this case a single point $P$ on the welded surface $S$ will open up into two points $P_1$ and $P_2$ on the two open ends. Thus the displacement at $P$ is double-valued.

In the example posed here, the strain components and their derivatives may be discontinuous on the two sides of the interface $S$ of Fig. 4.8:1(b). Hence, the value of a line integral around a contour in the body that crosses the interface $S$ will depend on where the crossing point is. In other words,

$$(1) \qquad u_i(P_2) - u_i(P_1) = \int_{\mathscr{L}(P)} U_{il}\,dx^l$$

depends on the location of $P$, where $\mathscr{L}(P)$ is a contour passing through the point $P$ as indicated in the figure and $U_{il}$ is given in Eq. (4.6:14).

On the other hand, if we have a ring as in Fig. 4.8:1(b) and apply some load on it and ask what is the corresponding deformation, we must first decide whether some crack is allowed to be opened or not. If the ring is to

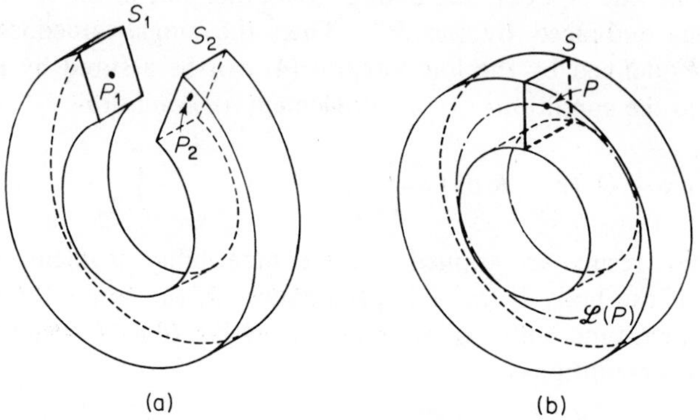


**Fig. 4.8:1.** Possibility and use of multivalued displacements.

remain integral, the deformation must be such that the displacement field is single-valued. In this case the strain due to the load must be such that

(2) $$\int_{\mathscr{L}(P)} U_{il}\, dx^l = 0$$

for a contour $\mathscr{L}(P)$ as shown in Fig. 4.8:1(b). Of course, Eq. (2) in conjunction with the compatibility conditions (4.6:4) assures that the line integral $\int U_{il}\, dx^l$ will vanish on any contour in the ring. Thus the point $P$ in Eq. (2) is arbitrary.

These examples show that both single-valued displacement fields and multi-valued displacement fields have proper use in answering appropriate questions.

## 4.9. PROPERTIES OF THE STRAIN TENSOR

The symmetric strain tensor $e_{ij}$ has many properties in common with the stress tensor. Thus, the existence of real-valued principal strains and principal planes, the representation of the Lamé strain ellipsoid, the Cauchy strain quadric, the Mohr circle for strain, the octahedral strain, strain deviations, and the strain invariants need no further discussion.

The first strain invariant,

(1) $$e = e_{11} + e_{22} + e_{33} = \text{sum of principal strains},$$

has a simple geometrical meaning in the case of infinitesimal strain. Let a volume element consist of a rectangular parallelepiped with edges parallel to the principal directions of strain. Let the length of the edges be $l_1$, $l_2$, $l_3$ in the unstrained state. Let $e_{11}$, $e_{22}$, $e_{33}$ be the principal strains. In the strained configuration, the edges become of lengths $l_1(1 + e_{11})$, $l_2(1 + e_{22})$, $l_3(1 + e_{33})$ and remain orthogonal to each other. Hence, for small strain the change of volume is

$$\Delta V = l_1 l_2 l_3 (1 + e_{11})(1 + e_{22})(1 + e_{33}) - l_1 l_2 l_3$$
$$\doteq l_1 l_2 l_3 (e_{11} + e_{22} + e_{33}).$$

Therefore,

$$e = e_{ii} = \Delta V / V. \tag{2}$$

*Thus, in the infinitesimal strain theory the first invariant represents the expansion in volume per unit volume.* For this reason, $e$ is called the *dilatation*.

If two-dimensional strain state (plane strain) is considered $u_3$ or $w \equiv 0$. the first invariant $e_{11} + e_{22}$ represents the change of area per unit area of the surface under strain.

For finite strain, the sum of principal strains does not have such a simple interpretation.

## PROBLEMS

**4.2.** A state of deformation in which all the strain components are constant throughout the body is called a *homogeneous deformation*. What is the equation, of the type $f(x, y, z) = 0$, of a surface which becomes a sphere $x^2 + y^2 + z^2 = r^2$ *after* a homogeneous deformation? What kind of surface is it? ($x, y, z$ are rectangular Cartesian coordinates.)

**4.3.** Show that a strain state described in rectangular Cartesian coordinates

$$e_{xx} = k(x^2 + y^2), \qquad e_{yy} = k(y^2 + z^2), \qquad e_{xy} = k'xyz,$$
$$e_{xz} = e_{yz} = e_{zz} = 0,$$

where $k$, $k'$ are small constants, is not a possible state of strain for a continuum.

**4.4.** A solid is heated nonuniformly to a temperature $T(x, y, z)$. If it is supposed that each element has unrestrained thermal expansion, the strain components will be

$$e_{xx} = e_{yy} = e_{zz} = \alpha T,$$
$$e_{xy} = e_{yz} = e_{zx} = 0,$$

where $x, y, z$, are rectangular Cartesian coordinates and $\alpha$ is the thermal expansion coefficient (a constant). Prove that this can only occur when $T$ is a linear function of $x, y, z$; i.e., $T = c_1 x + c_2 y + c_3 z + c_4$, where $c_1, \ldots, c_4$ are constants.

**4.5.** A soap-film like membrane stretched over a ring is deformed by uniform pressure into a hemisphere of the same radius (Fig. P4.5). In so doing, a point $P$ on the flat surface over the ring is deformed into a point $Q$ on the sphere. Determine a mathematical transformation from $P$ to $Q$. *Note:* A "soap-film like" membrane shall be defined as a membrane in which the tension is a constant, isotropic, and independent of the stretching of the membrane.

Let $P$ be referred to plane polar coordinates $(r, \theta)$ and $Q$ be referred to spherical polar coordinates $(R, \theta, \phi)$ with the same origin. We assume that $R = a$, $\theta = \theta$. The problem is to determine the angle $\phi$ as a function of $r$.

If the thickness of the film is uniform originally, it will not remain so after being blown into spherical shape. It will be interesting to determine the thickness distribution in the final configuration under the assumption of material incompressibility.

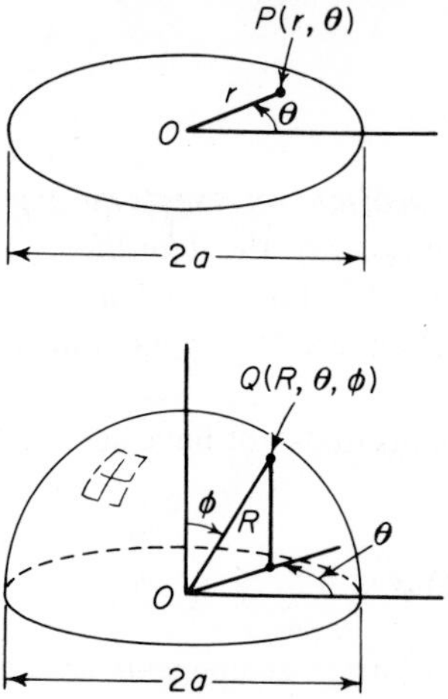


P4.5

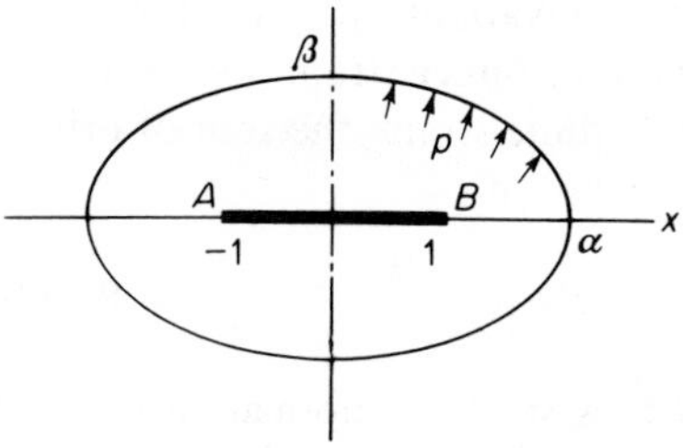


P4.6

**4.6.** Consider a two-dimensional inflatable structure which consists of an infinitely long tube. When folded flat, the cross section of the tube appears like a line segment $AB$ of length 2 in Fig. P4.6. When blown up with an internal gas pressure $p$, the tube assumes the form of an ellipse with major axis $\alpha$ and minor axis $\beta$.

*Questions:*

(a) Devise (arbitrarily, but specifically) a law of transformation which transforms the folded tube into the elliptic cylinder.

(b) Compute the components of membrane strain in the midsurface of the tube referred to the original folded tube (components of Green's strain tensor).

(c) Compute the corresponding components of strain referred to the blown up tube (components of Almansi's strain tensor).

(d) For equilibrium, what must be the distribution of membrane tension force $T$ in the blown-up tube? (The internal pressure is uniform.)

(e) Assume that $T = khe_1$, where $T$ is the membrane tension, $e_1$ is the strain component of Almansi on the midsurface of the tube wall, $h$ is the wall thickness of the blown up tube, and $k$ is a constant. Determine $h$.

(f) If the tube material is incompressible (like natural rubber), what would have to be the initial thickness distribution. ?

## 4.10. PHYSICAL COMPONENTS

In Sec. 2.16 we defined the physical components of a vector $\mathbf{v}$ as the components of $\mathbf{v}$ resolved in the directions of a set of unit vectors which are parallel either to the set of base vectors or to the set of reciprocal base vectors. Thus,

$$\begin{array}{llll} & \textit{Tensor components:} & v^r & v_r \\ (1) & \textit{Physical components:} & v^r\sqrt{g_{rr}} & v_r\sqrt{g^{rr}} \\ & \textit{Reference base vectors:} & \mathbf{g}_r & \mathbf{g}^r \end{array}$$

All the physical components have uniform physical dimensions; but they do not transform conveniently under coordinate transformations.

In Sec. 3.15 we defined the physical components of the stress tensors $\tau^{ij}$ as the components of the stress vector $\overset{i}{\mathbf{T}}$ resolved in the directions of unit vectors parallel to the base vectors. If the physical components of $\tau^{ij}$ are denoted by $\sigma^{ij}$, we have

$$\sigma^{ij} = \sqrt{\frac{g_{jj}}{g^{ii}}}\,\tau^{ij}, \qquad i, j \text{ not summed.} \tag{2}$$

Correspondingly, when $\overset{i}{\mathbf{T}}$ is resolved with respect to the reciprocal base vectors, the physical components are

$$\sigma^i_j = \sqrt{\frac{g^{jj}}{g^{ii}}}\,\tau^i_{\cdot j}, \qquad i, j \text{ not summed.} \tag{3}$$

All components $\sigma^{ij}$, $\sigma^i_j$ have uniform physical dimensions.

Now consider the strain tensor $e_{ij}$, which is covariant in the most natural form of definition (Sec. 4.1). If $ds$ is the length of an element $(d\theta^1, d\theta^2, d\theta^3)$ at a point $(\theta^1, \theta^2, \theta^3)$, and $ds_0$ is the length of the same element before the deformation takes place, then

$$ds^2 - ds_0^2 = 2e_{ij}\,d\theta^i\,d\theta^j. \tag{4}$$

The differential element $(d\theta^i)$ is a vector, whose physical components are $\sqrt{g_{ii}}\,d\theta^i$, ($i$ not summed). We may rewrite (4) as

$$ds^2 - ds_0^2 = \sum_{i=1}^{3}\sum_{j=1}^{3} e_{ij}\frac{1}{\sqrt{g_{ii}}\sqrt{g_{jj}}}(\sqrt{g_{ii}}\,d\theta^i)(\sqrt{g_{jj}}\,d\theta^j). \tag{5}$$

Since all the components $\sqrt{g_{ii}}\,d\theta^i$, $\sqrt{g_{jj}}\,d\theta^j$ have the dimension of length, we may define the *physical strain components*

$$\epsilon_{ij} = \frac{1}{\sqrt{g_{ii}}\sqrt{g_{jj}}}\,e_{ij}, \qquad i, j \text{ not summed,} \tag{6}$$

which are all dimensionless.

The rules of forming physical components now appear rather complicated. The appearance will become more systematic if we *limit ourselves to orthogonal curvilinear coordinates*. For any set of orthogonal curvilinear coordinates,

$$g_{ij} = g^{ij} = 0 \quad \text{if} \quad i \neq j, \tag{7}$$

and

$$g^{ii} = (g_{ii})^{-1}, \qquad i \text{ not summed.} \tag{8}$$

Then Eqs. (2), (3), and (6) may be written, in orthogonal curvilinear coordinates,

$$\sigma^{ij} = \sqrt{g_{ii}}\sqrt{g_{jj}}\,\tau^{ij}, \tag{9}$$

$$\sigma^{i}_{j} = \sqrt{g^{jj}}\sqrt{g_{ii}}\,\tau^{i}_{j}, \tag{10}$$

$$\epsilon_{ij} = \sqrt{g^{ii}}\sqrt{g^{jj}}\,e_{ij}, \tag{11}$$

where $i$, $j$ are not summed. These formulas, similar to Eq. (1), may serve as a pattern for defining the physical components of any tensor in any orthogonal curvilinear coordinates.

Let us repeat. The physical components and the tensor components of a physical quantity have the same geometric and physical interpretations, except for physical dimensions. The unification in dimensions is achieved by multiplying the tensor components with appropriate scale factors—the components of the metric tensor. The physical components do not transform conveniently under general transformation of coordinates. In practical applications, therefore, we generally write basic equations in the tensor form and then substitute the individual tensor components by their physical counterparts, if it is so desired.

We shall illustrate these applications in the following sections.

## 4.11. EXAMPLE—SPHERICAL COORDINATES

Let $x^1 = r$, $x^2 = \phi$, $x^3 = \theta$ (Fig. 4.11:1). Then

$$ds^2 = dr^2 + r^2\,d\phi^2 + r^2 \sin^2\phi\,d\theta^2,$$

$$g_{11} = 1, \qquad g_{22} = r^2,$$

$$g_{33} = r^2 \sin^2\phi, \qquad g_{ij} = 0 \quad \text{if} \quad i \neq j.$$

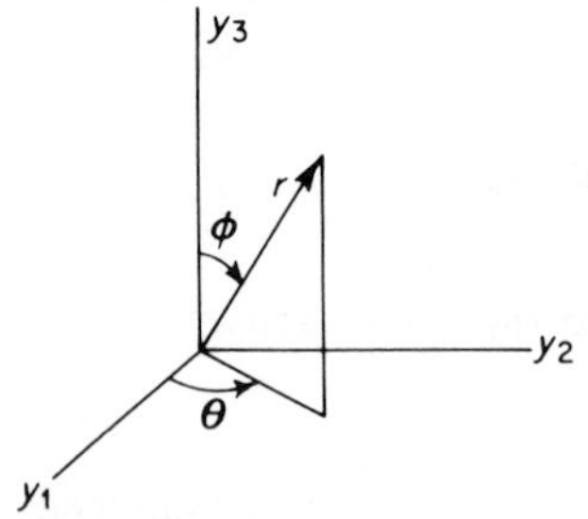

Fig. 4.11:1. Spherical polar coordinates.

Thus the coordinate system is orthogonal. From the definition of $g^{\alpha\beta}$, we have

$$g^{11} = 1, \qquad g^{22} = \frac{1}{r^2},$$

$$g^{33} = \frac{1}{r^2 \sin^2\phi}, \qquad g^{ij} = 0 \quad \text{if} \quad i \neq j.$$

The Euclidean Christoffel symbols are

$$\Gamma^1_{22} = -r, \qquad \Gamma^1_{33} = -r\sin^2\phi,$$

$$\Gamma^2_{12} = \Gamma^2_{21} = \frac{1}{r}, \qquad \Gamma^2_{33} = -\sin\phi\cos\phi,$$

$$\Gamma^3_{13} = \Gamma^3_{31} = \frac{1}{r}, \qquad \Gamma^3_{23} = \Gamma^3_{32} = \cot\phi,$$

all other $\Gamma^i_{jk} = 0$.

Let $u_i$ be the covariant components of the displacement vector. We have the infinitesimal strain tensor components

$$e_{ij} = \frac{1}{2}(u_i|_j + u_j|_i) = \frac{1}{2}\left(\frac{\partial u_i}{\partial x^j} + \frac{\partial u_j}{\partial x^i}\right) - \Gamma^\sigma_{ij}u_\sigma,$$

since $\Gamma^\sigma_{ij} = \Gamma^\sigma_{ji}$. Hence,

$$e_{11} = \frac{\partial u_1}{\partial r},$$

$$e_{22} = \frac{\partial u_2}{\partial \phi} + ru_1,$$

$$e_{33} = \frac{\partial u_3}{\partial \theta} + r\sin^2\phi\, u_1 + \sin^2\phi\cos^2\phi\, u_2,$$

$$e_{12} = \frac{1}{2}\left(\frac{\partial u_1}{\partial \phi} + \frac{\partial u_2}{\partial r}\right) - \frac{1}{r}u_2,$$

$$e_{23} = \frac{1}{2}\left(\frac{\partial u_2}{\partial \theta} + \frac{\partial u_3}{\partial \phi}\right) - \cot\phi\, u_3,$$

$$e_{31} = \frac{1}{2}\left(\frac{\partial u_3}{\partial r} + \frac{\partial u_1}{\partial \theta}\right) - \frac{1}{r}u_3.$$

The $u_i$ and $e_{ij}$ are tensor components. Let the corresponding physical components of the displacement vector be written as $\xi_r$, $\xi_\phi$, $\xi_\theta$ and that of the strain tensor be written as $\epsilon_{ij}$. Then, since the spherical coordinates are orthogonal,

$$\xi_r = \sqrt{g^{11}}\,u_1 = u_1, \qquad \xi_\phi = \sqrt{g^{22}}\,u_2 = \frac{u_2}{r},$$

$$\xi_\theta = \sqrt{g^{33}}\,u_3 = \frac{u_3}{r\sin\phi}, \qquad \epsilon_{ij} = \sqrt{g^{ii}g^{jj}}\,e_{ij}.$$

Therefore,

$$\epsilon_{11} = \frac{\partial \xi_r}{\partial r},$$

$$\epsilon_{22} = \frac{\partial \xi_\phi}{r\,\partial \phi} + \frac{\xi_r}{r},$$

$$\epsilon_{33} = \frac{1}{r \sin\phi}\frac{\partial \xi_\theta}{\partial \theta} + \frac{1}{r}\xi_r + \frac{\cot\phi}{r}\xi_\phi,$$

$$\epsilon_{12} = \frac{1}{2}\left(\frac{1}{r}\frac{\partial \xi_r}{\partial \phi} + \frac{1}{r}\frac{\partial (r\xi_\phi)}{\partial r}\right) - \frac{\xi_\phi}{r} = \frac{1}{2}\left(\frac{1}{r}\frac{\partial \xi_r}{\partial \phi} + \frac{\partial \xi_\phi}{\partial r} - \frac{\xi_\phi}{r}\right),$$

$$\epsilon_{23} = \frac{1}{2}\left(\frac{1}{r\sin\phi}\frac{\partial \xi_\phi}{\partial \theta} + \frac{1}{r}\frac{\partial \xi_\theta}{\partial \phi} - \frac{\cot\phi}{2r}\xi_\theta\right),$$

$$\epsilon_{31} = \frac{1}{2}\left(\frac{1}{r\sin\phi}\frac{\partial \xi_r}{\partial \theta} + \frac{\partial \xi_\theta}{\partial r} - \frac{\xi_\theta}{r}\right).$$

The equations of equilibrium now become, with $F_r$, $F_\theta$, $F_\phi$ denoting the physical components of the body force vector,

$$-F_r = \frac{1}{r}\frac{\partial}{\partial r}(r^2\sigma_{rr}) + \frac{1}{r\sin\theta}\frac{\partial}{\partial \theta}(\sin\theta\,\sigma_{r\theta}) + \frac{1}{r\sin\theta}\frac{\partial}{\partial \phi}(\sigma_{r\phi}) - \frac{1}{r}(\sigma_{\theta\theta} + \sigma_{\phi\phi}),$$

$$-F_\theta = \frac{1}{r^3}\frac{\partial}{\partial r}(r^3\sigma_{\theta r}) + \frac{1}{r\sin^2\phi}\frac{\partial}{\partial \phi}(\sin^2\phi\,\sigma_{\theta\phi}) + \frac{1}{r\sin\phi}\frac{\partial}{\partial \theta}(\sigma_{\theta\theta}),$$

$$-F_\phi = \frac{1}{r^3}\frac{\partial}{\partial r}(r^3\sigma_{r\phi}) + \frac{1}{r\sin\phi}\frac{\partial}{\partial \phi}(\sin\phi\,\sigma_{\phi\phi}) + \frac{1}{r\sin\phi}\frac{\partial}{\partial \theta}(\sigma_{\theta\phi}) - \frac{\cot\phi}{r}\sigma_{\theta\theta}.$$

## 4.12. EXAMPLE—CYLINDRICAL POLAR COORDINATES

Letting $\xi_r$, $\xi_\theta$, $\xi_z$ be the physical components of displacement and $\epsilon_{rr}$, $\epsilon_{r\theta}$, . . . , $\sigma_{rr}$, $\sigma_{r\theta}$, . . . , etc., be the physical components of the strain and stress, respectively, we obtain the following results.

$$x^1 = r, \qquad x^2 = \theta, \qquad x^3 = z,$$

$$ds^2 = dr^2 + r^2\,d\theta^2 + dz^2,$$

$$g_{11} = 1, \qquad g_{22} = r^2, \qquad g_{33} = 1, \quad \text{all other } g_{ij} = 0,$$

$$g^{11} = 1, \qquad g^{22} = \frac{1}{r^2}, \qquad g^{33} = 1, \quad \text{all other } g^{ij} = 0,$$

$$\Gamma^1_{22} = -r, \qquad \Gamma^2_{12} = \Gamma^2_{21} = \frac{1}{r}, \quad \text{all other } \Gamma^i_{jk} = 0.$$

Hence,

$$\epsilon_{rr} = \frac{\partial \xi_r}{\partial r}, \qquad \epsilon_{\theta\theta} = \frac{1}{r}\frac{\partial \xi_\theta}{\partial \theta} + \frac{\xi_r}{r}, \qquad \epsilon_{zz} = \frac{\partial \xi_z}{\partial z},$$

$$\epsilon_{\theta z} = \frac{1}{2}\left(\frac{\partial \xi_\theta}{\partial z} + \frac{1}{r}\frac{\partial \xi_z}{\partial \theta}\right), \qquad \epsilon_{zr} = \frac{1}{2}\left(\frac{\partial \xi_z}{\partial r} + \frac{\partial \xi_r}{\partial z}\right), \qquad \epsilon_{r\theta} = \frac{1}{2}\left(\frac{1}{r}\frac{\partial \xi_r}{\partial \theta} + \frac{\partial \xi_\theta}{\partial r} - \frac{\xi_\theta}{r}\right).$$

The equations of equilibrium are

$$-F_r = \frac{1}{r}\frac{\partial}{\partial r}(r\sigma_{rr}) + \frac{1}{r}\frac{\partial}{\partial \theta}\sigma_{r\theta} + \frac{\partial}{\partial z}\sigma_{rz} - \frac{\sigma_{\theta\theta}}{r},$$

$$-F_\theta = \frac{1}{r^2}\frac{\partial}{\partial r}(r^2\sigma_{\theta r}) + \frac{1}{r}\frac{\partial}{\partial \theta}\sigma_{\theta\theta} + \frac{\partial}{\partial z}\sigma_{\theta z},$$

$$-F_z = \frac{1}{r}\frac{\partial}{\partial r}(r\sigma_{zr}) + \frac{1}{r}\frac{\partial}{\partial \theta}\sigma_{z\theta} + \frac{\partial}{\partial z}\sigma_{zz}.$$

## PROBLEMS

**4.7.** Give proper definitions of strain tensors $e^{ij}$, $e^i_j$ in general frames of reference in terms of $e_{ij}$. Show that $e_{ij}$, $e^{ij}$ are symmetric tensors. Show that the principal strains $e_1$, $e_2$, $e_3$ are the roots of the characteristic equation

$$|e^i_j - e\delta^i_j| = 0$$

and that the first invariant is $e^i_i$.

**4.8.** The generalization of the expressions of strain tensors in terms of the displacement vector field **u**, which represents the displacement of any particle as the body deforms as explained in Sec. 4.2, can be done as follows. The displacement vector **u** can be resolved into covariant and contravariant components $u_i$ and $u^i$, respectively, along base vectors and reciprocal base vectors defined in the original coordinates $a_1, a_2, a_3$. Then Green's strain tensor is

$$E_{ij} = \tfrac{1}{2}[u_j|_i + u_i|_j + u^k|_i\, u_k|_j],$$

where $u_i|_j$ is the covariant derivative of $u_i$ with respect to $a_j$.

The displacement vector **u** can be resolved also into covariant and contravariant components $u_i$ and $u^i$ along base vectors and reciprocal base vectors defined in the Eulerian coordinates $x_1, x_2, x_3$. Then the Almansi strain tensor is

$$e_{ij} = \tfrac{1}{2}[u_j\|_i + u_i\|_j - u^k\|_i\, u_k\|_j],$$

where a double bar is used to indicate that $u_i$, $u^j$ and the covariant derivatives are referred to the coordinates $x_i$.

Prove these statements.

**4.9.** The condition of compatibility for finite strain can be derived easily if convected coordinates are used [see Eq. (4.1:15)]. By the fact that the space of the deformed body is Euclidean, derive the compatibility conditions. *Hint:* See comments on p. 101 and Prob. 2.31. Note: however, the properties $g^{ij}|_k = 0, g_{ij}|_k = 0$ are valid in Riemannian space as well as Euclidean space. The feature that distinguishes a Euclidean space is the vanishing of the Riemann-Christoffel curvature tensor $R^i_{pst}$ of Prob. 2.31. Expressing $R^i_{pst} = 0$ in terms of the metric tensors before and after deformation leads to the compatibility conditions.

# 5

# CONSERVATION LAWS

We shall discuss in this chapter the basic laws of conservation of mass and momentum. A rectangular Cartesian frame of reference will be used throughout. All tensors are Cartesian tensors.

## 5.1. GAUSS' THEOREM

Consider a convex region $V$ bounded by a surface $S$ that consists of a finite number of parts whose outer normals form a continuous vector field (Fig. 5.1:1). Let a function $A(x_1, x_2, x_3)$ be defined and differentiable in the region $V + S$. Then, by the usual process of integration, we obtain

$$(1) \qquad \iiint_V \frac{\partial A}{\partial x_1}\, dx_1\, dx_2\, dx_3 = \iint_S (A^* - A^{**})\, dx_2\, dx_3,$$

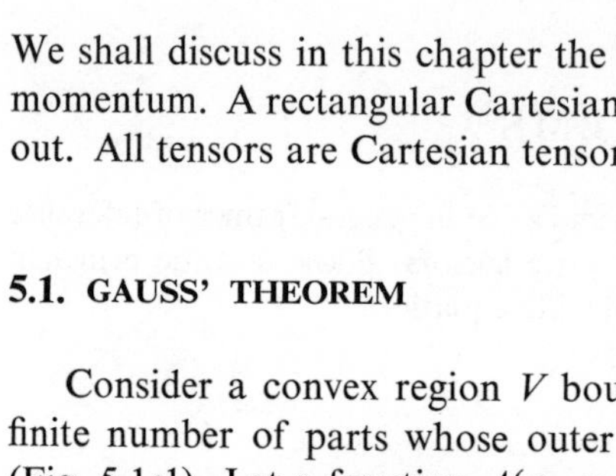

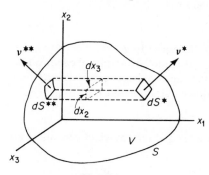

Fig. 5.1:1.

where $A^*$ and $A^{**}$ are the values of $A$ on the surface $S$ at the right and left ends of a line parallel to the $x_1$-axis, respectively. The factors $\pm dx_2\, dx_3$ in the surface integral in (1) are the projections of the $x_2$, $x_3$-plane of the areas $dS^*$ and $dS^{**}$ at the ends of a line parallel to the $x_1$-axis. Let $\boldsymbol{\nu} = (\nu_1, \nu_2, \nu_3)$ be the unit vector along the outer normal of $S$. Then $dx_2\, dx_3 = \nu_1^*\, dS^*$ at the right end and $dx_2\, dx_3 = -\nu_1^{**}\, dS^{**}$ at the left end. Therefore, the surface integral in Eq. (1) can be written as

$$\int_S (A\, \nu_1^*\ dS^* + A^{**}\nu_1^{**}\, dS^{**}) = \int_S A\nu_1\, dS.$$

Thus, Eq. (1) may be written as

$$\int_V \frac{\partial A}{\partial x_1}\, dV = \int_S A\nu_1\, dS,$$

where $dV$ and $dS$ denote the elements of $V$ and $S$, respectively. A similar argument applies to the volume integral of $\partial A/\partial x_2$, or $\partial A/\partial x_3$. In summary, we obtain Gauss' theorem

$$\int_V \frac{\partial A}{\partial x_i} dV = \int_S A\nu_i \, dS. \tag{2}$$

This formula holds for any convex regular region or for any region that can be decomposed into a finite number of convex regular regions, as can be seen by summing (2) over these component regions.

Now let us consider a tensor field $A_{jkl\ldots}$. Let the region $V$ with boundary surface $S$ be within the region of definition of $A_{jkl\ldots}$. Let every component of $A_{jkl\ldots}$ be continuously differentiable. Then Eq. (2) is applicable to every component of the tensor, and we may write

$$\int_V \frac{\partial}{\partial x_i} A_{jkl\ldots} \, dV = \int_S \nu_i A_{jkl\ldots} \, dS. \tag{3 ▲}$$

This is Gauss' theorem in a general form.

**Problem 5.1.** Show that

$$\int \phi_{,i} \, dV = \int \phi \nu_i \, dS \quad \text{or} \quad \int \text{grad}\, \phi \, dV = \int \boldsymbol{\nu} \phi \, dS,$$

$$\int u_{i,i} \, dV = \int u_i \nu_i \, dS \quad \text{or} \quad \int \text{div}\, \mathbf{u} \, dV = \int \boldsymbol{\nu} \cdot \mathbf{u} \, dS,$$

$$\int e_{ijk} u_{k,j} \, dV = e_{ijk} \int u_{k,j} \, dV = e_{ijk} \int u_k \nu_j \, dS = \int e_{ijk} u_k \nu_j \, dS,$$

or

$$\int \text{curl}\, \mathbf{u} \, dV = \int \boldsymbol{\nu} \times \mathbf{u} \, dS,$$

where $e_{ijk}$ is the permutation tensor. Verify that these formulas are also valid in two dimensions, in which case the range of indices is 1, 2 and the volume and surface integrals are replaced by surface and line integrals, respectively.

## 5.2. MATERIAL AND SPATIAL DESCRIPTION OF CHANGING CONFIGURATIONS

We shall speak of a *particle* in the sense of a material particle as we know it in Newtonian particle mechanics. The instantaneous geometric location of a particle will be spoken of as a *point*. A body is composed of particles. To label the particles of a body we choose a Cartesian frame of reference and identify the coordinates of the particles at a time $t = 0$. Let $(a_1, a_2, a_3)$ be the coordinates of a particle at $t = 0$ (Fig. 5.2:1). At a later time the particle

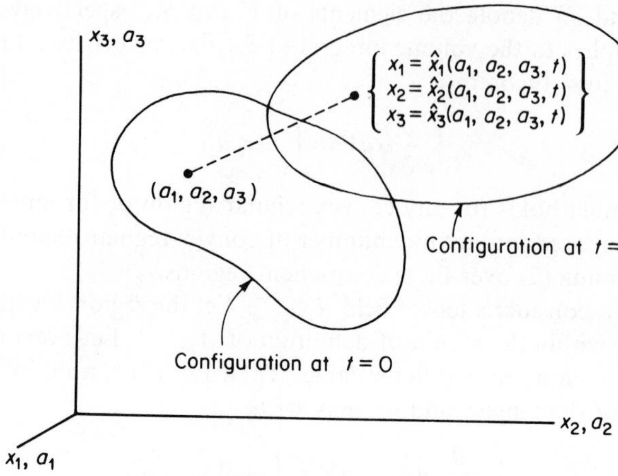


**Fig. 5.2:1.** Labeling of particles.

is moved to another point whose coordinates are $(x_1, x_2, x_3)$, *referred to the same coordinate system*. The relation

$$x_i = \hat{x}_i(a_1, a_2, a_3, t), \qquad i = 1, 2, 3, \tag{1}$$

links the configurations of the body at different instants of time. The functions $\hat{x}_i$ are single-valued continuous functions whose Jacobian does not vanish.

A basic property of bodies is that they have mass. In classical mechanics, mass is assumed to be conserved; i.e., the mass of a material body is the same at all times. In continuum mechanics it is further assumed that the mass is an absolutely continuous function of volume. In other words, it is assumed that a positive quantity $\rho$, called *density*, can be defined at every point in the body as the limit

$$\rho(\mathbf{x}) = \lim_{k\to\infty} \frac{\text{mass of } B_k}{\text{volume of } B_k}, \tag{2}$$

where $B_k$ is a suitably chosen infinite sequence of particle sets shrinking down upon the point $\mathbf{x}$; the symbol $\mathbf{x}$ stands for $(x_1, x_2, x_3)$. At time $t = 0$, the density at the point $(a_1, a_2, a_3)$ is denoted by $\rho_0(\mathbf{a})$.

The conservation of mass is expressed in the formula

$$\int \rho(\mathbf{x})\, dx_1\, dx_2\, dx_3 = \int \rho_0(\mathbf{a})\, da_1\, da_2\, da_3, \tag{3}$$

where the integrals extend over the same particles. Since

$$\int \rho(\mathbf{x})\, dx_1\, dx_2\, dx_3 = \int \rho(\mathbf{x}) \left| \frac{\partial x_i}{\partial a_j} \right| da_1\, da_2\, da_3,$$

and since this relation must hold for all bodies, we have

(4) ▲
$$\rho_0(\mathbf{a}) = \rho(\mathbf{x}) \left| \frac{\partial x_i}{\partial a_j} \right|,$$
$$\rho(\mathbf{x}) = \rho_0(\mathbf{a}) \left| \frac{\partial a_i}{\partial x_j} \right|,$$

where $|\partial a_i/\partial x_j|$ denotes the determinant of the matrix $\{\partial a_i/\partial x_j\}$. These equations relate the density in different configurations of the body to the transformation that leads from one configuration to the other.

For the particle $(a_1, a_2, a_3)$ whose trajectory is given by (1), the velocity is

(5) ▲
$$v_i(\mathbf{a}, t) = \frac{\partial}{\partial t} x_i(\mathbf{a}, t)$$

and the acceleration is

(6) ▲
$$\dot{v}_i(\mathbf{a}, t) = \frac{\partial^2}{\partial t^2} x_i(\mathbf{a}, t) = \frac{\partial}{\partial t} v_i(\mathbf{a}, t),$$

where **a** stands for $(a_1, a_2, a_3)$ and is held constant.

A description of mechanical evolution which uses $(a_1, a_2, a_3)$ and $t$ as independent variables is called a *material description*. In hydrodynamics, traditionally, a different description, called the *spatial description*, is used. In the spatial description, the location $(x_1, x_2, x_3)$ and time $t$ are taken as independent variables. This is convenient because measurements in many kinds of materials are more directly interpreted in terms of what happens at a certain place, rather than following the particles. These two methods of description are commonly designated as the *Lagrangian* and the *Eulerian descriptions*, respectively, although both are due to Euler. The variables $a_1, a_2, a_3, t$ are usually called the *Lagrangian variables*, whereas $x_1, x_2, x_3, t$ are called the *Eulerian variables*; they are related by Eq. (1). *For a given particle, it is convenient to speak of* $(a_1, a_2, a_3)$ *as the Lagrangian coordinates of the particle at* $(x_1, x_2, x_3)$.

In a spatial description, the instantaneous motion of the body is described by the velocity vector field $v_i(x_1, x_2, x_3, t)$ associated with the instantaneous location of each particle. The acceleration of the particle is given by the formula

(7) ▲
$$\dot{v}_i(\mathbf{x}, t) = \frac{\partial v_i}{\partial t}(\mathbf{x}, t) + v_j \frac{\partial v_i}{\partial x_j}(\mathbf{x}, t),$$

where **x** again stands for the variables $x_1, x_2, x_3$, and every quantity in this formula is evaluated at $(\mathbf{x}, t)$. The proof follows the fact that a particle located at $(x_1, x_2, x_3)$ at time $t$ is moved to a point with coordinates $x_i + v_i\, dt$ at the time $t + dt$; and that, according to Taylor's theorem,

$$\dot{v}_i(\mathbf{x}, t)\, dt = v_i(x_j + v_j\, dt, t + dt) - v_i(x, t)$$
$$= v_i + \frac{\partial v_i}{\partial t} dt + \frac{\partial v_i(x, t)}{\partial x_j} v_j\, dt - v_i,$$

which reduces to (7). The first term in (7) may be interpreted as arising from the time dependence of the velocity field; the second term as the contribution of the motion of the particle in the instantaneous velocity field. Accordingly, these terms are called the *local* and the *convective* parts of the acceleration, respectively.

The same reasoning that led to (7) is applicable to any function $F(x_1, x_2, x_3, t)$ that is attributable to the moving particles, such as the temperature. A convenient terminology is the *material derivative*, and it is denoted by a dot or the symbol $D/Dt$. Thus

$$(8) \quad \blacktriangle \qquad \dot{F} \equiv \frac{DF}{Dt} \equiv \left(\frac{\partial F}{\partial t}\right)_{\mathbf{x}=\text{const.}} + v_1 \frac{\partial F}{\partial x_1} + v_2 \frac{\partial F}{\partial x_2} + v_3 \frac{\partial F}{\partial x_3}$$
$$\equiv \left(\frac{\partial F}{\partial t}\right)_{\mathbf{a}=\text{const.}},$$

where $\mathbf{a} = (a_1, a_2, a_3)$ is the Lagrangian coordinate of the particle which is located at $\mathbf{x}$ at the time $t$, connected by Eq. (1).

## 5.3. MATERIAL DERIVATIVE OF VOLUME INTEGRAL

Consider a volume integral taken over the body

$$(1) \qquad I = \int_V A(x, t)\, dV,$$

where $A(x, t)$ denotes a property of the continuum and the integral is evaluated at an instant of time $t$. We may wish to know how fast the body itself sees the value of $I$ is changing, so it is of interest to know *the derivative of I with respect to time for a given set of particles*. Now the particle at $x_i$ at the instant $t$ will have the coordinates $x_i' = x_i + v_i\, dt$ at the time $t + dt$. The boundary $S$ of the body at the instant $t$ will have moved at time $t + dt$ to a neighboring surface $S'$, which bounds the volume $V'$ (Fig. 5.3:1). The material derivative of $I$ is defined as

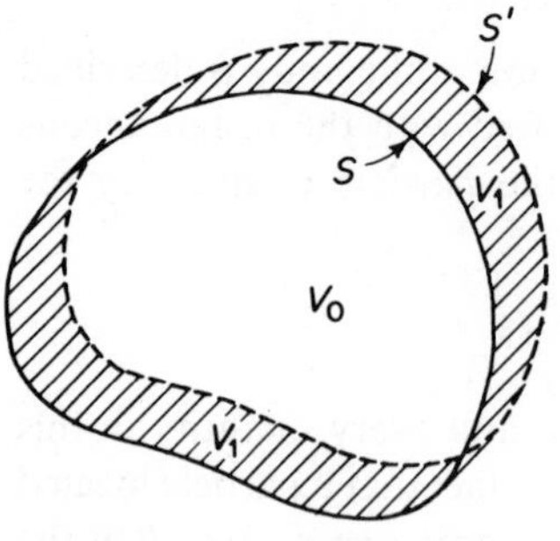

**Fig. 5.3:1.** Continuous change of the boundary of a region.

$$(2) \qquad \frac{DI}{Dt} = \lim_{dt \to 0} \frac{1}{dt} \left[ \int_{V'} A(\mathbf{x}', t + dt)\, dV' - \int_V A(\mathbf{x}, t)\, dV \right].$$

Now there are two contributions to the difference on the right-hand side of (2); one over the region $V_0$ where $V$ and $V'$ share in common, and another over the region $V_1$ where $V$ and $V'$ differ. The former contribution to $DI/Dt$ is evidently

$$\int_{V_0} \frac{\partial A}{\partial t}\, dV.$$

The latter contribution comes from the value of $A$ on the boundary multiplied by the volume swept by the particles on the boundary in the time interval $dt$. If $\nu_i$ is the unit vector along the exterior normal of $S$, then, since the displacement of a particle on the boundary is $v_i\, dt$, the volume swept by particles occupying an element of area $dS$ on the boundary $S$ is $dV = v_i \nu_i\, dS \cdot dt$. The contribution of this element to $DI/Dt$ is $Av_i\nu_i\, dS$. The total contribution is obtained by an integration over $S$. Therefore,

(3) ▲ $$\frac{D}{Dt}\int_V A\, dV = \int_V \frac{\partial A}{\partial t}\, dV + \int_S A v_j \nu_j\, dS.$$

Transforming the last integral by Gauss' theorem and using Eq. (5.2:8), we have

(4) ▲ $$\begin{aligned} \frac{D}{Dt}\int_V A\, dV &= \int_V \frac{\partial A}{\partial t}\, dV + \int_V \frac{\partial}{\partial x_j}(Av_j)\, dV \\ &= \int_V \left(\frac{\partial A}{\partial t} + v_j \frac{\partial A}{\partial x_j} + A\frac{\partial v_j}{\partial x_j}\right) dV \\ &= \int_V \left(\frac{DA}{Dt} + A\frac{\partial v_j}{\partial x_j}\right) dV. \end{aligned}$$

This important formula will be used over and over again below. It should be noted that *the operation of forming the material derivative and that of spatial integration is noncommutative* in general.

## 5.4. THE EQUATION OF CONTINUITY

The law of conservation of mass has been discussed in Sec. 5.2. With the results of Sec. 5.3, we can now give some alternative forms.

The mass contained in a region $V$ at a time $t$ is

(1) $$m = \int_V \rho\, dV,$$

where $\rho = \rho(x, t)$ is the density field of the continuum. Conservation of mass requires that $Dm/Dt = 0$. The derivative $Dm/Dt$ is given by Eqs. (5.3:3) and (5.3:4), when $A$ is replaced by $\rho$. Since the result must hold for arbitrary

$V$, the integrand must vanish. Hence, we obtain the following alternative forms of the law of conservation of mass.

(2) ▲
$$\int_V \frac{\partial \rho}{\partial t}\, dV + \int_S \rho v_j \nu_j\, dS = 0.$$

(3) ▲
$$\frac{\partial \rho}{\partial t} + \frac{\partial \rho v_j}{\partial x_j} = 0.$$

(4) ▲
$$\frac{D\rho}{Dt} + \rho \frac{\partial v_j}{\partial x_j} = 0.$$

These are called the *equations of continuity*. The integral form (2) is useful when the differentiability of $\rho v_j$ cannot be assumed.

In problems of statics, these equations are identically satisfied. Then the conservation of mass must be expressed by Eq. (5.2:3), or (5.2:4).

## 5.5. THE EQUATIONS OF MOTION

Newton's laws of motion state that, in an inertial frame of reference, the material rate of change of the linear momentum of a body is equal to the resultant of applied forces and that the material rate of change of the moment of momentum with respect to the coordinate origin is equal to the resultant moment of applied forces about the same origin.

At an instant of time $t$, a regular region $V$ of space contains the linear momentum

(1)
$$\mathscr{P}_i = \int_V \rho v_i\, dV.$$

If the body is subjected to surface tractions $\overset{\nu}{T}_i$ and body force per unit volume $X_i$, the resultant force is

(2)
$$\mathscr{F}_i = \int_S \overset{\nu}{T}_i\, dS + \int_V X_i\, dV.$$

According to the stress principle of Euler and Cauchy (Secs. 3.2 and 3.3), $\overset{\nu}{T}_i = \sigma_{ji}\nu_j$, where $\sigma_{ij}$ is the stress field and $\nu_j$ is the unit vector along the exterior normal to the boundary surface $S$ of the region $V$. Substituting into (2) and transforming into a volume integral by Gauss' theorem, we have

(3)
$$\mathscr{F}_i = \int_V \left(\frac{\partial \sigma_{ij}}{\partial x_j} + X_i\right) dV.$$

Newton's law states that

(4)
$$\frac{D}{Dt}\, \mathscr{P}_i = \mathscr{F}_i.$$

Hence, according to Eq. (5.3:4), with $A$ identified with $\rho v_i$, we have

(5) $$\int_V \left[\frac{\partial \rho v_i}{\partial t} + \frac{\partial}{\partial x_j}(\rho v_i v_j)\right] dV = \int_V \left(\frac{\partial \sigma_{ij}}{\partial x_j} + X_i\right) dV.$$

Since this equation must hold for an arbitrary region $V$, the integrand on the two sides must be equal. Thus

(6) $$\frac{\partial \rho v_i}{\partial t} + \frac{\partial}{\partial x_j}(\rho v_i v_j) = \frac{\partial \sigma_{ij}}{\partial x_j} + X_i.$$

The left-hand side of (6) is equal to

$$v_i\left(\frac{\partial \rho}{\partial t} + \frac{\partial \rho v_j}{\partial x_j}\right) + \rho\left(\frac{\partial v_i}{\partial t} + v_j \frac{\partial v_i}{\partial x_j}\right).$$

The first parentheses vanish by the equation of continuity (5.4:3), while the second is the acceleration $Dv_i/Dt$. Hence,

(7) ▲ $$\rho \frac{Dv_i}{Dt} = \frac{\partial \sigma_{ij}}{\partial x_j} + X_i.$$

This is the *Eulerian equation of motion* of a continuum. The equation of equilibrium discussed in Sec. 3.4 is obtained by setting all velocity components $v_i$ equal to zero.

If differentiability of the stress field or the momentum field cannot be assumed, we may use Eq. (5.3:3) to compute $D\mathscr{P}/Dt$. Then Eqs. (2) and (4) give *Euler's equation in the integral form*,

(8) $$\int_V \frac{\partial \rho v_i}{\partial t}\, dV = \int_S (\sigma_{ij} - \rho v_i v_j)\nu_j\, dS + \int_V X_i\, dV.$$

The corresponding equations of static equilibrium are obtained, of course, by setting all velocity components to zero.

## 5.6. MOMENT OF MOMENTUM

An application of the law of balance of angular momentum to the particular case of *static equilibrium* leads to the conclusion that stress tensors are symmetric tensors (see Sec. 3.4). We shall now show that no additional restriction to the motion of a continuum is introduced in dynamics by the angular momentum postulate.

At an instant of time $t$, a body occupying a regular region $V$ of space with boundary $S$ has the moment of momentum [Eq. (3.2:2)]

(1) $$\mathscr{H}_i = \int_V e_{ijk} x_j \rho v_k\, dV,$$

with respect to the origin. If the body is subjected to surface traction $\overset{\nu}{T}_i$ and body force per unit volume $X_i$, the resultant moment about the origin is

$$\mathscr{L}_i = \int_V e_{ijk} x_j X_k \, dV + \int_S e_{ijk} x_j \overset{\nu}{T}_k \, dS. \tag{2}$$

Introducing Cauchy's formula $\overset{\nu}{T}_k = \sigma_{lk}\nu_l$ into the last integral and transforming the result into a volume integral by Gauss' theorem, we obtain

$$\mathscr{L}_i = \int_V e_{ijk} x_j X_k \, dV + \int_V (e_{ijk} x_j \sigma_{lk})_{,l} \, dV. \tag{3}$$

Euler's law states that for any region $V$

$$\frac{D}{Dt} \mathscr{H}_i = \mathscr{L}_i. \tag{4}$$

Evaluating the material derivative of $\mathscr{H}_i$ according to (5.3:4), and using (3), we obtain

$$e_{ijk} x_j \frac{\partial}{\partial t}(\rho v_k) + \frac{\partial}{\partial x_l}(e_{ijk} x_j \rho v_k v_l) = e_{ijk} x_j X_k + e_{ijk}(x_j \sigma_{lk})_{,l}. \tag{5}$$

The second term in (5) can be written as

$$e_{ijk} \rho v_j v_k + e_{ijk} x_j \frac{\partial}{\partial x_l}(\rho v_k v_l) = e_{ijk} x_j \frac{\partial}{\partial x_l}(\rho v_k v_l).$$

The last term in (5) can be written as $e_{ijk}\sigma_{jk} + e_{ijk} x_j \sigma_{lk,l}$. Hence, Eq. (5) becomes

$$e_{ijk} x_j \left[ \frac{\partial}{\partial t} \rho v_k + \frac{\partial}{\partial x_l}(\rho v_k v_l) - X_k - \sigma_{lk,l} \right] - e_{ijk}\sigma_{jk} = 0. \tag{6}$$

The sum in the square brackets vanishes by the equation of motion (5.5:6). Hence, Eq. (6) is reduced to

$$e_{ijk}\sigma_{jk} = 0; \tag{7}$$

i.e., $\sigma_{jk} = \sigma_{kj}$. Thus, if the stress tensor is symmetric, the law of balance of moment of momentum is identically satisfied.

## 5.7. OTHER FIELD EQUATIONS

The motion of a continuum must be governed further by the law of conservation of energy. If mechanical energy is the only form of energy of interest in a problem, then the energy equation is merely the first integral of the equation of motion. If the interaction of thermal process and mechanical process is significant, then the equation of energy contains a thermal energy term and is an independent equation to be satisfied. We shall discuss the energy equation in Sec. 12.2.

The equations of continuity and motion constitute four equations for ten unknown functions of time and position; namely, the density $\rho$, the three velocity components $v_i$ (or displacements $u_i$) and the six independent stress components $\sigma_{ij}$. Further restrictions would have to be introduced before the motion of a continuum can be determined. One group of such additional restrictions comes from a statement about the mechanical property of the medium, in the form of a specification of stress-strain relationship. These specifications are called *constitutive equations.*

One of the most ambitious programs of research in mathematical physics is to derive the relationship between stress, strain, and other physical characteristics of a material from the properties of the constituent elementary particles. The field of endeavor in this respect is statistical mechanics; in particular, the newer statistical mechanics of transport phenomena that deals with nonequilibrium states. At present, however, this program is still in its initial stages of development.

A different approach is to determine the physical relations experimentally. From the experimental results, certain general laws characterizing a material are abstracted. Once such a set of laws is formulated, new problems can be proposed and solved mathematically, new experiments can be suggested and performed to check the accuracy of the basic laws, and, if it is satisfactory, to be used in solving technical problems of engineering importance.

Since experimental observation as a human endeavor is always limited in scope, the abstraction of observed facts into a mathematical law can only be a mental process guided by logic and consistency with the general principles of physics as a whole. This abstraction and deduction is an important function of continuum mechanics.

In the next chapter we shall discuss the simplest constitutive equation for solids—that of linear elastic body. In the theory of linear elasticity, the stress-strain relationship provides six additional equations relating the variables named above, making the motion of the continuum deterministic. In later chapters we shall discuss the cases in which the constitutive equations involve new variables so that additional equations from thermodynamics or other physical principles must be considered.

## PROBLEMS

**5.2.** Express the following statements in tensor equations. Define your symbols.

(a) The force of gravitational attraction between two particles of inertial masses $m_1$ and $m_2$, respectively, and separated by a distance $r$, is equal to $Gm_1m_2/r^2$ and is directed toward each other.

(b) The components of stress is a linear function of the components of strain.

(c) A normal vector of a surface is perpendicular to any two line elements tangent to the surface (in particular, to tangents of the parametric curves).

**5.3.** If the stress-strain law in rectangular Cartesian coordinates is

$$\tau_{ij} = \lambda\theta\, \delta_{ij} + 2Ge_{ij}, \qquad \theta = e_{\alpha\alpha},$$

what is the proper form of tensor relation between $\tau^{ij}$ and $e_{ij}$ in general coordinates of reference?

**5.4.** Discuss the appropriateness, from the tensorial point of view, of the following proposals for the constitutive equations in Cartesian tensors for certain nonlinear, isotropic, elastic materials.

(a) $\tau_{ij} = P(e_{mn})e_{ij}$, where $P(e_{mn}) = ae_{11} + be_{11}^2$ ($a$, $b$ constants).

(b) $\tau_{ij} = P(e_{mn})e_{ij}$, where $P(e_{mn}) = a + b\,|e_1|^2$ ($a$, $b$ constants), and $e_1$, $e_2$, $e_3$ are the three principal strains satisfying the relation $e_1 \geqslant e_2 \geqslant e_3$.

(c) $\tau_{ij} = Q(I_1, I_2, I_3)e_{ij}$, where $I_1$, $I_2$, $I_3$ are the first, second, and third invariants of $e_{ij}$, respectively.

(d) $\tau_{ij} = \alpha\delta_{ij} + \beta e_{ij} + \gamma e_{ik}e_{kj} + \lambda e_{ik}e_{km}e_{mj}$, where $\alpha$, $\beta$, $\gamma$, $\lambda$ are constants.

(e) If in (d), $\alpha$, $\beta$, $\gamma$, $\lambda$ are permitted to depend on the stress components, what kind of combinations of the stress components would be allowed for $\alpha$, $\beta$, $\gamma$, $\lambda$ to be functionally dependent upon?

# 6

# ELASTIC AND PLASTIC BEHAVIOR OF MATERIALS

In this chapter, some simple constitutive laws are considered. These laws correspond to material behavior which is isothermal, in the range of relatively small strains, with slow rates of flow. Thus, the plasticity considered here is appropriate to most problems in structural engineering, in which excessive plastic flow is generally undesirable; but it does not meet the needs of solving problems in metal forming, wire drawing, rolling, etc.

We shall use Cartesian tensors in the text. In rectangular Cartesian coordinates the physical components of a stress tensor are the same as the tensor components, so stresses will be denoted by $\sigma_{ij}$. The reader should try to put all of the equations in this chapter into general tensor equations in curvilinear coordinates.

## 6.1. GENERALIZED HOOKE'S LAW

With the introduction of the concepts of stress and strain, Cauchy generalized Hooke's law into the statement that the components of stress are linearly related to the components of strain. As a tensor equation, the generalized Hooke's law may be written in the form

(1) ▲ $$\sigma^{ij} = C^{ijkl} e_{kl},$$

where $\sigma^{ij}$ is the stress tensor, $e_{kl}$ is the strain tensor,† and $C^{ijkl}$ is the tensor of the *elastic constants*, or *moduli*, of the material. Inasmuch as $\sigma^{ij} = \sigma^{ji}$, we must have

(2) $$C^{ijkl} = C^{jikl}.$$

Furthermore, since $e_{kl} = e_{lk}$, and in Eq. (1) the indices $k$ and $l$ are dummys, we can always symmetrize $C^{ijkl}$ with respect to $k$ and $l$ without altering the sum. Hence, without loss of generality, we may assume that

(3) $$C^{ijkl} = C^{ijlk}.$$

According to these symmetry properties, the maximum number of the independent elastic constants is 36.

† If the displacement is infinitesimal, $e_{ij} = \frac{1}{2}(u_{i,j} + u_{j,i})$. If $u_i$ is finite, $e_{ij}$ is the Almansi strain tensor of Sec. 4.1. See also Sec. 16.6, 16.7.

If there exists a strain energy function $W$,

(4) $$W = \tfrac{1}{2}C^{ijkl}e_{ij}e_{kl},$$

with the property

(5) ▲ $$\frac{\partial W}{\partial e_{ij}} = \sigma^{ij},$$

then we can always suppose that the quadratic form (4) is symmetric, and it follows that

(6) $$C^{ijkl} = C^{klij}.$$

Under the symmetry condition (6), the number of independent elastic constants is reduced further to 21.

The question of existence of strain energy function will be discussed later in Chap. 12. It is now generally accepted that the number of independent elastic constants in the generalized Hooke's law is 21 in the most general form of anisotropy.

If a material possesses further symmetry in its elastic property in certain directions, the number of independent elastic constants will be reduced further. For example, if the material exhibits symmetry with respect to a plane, the number of independent elastic constants becomes 13. If there is symmetry with respect to three mutually perpendicular planes, the number becomes 9. However, if the material is elastically *isotropic*, i.e., if its elastic properties are identical in all directions, then the number of independent elastic constants reduces to 2.

The study of crystal symmetry is a very interesting subject. Many excellent references exist. See Love,[1.2] Green and Adkins,[1.2] Sokolnikoff.[1.2]

In the following discussion, we shall limit our attention to *isotropic* materials.

## 6.2. STRESS-STRAIN RELATIONSHIP FOR AN ISOTROPIC ELASTIC MATERIAL

For an *isotropic elastic material* in which there is no change of temperature, Hooke's law referred to a set of rectangular Cartesian coordinates may be stated in the form

(1) ▲ $$\sigma_{\alpha\alpha} = 3Ke_{\alpha\alpha},$$

(2) ▲ $$\sigma'_{ij} = 2Ge'_{ij},$$

where $K$ and $G$ are constants and $\sigma'_{ij}$ and $e'_{ij}$ are the stress deviation and strain deviation, respectively; i.e.,

(3) ▲ $$\sigma'_{ij} = \sigma_{ij} - \tfrac{1}{3}\sigma_{\alpha\alpha}\,\delta_{ij},$$

(4) ▲ $$e'_{ij} = e_{ij} - \tfrac{1}{3}e_{\alpha\alpha}\,\delta_{ij}.$$

We have seen before that $\frac{1}{3}\sigma_{\alpha\alpha}$ is the mean stress at a point and that, if the strain were infinitesimal, $e_{\alpha\alpha}$ is the change of volume per unit volume: both are invariants. Thus, (1) states that the change of volume of the material is proportional to the mean stress. In the special case of hydrostatic compression,

$$\sigma_{xx} = \sigma_{yy} = \sigma_{zz} = -p, \qquad \sigma_{xy} = \sigma_{yz} = \sigma_{zx} = 0,$$

we have $\sigma_{\alpha\alpha} = -3p$, and Eq. (1) may be written, in the case of infinitesimal strain, with $v$ and $\Delta v$ denoting volume and change of volume, respectively,

$$(5) \blacktriangle \qquad \frac{\Delta v}{v} = -\frac{p}{K}.$$

Thus, the coefficient $K$ is appropriately called the *bulk modulus* of the material.

The strain deviation $e'_{ij}$ describes a deformation without volume change. Equation (2) states that the stress deviation is simply proportional to the strain deviation. The constant $G$ is called the *modulus of elasticity in shear*, or *shear modulus*, or the *modulus of rigidity*. In the special case in which $e_{xy} \neq 0$ but all other strain components vanish, we have

$$(6) \blacktriangle \qquad \sigma_{xy} = 2Ge_{xy},$$

whereas all other stress components vanish. The coefficient 2 is included because, before the tensor concept was introduced, it was customary to define the shear strain as $\gamma_{xy} = 2e_{xy}$.

If we substitute (3) and (4) into (2) and make use of (1), the result may be written in the form

$$(7) \blacktriangle \qquad \sigma_{ij} = \lambda e_{\alpha\alpha}\,\delta_{ij} + 2Ge_{ij},$$

or

$$(8) \blacktriangle \qquad e_{ij} = \frac{1+\nu}{E}\,\sigma_{ij} - \frac{\nu}{E}\,\sigma_{\alpha\alpha}\,\delta_{ij}.$$

The constants $\lambda$ and $G$ are called *Lamé's constants* (G. Lamé, 1852). In many books $\mu$ is used in place of $G$. The constant $E$ is called the modulus of elasticity, or Young's modulus (Thomas Young, 1807). The constant $\nu$ is called Poisson's ratio. The relationships between these constants are

$$\lambda = \frac{2G\nu}{1-2\nu} = \frac{G(E-2G)}{3G-E} = K - \frac{2}{3}G = \frac{E\nu}{(1+\nu)(1-2\nu)}$$
$$= \frac{3K\nu}{1+\nu} = \frac{3K(3K-E)}{9K-E},$$

$$G = \frac{\lambda(1-2\nu)}{2\nu} = \frac{3}{2}(K-\lambda) = \frac{E}{2(1+\nu)} = \frac{3K(1-2\nu)}{2(1+\nu)} = \frac{3KE}{9K-E},$$

(9)
$$\nu = \frac{\lambda}{2(\lambda + G)} = \frac{\lambda}{(3K - \lambda)} = \frac{E}{2G} - 1 = \frac{3K - 2G}{2(3K + G)} = \frac{3K - E}{6K},$$

$$E = \frac{G(3\lambda + 2G)}{\lambda + G} = \frac{\lambda(1 + \nu)(1 - 2\nu)}{\nu} = \frac{9K(K - \lambda)}{3K - \lambda}$$

$$= 2G(1 + \nu) = \frac{9KG}{3K + G} = 3K(1 - 2\nu),$$

$$K = \lambda + \frac{2}{3}G = \frac{\lambda(1 + \nu)}{3\nu} = \frac{2G(1 + \nu)}{3(1 - 2\nu)} = \frac{GE}{3(3G - E)} = \frac{E}{3(1 - 2\nu)}.$$

To these we may add the following combinations that appear frequently.

$$\frac{G}{\lambda + G} = 1 - 2\nu, \qquad \frac{\lambda}{\lambda + 2G} = \frac{\nu}{1 - \nu}.$$

In unabridged notation, Eq. (8) reads

(10)
$$\begin{aligned}
e_{xx} &= \frac{1}{E}\left[\sigma_{xx} - \nu(\sigma_{yy} + \sigma_{zz})\right],\\
e_{yy} &= \frac{1}{E}\left[\sigma_{yy} - \nu(\sigma_{xx} + \sigma_{zz})\right],\\
e_{zz} &= \frac{1}{E}\left[\sigma_{zz} - \nu(\sigma_{xx} + \sigma_{yy})\right],\\
e_{xy} &= \frac{1 + \nu}{E}\sigma_{xy} = \frac{1}{2G}\sigma_{xy},\\
e_{yz} &= \frac{1 + \nu}{E}\sigma_{yz} = \frac{1}{2G}\sigma_{yz},\\
e_{zx} &= \frac{1 + \nu}{E}\sigma_{zx} = \frac{1}{2G}\sigma_{zx}.
\end{aligned}$$

Table 6.2:1 gives the average values of $E$, $G$, and $\nu$ at toom temperature for several engineering materials which are approximately isotropic.

In 1829, Poisson advanced arguments, later found untenable, that the value of $\nu$ should be $\frac{1}{4}$. The special value of Poisson's ratio $\nu = \frac{1}{4}$ makes

(11)
$$\lambda = G$$

and simplifies the equations of elasticity considerably. Consequently, this assumption is often used, particularly in geophysics, in the study of

Table 6.2:1

| | $E$, $10^6$ lb/sq in. | $G$, $10^6$ lb/sq in. | $\nu$ | Speed of Sound (Dilatational Wave), $10^3$ ft per sec |
|---|---|---|---|---|
| Metals: | | | | |
| Steels | 30 | 11.5 | 0.29 | 16.3 |
| Aluminum alloys | 10 | 2.4 | 0.31 | 16.5 |
| Magnesium alloys | 6.5 | 2.4 | 0.35 | 16.6 |
| Copper (hot rolled) | 15.0 | 5.6 | 0.33 | – |
| Plastics: | | | | |
| Cellulose acetate | 0.22 | | | 3.6 |
| Vinylchloride acetate | 0.46 | | | 5.1 |
| Phenolic laminates | 1.23 | | 0.25 | 8.2 |
| Glass | 8 | 3.2 | 0.25 | |
| Concrete | 4 | | 0.2 | |

complicated wave-propagation problems. The special value $\nu = \frac{1}{2}$ implies that

$$G = \frac{1}{3} E, \qquad \frac{1}{K} = 0, \tag{12}$$

and

$$\frac{\Delta v}{v} = e_{\alpha\alpha} = 0. \tag{13}$$

Inasmuch as experience shows that hydrostatic compression induces reduction of volume, $\nu = \frac{1}{2}$ is an upper limit of the Poisson's ratio (cf. Sec. 12.5).

## 6.3. IDEAL PLASTIC SOLIDS

Metals obey Hooke's law only in a certain range of small strain. When a metal is strained beyond an *elastic limit*, Hooke's law no longer applies. The behavior of metals beyond their elastic limit is rather complicated. We shall give a very brief outline of some experimental facts in the next section, and we will discuss the formulation of the constitutive relations in the plastic regime later. However, before engaging in such a long and involved discussion, let us present a set of laws which defines the simplest of plastic materials—an *ideal plastic solid obeying von Mises' yielding criterion*. That such a set of laws is a reasonable abstraction of the behavior of certain real materials will be discussed later. At the moment, we shall just exhibit the minimum ingredients that constitute a theory of plasticity.

In Secs. 3.12 and 3.13 we defined the stress deviation $\sigma'_{ij}$, which has a second invariant $J_2$

(1) $$J_2 = \tfrac{1}{2}\sigma'_{ij}\sigma'_{ij}.$$

In a similar way, the strain deviation $e'_{ij}$ is defined by subtracting the mean strain from the strain tensor $e_{ij}$,

(2) ▲ $$e'_{ij} = e_{ij} - \tfrac{1}{3}e_{\alpha\alpha}\,\delta_{ij}.$$

If the material is isotropic and obeys Hooke's law, then

(3) $$e'_{ij} = \frac{1}{2G}\,\sigma'_{ij}.$$

When the elastic limit is reached, $e'_{ij}$ is no longer given by Hooke's law. In this case we define the *plastic strain deviation* as the actual strain deviation minus the strain deviation that would be computed if Hooke's law still applied. Let the plastic strain deviation be denoted by $e_{ij}^{(p)}$,

(4) $$e_{ij}^{(p)} = e'_{ij} - \frac{1}{2G}\,\sigma'_{ij}.$$

Then the total strain deviation may be written as

(5) ▲ $$e'_{ij} = e_{ij}^{(p)} + \frac{1}{2G}\,\sigma'_{ij}.$$

If at each instant of time the rate of plastic deformation is $\dot{e}_{ij}^{(p)}$, then the increment of plastic strain in the time interval $dt$ is $\dot{e}_{ij}^{(p)}\,dt$, and the total plastic strain deviation after successive stages of yielding will be the algebraic sum of the deformations that occur at all stages:

(6) $$e_{ij}^{(p)} = e_{ij}^{(p)}(0) + \int_0^t \dot{e}_{ij}^{(p)}(t)\,dt$$

where $e_{ij}^{(p)}(0)$ is the initial value of $e_{ij}^{(p)}$ at time $t = 0$. A theory of plasticity is formulated by specifying how $\dot{e}_{ij}^{(p)}$ can be computed.

For an *ideal plastic solid obeying von Mises' yielding criterion and flow rule* our specifications are as follows.

(a) Hooke's law holds for the mean stress and the mean strain at all times.

$$\sigma_{\alpha\alpha} = 3Ke_{\alpha\alpha}.$$

Hence, the plastic strain is incompressible ($e_{ii}^{(p)} = 0$) and the plastic strain deviation tensor is the same as the plastic strain tensor. This justifies the notation $e_{ij}^{(p)}$ in places where normally we would write $e_{ij}'^{(p)}$, to indicate a strain deviation.

(b) The material is elastic and obeys Hooke's law as long as the second invariant $J_2$ is less than a constant $k^2$. In other words, no change in plastic strain can occur as long as $J_2 < k^2$.

$$\dot{e}_{ij}^{(p)} = 0 \quad \text{when} \quad J_2 < k^2.$$

(c) Yielding can occur (elastic limit is reached) when and only when $J_2 = k^2$. When the yielding condition $J_2 - k^2 = 0$ prevails, the rate of change of the plastic strain is proportional to the stress deviation.

$$\dot{e}_{ij}^{(p)} = \frac{1}{\mu}\,\sigma'_{ij}, \qquad \mu > 0$$

where $\mu$ is a *positive* factor of proportionality which has the dimensions of the coefficient of viscosity of a fluid.

(d) Any stress state corresponding to $J_2 > k^2$ cannot be realized in the material.

The set of laws above contains two essential parts: the criterion for yielding, and the stress-strain relation in the elastic and plastic regimes. In the specifications above, the yielding condition is based on the second invariant of the stress deviation tensor. Such a yielding criterion was first proposed by von Mises. The constant $k$ can be identified with the yield stress in simple shear.

We notice that only *the rate* of plastic strain is specified by our laws. Under a varying loading program, successive plastic strain increments must be added together algebraically according to Eq. (6). Such a theory is called an *incremental theory*. Since the material behaves like a viscous fluid when it is in the plastic state, it is called an *ideal viscoplastic material*.

In many applications of the theory of plasticity the rate at which plastic flow occurs is of little interest. Sometimes an engineer is concerned only with the total amount of plastic flow. In these applications we often replace Eq. (c) by the incremental law

$$\Delta e_{ij}^{(p)} = \lambda \sigma'_{ij}, \qquad \lambda > 0,$$

where $\lambda$ is an arbitrary factor of proportionality and not a characteristic constant of the material. The sign of $\lambda$ is determined by the fact that plastic flow involves dissipation of energy. Under a given set of loading conditions the value of $\lambda$, and hence the total plastic flow, is determined by the total work done by the external load.

Although the bulk of the mathematical theory of plasticity deals with ideal plastic materials as specified above, actual materials may exhibit much more complicated plastic behavior. For this reason, other yielding conditions and other plastic flow rules have been proposed. As a preparatory measure for the study of these formulations, we shall review briefly below some basic experimental facts.

## 6.4. SOME EXPERIMENTAL INFORMATION

***Simple Tension Tests.*** If a rod of a ductile metal is pulled in a testing machine at room temperature, the load applied on the test specimen may be plotted against the elongation,

$$\epsilon = \frac{l - l_0}{l_0},$$

where $l_0$ is original length of the rod and $l$ is the length under load. Numerous experiments show typical load-elongation relationships, as indicated in the

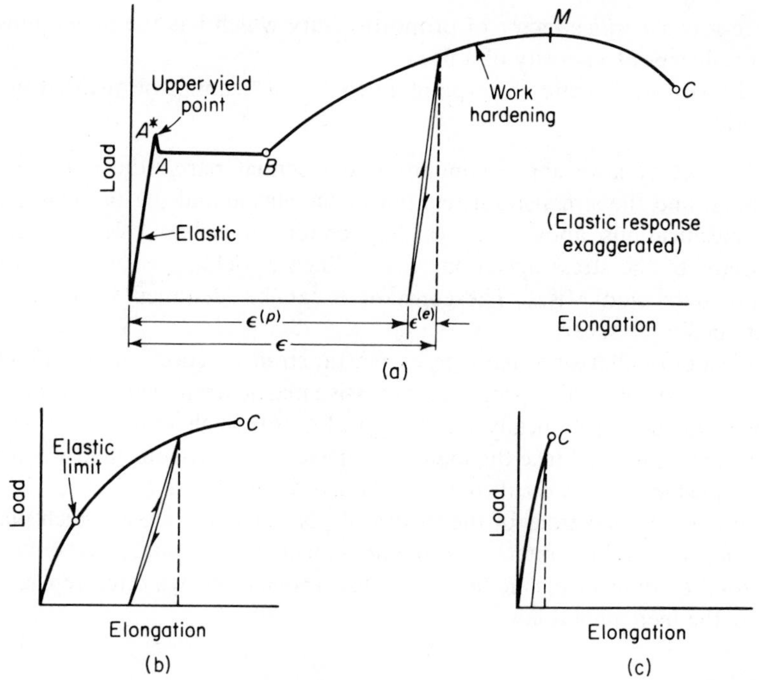


**Fig. 6.4:1.** Typical load-elongation curves in simple tension tests.

diagrams of Fig. 6.4:1. The initial region appears as a straight line. This is the region in which the law of linear elasticity is expected to hold. Mild steel shows an upper yield point and a flat yield region which is caused by many microscopic discontinuous small steps of slip along slip planes of the crystals [Fig. 6.4:1(a)]. Most other metals do not have such a flat yield region [Fig. 6.4:1(b)].

Upon unloading at any stage in the deformation, the strain is reduced along an elastic unloading line, and reloading retraces the unloading curve with relatively minor deviation and then produces further plastic deformation when approximately the previous maximum stress is exceeded. The test specimen may "neck" at a certain strain, so that its cross-sectional area is reduced in a small region. When necking occurs under continued elongation, the load reaches a maximum and then drops down, although the actual average stress in the neck region (load divided by the true area of the neck) continues to increase. The maximum $M$ is the ultimate load. Beyond the ultimate load the metal flows. At the point $C$ in the curves of Fig. 6.4:1 the specimen breaks.

Materials like cast iron, titanium carbide, beryllium, or any rock material which allow very little plastic deformation before reaching the breaking

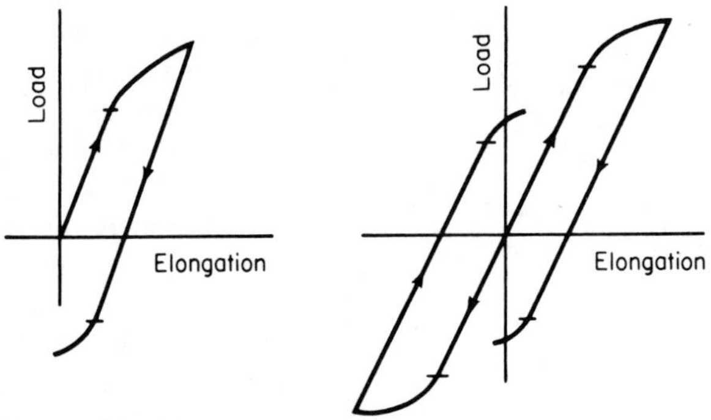


**Fig. 6.4:2.** Bauschinger effect in a simple tension-compression test.

point, are called *brittle* materials. The load-strain curve for a brittle material will appear as Fig. 6.4:1(c). The point $C$ is the breaking point.

A fact of great importance for geology is that brittle materials such as rocks tend to become ductile when subjected to large hydrostatic pressure (large negative mean stress). This was demonstrated by Theodore von Kármán (1911) in his classical experiments on marbles.

Tests of specimens subjected to simple compression or simple shear lead to load-strain diagrams similar to those of Fig. 6.4:1.

***Bauschinger Effect.*** When a metal specimen is subjected to repeated tension-compression tests, the load-deflection curve sometimes appears as in Fig. 6.4:2. The tension stroke and the compression stroke are dissimilar. This is referred to as *Bauschinger effect*, after J. Bauschinger's basic paper on strain hardening published in 1886.

***Anisotropy.*** Plastic deformation is physically anisotropic. The process of slip on a crystal plane is clearly directional. As a consequence, any initial isotropy which may have been present is usually destroyed by plastic deformation. From the point of view of the dislocation theory, slip is an irreversible process; every slip produces a new material. These changes are revealed in the Bauschinger effect and in the anisotropy of materials after plastic deformation.

***Time and Temperature Effects.*** The results described above are typical of those which would be obtained by slow application of the load in a testing machine. Under rapid loading, the hereditary nature of the material usually reveals itself. If the hereditary stress-strain law is linear, it is often described as *anelasticity* or *viscoelasticity* (see Chaps. 1 and 15). If the strain is large, the hereditary stress-strain laws for metals are generally nonlinear. The time-dependent plastic flow is often described as *creep*.

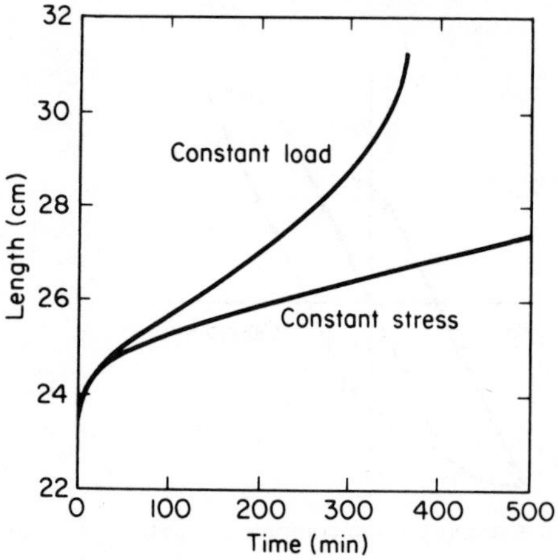


**Fig. 6.4:3.** Creep test of a lead wire in tension. The initial lengths and initial loads were the same in both tests. (After Andrade, 1910; courtesy of the Royal Society, London.)

A simple tension specimen of lead wire under a constant tension load shows a creep curve as illustrated in Fig. 6.4:3. Following an initial extension, the rate of strain first decreases gradually, then remains nearly constant for a while, and finally accelerates until the specimen breaks. The three stages of creep are called, respectively, the primary, the secondary, and the tertiary creep.

Andrade has pointed out, however, that if a creep test is performed at constant stress, the final stage of accelerated elongation generally does not appear. As shown in Fig. 6.4:3, the creep curve at constant load differs considerably from that at constant stress.

The creep rate and the length of the primary and secondary stages are dependent on the temperature and the stress level. At high temperature and high stress levels the secondary creep stage may not be appreciable.

The phenomenon of creep is important in geophysics and in those engineering problems in which accurate dimensions must be maintained over a long period of time. It is a controlling factor for the design of machinery that operates at high temperature. On the other hand, plastic deformation generates heat, and a very rapid application of load is usually associated with a large temperature gradient, localized plastic flow and anisotropic effects. Continuous metal cutting is a good example of large localized plastic deformation.

***Size Effect and Stress Inhomogeneity.*** Since practically all materials have a nonhomogeneous microscopic structure, plastic properties of a small region may be different from those of a larger region. This consideration becomes important in the question of stress concentration in a notched specimen.

***Combined Stress.*** One of the most important tasks of mathematical theory is to extend and generalize the experiences of simple experiments and to suggest crucial tests to verify the basic assumptions of the theory. In plasticity, this function of mathematics is particularly evident, because here one must consider the stress and strain tensors; yet no direct observation of all components of these tensors is possible.

A relatively simple combined-stress experiment employs a thin-walled circular cylinder. With the combination of an axial tension, a twisting moment about the cylinder axis, and an internal or external pressure, a good approximation to a biaxial state of stress (involving $\sigma_{11}$, $\sigma_{12}$, $\sigma_{22}$, with $\sigma_{31} = \sigma_{32} = \sigma_{33} = 0$) can be produced. Much information has been gathered from such experiments. Among other things, it is found that the shear stress is by far the major cause of yielding.

P. W. Bridgman's renowned experiments on the influence of hydrostatic pressure on yielding and other mechanical and metallographical characteristics on metals have shown that hydrostatic pressures of the order of the yield stress have practically no influence on yielding of metals. If a tensile test or shear test is run at atmospheric pressure in the standard manner, with load and deformation recorded, immersion of the whole setup in a chamber under high hydrostatic pressure will hardly alter the stress-strain curves in the small strain range. The major effect is to increase greatly the ductility of the material and thus permit much larger deformation prior to fracture.

Conversely, very small changes in material density are found when a metal is subjected to repeated plastic deformation, indicating that the plastic volume change is small.

## 6.5. A BASIC ASSUMPTION OF THE MATHEMATICAL THEORY OF PLASTICITY

If the laws of linear elasticity and ideal plasticity, as described in Secs. 6.1, 6.2, and 6.3, are compared with the typical experimental features described in 6.4, we can determine the applicability and limitations of the mathematical formulations. We see that near the origin of the stress-strain diagrams of Fig. 6.4:1 the linear relationship is good in nearly all cases. For the case of mild steel represented by Fig. 6.4:1(a), the large, flat, yielding region *A*-*B* has the features of the ideal plasticity. The slightly unstable region of the yield curve at the peak $A^*$ in Fig. 6.4:1(a) is, of course, ignored by the ideal plasticity law. The rise beyond the point *B* in Fig. 6.4:1(a), as well as the curved portion of the curves in Fig. 6.4:1(b) and (c), obviously can be represented neither by the linear elasticity law nor by the ideal plasticity law. The features of these curved portions of stress-strain curves are said to be features of *strain hardening*.† *It is evident that the stress-strain laws of strain-hardening will be basically different from those of ideal plasticity.*

It is the objective of the theory of plasticity to offer a mathematical description of the mechanical behavior of materials in the plastic range. Since the foundation of the theory of plasticity is not yet firmly established, we shall present below only an outline of some of the best known theories.

For the purpose of describing the state of stress at any point in the material, it is convenient to represent each state of stress by a point in a nine-dimensional *stress space*, with axes $\sigma_{ij}\,(i, j = 1, 2, 3)$. Similarly, a state of strain may be spoken of as a point in a nine-dimensional *strain space*, with components $e_{ij}$. In particular, a state of plastic strain $e_{ij}^{(p)}$ may be so represented. A program of loading may be regarded as a path in the stress space and the corresponding deformation history as a path in the strain space.

A basic assumption is made that there exists a scalar function, called a *yield function* or *loading function*, and denoted by $f(\sigma_{ij}, e_{ij}^{(p)}, \kappa)$, which depends on the state of stress and strain and the history of loading, and which characterizes the yielding of the material as follows. The equation $f = 0$ represents a closed surface in the stress space. No change in plastic deformation occurs as long as $f < 0$. Change in plastic deformation occurs when $f = 0$. No meaning is associated with $f > 0$. The parameter $\kappa$ is called a *work-hardening parameter*, which is assumed to depend on the plastic deformation history of the material.

Work hardening will be considered later. Here let us consider two simple examples to see the plausibility of the yielding assumption.

*Example 1. Yielding Condition of von Mises*

The yielding condition of von Mises is defined by the yield function

$$f(\sigma_{ij}) = J_2 - k^2, \tag{1}$$

† The material that is most qualified to be called *elastic* is probably natural rubber. But if the stress-strain relationship of natural rubber is plotted in the manner of Fig. 6.4:1, the resulting curve will not be a straight line. In some rubbers this can be corrected by plotting the stress components versus the (finite) Eulerian (Almansi) strains or, equivalently, by plotting the Kirchhoff stress components versus Green's strain components. In such cases the material is said to be *linearly elastic* with respect to finite strain. In other cases the nonlinearity in the stress-strain relationship remains, although the material returns to its original state along the same stress-strain curve when all the loads are removed. In this case the material is said to be *nonlinearly elastic*. In this chapter we discuss the strain hardening only under the assumption of small strain. Finite deformation is discussed in Chap. 16.

where $J_2$ is the second invariant of the stress deviation,

$$J_2 = \tfrac{1}{2}\sigma'_{ij}\sigma'_{ij}. \tag{2}$$

For an ideal plastic material obeying von Mises' condition, $k$ is a constant independent of strain history. The stress state is characterized by the condition $J_2 \leqslant k^2$, with plastic flow possible only when $J_2 = k^2$, whereas the condition $J_2 > k^2$ can never be realized. These are the conditions specified in Sec. 6.3. For a work-hardening material, $k$ will be allowed to change with strain history.

If a material obeying von Mises' yield condition is subjected to a simple shear $\sigma_{12}$ while all other stress components vanish (a state of stress realizable in torsion of a thin-walled tube), then $J_2 = \sigma_{12}^2$, and yielding should occur when

$$\sigma_{12} = k.$$

Hence *the number k means the yield stress in simple shear.*

*Example 2. Yielding Condition of Tresca*

Tresca first advanced the idea of yielding criterion. Through his work on metal forming in an armory, he concluded that the decisive factor for yielding is the maximum shear stress in the material. Tresca's criterion stipulates that the maximum shear stress must have the constant value $k$ during plastic flow.

To express Tresca's idea analytically, it is the simplest to use the principal stresses $\sigma_1$, $\sigma_2$, $\sigma_3$. If it is known that

$$\sigma_1 \geqslant \sigma_2 \geqslant \sigma_3, \tag{3}$$

then Tresca's yielding condition is

$$f \equiv \sigma_1 - \sigma_3 - 2k = 0. \tag{4}$$

However, $f$ in this form is not analytic; it violates the rule that the manner in which the principal axes are labeled 1, 2, 3 should not affect the form of the yield function. To obey this rule, we observe that Tresca's condition states that during plastic flow one of the differences $|\sigma_1 - \sigma_2|$, $|\sigma_2 - \sigma_3|$, $|\sigma_3 - \sigma_1|$ has the value $2k$. Hence we may write

$$f \equiv [(\sigma_1 - \sigma_2)^2 - 4k^2][(\sigma_2 - \sigma_3)^2 - 4k^2][(\sigma_3 - \sigma_1)^2 - 4k^2] = 0. \tag{5}$$

This equation is now symmetrical with respect to the principal stresses, and it can be put into an invariant form which is due to Reuss:

$$4J_2^3 - 27J_3^2 - 36k^2J_2^2 + 96k^4J_2 - 64k^6 = 0, \tag{6}$$

where $J_2$, $J_3$ are the second and third invariant of the stress deviation tensor.

In certain problems, the direction of the principal axes and the relative magnitudes of the principal stresses are known by symmetry considerations or by intuition. Then it is possible to use the simple form (4) for Tresca's criterion. In general, however, we cannot tell a priori the relative magnitude of the principal stresses, and we would be obliged to use the general form (6).

**Problem 6.1.** Apply the yielding criteria of von Mises and Tresca to simple tension and to simple shear. Show that, although they give the same yielding stress in simple shear, the tensile yielding stress predicted by von Mises' criterion is smaller than that predicted by Tresca's criterion by a factor $\sqrt{3}/2$.

**Problem 6.2.** Show that in no other type of stress is the discrepancy between the predictions of the yielding stresses by von Mises' and by Tresca's criteria as large as it is in simple tension (von Mises,[6.3] 1913).

## 6.6. LOADING AND UNLOADING CRITERIA

Let us first clarify what is meant by loading and unloading in a plastic state. Consider a plastic state at which the yield function vanishes,

$$f(\sigma_{ij}, e_{ij}^{(p)}, \kappa) = 0. \tag{1}$$

If we consider the time rate of $f$, i.e.,

$$\dot{f} = \frac{\partial f}{\partial \sigma_{ij}}\,\dot{\sigma}_{ij} + \frac{\partial f}{\partial e_{ij}^{(p)}}\,\dot{e}_{ij}^{(p)} + \frac{\partial f}{\partial \kappa}\,\dot{\kappa}, \tag{2}$$

then, obviously, $f = 0$ and $\dot{f} < 0$ at a time $t$ would imply $f < 0$ the next instant of time. Such a change leads to an elastic state and is a natural attribute to the term *unloading*. However, we also require that in an unloading process no plastic strain occurs, $\dot{e}_{ij}^{(p)} = 0$, and that by its very nature, the rate of change of the strain-hardening parameter, $\dot{\kappa}$, must also vanish. Hence, by Eq. (2) we stipulate that the criterion for unloading from a plastic state is

$$\frac{\partial f}{\partial \sigma_{ij}}\,\dot{\sigma}_{ij} < 0, \qquad f = 0, \qquad \text{during } \textit{unloading}. \tag{3}$$

Otherwise it is said to be *loading* or *neutral loading*. Thus,

$$\frac{\partial f}{\partial \sigma_{ij}}\,\dot{\sigma}_{ij} = 0, \qquad f = 0, \qquad \text{during neutral loading;} \tag{4}$$

$$\frac{\partial f}{\partial \sigma_{ij}}\,\dot{\sigma}_{ij} > 0, \qquad f = 0, \qquad \text{during } \textit{loading}. \tag{5}$$

The function $f$ is called a loading function because of its prominence in these loading criteria.

A simple geometric interpretation of these criteria exists. Since the yield surface, i.e., the loading surface, is assumed to be a closed surface, we can speak of its inside and outside. Then, for a state of stress on the loading surface, loading, unloading, or neutral loading takes place, according to whether the stress increment vector is directed outward, inward, or along the tangent to the loading surface, respectively. Because of this geometric interpretation (also for reasons to be discussed in Sec. 6.9), it is important to obey the sign convention in writing Eq. (1) so that $\partial f/\partial \sigma_{ij}$ be directed along the outer normal to the surface $f = 0$.

## 6.7. ISOTROPIC STRESS THEORIES

A material is said to be *isotropic* if there is no orientation effect in the material. If the yield function $f$ depends only on the *invariants* (with respect to rotation of coordinates) of stress, strain, and strain history, then the plastic characteristics of the material is isotropic. If the yield function $f$ is an isotropic function of the stress alone, then the theory of plasticity is called an *isotropic stress theory*. In such theories,

(1) $$f = f(I_1, I_2, I_3),$$

where $I_1, I_2, I_3$ are the three invariants of the stress tensor $\sigma_{ij}$. Equivalently, we may write (1) in terms of principal stresses,

(2) $$f = f(\sigma_1, \sigma_2, \sigma_3).$$

If the principal stresses $\sigma_1$, $\sigma_2$, $\sigma_3$ are taken as the coordinate axes, the surface $f(\sigma_1, \sigma_2, \sigma_3) = 0$ can be plotted in a three-dimensional stress space. For example, if we take $f = J_2 - k^2$, the surface $f = 0$ will appear as a circular cylinder whose axis is equally inclined to the coordinate axes, as shown in Fig. 6.7:1.

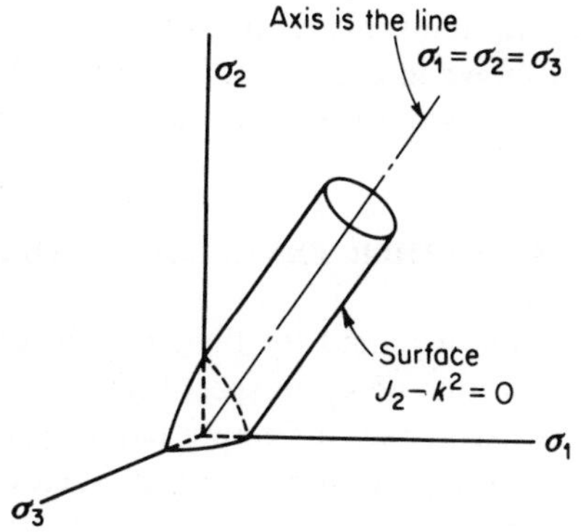


**Fig. 6.7:1.** A yield surface in the principal stress space.

According to P. W. Bridgman[6.3] (1923), the plastic deformation of metals essentially is independent of hydrostatic pressure; Crossland[6.3] (1954) and many others have verified this result. If this conclusion is accepted, the yield function will be independent of $I_1 = \sigma_1 + \sigma_2 + \sigma_3$. Then it is advantageous to introduce the stress deviations $\sigma'_{ij} = \sigma_{ij} - \frac{1}{3}\sigma_{kk}\,\delta_{ij}$. Let $J_2$, $J_3$ be the second and third invariants of the stress deviation $\sigma'_{ij}$. Then $f = f(J_2, J_3)$. von Mises took the simplest of such functions, assuming further that $J_3$ does not appear, and in this way obtained his famous criterion $f = J_2 -$ const. The surface $f(J_2) = 0$, when plotted in the space of principal stresses, will be a cylinder perpendicular to the plane

(3) $$I_1 = \sigma_1 + \sigma_2 + \sigma_3 = 0.$$

Its axis is equally inclined to the coordinate axes $\sigma_1, \sigma_2, \sigma_3$, but its cross section is no longer circular if $J_3$ participates in $f$.

When $f$ depends on $J_2, J_3$ alone, it can be written in the form $f(\sigma_1 - \sigma_3, \sigma_2 - \sigma_3)$. Then a two-dimensional plot of the surface $f = 0$ is possible, with $\sigma_1 - \sigma_3, \sigma_2 - \sigma_3$ as coordinate axes (see Fig. 6.7:2).

Another way of representing a yield surface when it is unaffected by hydrostatic pressure is to *project the yield surface on the so-called $\pi$-plane*,

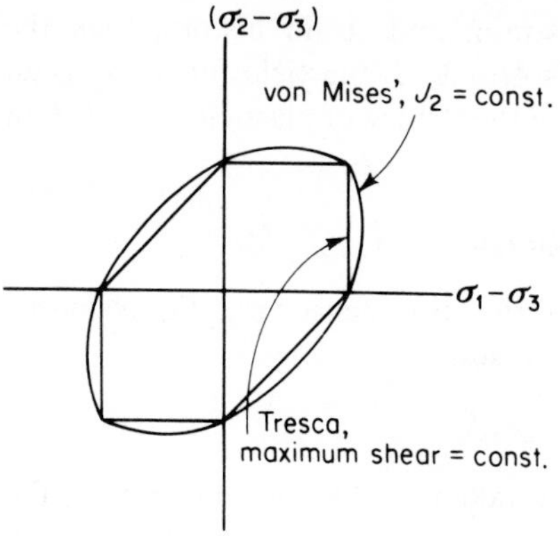


**Fig. 6.7:2.** Yield surface plotted on the plane of $(\sigma_1 - \sigma_3)$, $(\sigma_2 - \sigma_3)$.

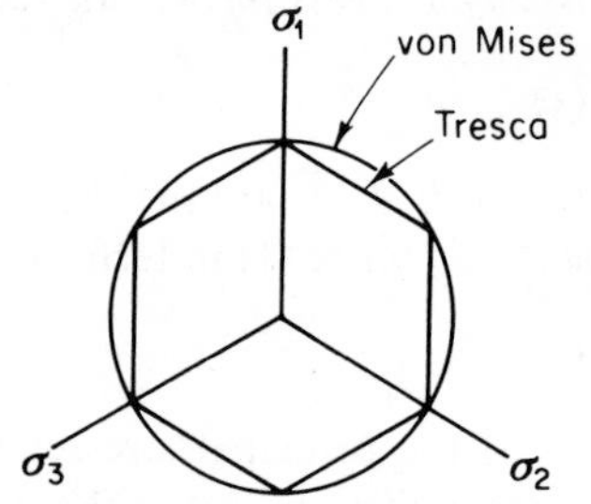


**Fig. 6.7:3.** Projection of yield surfaces on the $\pi$-plane.

$\sigma_1 + \sigma_2 + \sigma_3 = 0$. For example, von Mises' criterion would appear as a circle on this plane, and Tresca's criterion would appear as a regular hexagon, as shown in Fig. 6.7:3. The $\pi$-plane projection will be used extensively below in discussing the flow and the hardening rules in plasticity.

## 6.8. FURTHER EXAMPLES OF YIELD FUNCTIONS

Let $\sigma'_{ij}$ stand for the stress deviation and $J_2, J_3$ for the second and third invariants of $\sigma'_{ij}$, respectively. Von Mises' and Tresca's yield function discussed in Sec. 6.5 imply initial isotropy, and equality of tensile and compressive yield stresses at all stages of the deformation. The more general expression

(1) $$f = F(J_2, J_3) - k^2$$

still contains no Bauschinger effect. A simple example is

(2) $$F = J_2^3 - cJ_3^2.$$

A good correlation with Osgood's experimental data[6.3] (1947) is obtained by taking $c = 2.25$.

To include Bauschinger effect, but preserve isotropy, the following yield functions have been suggested.

(3) $$f = F(J_2) - m\sigma'_{ij}e^{(p)}_{ij} - k^2, \qquad m, \text{ a constant,}$$

(4) $$f = F(J_2, J_3) - m\sigma'_{ij}e^{(p)}_{ij} - k^2,$$

(5) $$f = F(J_2, J_3) - H(J_2, J_3)\sigma'_{ij}e^{(p)}_{ij} - k^2,$$

(6) $$f = F(J_2, J_3) - [P(J_2, J_3)\sigma'_{ij} + Q(J_2, J_3)t_{ij}]e^{(p)}_{ij} - k^2,$$

where

(7) $$t_{ij} = \sigma'_{ik}\sigma'_{kj} - \tfrac{2}{3}J_2\delta_{ij}.$$

To reveal and preserve *initial anisotropy*, we may use

(8) $$f = D_{ijkl}\sigma_{ij}\sigma_{kl} - k^2,$$

which contains no Bauschinger effect. Here $D_{ijkl}$ is composed of a set of 21 independent constants, due to the symmetry conditions

$$D_{ijkl} = D_{klij} = D_{jikl} = D_{ijlk}$$

If incompressibility is assumed, the number of independent constants is reduced to 15.

Finally, the expression

(9) $$f = D_{ijkl}(\sigma'_{ij} - me^{(p)}_{ij})(\sigma'_{kl} - me^{(p)}_{kl}) - k^2$$

yields the Bauschinger effect which is controlled by one constant. The initial anisotropy is not preserved during deformation. Yield functions such as this will be discussed more fully in Sec. 6.12.

These examples all use a single analytic function to represent the entire yield surface. If a yield surface is composed of piecewise smooth surfaces which meet to form corners, it would be convenient to use a separate expression for each of these piecewise smooth surfaces. This concept leads to Koiter's generalization (see Sec. 6.10).

## 6.9. WORK HARDENING—DRUCKER'S DEFINITION

Work hardening in a simple tension experiment means that the stress is a monotonically increasing function of increasing strain. To generalize this concept, D. C. Drucker[6.3] (1951) considers the work done by an external agency which slowly applies a set of self-equilibrating forces and then slowly removes them. This external agency is to be understood as entirely separate and distinct from the agency which causes the existing state of stress. The original configuration may or may not be restored after the cycle. *Work hardening is then defined to mean that for all such added sets of stresses a positive work is done by the external agency during the application of the stresses, and the net work performed by it over the cycle of application and removal is either zero or positive.*

Rephrased, Drucker's definition of work hardening means that useful net energy over and above the elastic energy cannot be extracted from the material and the system of forces acting upon it.

Consider a volume of material in which there is a homogeneous state of stress $\sigma_{ij}$ and strain $e_{ij}$. Suppose that an external agency applies small surface tractions which alter the stresses at each point by $d\sigma_{ij}$, and produces small strain increment $de_{ij}$. On removing the stresses $d\sigma_{ij}$, a strain $de_{ij}^{(e)}$ is recovered. Then, according to the definition above, *the material is said to be work-hardening if the following two conditions hold true.*

(1) $$d\sigma_{ij}\, de_{ij} > 0, \qquad \textit{upon loading};$$

(2) $$d\sigma_{ij}(de_{ij} - de_{ij}^{(e)}) \geqslant 0, \qquad \textit{on completing a cycle.}$$

Let us denote by $de_{ij}^{(p)}$ the *plastic strain increment* which is not recovered by the process above. Then, since $de_{ij}^{(p)} = de_{ij} - de_{ij}^{(e)}$, the second condition may be written as

(3) $$d\sigma_{ij}\, de_{ij}^{(p)} \geqslant 0.$$

## 6.10. IDEAL PLASTICITY

If we accept Drucker's definition of strain hardening, we can define ideal plasticity as a plastic deformation without strain hardening; mathematically specified by the condition that when plastic deformation occurs the equality sign prevails in Eq. (6.9:3):

(1) $$d\sigma_{ij}\, de_{ij}^{(p)} = 0,$$

and that the yield function is unaffected by $e_{ij}^{(p)}$. The differentials $d\sigma_{ij}$, $de_{ij}^{(p)}$ must be interpreted as in Sec. 6.9.

As an application of this definition, we shall derive the *flow rule* during plastic deformation. We notice that the yield function furnishes a criterion to tell whether yielding occurs or not. If yielding does occur, we need further information concerning the increment or rate of deformation in order to complete our description of the material behavior. In other words, we need the flow rule.

Now, for an ideal plastic material we assume that a yield function $f(\sigma_{ij})$ exists, which is a function of the stresses $\sigma_{ij}$ and not of the strains $e_{ij}^{(p)}$, such that $f(\sigma_{ij}) \leqslant 0$ prevails; and

(2) $$\dot{e}_{ij}^{(p)} \neq 0 \quad \text{only if} \quad f(\sigma_{ij}) = 0.$$

Since $f$ is assumed to be a function of $\sigma_{ij}$ only, any change in stresses during plastic flow must satisfy the relation

(3) $$df = \frac{\partial f}{\partial \sigma_{ij}}\, d\sigma_{ij} = 0.$$

But, when plastic flow occurs, Eq. (1) holds. A comparison of (1) and (3) shows that

(4) $$de_{ij}^{(p)} = \lambda \frac{\partial f}{\partial \sigma_{ij}},$$

where $\lambda$ is an arbitrary constant of proportionality. We shall write (4) in terms of the rate of strain

(5) $$\dot{e}_{ij}^{(p)} = \frac{1}{\mu} \frac{\partial f}{\partial \sigma_{ij}}$$

If $\partial f/\partial \sigma_{ij}$ has the dimension of stress, then $\mu$ has the significance of the coefficient of viscosity. The sign of $\lambda$ or $\mu$ is restricted by the condition that plastic flow always involves dissipation of mechanical energy, a condition which may be written as

(6) $$\dot{W} = \sigma_{ij}\dot{e}_{ij}^{(p)} > 0.$$

The formula (4) gives the rule of plastic flow (plastic strain increments). The connection between yield conditions and flow rule as expressed in (4) was first stipulated by Richard von Mises[6.3] (1928) and is known as the *theory of the plastic potential.*

W. T. Koiter[6.3] (1953) generalized this theory by allowing the yield limit to be specified by several yield functions,

$$f_1(\sigma_{ij}), \qquad f_2(\sigma_{ij}), \qquad \ldots, \qquad f_n(\sigma_{ij}).$$

A state of stress is below the yield limit if all these functions are negative. For a state of stress at the yield limit, at least one yield function vanishes, while none has a value greater than zero.

If the functions $f_h = \ldots = f_m = 0$, whereas all other $f$'s are negative, the Koiter generalization of the flow rule (5) is

(7) $$\dot{e}_{ij}^{(p)} = \lambda_h \frac{\partial f_h}{\partial \sigma_{ij}} + \ldots + \lambda_m \frac{\partial f_m}{\partial \sigma_{ij}},$$

where $\lambda_h, \ldots, \lambda_m$ are arbitrary nonnegative factors of portionality. Thus, in the case of ideal plasticity, the basic concept leads at once to a general incremental stress-strain relationship.

An interesting geometric interpretation of the plastic potential is given by Prager. The formula $f(\sigma_{ij}) = 0$ defines a surface in the nine-dimensional stress space with $\sigma_{ij}(i, j = 1, 2, 3)$ as coordinates. The normal vector to this surface has the components $\partial f/\partial\sigma_{ij}$. Equation (5) states thus that the vector of plastic deformation rate $\dot{e}_{ij}^{(p)}$ is normal to the surface $f = 0$ in the stress space.

*Example 1*

Under the yielding condition of von Mises, Eq. (6.5:1), and assuming that the constant $k$ is independent of plastic deformation, we derive the flow rule according to Eq. (5)

$$\dot{e}_{ij}^{(p)} = \lambda\sigma'_{ij}, \tag{8}$$

which is the rule presented in Sec. 6.3.

*Example 2*

Tresca's yielding condition can be expressed in terms of Koiter's generalized plastic potential by defining the functions

$$\begin{aligned} &f_1 = \sigma_2 - \sigma_3 - 2k, \qquad f_2 = \sigma_3 - \sigma_1 - 2k, \qquad f_3 = \sigma_1 - \sigma_2 - 2k,\\ &f_4 = -(\sigma_2 - \sigma_3) - 2k, \qquad f_5 = -(\sigma_3 - \sigma_1) - 2k,\\ &f_6 = -(\sigma_1 - \sigma_2) - 2k. \end{aligned} \tag{9}$$

where $\sigma_1, \sigma_2, \sigma_3$ are the principal stresses. A state of stress is below the yield limit if $f_1, \ldots, f_6$ are all negative. Yielding occurs when one or more of these $f$'s are equal to zero. None of the $f$'s can have a positive value. When yielding occurs, the flow rule is given by Eq. (7). For example, if $f_1 = 0$, while all the other $f$'s are negative, then

$$\dot{e}_2^{(p)} = \lambda_1, \qquad \dot{e}_3^{(p)} = -\lambda_1, \qquad \dot{e}_1^{(p)} = 0, \tag{10}$$

where $\lambda_1 > 0$, and $e_1$, $e_2$, $e_3$ are the principal strains corresponding to $\sigma_1, \sigma_2, \sigma_3$. The principal axes of the strain tensor coincide with those of the stress tensor under Tresca's condition.

*Remark*: Formal application of our method of derivation to a simple load-deflection experiment may appear difficult. If we twist a tube of ideal plastic material in torsion, the yield condition according to von Mises' criterion is reached when the shearing stress $\tau = k$, the yielding stress. Plastic flow will continue with no possibility of increasing $\tau$ beyond $k$. Hence, if we limit ourselves to torsion and apply Eqs. (1) or (2) we would have obtained $d\tau = 0$, which yields no useful information. To deduce anything significant we would have to consider adding other loads, for example, tension or internal pressure in the tube. These additional varieties of loads will alter the plastic flow, thus providing nontrivial changes $d\sigma_{ij}$ and $de_{ij}^{(p)}$, to which the derivation above applies.

## 6.11. FLOW RULE FOR WORK-HARDENING MATERIALS

Under Drucker's hypothesis, as incorporated in his definition of work-hardening materials, the following consequences will be proved.

A. The yield surface and all subsequent loading surfaces must be convex.
B. The plastic strain increment vector must be normal to the loading surface at a regular point, and it must lie between adjacent normals to the loading surface at a corner of the surface.
C. The rate of change of plastic strain must be a linear function of the rate of change of the stress.

To facilitate the proof,† let us think of an increment of stress $d\sigma_{ij}$ as a vector $d\mathbf{S}$ in the nine-dimensional stress space, and the corresponding plastic strains $de_{ij}^{(p)}$ as the components of a vector $d\mathbf{E}$ in the same (stress) space. Then, by Eq. (6.9:3), we have

$$d\mathbf{S} \cdot d\mathbf{E} = |d\mathbf{S}|\,|d\mathbf{E}| \cos \psi \geqslant 0, \tag{1}$$

This implies that

$$-\frac{\pi}{2} \leqslant \psi \leqslant \frac{\pi}{2}, \tag{2}$$

which means that the angle between $d\mathbf{S}$ and $d\mathbf{E}$ is acute.

Now let $P$ be a regular (smooth) point on the yield surface. Let $d\mathbf{E}$ be a plastic strain vector at $P$. According to Eqs. (1) and (2), all stress increments $d\mathbf{S}$ that will produce $d\mathbf{E}$, i.e., vectors that represent loading in the stress space and end at $P$, and corresponding to $d\mathbf{E}$, must form an acute angle with $d\mathbf{E}$. If we represent a hyperplane normal to $d\mathbf{E}$ by $AB$ in Fig. 6.11:1, then the vectors $d\mathbf{S}(= d\sigma_{ij})$ must all originate in one side of the hyperplane $AB$. The initial points of the $d\mathbf{S}$ vectors, represented by stress states inside or on the yield surface, must thus all lie on one side on the hyperplane $AB$. However, $d\mathbf{S}$ are loading vectors (since $d\mathbf{E}$ exists); they are outward vectors whose directions are bounded by the tangent plane of the yield surface (see Sec. 6.6). Hence, the hyperplane $AB$ must be tangent to the loading surface. Since $d\mathbf{E}$ is normal to the hyperplane, it is also normal to the yield surface at the point of tangency. Furthermore, since the yield surface lies on one side of the tangent plane, the yield surface is convex

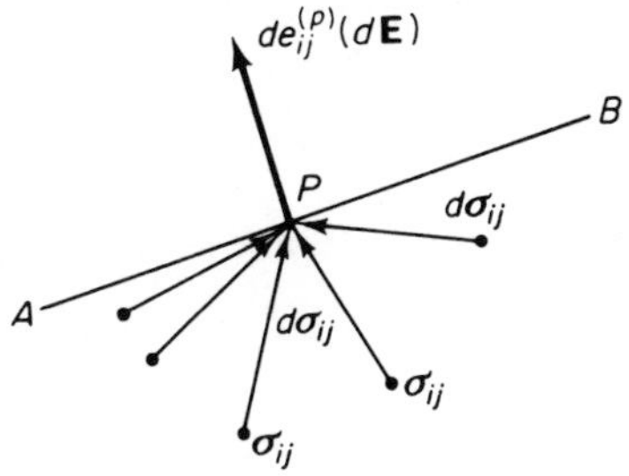


**Fig. 6.11:1.** Possible stress increments corresponding to a plastic strain increment $d\mathbf{E}$.

† The following explanation follows that of P. M. Naghdi[6.3] (1960).

at $P$. Finally, since $P$ is an arbitrary point on the yield surface, the convexity of the entire yield surface is established.

Since at a regular point on a surface there is a unique tangent plane, the hyperplane $AB$ is unique at a regular point of the yield surface. Thus, the direction of $d\mathbf{E}$ normal to the hyperplane is also unique. In other words, at a smooth point of the yield surface the direction of $d\mathbf{E}$ (i.e., $de_{ij}^{(p)}$) is independent of the direction of $d\sigma_{ij}$.

At a corner on the yield surface, there may have been more than one limiting tangent plane; convexity must still hold, but the direction of strain increment $de_{ij}^{(p)}$ may depend on the direction of the loading vector $d\sigma_{ij}$.

Now the normality of the plastic strain rate vector at a smooth point of the loading surface requires that

(3) ▲ $$\dot{e}_{ij}^{(p)} = \Lambda \frac{\partial f}{\partial \sigma_{ij}},$$

where $\Lambda$ is a function which may depend on stress, stress rate, strain, and strain history. Since the work done by an external agency during loading must be positive, it is easy to show that $\Lambda$ must be nonnegative. Equation (3) here is similar to Eq. (6.10:4); thus, the loading function $f$ plays the role of a *plastic potential.*

Since subsequent loading surfaces pass through the loading point, we must have

(4) $$f(\sigma_{ij}, e_{ij}^{(p)}, \kappa) = 0.$$

This equation is called the *consistency* condition by Prager. It means that loading from a plastic state must lead to another plastic state. Assuming $\kappa$ to be a function of $e_{ij}^{(p)}$, then during loading we must have

(5) $$\dot{f} = \frac{\partial f}{\partial \sigma_{ij}} \dot{\sigma}_{ij} + \frac{\partial f}{\partial e_{ij}^{(p)}} \dot{e}_{ij}^{(p)} + \frac{\partial f}{\partial \kappa} \frac{\partial \kappa}{\partial e_{ij}^{(p)}} \dot{e}_{ij}^{(p)} = 0.$$

Substituting (3) in (5) and solving for $\Lambda$, we obtain

(6) $$\Lambda = -\frac{\dfrac{\partial f}{\partial \sigma_{kl}} \dot{\sigma}_{kl}}{\left(\dfrac{\partial f}{\partial e_{ij}^{(p)}} + \dfrac{\partial f}{\partial \kappa} \dfrac{\partial \kappa}{\partial e_{ij}^{(p)}}\right) \dfrac{\partial f}{\partial \sigma_{ij}}}.$$

Combining (6) with (3), we have

(7) ▲ $$\dot{e}_{ij}^{(p)} = \hat{G} \frac{\partial f}{\partial \sigma_{ij}} \frac{\partial f}{\partial \sigma_{kl}} \dot{\sigma}_{kl},$$

where

(8) $$\hat{G} = -\frac{1}{\left(\dfrac{\partial f}{\partial e_{mn}^{(p)}} + \dfrac{\partial f}{\partial \kappa} \dfrac{\partial \kappa}{\partial e_{mn}^{(p)}}\right) \dfrac{\partial f}{\partial \sigma_{mn}}}.$$

These results were first given by Prager[6.3] (1948) and Drucker[6.3] (1959). Equation (7) proves the linearity statement (C) given above.

For a loading surface that has corners, Koiter's generalization can be used. Such a surface is composed of a number of individual smooth loading surfaces $f_p$ which meet to form corners. Koiter[6.3] (1953) has shown that if the loading surfaces described by $f_p = 0$ act independently, the total plastic deformation can be written as a sum of contributions from certain of the $f_p$'s, as follows.

$$\dot{e}_{ij}^{(p)} = \sum_{p=1}^{n} C_p h_p \frac{\partial f_p}{\partial \sigma_{ij}} \frac{\partial f_p}{\partial \sigma_{kl}} \dot{\sigma}_{kl}, \tag{9}$$

where

$$\begin{aligned} C_p &= 0 \quad \text{if} \quad f_p < 0, \quad \text{or} \quad (\partial f_p/\partial \sigma_{ij})\dot{\sigma}_{ij} < 0, \\ C_p &= 1 \quad \text{if} \quad f_p = 0, \quad \text{and} \quad (\partial f_p/\partial \sigma_{ij})\dot{\sigma}_{ij} \geqslant 0, \end{aligned} \tag{10}$$

and $h_p$'s are positive functions of stress, strain, and strain history. Equations (10) specify, of course, the condition of yielding and loading.

It should be remarked that the properties deduced above, namely, the convexity, normality, and linearity, followed Drucker's hypothesis—which is often interpreted as a statement that the material is stable. Hence, these properties may hold only for the class of stable materials. Unstable materials do exist. Mild steel at its upper yield point (see the point $A^*$ in Fig. 6.4:1) is a well-known example. However, no general stress-strain relations have been proposed as yet for unstable materials.

*Example*

If we choose $f = J_2 - \kappa$, where $J_2$ is the second invariant of the stress deviation, and $\kappa$ is a hardening parameter which depends on plastic deformation then the flow rule at yielding condition is

$$\begin{aligned} \dot{e}_{ij}^{(p)} &= \hat{G} \frac{\partial f}{\partial \sigma_{ij}} \frac{\partial f}{\partial \sigma_{kl}} \dot{\sigma}_{kl} = \hat{G}\left(\frac{\partial f}{\partial J_2}\frac{\partial J_2}{\partial \sigma'_{ij}}\right)\left(\frac{\partial J_2}{\partial \sigma'_{kl}}\frac{d\sigma'_{kl}}{dt}\right) \\ &= \hat{G}\,\sigma'_{ij}\,\dot{J}_2, \qquad \dot{J}_2 \geqslant 0. \end{aligned}$$

If we set $f = F(J_2, J_3) - \kappa$, and assuming $\partial\kappa/\partial e_{mn}^{(p)} \neq 0$ for some $m$, $n$, then the flow rule is

$$\dot{e}_{ij}^{(p)} = \hat{G}\left[\frac{\partial F}{\partial J_2}\sigma'_{ij} + \frac{\partial F}{\partial J_3}t_{ij}\right]\dot{F}, \qquad \dot{F} \geqslant 0,$$

where $t_{ij} = \sigma'_{ik}\sigma'_{kj} - \frac{2}{3}J_2\delta_{ij}$.

Note that the yield-function of a work-hardening material depends on the plastic strain $e_{ij}^{(p)}$ in some significant manner, in contrast to an ideal plastic material the yield function of which is independent of the plastic strain.

## 6.12. SUBSEQUENT LOADING SURFACES—HARDENING RULES

We have discussed the yield surface and the flow rule. Now we must consider the third aspect: the determination of subsequent loading surfaces as plastic flow proceeds. In other words, we must now consider how the plastic deformation $e_{ij}^{(p)}$ enters into the loading surface

$$f(\sigma_{ij}, e_{ij}^{(p)}, \kappa) = 0. \tag{1}$$

Laws governing this aspect are called *hardening rules*. Researches in this area are much discussed in the current literature, but a firm experimental foundation is still lacking.

If we assume that plastic deformation is independent of hydrostatic pressure, then the loading surfaces in the principal stress space are cylinders of infinite length perpendicular to the $\pi$-plane $\sigma_1 + \sigma_2 + \sigma_3 = 0$. Hence, they can be represented by their cross sections on this plane (see Sec. 6.7). We know that these cross-sectional curves are closed, convex, and piecewise smooth; but they change in size and shape during plastic deformation.

We shall illustrate several proposed hardening rules in Fig. 6.12:1. In Fig. 6.12:1(a) is shown Tresca's initial yield surface, a regular hexagon on

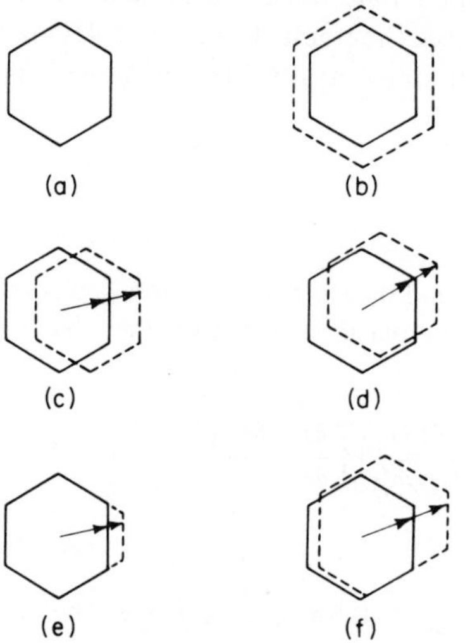


**Fig. 6.12:1.** Several hardening rules. (a) Initial yield condition (Tresca), (b) isotropic hardening, (c) kinematic hardening (Prager), (d) kinematic hardening (Ziegler's modification), (e) independently acting plane loading surfaces, (f) interdependent plane loading surfaces. (From Naghdi,[6.3] 1960.)

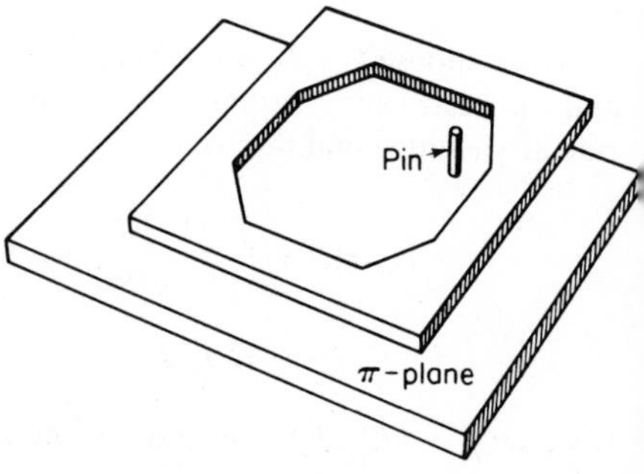


**Fig. 6.12:2.** A mechanical model used in explaining the kinematic hardening rule.

the $\pi$-plane $\sigma_1 + \sigma_2 + \sigma_3 = 0$. In Fig. 6.12:1(b) is shown the so-called *isotropic hardening*, which assumes a uniform expansion of the initial yield surface. The subsequent yield surfaces may be written as

(2) $$f = f^*(J_2, J_3) - \kappa = 0,$$

where $\kappa$ depends on the plastic strain history. Isotropic hardening, though widely used in analysis, has little experimental support.

Figure 6.12:1(c) illustrates Prager's kinematic hardening. The initial yield surface translates in the $\pi$-plane without rotation and without change in size. To explain this rule, Prager[6.3] (1954) used a mechanical model which may be represented as in Fig. 6.12:2. The initial yield surface is regarded as a planar rigid frame lying on the $\pi$-plane. The loading point on the $\pi$-plane is represented by a small, frictionless pin. If the pin engages the frame, it may push the frame about. Under the assumption of frictionlessness, any motion that can be imparted by the pin to the frame must be normal to the edge in contact. However, when a corner of the frame is caught by the pin, the pin may carry the frame if the direction of motion lies within a certain angle. Rotation of the frame is supposed to be prevented by some mechanism. If the pin disengages and moves away from the frame, the frame stays put, and the change represents an unloading. It is obvious that none of the flow rules deduced from Drucker's hypothesis is violated, and no theoretical objection can be raised against interpreting the motion of the rigid frame as a hardening rule. In fact, the Bauschinger effect is represented very simply, and the development of anistropy due to plastic deformation appears most naturally.

With some variation in the model, Prager[6.3] (1954) was able to represent various models of plasticity: rigid, perfectly plastic, rigid work-hardening elastic, etc. Almost simultaneously, a similar concept was introduced independently by Ishlinski[6.3] (1954) in the Soviet Union. Further developments were made by Prager,[6.3] Boyce,[6.3] Hodge,[6.3] Novozhilov,[6.3] Kadasevich,[6.3] and others. Hodge[6.3] (1956) points out that the concept of kinematic hardening must be applied in nine-dimensional stress space.

An example of Prager's kinematic hardening is given in Eq. (6.8:9). In general, Prager's loading surface may be expressed as

(3) $$f(\sigma_{ij} - \alpha_{ij}) = 0,$$

where $\alpha_{ij}$ represents the translation of the center of the initial yield surface. If linear hardening is assumed, we have

(4) $$\dot{\alpha}_{ij} = c\dot{e}_{ij}^{(p)},$$

where $c$ is a constant (Shield and Zielger,[6.3] 1958). From (3) and (4), we obtain easily, according to Eq. (6.11:6),

(5) $$\Lambda = \frac{1}{c} \frac{(\partial f/\partial \sigma_{ij})\dot{\sigma}_{ij}}{(\partial f/\partial \sigma_{kl})(\partial f/\partial \sigma_{kl})}.$$

Ziegler[6.3] (1959) modified Prager's rule by suggesting that Eq. (4) be replaced by the relation

$$\dot{\alpha}_{ij} = \dot{\mu}(\sigma_{ij} - \alpha_{ij}), \tag{6}$$

where $\dot{\mu} > 0$. Geometrically, this means that the direction of motion of the center of the initial yield surface agrees with the radius vector that joins the instantaneous center $\alpha_{ij}$ with the load point $\sigma_{ij}$.

Ziegler's modified kinematic hardening is illustrated in Fig. 6.12:1(d).

Figures 6.12:1(e) and (f) show two other hardening rules. In Fig. 6.12:1(e) the plastic deformation causes a linear segment to move. In Fig. 6.12:1(f) the plane loading surfaces changes with plastic loading in some interdependent manner.

More complicated hardening rules have been proposed from time to time. For example, Hodge[6.3] (1957) has extended the kinematic hardening to include an expansion of the yield surface simultaneously with its translation.

A significant, new line of thinking was introduced independently by Budiansky[6.3] and Kliushnikov[6.3] in 1959. They considered the possibility of creating corners in subsequent yield surfaces at the point of loading, and both achieved a compromise with the Hencky-Nadai "deformation" theory or the "total strain" theory. For an account of these theories the reader is referred to Notes 6.1, p. 476. See Hill,[6.2] Nadai,[6.2] and papers by Budiansky, Kliushnikov, Naghdi, Sanders, Phillips, etc., in Bibliography 6.2, and 6.4–6.6.

## PROBLEMS

**6.3.** A simple torsion test of certain material, using a hollow cylinder specimen as shown in Fig. P6.3(a), shows that the load-deflection curve is linear for a shearing stress below 25,000 lb/sq in. and that at the stress 25,000 lb/sq in. yielding occurs. If the criterion of yielding for this material is von Mises' $J_2 = k^2$, what is the expected value of the tensile stress at which yielding would occur in a tension test specimen as shown in Fig. P6.3(b).

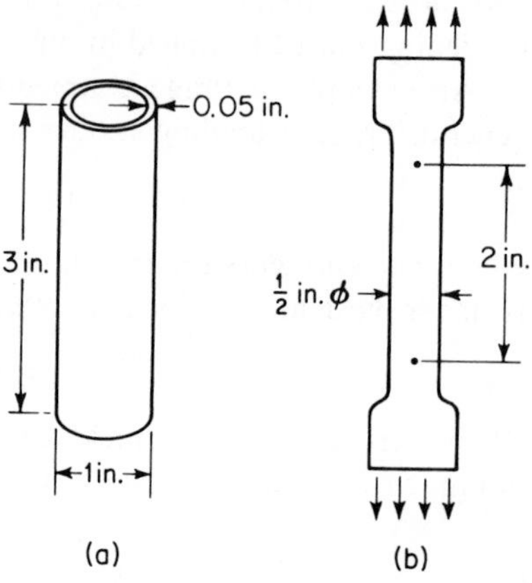


**P6.3**

**6.4.** Consider an isotropic, incompressible, but nonlinearly elastic material. Let $e'_{ij}$, $\sigma'_{ij}$ be the strain deviation and stress deviation, respectively. If the material were elastic,

$$e'_{ij} = \frac{1}{2G}\sigma'_{ij}.$$

This can be written, of course, as a matrix equation,

$$\begin{bmatrix} e'_{11} & e'_{12} & e'_{13} \\ e'_{21} & e'_{22} & e'_{23} \\ e'_{31} & e'_{32} & e'_{33} \end{bmatrix} = \frac{1}{2G} \begin{bmatrix} \sigma'_{11} & \sigma'_{12} & \sigma'_{13} \\ \sigma'_{21} & \sigma'_{22} & \sigma'_{23} \\ \sigma'_{31} & \sigma'_{32} & \sigma'_{33} \end{bmatrix},$$

or, in short,

$$\mathbf{e}' = \frac{1}{2G}\boldsymbol{\sigma}'.$$

To generalize such a relation to a nonlinear elastic material, we assume that the matrix $\mathbf{e}'$ can be represented as a power series in $\boldsymbol{\sigma}'$,

$$\mathbf{e}' = \mathbf{C}_1\boldsymbol{\sigma}' + \mathbf{C}_2\boldsymbol{\sigma}'^2 + \mathbf{C}_3\boldsymbol{\sigma}'^3 + \dots.$$

Prove, by an application of the Cayley-Hamilton theorem, that the most general nonlinear elasticity law of this type can be reduced to the form

$$\mathbf{e}' = \mathbf{P}\boldsymbol{\sigma}' + \mathbf{Q}\boldsymbol{\sigma}'^2. \qquad \text{(Prager)}$$

**6.5.** In soil mechanics, Coulomb's yield condition is widely used. Consider a two-dimensional problem. If $\sigma_1$ and $\sigma_2$ are the principal stresses in the plane of the problem, Coulomb's condition is specified by the yield functions

$$f_1 = \sigma_1(1 + \sin\phi) - \sigma_2(1 - \sin\phi) - 2c\cos\phi,$$
$$f_2 = -\sigma_1(1 - \sin\phi) + \sigma_2(1 + \sin\phi) - 2c\cos\phi,$$

where $c$ = cohesion and $\phi$ = angle of internal friction. The soil is elastic if both $f_1$ and $f_2$ are negative, and it is plastic when $f_1$, or $f_2$, or both, vanish.

(a) Show that the flow rule derived from this condition by means of Koiter's theory of generalized plastic potential requires that the rate of the planar dilatation $\dot{e}_1^{(p)} + \dot{e}_2^{(p)}$ and the maximum shear rate $|\dot{e}_1^{(p)} - \dot{e}_2^{(p)}|$ have the constant ratio $\sin\phi$ for all states of stress at the yield limit, except the state $\sigma_1 = \sigma_2 = c\cot\phi$.

(b) Generalize the results above to obtain the generalized plastic potentials for three-dimensional problems.

(c) According to your generalization (b), is there any general relationship between the three-dimensional dilatation and the maximum shear rate for stress states at the yield limit?

*Ref.* W. Prager, "On the Kinematics of Soils. *Memoires des Sciences* **28** (1954), Ac. Roy. Belgique, pp. 3–8.

# 7

# LINEAR ELASTICITY

In the following chapter we shall discuss the classical theory of elasticity—a generalization of the method of Chap. 1 to a continuum. We shall discuss the general structure of the theory and illustrate the applications of the theory by a few examples.

*Rectangular Cartesian* coordinates of reference will be used throughout. The coordinates will be denoted by $x_1, x_2, x_3$ or $x, y, z$ unless stated otherwise.

## 7.1. BASIC EQUATIONS OF ELASTICITY FOR HOMOGENEOUS ISOTROPIC BODIES

An elastic body has a unique natural state, to which the body returns when all external loads are removed. All stresses, strains, and particle displacements are measured from this natural state: their values are counted as zero in that state.

There are two ways to describe a deformed body: the *material* and the *spatial* (see Sec. 5.2). Consider the spatial description. The motion of a continuum is described by the instantaneous velocity field $v_i(x_1, x_2, x_3, t)$. To describe the strain in the body, a displacement field $u_i(x_1, x_2, x_3, t)$ is specified which describes the displacement of a particle located at $x_1x_2x_3$ at time $t$ from its position in the natural state. Various strain measures can be defined for the displacement field. The Almansi strain tensor is expressed in terms of $u_i(x_1, x_2, x_3, t)$ according to Eq. (4.2:4),

(1) $$e_{ij} = \frac{1}{2}\left[\frac{\partial u_j}{\partial x_i} + \frac{\partial u_i}{\partial x_j} - \frac{\partial u_k}{\partial x_i}\frac{\partial u_k}{\partial x_j}\right].$$

The particle displacements $u_i$ are functions of time and position. The particle velocity is given by the material derivative of the displacement,

(2) $$v_i = \frac{\partial u_i}{\partial t} + v_j\frac{\partial u_i}{\partial x_j}.$$

The particle acceleration is given by the material derivative of the velocity (5.2:7),

(3) $$\alpha_i = \frac{\partial v_i}{\partial t} + v_j\frac{\partial v_i}{\partial x_j}$$

The motion of the body must obey the equation of continuity (5.4:3)

$$\frac{\partial \rho}{\partial t} + \frac{\partial(\rho v_i)}{\partial x_i} = 0 \tag{4}$$

and the equation of motion (5.5:7)

$$\rho \alpha_i = \frac{\partial \sigma_{ij}}{\partial x_j} + X_i. \tag{5}$$

In addition to the field Eqs. (4) and (5), the theory of linear elasticity is based on Hooke's law. For a homogeneous isotropic material, this is (6.2:7).

$$\sigma_{ij} = \lambda e_{kk}\delta_{ij} + 2Ge_{ij}, \tag{6}$$

where $\lambda$ and $G$ are constants independent of the spatial coordinates.†

The famous nonlinear terms in (1), (2), and (3) are sources of major difficulty in the theory of elasticity. To make some progress, we are forced to *linearize* by considering small displacements and small velocities, i.e., by restricting ourselves to values of $u_i$, $v_i$ so small that the nonlinear terms in (1)–(3) may be neglected. In such a linearized theory, we have

$$e_{ij} = \tfrac{1}{2}(u_{i,j} + u_{j,i}), \tag{7}$$

$$v_i = \frac{\partial u_i}{\partial t}, \qquad \alpha_i = \frac{\partial v_i}{\partial t}. \tag{8}$$

*Unless stated otherwise, all that is discussed below is subjected to this restriction of linearization.* Fortunately, many useful results can be obtained from this linearized theory.

Equations (1)–(6) or (4)–(8) together are 22 equations for the 22 unknowns $\rho$, $u_i$, $v_i$, $e_{ij}$, $\sigma_{ij}$. In the infinitesimal displacement theory we may eliminate $\sigma_{ij}$ by substituting Eq. (6) into (5) and using (7) to obtain the well-known *Navier's equation*,

(9) ▲
$$Gu_{i,jj} + (\lambda + G)u_{j,ji} + X_i = \rho \frac{\partial^2 u_i}{\partial t^2}.$$

This can be written in the form

(10) ▲
$$G\,\nabla^2 u_i + (\lambda + G)e_{,i} + X_i = \rho \frac{\partial^2 u_i}{\partial t^2},$$

where

$$e = u_{j,j} \tag{11}$$

$$\nabla^2 u_i = (u_i)_{,jj}. \tag{12}$$

† The corresponding equations based on the material description are the following: velocity and acceleration, Eqs. (5.2:5), (5.2:6); strain measure, the Green's strain tensor, Eq. (4.2:5); the equation of continuity, Eq. (5.2:1); stress tensors, Sec. 16.2; the equations of motion, Eqs. (16.3:6)–(16.3:9); the stress-strain laws, see Sec. 16.6, 16.7; in particular, Eqs. (16.7:1) and (16.7:2). It can be seen that the kinematical relations appear simpler in the material description, but the equations of motion appear more complicated.

The quantity $e$ is the *divergence* of the displacement vector $u_i$. $\nabla^2$ is the *Laplace operator*. If we write $x$, $y$, $z$ instead of $x_1$, $x_2$, $x_3$, we have

$$e = \frac{\partial u}{\partial x} + \frac{\partial v}{\partial y} + \frac{\partial w}{\partial z}, \tag{13}$$

$$\nabla^2 = \frac{\partial^2}{\partial x^2} + \frac{\partial^2}{\partial y^2} + \frac{\partial^2}{\partial z^2}, \tag{14}$$

Love[1.2] writes Eq. (10) in the form,

$$G\,\nabla^2(u, v, w) + (\lambda + G)\left(\frac{\partial}{\partial x}, \frac{\partial}{\partial y}, \frac{\partial}{\partial z}\right)e + (X, Y, Z) = \rho\frac{\partial^2}{\partial t^2}(u, v, w), \tag{15}$$

which is a shorthand for three equations of the type

(16) ▲
$$G\,\nabla^2 u + (\lambda + G)\frac{\partial e}{\partial x} + X = \rho\frac{\partial^2 u}{\partial t^2}.$$

This can also be written as

(17) ▲
$$G\left(\nabla^2 u + \frac{1}{1 - 2\nu}\frac{\partial e}{\partial x}\right) + X = \rho\frac{\partial^2 u}{\partial t^2}.$$

If we introduce the rotation vector

$$\begin{aligned}(\omega_x, \omega_y, \omega_z) &\equiv \tfrac{1}{2}\,\mathrm{curl}\,(u, v, w)\\ &\equiv \frac{1}{2}\left(\frac{\partial w}{\partial y} - \frac{\partial v}{\partial z}, \frac{\partial u}{\partial z} - \frac{\partial w}{\partial x}, \frac{\partial v}{\partial x} - \frac{\partial u}{\partial y}\right)\end{aligned} \tag{18}$$

and use the identity

$$\nabla^2(u, v, w) = \left(\frac{\partial}{\partial x}, \frac{\partial}{\partial y}, \frac{\partial}{\partial z}\right)e - 2\,\mathrm{curl}\,(\omega_x, \omega_y, \omega_z), \tag{19}$$

then (10) may be written as

(20)
$$(\lambda + 2G)\left(\frac{\partial}{\partial x}, \frac{\partial}{\partial y}, \frac{\partial}{\partial z}\right)e - 2G\,\mathrm{curl}\,(\omega_x, \omega_y, \omega_z) + (X, Y, Z) = \rho\frac{\partial^2}{\partial t^2}(u, v, w).$$

## 7.2. EQUILIBRIUM OF AN ELASTIC BODY UNDER ZERO BODY FORCE

Consider the conditions of static equilibrium of an elastic body. If the body force vanishes, $X_i = 0$, then by taking divergence of Eq. (7.1:9), we have

$$Gu_{i,jji} + (\lambda + G)u_{j,jii} = 0,$$

or

$$(u_{j,j})_{,ii} = 0. \tag{1}$$

In unabridged notations for rectangular Cartesian coordinates, this is

$$\left(\frac{\partial^2}{\partial x^2} + \frac{\partial^2}{\partial y^2} + \frac{\partial^2}{\partial z^2}\right) e = 0, \quad \text{or} \quad \nabla^2 e = 0. \tag{2}$$

Equation (2) is a Laplace equation. A function satisfying Eq. (2) is called a *harmonic function*. Thus, *the dilation e is a harmonic function when the body force vanishes.*

But

$$(3\lambda + 2G)e = \sigma_{xx} + \sigma_{yy} + \sigma_{zz} = 3\sigma,$$

where $\sigma$ is the mean stress. Hence, *the mean stress is also a harmonic function:*

$$\nabla^2 \sigma = 0. \tag{3}$$

If we put $X = 0$, $\partial^2 u/\partial t^2 = 0$, and operate on Eq. (7.1:16) with the Laplacian $\nabla^2$, we have

$$(\lambda + G)\frac{\partial}{\partial x}\nabla^2 e + G\nabla^2\nabla^2 u = 0.$$

With Eq. (2), this implies that

$$\nabla^4 u = 0, \tag{4}$$

where, in rectangular Cartesian coordinates,

$$\nabla^4 = \frac{\partial^4}{\partial x^4} + \frac{\partial^4}{\partial y^4} + \frac{\partial^4}{\partial z^4} + 2\frac{\partial^4}{\partial x^2\,\partial y^2} + 2\frac{\partial^4}{\partial y^2\,\partial z^2} + 2\frac{\partial^4}{\partial z^2\,\partial x^2}. \tag{5}$$

Equation (4) is called a *biharmonic* equation, and its solution is called a *biharmonic function*. Hence, *the displacement component u is biharmonic. Similarly, the components v, w are biharmonic.* It follows that *when the body force is zero, each of the strain components and each of the stress components, being linear combination of the first derivatives of u, v, w, are all biharmonic functions:*

$$\nabla^4 \sigma_{ij} = 0, \tag{6}$$

$$\nabla^4 e_{ij} = 0. \tag{7}$$

## 7.3. BOUNDARY VALUE PROBLEMS

Navier's equation (7.1:9), combines Hooke's law and the equation of motion. It is to be solved for appropriate boundary and initial conditions. The boundary conditions that occur are usually one of two kinds:

A. *Specified displacements*. The components of displacement $u_i$ are prescribed on the boundary.
B. *Specified surface tractions*. The components of surface traction $\overset{\nu}{T}_i$ are assigned on the boundary.

In most problems of elasticity, the boundary conditions are such that over part of the boundary displacements are specified, whereas over another part the surface tractions are specified. Let the region occupied by an elastic body be denoted by $V$. Let the boundary surface of $V$ be denoted by $S$. We separate $S$ into two parts: $S_u$, where displacements are specified; and $S_\sigma$, where surface tractions are specified. Therefore, on $S_\sigma$,

$$\overset{\nu}{T}_i = \sigma_{ij}\nu_j = \text{a prescribed function},$$

where $\nu_j$ is a unit vector along the outer normal to the surface $S_\sigma$. By Hooke's law, this may be written as

$$[\lambda u_{k,k}\delta_{ij} + G(u_{i,j} + u_{j,i})]\nu_j = \text{prescribed function}.$$

Hence, over the entire surface, the boundary conditions are that either $u_i$, or a combination of the first derivatives of $u_i$, are prescribed.

In dynamic problems, a set of initial conditions on $u_i$ or $\sigma_{ij}$ must be specified in the region $V$ and on the boundary $S$.

The question arises whether a boundary-value problem posed in this way has a solution, and whether the solution is unique or not. The question has two parts. First, do we expect a unique solution on physical grounds? Second, does the specific mathematical problem have a unique solution? In continuum mechanics, there are many occasions in which we do not expect a unique solution to exist. For example, when a thin-walled spherical shell is subjected to a uniform external pressure, the phenomenon of buckling may occur when the pressure exceeds certain specific value. At the buckling load, the shell may assume several different forms of deformation, some stable, some unstable. On the other hand, our everyday experience about the physical world tells us that the vast majority of mechanical cause-and-effect relationships are unique. Theoretically, the physical question is partly answered by thermodynamics. But the mathematical question must be answered by the theory of partial differential equations. A satisfactory theory must bring harmony between the mathematical formulation and the physical world. In the next section we shall present a restricted proof of uniqueness due to Kirchhoff. Further remarks are given in Notes 7.1, p. 478.

In the preceding discussions we have taken the displacements $u_i$ as the basic unknown variables. In problems of static equilibrium, however, it is customary to use an alternate procedure. The equations of equilibrium are first solved for the stresses $\sigma_{ij}$. We then use Hooke's law to obtain the strain $e_{ij}$. This solution will not be unique. In fact, an infinite set of solutions will be found. The correct one is then singled out by the conditions of compatibility. Only the one solution that satisfies both the equation of equilibrium and the equations of compatibility corresponds to a continuous displacement field. This procedure becomes very attractive when stress functions are introduced, which yield general solutions of the equations of equilibrium. (see Sec. 9.2).

By means of Hooke's law, the compatibility equation

(1) $$e_{ij,kl} + e_{kl,ij} - e_{ik,jl} - e_{jl,ik} = 0$$

can be expressed directly in terms of stress components. On substituting

$$e_{ij} = \frac{1+\nu}{E}\sigma_{ij} - \frac{\nu}{E}\theta\delta_{ij}, \qquad \theta = \sigma_{kk},$$

into (1), we obtain

$$(2) \qquad \sigma_{ij,kl} + \sigma_{kl,ij} - \sigma_{ik,jl} - \sigma_{jl,ik} = \frac{\nu}{1+\nu}(\delta_{ij}\theta_{,kl} + \delta_{kl}\theta_{,ij} - \delta_{ik}\theta_{,jl} - \delta_{jl}\theta_{,ik}).$$

Since only six of the 81 equations represented by (1) are linearly independent, the same must be true for (2). If we combine Eqs. (2) linearly by setting $k = l$ and summing, we obtain

$$(3) \qquad \sigma_{ij\,kk} + \sigma_{kk,ij} - \sigma_{ik.jk} - \sigma_{jk,ik} = \frac{\nu}{1+\nu}(\delta_{ij}\theta_{,kk} + \delta_{kk}\theta_{,ij} - \delta_{ik}\theta_{,jk} - \delta_{jk}\theta_{,ik}),$$

which is a set of nine equations of which six are independent because of the symmetry in $i$ and $j$. Hence, the number of independent equations is not reduced, and (2) and (3) are equivalent. Since $\sigma_{kk} = \theta$ and $\sigma_{ij,kk} = \nabla^2\sigma_{ij}$, if we use the equation of equilibrium to replace, say, $\sigma_{ik,kj}$ by $-X_{i,j}$, we can write (3) as

$$(4) \qquad \nabla^2\sigma_{ij} + \frac{1}{1+\nu}\theta_{,ij} - \frac{\nu}{1+\nu}\delta_{ij}\nabla^2\theta = -(X_{i,j} + X_{j,i}),$$

where $X_i$ is the body force per unit volume. In dynamic problems the inertia force should be included in $X_i$. With a contraction $i = j$, Eq. (4) furnishes a relation between $\nabla^2\theta$ and $X_{i,i}$. If this is used to transform the third term in (4), we obtain

$$(5) \ \blacktriangle \qquad \nabla^2\sigma_{ij} + \frac{1}{1+\nu}\theta_{,ij} = -\frac{\nu}{1-\nu}\delta_{ij}X_{k,k} - (X_{i,j} + X_{j,i}).$$

Written out *in extenso*, these are

$$(6) \qquad \begin{aligned}
\nabla^2\sigma_{xx} + \frac{1}{1+\nu}\frac{\partial^2\theta}{\partial x^2} &= -\frac{\nu}{1-\nu}\left(\frac{\partial X}{\partial x} + \frac{\partial Y}{\partial y} + \frac{\partial Z}{\partial z}\right) - 2\frac{\partial X}{\partial x}, \\
\nabla^2\sigma_{yy} + \frac{1}{1+\nu}\frac{\partial^2\theta}{\partial y^2} &= -\frac{\nu}{1-\nu}\left(\frac{\partial X}{\partial x} + \frac{\partial Y}{\partial y} + \frac{\partial Z}{\partial z}\right) - 2\frac{\partial Y}{\partial y}, \\
\nabla^2\sigma_{zz} + \frac{1}{1+\nu}\frac{\partial^2\theta}{\partial z^2} &= -\frac{\nu}{1-\nu}\left(\frac{\partial X}{\partial x} + \frac{\partial Y}{\partial y} + \frac{\partial Z}{\partial z}\right) - 2\frac{\partial Z}{\partial z}, \\
\nabla^2\sigma_{yz} + \frac{1}{1+\nu}\frac{\partial^2\theta}{\partial y\,\partial z} &= -\left(\frac{\partial Y}{\partial z} + \frac{\partial Z}{\partial y}\right), \\
\nabla^2\sigma_{zx} + \frac{1}{1+\nu}\frac{\partial^2\theta}{\partial z\,\partial x} &= -\left(\frac{\partial Z}{\partial x} + \frac{\partial X}{\partial z}\right), \\
\nabla^2\sigma_{xy} + \frac{1}{1+\nu}\frac{\partial^2\theta}{\partial x\,\partial y} &= -\left(\frac{\partial X}{\partial y} + \frac{\partial Y}{\partial x}\right).
\end{aligned}$$

These equations were obtained by Michell in 1900, and, for the case in which body forces are absent, by Beltrami in 1892. They are known as the *Beltrami-Michell compatibility equations.*

For a simply connected region, satisfaction of the Beltrami-Michell equations implies that the stress system $\sigma_{ij}$ is derivable from a continuous displacement field. If the region concerned is multiply connected, additional conditions in the form of certain line integrals must be satisfied (see Sec. 4.7).

## 7.4. THE PROBLEM OF EQUILIBRIUM AND THE UNIQUENESS OF SOLUTIONS IN ELASTICITY

Consider the problem of determining the state of stress and strain in a body of a given shape which is held strained by body forces $X_i$ and surface tractions $\overset{\nu}{T}_i$. Let us assume that a function $W(e_{11}, e_{12}, \ldots, e_{33})$, called *the strain energy function of the elastic material*, exists and has the property

$$\frac{\partial W}{\partial e_{ij}} = \sigma_{ij}. \tag{1}$$

For example, if the material obeys Hooke's law [Eq. (6.1:1)], then

$$W = \tfrac{1}{2} C_{ijkl} e_{ij} e_{kl}. \tag{2}$$

The existence and the positive definiteness of the strain energy function for an elastic body are discussed in Chap. 12. In Sec. 12.4 it is shown that $W$ is positive definite in the neighborhood of the natural state. A natural state of a material is a stable state in which the material can exist by itself in thermodynamic equilibrium. Limiting ourselves to consider materials which have a unique natural state in the theory of elasticity, we are assured of a positive definite strain energy function in the neighborhood of the natural state.

The equations of equilibrium $\sigma_{ij,j} + X_i = 0$ may be written in terms of $W$ as

$$\left(\frac{\partial W}{\partial e_{ij}}\right)_{,j} + X_i = 0. \tag{3}$$

The boundary conditions over the boundary surface $S_u + S_\sigma$ are:

(4a) Over $S_u$, the values of $u_i$ are given,

(4b) Over $S_\sigma$, the tractions $\overset{\nu}{T}_i = \dfrac{\partial W}{\partial e_{ij}} \nu_j$ are specified

(see Sec. 7.3). If $S_\sigma$ constitutes the entire surface of the body, it is obvious that equilibrium would be impossible unless the system of body forces and surface tractions satisfy the conditions of static equilibrium for the body as a whole. In the same case, the displacement will be indeterminate to the extent of a possible rigid-body motion.

We shall prove the following theorem due to Kirchhoff. *If either the surface displacements or the surface tractions are given, the solution of the problem of equilibrium of an elastic body as specified by* Eqs. (1)–(4) *is unique in the sense that the state of stress (and strain) is determinate without ambiguity, provided that the magnitude of the stress (and strain) is so small that the strain energy function exists and remains positive definite.*

*Proof.* Since a strained state must be either unique or nonunique, a proof can be constructed by showing that an assumption of nonuniqueness leads to absurdity. Assume that there exist two systems of displacements $u_i'$ and $u_i''$ which define two states of strain, both satisfying Eq. (3) and the boundary conditions (4a) and (4b). Then the difference $u_i \equiv u_i' - u_i''$ satisfies the equation

$$\left(\frac{\partial W}{\partial e_{ij}}\right)_{,j} = 0 \tag{5}$$

and the boundary conditions that

$$u_i = 0 \quad \text{on} \quad S_u \qquad \text{and} \qquad \frac{\partial W}{\partial e_{ij}}\nu_j = 0 \quad \text{on} \quad S_\sigma. \tag{6}$$

From (5) we have

$$\int_V u_i \left(\frac{\partial W}{\partial e_{ij}}\right)_{,j} dV = 0,$$

which, on integrating by parts, becomes

$$\int_S u_i \frac{\partial W}{\partial e_{ij}} \nu_j \, dS - \int_V \frac{\partial W}{\partial e_{ij}} u_{i,j} \, dV = 0. \tag{7}$$

The first surface integral vanishes because of the boundary conditions (6). The second volume integral may be written as

$$\int_V \frac{\partial W}{\partial e_{ij}} e_{ij} \, dV.$$

Now, when $W$ is a homogeneous quadratic function of $e_{ij}$ [Eq. (2)], the integral above is equal to $\int 2W\,dV$. Since $W$ is assumed to be positive definite, the integral $\int W\,dV$ cannot vanish unless $W$ vanishes, which in turn implies that $e_{ij} = 0$ everywhere. Hence, $u_i = u_i' - u_i''$ corresponds to the natural, unstrained state of the body. Therefore, the states of strain (and stress) defined by $u_i'$ and $u_i''$ are the same, contrary to the assumption. Hence, the state of strain (and stress) is unique. Q.E.D.

We must note that the uniqueness theorem is proved only in the neighborhood of the natural state. In fact when the strain energy function fails to remain positive definite, a multi-valued solution or several solutions may be possible. See further remarks in Notes 7.1, p. 478.

**Problem 7.1.** Prove Neumann's theorem that the solution $u_i(x, t)$, for $x$ in $V + S_u + S_\sigma$ and $t \geqslant 0$, of the following system of equations, is unique.

(8) $$\frac{\partial}{\partial x_j}\left(\frac{\partial W}{\partial e_{ij}}\right) + X_i - \rho \frac{\partial^2 u_i}{\partial t^2} = 0, \qquad \text{for } x \text{ in } V,\ t \geqslant 0,$$

(9) $$u_i = f_i(x, t), \qquad \text{for } x \text{ on } S_u,\ t \geqslant 0,$$

(10) $$\frac{\partial W}{\partial e_{ij}} \nu_j = g_i(x, t), \qquad \text{for } x \text{ on } S_\sigma,\ t \geqslant 0,$$

(11) $$u_i = h_i(x), \qquad \dot{u}_i = k_i(x), \text{ when } t = 0,\ x \text{ in } V + S,$$

(12) $$e_{ij} = \tfrac{1}{2}(u_{i,j} + u_{j,i}),$$

where $X_i$, $f_i$, $g_i$, $h_i$, $k_i$ are preassigned functions and $W(e_{ij})$ is a positive definite quadratic form. *Hint:* Multiply Eq. (8) by $\partial u_i/\partial t$ and integrate over $V$ and $(0, t)$. Note that the kinetic energy is positive definite. (Ref., Love, Elasticity,[1.2] p. 176.)

## 7.5. SAINT VENANT'S THEORY OF TORSION

To illustrate the applications of the theory of elasticity, we shall consider the problem of torsion of a cylindrical body. A cylindrical shaft, with an axis $z$, is acted on at its ends by a distribution of shearing stresses whose resultant force is zero but whose resultant moment is a torque $T$. The lateral surface of the shaft is stress free (see Fig. 7.5:1).

If the shaft is a circular cylinder, it is very easy to show that all plane cross sections normal to the $z$-axis remain plane, and the deformation consists of relative rotation $\theta$ of the cross sections. The rate of rotation per unit axial length $d\theta/dz$ is proportional to the torque $T$, with a proportionality constant equal to the product of the shear modulus $G$ of the shaft material, and the polar moment of inertia $J$ of the shaft cross-sectional area:

(1) $$GJ \frac{d\theta}{dz} = T.$$

The only nonvanishing component of stress is the shear in cross sections perpendicular to $z$, whose magnitude is

(2) $$\tau = \frac{Tr}{J},$$

where $r$ is the radius vector from the central axis $z$.

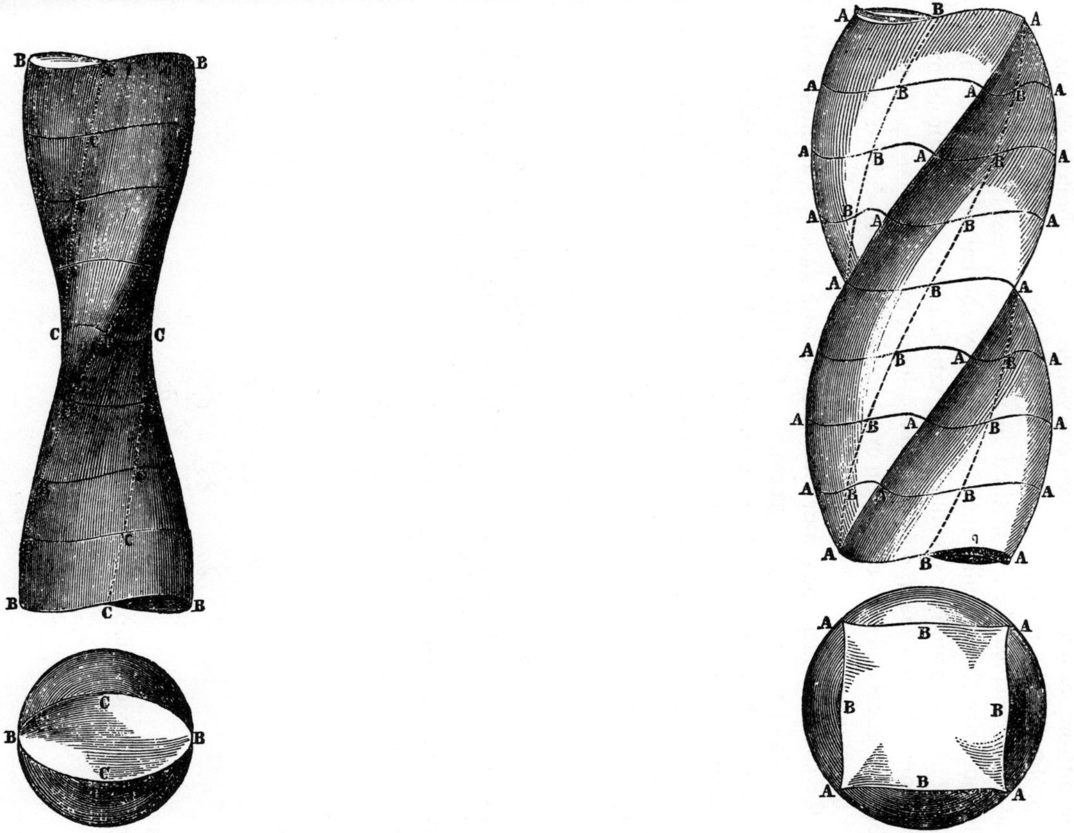


Fig. 7.5:1. Torsion of bars of elliptic and square cross section, as drawn by St. Venant.

If the cross section of the shaft is not circular, a plane cross section does not remain plane; it warps, as is shown in Fig. 7.5:1. The problem is to calculate the stress distribution and the deformation of the shaft.

This is an important problem in engineering; for shafts are used to transmit torques and they are seen everywhere. The celebrated solution to the problem is due to Barre de Saint-Venant (1855), who used the so-called *semi-inverse method*, i.e., a method in which one guesses at part of the solution, tries to determine the rest rationally so that all the differential equations and boundary conditions are satisfied. The torsion problem is not simple. Saint-Venant, guided by the solution of the circular shaft, made a brilliant guess and showed that an exact solution to a well-defined problem can be obtained.

We shall consider, then, a cylindrical shaft with an axis $z$, with the ends at $z = 0$ and $z = L$. A set of rectangular Cartesian coordinates $x, y, z$ will be used, with the $x, y$-plane perpendicular to the axis of the shaft. The displacement components in the $x, y, z$-direction will be written as $u, v, w$, respectively. Saint-Venant assumed that as the shaft twists the plane cross sections are warped but the *projections* on the $x$, $y$-plane rotate as a rigid body; i.e.,

$$u = -\alpha z y, \qquad v = \alpha z x, \qquad w = \alpha \varphi(x, y), \tag{3}$$

where $\varphi(x, y)$ is some function of $x$ and $y$, called the *warping function*, and $\alpha$ is the angle of twist per unit length of the bar and is assumed to be very small ($\ll 1$). We rely on the function $\varphi(x, y)$ to satisfy the differential equations of equilibrium (without body force)

$$\begin{aligned} \frac{\partial \sigma_{xx}}{\partial x} + \frac{\partial \sigma_{xy}}{\partial y} + \frac{\partial \sigma_{xz}}{\partial z} &= 0, \\ \frac{\partial \sigma_{yx}}{\partial x} + \frac{\partial \sigma_{yy}}{\partial y} + \frac{\partial \sigma_{yz}}{\partial z} &= 0, \\ \frac{\partial \sigma_{zx}}{\partial x} + \frac{\partial \sigma_{zy}}{\partial y} + \frac{\partial \sigma_{zz}}{\partial z} &= 0, \end{aligned} \tag{4}$$

the boundary conditions on the lateral surface of the cylinder

$$\begin{aligned} \sigma_{xx}\nu_x + \sigma_{xy}\nu_y &= 0, \\ \sigma_{yx}\nu_x + \sigma_{yy}\nu_y &= 0, \\ \sigma_{zx}\nu_x + \sigma_{zy}\nu_y &= 0, \end{aligned} \tag{5}$$

and the boundary conditions at the ends $z = 0$ and $z = L$:

$$\begin{gathered} \sigma_{zz} = 0 \\ \sigma_{zx}, \sigma_{zy} \text{ equipollent to a torque } T. \end{gathered} \tag{6}$$

The constants $\nu_x$, $\nu_y$ are the direction cosines of the exterior normal to the lateral surface ($\nu_z = 0$).

From (3) we obtain the stresses according to Hooke's law.

$$\sigma_{yz} = \alpha G\left(\frac{\partial \varphi}{\partial y} + x\right), \qquad \sigma_{zx} = \alpha G\left(\frac{\partial \varphi}{\partial x} - y\right),$$
(7)
$$\sigma_{xy} = \sigma_{xx} = \sigma_{yy} = \sigma_{zz} = 0.$$

A substitution of these values into Eqs. (4) shows that the equilibrium equations will be satisfied if $\varphi(x, y)$ satisfies the equation

(8) $$\frac{\partial^2 \varphi}{\partial x^2} + \frac{\partial^2 \varphi}{\partial y^2} = 0$$

throughout the cross section of the cylinder. To satisfy the boundary conditions (5), we must have

(9) $$\left(\frac{\partial \varphi}{\partial x} - y\right) \cos(\nu, x) + \left(\frac{\partial \varphi}{\partial y} + x\right) \cos(\nu, y) = 0, \quad \text{on } C,$$

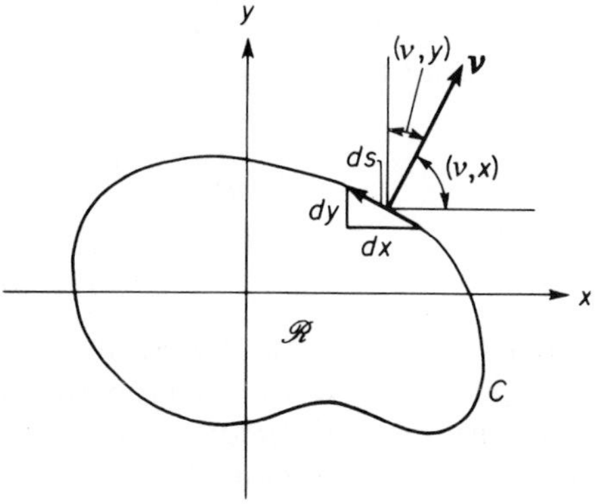


**Fig. 7.5:2.** Notations.

where $C$ is the boundary of the cross section $R$ (Fig. 7.5:2). But

$$\frac{\partial \varphi}{\partial x} \cos(x, \nu) + \frac{\partial \varphi}{\partial y} \cos(y, \nu) \equiv \frac{\partial \varphi}{\partial \nu};$$

hence, the boundary condition (9) can be written as

(10) $$\frac{\partial \varphi}{\partial \nu} = y \cos(x, \nu) - x \cos(y, \nu) \quad \text{on } C.$$

The boundary conditions (6) are satisfied if

(11) $$\iint_R \sigma_{zx}\, dx\, dy = 0, \qquad \iint_R \sigma_{zy}\, dx\, dy = 0,$$

(12) $$\iint_R (x\sigma_{zy} - y\sigma_{zx})\, dx\, dy = T.$$

We can show that Eqs. (11) are readily satisfied if $\varphi$ satisfies (8) and (10); because

$$\iint_R \sigma_{zx}\, dx\, dy = \alpha G \iint_R \left(\frac{\partial \varphi}{\partial x} - y\right) dx\, dy$$
$$= \alpha G \iint_R \left\{\frac{\partial}{\partial x}\left[x\left(\frac{\partial \varphi}{\partial x} - y\right)\right] + \frac{\partial}{\partial y}\left[x\left(\frac{\partial \varphi}{\partial y} + x\right)\right]\right\} dx\, dy,$$

since $\varphi$ satisfies (8). On applying Gauss' theorem to the last integral, it becomes a line integral on the boundary $C$ of the region $R$:

$$\alpha G \int_C x \left[ \frac{\partial \varphi}{\partial \nu} - y \cos (x, \nu) + x \cos (y, \nu) \right] ds,$$

which vanishes on account of (10). Similarly, the second of Eqs. (11) is satisfied. Finally, the last condition (12) requires that

$$T = \alpha G \iint_R \left( x^2 + y^2 + x \frac{\partial \varphi}{\partial y} - y \frac{\partial \varphi}{\partial x} \right) dx\, dy. \tag{13}$$

Writing $J$ for the integral

$$J \equiv \iint_R \left( x^2 + y^2 + x \frac{\partial \varphi}{\partial y} - y \frac{\partial \varphi}{\partial x} \right) dx\, dy, \tag{14}$$

we have

$$T = \alpha G J. \tag{15}$$

This merely shows that the torque $T$ is proportional to the angle of twist per unit length $\alpha$, with a proportionality constant $GJ$, which is usually called the *torsional rigidity* of the shaft. The $J$ represents the polar moment of inertia of the section when the cross section is circular. However, it is conventional to retain the notation $GJ$ for torsional rigidity, even for noncircular cylinders.

Thus, we see that the problem of torsion is reduced to the solution of Eqs. (8) and (10). The solution will yield a stress system $\sigma_{zx}$, $\sigma_{zy}$. If the end sections of the shaft are free to warp, and if the stresses prescribed on the end sections are exactly the same as those given by the solution, then an exact solution is obtained, and the solution is unique (see Sec. 7.4). If the distribution of stresses acting on the end sections, while equipollent to a torque $T$, does not agree exactly with that given by Eq. (7), then only an approximate solution is obtained. According to a principle proposed by Saint-Venant, the error in the approximation is significant only in the neighborhood of the end section (see Secs. 10.11–10.13).

Equation (8) is a *potential* equation; its solutions are called *harmonic functions*. The same equation appears in hydrodynamics. The boundary condition (10) is similar to that for the velocity potential in hydrodynamics with prescribed velocity efflux over the boundary. In the hydrodynamics problem, the condition for the existence of a solution $\varphi$ is that the total flux of fluid across the boundary must vanish. Translated to our problem, the condition for the existence of a solution $\varphi$ is that the integral of the normal derivative of the function $\varphi$, calculated over the entire boundary $C$, must vanish. This follows from the identity

$$\int_C \frac{\partial \varphi}{\partial \nu} ds = \iint_R \operatorname{div} (\operatorname{grad} \varphi)\, dx\, dy = \iint_R \nabla^2 \varphi\, dx\, dy \tag{16}$$

and from the fact that $\nabla^2\varphi = 0$. This condition is satisfied in our case by (10), as can be easily shown. Therefore, our problem is reduced to the solution of a potential problem (called Neumann's problem) subjected to the boundary condition (10).

An alternate approach was proposed by Prandtl, who takes the stress components as the principal unknowns. If we assume that only $\sigma_{xz}$, $\sigma_{yz}$ differ from zero, then all the equations of equilibrium (4) are satisfied if

$$\frac{\partial \sigma_{xz}}{\partial x} + \frac{\partial \sigma_{yz}}{\partial y} = 0. \tag{17}$$

Prandtl observes that this equation is identically satisfied if $\sigma_{xz}$ and $\sigma_{yz}$ are derived from a *stress function* $\psi(x, y)$ so that

$$\sigma_{xz} = \frac{\partial \psi}{\partial y}, \qquad \sigma_{yz} = -\frac{\partial \psi}{\partial x}. \tag{18}$$

This corresponds to the stream function in hydrodynamics, if $\sigma_{xz}$ and $\sigma_{yz}$ were identified with velocity components. Although $\psi$ can be arbitrary as far as equilibrium conditions are concerned, the stress system (18) must satisfy the boundary conditions (5) and (6), and the compatibility conditions. From Eq. (7.3:6), we see that compatibility requires that (in the absence of body force),

$$\nabla^2\sigma_{yz} = 0; \qquad \nabla^2\sigma_{zx} = 0,$$

where $\nabla^2$ denotes $\left(\frac{\partial^2}{\partial x^2} + \frac{\partial^2}{\partial y^2}\right)$. Hence,

$$\frac{\partial}{\partial x}\nabla^2\psi = 0, \qquad \frac{\partial}{\partial y}\nabla^2\psi = 0. \tag{19}$$

It follows that

$$\nabla^2\psi = \text{const.} \tag{20}$$

Of the boundary conditions (5), only the last equation is not identically satisfied. If we note from Fig. 7.5:2, that

$$\nu_x = \cos(\nu, x) = \frac{dy}{ds}, \qquad \nu_y = \cos(\nu, y) = -\frac{dx}{ds}, \tag{21}$$

we can write the last of Eq. (5) as

$$\frac{\partial \psi}{\partial y}\frac{dy}{ds} + \frac{\partial \psi}{\partial x}\frac{dx}{ds} = \frac{d\psi}{ds} = 0, \quad \text{on } C. \tag{22}$$

Hence $\psi$ must be a constant along the boundary curve $C$. For a simply connected region, no loss of generality is involved in setting

$$\psi = 0, \quad \text{on } C. \tag{23}$$

If the cross section occupies a region $R$ that is multi-connected, additional conditions of compatibility must be imposed (see Sec. 4.7).

It remains to examine the boundary conditions (6). The first, $\sigma_{zz} = 0$, follows the starting assumption. The other conditions are stated in Eqs. (11) and (12). Now,

$$\iint_R \sigma_{zx}\, dx\, dy = \iint_R \frac{\partial \psi}{\partial y}\, dx\, dy.$$

By Gauss' theorem, this is $\int_C \psi \nu_y\, ds$, and it vanishes on account of (23). Similarly, the resultant force in the $y$-direction vanishes. Thus, Eqs. (11) are satisfied. Finally, Eq. (12) requires that

$$T = -\iint_R \left( x \frac{\partial \psi}{\partial x} + y \frac{\partial \psi}{\partial y} \right) dx\, dy,$$

which can be transformed by Gauss' theorem as follows:

$$\begin{aligned} T &= -\iint_R \left\{ \frac{\partial}{\partial x}(x\psi) + \frac{\partial}{\partial y}(y\psi) - 2\psi \right\} dx\, dy \\ &= -\int_C \{x\psi \cos(\nu, x) + y\psi \cos(\nu, y)\}\, ds + \iint_R 2\psi\, dx\, dy. \end{aligned} \tag{24}$$

If $R$ is a simply connected region, the line integral vanishes by the boundary condition (23). Hence,

$$T = 2 \iint_R \psi\, dx\, dy. \tag{24a}$$

Thus, all differential equations and boundary conditions concerning stresses are satisfied if $\psi$ obeys Eqs. (20), (23), and (24). But there remains an indeterminate constant in Eq. (20). This constant has to be determined by boundary conditions on displacements. We have, from Eqs. (3) and (7),

$$\frac{\partial w}{\partial x} = \frac{\sigma_{zx}}{G} + \alpha y, \qquad \frac{\partial w}{\partial y} = \frac{\sigma_{zy}}{G} - \alpha x. \tag{25}$$

Differentiating with respect to $y$ and $x$, respectively, and subtracting, we get

$$\frac{1}{G}\left( \frac{\partial \sigma_{zx}}{\partial y} - \frac{\partial \sigma_{zy}}{\partial x} \right) = -2\alpha. \tag{26}$$

Hence, a substitution from (18) gives

$$\frac{\partial^2 \psi}{\partial x^2} + \frac{\partial^2 \psi}{\partial y^2} = -2G\alpha. \tag{27}$$

In this way, the problem of torsion is reduced to the solution of the Poisson Eq. (27) with boundary condition (23).

With either of the two approaches outlined above, the problem of torsion is reduced to standard problems in the theory of potentials in two dimensions. Such potential problems occur also in the theory of hydrodynamics, gravitation, static electricity, steady flow of heat, etc. A great deal is known about these potential problems; many special solutions have been worked out in detail and general methods of solution are available. The most powerful tool for potential theory in two dimensions comes from the theory of functions of a complex variable. Since this branch of applied mathematics is probably well-known to the reader in his study of other branches of physics, we shall not elaborate further. In any case, a detailed account of this beautiful field will require a book in its own name, and indeed many excellent books exist (see, for example, Courant,[5.1] Courant and Hilbert,[10.1] Kellogg,[5.1] etc.) The complex variable method and the associated singular integral equations approach are developed in the monumental works of Muskhelishvili;[1.2] a shorter account is given by Sokolnikoff.[1.2] Other methods of solution and detailed examples can be found in the classical books[1.2] of Love, Sokolnikoff, Southwell, Timoshenko and Goodier, Treffitz, Sechler, Green and Zerna. Goodier's article in Flügge's[1.4] *Handbook of Engineering Mechanics* contains results for various cross sections commonly used in engineering.

*Example. Bars with Elliptical Cross Section*

Let the boundary of the cross section (Fig. 7.5:3) be given by the equation

$$\frac{x^2}{a^2} + \frac{y^2}{b^2} - 1 = 0.$$

Then Eqs. (27) and (23) are satisfied by

$$\psi = -\frac{a^2 b^2 G\alpha}{(a^2 + b^2)}\left(\frac{x^2}{a^2} + \frac{y^2}{b^2} - 1\right).$$

Equation (24) gives the relation between the torque and the rate of twist $\alpha = d\theta/dz$.

$$T = \frac{\pi a^3 b^3}{(a^2 + b^2)} G \frac{d\theta}{dz}.$$

The stresses are given by Eq. (18). Note that the curves $\psi(x, y) =$ const. have an interesting meaning. The slope $dy/dx$ of the tangent to such a curve is determined by the formula

$$\frac{\partial \psi}{\partial x} + \frac{\partial \psi}{\partial y}\frac{dy}{dx} = 0.$$

Hence, according to Eq. (18), we have

$$\frac{dy}{dx} = \frac{\sigma_{zy}}{\sigma_{zx}}. \tag{28}$$

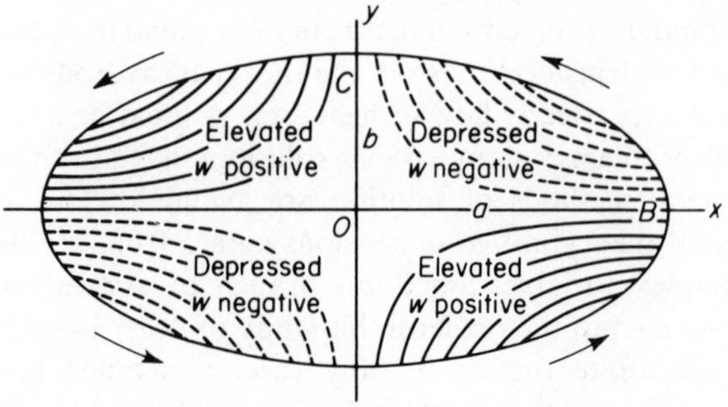


**Fig. 7.5:3.** Lines of constant warping, according to St. Venant.

Thus, at each point of the curve $\psi(x, y) =$ const., the stress vector $(\sigma_{zx}, \sigma_{zy})$ is directed along the tangent to the curve. The curve $\psi(x, y) =$ const. are called the *lines of shearing stress*. The magnitude of the tangential stress is

$$\tau = \sqrt{\sigma_{zx}^2 + \sigma_{zy}^2} = \sqrt{\left(\frac{\partial\psi}{\partial x}\right)^2 + \left(\frac{\partial\psi}{\partial y}\right)^2}. \tag{29}$$

Hence, $\tau$ is proportional to the absolute value of the gradient of the surface $z = \psi(x, y)$. The maximum stress occurs where the gradient is the largest.

In the present example, the lines of shearing stress are concentric ellipses. It is easy to see that the spacing of the $\psi =$ const. lines are closest at the end of the minor axis. The maximum shearing stress occurs there and is given by Eq. (29) to be

$$\tau_{\max} = 2G\alpha \frac{a^2 b}{a^2 + b^2}.$$

A general theorem can be proved that the points at which the maximum shearing stress occurs lie on the boundary curve of the cross section.

The warping function $\varphi(x, y)$ is easily shown to be

$$\varphi = -\frac{a^2 - b^2}{a^2 + b^2} xy.$$

Contour lines of constant displacement along the $z$-axis, $w = \alpha\varphi(x, y) =$ const., are hyperbolas as shown in Figs. 7.5:3, which were taken from St. Venant's original publication. The solid lines in the figure indicate where $w$ is positive, the dotted lines where $w$ is negative.

## 7.6. SOAP FILM ANALOGY

As remarked before, Eqs. (7.5:8) and (7.5:27) occur in many other physical theories entirely unrelated to the torsion problem. For example, if we consider a thin film of liquid, such as that of a soap bubble, we see that the predominant force acting in the film is the surface tension, which may be considered to be constant. The equation of equilibrium of an element of soap film is

$$\frac{T}{R_1} + \frac{T}{R_2} = p,$$

where $R_1$, $R_2$ are the principal radii of curvature and $p$ is the pressure per unit area normal to the film. The derivation of this equation is simple and can be understood from elementary considerations as illustrated in Fig. 7.6:1. Now if we take a tube whose cross-sectional shape is the same as that of the shaft whose torsional property is questioned, cut a plane section, spread a

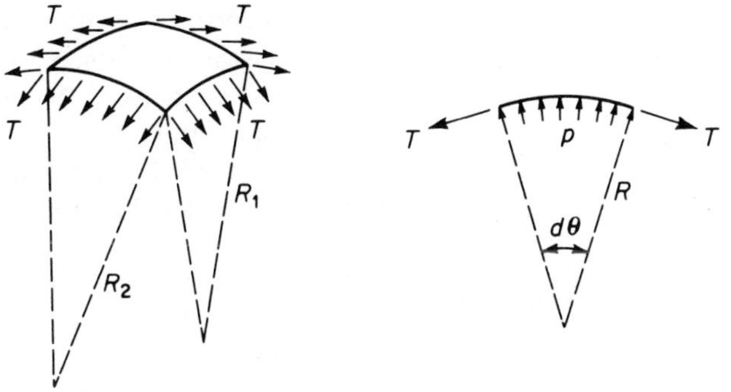


**Fig. 7.6:1.** Equilibrium of a soap film.

soap film over it under a small pressure $p$. If the film deflection is sufficiently small, the mean curvature of the film is given by the sum of the second derivatives of the deflection surface. Thus,

$$T\left(\frac{\partial^2 w}{\partial x^2} + \frac{\partial^2 w}{\partial y^2}\right) = p, \qquad w = 0 \quad \text{on boundary},$$

where $w$ denotes the deflection of the film measured from the plane of the cross section and $x$, $y$ are a set of rectangular coordinates. These equations are identical with Eqs. (7.5:23) and (7.5:27). Thus, we obtain Prandtl's *soap film analogy*. The gradient of the soap film is proportional to the shear stress in torsion. The volume under the film and above the cross section is proportional to the total torque.

The value of an analogy lies in its power of suggestion. Most people can visualize the shape of a soap film, perhaps because of their long experience with it. Thus, with the soap film analogy it is very easy to explain the stress concentration at an re-entrant corner in a cross section, such as the points marked by $P$ in Fig. 7.6:2.

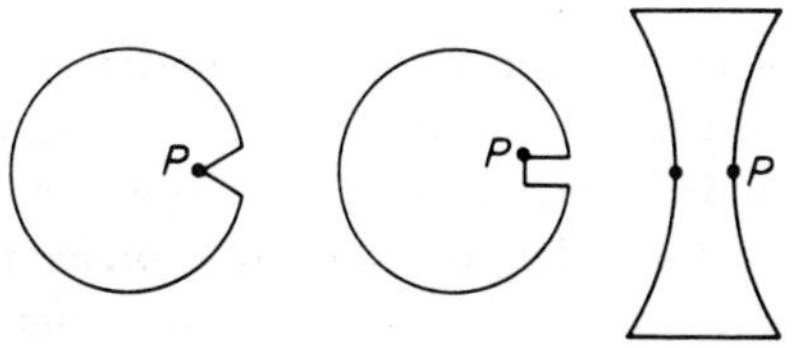


**Fig. 7.6:2.** Re-entrant corners suggest stress concentration.

Economical and efficient use of materials to transmit forces is an important objective in engineering; and the problem of avoiding the weakest links—points where stress concentrations occur—is obviously of great interest.

### 7.7. BENDING OF BEAMS

When a shaft is used to transmit bending moments and transverse shear, it is called a *beam*. Since beams are used in every engineering structure, the theory of beams is of great importance. The long history of the development of man's understanding of the action of the beam is a fascinating subject well-recorded in Timoshenko's book.[1.1] Modern investigation began with Galileo, but it is again to the credit of Saint-Venant that the problem is solved within the general theory of elasticity.

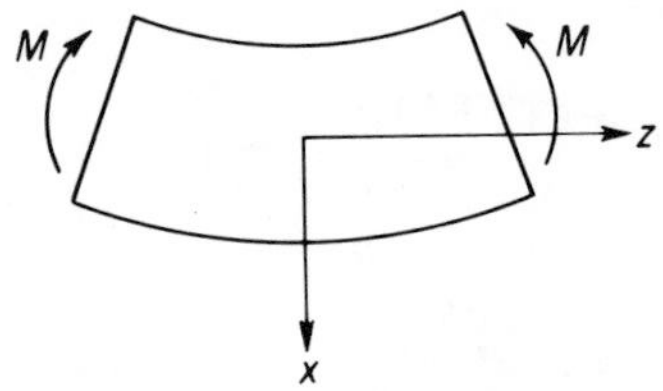

**Fig. 7.7:1.** Pure bending of a prismatic beam.

Consider first the pure bending of a prismatic beam (Fig. 7.7:1). Let the beam be subjected to two equal and opposite couples $M$ acting in one of its principal planes.† Let the origin of the coordinates be taken at the centroid of a cross section, and let the $x$, $z$-plane be the principal plane of bending. The usual elementary theory of bending assumes that the stress components are

$$\sigma_{zz} = \frac{Ex}{R}, \qquad \sigma_{xx} = \sigma_{yy} = \sigma_{xy} = \sigma_{xz} = \sigma_{yz} = 0, \tag{1}$$

in which $R$ is the radius of curvature of the beam after bending. It is easily verified that the stress system (1) satisfies all the equations of equilibrium (7.5:4) and compatibility (7.3:6). The boundary conditions on the lateral surface of the beam are also satisfied. If the normal stress at the ends of the beam is linearly distributed as in Eq. (1), and if the bending moment is

$$M = \iint \sigma_{zz} x \, dx \, dy = \iint \frac{1}{R} Ex^2 \, dx \, dy = \frac{EI}{R}, \tag{2}$$

where $I$ is the moment of inertia of the cross section of the beam with respect to the neutral axis parallel to the $y$-axis, then every condition is satisfied, and Eq. (1) gives an exact solution.

Consider next the same prismatic beam loaded by a lateral force $P$ at the end $z = 0$, and clamped at the end $z = L$ (Fig. 7.7:2). For a beam so

† A principal plane is one that contains the principal axes of the moment of area of the cross sections of the beam.

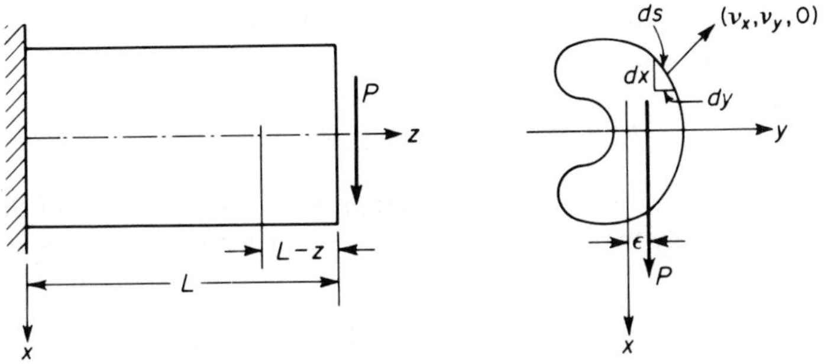


**Fig. 7.7:2.** Cantilever beam loaded at one end.

loaded, the resultant shear $P$ has to be resisted by the shearing stresses $\sigma_{zx}, \sigma_{zy}$. Hence, the stress system (1) will not suffice.

Let the force $P$ be parallel to one of the principal axes of the cross section of the beam. (An arbitrary force can be resolved into components parallel to the principal axes, and the action of each component may be considered separately.) Let $P$ be parallel to the $x$-axis and let the $x, z$-plane be a principal plane. Saint-Venant, using his semi-inverse method, assumes that

$$\sigma_{zz} = -\frac{P(L - z)x}{I}, \qquad \sigma_{xx} = \sigma_{yy} = \sigma_{xy} = 0, \tag{3}$$

leaving $\sigma_{zx}, \sigma_{zy}$ undetermined. The first term is given by the elementary theory, $P(L - z)$ being the bending moment at section $z$. Now we must see how the equilibrium, compatibility, and boundary conditions can be satisfied. In the absence of body force, the equilibrium equations (7.5:4) require that

$$\frac{\partial \sigma_{zx}}{\partial z} = 0, \qquad \frac{\partial \sigma_{yz}}{\partial z} = 0, \tag{4}$$

$$\frac{\partial \sigma_{xz}}{\partial x} + \frac{\partial \sigma_{yz}}{\partial y} + \frac{Px}{I} = 0. \tag{5}$$

The Beltrami-Michell compatibility conditions (7.3:6) require that

$$\nabla^2 \sigma_{yz} = 0, \qquad \nabla^2 \sigma_{xz} + \frac{1}{1 + \nu} \frac{P}{I} = 0, \tag{6}$$

where $\nabla^2 = \partial^2/\partial x^2 + \partial^2/\partial y^2$. The stress-free boundary condition on the lateral surface requires that [Eq. (7.5:5)]

$$\sigma_{zx} \cos(\nu, x) + \sigma_{zy} \cos(\nu, y) = 0, \quad \text{on } C. \tag{7}$$

The end condition at $z = L$ requires that

$$\sigma_{zz} = 0, \qquad \iint \sigma_{zy}\, dx\, dy = 0, \tag{8}$$

$$\iint \sigma_{zx}\, dx\, dy = P. \tag{9}$$

The end condition at $z = 0$ is concerned with the conditions of clamping, and it is usually stated in the form

$$u = \frac{\partial u}{\partial z} = 0 \quad \text{at} \quad x = y = z = 0. \tag{10}$$

The method of solving Eqs. (4) through (9) is similar to that of Sec. 7.5. Equations (4) imply that both $\sigma_{xz}$ and $\sigma_{yz}$ are independent of $z$. Equation (5) may be written as

$$\frac{\partial}{\partial x}\left(\sigma_{xz} + \frac{Px^2}{2I} - f(y)\right) + \frac{\partial \sigma_{yz}}{\partial y} = 0, \tag{11}$$

where $f(y)$ is a function of $y$ only. Equation (11) can be satisfied identically if the stresses $\sigma_{xz}$, $\sigma_{yz}$ are derived from a stress function $\psi(x, y)$ such that

$$\sigma_{xz} = \frac{\partial \psi}{\partial y} - \frac{Px^2}{2I} + f(y), \qquad \sigma_{yz} = -\frac{\partial \psi}{\partial x}. \tag{12}$$

Equations (6) imply that

$$\frac{\partial}{\partial x}(\nabla^2 \psi) = 0, \qquad \frac{\partial}{\partial y}(\nabla^2 \psi) = \frac{\nu}{1 + \nu}\frac{P}{I} - \frac{d^2 f}{dy^2}. \tag{13}$$

Hence,

$$\frac{\partial^2 \psi}{\partial x^2} + \frac{\partial^2 \psi}{\partial y^2} = \frac{\nu}{1 + \nu}\frac{Py}{I} - \frac{df}{dy} + c. \tag{14}$$

The integration constant $c$ has a very simple physical meaning. Consider the rotation $\omega$ of an element of area in the plane of a cross section.

$$\omega = \frac{1}{2}\left(\frac{\partial v}{\partial x} - \frac{\partial u}{\partial y}\right). \tag{15}$$

The rate of change of this rotation in the $z$-axis direction is

$$\begin{aligned} \frac{\partial \omega}{\partial z} &= \frac{1}{2}\frac{\partial}{\partial z}\left(\frac{\partial v}{\partial x} - \frac{\partial u}{\partial y}\right) \\ &= \frac{1}{2}\frac{\partial}{\partial x}\left(\frac{\partial v}{\partial z} + \frac{\partial w}{\partial y}\right) - \frac{1}{2}\frac{\partial}{\partial y}\left(\frac{\partial u}{\partial z} + \frac{\partial w}{\partial x}\right) = \frac{\partial e_{yz}}{\partial x} - \frac{\partial e_{xz}}{\partial y} \\ &= \frac{1}{2G}\left(\frac{\partial \sigma_{yz}}{\partial x} - \frac{\partial \sigma_{xz}}{\partial y}\right) = -\frac{1}{2G}\left(\frac{\partial^2 \psi}{\partial x^2} + \frac{\partial^2 \psi}{\partial y^2} + \frac{df}{dy}\right). \end{aligned} \tag{16}$$

In deriving the last line, Hooke's law and Eq. (12) are used. Hence, Eq. (14) leads to

$$-2G\frac{\partial \omega}{\partial z} = \frac{\nu}{1 + \nu}\frac{Py}{I} + c. \tag{17}$$

This shows that $c$ represents a constant rate of rotation, i.e., it corresponds to a rigid-body rotation of a cross section (the same kind as in the torsion problem). It can be shown that by a proper shifting of the load $P$ parallel to itself in the plane $z = L$, the torsional deformation can be eliminated so that $c = 0$. (This leads to the concept of *shear center*, the point through which $P$ must act so that $c = 0$.) In the following discussion we shall assume that $P$ acts through the shear center. The more general problem can be solved obviously by a linear superposition of the solutions of bending and of torsion.

On setting $c = 0$, Eq. (14) becomes

$$\frac{\partial^2 \psi}{\partial x^2} + \frac{\partial^2 \psi}{\partial y^2} = \frac{\nu}{1+\nu}\frac{Py}{I} - \frac{df}{dy}. \tag{18}$$

The boundary condition (7) now requires that [see Fig. 7.5:2 and Eqs. (7.5:21)]

$$\frac{\partial \psi}{\partial y}\frac{dy}{ds} + \frac{\partial \psi}{\partial x}\frac{dx}{ds} = \frac{\partial \psi}{\partial s} = \left[\frac{Px^2}{2I} - f(y)\right]\frac{dy}{ds}. \tag{19}$$

From these equations, the value of $\psi$ can be determined up to an integration constant which does not contribute anything to the stress system. The function $f(y)$ is arbitrary; it was introduced by Timoshenko to simplify the solution in case the boundary curve of the cross section can be written in the form

$$C: \quad \frac{Px^2}{2I} - f(y) = 0. \tag{20}$$

This would be the case, for example, if $C$ is a circle or an ellipse. In such a case, we choose $f(y)$ according to Eq. (20). Then the boundary condition may be written as

$$\psi = 0 \quad \text{on } C. \tag{21}$$

It remains to show that the load at the end $z = L$ is equipollent to a shear $P$, i.e. that Eqs. (8) and (9) are satisfied. This is easily done. For example, using Eqs. (5) and (7), we have

$$\begin{aligned}
\iint \sigma_{xz}\,dx\,dy &= \iint \left[\sigma_{xz} + x\left(\frac{\partial \sigma_{xz}}{\partial x} + \frac{\partial \sigma_{yz}}{\partial y}\right) + \frac{Px^2}{I}\right] dx\,dy \\
&= P + \int_C x[\sigma_{xz}\cos(\nu, x) + \sigma_{yz}\cos(\nu, y)]\,ds = P, \\
\iint \sigma_{yz}\,dx\,dy &= \iint \left[\sigma_{yz} + y\left(\frac{\partial \sigma_{xz}}{\partial x} + \frac{\partial \sigma_{yz}}{\partial y}\right) + \frac{Pxy}{I}\right] dx\,dy \\
&= \iint y[\sigma_{xz}\cos(\nu, x) + \sigma_{yz}\cos(\nu, y)]\,ds = 0,
\end{aligned}$$

since $\iint xy\,dx\,dy$ vanishes because the $x$-axis is assumed to be a principal axis.

Thus, all the equations are satisfied if a solution $\psi(x, y)$ is found from Eqs. (18) and (19) or (21). This reduces the beam problem to a standard problem in two-dimensional potential theory.

If a solution is obtained and the moment of the shearing stresses is computed and set equal to $P\epsilon$

$$\iint (x\sigma_{yz} - y\sigma_{xz})\, dx\, dy = P\epsilon, \tag{22}$$

the constant $\epsilon$ will give the location of the shear center. The applied load must have an eccentricity $\epsilon$ in order to obtain a bending without twisting (see discussion about the constant $c$ above).

*Example. Beam of Circular Cross Section of Radius r*

The boundary curve $C$ is given by the equation

$$x^2 + y^2 = r^2.$$

Hence, we take

$$f(y) = \frac{P}{2I}(r^2 - y^2).$$

Equation (18) becomes

$$\frac{\partial^2\psi}{\partial x^2} + \frac{\partial^2\psi}{\partial y^2} = \frac{1+2\nu}{1+\nu}\frac{Py}{I}.$$

It is easily verified that the solution that satisfies the boundary condition (21) is

$$\psi = -\frac{(1+2\nu)}{8(1+\nu)}\frac{P}{I}(x^2 + y^2 - r^2)y.$$

The stress components are

$$\sigma_{xz} = \frac{(3+2\nu)}{8(1+\nu)}\frac{P}{I}\left(r^2 - x^2 - \frac{1-2\nu}{3+2\nu}y^2\right),$$

$$\sigma_{yz} = -\frac{(1+2\nu)}{4(1+\nu)}\frac{Pxy}{I}.$$

A large number of examples can be found in the books[1.2] of Love, Sokolnikoff, Timoshenko, etc. An extensive discussion of the shear center can be found in Sechler's and Sokolnikoff's books.[1.2] There are several possible definitions of shear center and center of twist [Trefftz (1935), Goodier (1944) and Weinstein (1947)]; an outline and comparison of various definitions can be found in Fung's *Aeroelasticity*[10.5] (1957) pp. 471–475. See Bibliography 7.2, p. 491.

## 7.8. PLANE ELASTIC WAVES

As a further illustration of the theory of elasticity, let us consider some simple types of waves in an elastic medium. The displacement components $u_1, u_2, u_3$ (or in unabridged notations, $u, v, w$) will be assumed to be infinitesimal so that all equations are linearized. The basic field equation, in the absence of body force, is Navier's Eq. (7.1:9)

$$\rho\frac{\partial^2 u_i}{\partial t^2} = Gu_{i,jj} + (\lambda + G)u_{j,ji}. \tag{1}$$

We shall first verify that the motion

$$(2) \qquad u = A \sin \frac{2\pi}{l}(x \pm ct), \qquad v = w = 0,$$

where $A$, $l$, $c$ are constants, is possible if $c$ assumes the special value $c_L$,

$$(3) \ ▲ \qquad c_L = \sqrt{\frac{\lambda + 2G}{\rho}} = \sqrt{\frac{E(1 - \nu)}{(1 + \nu)(1 - 2\nu)\rho}}.$$

This can be verified at once by substituting Eqs. (2) into (1). The pattern of motion expressed by Eqs. (2) is unchanged when $x \pm c_L t$ remains constant. Hence, if the negative sign were taken, the pattern would move to the right with a velocity $c_L$ as the time $t$ increases. The constant $c_L$ is called the *phase velocity* of the wave motion. In Eqs. (2), $l$ is the *wave length*, as can be seen from the sinusoidal pattern of $u$ as a function of $x$, at any instant of time. The particle velocity represented by Eqs. (2) is in the same direction as that of the wave propagation (namely, the $x$-axis). Such a motion is said to constitute a train of *longitudinal waves*. Since at any instant of time the wave crests lie in parallel planes, the motion represented by (2) is called a train of *plane waves*.

Next, let us consider the motion

$$(4) \qquad u = 0, \qquad v = A \sin \frac{2\pi}{l}(x \pm ct), \qquad w = 0,$$

which represents a train of plane waves of wave length $l$ propagating in the $x$-axis direction with a phase velocity $c$. When Eqs. (4) are substituted into Eqs. (1), it is seen that $c$ must assume the value $c_T$,

$$(5) \ ▲ \qquad c_T = \sqrt{\frac{G}{\rho}}.$$

The particle velocity (in the $y$-direction) represented by Eqs. (4) is perpendicular to the direction of wave propagation ($x$-direction). Hence, it is said to be a *transverse wave*. The speeds $c_L$ and $c_T$ are called the characteristic *longitudinal wave speed* and *transverse wave speed*, respectively. They depend on the elastic constants and the density of the material. The ratio $c_T/c_L$ depends on Poisson's ratio,

$$(6) \ ▲ \qquad c_T = c_L \sqrt{\frac{1 - 2\nu}{2(1 - \nu)}}.$$

If $\nu = 0.25$, then $c_L = \sqrt{3}\, c_T$.

Similar to Eqs. (4), the following example represents a transverse wave in which the particles move in the $z$-axis direction.

$$(7) \qquad u = 0, \qquad v = 0, \qquad w = A \sin \frac{2\pi}{l}(x \pm c_T t).$$

The plane parallel to which the particles move [such as the $x, y$-plane in Eqs. (4), or the $x, z$-plane in Eqs. (7)], is called the *plane of polarization*.

Table 6.2:1, p. 131, gives a brief list of the longitudinal wave velocities of some common media. It is very interesting to see that most metals and alloys have approximately the same wave velocities.

Plane waves as described above may exist only in an unbounded elastic continuum. In a finite body, a plane wave will be reflected when it hits a boundary. If there is another elastic medium beyond the boundary, refracted waves occur in the second medium. The features of reflection and refraction are similar to those in acoustics and optics; the main difference is that, in general, an incident longitudinal wave will be reflected and refracted in a combination of longitudinal and transverse waves, and an incident transverse wave will also be reflected in a combination of both types of waves. The details can be worked out by a proper combination of these waves so that the boundary conditions are satisfied. See Sec. 8.14.

## 7.9. RAYLEIGH SURFACE WAVES

In an elastic body, it is possible to have another type of wave, which is propagated over the surface and which penetrates only a little into the interior of the body. These waves are similar to waves produced on a smooth surface of water when a stone is thrown into it. They are called *surface waves*. The simplest is the *Rayleigh wave* that occurs on the free surface of a homogeneous, isotropic, semi-infinite solid. It is an important type of wave because the largest disturbances caused by an earthquake recorded on a distant seismogram are usually those of Rayleigh waves.

The criterion for surface waves is that the amplitude of the displacement in the medium diminishes exponentially with increasing distance from the boundary.

Let us demonstrate the existence of Rayleigh waves in the simple two-dimensional case. Consider an elastic half-space $y \geqslant 0$. The surface $y = 0$ is stress-free. Let us consider displacements represented by the real part of the following expressions:

$$\begin{aligned} u &= A\, e^{-by} \exp\,[ik(x - ct)], \\ v &= B\, e^{-by} \exp\,[ik(x - ct)], \\ w &= 0, \end{aligned} \tag{1}$$

where $i$ is the imaginary unit $\sqrt{-1}$ and $A$ and $B$ are complex constants. The coefficient $b$ is supposed to be real and positive so that the amplitude of the waves decreases exponentially with increasing $y$, and tends to zero as $y \to \infty$.

We would first see if the displacements given by (1) can satisfy the equations of motion, which, in view of the definitions of $c_L$ and $c_T$ by Eqs. (7.8:3) and (7.8:5), can be written as

$$\frac{\partial^2 u_i}{\partial t^2} = c_T^2 u_{i,jj} + (c_L^2 - c_T^2) u_{j,ji}. \tag{2}$$

Substituting (1) into (2), cancelling the common exponential factor and rearranging terms, we obtain the equations

$$\begin{aligned} [c_T^2 b^2 + (c^2 - c_L^2)k^2]A - i(c_L^2 - c_T^2)bkB &= 0, \\ -i(c_L^2 - c_T^2)bkA + [c_L^2 b^2 + (c^2 - c_T^2)k^2]B &= 0. \end{aligned} \tag{3}$$

The condition for the existence of a nontrivial solution is the vanishing of the determinant of the coefficients, which may be written in the form

$$[c_L^2 b^2 - (c_L^2 - c^2)k^2][c_T^2 b^2 - (c_T^2 - c^2)k^2] = 0. \tag{4}$$

This gives the following roots for $b$.

$$b' = k\left(1 - \frac{c^2}{c_L^2}\right)^{1/2}, \qquad b'' = k\left(1 - \frac{c^2}{c_T^2}\right)^{1/2}. \tag{5}$$

The assumption that $b$ is real requires that $c < c_T < c_L$. Corresponding to $b'$ and $b''$, respectively, the ratio $B/A$ can be solved from Eq. (3).

$$\left(\frac{B}{A}\right)' = -\frac{b'}{ik}, \qquad \left(\frac{B}{A}\right)'' = \frac{ik}{b''}. \tag{6}$$

Hence, a general solution of the type (1), satisfying the equations of motion, may be written as

$$\begin{aligned} u &= A'e^{-b'y} \exp[ik(x - ct)] + A''e^{-b''y} \exp[ik(x - ct)], \\ v &= -\frac{b'}{ik} A'e^{-b'y} \exp[ik(x - ct)] + \frac{ik}{b''} A''e^{-b''y} \exp[ik(x - ct)], \\ w &= 0. \end{aligned} \tag{7}$$

Now, we would like to show that the constants $A'$, $A''$, $k$, and $c$ can be so chosen that the boundary conditions on the free surface $y = 0$ can be satisfied:

$$\sigma_{yx} = \sigma_{yy} = \sigma_{yz} = 0, \qquad \text{on } y = 0. \tag{8}$$

By Hooke's law, and in view of (7), conditions (8) are equivalent to

$$\begin{aligned} \frac{\partial u}{\partial y} + \frac{\partial v}{\partial x} &= 0, & \text{on } y = 0, \\ \lambda\left(\frac{\partial u}{\partial x} + \frac{\partial v}{\partial y}\right) + 2G\frac{\partial v}{\partial y} &= 0, & \text{on } y = 0. \end{aligned} \tag{9}$$

On substituting Eqs. (7) into (9), setting $y = 0$, omitting the common factor $\exp[ik(x - ct)]$, and writing

$$G = \rho c_T^2, \qquad \lambda = \rho(c_L^2 - 2c_T^2),$$

we obtain the results

$$(10)\qquad \begin{aligned} -2b'A' - \left(b'' + \frac{k^2}{b''}\right)A'' &= 0, \\ \left((c_L^2 - 2c_T^2) - c_L^2\frac{b'^2}{k^2}\right)A' - 2c_T^2A'' &= 0. \end{aligned}$$

This can be written more symmetrically, by Eqs. (5), as

$$(11)\qquad \begin{aligned} 2b'A' + \left(2 - \frac{c^2}{c_T^2}\right)k^2\frac{A''}{b''} &= 0, \\ \left(2 - \frac{c^2}{c_T^2}\right)A' + 2b''\frac{A''}{b''} &= 0. \end{aligned}$$

For a nontrivial solution, the determinant of the coefficients of $A'$, $A''$ must vanish, yielding the characteristic equation for $c$

$$(12)\qquad \left(2 - \frac{c^2}{c_T^2}\right)^2 = 4\left(1 - \frac{c^2}{c_L^2}\right)^{1/2}\left(1 - \frac{c^2}{c_T^2}\right)^{1/2}.$$

The quantity $c^2/c_T^2$ can be factored out after rationalization, and Eq. (12), called the *Rayleigh equation*, takes the form

$$(13)\qquad \frac{c^2}{c_T^2}\left[\frac{c^6}{c_T^6} - 8\frac{c^4}{c_T^4} + c^2\left(\frac{24}{c_T^2} - \frac{16}{c_L^2}\right) - 16\left(1 - \frac{c_T^2}{c_L^2}\right)\right] = 0.$$

If $c = 0$, Eqs. (7) are independent of time, and from (11) we have $A'' = -A'$ and $u = v = 0$. Hence, this solution is of no interest. The second factor in (13) is negative for $c = 0$, $c_T < c_L$, and is positive for $c = c_T$. There is always a root $c$ of Eq. (13) in the range $(0, c_T)$. Hence, surface waves can exist with a speed less than $c_T$.

For an incompressible solid $c_L \to \infty$, Eq. (13) becomes

$$(14)\qquad \frac{c^6}{c_T^6} - 8\frac{c^4}{c_T^4} + 24\frac{c^2}{c_T^2} - 16 = 0.$$

This cubic equation in $c^2$ has a real root at $c^2 = 0.91275c_T^2$, corresponding to surface waves with speed $c \simeq 0.95538c_T$. The other two roots for this case are complex and do not represent surface waves.

If the Poisson ratio is $\frac{1}{4}$, so that $\lambda = G$ and $c_L = \sqrt{3}\,c_T$, Eq. (13) becomes

$$(15)\qquad \frac{c^6}{c_T^6} - 8\frac{c^4}{c_T^4} + \frac{56}{3}\frac{c^2}{c_T^2} - \frac{32}{3} = 0.$$

This equation has three real roots: $c^2/c_T^2 = 4,\ 2 + 2/\sqrt{3},\ 2 - 2/\sqrt{3}$. The last root alone can satisfy the condition that $b'$ and $b''$ be real for surface waves. The other two roots correspond to complex $b'$ and $b''$ and do not represent surface waves. In fact, these extraneous roots do not satisfy Eq. (12); they

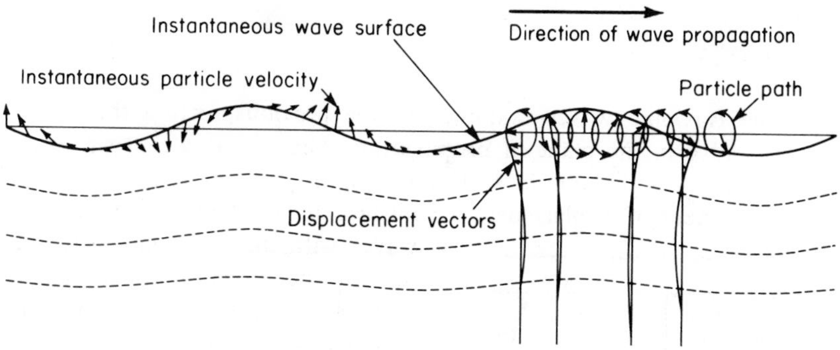


**Fig. 7.9:1.** Schematic drawing for Rayleigh surface waves.

arise from the rationalization process of squaring. The last root corresponds to the velocity

$$c_R = 0.9194 c_T, \tag{16}$$

which corresponds, in turn, to the displacements

$$\begin{aligned} u &= A'(e^{-0.8475ky} - 0.5773\, e^{-0.3933ky}) \cos k(x - c_R t), \\ v &= A'(-0.8475\, e^{-0.8475ky} + 1.4679\, e^{-0.3933ky}) \sin k(x - c_R t), \end{aligned} \tag{17}$$

where $A'$ is taken to be a real number, which may depend on $k$. From (17) it is seen the particle motion for Rayleigh waves is elliptical retrograde in contrast to the elliptical direct orbit for surface waves on water (see Fig. 7.9:1). The vertical displacement is about 1.5 times the horizontal displacement at the surface. Horizontal motion vanishes at a depth of 0.192 of a wavelength and reverses sign below this.

Figure 7.9:2 shows Knopoff's calculated results for the ratios $c_R/c_T$, $c_R/c_L$ for Rayleigh waves as functions of Poisson's ratio $\nu$.

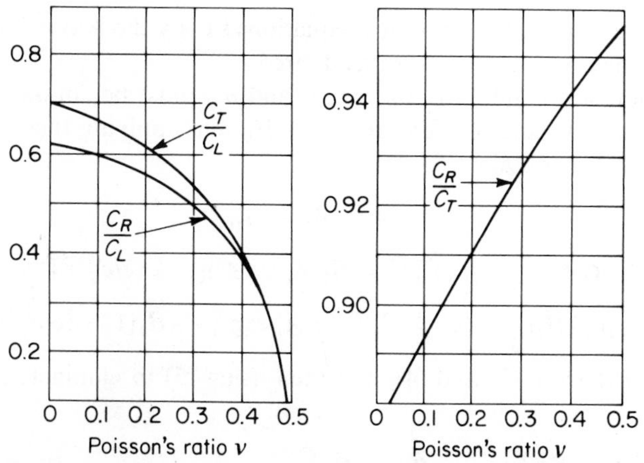


**Fig. 7.9:2.** Ratios $c_R/c_L$, $c_R/c_T$, and $c_T/c_L$, where $c_R$ = Rayleigh wave speed, $c_L$ = the longitudinal wave speed, $c_T$ = the transverse wave speed.

## 7.10. LOVE WAVES

In the Rayleigh waves examined in the previous section the material particles move in the plane of propagation. Thus, in Rayleigh waves over the half-space $y \geqslant 0$ along the surface $y = 0$, propagating in the $x$-direction, the $z$-component of displacement $w$ vanishes. It may be shown that surface waves with displacements perpendicular to the direction of propagation (the so-called *SH waves*) is impossible in a homogeneous half-space. However, *SH* surface waves are observed as prominently on the Earth's surface as other surface waves. Love showed that a theory sufficient to include *SH* surface waves can be constructed by having a homogeneous layer of a medium $M_1$ of uniform thickness $H_1$, overlying a homogeneous half-space of another medium $M$.

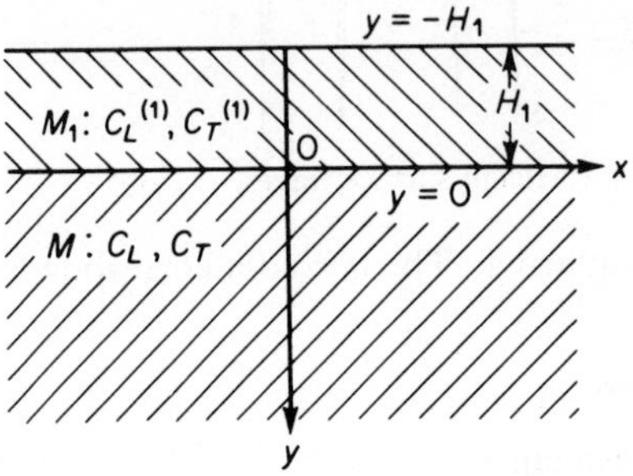


**Fig. 7.10:1.** A layered half-space.

Using axes as in Fig. 7.10:1, we take $u = v = 0$, and

$$(1) \qquad w = A \exp\left[-k\sqrt{1 - \frac{c^2}{c_T^2}}\, y\right] \exp\left[ik(x - ct)\right]$$

in $M$, and

$$(2) \qquad w = \left\{A_1 \exp\left[-k\sqrt{1 - \left(\frac{c}{c_T^{(1)}}\right)^2}\, y\right] + A_1' \exp\left[k\sqrt{1 - \left(\frac{c}{c_T^{(1)}}\right)^2}\, y\right]\right\} \times \exp\left[ik(x - ct)\right]$$

in $M_1$. It is easily verified that these equations satisfy the Navier's equations. If $c < c_T$, then $w \to 0$ as $y \to \infty$, as desired.

The boundary conditions are that $w$ and $\sigma_{zy}$ must be continuous across the surface $y = 0$, and $\sigma_{zy}$ zero at $y = -H_1$. On applying these conditions to (1) and (2), we obtain

$$(3) \qquad A = A_1 + A_1',$$

$$(4) \qquad GA[1 - (c/c_T)^2]^{1/2} = G_1(A_1 - A_1')[1 - (c/c_T^{(1)})^2]^{1/2},$$

$$(5) \qquad A_1 \exp\{kH_1[1 - (c/c_T^{(1)})^2]^{1/2}\} = A_1' \exp\{-kH_1[1 - (c/c_T^{(1)})^2]^{1/2}\}.$$

Eliminating $A$ from (3) and (4), and then using (5) to eliminate $A_1$ and $A_1'$, we have

$$\frac{G[1 - (c/c_T)^2]^{1/2}}{G_1[1 - (c/c_T^{(1)})^2]^{1/2}} = \frac{A_1 - A_1'}{A_1 + A_1'} = i \tan\{ikH_1[1 - (c/c_T^{(1)})^2]^{1/2}\}.$$

Hence, we have

$$(6) \qquad G[1-(c/c_T)^2]^{1/2} - G_1[(c/c_T^{(1)})^2 - 1]^{1/2} \tan \{kH_1[(c/c_T^{(1)})^2 - 1]^{1/2}\} = 0$$

as the equation to give the *SH* surface wave velocity $c$ in the present conditions.

If $c_T^{(1)} < c_T$, Eq. (6) yields a real value of $c$ which lies in the range $c_T^{(1)} < c < c_T$ and depends on $k$ and $H_1$ (as well as on $G$, $G_1$, $c_T$, and $c_T^{(1)}$), because for $c$ in this range the values of the left-hand-side terms in (6) are real and opposite in sign. Thus, *SH* surface waves can occur under the stated boundary conditions, provided the shear velocity $c_T^{(1)}$ in the upper layer is less than that in the medium $M$. These waves are called *Love waves.*

Love waves of general shape may be derived by superposing harmonic Love waves of the type (2) with different $k$. The dependence of the wave speed $c$ on the wave number $k$ introduces a dispersion phenomenon which will be considered later.

## PROBLEMS

**7.2.** Derive Navier's equation in spherical polar coordinates.

**7.3.** From data given in various handbooks, determine the longitudinal and shear wave speeds in the following materials:

(a) Gases: air at sea level, and at 100,000 ft altitude.

(b) Metals: iron, a carbon steel, a stainless steel, copper, bronze, brass, nickle, aluminium, an aluminium alloy, titanium, titanium carbide, berylium, berylium oxide.

(c) Rocks and soils: a granite, a sandy loam.

(d) Wood: spruce, mahogany, balsa.

(e) Plastics: lucite, a foam rubber.

**7.4.** Sketch the instantaneous wave surface, particle velocities, and particle paths of a Love wave.

**7.5.** Investigate plane wave propagations in an anisotropic elastic material. Apply the results to a cubic crystal. *Note:*

$$\rho \frac{\partial^2 u_i}{\partial t^2} = C_{ijkl} \frac{\partial u_l}{\partial x_j \partial x_k}, \qquad u_l = A_l\, e^{-i(\omega t - k_j x_j)}$$

where $\mathbf{k}(k_1, k_2, k_2)$ is the wave vector normal to the wave front.

**7.6.** Determine the stress field in a rotating, gravitating sphere of uniform density.

# 8

# SOLUTION OF PROBLEMS IN ELASTICITY BY POTENTIALS

In this chapter we shall consider the use of the method of potentials in treating static and dynamic problems of an isotropic elastic body subjected to forces acting on the boundary surface of the body, the forces being independent of the deformation of the body. The main problem is determining the stresses and displacements at every point in the body under appropriate boundary conditions. The displacements will be assumed to be infinitesimal and Hooke's law is assumed to hold, so that the basic equations are those listed in Sec. 7.1.

In both static and dynamic problems, we may start with Navier's equation and try to determine a continuous and twice differentiable solution $u_i$ under appropriate boundary conditions. In the case of static equilibrium, we may also start with a general solution of the equation of equilibrium and use the compatibility and boundary conditions to determine the unique answer.

To simplify the solution, a number of potentials have been introduced. The most powerful ones are presented below. Potentials related to displacements are the scalar and vector potentials, the Galerkin vectors, and the Papkovich-Neuber functions. These will be discussed in this chapter. Potentials that generate systems of equilibrating stresses are the Airy stress-functions, the Maxwell-Morera stress functions, etc. (see Chap. 9). Some applications will be illustrated in this and subsequent chapters.

## 8.1. THE SCALAR AND VECTOR POTENTIALS FOR THE DISPLACEMENT VECTOR FIELDS

It is well-known (sometimes referred to as Helmholtz' theorem) that any analytic vector field $\mathbf{u}(u_1, u_2, u_3)$ can be expressed in the form

(1) ▲
$$u_i = \phi_{,i} + e_{ijk}\psi_{k,j}$$

i.e.,

(2a) ▲
$$\mathbf{u} = \text{grad}\ \phi + \text{curl}\ \boldsymbol{\psi}$$

involving three equations of the type

$$u_1 = \frac{\partial \phi}{\partial x_1} + \frac{\partial \psi_3}{\partial x_2} - \frac{\partial \psi_2}{\partial x_3}, \tag{2b}$$

where $\phi$ is a scalar function and $\boldsymbol{\psi}$ is a vector field with three components $\psi_1, \psi_2, \psi_3$. The requirement (1) leaves $\psi_i$ indeterminate to the extent that its divergence is arbitrary. For definiteness, we may impose the condition

$$\psi_{i,i} = 0. \tag{3}$$

Then $\phi$ is called the *scalar potential* and $\boldsymbol{\psi}$ the *vector potential* of the vector field **u**. We shall prove this theorem presently by showing that the functions $\phi$ and $\psi_1, \psi_2, \psi_3$ so proposed can be found.

Applying Eq. (1) to the elastic displacement vector $u_i$, we shall see that the dilatation $e$ can be derived from the scalar potential $\phi$ and that the rotation can be derived from the vector potential $\boldsymbol{\psi}$. Note first that if (1) is effected, then

$$u_{i,i} = \phi_{,ii} + e_{ijk}\psi_{k,ji}. \tag{4}$$

The last term vanishes when summed in pairs since $\psi_{k,ji} = \psi_{k,ij}$, but $e_{ijk} = -e_{jik}$. Hence,

$$e = u_{i,i} = \phi_{,ii}. \tag{5}$$

Furthermore,

$$\begin{aligned} e_{ijk}u_{k,j} &= e_{ijk}\phi_{,kj} + e_{ijk}e_{klm}\psi_{m,lj} \\ &= e_{kij}e_{klm}\psi_{m,lj} && \text{[by symmetry of } \phi_{,kj}] \\ &= (\delta_{il}\delta_{jm} - \delta_{im}\delta_{jl})\psi_{m,lj} && \text{[by the } e\text{-}\delta \text{ identity, Eq. (2.1:11)]} \\ &= -\psi_{i,jj} + \psi_{j,ij} \\ &= -\psi_{i,jj}, && \text{[by virtue of Eq. (3)].} \end{aligned} \tag{6}$$

The curl of $u_i$, given by the left-hand side of (6), is twice the rotation vector $\omega_i$:

$$\omega_i = \tfrac{1}{2}e_{ijk}u_{k,j}.$$

Hence, we may write

$$\psi_{i,jj} = -2\omega_i. \tag{7}$$

This shows that each component of the vector $\psi_i$ is related to a component of the rotation vector field.

Now, particular solutions of (5) and (7) are known in the theory of Newtonian potentials. They are

$$\phi(x_1, x_2, x_3) = -\frac{1}{4\pi}\iiint \frac{e(x_1', x_2', x_3')}{r}\, dx_1'\, dx_2'\, dx_3', \tag{8}$$

$$\bar{\psi}_i(x_1, x_2, x_3) = \frac{1}{2\pi}\iiint \frac{\omega_i(x_1', x_2', x_3')}{r}\, dx_1'\, dx_2'\, dx_3', \tag{9}$$

where

$$r^2 = (x_1 - x_1')^2 + (x_2 - x_2')^2 + (x_3 - x_3')^2$$

and the integration extends over the entire body. But $\bar{\psi}_i$ given in (9) does not always satisfy Eq. (3). To overcome this difficulty, we note that the curl of the gradient of a scalar function vanishes, and a general vector potential is

$$\psi_i = \bar{\psi}_i + \theta_{,i}, \tag{10}$$

where $\theta$ is a scalar function. The added term $\theta_{,i}$ contributes nothing to $u_i$. However, $\theta$ can be determined in such a manner that

$$\theta_{,ii} = -\bar{\psi}_{i,i} \tag{11}$$

by an integral analogous to (8). With $\theta$ so determined, the function $\psi_i$ given by (10) will satisfy the requirement (3).

The general solution of Eq. (5) is the sum of (8) and an arbitrary harmonic function $\phi^*$. The general solution of Eq. (7) is the sum of (10) and a set of harmonic functions $\psi_i^*$. Thus, the equation (1), considered as a set of partial differential equations for the unknown functions $\phi$ and $\psi$, can be solved. Q.E.D.

Love shows (Mathematical Theory of Elasticity,[1.2] p. 48), through integration by parts of (8) and (9), that the following resolution meets all the potentials requirements:

$$\phi = -\frac{1}{4\pi}\iiint u_i(x_1', x_2', x_3')\frac{\partial}{\partial x_i}\left(\frac{1}{r}\right) dx_1'\, dx_2'\, dx_3', \tag{12}$$

$$\psi_i = \frac{1}{4\pi}\iiint e_{ijk}u_k(x_1', x_2', x_3')\frac{\partial}{\partial x_j}\left(\frac{1}{r}\right) dx_1'\, dx_2'\, dx_3'. \tag{13}$$

Pendse[8.1] (1948) pointed out, however, that the use of scalar and vector potentials is not always free from ambiguity.

## 8.2. EQUATIONS OF MOTION IN TERMS OF DISPLACEMENT POTENTIALS

Let the displacement vector field $u_1, u_2, u_3$ be represented by a scalar potential $\phi(x_1, x_2, x_3, t)$ and a triple of vector potentials $\psi_i(x_1, x_2, x_3, t)$, $i = 1, 2, 3$, so that

$$u_i = \frac{\partial \phi}{\partial x_i} + e_{ijk}\frac{\partial \psi_k}{\partial x_j}, \qquad \psi_{i,i} = 0. \tag{1}$$

We shall assume that the body force is zero. [If this is not the case, the body force term can be removed by a particular integral such as Kelvin's, see Eq.

(8.8:9). See also Sec. 8.18.] We can write the Navier's equation of linear elasticity for a homogeneous, isotropic body,

$$\rho \frac{\partial^2 u_i}{\partial t^2} = G u_{i,jj} + (\lambda + G) u_{j,ji}, \tag{2a}$$

in the form

$$\rho \frac{\partial}{\partial x_i}\left(\frac{\partial^2 \phi}{\partial t^2}\right) + \rho e_{ijk} \frac{\partial}{\partial x_j}\left(\frac{\partial^2 \psi_k}{\partial t^2}\right) = (\lambda + G) \frac{\partial}{\partial x_i} \nabla^2 \phi + G \frac{\partial}{\partial x_i} \nabla^2 \phi + G e_{ijk} \frac{\partial}{\partial x_j} \nabla^2 \psi_k, \tag{2}$$

where $\nabla^2$ is the Laplacian operator

$$\nabla^2 = \frac{\partial^2}{\partial x_1^2} + \frac{\partial^2}{\partial x_2^2} + \frac{\partial^2}{\partial x_3^2}.$$

The fact that $\psi_{i,i} = 0$, [Eq. (8.1:3)], is used here. Furthermore, since the displacements are assumed to be infinitesimal, any change in the density $\rho$ is an infinitesimal of the first order, as can be seen easily from the equation of the continuity [Eq. (5.4:4)]. Since in Eq. (2a) $\rho$ is multiplied by small quantities of the first order, $u_i$, it can be treated as a constant. Now, it is easy to see that Eq. (2) will be satisfied if the functions $\phi$ and $\psi_i$ are solutions of the equations

$$\blacktriangle \qquad \nabla^2 \phi = \frac{1}{c_L^2} \frac{\partial^2 \phi}{\partial t^2}, \qquad \nabla^2 \psi_k = \frac{1}{c_T^2} \frac{\partial^2 \psi_k}{\partial t^2}, \tag{3}$$

where

$$c_L = \sqrt{\frac{\lambda + 2G}{\rho}}, \qquad c_T = \sqrt{\frac{G}{\rho}}. \tag{4}$$

These are wave equations. They indicate that two types of disturbances with velocities $c_L$ and $c_T$ may be propagated through an elastic solid. A comparison with Eqs. (7.8:3) and (7.8:5) shows that $c_L$ and $c_T$ are the speeds of plane longitudinal and transverse waves, respectively. These waves are also referred to as *dilatational* and *distorsional* waves. The latter are also known as *shear*, *equivoluminal*, or *rotational* waves. If we write $e$ for the dilatation $u_{i,i}$ and $\boldsymbol{\omega}(\omega_1, \omega_2, \omega_3)$ for the rotation, then, by Eqs. (8.1:5) and (8.1:7), we have, from (3),

$$\nabla^2 e = \frac{1}{c_L^2} \frac{\partial^2 e}{\partial t^2}, \qquad \nabla^2 \omega_i = \frac{1}{c_T^2} \frac{\partial^2 \omega_i}{\partial t^2}. \tag{5}$$

Thus, by introducing the potentials $\phi$ and $\psi_1, \psi_2, \psi_3$, we have reduced the problems of linear elasticity to that of solving the wave equations.

If static equilibrium is considered, then all derivatives with respect to time vanish, and we see that Eq. (2) is satisfied if

$$\nabla^2\phi = \text{const.}, \qquad \nabla^2\psi_i = \text{const.} \tag{6}$$

In the following sections we shall consider some applications to statics based on Eqs. (6).

Of course, all solutions of Eq. (2) are not given by (3) and (6). We shall now consider a more general solution. Such a generalization often becomes important when we wish to examine the limiting case of a dynamic problem as the loading tends to a steady state. For example, if we consider a load suddenly applied at $t = 0$ on a body and then kept constant afterwards, we may question whether the steady-state solution is given by Eqs. (3) with $\partial\phi/\partial t$, $\partial\psi_k/\partial t$ terms set to zero. A quick comparison with Eq. (6) shows that this may not be the case if the constants in Eq. (6) do not vanish.

Let us differentiate Eq. (2) with respect to $x_m$.

$$\frac{\partial^2}{\partial x_m\,\partial x_i}\left(\rho\,\frac{\partial^2\phi}{\partial t^2}\right) + e_{ijk}\frac{\partial^2}{\partial x_m\,\partial x_j}\left(\rho\,\frac{\partial^2\psi_k}{\partial t^2}\right) \tag{7}$$

$$= (\lambda + G)\frac{\partial^2}{\partial x_i\,\partial x_m}\nabla^2\phi + G\frac{\partial^2}{\partial x_i\,\partial x_m}\nabla^2\phi + Ge_{ijk}\frac{\partial^2}{\partial x_m\,\partial x_j}\nabla^2\psi_k.$$

A contraction of $m$ with $i$ yields

$$\nabla^2\left(\nabla^2\phi - \frac{1}{c_L^2}\frac{\partial^2\phi}{\partial t^2}\right) = 0. \tag{8}$$

A multiplication of (7) with $e_{iml}$, summing over $m$ and summing over $i$, and using Eq. (8.1:3), yields, on the other hand,

$$\nabla^2\left(\nabla^2\psi_k - \frac{1}{c_T^2}\frac{\partial^2\psi_k}{\partial t^2}\right) = 0. \tag{9}$$

Hence,

$$\nabla^2\phi - \frac{1}{c_L^2}\frac{\partial^2\phi}{\partial t^2} = \Phi, \qquad \nabla^2\Phi = 0, \tag{10 ▲}$$

$$\nabla^2\psi_k - \frac{1}{c_T^2}\frac{\partial^2\psi_k}{\partial t^2} = \Psi_k, \qquad \nabla^2\Psi_k = 0, \tag{11 ▲}$$

where $\Phi$, $\Psi_1$, $\Psi_2$, $\Psi_3$ are harmonic functions. The $\Phi$ and $\Psi$'s are not entirely independent, but are connected through Eq. (2).

$$\frac{c_L^2}{c_T^2}\frac{\partial\Phi}{\partial x_i} + e_{ijk}\frac{\partial}{\partial x_j}\Psi_k = 0. \tag{12 ▲}$$

Equations (10)–(12) give the general solution of (2). Equations (3) are obtained by setting $\Phi = \Psi_k = 0$; Eqs. (6), by taking $\Phi$ and $\Psi_k$ to be constants and by letting time derivatives vanish.

## 8.3. STRAIN POTENTIAL

In this section we shall consider the static equilibrium of elastic bodies in the restricted case in which the components of displacement can be derived from a scalar function $\phi(x_1, x_2, x_3)$ so that

$$2Gu_i = \frac{\partial \phi}{\partial x_i}. \tag{1}$$

The function $\phi$ is called Lamé's *strain potential.* With Eq. (1), the dilatation $e$, the strain tensor $e_{ij}$, and the stress tensor $\sigma_{ij}$ are

$$2Ge = 2Gu_{i,i} = \phi_{,ii}, \tag{2}$$

$$e_{ij} = \frac{1}{2}(u_{i,j} + u_{j,i}) = \frac{1}{2G}\phi_{,ij}, \tag{3}$$

$$\sigma_{ij} = \lambda\delta_{ij}e + 2Ge_{ij} = \lambda\delta_{ij}e + \phi_{,ij}. \tag{4}$$

In the absence of body force, the function $\phi$ must satisfy Eqs. (8.2:6). Since our objective is to obtain some solution, not necessarily general, the constant in (8.2:6) will be chosen as zero. Then $\phi$ satisfies the Laplace equation

$$\nabla^2\phi = 0. \tag{6}$$

Hence, it is a *harmonic function* by definition. Under this choice, we have

$$\sigma_{ij} = \phi_{,ij}. \tag{7}$$

In unabridged notations with respect to rectangular Cartesian coordinates $x, y, z$, Eqs. (2)–(7) are

$$2Gu = \frac{\partial \phi}{\partial x}, \qquad 2Gv = \frac{\partial \phi}{\partial y}, \qquad 2Gw = \frac{\partial \phi}{\partial z}, \tag{8}$$

$$2Ge = \nabla^2\phi = \left(\frac{\partial^2}{\partial x^2} + \frac{\partial^2}{\partial y^2} + \frac{\partial^2}{\partial z^2}\right)\phi = 0, \tag{9}$$

$$\sigma_{xx} = \frac{\partial^2 \phi}{\partial x^2}, \qquad \sigma_{xy} = \frac{\partial^2 \phi}{\partial x\, \partial y}, \qquad \text{etc.} \tag{10}$$

Corresponding formulas for curvilinear coordinates can be obtained by converting Eqs. (1)–(7) into general tensor equations and then specializing. The results for cylindrical coordinates $r, \theta, z$ are as follows, where $(\xi_r, \xi_\theta, \xi_z)$

and ($\sigma_{rr}$, $\sigma_{r\theta}$, etc.) denote the physical components of displacement and stresses, respectively.

$$2G\xi_r = \frac{\partial \phi}{\partial r}, \qquad 2G\xi_\theta = \frac{1}{r}\frac{\partial \phi}{\partial \theta}, \qquad 2G\xi_z = \frac{\partial \phi}{\partial z}, \tag{11}$$

$$\nabla^2 \phi \equiv \left(\frac{\partial^2}{\partial r^2} + \frac{1}{r}\frac{\partial}{\partial r} + \frac{1}{r^2}\frac{\partial^2}{\partial \theta^2} + \frac{\partial^2}{\partial z^2}\right)\phi = 0, \tag{12}$$

$$\sigma_{rr} = \frac{\partial^2 \phi}{\partial r^2}, \qquad \sigma_{\theta\theta} = \frac{1}{r}\frac{\partial \phi}{\partial r} + \frac{1}{r^2}\frac{\partial^2 \phi}{\partial \theta^2}, \qquad \sigma_{zz} = \frac{\partial^2 \phi}{\partial z^2}, \tag{13}$$

$$\sigma_{r\theta} = \frac{\partial}{\partial r}\left(\frac{1}{r}\frac{\partial \phi}{\partial \theta}\right) = \frac{\partial^2}{\partial r\,\partial \theta}\left(\frac{\phi}{r}\right), \tag{14}$$

$$\sigma_{\theta z} = \frac{1}{r}\frac{\partial^2 \phi}{\partial \theta\,\partial z}, \qquad \sigma_{zr} = \frac{\partial^2 \phi}{\partial z\,\partial r}. \tag{15}$$

Harmonic functions are well-known. For example, it can be easily verified that the functions

$$A(x^2 - y^2) + 2Bxy, \tag{16}$$

$$Cr^n \cos n\theta, \qquad r^2 = x^2 + y^2, \tag{17}$$

$$C \log (r/a), \qquad r^2 = x^2 + y^2, \tag{18}$$

$$C\theta, \qquad \theta = \tan^{-1}(y/x), \tag{19}$$

$$\frac{C}{R}, \qquad R^2 = x^2 + y^2 + z^2, \tag{20}$$

$$C \log (R + z), \tag{21}$$

$$C \log \left\{\frac{(R_1 + z - c)(R_2 - z - c)}{r^2}\right\}, \qquad R_1^2 = r^2 + (z - c)^2, \quad R_2^2 = r^2 + (z + c)^2, \tag{22}$$

are harmonic functions, and combinations of them may solve some important practical problems in elasticity. In (16)–(19) the coordinate $z$ does not appear. Hence, the component of displacement in the $z$-direction, $\partial\phi/\partial z$, vanishes, and any combination of these functions can only correspond to a plane strain condition. It is obvious that the solutions (17)–(19) may be useful for cylindrical bodies, (20) may be useful for spheres, and (21) and (22) may be useful for bodies of revolution about the $z$-axis.

To the stress systems derived from these potentials, we may add the uniform stress distributions given by

$$\phi = Cr^2 = C(x^2 + y^2) \tag{23}$$

$$\phi = CR^2 = C(x^2 + y^2 + z^2), \tag{24}$$

which are special solutions of Eq. (8.2:6). With (23) or (24), Eqs. (9)–(10) and (12)–(15) must be modified because $e \neq 0$.

*Example. Hollow Spheres Subjected to Internal and External Pressure*

Let a hollow sphere of inner radius $a$ and outer radius $b$ be subjected to an internal pressure $p$ and an external pressure $q$. Because of the spherical symmetry of the problem, it will be advantageous to use spherical coordinates $R, \theta, \varphi$. But it is also possible to use rectangular coordinates and note that at a point $(x, 0, 0)$ on the $x$-axis the radial stress $\sigma_{RR}$ can be identified with the normal stress $\sigma_{xx}$, and the tangential stresses $\sigma_{\theta\theta}$ can be identified with $\sigma_{yy}$. All shearing stresses $\sigma_{R\theta}, \sigma_{\varphi\theta}$, etc., vanish on account of symmetry.

The solution is furnished by the strain potential (20) and (24),

$$\phi = \frac{C}{R} + DR^2, \qquad R^2 = x^2 + y^2 + z^2,$$

from which

$$\frac{\partial^2\phi}{\partial x^2} = \frac{3Cx^2}{R^5} - \frac{C}{R^3} + 2D, \qquad \frac{\partial^2\phi}{\partial y^2} = \frac{3Cy^2}{R^5} - \frac{C}{R^3} + 2D.$$

Hence, at the point $(R, 0, 0)$,

$$\sigma_{RR} = \sigma_{xx} = \frac{2C}{R^3} + 2D,$$

$$\sigma_{\theta\theta} = \sigma_{yy} = -\frac{C}{R^3} + 2D.$$

The constants $C$ and $D$ are easily determined from the boundary conditions,

$$\sigma_{RR} = -p \quad \text{when} \quad R = a, \qquad \sigma_{RR} = -q \quad \text{when} \quad R = b.$$

The results are

$$\sigma_{RR} = -p\frac{(b/R)^3 - 1}{(b/a)^3 - 1} - q\frac{1 - (a/R)^3}{1 - (a/b)^3},$$

$$\sigma_{\theta\theta} = \frac{p}{2}\frac{(b/R)^3 + 2}{(b/a)^3 - 1} - \frac{q}{2}\frac{(a/R)^3 + 2}{1 - (a/b)^3}.$$

**Problem 8.1.** Find the solution for a solid sphere subjected to a uniform external pressure $q$.

**Problem 8.2.** Let the inner wall of the hollow spherical shell, at $R = a$, be rigid (a condition realized by filling the hole with some incompressible material). When the outer surface $R = b$ is subjected to a uniform pressure $q$, what are the stresses at the inner wall?

**Problem 8.3.** Show that the function $\phi = C\theta$ solves the problem of a circular disk subjected to uniformly distributed tangential shear on the circumference.

## 8.4. GALERKIN VECTOR

We have shown in Sec. 8.2 that when the displacement field $u_i$ is expressed in terms of scalar and vector potentials, $\phi$ and $\psi_1, \psi_2, \psi_3$, respectively,

$$(1) \qquad 2Gu_i = \phi_{,i} + e_{ijk}\psi_{k,j}, \qquad \psi_{i,i} = 0$$

the basic equations of static equilibrium for a homogeneous isotropic linear elastic medium are satisfied if

$$\nabla^2\phi = \text{const.}, \qquad \nabla^2\psi_k = \text{const.} \tag{2}$$

Thus, a broad class of problems in elastic equilibrium is solved by determining the four functions $\phi$, $\psi_1$, $\psi_2$, $\psi_3$. However, this may not be the general solution. The general solution (8.2:10)–(8.2:12) is more complex.

In searching for other solutions of equal generality, Galerkin introduced, in papers published in 1930, displacement potential functions which satisfy biharmonic equations. Papkovich noted in 1932 that Galerkin's functions are components of a vector. It is perhaps simpler to introduce the Galerkin vector by considering the vector potential $\psi_k$ itself as being generated by another vector field $\tilde{F}_i$,

$$\psi_k = -e_{klm}c\tilde{F}_{m,l}, \tag{3}$$

where $c$ is a constant. Then Eq. (1) becomes

$$2Gu_i = \phi_{,i} - e_{ijk}e_{klm}c\tilde{F}_{m,lj}. \tag{4}$$

This can be simplified by means of the $e$-$\delta$ identity [Eq. 2.1:11)] into

$$\begin{aligned} 2Gu_i &= \phi_{,i} - (\delta_{il}\delta_{jm} - \delta_{im}\delta_{jl})c\tilde{F}_{m,lj} \\ &= \phi_{,i} + c\tilde{F}_{i,jj} - c\tilde{F}_{j,ji}. \end{aligned} \tag{5}$$

But $c\tilde{F}_{j,j}$ is a scalar function, and can be specified arbitrarily without disturbing the definition (3). This suggests the representation†

$$2Gu_i = cF_{i,jj} - F_{j,ji}. \tag{6}$$

The hope now arises that, by a judicious choice of the constant $c$, the Navier Eq. (7.1:9), namely,

$$(\lambda + G)u_{j,jk} + Gu_{k,mm} + X_k = 0,$$

where $X_k$ represents the body force per unit volume, can be simplified. Indeed, on substituting Eq. (6) into the Navier equation, and remembering that $\lambda + G = G/(1 - 2\nu)$, we obtain

$$\frac{1}{2(1-2\nu)}[cF_{i,jji} - F_{j,jii}]_{,k} + \frac{1}{2}[cF_{k,jj} - F_{j,jk}]_{,mm} + X_k = 0.$$

With a proper change of dummy indices, this is

$$\left[\frac{c-1}{2(1-2\nu)} - \frac{1}{2}\right]F_{j,jiik} + \frac{c}{2}F_{k,jjii} + X_k = 0.$$

The coefficient of the first term vanishes if

$$c = 2(1 - \nu).$$

† In order that (5) and (6) both represent the same displacement field, we may choose $F_i = \tilde{F}_i + H_i$, where $H_i$ are arbitrary harmonic functions ($H_{i,jj} = 0$) and demand that $\tilde{F}_{j,j} = (\phi + H_{j,j})/(c - 1)$.

Therefore, we conclude that the basic equation of elasticity is satisfied when

(7) ▲ $$2Gu_i = 2(1-\nu)F_{i,jj} - F_{j,ji},$$

if $F_i$ satisfies the equation

(8) ▲ $$F_{i,jjmm} = -\frac{X_i}{1-\nu}.$$

The vector $F_i$ so defined is called the *Galerkin vector*. If $X_i = 0$, Eq. (8) is said to be biharmonic and its solutions are called biharmonic functions. Thus, to solve a problem in static equilibrium is to determine the three functions $F_1$, $F_2$, $F_3$.

From (7) we can derive expressions for the stresses in terms of $F_i$. In unabridged notation, we have, with

$$\nabla^2 \equiv \frac{\partial^2}{\partial x^2} + \frac{\partial^2}{\partial y^2} + \frac{\partial^2}{\partial z^2},$$

the following results.

(9) $$\nabla^4 F_1 = -\frac{X_1}{1-\nu},$$

(10) $$2Gu_1 = 2(1-\nu)\nabla^2 F_1 - \frac{\partial}{\partial x_1}\left(\frac{\partial F_1}{\partial x_1} + \frac{\partial F_2}{\partial x_2} + \frac{\partial F_3}{\partial x_3}\right),$$

(11) $$2G\left(\frac{\partial u_1}{\partial x_1} + \frac{\partial u_2}{\partial x_2} + \frac{\partial u_3}{\partial x_3}\right) = (1-2\nu)\nabla^2\left(\frac{\partial F_1}{\partial x_1} + \frac{\partial F_2}{\partial x_2} + \frac{\partial F_3}{\partial x_3}\right),$$

(12) $$\sigma_{11} = 2(1-\nu)\frac{\partial}{\partial x_1}\nabla^2 F_1 + \left(\nu\nabla^2 - \frac{\partial^2}{\partial x_1^2}\right)\left(\frac{\partial F_1}{\partial x_1} + \frac{\partial F_2}{\partial x_2} + \frac{\partial F_3}{\partial x_3}\right),$$

(13) $$\sigma_{12} = (1-\nu)\left(\frac{\partial}{\partial x_2}\nabla^2 F_1 + \frac{\partial}{\partial x_1}\nabla^2 F_2\right) - \frac{\partial^2}{\partial x_1\,\partial x_2}\left(\frac{\partial F_1}{\partial x_1} + \frac{\partial F_2}{\partial x_2} + \frac{\partial F_3}{\partial x_3}\right),$$

(14) $$(\sigma_{11} + \sigma_{22} + \sigma_{33}) = (1+\nu)\nabla^2\left(\frac{\partial F_1}{\partial x_1} + \frac{\partial F_2}{\partial x_2} + \frac{\partial F_3}{\partial x_3}\right).$$

Other components $u_2$, $u_3$, $\sigma_{22}$, $\sigma_{33}$, etc., are obtained by cyclic permutation of the subscripts 1, 2, 3.

**Problem 8.4.** Determine the body forces, stresses, and displacements defined by the following Galerkin vectors in rectangular Cartesian coordinates:

(a) $F_1 = F_2 = 0, \quad F_3 = R^2,$

(b) $F_1 = yR^2, \quad F_2 = -xR^2, \quad F_3 = 0,$

(c) $F_1 = F_2 = 0, \quad F_3 = z^4,$

where $R^2 = x^2 + y^2 + z^2$.

**Problem 8.5.** Show that the Lamé strain potential $\phi$ can be identified with the divergence $-F_{i,i}$, if the body force vanishes and if $F_i$ is harmonic.

## 8.5. EQUIVALENT GALERKIN VECTORS

If two Galerkin vectors $\bar{F}_i$ and $\bar{\bar{F}}_i$ define the same displacements, they are equivalent. Let $\bar{F}_i$ and $\bar{\bar{F}}_i$ be two equivalent Galerkin vectors. Then their difference

$$F_i = \bar{F}_i - \bar{\bar{F}}_i \tag{1}$$

is a biharmonic vector and corresponds to zero displacement, $u_i = 0$. A general form of $F_i$ can be obtained if we set, as in Sec. 8.4,

$$F_i = cf_{i,jj} + f_{j,ji} \tag{2}$$

and determine $c$ such that $u_i = 0$, where

$$2Gu_i = 2(1 - \nu)F_{i,jj} - F_{j,ji}.$$

The result is

$$F_i = (1 - 2\nu)f_{i,jj} + f_{j,ji}, \tag{3}$$

provided that

$$f_{i,jjkk} = 0. \tag{4}$$

As an application, let us consider a Galerkin vector the third component of which does not vanish,

$$\bar{F}_1, \quad \bar{F}_2, \quad \bar{F}_3 \neq 0. \tag{5}$$

It will be shown that an equivalent $\bar{\bar{F}}_i$ with $\bar{\bar{F}}_3 = 0$ can be formed if the body force $X_i = 0$. A particular solution is obtained by taking a special vector $f_i$

$$f_1 = f, \qquad f_2 = f_3 = 0. \tag{6}$$

Then, from (3) and (5),

$$F_1 = \bar{F}_1 - \bar{\bar{F}}_1 = \left[(1 - 2\nu)\nabla^2 + \frac{\partial^2}{\partial x^2}\right]f, \tag{7}$$

$$F_2 = \bar{F}_2 - \bar{\bar{F}}_2 = \frac{\partial^2 f}{\partial x\, \partial y}, \tag{8}$$

$$F_3 = \bar{F}_3 - \bar{\bar{F}}_3 = \bar{F}_3 = \frac{\partial^2 f}{\partial x\, \partial z}, \qquad \bar{\bar{F}}_3 = 0. \tag{9}$$

Now, since $X_i = 0$ implies that $\bar{F}_3$ is biharmonic, it will be possible to determine a biharmonic function $f$ that satisfies Eq. (9). When $f$ is so determined, we could compute $\bar{\bar{F}}_1$ and $\bar{\bar{F}}_2$ according to Eqs. (7) and (8). Thus, when the body forces are zero, any Galerkin vector has an equivalent Galerkin vector with a zero component in the direction of $z$.

## 8.6. EXAMPLE—VERTICAL LOAD ON THE HORIZONTAL SURFACE OF A SEMI-INFINITE SOLID

Consider a semi-infinite homogeneous, isotropic, linear elastic solid which occupies the space $z \geqslant 0$ as shown in Fig. 8.6:1, and subjected to a sinusoidally distributed vertical load on the boundary surface $z = 0$, so that the boundary conditions are

$$(1) \qquad z = 0: \qquad \begin{aligned} &\text{(a)} \quad \sigma_{zz} = \mathscr{A} \cos \frac{\pi x}{l} \cos \frac{\pi y}{L}, \\ &\text{(b)} \quad \sigma_{zx} = \sigma_{zy} = 0. \end{aligned}$$

The boundary condition at $z = \infty$ shall be assumed to be the vanishing of all stress components.

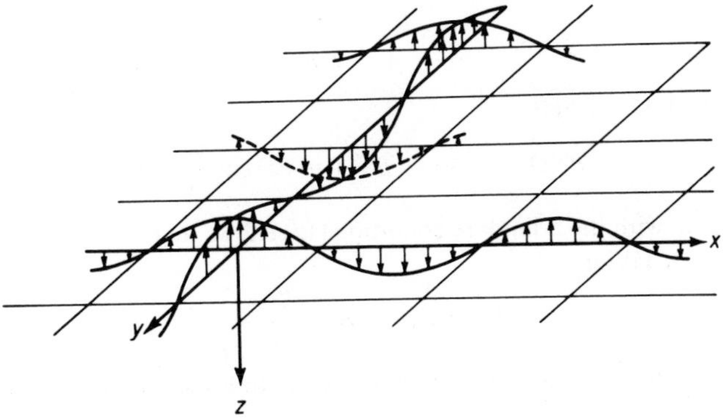


**Fig. 8.6:1.** Sinusoidally distributed load acting on a semi-infinite solid.

The problem can be solved by taking a Galerkin vector with one nonvanishing component,

$$(2) \qquad F_1 = F_2 = 0, \qquad F_3 = Z.$$

The function $Z$ must be biharmonic. It can be easily verified that, if $\psi$ is a harmonic function, then the following function is biharmonic (cf. Sec. 8.11):

$$(3) \qquad Z = (A + Bcz)\psi.$$

The form of the boundary condition (1$a$) suggests that we should try

$$(4) \qquad \psi = \cos \frac{\pi x}{l} \cos \frac{\pi y}{L} f(z).$$

On substituting (4) into the Laplace equation $\nabla^2 \psi = 0$, one finds that

$$f(z) = e^{-cz},$$

where

$$(5) \qquad c = \sqrt{\left(\frac{\pi}{l}\right)^2 + \left(\frac{\pi}{L}\right)^2}.$$

The other solution $e^{cz}$ must be rejected by the boundary condition at infinity.

The stresses corresponding to the Galerkin vector above can be easily derived. Note that

$$
\begin{aligned}
\nabla^2 Z &= 2Bc\frac{\partial \psi}{\partial z} = -2Bc^2\psi, \\
\frac{\partial Z}{\partial z} &= -cA\psi + Bc(1 - cz)\psi, \\
\left(\frac{\partial^2}{\partial x^2} + \frac{\partial^2}{\partial y^2}\right)Z &= -(A + Bcz)c^2\psi.
\end{aligned} \tag{6}
$$

Hence, by Eq. (8.4:13),

$$\sigma_{zx} = \frac{\partial}{\partial x}[2\nu Bc^2 - (A + Bcz)c^2]\psi. \tag{7}$$

The boundary conditions (1b), that $\sigma_{zx}$ and $\sigma_{zy}$ vanish at the surface $z = 0$, are satisfied if

$$A = 2\nu B. \tag{8}$$

Then, Eq. (8.4:12) yields

$$\sigma_{zz} = Bc^3(1 + cz)\psi = Bc^3(1 + cz)\cos\frac{\pi x}{l}\cos\frac{\pi y}{L}e^{-cz}. \tag{9}$$

A comparison with the boundary condition (1a) shows that all boundary conditions are satisfied by taking

$$B = \frac{\mathscr{A}}{c^3}, \qquad A = \frac{2\nu\mathscr{A}}{c^3}. \tag{10}$$

The resulting displacements and stresses are as follows, where we write, for convenience,

$$\alpha = \frac{\pi}{l}, \qquad \beta = \frac{\pi}{L}, \tag{11}$$

$$
\begin{aligned}
u_x &= \frac{\mathscr{A}\alpha}{2Gc^2}(-1 + 2\nu + cz)\sin\alpha x\cos\beta y\, e^{-cz}, \\
u_y &= \frac{\mathscr{A}\beta}{2Gc^2}(-1 + 2\nu + cz)\cos\alpha x\sin\beta y\, e^{-cz}, \\
u_z &= \frac{\mathscr{A}}{2Gc}[2(1 - \nu) + cz]\cos\alpha x\cos\beta y\, e^{-cz}, \\
\sigma_{xx} &= \frac{\mathscr{A}}{c^2}(-\alpha^2 - 2\nu\beta^2 + \alpha^2 cz)\,\psi, \\
\sigma_{yy} &= \frac{\mathscr{A}}{c^2}(-\beta^2 - 2\nu\alpha^2 + \beta^2 cz)\,\psi, \\
\sigma_{zz} &= -\mathscr{A}(1 + cz)\,\psi, \\
\sigma_{xy} &= \frac{\mathscr{A}\alpha\beta}{c^2}(1 - 2\nu - cz)\sin\alpha x\sin\beta y\, e^{-cz}, \\
\sigma_{yz} &= -\mathscr{A}\beta z\cos\alpha x\sin\beta y\, e^{-cz}, \\
\sigma_{zx} &= -\mathscr{A}\alpha z\sin\alpha x\cos\beta y\, e^{-cz}.
\end{aligned} \tag{12}
$$

Thus, stresses are attenuated exponentially as the depth $z$ is increased; the rate of attenuation $c$ depends on the wave lengths $l$ and $L$.

Solutions of this form may be superposed together to produce further solutions. The method of Fourier series may be used to obtain periodic loadings on the surface $z = 0$, and the method of Fourier integral may be used to obtain more general loadings.

## 8.7. LOVE'S STRAIN FUNCTION

A Galerkin vector that has only one nonvanishing component $F_3$ is called *Love's strain function* and shall be denoted by

$$F_3 = Z.$$

In many applications it may be desired to express $Z$ both in rectangular coordinates $x$, $y$, $z$ and in cylindrical coordinates $r$, $\theta$, $z$. In both cases,

$$\nabla^4 Z = -\frac{X_z}{1-\nu},$$

where

$$\nabla^2 \equiv \frac{\partial^2}{\partial x^2} + \frac{\partial^2}{\partial y^2} + \frac{\partial^2}{\partial z^2} = \frac{\partial^2}{\partial r^2} + \frac{1}{r}\frac{\partial}{\partial r} + \frac{1}{r^2}\frac{\partial^2}{\partial \theta^2} + \frac{\partial^2}{\partial z^2}$$

and $X_z$ is the body force per unit volume in the $z$-direction—the only non-vanishing component that can be treated by such a strain function.

On putting $F_1 = F_2 = 0$ and $F_3 = Z$ in Eqs. (8.4:10)–(8.4:14) we obtain the following expressions for the physical components of displacements and stresses:

*In rectangular coordinates:*

$$2Gu_x = -\frac{\partial^2 Z}{\partial x\,\partial z}, \qquad 2Gu_y = -\frac{\partial^2 Z}{\partial y\,\partial z}, \qquad 2Gu_z = \left[2(1-\nu)\nabla^2 - \frac{\partial^2}{\partial z^2}\right] Z,$$

$$\theta = \sigma_{xx} + \sigma_{yy} + \sigma_{zz} = (1+\nu)\frac{\partial \nabla^2 Z}{\partial z},$$

$$\sigma_{xx} = \frac{\partial}{\partial z}\left(\nu\nabla^2 - \frac{\partial^2}{\partial x^2}\right) Z,$$

$$\sigma_{yy} = \frac{\partial}{\partial z}\left(\nu\nabla^2 - \frac{\partial^2}{\partial y^2}\right) Z,$$

$$\sigma_{zz} = \frac{\partial}{\partial z}\left[(2-\nu)\nabla^2 - \frac{\partial^2}{\partial z^2}\right] Z,$$

$$\sigma_{zx} = \frac{\partial}{\partial y}\left[(1-\nu)\nabla^2 - \frac{\partial^2}{\partial z^2}\right] Z, \qquad \sigma_{xy} = -\frac{\partial^3 Z}{\partial x\,\partial y\,\partial z},$$

$$\sigma_{zy} = \frac{\partial}{\partial x}\left[(1-\nu)\nabla^2 - \frac{\partial^2}{\partial z^2}\right] Z.$$

*In cylindrical coordinates:*

$$2G\xi_r = -\frac{\partial^2 Z}{\partial r\,\partial z}, \qquad 2G\xi_\theta = -\frac{1}{r}\frac{\partial^2 Z}{\partial\theta\,\partial z}, \qquad 2G\xi_z = \left[2(1-\nu)\nabla^2 - \frac{\partial^2}{\partial z^2}\right]Z,$$

$$\theta = \sigma_{rr} + \sigma_{\theta\theta} + \sigma_{zz} = (1+\nu)\frac{\partial\,\nabla^2 Z}{\partial z}$$

$$\sigma_{rr} = \frac{\partial}{\partial z}\left(\nu\,\nabla^2 - \frac{\partial^2}{\partial r^2}\right)Z,$$

$$\sigma_{\theta\theta} = \frac{\partial}{\partial z}\left(\nu\,\nabla^2 - \frac{1}{r}\frac{\partial}{\partial r} - \frac{1}{r^2}\frac{\partial}{\partial\theta^2}\right)Z,$$

$$\sigma_{zz} = \frac{\partial}{\partial z}\left[(2-\nu)\nabla^2 - \frac{\partial^2}{\partial z^2}\right]Z,$$

$$\sigma_{r\theta} = -\frac{\partial^3}{\partial r\,\partial\theta\,\partial z}\left(\frac{Z}{r}\right),$$

$$\sigma_{\theta z} = \frac{1}{r}\frac{\partial}{\partial\theta}\left[(1-\nu)\nabla^2 - \frac{\partial^2}{\partial z^2}\right]Z,$$

$$\sigma_{zr} = \frac{\partial}{\partial r}\left[(1-\nu)\nabla^2 - \frac{\partial^2}{\partial z^2}\right]Z.$$

Love introduced, in 1906, the strain function $Z$ as a function of $r$ and $z$ only in treating solids of revolution under axi-symmetric loading.

## 8.8. KELVIN'S PROBLEM—A SINGLE FORCE ACTING IN THE INTERIOR OF AN INFINITE SOLID

Let a force $2P$ be applied at the origin in the direction of $z$ (Fig. 8.8:1). This concentrated force may be regarded as the limit of a system of loads applied on the surface of a small cavity at the origin. The boundary conditions of the problem are: (1) At infinity, all stresses vanish. (2) At the origin, the stress singularity is equivalent to a concentrated force of magnitude $2P$ in the $z$-direction.

The symmetry of the problem suggests the use of cylindrical coordinates, and Love's strain function $Z(r, z)$ suggests itself. Since body force is absent, $Z$ must be a biharmonic function whose third partial derivatives should define stresses that vanish at infinity, but have a singularity at the origin. One such function is

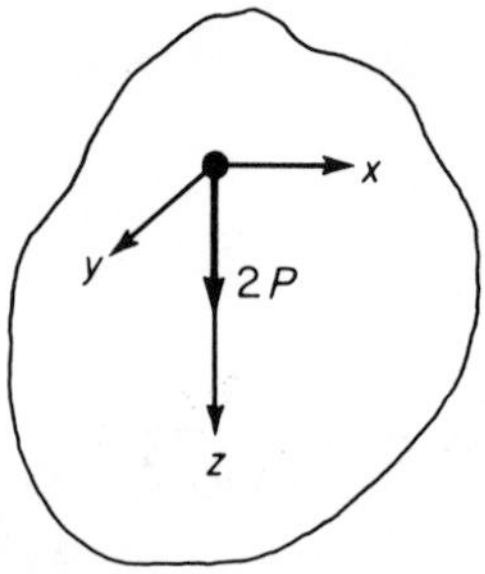


**Fig. 8.8:1.** Kelvin's problem.

(1) $$Z = BR = B(z^2 + r^2)^{1/2}$$

(the function $\frac{1}{R}$ is harmonic, hence $R = R^2 \frac{1}{R}$ is biharmonic see Theorem 2, p. 208), for which

$$\frac{\partial Z}{\partial r} = \frac{Br}{R}, \qquad \frac{\partial^2 Z}{\partial r^2} = B\left(\frac{1}{R} - \frac{r^2}{R^3}\right) = \frac{Bz^2}{R^3},$$

$$\frac{\partial Z}{\partial z} = \frac{Bz}{R}, \qquad \frac{\partial^2 Z}{\partial z^2} = B\left(\frac{1}{R} - \frac{z^2}{R^3}\right) = \frac{Br^2}{R^3},$$

$$\nabla^2 Z = \frac{2B}{R}.$$

Therefore,

$$2G\xi_r = \frac{Brz}{R^3}, \qquad 2G\xi_\theta = 0, \qquad 2G\xi_z = B\left[\frac{2(1-2\nu)}{R} + \frac{1}{R} + \frac{z^2}{R^3}\right],$$

$$\sigma_{rr} = B\left[\frac{(1-2\nu)z}{R^3} - \frac{3r^2 z}{R^5}\right],$$

$$\sigma_{\theta\theta} = \frac{(1-2\nu)Bz}{R^3},$$

$$\sigma_{zz} = -B\left[\frac{(1-2\nu)z}{R^3} + \frac{3z^3}{R^5}\right],$$

$$\sigma_{rz} = -B\left[\frac{(1-2\nu)r}{R^3} + \frac{3rz^2}{R^5}\right],$$

$$\sigma_{z\theta} = \sigma_{r\theta} = 0.$$

These stresses are singular at the origin and vanish at infinity, and they have the correct symmetry. Therefore, the stress singularity can only be equivalent to a vertical force. Thus, (1) is indeed the desired solution if this force can be made equal to $2P$ by a proper choice of the constant $B$. To determine $B$, we consider a cylinder with a cavity at its center at the origin and with bases at $z = \pm a$ (Fig. 8.8:2). Since this cylinder is in equilibrium, the resultant of surface tractions on the outer surface must balance the load $2P$ acting on the surface of the cavity at the origin. Therefore, we must have, when the radius of this cylinder tends to infinity,

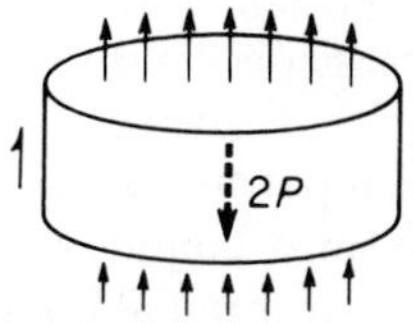


**Fig. 8.8:2.** Boundary used in evaluating the integration constant.

$$2P = \int_0^\infty 2\pi r\, dr\, (-\sigma_{zz})_{z=a} + \int_0^\infty 2\pi r\, dr\, (\sigma_{zz})_{z=-a} + \lim_{r\to\infty} \int_{-a}^{a} 2\pi r\, dz\, (\sigma_{rz}).$$

The values of the first and the second integrals are seen to be the same, and the third integral vanishes in the limit. Noting that $r\,dr = R\,dR$, we have

$$\begin{aligned} 2P &= 2\int_0^\infty 2\pi R\,dR\,(-\sigma_z)_{z=a} \\ &= 4\pi B\left[(1-2\nu)a\int_a^\infty \frac{R\,dR}{R^3} + 3a^3\int_a^\infty \frac{R\,dR}{R^5}\right] \\ &= 8\pi(1-\nu)B. \end{aligned}$$

Hence,

$$B = \frac{P}{4\pi(1-\nu)} \tag{4}$$

and the solution is completed.

An especially simple situation results when the Poisson's ratio is $\frac{1}{2}$. Then the factor $(1 - 2\nu)$ vanishes, and the stresses become

$$\begin{aligned} &\sigma_{rr} = -\frac{3Pr^2z}{2\pi R^5}, \qquad \sigma_{\theta\theta} = 0, \qquad \sigma_{r\theta} = \sigma_{z\theta} = 0, \\ &\sigma_{zz} = -\frac{3Pz^3}{2\pi R^5}, \qquad \sigma_{rz} = -\frac{3Prz^2}{2\pi R^5}. \end{aligned} \tag{5}$$

These results appear even simpler if the stresses are resolved in directions of spherical coordinates. In any meridional plane (see Fig. 8.8:3), at a point whose coordinates are $(r, z)$ in cylindrical and $(R, \varphi)$ in spherical coordinates, we find

$$\begin{aligned} \sigma_{RR} &= \sigma_{rr}\sin^2\varphi + \sigma_{zz}\cos^2\varphi \\ &\quad + 2\sigma_{rz}\sin\varphi\cos\varphi = \sigma_{rr}\frac{r^2}{R^2} \\ &\quad + \sigma_{zz}\frac{z^2}{R^2} + 2\sigma_{rz}\frac{rz}{R^2} = -\frac{3Pz}{2\pi R^3}. \end{aligned} \tag{6}$$

Similarly, we have

$$\sigma_{\varphi\varphi} = \sigma_{R\varphi} = \sigma_{\theta\theta} = 0, \tag{7}$$

and, by symmetry,

$$\sigma_{R\theta} = \sigma_{\varphi\theta} = 0. \tag{8}$$

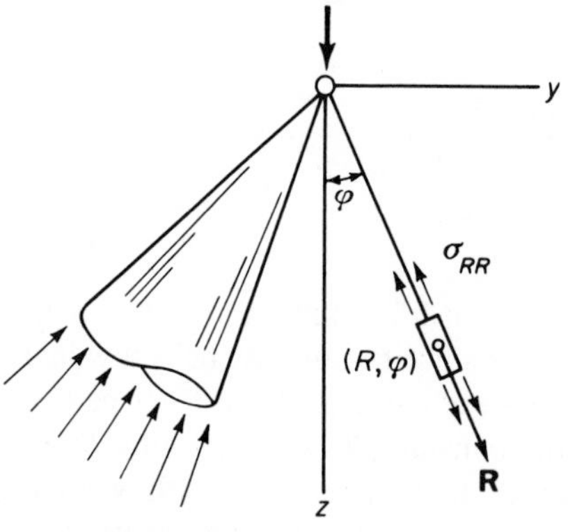


**Fig. 8.8:3.** Simple interpretation of the solution to Kelvin's problem when the Poisson's ratio is 1/2.

Hence, $\sigma_{RR}$, $\sigma_{\varphi\varphi}$, $\sigma_{\theta\theta}$ are principal stresses, and the only nonvanishing component is $\sigma_{RR}$. If the solid were divided into many cones extending from a common vertex at the origin, each of these cones would transmit its own radial force without reaction from the adjacent cones.

In the special case of Poisson's ratio $\nu = \frac{1}{2}$, it is evident that the solution of the Kelvin problem also furnishes the solution to Boussinesq's problem of a normal force and Cerruti's problem of a tangential force acting on the boundary of a semi-infinite solid (see Fig. 8.8:4). When $\nu = \frac{1}{2}$, Eqs. (5)–(8) hold for Boussinesq's problem when the solid occupies the space $z \geqslant 0$, and for Cerruti's problem when the solid occupies the space $x \geqslant 0$.

The observation that the solutions of Boussinesq's and Cerruti's problems take on such simple form when Poisson's ratio is $\frac{1}{2}$ led Westergaard to consider a method of solution for these problems by perturbation of Poisson's ratio, a method to be described in the next section.

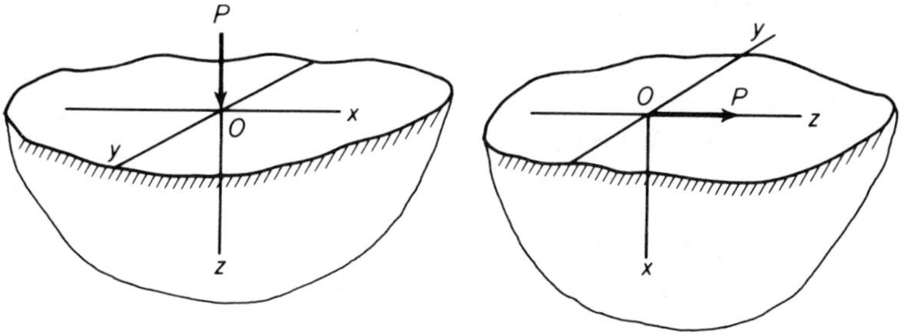


**Fig. 8.8:4.** Boussinesq and Cerruti's problems.

Originally, Lord Kelvin, Boussinesq, and Cerruti obtained solutions to the problems which now are adorned by their names (in 1848, 1878, 1882, respectively) by an extension of the method of singularities in the theory of Newtonian potentials. Lord Kelvin (1848) discovered that a particular solution of Navier's Eq. (7.1:9) is

$$(9) \qquad u_i(x) = A\iiint\left[B\frac{X_i(\xi)}{r} - (x_j - \xi_j)X_j(\xi)\frac{\partial}{\partial x_i}\left(\frac{1}{r}\right)\right] d\xi_1\, d\xi_2\, d\xi_3,$$

where

$$r = [(x_1 - \xi_1)^2 + (x_2 - \xi_2)^2 + (x_3 - \xi_3)^2]^{1/2},$$

$$A = \frac{\lambda + G}{8\pi G(\lambda + 2G)}, \qquad B = \frac{\lambda + 3G}{\lambda + G}.$$

In this formula $x$ stands for $(x_1, x_2, x_3)$, and $\xi$ for $(\xi_1, \xi_2, \xi_3)$. The quantity $r$ is the distance between the field point $x$ to the variable point $\xi$. The functions $X_i(\xi)$ are the components of the body force $X_i$ expressed in terms of the variables of integration $\xi_i$. If an idealized case is considered in which a small sphere is isolated in the body, in which the body force concentrates, the limiting case will provide a solution of the Kelvin problem named above. A combination of such a solution with some other singular solutions of Navier's equation yields solutions to the Boussinesq and Cerruti problems. A general theory for this adaptation was given by Betti (1872), who showed how to express the dilatation and rotation at every point in a solid body by surface integrals containing explicitly the surface tractions and surface displacements. A lucid exposition of this approach can be found in Love, *Elasticity*,[1.2] Chapter X.

The corresponding problems of specific normal and tangential displacements at the origin on a semi-infinite solid were also solved by Boussinesq (1888). There are numerous ways of arriving at these results. For references, see Love,[1.2] p. 243.

**Problem 8.6.** Investigate the system of stresses corresponding to

(a) Strain potential $\phi = \log(R + z)$, $\qquad R^2 = x^2 + y^2 + z^2$.

(b) Love's strain functions $Z = \log(R + z)$,

(c) $Z = x \log(R + z)$.

Find a strain potential that is equivalent to (b).

## 8.9. PERTURBATION OF ELASTICITY SOLUTIONS BY A CHANGE OF POISSON'S RATIO

In many problems of elasticity, the effect of Poisson's ratio is relatively unimportant. In two-dimensional problems without body force, Poisson's ratio has no effect on the stress distribution at all. In fact, this is the basic justification of most photoelasticity studies, in which the selection of materials for the models can be made regardless of elastic constants.

In three-dimensional problems, the effect of Poisson's ratio generally cannot be disregarded, but the examples of Boussinesq's and Cerruti's problems in Sec. 8.8 show that particularly simple solutions can be obtained when Poisson's ratio assumes a certain particular value. This leads one to consider the effect of Poisson's ratio. Westergaard proposes to obtain a solution by first solving a problem for a specific value of Poisson's ratio $m$, and then determining the necessary corrections when the actual value of $\nu$ is used. The value of the shear modulus $G$ is considered fixed.

Consider a body bounded by a surface $S$. Let $u_i$ be the solution of a certain problem, i.e., it satisfies the Navier equation (in rectangular Cartesian coordinates).

(1) $$\frac{1}{1-2\nu} e_{,i} + u_{i,\alpha\alpha} + \frac{X_i}{G} = 0, \qquad e = u_{\alpha,\alpha},$$

and assumes on the boundary $S$ surface traction $\overset{\nu}{T}_i$ and surface displacement $u_i$.

Let $u_i^*$ be a solution of another problem, satisfying, for a different value of Poisson's ratio $m$, but the same value of $G$, the equation

(2) $$\frac{1}{1-2m} e^*_{,i} + u^*_{i,\alpha\alpha} + \frac{X_i^*}{G} = 0, \qquad e^* = u^*_{\alpha,\alpha},$$

$$\overset{\nu}{T}{}_i^*, \quad u_i^* \quad \text{on } S.$$

This problem shall be assumed to have been solved, and is the starting point for further investigation on the effect of variation of Poisson's ratio.

Subtracting (2) from (1), and writing

$$(3) \qquad u_i^{**} = u_i - u_i^*, \qquad e_{ij}^{**} = e_{ij} - e_{ij}^*, \qquad X_i^{**} = X_i - X_i^*,$$

etc., we obtain

$$(4) \qquad \left(\frac{1}{1-2\nu} - \frac{1}{1-2m}\right)e_{,i}^* + \frac{1}{1-2\nu}\,e_{,i}^{**} + u_{i,\alpha\alpha}^{**} + \frac{X_i^{**}}{G} = 0$$

and

$$(5) \qquad \overset{\nu}{T}{}_i^{**} = \overset{\nu}{T}_i - \overset{\nu}{T}{}_i^*, \qquad u_i^{**} = u_i - u_i^* \quad \text{on } S.$$

Equations (4) and (5) define another problem in the linear theory of elastic equilibrium. Any solution of (4) and (5), added to (2), which corresponds to Poisson's ratio $m$, will yield a solution of (1) with the value of Poisson's ratio equal to $\nu$.

To make this concept useful, let us consider a class of problems such as Boussinesq's, in which the boundary conditions take the form of specified surface tractions; e.g.,

$$(6) \qquad \sigma_{33}^*, \sigma_{31}^*, \sigma_{32}^* \quad \text{are specified on } S\colon \quad x_3 = \text{const.}$$

Then it is expedient to consider a problem (4) with

$$(7) \qquad \sigma_{33}^{**} = \sigma_{31}^{**} = \sigma_{32}^{**} = 0 \quad \text{on } S\colon \qquad x_3 = \text{const.}$$

When such a particular solution is found, it can be added to the solution of (2) to form a solution of (1) with the same specified boundary values of $\sigma_{33}, \sigma_{31}, \sigma_{32}$ as in (6).

Westergaard shows how to construct a general solution in which $\sigma_{33}^{**}$, $\sigma_{31}^{**}$, and $\sigma_{32}^{**}$ vanish identically throughout the elastic body. The method is based on the assumption that the components of displacement are derived from a scalar function $\varphi$ according to the rule

$$(8) \qquad 2Gu_1^{**} = \varphi_{,1}, \qquad 2Gu_2^{**} = \varphi_{,2}, \qquad 2Gu_3^{**} = -\varphi_{,3}.$$

Westergaard calls this a "twinned gradient," the third component $-\varphi_{,3}$ being regarded as the "twin" of the ordinary gradient $\varphi_{,3}$. Now

$$(9) \qquad \begin{aligned} \sigma_{ij} &= \frac{2\nu}{1-2\nu} Ge_{\alpha\alpha}\delta_{ij} + 2Ge_{ij}, \\ \sigma_{ij}^* &= \frac{2m}{1-2m} Ge_{\alpha\alpha}^*\delta_{ij} + 2Ge_{ij}^*. \end{aligned}$$

Hence,

$$(10) \qquad \begin{aligned} \sigma_{ij}^{**} &= \sigma_{ij} - \sigma_{ij}^* \\ &= \frac{2G(\nu - m)}{(1-2\nu)(1-2m)}\, e_{\alpha\alpha}^*\delta_{ij} + \frac{2G\nu}{1-2\nu}\, e_{\alpha\alpha}^{**}\delta_{ij} + 2Ge_{ij}^{**}. \end{aligned}$$

From (8), we have

$$e_{23}^{**} \equiv 0, \qquad e_{31}^{**} \equiv 0. \tag{11}$$

Hence,

$$\sigma_{23}^{**} = 0, \qquad \sigma_{31}^{**} = 0. \tag{12}$$

Furthermore,

$$2Ge_{\alpha\alpha}^{**} = \varphi_{,11} + \varphi_{,22} - \varphi_{,33} = \nabla^2\varphi - 2\varphi_{,33}, \tag{13}$$

and, from (10),

$$\sigma_{33}^{**} = \frac{2G(\nu - m)}{(1 - 2\nu)(1 - 2m)}\, e_{\alpha\alpha}^{*} + \frac{\nu}{1 - 2\nu}(\nabla^2\varphi - 2\varphi_{,33}) - \varphi_{,33}. \tag{14}$$

Hence, if we choose to have

$$\sigma_{33}^{**} \equiv 0, \tag{15}$$

then $\varphi$ must satisfy the equation

$$\nu\,\nabla^2\varphi - \varphi_{,33} = -\frac{2G(\nu - m)}{(1 - 2m)}\, e_{\alpha\alpha}^{*}. \tag{16}$$

With $\varphi$ so chosen, Eq. (4) will yield the required value of the body force $X_i^{**}$ in order to keep the body in equilibrium. However, it is simpler to proceed as follows. Compute first the stress $\sigma_{11}^{**}$,

$$\sigma_{11}^{**} = \sigma_{11}^{**} - \sigma_{33}^{**} = 2G(e_{11}^{**} - e_{33}^{**}) = \varphi_{,11} + \varphi_{,33}. \tag{17}$$

Then the only nonvanishing components of stress are

$$\sigma_{11}^{**} = \nabla^2\varphi - \varphi_{,22}, \qquad \sigma_{22}^{**} = \nabla^2\varphi - \varphi_{,11}, \qquad \sigma_{12}^{**} = \varphi_{,12}. \tag{18}$$

The equations of equilibrium expressed in stresses show at once that the components of the required body force are

$$-X_1^{**} = (\nabla^2\varphi)_{,1}, \quad -X_2^{**} = (\nabla^2\varphi)_{,2}. \tag{19}$$

The solution is established if a function $\varphi$ can be found that satisfies (16) and (19) simultaneously.

In a majority of significant problems, both the original and the final body forces are zero; i.e.,

$$X_i = X_i^{*} = X_i^{**} = 0. \tag{20}$$

Then (16) and (19) can be satisfied if

$$\nabla^2\varphi = 0 \quad \text{and} \quad \varphi_{,33} = \frac{2G(\nu - m)}{(1 - 2m)}\, e_{\alpha\alpha}^{*} = \frac{\nu - m}{1 + m}\, \sigma_{\alpha\alpha}^{*}. \tag{21}$$

But $e_{\alpha\alpha}^{*}$ is a harmonic function. Hence, these two equations can be satisfied simultaneously, and the method is established.

The formulas given above are valid as long as $\nu$ and $m$ are not exactly equal to $\frac{1}{2}$. In case $m \to \frac{1}{2}$, the dilatation $e^{*}$ tends to zero, but the bulk modulus tends to infinity. The sum of normal stresses

$$\sigma_{ii}^{*} = \frac{2G(1 + m)}{(1 - 2m)}\, e_{ii}^{*} \tag{22}$$

remains finite as $m \to \frac{1}{2}$. It is easy to show that all the formulas in this section remain valid when $m = \frac{1}{2}$, provided that the $e_{ii}^*$ term is replaced by the $\sigma_{ii}^*$ term according to (22), as in the last equation of (21).

## 8.10. BOUSSINESQ'S PROBLEM

A load $P$ acts at the origin of coordinates and perpendicular to the plane surface of a semi-infinite solid occupying the space $z \geqslant 0$, (Fig. 8.8:4). When Poisson's ratio is $\frac{1}{2}$, the problem has the simple solution as stated in Eqs. (8.8:5) which gives

$$(1)\quad \theta^* = \sigma_{11} + \sigma_{22} + \sigma_{33} = -\frac{3Pz}{2\pi R^3}, \qquad R^2 = z^2 + r^2 = x^2 + y^2 + z^2.$$

If Poisson's ratio is $\nu$, Westergaard's Eq. (8.9:21) becomes, in this case,

$$(2)\qquad \nabla^2\varphi = 0, \qquad \frac{\partial^2\varphi}{\partial z^2} = \frac{(1-2\nu)P}{2\pi}\frac{z}{R^3}.$$

Integrating, we have

$$(3)\qquad \frac{\partial\varphi}{\partial z} = -\frac{(1-2\nu)P}{2\pi R}, \qquad \varphi = -\frac{(1-2\nu)P}{2\pi}\log(R+z).$$

It can be shown that $\varphi$ is harmonic; therefore the problem is solved.

On substituting (3) into Eqs. (8.9:8)–(8.9:14), and adding to the solution for $m = \frac{1}{2}$, we obtain the final results, in which the only nonvanishing components are

$$(4)\qquad \xi_r = \frac{P}{4\pi GR}\left[\frac{rz}{R^2} - \frac{(1-2\nu)r}{R+z}\right],$$

$$(5)\qquad \xi_z = \frac{P}{4\pi GR}\left[2(1-\nu) + \frac{z^2}{R^2}\right],$$

$$(6)\qquad \theta = \sigma_{rr} + \sigma_{\theta\theta} + \sigma_{zz} = -\frac{(1+\nu)}{\pi}P\frac{z}{R^3},$$

$$(7)\qquad \sigma_{rr} = \frac{P}{2\pi R^2}\left[-\frac{3r^2z}{R^3} + \frac{(1-2\nu)R}{R+z}\right],$$

$$(8)\qquad \sigma_{\theta\theta} = \frac{(1-2\nu)P}{2\pi R^2}\left[\frac{z}{R} - \frac{R}{R+z}\right],$$

$$(9)\qquad \sigma_{zz} = -\frac{3Pz^3}{2\pi R^5},$$

$$(10)\qquad \sigma_{rz} = -\frac{3Prz^2}{2\pi R^5}.$$

where $r^2 = x^2 + y^2$, $\quad R^2 = r^2 + z^2 = x^2 + y^2 + z^2$.

**Problem 8.7.** Solve the Boussinesq problem by a combination of a Galerkin vector

$$F_1 = F_2 = 0, \qquad F_3 = BR, \qquad R = (r^2 + z^2)^{1/2}$$

and a Lamé strain potential $\Phi = c \log (R + z)$. Show that $c = -(1 - 2\nu)B$, $B = P/2\pi$.

**Problem 8.8.** Solve Cerruti's problem by the method of "twinned gradient" (Ref., Westergaard,[1.2] p. 142).

## 8.11. ON BIHARMONIC FUNCTIONS

We have seen that many problems in elasticity are reduced to the solution of biharmonic equations with appropriate boundary conditions. It will be useful to consider the mathematical problem in some detail.

We shall consider the equation

$$\nabla^2 \nabla^2 u = 0, \tag{1}$$

where the operator $\nabla^2$ is, in rectangular Cartesian coordinates $(x, y, z)$,

$$\nabla^2 \equiv \frac{\partial^2}{\partial x^2} + \frac{\partial^2}{\partial y^2} + \frac{\partial^2}{\partial z^2}, \tag{2}$$

in cylindrical polar coordinates $(r, \theta, z)$,

$$x = r\cos\theta, \qquad y = r\sin\theta, \qquad z = z,$$

$$\nabla^2 \equiv \frac{\partial^2}{\partial r^2} + \frac{1}{r}\frac{\partial}{\partial r} + \frac{1}{r^2}\frac{\partial^2}{\partial \theta^2} + \frac{\partial^2}{\partial z^2} \tag{3}$$

and, in spherical polar coordinates $(R, \varphi, \theta)$,

$$x = R\sin\varphi\cos\theta, \qquad y = R\sin\varphi\sin\theta, \qquad z = R\cos\varphi,$$

$$\begin{aligned} \nabla^2 &\equiv \frac{1}{R^2}\frac{\partial}{\partial R}\left(R^2\frac{\partial}{\partial R}\right) + \frac{1}{R^2\sin\varphi}\frac{\partial}{\partial\varphi}\left(\sin\varphi\frac{\partial}{\partial\varphi}\right) + \frac{1}{R^2\sin^2\varphi}\frac{\partial^2}{\partial\theta^2} \\ &= \frac{\partial^2}{\partial R^2} + \frac{2}{R}\frac{\partial}{\partial R} + \frac{1}{R^2}\frac{\partial^2}{\partial\varphi^2} + \frac{\cot\varphi}{R^2}\frac{\partial}{\partial\varphi} + \frac{1}{R^2\sin^2\varphi}\frac{\partial^2}{\partial\theta^2}. \end{aligned} \tag{4}$$

The $\nabla^4$ operator is obtained by repeated operation of the above. Thus,

$$\begin{aligned} \nabla^4 &\equiv \left(\frac{\partial^2}{\partial x^2} + \frac{\partial^2}{\partial y^2} + \frac{\partial^2}{\partial z^2}\right)\left(\frac{\partial^2}{\partial x^2} + \frac{\partial^2}{\partial y^2} + \frac{\partial^2}{\partial z^2}\right) \\ &= \frac{\partial^4}{\partial x^4} + \frac{\partial^4}{\partial y^4} + \frac{\partial^4}{\partial z^4} + 2\frac{\partial^4}{\partial x^2\,\partial y^2} + 2\frac{\partial^4}{\partial x^2\,\partial z^2} + 2\frac{\partial^4}{\partial y^2\,\partial z^2}. \end{aligned} \tag{5}$$

A regular solution of Eq. (1) in a region $\mathscr{R}$ is one that is four times continuously differentiable in $\mathscr{R}$. A regular solution of (1) is called a *biharmonic function*. Since Eq. (1) is obtained by repeated operation of the Laplace operator (2), and the regular solution of the equation $\nabla^2 u = 0$ is called a *harmonic function*, it is expected that biharmonic functions are closely connected with harmonic functions. In fact, we have the following theorems due to Almansi.[8.1]

THEOREM 1. *If $u_1$, $u_2$ are two functions, harmonic in a region $\mathscr{R}(x, y, z)$, then*

$$u = xu_1 + u_2 \tag{6}$$

*is biharmonic in $\mathscr{R}$. Conversely, if $u$ is a given biharmonic function in a region $\mathscr{R}$, and if every line parallel to the x-axis intersects the boundary of $\mathscr{R}$ in at most two points, then there exist two harmonic functions $u_1$ and $u_2$ in $\mathscr{R}$, so that $u$ can be represented in the form of Eq.* (6).

*Proof.* The first part of the theorem can be verified directly according to the identity

$$\nabla^2(\phi\psi) = \phi\,\nabla^2\psi + \psi\,\nabla^2\phi + 2\left(\frac{\partial\phi}{\partial x}\frac{\partial\psi}{\partial x} + \frac{\partial\phi}{\partial y}\frac{\partial\psi}{\partial y} + \frac{\partial\phi}{\partial z}\frac{\partial\psi}{\partial z}\right). \tag{7}$$

To prove the converse theorem, we note that the theorem is established if we can show that there exist a function $u_1$ such that

$$\text{(a)}\quad \nabla^2 u_1 = 0, \qquad \text{(b)}\quad \nabla^2(xu_1 - u) = 0. \tag{8}$$

By virtue of (8a), the second equation can be written as

$$\nabla^2 u = \nabla^2(xu_1) = 2\,\frac{\partial u_1}{\partial x}, \tag{9}$$

which has a particular solution

$$\bar{u}_1(x, y, z) = \int_{x_0}^{x} \tfrac{1}{2}\,\nabla^2 u(\xi, y, z)\, d\xi, \tag{10}$$

where $x_0$ is an arbitrary point in the region $\mathscr{R}$. This particular solution does not necessarily satisfy Eq. (8a). However, since $u$ is biharmonic, we have

$$\frac{\partial}{\partial x}\nabla^2\bar{u}_1 = \nabla^2\frac{\partial\bar{u}_1}{\partial x} = \frac{1}{2}\nabla^4 u = 0. \tag{11}$$

Hence $\nabla^2\bar{u}_1$ is a function $v(y, z)$ of the variables $y$, $z$ only. Now let us determine a function $\bar{\bar{u}}_1(y, z)$ so that

$$\left(\frac{\partial^2}{\partial y^2} + \frac{\partial^2}{\partial z^2}\right)\bar{\bar{u}}_1 = -v(y, z); \tag{12}$$

for example, by

$$\bar{\bar{u}}_1(y, z) = -\iint (\log r)\cdot v(\eta, \zeta)\, d\eta\, d\zeta, \tag{13}$$

where $r^2 = (y - \eta)^2 + (z - \zeta)^2$ and the integral extends through the region $\mathscr{R}$. Then the function $u_1 = \bar{u}_1 + \bar{\bar{u}}_1$ satisfies both conditions (9) and (8a) and the theorem is proved.

By a slight change in the proof it can be shown that Theorem 1 holds as well in the two-dimensional case.

Similarly, we have the following

THEOREM 2. *If $u_1, u_2$ are two harmonic functions in a three-dimensional region $\mathscr{R}$, then*

$$u = (R^2 - R_0^2)u_1 + u_2, \tag{14}$$

*is biharmonic in $\mathscr{R}$, where $R^2 = x^2 + y^2 + z^2$ and $R_0$ is an arbitrary constant. Conversely, if $u$ is a given biharmonic function in a region $\mathscr{R}$, and if $\mathscr{R}$ is such that, with an origin inside $\mathscr{R}$, each radius vector intersects the boundary of $\mathscr{R}$ in at most one point, then two harmonic functions $u_1, u_2$ can be determined so that* (14) *holds.*

*Proof.* The proof of the first part again follows by direct calculation. Since $u_1, u_2$ are harmonic, an application of the identity (7) yields

$$\begin{aligned} \nabla^2 u &= u_1 \nabla^2 R^2 + 4\left(x\frac{\partial u_1}{\partial x} + y\frac{\partial u_1}{\partial y} + z\frac{\partial u_1}{\partial z}\right) = 6u_1 + 4R\frac{\partial u_1}{\partial R}, \\ \nabla^2 \nabla^2 u &= 6\nabla^2 u_1 + 8\left(\frac{\partial^2 u_1}{\partial x^2} + \frac{\partial^2 u_1}{\partial y^2} + \frac{\partial^2 u_1}{\partial z^2}\right) = 0. \end{aligned} \tag{15}$$

To prove the converse theorem, we note that the theorem is established if we can determine a function $u_1$ with the properties

$$\text{(a)}\ \ \nabla^2 u_1 = 0, \qquad \text{(b)}\ \ \nabla^2[u - (R^2 - R_0^2)u_1] = 0. \tag{16}$$

Equation (16b) can be simplified, by virtue of (16a), into

$$\nabla^2 u = 6u_1 + 4R\frac{\partial u_1}{\partial R}. \tag{17}$$

An integral of this differential equation is

$$u_1 = R^{-3/2}\int_0^R \tfrac{1}{4}\rho^{1/2}\,\nabla^2 u\,d\rho. \tag{18}$$

It will now be shown that this integral indeed satisfies the condition (16a) and, hence, is the desired function. The demonstration will be simpler if the spherical coordinates are used. From (18),

$$\begin{aligned} \nabla^2 u_1 \equiv \Bigg\{&\frac{1}{R^2}\frac{\partial}{\partial R}\left(R^2\frac{\partial}{\partial R}\right) \\ &+ \left[\frac{1}{R^2 \sin\varphi}\frac{\partial}{\partial\varphi}\left(\sin\varphi\frac{\partial}{\partial\varphi}\right) + \frac{1}{R^2\sin^2\varphi}\frac{\partial^2}{\partial\theta^2}\right]\Bigg\} R^{-3/2}\int_0^R \frac{1}{4}\rho^{1/2}\,\nabla^2 u\,d\rho. \end{aligned} \tag{19}$$

The operator in the square bracket can be taken under the sign of integration. But, since $u$ is biharmonic $\nabla^2 \nabla^2 u = 0$, we have

$$\left[\frac{1}{\sin\varphi}\frac{\partial}{\partial\varphi}\left(\sin\varphi\frac{\partial}{\partial\varphi}\right)+\frac{1}{\sin^2\varphi}\frac{\partial^2}{\partial\theta^2}\right]\nabla^2 u = -\frac{\partial}{\partial\rho}\left(\rho^2\frac{\partial}{\partial\rho}\right)\nabla^2 u.$$

Therefore, (19) may be written as

$$\nabla^2 u_1 = \frac{1}{R^2}\frac{\partial}{\partial R}\left(R^2\frac{\partial}{\partial R}\right)R^{-3/2}\int_0^R \frac{1}{4}\rho^{1/2}\nabla^2 u\,d\rho \\ - R^{-7/2}\int_0^R \frac{1}{4}\rho^{1/2}\frac{\partial}{\partial\rho}\left(\rho^2\frac{\partial}{\partial\rho}\right)\nabla^2 u\,d\rho.$$

On carrying out the indicated differentiation in the first term and integrating the second term twice by parts, we obtain finally

$$\begin{aligned}\nabla^2 u_1 &= \frac{1}{R^2}\frac{\partial}{\partial R}\left\{-\frac{3}{8}R^{-1/2}\int_0^R \rho^{1/2}\nabla^2 u\,d\rho + \frac{R}{4}\nabla^2 u\right\} \\ &\qquad - \frac{1}{4}\frac{1}{R}\frac{\partial\nabla^2 u}{\partial R} + \frac{1}{8}R^{-7/2}\int_0^R \rho^{3/2}\frac{\partial\nabla^2 u}{\partial\rho}\,d\rho \\ &= \frac{1}{R^2}\left\{\frac{3}{8}\frac{1}{2}R^{-3/2}\int_0^R \rho^{1/2}\nabla^2 u\,d\rho - \frac{3}{8}\nabla^2 u + \frac{1}{4}\nabla^2 u + \frac{R}{4}\frac{\partial\nabla^2 u}{\partial R}\right\} \\ &\qquad - \frac{1}{4}\frac{1}{R}\frac{\partial\nabla^2 u}{\partial R} + \frac{1}{8}\frac{1}{R^2}\nabla^2 u - \frac{1}{8}\frac{3}{2}R^{-7/2}\int_0^R \rho^{1/2}\nabla^2 u\,d\rho \\ &= 0.\end{aligned}$$

Hence, $u_1$ given by (18) satisfies all the requirements, and the theorem is proved.

Theorem 2 holds also in the two-dimensional case when $R^2$ is replaced by $r^2 = x^2 + y^2$. The proof is analogous to the above. It is also evident that the choice of $x$ in Theorem 1 is incidental. The theorem holds when $x$ is replaced by $y$ or $z$.

Special representation of biharmonic functions in two-dimensions by means of analytic functions of a complex variable will be discussed in Sec. 9.5.

**Problem 8.9.** *Yih's solution of multiple Bessel equations.* Harmonic functions in cylindrical polar coordinates may assume the form $Z(r)e^{i\alpha x}e^{i\beta\theta}$, where $Z(r)$ is a Bessel function. Now consider the hyper-Bessel equation:

$$\left(\frac{d^2}{dr^2}+\frac{1}{r}\frac{d}{dr}-\frac{p^2}{r^2}+k^2\right)^n f = 0, \qquad n\text{, positive integer.}$$

With the help of Almansi's theorems discussed above, show that if $p$ (taken to be positive for convenience) is not an integer, the solutions are $r^m J_{\pm(p+m)}(kr)$, in which $m = 0, 1, 2, \ldots, n-1$; otherwise they are $r^m J_{p+m}(kr)$ and $r^m N_{p+m}(kr)$, with $m$

ranging over the same integers. The symbols $J$ and $N$ stand for the Bessel function and the Neumann function, respectively. (Chia Shun Yih, *Quart. Appl. Math.*, **13**, 4, 462–463, 1956.)

**Problem 8.10.** *Generation of biharmonic functions in cylindrical coordinates.* If $u_1(z, x)$ is a harmonic function in a space $\mathscr{R}(x, y, z)$, then

$$v_1 = \frac{1}{2\pi}\int_0^{2\pi} u_1(z, r\cos\theta)\, d\theta,$$

obtained by turning the space around the $z$-axis, is also harmonic. By selecting $u(z, x)$ as the real part of $(z + ix)^n$, in which $n$ is a positive integer, show that the following functions are harmonic in cylindrical coordinates $(r, \theta, z)$, where $r^2 = x^2 + y^2$.

$$\begin{aligned}
\psi_2 &= z^2 - \tfrac{1}{2}r^2,\\
\psi_3 &= z^3 - \tfrac{3}{2}zr^2,\\
\psi_4 &= z^4 - 3z^2r^2 + \tfrac{3}{8}r^4,\\
\psi_5 &= z^5 - 5z^3r^2 + \tfrac{15}{8}zr^4,\\
\psi_6 &= z^6 - \tfrac{15}{2}z^4r^2 + \tfrac{45}{8}z^2r^4 - \tfrac{5}{16}r^6.
\end{aligned}$$

Show that $z\psi_n$ and $(z^2 + r^2)\psi_n$ are biharmonic $(n = 2, 3, \ldots,)$. *Note:* If $u(z, x)$ is a biharmonic function, then the process indicated above generates a biharmonic function $v(z, r)$.

## 8.12. NEUBER-PAPKOVICH REPRESENTATION

In Sec. 8.4, the displacement field $u_i$ is represented by the Galerkin vector $(F_1, F_2, F_3)$, in the form

$$(1) \qquad 2Gu_i = 2(1 - \nu)\,\nabla^2 F_i - F_{j,ji} \qquad i = 1, 2, 3.$$

If we set

$$(2) \qquad \nabla^2 F_i = \frac{1}{2(1-\nu)}\,\Phi_i, \qquad F_{j,j} = \Psi,$$

Eq. (1) becomes

$$(3) \qquad 2Gu_i = \Phi_i - \Psi_{,i}.$$

Consider first the case in which the body force is absent. Since in the absence of body force the Galerkin vector must satisfy the biharmonic equation, we see that $\Phi_i$ and $\Psi$ satisfy the equations

$$(4) \qquad \nabla^2\Phi_i = 0, \qquad \nabla^4\Psi = 0.$$

Hence the $\Phi$'s are harmonic and the $\Psi$ is biharmonic. They are related through (2) by the equation

$$(5) \qquad \nabla^2\Psi = \frac{1}{2(1-\nu)}\,\Phi_{j,j}.$$

On noting that $\nabla^2(x_j\Phi_j) = 2\Phi_{j,j}$, we see that a particular solution of this equation is $\frac{1}{4(1-\nu)} x_j\Phi_j$. Hence, the general solution can be written as

$$\Psi = \frac{1}{4(1-\nu)} x_j\Phi_j + \phi_0, \tag{6}$$

where $\phi_0$ is an arbitrary harmonic function. On substituting into Eq. (3), we obtain

$$2Gu_i = \frac{3-4\nu}{4(1-\nu)}\Phi_i - \frac{1}{4(1-\nu)} x_j\Phi_{j,i} - \phi_{0,i}. \tag{7}$$

If we define

$$\phi_i = \frac{\Phi_i}{4(1-\nu)}, \tag{8}$$

$$\kappa = 3 - 4\nu, \tag{9}$$

we get

$$2Gu_i = \kappa\phi_i - x_j\phi_{j,i} - \phi_{0,i}. \tag{10}$$

This formula expresses $u_i$ in terms of four harmonic functions $\phi_0$, $\phi_1$, $\phi_2$, $\phi_3$. It was given independently by P. F. Papkovich (1932) and H. Neuber (1934) by different methods. The connection with the Galerkin vector was pointed out by Mindlin (1936). The special importance of the Neuber-Papkovich solution lies in its strict similarity to a general solution in two dimensions (Sec. 9.6), in which case a well-known procedure exists for the determination of the harmonic functions involved from specified boundary conditions.

The question of whether all four of the harmonic functions are independent, or whether one of them may be eliminated so that the general solution of the three-dimensional Navier's equation involves only three independent harmonic functions, has been a subject of much discussion. See Sokolnikoff, *Elasticity*,[1.2] 2nd ed. (1956), p. 331, and Naghdi[8.1] (1960).

If the body force does not vanish, a general solution of Navier's equation can be obtained by adding a particular integral to the Neuber-Papkovich solution [see Eq. (8.8:9)].

## 8.13. OTHER METHODS OF SOLUTION OF ELASTOSTATIC PROBLEMS

Two other classical methods of solving Navier's equations for elastostatic problems must be mentioned. The first is Betti's method, which was referred to in Sec. 8.8. The second is the method of integral transformation (Fourier, Laplace, Hankel, Mellin, Stieltjes, etc.). For the former, see Love, *Mathematical Theory*,[1.2] Chap. 10, and for the latter, Sneddon, *Fourier Transforms*[9.2] (1951), Chap. 10, and Flügge, *Encyclopedia of Physics*,[1.4] Vol. 6 (1962).

## 8.14. REFLECTION AND REFRACTION OF PLANE *P* AND *S* WAVES

So far we have considered only static problems. We shall now consider some dynamic problems in order to illustrate the use of displacement potentials in dynamics.

According to Sec. 8.2, when the displacements are represented by a scalar potential $\phi$ and vector potentials $\psi_1, \psi_2, \psi_3$ through the expression

$$u_i = \frac{\partial \phi}{\partial x_i} + e_{ijk} \frac{\partial \psi_k}{\partial x_j},$$

a broad class of solution is obtained if $\phi$ and $\psi_k$ satisfy the wave equations

$$\frac{\partial^2 \phi}{\partial x^2} + \frac{\partial^2 \phi}{\partial y^2} + \frac{\partial^2 \phi}{\partial z^2} = \frac{1}{c_L^2} \frac{\partial^2 \phi}{\partial t^2},$$

$$\frac{\partial^2 \psi_k}{\partial x^2} + \frac{\partial^2 \psi_k}{\partial y^2} + \frac{\partial^2 \psi_k}{\partial z^2} = \frac{1}{c_T^2} \frac{\partial^2 \psi_k}{\partial t^2}.$$

The functions $\phi$ and $\psi_1, \psi_2, \psi_3$ define dilatational and distorsional waves, which are called, in seismology, the *primary*, or *P* (or "push"), waves, and the secondary, or *S* (or "shake"), waves. If we consider plane waves, as in Sec. 7.8, we see that the *S* waves are polarized. If an *S*-wave train propagates along the *x*-axis in the *x*, *z*-plane, (*z* "vertical," *x* "horizontal"), and the material particles move in the *z*-direction (vertical), then we speak of *SV* waves. If the *S* waves propagate along *x* in the *x*, *z*-plane but the particles move in the *y*-direction ("horizontal"), then we speak of *SH* waves.

Consider a homogeneous isotropic elastic medium occupying a half-space $z \geqslant 0$. Since an elastic medium has two characteristic wave speeds, plane *P* waves hitting the free boundary $z = 0$ are reflected into plane *P* waves and plane *S* waves. Similarly, incident *SV* waves are reflected as both *P* and *SV* waves. If two elastic media are in contact with a "welded" interface, then incident *P* waves will be reflected in the first medium into *P* and *S* waves, and also refracted in the second medium in *P* and *S* waves. A similar statement holds for incident *SV* waves. The *SH* waves behave simpler. A train of incident *SH* waves will not generate *P* waves at the interface, so it is reflected and refracted in *SH* waves.

We shall show that the laws of reflection and refraction are the same Snell's law as in optics. Thus, if we have two homogeneous isotropic elastic media $M$ and $M_1$, of infinite extent and in "welded" contact at the plane $z = 0$, as shown in Fig. 8.14:1, and if the directions of advance of the waves are all parallel to the *x*, *z*-plane as illustrated by rays in the figure, then for incident *SV* waves,

$$(1) \qquad \frac{c_T}{\cos f_0} = \frac{c_T}{\cos f} = \frac{c_L}{\cos e} = \frac{c_T^{(1)}}{\cos f_1} = \frac{c_L^{(1)}}{\cos e_1}, \qquad (\therefore f = f_0).$$

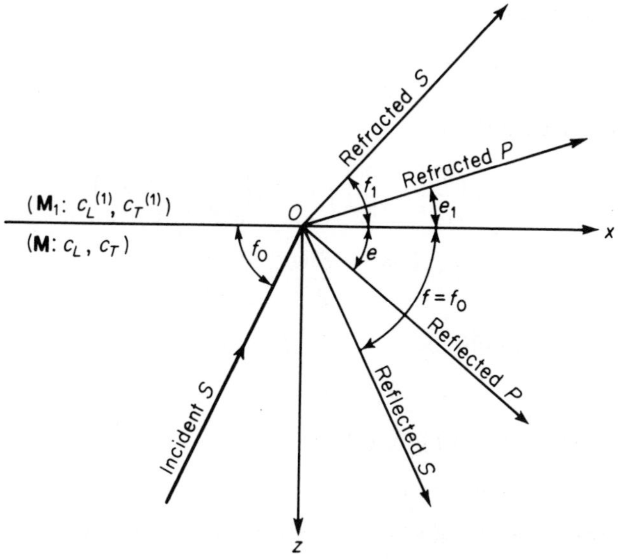


**Fig. 8.14:1.** Reflection of a $SV$ ray incident against a plane boundary.

In this equation, $c_L$, $c_T$ are, respectively, the longitudinal and transverse wave speeds of the medium $M$, and $c_L^{(1)}$, $c_L^{(1)}$ are the corresponding speeds of the medium $M_1$. The angle $f_0$ between the ray of the incident waves and the plane boundary is called the *angle of emergence* of the waves. Its complement is called the *angle of incidence*. Similarly, for incident $SH$ waves, we have

$$\frac{c_T}{\cos f_0} = \frac{c_T}{\cos f} = \frac{c_T^{(1)}}{\cos f_1}, \qquad (\therefore f = f_0), \tag{2}$$

and for incident $P$ waves,

$$\frac{c_L}{\cos e_0} = \frac{c_L}{\cos e} = \frac{c_T}{\cos f} = \frac{c_L^{(1)}}{\cos e_1} = \frac{c_T^{(1)}}{\cos f_1} \tag{3}$$

(so that $e = e_0$). If we consider a half-space $M$ with a free surface $z = 0$, we have these same equations, with the $c_L^{(1)}$, $c_T^{(1)}$ terms, which are now irrelevant, omitted, of course.

These results are easily proved. Let us work out the case of $SV$ waves emerging against a free boundary. Other cases are similar.

In $SV$ waves emerging at an angle $f_0$ from a free plane boundary, the wave front has a normal in the direction of a unit vector whose direction cosines are $(\cos f_0, 0, \sin f_0)$; but in the incident $SV$ waves the normal to the wave front has a different direction, with direction cosines $(\cos f_0, 0, -\sin f_0)$. This change of direction excites a reflected $P$ wave. We assume, therefore, that

$$\begin{gathered} \phi = \Phi(x \cos e + z \sin e - c_L t), \qquad \psi_1 = \psi_3 = 0, \\ \psi_2 = \psi = \Psi_0(x \cos f_0 - z \sin f_0 - c_T t) + \Psi(x \cos f + z \sin f - c_T t). \end{gathered} \tag{4}$$

The displacements are

(5) $$u = \frac{\partial \phi}{\partial x} - \frac{\partial \psi}{\partial z}, \qquad w = \frac{\partial \phi}{\partial z} + \frac{\partial \psi}{\partial x},$$

and the stresses are given by

(6) $$\sigma_{zz} = \lambda\left(\frac{\partial^2 \phi}{\partial x^2} + \frac{\partial^2 \phi}{\partial z^2}\right) + 2G\left(\frac{\partial^2 \phi}{\partial z^2} + \frac{\partial^2 \psi}{\partial x\, \partial z}\right),$$

$$\sigma_{zx} = G\left(2\frac{\partial^2 \phi}{\partial x\, \partial z} + \frac{\partial^2 \psi}{\partial x^2} - \frac{\partial^2 \psi}{\partial z^2}\right).$$

The boundary conditions are

(7) $$\text{at } z = 0, \qquad \sigma_{zz} = \sigma_{zx} = 0.$$

On substituting (4) and (6) into (7), we have

(8) $$(\lambda + 2G \sin^2 e)\Phi''(x \cos e - c_L t)$$
$$- 2G[\cos f_0 \sin f_0 \Psi_0''(x \cos f_0 - c_T t) - \cos f \sin f \Psi''(x \cos f - c_T t)] = 0,$$

(9) $$\cos e \sin e\, \Phi''(x \cos e - c_L t) + (\cos^2 f_0 - \sin^2 f_0)\Psi_0''(x \cos f_0 - c_T t)$$
$$+ (\cos^2 f - \sin^2 f)\Psi''(x \cos f - c_T t) = 0.$$

These equations can be satisfied for all values of $x$ and $t$ only if the arguments of the various $\Phi$ and $\Psi$ functions are in a constant ratio. Hence,

(10) $$\frac{c_T}{\cos f_0} = \frac{c_L}{\cos e} = \frac{c_T}{\cos f}. \qquad \text{Q.E.D.}$$

When Eqs. (10) are satisfied, the functional relationships between $\Psi_0$, $\Psi$, and $\Phi$ are given by Eqs. (8) and (9). A detailed study of such functional relationships yields information about the partitioning of energy among the various components of reflected and refracted waves—an important subject whose details can be found in Ewing, Jardetzky, and Press.[7.4]

## 8.15. LAMB'S PROBLEM—LINE LOAD SUDDENLY APPLIED ON ELASTIC HALF-SPACE

Lamb,[7.4] in a classic paper published in 1904, considered the disturbance generated in a semi-infinite medium by an impulsive force applied along a line or at a point on the surface or inside the medium. Lamb's solution, as well as the extensions thereof, has been studied by Nakano, Lapwood, Pekeris, Cagniard, Garvin, Chao, and others. In this section we shall consider only the problem of a line load suddenly applied on the surface. (Fig. 8.15:1.)

Consider a semi-infinite body of homogeneous isotropic linear elastic material occupying the space $z \geqslant 0$. For time $t < 0$, the medium is stationary.

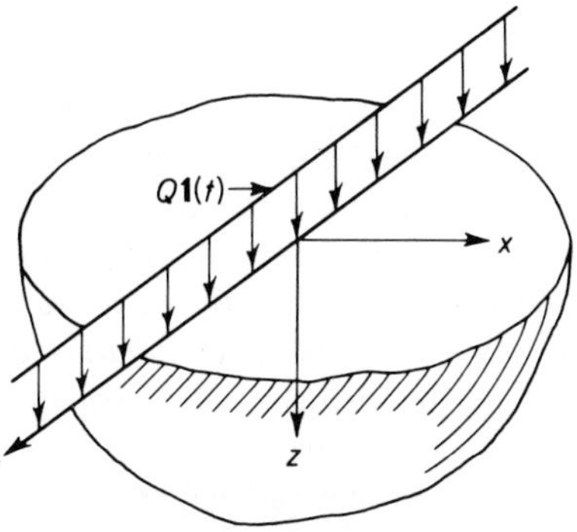


**Fig. 8.15:1.** A suddenly applied line load on an elastic half-space.

At $t = 0$, a concentrated load is suddenly applied normal to the free surface $z = 0$, along a line coincident with the $y$-axis. The boundary conditions are therefore two-dimensional. We may assume the deformation state to be plane strain. The displacement $v$ vanishes, and $u$ and $w$ are independent of $y$. Under this assumption, only one component of the vector potential is required. According to Sec. 8.2, the displacements are represented by

(1) $$u = \frac{\partial \phi}{\partial x} - \frac{\partial \psi}{\partial z}, \qquad w = \frac{\partial \phi}{\partial z} + \frac{\partial \psi}{\partial x}.$$

Navier's equations of motion are satisfied if $\phi$ and $\psi$ satisfy the wave equations

(2) $$\frac{\partial^2 \phi}{\partial x^2} + \frac{\partial^2 \phi}{\partial z^2} = \frac{1}{c_L^2}\frac{\partial^2 \phi}{\partial t^2}, \qquad \frac{\partial^2 \psi}{\partial x^2} + \frac{\partial^2 \psi}{\partial z^2} = \frac{1}{c_T^2}\frac{\partial^2 \psi}{\partial t^2}.$$

The stress components are

(3) $$\sigma_{yz} = \sigma_{xy} = 0, \qquad \sigma_{yy} = \nu(\sigma_{xx} + \sigma_{zz}).$$

(4) $$\sigma_{zx} = G\left(\frac{\partial w}{\partial x} + \frac{\partial u}{\partial z}\right) = G\left(2\frac{\partial^2 \phi}{\partial x\,\partial z} + \frac{\partial^2 \psi}{\partial x^2} - \frac{\partial^2 \psi}{\partial z^2}\right),$$
$$\sigma_{zz} = \lambda e + 2G\frac{\partial w}{\partial z} = \lambda\left(\frac{\partial^2 \phi}{\partial x^2} + \frac{\partial^2 \phi}{\partial z^2}\right) + 2G\left(\frac{\partial^2 \phi}{\partial z^2} + \frac{\partial^2 \psi}{\partial x\,\partial z}\right),$$
$$\sigma_{xx} = \lambda e + 2G\frac{\partial u}{\partial x} = \lambda\left(\frac{\partial^2 \phi}{\partial x^2} + \frac{\partial^2 \phi}{\partial z^2}\right) + 2G\left(\frac{\partial^2 \phi}{\partial x^2} - \frac{\partial^2 \psi}{\partial x\,\partial z}\right),$$

The boundary conditions on the surface $z = 0$ are

(5) $$(\sigma_{zx})_{z=0} = (\sigma_{zy})_{z=0} = 0, \qquad (\sigma_{zz})_{z=0} = -Q\delta(x)\mathbf{1}(t),$$

where $\mathbf{1}(t)$ is the unit-step function

(6) $$\mathbf{1}(t) = 0 \quad \text{for} \quad t < 0, \qquad \mathbf{1}(t) = 1 \quad \text{for} \quad t > 0,$$

while $\delta(x)$ is the Dirac delta function, i.e., one whose value is zero everywhere except in the neighborhood of $x = 0$ where it becomes infinitely large in such a way that $\int_{-\infty}^{\infty} \delta(x)\, dx = 1$. It is seen that the surface stress given by (6) is equivalent to a concentrated load $Q$ per unit length suddenly applied on the line $x = 0$, $z = 0$ and maintained constant afterwards.

For the conditions at infinity, it is reasonable to demand that (a) all displacements and stresses remain finite at infinity, and (b) at large distances from the point of application of the load the disturbance consists of outgoing waves. These are called the *finiteness* and *radiation conditions*, respectively.

In the present problem, the disturbance is propagated outward at a finite velocity, so conditions (a) and (b) are equivalent to the statement that there exists an outgoing wave front, beyond which the medium is undisturbed. The question arises whether the boundary conditions (5) and the finiteness and radiation conditions, in the absence of any other disturbances in the medium, will determine a unique solution of our problem. For a *point* load, the answer is obviously affirmative, because for a suddenly applied point load, the wave front will be at a finite distance from the point of application of the load at any finite time. Hence, if we take a volume sufficiently large so that it includes the wave front in its interior, we have a finite body over whose entire surface the surface tractions are specified. Neumann's uniqueness theorem (Prob. 7.1, p. 162) then guarantees a unique solution. For a *line* load, we do not have such a simple and general proof. In fact, difficulty may arise in two-dimensional problems. (For example, in the corresponding static problem—a static line load acting on the surface—the displacement at infinity is logarithmically divergent. See Sec. 9.4, Example 2.) However, for the present problem, a unique solution can be determined if we assume the deformation to be truly two-dimensional.

(The significance of the last assumption is made clear by the following remarks. Note that a cylindrical body subjected to surface forces uniform along the generators *may* have an internal stress state that is not uniform along the axis. For example, transient axial waves may be superposed without disturbing the lateral boundary conditions. In other words, a seemingly two-dimensional problem may actually be three-dimensional. Such an occasion also occurs in fluid mechanics. A nontrivial example in hydrodynamics is the flow around a circular cylinder, with a velocity field uniform at infinity and normal to the cylinder axis. At supercritical Reynolds numbers, the three-dimensionality of the flow in the wake, i.e., variation along the cylinder axis, is very pronounced and becomes a predominant feature.)

The boundary conditions (5) can be written in a different form. The unit-step function $\mathbf{1}(t)$ has no Fourier transform. But if we consider it to be the limiting case of the function $e^{-\beta t}\,\mathbf{1}(t)$, which has a Fourier transform, then the Fourier integral theorem

$$f(x) = \frac{1}{2\pi}\int_{-\infty}^{\infty} e^{ikx}\, dk \int_{-\infty}^{\infty} f(\xi)\, e^{-ik\xi}\, d\xi, \tag{7}$$

which is valid for an arbitrary function $f(x)$ that is square-integrable in the Lebesgue sense, yields the representation

$$\mathbf{1}(t) = \frac{1}{2\pi} \lim_{\beta \to 0} \int_{-\infty}^{\infty} \frac{e^{i\omega t}}{\beta + i\omega} \, d\omega \,. \tag{8}$$

Equation (8) is obtained on substituting $e^{-\beta\xi}\,\mathbf{1}(\xi)$ for $f(\xi)$ in Eq. (7) and changing $k$ to $\omega$. Similarly, $\delta(x)$ has no Fourier transform. But, considering the delta function as the limit of a square wave

$$\frac{1}{\varepsilon}\left[\mathbf{1}\left(x + \frac{\varepsilon}{2}\right) - \mathbf{1}\left(x - \frac{\varepsilon}{2}\right)\right]$$

as $\varepsilon \to 0$, we can use (7) to obtain the representation

$$\delta(x) = \lim_{\varepsilon \to 0} \frac{1}{2\pi\varepsilon} \int_{-\infty}^{\infty} \frac{\sin k\varepsilon}{k} \, e^{ikx} \, dk. \tag{9}$$

Therefore, the second condition in (5) may be written

$$[\sigma_{zz}]_{z=0} = \frac{-Q}{4\pi^2} \lim_{\substack{\varepsilon \to 0 \\ \beta \to 0}} \int_{-\infty}^{\infty} \int_{-\infty}^{\infty} \frac{\sin k\varepsilon}{k\varepsilon} \, \frac{1}{\beta + i\omega} \, e^{i(\omega t + kx)} \, d\omega \, dk. \tag{10}$$

From this, we see that if we can obtain an elementary solution of Eqs. (2) satisfying the boundary conditions

$$[\sigma_{xz}]_{z=0} = 0, \qquad [\sigma_{zz}]_{z=0} = Ze^{i(\omega t + kx)}, \tag{11}$$

then by the principle of superposition the solution to the original problem with boundary conditions (5) can be obtained by setting

$$Z(k, \omega) = \frac{-Q}{4\pi^2} \frac{\sin k\varepsilon}{k\varepsilon} \frac{1}{\beta + i\omega}, \tag{12}$$

integrating with respect to $k$ and $\omega$ both from $-\infty$ to $\infty$, and then passing to the limit $\beta \to 0$, $\varepsilon \to 0$.

The solution of the elementary problem is obtained by assuming

$$\phi = Ae^{-\nu z + ikx + i\omega t}, \qquad \psi = Be^{-\nu' z + ikx + i\omega t}, \tag{13}$$

which satisfy the wave Eqs. (2) if

$$\nu^2 = k^2 - k_\alpha^2, \qquad \nu'^2 = k^2 - k_\beta^2, \qquad k_\alpha = \frac{\omega}{c_L}, \qquad k_\beta = \frac{\omega}{c_T}. \tag{14}$$

On substituting (13) into (4) and (11), we obtain

$$\begin{aligned} -2Ai\nu k - (2k^2 - k_\beta^2)B &= 0, \\ (2k^2 - k^2)A - 2Bik\nu' &= \frac{1}{G} Z(k, \omega). \end{aligned} \tag{15}$$

Solving these equations for $A$ and $B$, we obtain

$$A = \frac{2k^2 - k_\beta^2}{F(k)} \frac{Z(k, \omega)}{G}, \qquad B = \frac{-2ik\nu}{F(k)} \frac{Z(k, \omega)}{G}, \tag{16}$$

where

$$F(k) \equiv (2k^2 - k_\beta^2)^2 - 4k^2\nu\nu' \tag{17}$$

is Rayleigh's function (see Sec. 7.9).

A formal solution of our problem is obtained by substituting (16) and (12) into (13) and integrating with respect to $\omega$ and $k$ from $-\infty$ to $\infty$ as indicated before. In so doing, an appropriate branch of the multi-valued functions $\nu$ and $\nu'$ must be chosen so that the conditions at infinity are satisfied.

The evaluation of these integrals is a formidable task. Direct integration or numerical integration are exceedingly difficult. Lamb uses the method of contour integration. The variables of integration $k$ and $\omega$ are replaced by complex variables, and the contours of integration are deformed in such a way that some explicit results are obtained. However, we shall not pursue this method here. Details can be found in Lamb's paper, or in Ewing, Jardetzky, and Press, *Elastic Waves in Layered Media*[7.4] pp. 44–64. In the following section, we shall explain a very elegant exact solution given by Cagniard's method.

## 8.16. SOLUTION BY CAGNIARD'S METHOD

Let Eqs. (8.15:2) and (8.15:5) be transformed by Laplace transformation with respect to time. Let the Laplace transforms of $\phi$ and $\psi$ be written as

$$\bar{\phi} = \int_0^\infty e^{-st}\phi(x, z; t)\, dt, \qquad \bar{\psi} = \int_0^\infty e^{-st}\psi(x, z; t)\, dt, \qquad \text{Rl}\, s > \rho, \tag{1}$$

where $\rho$ is the abscissa of convergence. Since the initial conditions are $\phi = \psi = 0$ for $t < 0$, Eqs. (8.15:2) are transformed into

$$\frac{\partial^2\bar{\phi}}{\partial x^2} + \frac{\partial^2\bar{\phi}}{\partial z^2} = \frac{s^2}{c_L^2}\bar{\phi}, \qquad \frac{\partial^2\bar{\psi}}{\partial x^2} + \frac{\partial^2\bar{\psi}}{\partial z^2} = \frac{s^2}{c_T^2}\bar{\psi}. \tag{2}$$

The boundary conditions (8.15:5), together with (8.15:4), are now expressed as

$$2\frac{\partial^2\bar{\phi}}{\partial x\, \partial z} + \frac{\partial^2\bar{\psi}}{\partial x^2} - \frac{\partial^2\bar{\psi}}{\partial z^2} = 0, \quad \text{at } z = 0, \tag{3}$$

$$\lambda\left(\frac{\partial^2\bar{\phi}}{\partial x^2} + \frac{\partial^2\bar{\phi}}{\partial z^2}\right) + 2G\left(\frac{\partial^2\bar{\phi}}{\partial z^2} + \frac{\partial^2\bar{\psi}}{\partial x\, \partial z}\right) = -\frac{Q\delta(x)}{s}, \quad \text{at } z = 0. \tag{4}$$

An elementary solution of $\bar{\phi}$ from Eq. (2) may be posed in the form $e^{iksx-\nu sz}$, provided that $\nu = \sqrt{k^2 + c_L^{-2}}$. Henceforth, we shall write, for simplicity,

$$v_1 = \frac{1}{c_L}, \qquad v_2 = \frac{1}{c_T}. \tag{5}$$

Then the elementary solutions of Eqs. (2) may be written as

$$\bar{\phi} = e^{iksx - \sqrt{k^2+v_1{}^2}\,sz}, \qquad \bar{\psi} = e^{iksx - \sqrt{k^2+v_2{}^2}\,sz}.$$

By the principle of superposition, we see that a general solution of Eqs. (2) can be written as

$$\bar{\phi} = \int_\Gamma P_1(k)\, e^{-s[\sqrt{v_1{}^2+k^2}\, z - ikx]}\, dk. \tag{6}$$

$$\bar{\psi} = \int_\Gamma P_2(k)\, e^{-s[\sqrt{v_2{}^2+k^2}\, z - ikx]}\, dk, \tag{7}$$

where $P_1(k)$ and $P_2(k)$ are two arbitrary functions and $\Gamma$ is chosen to be a path of integration in the positive direction along the axis of real $k$. To insure the existence of Laplace transforms, the real part of $s$ must be greater than the abscissa of convergence $\rho$. It is sufficient to consider $s$ real and positive and $>\rho$. The quantities $\sqrt{v_1^2 + k^2}$ and $\sqrt{v_2^2 + k^2}$ are given positive real parts along the path of integration so that $\bar{\phi}$ and $\bar{\psi}$ tend to zero as $z \to \infty$. On substituting (6) and (7) into (3) and (4), we obtain†

$$\int_\Gamma [-2ik\sqrt{v_1^2 + k^2}\, P_1(k) - (2k^2 + v_2^2)P_2(k)]s^2\, e^{iksx}\, dk = 0, \tag{8}$$

$$\int_\Gamma [(2k^2 + v_2^2)P_1(k) - 2ik\sqrt{v_2^2 + k^2}\, P_2(k)]Gs^2\, e^{iksx}\, dk = -\frac{Q\delta(x)}{s}. \tag{9}$$

Equation (8) is satisfied by

$$P_1(k) = (2k^2 + v_2^2)R(k), \qquad P_2(k) = -2ik\sqrt{v_1^2 + k^2}\, R(k), \tag{10}$$

where $R(k)$ is a new unknown. Substituting (10) into (9) and using the Fourier representation of $\delta(x)$, Eq. (8.15:9), we obtain

(11)

$$\int_\Gamma \left[(2k^2 + v_2^2)^2 - 4k^2\sqrt{(v_1^2 + k^2)(v_2^2 + k^2)}\right] R(k) G s^2\, e^{iksx}\, dk = -\frac{Q}{2\pi}\int_\Gamma e^{iksx}\, dk,$$

which is satisfied by

$$R(k) = -\frac{Q}{2\pi G}\frac{1}{s^2}\left[(2k^2 + v_2^2)^2 - 4k^2\sqrt{(k^2 + v_1^2)(k^2 + v_2^2)}\right]^{-1}. \tag{12}$$

† Note that from Eqs. (5) and (8.2:4) we have $(\lambda + 2G)v_1^2 = Gv_2^2$.

Recalling Eqs. (6), (7), and (10), we see that

$$(13)\qquad \bar{\phi} = -\frac{Q}{2\pi G s^2}\int_\Gamma \frac{(2k^2+v_2^2)\,e^{-s[\sqrt{v_1^2+k^2}\,z-ikx]}}{(2k^2+v_2^2)^2-4k^2\sqrt{(k^2+v_1^2)(k^2+v_2^2)}}\,dk,$$

$$(14)\qquad \bar{\psi} = \frac{Qi}{\pi G s^2}\int_\Gamma \frac{k\sqrt{v_1^2+k^2}\,e^{-s[\sqrt{v_2^2+k^2}\,z-ikx]}}{(2k^2+v_2^2)^2-4k^2\sqrt{(k^2+v_1^2)(k^2+v_2^2)}}\,dk.$$

The Laplace transforms of the stresses can be expressed, according to Eq. (8.15:4), as

$$(15)\qquad \bar{\sigma}_{zz} = -\frac{Q}{2\pi}\int_\Gamma \frac{(2k^2+v_2^2)^2}{\Delta}e^{\xi}\,dk + \frac{2Q}{\pi}\int_\Gamma \frac{k^2\sqrt{(v_1^2+k^2)(v_2^2+k^2)}}{\Delta}e^{\eta}\,dk,$$

$$(16)\qquad \bar{\sigma}_{zx} = \frac{Qi}{\pi}\int_\Gamma \frac{(2k^2+v_2^2)k\sqrt{v_1^2+k^2}}{\Delta}e^{\xi}\,dk - \frac{Qi}{\pi}\int_\Gamma \frac{(2k^2+v_2^2)k\sqrt{v_1^2+k^2}}{\Delta}e^{\eta}\,dk,$$

$$(17)\qquad \bar{\sigma}_{xx} = -\frac{Q}{2\pi}\int_\Gamma \frac{(v_2^2-2v_1^2-2k^2)(2k^2+v_2^2)}{\Delta}e^{\xi}\,dk - \frac{2Q}{\pi}\int_\Gamma \frac{k^2\sqrt{(v_1^2+k^2)(v_2^2+k^2)}}{\Delta}e^{\eta}\,dk,$$

where

$$(18)\qquad \Delta = \left[(2k^2+v_2^2)^2-4k^2\sqrt{(k^2+v_1^2)(k^2+v_2^2)}\right],$$

$$(19)\qquad \xi = -s\sqrt{v_1^2+k^2}\,z + iskx, \qquad \eta = -s\sqrt{v_2^2+k^2}\,z + iskx.$$

To find the inverse of these transforms, we use a method originally due to Cagniard[7.4] (1935) and later modified by De Hoop[7.4] (1958). The details of our solution are very similar to those given by Ang[9.2] (1960) in his analysis of suddenly started moving line loads. The method consists in considering $k$ as a complex number and so deforming the path of integration $\Gamma$ that the integrals in Eqs. (13) and (14) can be recognized as the Laplace transforms of certain explicit functions of time, thus allowing us to write down the inverse transforms by inspection. This is done, for example, for the integral in Eq. (13), by introducing a transformation of variable from $k$ to $t$ in such a way that

$$(20)\qquad \sqrt{v_1^2+k^2}\,z - ikx = t$$

and deforming $\Gamma$ into another curve $\Gamma_1$ in such a way that the corresponding path of integration on the $t$-plane is the positive real axis from 0 to $\infty$, so that the integral assumes the form

$$\int_0^\infty f(x, z, t)\, e^{-st}\, dt.$$

If this scheme is successful, then the inverse of that integral is exactly $f(x, z, t)$.

Since the entire complex $k$-plane will be considered, we shall make the functions $\sqrt{v_1^2 + k^2}$ and $\sqrt{v_2^2 + k^2}$ single-valued and having their real parts positive throughout the $k$-plane. This is achieved by introducing branch cuts along the imaginary axis from the branch points $\pm iv_1$ to $\pm i\infty$ and from $\pm iv_2$ to $\pm i\infty$ (see Fig. 8.16:1). Note that $v_1 < v_2$, since $c_L > c_T$. The integrands in (13)–(17) have singular points where the denominator $\Delta$ vanishes. Now, if we write $k = i/c$, we see readily that the denominator $\Delta$ is exactly the Rayleigh function which has two real roots at $c = \pm c_R$, where $c_R$ is the Rayleigh surface wave speed [see Eq. (7.9:12) *et seq.*]. Hence, the integrand has two poles at the points $\pm ik_R$, on the imaginary axis, where $k_R = 1/c_R$. Since $c_R < c_T < c_L$, the relative position of the poles and the branch points are as shown in Fig. 8.16:1. This exhausts all singularities. The integrands of (13)–(17) are analytic on the entire Argand plane of $k$ cut in this manner.

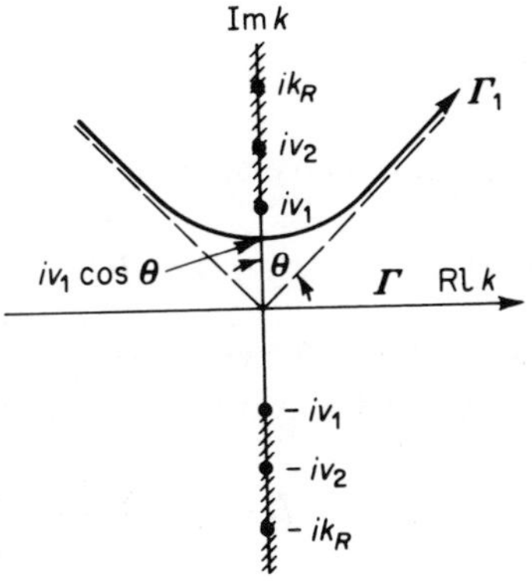


**Fig. 8.16:1.** Path of integration and location of branch points.

To illustrate the details, let us consider the integral

$$\bar{I}_1 = \int_\Gamma \frac{(2k^2 + v_2^2)^2}{\Delta}\, e^{-s(\sqrt{v_1{}^2+k^2}\, z - ikx)}\, dk. \tag{21}$$

Introducing the transformation (20) and solving for $k$, we obtain

$$k = \pm \sqrt{\frac{t^2}{r^2} - v_1^2}\, \sin\theta + \frac{it}{r}\cos\theta, \tag{22}$$

where

$$r = \sqrt{x^2 + z^2}, \qquad \theta = \tan^{-1}\frac{z}{x}, \qquad 0 \leqslant \theta \leqslant \pi. \tag{23}$$

Let $t$ be real and positive. As $t$ varies from $v_1 r$ to $\infty$, $k$ traces on the Argand plane a hyperbola whose asymptotes make an angle $\theta$ with the imaginary-$k$ axis. This is shown as curve $\Gamma_1$ in Fig. 8.16:1. Hence, our scheme will work if the original path of integration $\Gamma$ can be deformed into the path $\Gamma_1$ on the $k$-plane. Now, we have seen before that all the singularities of the integrand lie on the imaginary axis, and all have moduli $\geqslant v_1$. Hence, for

$0 \leqslant \theta \leqslant \pi$ there is no singularity between $\Gamma$ and $\Gamma_1$. Furthermore, the integrand $\to 0$ exponentially as $k \to \infty$, so that an integral along the circular arcs connecting the ends of $\Gamma$ and $\Gamma_1$ gives no contribution as $|k| \to \infty$. Hence, the paths $\Gamma$ and $\Gamma_1$ are equivalent. Therefore, we can write $\bar{I}_1$ as

$$\bar{I}_1 = \int_{v_1 r}^{\infty} \left[ M_1(k_+^{(1)}) \frac{\partial k_+^{(1)}}{\partial t} - M_1(k_-^{(1)}) \frac{\partial k_-^{(1)}}{\partial t} \right] e^{-st}\, dt, \tag{24}$$

where

$$M_1(k) = (2k^2 + v_2^2)^2[(2k^2 + v_2^2)^2 - 4k^2\sqrt{(k^2 + v_1^2)(k^2 + v_2^2)}]^{-1}, \tag{25}$$

$$k_\pm^{(1)} = \pm\sqrt{\frac{t^2}{r^2} - v_1^2}\,\sin\theta + i\frac{t}{r}\cos\theta, \qquad 0 \leqslant \theta \leqslant \pi. \tag{26}$$

Hence, the inverse transform of $\bar{I}$ is

$$I(r, \theta, t) = \mathbf{1}(t - v_1 r)\left[ M_1(k_+^{(1)}) \frac{\partial k_+^{(1)}}{\partial t} - M_1(k_-^{(1)}) \frac{\partial k_-^{(1)}}{\partial t} \right], \tag{27}$$

where $k_\pm^{(1)}(r, \theta, t)$ are given by (26), and $\mathbf{1}(t)$ is the unit-step function.

Consider next the integral $\bar{I}_2$,

$$\bar{I}_2 = \int_\Gamma \frac{k^2\sqrt{(v_1^2 + k^2)(v_2^2 + k^2)}}{\Delta} e^{-s[\sqrt{v_2{}^2 + k^2}\, z - ikx]}\, dk. \tag{28}$$

Here we let

$$t = \sqrt{v_2^2 + k^2}\, z - ikx, \tag{29}$$

so that

$$k = \pm\sqrt{\frac{t^2}{r^2} - v_2^2}\,\sin\theta + \frac{it}{r}\cos\theta, \qquad (k_\pm^{(2)}) \tag{30}$$

with $r$, $\theta$ given by (23). For $t$ real and varying from $v_2 r$ to $\infty$, the curve $\Gamma_2$ on the $k$-plane is again a hyperbola with asymptotes making an angle $\theta$ with the imaginary axis. Now, depending on $\theta$, the intercept of the hyperbola on the imaginary axis, namely, $iv_2\cos\theta$, may or may not be above the point $iv_1$. If $\theta$ lies in the range

$$\cos^{-1}\frac{v_1}{v_2} \leqslant \theta \leqslant \pi - \cos^{-1}\frac{v_1}{v_2}, \tag{31}$$

the situation of Fig. 8.16:2 prevails, and $\Gamma$ and $\Gamma_2$ are equivalent. Hence, in this range of $\theta$ we have

$$I_2(r, \theta, t) = \mathbf{1}(t - v_2 r)\left[ M_2(k_+^{(2)}) \frac{\partial k_+^{(2)}}{\partial t} - M_2(k_-^{(2)}) \frac{\partial k_-^{(2)}}{\partial t} \right], \tag{32}$$

where $k_\pm^{(2)}(r, \theta, t)$ are given by (30) with corresponding $\pm$ signs, with $\theta$ limited by (31), and

$$M_2(k) = k^2\sqrt{(v_1^2 + k^2)(v_2^2 + k^2)}[(2k^2 + v_2^2)^2 - 4k^2\sqrt{(v_1^2 + k^2)(v_2^2 + k^2)}]^{-1}. \tag{33}$$

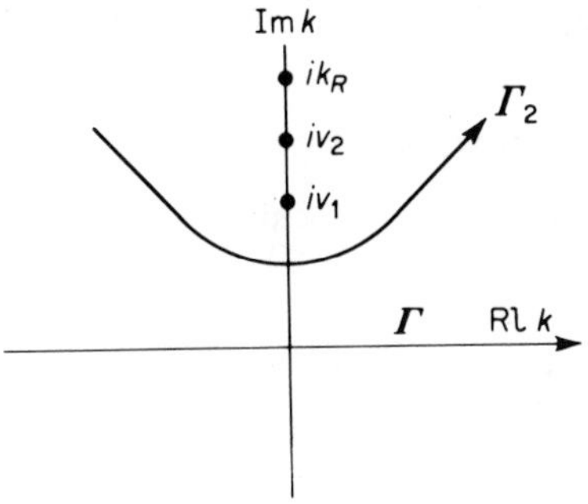


**Fig. 8.16:2.** Path of integration when Eq. (31) prevails.

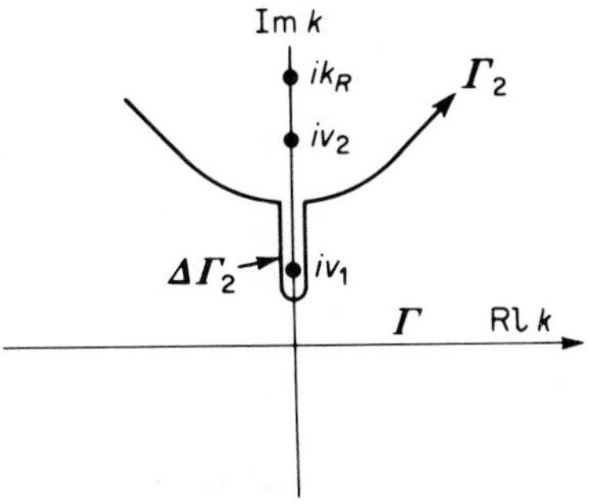


**Fig. 8.16:3.** Path of integration when Eq. (31) is not satisfied.

If $\theta$ does not belong to the range indicated in (31), then the hyperbola crosses the portion of the branch cuts given by Rl $k = 0$, $v_1 < |\text{Im } k| < v_2$. In this case, integration along $\Gamma_2$ does not give the same result as that along the original path $\Gamma$. But the path consisting of $\Gamma_2$ and $\Delta\Gamma_2$ shown in Fig. 8.16:3 is equivalent to $\Gamma$. The additional path $\Delta\Gamma_2$ consists of a circle of radius $\delta(\delta \to 0)$ centered at $k = iv_1$ and of two segments represented by

$$k = i\left(-\sqrt{v_2^2 - \frac{t^2}{r^2}} \sin\theta + \frac{t}{r}\cos\theta\right) \mp \delta, \qquad (k_{\mp}^{(3)}) \tag{34}$$

in the range

$$0 \leqslant \theta < \cos^{-1}\frac{v_1}{v_2}. \tag{35}$$

In (34), the range of $t$ is

$$v_2 r \geqslant t \geqslant v_1 r \cos\theta + r\sqrt{v_2^2 - v_1^2}\sin\theta. \tag{36}$$

While the contribution of the $\delta$-circular arc is nil as $\delta \to 0$, the path (34) together with the hyperbolic path $\Gamma_2$ gives

$$\begin{aligned} I_2(r, \theta, t) = \mathbf{1}(t - v_2 r)\left[M_2(k_+^{(2)})\frac{\partial k_+^{(2)}}{\partial t} - M_2(k_-^{(2)})\frac{\partial k_-^{(2)}}{\partial t}\right] \\ + f_\theta^{(1)} f_t^{(1)}\left[M_2(k_+^{(3)}) - M_2(k_-^{(3)})\right]\frac{\partial k_+^{(3)}}{\partial t}, \end{aligned} \tag{37}$$

which is valid for $\theta$ in (35), and where $k_{\mp}^{(3)}$ is given by Eq. (34) with the corresponding $\mp$ signs. Moreover, the functions $f_\theta^{(1)}$ and $f_t^{(1)}$ are

$$f_\theta^{(1)} = 1 \quad \text{for } \theta \text{ in the range (35)}; \quad f_\theta^{(1)} = 0 \text{ otherwise}; \tag{38}$$

$$f_t^{(1)} = 1 \quad \text{for } t \text{ in the range (36)}; \quad f_t^{(1)} = 0 \text{ otherwise}. \tag{39}$$

Finally, for $\theta$ in the range

$$\pi - \cos^{-1}\frac{v_1}{v_2} < \theta \leqslant \pi, \tag{40}$$

we have results similar to (37) except that $f_\theta^{(1)}, f_t^{(1)}$, and $k^{(3)}$ must be replaced by $f_\theta^{(2)}, f_t^{(2)}, k^{(4)}$.

(41) $$f_\theta^{(2)} = 1 \quad \text{for } \theta \text{ in (40)}; \quad f_\theta^{(2)} = 0 \text{ otherwise};$$

(42) $$f_t^{(2)} = 1 \quad \text{for } v_2 r \geq t \geq v_2 r\,|\cos\theta| + r\sqrt{v_2^2 - v_1^2}\sin\theta;$$
$$= 0 \quad \text{otherwise.}$$

(43) $$k_\pm^{(4)} = \pm\delta + i\left(\sqrt{v_2^2 - \frac{t^2}{r^2}}\sin\theta + \frac{t}{r}\cos\theta\right).$$

To summarize, the stress $\sigma_{zz}$ is given by

(44) $$\begin{aligned}\sigma_{zz} = &-\frac{Q}{2\pi}\mathbf{1}(t - v_1 r)\left[M_1(k_+^{(1)})\frac{\partial k_+^{(1)}}{\partial t} - M_1(k_-^{(1)})\frac{\partial k_-^{(1)}}{\partial t}\right] \\ &+ \frac{2Q}{\pi}\mathbf{1}(t - v_2 r)\left[M_2(k_+^{(2)})\frac{\partial k_+^{(2)}}{\partial t} - M_2(k_-^{(2)})\frac{\partial k_-^{(2)}}{\partial t}\right] \\ &+ \frac{2Q}{\pi}f_\theta^{(1)}f_t^{(1)}\left[M_2(k_+^{(3)}) - M_2(k_-^{(3)})\right]\frac{\partial k_+^{(3)}}{\partial t} \\ &+ \frac{2Q}{\pi}f_\theta^{(2)}f_t^{(2)}\left[M_2(k_+^{(4)}) - M_2(k_-^{(4)})\right]\frac{\partial k_+^{(4)}}{\partial t}, \qquad 0 \leq \theta \leq \pi.\end{aligned}$$

Other stress components can be obtained in a similar manner. This gives the exact solution in a closed form.

Let us examine the wave pattern of the stress response $\sigma_{zz}$ as revealed by Eq. (44). At a time $t$ after the application of the force $Q$ at the origin, the first term in (44) is nonvanishing only for $r \leq t/v_1 = c_L t$, whereas the second term is nonvanishing for $r \leq t/v_2 = c_T t$. These are the zones disturbed by the dilatational and the distorsional waves, respectively, as regions I and II in Fig. 8.16:4. The third and the fourth terms are nonvanishing in the shaded

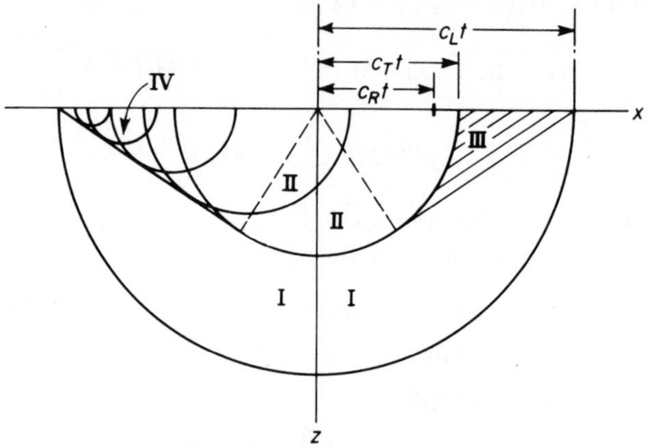


**Fig. 8.16:4.** Different regions of stress waves.

curvilinear triangle regions III and IV, respectively, as marked in Fig. 8.16:4. Waves in III and IV are called *head waves* or *von Schmidt* waves. The existence of the head waves can be understood from Huygens' principle simply by noting that as the wave front of the longitudinal waves intersects the free surface, the point of intersection may be regarded as a new source of disturbance which causes both dilatational and distorsional waves. The former is included in the first term in Eq. (44); the latter in the third or the fourth term. Such successive wave fronts and their envelope are sketched in Fig. 8.16:4.

A careful examination of the terms $M_1(k_+^{(1)})$, $M_2(k_+^{(2)})$, etc., will reveal also the Rayleigh surface waves, corresponding to the poles of the integrands in Eqs. (13)–(17), i.e., the roots of $\Delta(k)$ in Eq. (18). These poles are marked as $\pm ik_R$ in Figs. 8.16:1, 8.16:2, and 8.16:3.

## 8.17. MOTION OF THE SURFACE DUE TO A BURIED PULSE

The methods illustrated in the preceding sections can be used to solve Lamb's problems of the more general kind—such as a line load applied internally or a point load applied on the surface or internally. They can be extended also to layered media. Because of its importance to geophysics and seismology, a very extensive literature exists on this subject [see, for example, the bibliography in Ewing, Jardetzky, and Press[7.4] (1957)].

The problem of a concentrated vertical point load applied internally to a homogeneous elastic half-space at a distance $H$ below the horizontal free surface is solved by Pekeris[7.4] (1955), whose analytical results were evaluated by Pekeris and Lifson[7.4] (1957). Because of their general interest, we shall present here, without going into detail, some of Pekeris' numerical results.

Since a point load is considered, it is appropriate to use a system of cylindrical polar coordinates $(z, r, \theta)$. The load is applied at the point $z = H$, $r = 0$. The solution is independent of $\theta$. The surface $z = 0$ is free, on which $\sigma_{zz} = \sigma_{zr} = 0$. The load is of magnitude $Z$, pointing downward, and is a step function of time. The elastic constants $\lambda$ and $G$ are assumed to be equal, corresponding to Poisson's ratio $\nu = 0.25$. The ratio of the

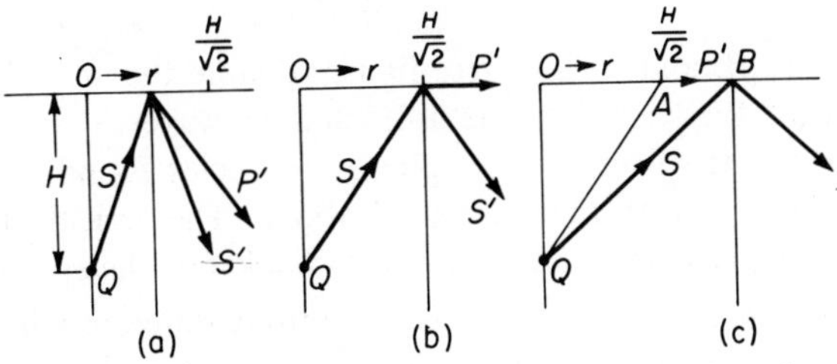


**Fig. 8.17:1.** Ray diagram of the reflection of a $SV$ wave at a free surface. (Courtesy Pekeris and Lifson.)

longitudinal and transverse wave velocities $c_L/c_T$ is therefore $\sqrt{3}$. We shall examine the motion of the free surface $z = 0$.

The high-speed compression wave which is the first to reach the surface varies, generally, in a monotone fashion. The slow-speed shear wave is of more complicated nature due to diffraction effects. In Fig. 8.17:1(a), $r < H/\sqrt{2}$, the $S$-wave incident on the surface is reflected partially as an $S'$-wave and partially as a derived $P'$ wave. At $r = H/\sqrt{2}$, $P'$ propagates in the horizontal direction, as shown in Fig. 8.17:1(b). At this point the shear wave is totally reflected. At $r > H/\sqrt{2}$, such as the point $B$ in Fig. 8.17:1(c), there is, in addition to the direct shear wave $S$, another wave which travels along the path $QAB$, covering the leg $QA$ with the shear wave speed $c_T$ and the horizontal leg $AB$ with the longitudinal wave speed $c_L$. This diffracted wave is denoted by $SP$. It can be shown that the travel time for $SP$ along $QAB$ is less than that for $S$ along $QB$, in spite of its longer trajectory. The order of arrival of the various phases is therefore, $P$, $S$, when $r < H/\sqrt{2}$, and $P$, $SP$, $S$ when $r > H/\sqrt{2}$.

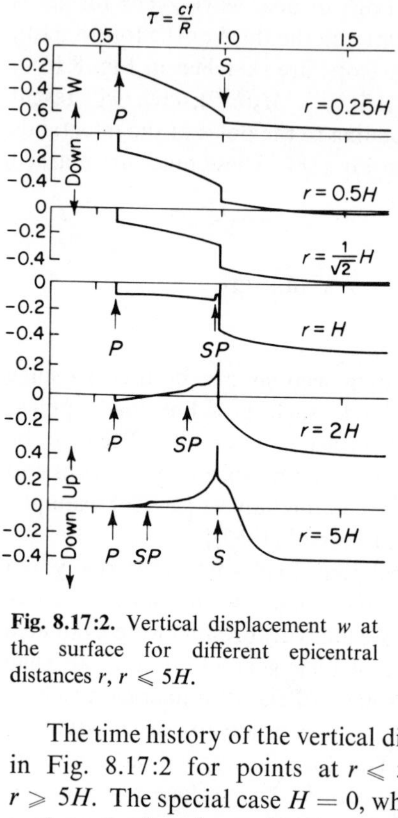


**Fig. 8.17:2.** Vertical displacement $w$ at the surface for different epicentral distances $r$, $r \leqslant 5H$.

The time history of the vertical displacement $w$ at the free surface is shown in Fig. 8.17:2 for points at $r \leqslant 5H$, and in Fig. 8.17:3 for points at $r \geqslant 5H$. The special case $H = 0$, which corresponds to a pulse applied at the surface, is also shown in Fig. 8.17:3. The coordinates $\tau$ and $W$ are non-dimensional and are related to physical quantities $w$, $r$, $t$ as follows:

$$R = (r^2 + H^2)^{1/2}, \qquad w = \frac{3Z}{\pi^2 GR} W, \qquad \tau = \frac{ct}{R}.$$

The points $P$, $S$, $R$, and $SP$ in these figures denote the arrival of the compressional, shear, Rayleigh, and the diffracted $SP$ waves.

The order of arrival of various phases of waves is indicated on these figures, in agreement with that discussed above. The arrival of the $P$-wave signals a finite displacement, which becomes weaker as the epicenter distance increases. The character of the $S$ phase is entirely different when $r < H/\sqrt{2}$ from the case when $r > H/\sqrt{2}$. In the former case, the arrival of the $S$-wave is marked by a finite jump in the displacement, whereas in the latter case the

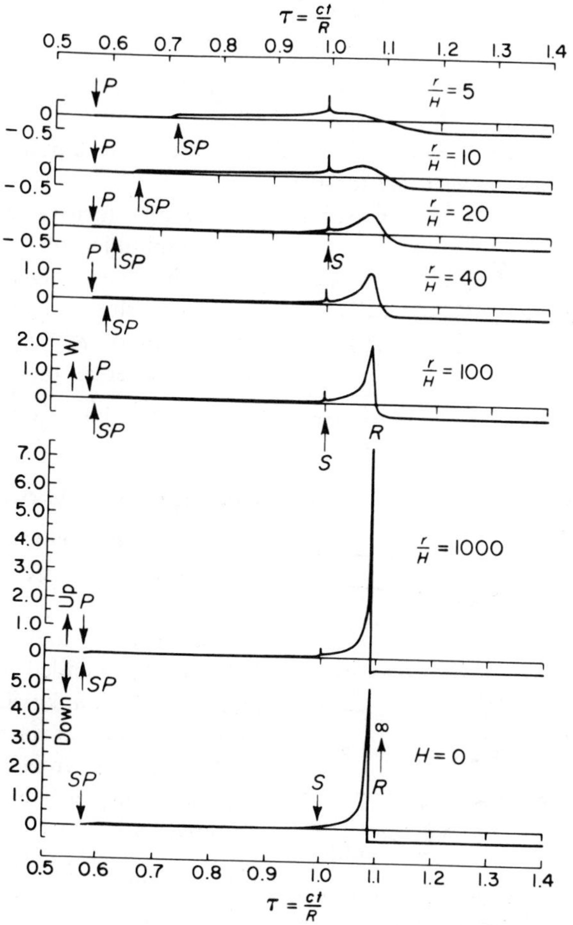


**Fig. 8.17:3.** Vertical displacement $w$ at the surface for different epicentral distances $r$, $r \geqslant 5H$. (Courtesy Pekeris and Lifson.)

$S$-phase is marked by a *logarithmic infinity* in the displacement directed opposite to the former. The $SP$-phase is rather weak, its strength increases as $r$ increases. The displacement following the arrival of $SP$-phase is upward and outward.

Referring to Fig. 8.17:2, we see that there is no sign of a surface wave for ranges $r < H/\sqrt{2}$. At $r = 5H$, the Rayleigh wave begins to emerge. At large epicentral distances, the Rayleigh surface wave is obviously the most important phase.

The time history of the horizontal (radial) displacement $q$ at the free surface is shown in Figs. 8.17:4 and 8.17:5. As before the points $P$, $S$, $R$, and $SP$ denote the arrival of the compressional, shear, Rayleigh, and the diffracted $SP$ waves, respectively. The dimensionless displacement $Q'$ is related to the physical quantity $q$ by the equation

$$q = -\frac{3Z}{\pi^2 GR} Q'.$$

The Rayleigh waves are more clearly discernible in the horizontal displacement when $r$ lies between $5H$ to $10H$.

The coordinates used in Figs. 8.17:2–8.17:4 are dimensionless quantities. The amplitude of the Rayleigh waves in these figures increases with increasing range because an extra factor of $R$ is applied for the ordinate, whereas the surface waves decrease only like $1/\sqrt{R}$.

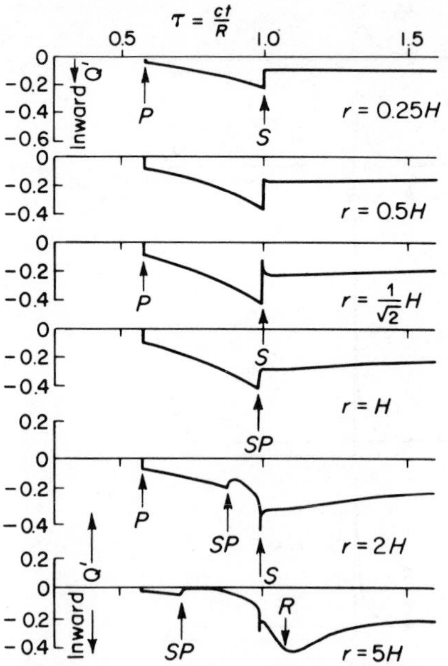


**Fig. 8.17:4.** Horizontal displacement $q$ at the surface for different epicentral distances $r$ ($r \leqslant 5H$).

## 8.18. GALERKIN VECTOR AND NEUBER-PAPKOVICH FUNCTIONS IN DYNAMICS

Iacovache,[8.1] and Sternberg and Eubanks[8.1] extended the Galerkin and Neuber-Papkovich representation of statics to wave equations.

We have shown in Sec. 8.2 that if the elastic displacements $u_i$ are represented by

(1) $$2Gu_i = \phi_{,i} + e_{ijk}\psi_{k,j}$$

$$\psi_{i,i} = 0, \quad i = 1, 2, 3,$$

then Navier's Eqs. (7.1:9) are satisfied by arbitrary functions $\phi$ and $\psi_k$ which satisfy the equations

(2) $$\Box_1 \phi = \Phi, \qquad \Box_2 \psi_k = \Psi_k, \qquad k = 1, 2, 3$$

(3) $$\Delta\Phi = -2\frac{c_T^2}{c_L^2} X_{i,i}, \qquad \Delta\Psi_k = -2e_{kml}X_{m,l},$$

(4) $$\frac{c_L^2}{c_T^2}\Phi_{,i} + e_{ijk}\Psi_{k,j} + 2X_i = 0,$$

where

(5) $$\Delta \equiv \frac{\partial^2}{\partial x_1^2} + \frac{\partial^2}{\partial x_2^2} + \frac{\partial^2}{\partial x_3^2},$$

(6) $$\Box_1 \equiv \Delta - \frac{1}{c_L^2}\frac{\partial^2}{\partial t^2}, \qquad \Box_2 \equiv \Delta - \frac{1}{c_T^2}\frac{\partial^2}{\partial t^2}.$$

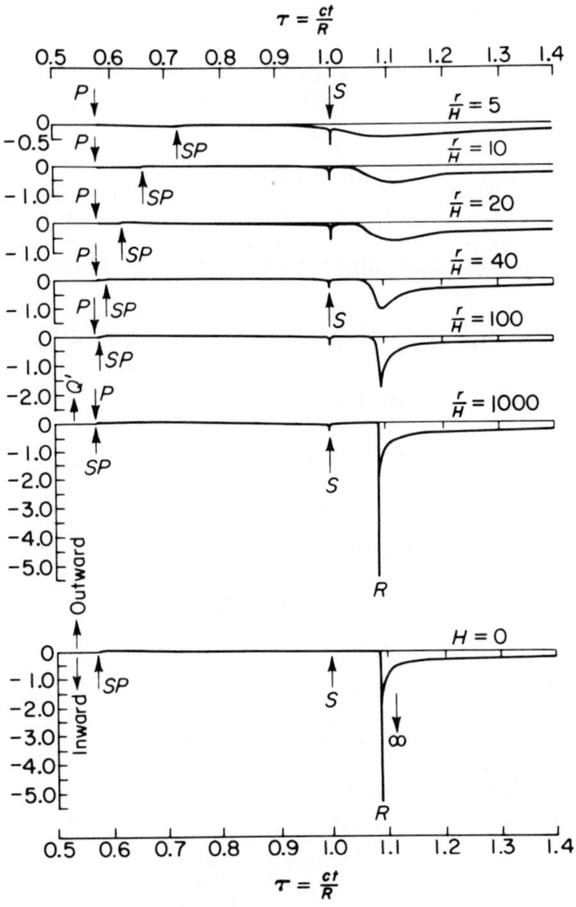


**Fig. 8.17:5.** Horizontal displacement $q$ at the surface for different epicentral distances $r$ ($r \geqslant 5H$). (Courtesy Pekeris and Lifson.)

In Eq. (4), we have retained the body force term $X_i$, which was omitted in Sec. 8.2. We have also added the constant factor $2G$ in Eq. (1) so that the final result may be directly compared with those of Secs. 8.4 and 8.12.

Now let us define

$$F_i = \frac{1}{2(1-\nu)}\left(\frac{c_L^2}{c_T^2}\,\phi'_{,i} + e_{ijk}\psi'_{k,j}\right), \tag{7}$$

where

$$\Box_2\phi' = \phi, \qquad \Box_1\psi'_k = \psi_k. \tag{8}$$

Then by (8), (2), and (4),

$$\Box_1\Box_2 F_i = \frac{1}{2(1-\nu)}\left[\frac{c_L^2}{c_T^2}\Phi_{,i} + e_{ijk}\Psi_{k,j}\right] = -\frac{X_i}{1-\nu}. \tag{9}$$

Furthermore, from (7), we have

$$(10) \qquad F_{i,i} = \frac{1}{2(1-\nu)} \frac{c_L^2}{c_T^2} \phi'_{,ii}$$

and

$$(11) \qquad \Box_1 F_i = \frac{1}{2(1-\nu)} \left[ \frac{c_L^2}{c_T^2} \Box_1 \phi'_{,i} + e_{ijk} \psi_{k,j} \right].$$

Solving (11) for $e_{ijk}\psi_{k,j}$ and substituting into (1), we obtain

$$2Gu_i = \phi_{,i} + 2(1-\nu)\Box_1 F_i - \frac{c_L^2}{c_T^2} \Box_1 \phi'_{,i}.$$

Expressing $\phi$ by $\phi'$ by (8), so that

$$2Gu_i = \left(1 - \frac{c_L^2}{c_T^2}\right) \Delta \phi'_{,i} + 2(1-\nu)\Box_1 F_i.$$

Using (10), and remembering that

$$\frac{c_T^2}{c_L^2} = \frac{1-2\nu}{2(1-\nu)},$$

we obtain finally

$$(12) \ \blacktriangle \qquad 2Gu_i = -F_{j,ji} + 2(1-\nu)\Box_1 F_i.$$

This is a general representation for $u_i$ which satisfies the Navier's equation if $F_i$ satisfies Eq. (9),

$$(9) \ \blacktriangle \qquad \Box_1 \Box_2 F_i = -\frac{X_i}{1-\nu}.$$

Equation (12) may be compared with Eq. (8.4:8). It is seen that the operator $\Box_1$ has replaced $\Delta$, and that $F_i$ satisfies Eq. (9) rather than the condition of biharmonicity. If the body force term is removed, the reader may prove the interesting fact that when $c_L \neq c_T$, any quantity $F_i$ satisfying Eq. (9) can be written as a sum $F'_i + F''_i$, where $\Box_1 F'_i = 0$, $\Box_2 F''_i = 0$.

To extend the Neuber-Papkovich representation (Sec. 8.12), we define

$$(13) \qquad \eta_i = 2(1-\nu)\Box_1 F_i.$$

$$(14) \qquad \omega = -F_{j,j} + \frac{1}{4(1-\nu)} x_j \eta_j.$$

Then Eq. (12) gives

$$(15) \qquad 2Gu_i = -\frac{1}{4(1-\nu)} (x_j \eta_j)_{,i} + \eta_i + \omega_{,i}.$$

Applying the operators $\Box_2$ and $\Box_1$ on Eqs. (13) and (14), respectively, we see that $\eta_i$ and $\omega$ must obey the following equations instead of being harmonic:

$$\Box_2 \eta_i = -2X_i, \tag{16}$$

$$\Box_1 \omega = \frac{1}{4(1-\nu)} x_j \Box_1 \eta_j. \tag{17}$$

## PROBLEMS

**8.11.** Derive potentials to solve the following equations:
Example:

$$\frac{\partial u}{\partial y} - \frac{\partial v}{\partial x} = 0 \quad \text{is solved by taking } u = \frac{\partial \phi}{\partial x}, \ v = \frac{\partial \phi}{\partial y}.$$

(a) $\dfrac{\partial u}{\partial x} + \dfrac{\partial v}{\partial y} = 0.$

(b) Plane stress:

$$\frac{\partial \sigma_x}{\partial x} + \frac{\partial \tau_{xy}}{\partial y} = 0, \qquad \frac{\partial \tau_{xy}}{\partial x} + \frac{\partial \sigma_y}{\partial y} = 0.$$

(c) In the theory of membrane stresses in a flat plate:

$$\frac{\partial N_x}{\partial x} + \frac{\partial N_{xy}}{\partial y} = 0, \qquad \frac{\partial N_{xy}}{\partial x} + \frac{\partial N_y}{\partial y} = 0.$$

(d) In theory of bending of plates:

$$\frac{\partial Q_x}{\partial x} + \frac{\partial Q_y}{\partial y} = 0, \qquad \frac{\partial M_x}{\partial x} + \frac{\partial M_{xy}}{\partial y} = Q_x, \qquad \frac{\partial M_{xy}}{\partial x} + \frac{\partial M_y}{\partial y} = Q_y.$$

**8.12.** A train of plane wave of wave length $L$ and phase velocity $c$ can be represented as

$$\phi = A \exp\left[i\frac{2\pi}{L}(\nu_1 x + \nu_2 y + \nu_3 z \pm ct)\right]$$

where $A$ is a constant and $\nu_1, \nu_2, \nu_3$ are the direction cosines of a vector normal to the wave front. Consider suitable superposition of these waves, derive expressions for trains of

(a) Cylindrical waves,

(b) Spherical waves, such as those generated by a point source.

*Ans.* (a) $BJ_0(kr)\, e^{-\nu|z|}\, e^{i\omega t}$, $\qquad \nu^2 = k^2 - k_\alpha^2,\ r^2 = x^2 + y^2.$

(b) $C\dfrac{1}{R}\exp\left[\pm i(k_\alpha R - \omega t)\right]$, $\qquad R^2 = x^2 + y^2 + z^2.$

Here $B$, $C$ are constants, $k_\alpha = \omega/c$, and $k$ is a parameter. The analysis may be simplified by means of contour integrations, regarding some variable of integration as complex numbers.

**8.13.** High pressure vessels of steel will be designed on the basis of von Mises' yield criterion. Consider spherical and cylindrical tanks of outer radius $b$ and inner radius $a$. Compare the maximum internal pressure $p$ at which yielding occurs.

*Ans.* $J_2 - k^2 = 0$, $k$ = yield stress in simple shear.

$$\text{Sphere: } p_{\text{yield}} = \sqrt{3}\left[\left(\frac{b}{a}\right)^3 - 1\right]\left[\frac{5}{2}\left(\frac{b}{a}\right)^6 + \left(\frac{b}{a}\right)^3 + 1\right]^{-1/2} k\,.$$

$$\text{Cylinder: } p_{\text{yield}} = \left(1 - \frac{a^2}{b^2}\right)k\,, \qquad \text{(far away from the ends).}$$

If $b/a = 1.1$, $p_{\text{yield}} = 0.523k$ for sphere, $= 0.222k$ for cylinder.

**8.14.** Consider a spherical fluid gyroscope which consists of a hollow metallic sphere filled with a dense fluid. This sphere is rotated at an angular velocity $\omega$ about its polar axis. In a steady-state rotation, ($\omega$ = constant), what are the stresses and displacements in the shell due to the fluid pressure and the centrifugal forces acting on the shell?

**8.15.** Determine the frequencies of radial vibrations for a hollow sphere whose inner wall is rigid and the outer wall is free.

**8.16.** An infinitely long circular cylindrical hole of radius $a$ is drilled in an infinite elastic medium. A pressure load is suddenly applied in the hole and starts to travel at a constant speed, so that the boundary conditions on the surface of the hole are, at $r = a$,

$$\sigma_{rr} = p\mathbf{1}\left(t - \frac{|z|}{c}\right), \qquad \sigma_{rz} = \sigma_{r\theta} = 0.$$

The medium is initially quiescent. Determine the response of the medium.

**8.17.** Shock tubes are common tools used in aerodynamic research. A shock tube consists of a long cylindrical shell, closed at both ends. Near one end a thin diaphragm is inserted normal to the tube axis. On one side of the diaphragm the air is evacuated; on the other side, gas at high pressure is stored. In operation, the diaphragm is suddenly split, the onrushing gas from the high pressure side creates a shock front that travels down the evacuated tube.

Elastic waves are generated in the tube wall by the bursting of the diaphragm and by the moving shock wave. These elastic waves have some effect on the instrumentation and measurements. Discuss the transient elastic response of the shock tube wall.

**8.18.** An infinite elastic medium contains a spherical hole of radius $a$. A sinusoidally fluctuating pressure acts on the surface of the hole. Determine the displacement field in the medium.

# 9

# TWO-DIMENSIONAL PROBLEMS IN ELASTICITY

The application of the Airy stress function reduces elastostatic problems in plane stress and plane strain to boundary-value problems of a biharmonic equation. A general method of solution using the theory of functions of a complex variable is available. We shall discuss this method briefly and illustrate its utility in solving a few important problems.

In the latter part of this chapter we shall return to dynamics. The problem of a load moving at constant speed over an elastic half-space will be discussed.

Throughout this chapter $x, y, z$ represent a set of rectangular Cartesian coordinates, with respect to which the displacement components are written as $u$, $v$, $w$, the strain components are $e_{xx}$, $e_{xy}$, etc., and the stress components are $\sigma_{xx}$, $\sigma_{xy}$, etc. We recall the factor $\frac{1}{2}$ in our definition of the strain components:

$$e_{ij} = \frac{1}{2}\left(\frac{\partial u_i}{\partial x_j} + \frac{\partial u_j}{\partial x_i}\right).$$

When curvilinear coordinates are used, we retain the notations of Chapter 4, in which $u_i$ and $e_{ij}$ denote the *tensor* components of the displacement and the strain, respectively; whereas $\xi_i$, $\epsilon_{ij}$ denote the *physical* components of these tensors. See Secs. 4.10–4.12.

## 9.1. PLANE STATE OF STRESS OR STRAIN

If the stress components $\sigma_{zz}$, $\sigma_{zx}$, $\sigma_{zy}$ vanish everywhere,

$$\sigma_{zz} = \sigma_{zx} = \sigma_{zy} = 0, \tag{1}$$

the state of stress is said to be *plane stress* parallel to the $x, y$-plane. In this case,

$$\begin{aligned} e_{xx} &= \frac{1}{E}(\sigma_{xx} - \nu\sigma_{yy}), \qquad e_{yy} = \frac{1}{E}(\sigma_{yy} - \nu\sigma_{xx}), \\ e_{zz} &= -\frac{\nu}{E}(\sigma_{xx} + \sigma_{yy}), \qquad e_{xy} = \frac{1}{2G}\sigma_{xy}, \qquad e_{xz} = e_{yz} = 0, \end{aligned} \tag{2}$$

$$\sigma_{xx} = \frac{E}{1-\nu^2}(e_{xx} + \nu e_{yy}), \qquad \sigma_{yy} = \frac{E}{1-\nu^2}(e_{yy} + \nu e_{xx}),$$

(3)

$$\sigma_{xy} = \frac{E}{1+\nu} e_{xy},$$

(4)
$$\sigma_{xx} + \sigma_{yy} = \frac{E}{1-\nu}(e_{xx} + e_{yy}),$$

(5)
$$e_{xx} + e_{yy} = \frac{\partial u}{\partial x} + \frac{\partial v}{\partial y}.$$

Substituting (3) into the equation of equilibrium (7.1:5), we obtain the basic equations for plane stress,

(6) ▲
$$G\left(\frac{\partial^2 u}{\partial x^2} + \frac{\partial^2 u}{\partial y^2}\right) + G\frac{1+\nu}{1-\nu}\frac{\partial}{\partial x}\left(\frac{\partial u}{\partial x} + \frac{\partial v}{\partial y}\right) + X = \rho\frac{\partial^2 u}{\partial t^2},$$
$$G\left(\frac{\partial^2 v}{\partial x^2} + \frac{\partial^2 v}{\partial y^2}\right) + G\frac{1+\nu}{1-\nu}\frac{\partial}{\partial y}\left(\frac{\partial u}{\partial x} + \frac{\partial v}{\partial y}\right) + Y = \rho\frac{\partial^2 v}{\partial t^2}.$$

If the $z$-component of displacement $w$ vanishes everywhere, and if the displacements $u$, $v$ are functions of $x$, $y$ only, and not of $z$, the body is said to be in *plane strain* state parallel to the $x$, $y$-plane. In plane strain we must have,

(7)
$$\frac{\partial u}{\partial z} = \frac{\partial v}{\partial z} = w = 0, \quad \text{and} \quad \sigma_{zz} = \nu(\sigma_{xx} + \sigma_{yy}), \quad (\text{since } e_{zz} = 0).$$

The basic equation (7.1:9) becomes, in plane strain,

(8) ▲
$$G\left(\frac{\partial^2 u}{\partial x^2} + \frac{\partial^2 u}{\partial y^2}\right) + \frac{1}{1-2\nu}G\frac{\partial}{\partial x}\left(\frac{\partial u}{\partial x} + \frac{\partial v}{\partial y}\right) + X = \rho\frac{\partial^2 u}{\partial t^2},$$
$$G\left(\frac{\partial^2 v}{\partial x^2} + \frac{\partial^2 v}{\partial y^2}\right) + \frac{1}{1-2\nu}G\frac{\partial}{\partial y}\left(\frac{\partial u}{\partial x} + \frac{\partial v}{\partial y}\right) + Y = \rho\frac{\partial^2 v}{\partial t^2}.$$

If $\nu$ is replaced by $\nu/(1+\nu)$ in Eq. (8), then it assumes the form (6). Hence, any problem of a plane state of strain may be solved as a problem of a plane state of stress after replacing the true value of $\nu$ by the "apparent value" $\nu/(1+\nu)$.† Conversely, any plane stress problem may be solved as a problem of plane strain by replacing the true value of $\nu$ by an apparent value $\nu/(1-\nu)$.†

The strain state in a long cylindrical body acted on by loads that are normal to the axis of the cylinder and uniform in the axial direction often can be approximated by a plane strain state. A constant axial strain $e_{zz}$ may be imposed on a plane strain state without any change in stresses in the $x$, $y$-plane. Hence a minor extension of the definition of plane strain can be

† This substitution refers only to the field equations (6) and (8). The boundary conditions, the stress-strain relationship, and the shear modulus $G$ are not to be changed.

formulated by requiring that $e_{zz}$ be a constant, that $u$ and $v$ be functions of $x, y$ only, and that $w$ be a linear function of $z$ only.

The state of stress in a thin flat plate acted on by forces parallel to the midplane of the plate is approximately plane stress. However, since in general $e_{zz}$ does not vanish, the displacements $u, v, w$ are functions of $z$, and the problem is not truly two-dimensional. In fact, it can be shown that the general state of plane stress, satisfying Eqs. (1) and the equations of equilibrium and the Beltrami-Michell compatibility conditions (7.3:6), is one in which the stresses $\sigma_{xx}, \sigma_{yy}, \sigma_{xy}$ are parabolically distributed throughout the thickness of the plate, i.e., of the form $f(x, y) + g(x, y)z^2$. (See Timoshenko and Goodier,[1.2] p. 241.) However, the part proportional to $z^2$ can be made as small as we please compared with the first term, by restricting ourselves to plates which are sufficiently thin (with the ratio $h/L \to 0$, where $h$ is the plate thickness and $L$ is a characteristic dimension of the plate).

## 9.2. AIRY STRESS FUNCTIONS FOR TWO-DIMENSIONAL PROBLEMS

For plane stress or plane strain problems, we may try to find general stress systems that satisfy the equations of equilibrium and compatibility and then determine the solution to a particular problem by the boundary conditions.

Let $x, y$ be a set of rectangular Cartesian coordinates. For plane stress and plane strain problems in the $x, y$-plane, the equations of equilibrium (3.4:2) are specialized into

$$\frac{\partial \sigma_{xx}}{\partial x} + \frac{\partial \sigma_{xy}}{\partial y} = -X, \tag{1}$$

$$\frac{\partial \sigma_{yy}}{\partial y} + \frac{\partial \sigma_{yx}}{\partial x} = -Y, \tag{2}$$

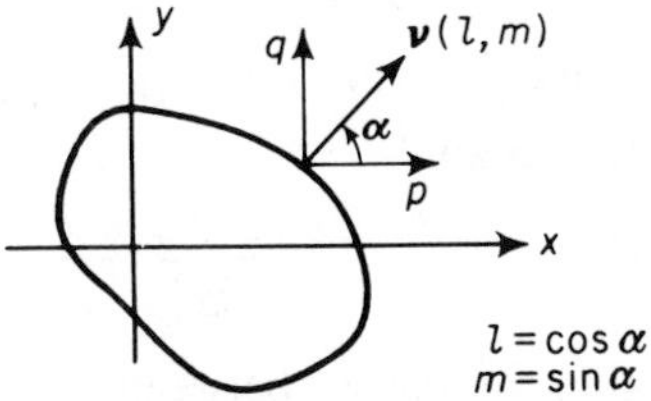


**Fig. 9.2:1.** Notations.

with the boundary conditions

$$l\sigma_{xx} + m\sigma_{xy} = p, \qquad m\sigma_{yy} + l\sigma_{xy} = q, \tag{3}$$

where $l, m$ are the direction cosines of the outer normal to the boundary curve and where $p, q$ are surface tractions acting on the boundary surface.

The strain components are,

(a) *In the plane stress case:*

$$e_{xx} = \frac{1}{E}(\sigma_{xx} - \nu\sigma_{yy}), \qquad e_{yy} = \frac{1}{E}(\sigma_{yy} - \nu\sigma_{xx}),$$
$$e_{xy} = \frac{1}{2G}\sigma_{xy} = \frac{(1+\nu)}{E}\sigma_{xy}; \tag{4}$$

(b) *In the plane strain case:*

$$e_{xx} = \frac{1}{E}[(1-\nu^2)\sigma_{xx} - \nu(1+\nu)\sigma_{yy}],$$

(5) $$e_{yy} = \frac{1}{E}[(1-\nu^2)\sigma_{yy} - \nu(1+\nu)\sigma_{xx}],$$

$$e_{xy} = \frac{1}{E}(1+\nu)\sigma_{xy}.$$

In view of what was discussed in the preceding section, for very thin plates we may assume $\sigma_{xx}, \sigma_{yy}, \sigma_{xy}$ to be independent of $z$. Then the plane stress problem becomes truly two-dimensional, as well as the plane strain problem.

The compatibility conditions are as follows (see Sec. 4.6):

(6)
$$\frac{\partial^2 e_{xx}}{\partial y^2} + \frac{\partial^2 e_{yy}}{\partial x^2} = 2\frac{\partial^2 e_{xy}}{\partial x\,\partial y}, \qquad \frac{\partial^2 e_{xx}}{\partial y\,\partial z} = \frac{\partial}{\partial x}\left(-\frac{\partial e_{yz}}{\partial x} + \frac{\partial e_{xz}}{\partial y} + \frac{\partial e_{xy}}{\partial z}\right),$$
$$\frac{\partial^2 e_{yy}}{\partial z^2} + \frac{\partial^2 e_{zz}}{\partial y^2} = 2\frac{\partial^2 e_{yz}}{\partial y\,\partial z}, \qquad \frac{\partial^2 e_{yy}}{\partial x\, dz} = \frac{\partial}{\partial y}\left(\frac{\partial e_{yz}}{\partial x} - \frac{\partial e_{xz}}{\partial y} + \frac{\partial e_{xy}}{\partial z}\right),$$
$$\frac{\partial^2 e_{zz}}{\partial x^2} + \frac{\partial^2 e_{xx}}{\partial z^2} = 2\frac{\partial^2 e_{xz}}{\partial z\,\partial x}, \qquad \frac{\partial^2 e_{zz}}{\partial x\,\partial y} = \frac{\partial}{\partial z}\left(\frac{\partial e_{yz}}{\partial x} + \frac{\partial e_{xz}}{\partial y} - \frac{\partial e_{xy}}{\partial z}\right).$$

On substituting (4) into the first equation of (6), we obtain, in the plane stress case,

(7) $$\frac{\partial^2}{\partial y^2}(\sigma_{xx} - \nu\sigma_{yy}) + \frac{\partial^2}{\partial x^2}(\sigma_{yy} - \nu\sigma_{xx}) = 2(1+\nu)\frac{\partial^2 \tau_{xy}}{\partial x\,\partial y}.$$

Differentiating (1) with respect to $x$ and (2) with respect to $y$ and adding, we obtain

(8) $$\frac{\partial^2 \sigma_{xx}}{\partial x^2} + \frac{\partial^2 \sigma_{yy}}{\partial y^2} + \frac{\partial X}{\partial x} + \frac{\partial Y}{\partial y} = -2\frac{\partial^2 \tau_{xy}}{\partial x\,\partial y}.$$

Eliminating $\tau_{xy}$ between (7) and (8), we obtain

(9) $$\left(\frac{\partial^2}{\partial x^2} + \frac{\partial^2}{\partial y^2}\right)(\sigma_{xx} + \sigma_{yy}) = -(1+\nu)\left(\frac{\partial X}{\partial x} + \frac{\partial Y}{\partial y}\right).$$

Similarly, in the plane strain case, we have

(10) $$\left(\frac{\partial^2}{\partial x^2} + \frac{\partial^2}{\partial y^2}\right)(\sigma_{xx} + \sigma_{yy}) = -\frac{1}{(1-\nu)}\left(\frac{\partial X}{\partial x} + \frac{\partial Y}{\partial y}\right).$$

Equations (1), (2), (3), and (9) or (10) define the plane problems in terms of the stress components $\sigma_{xx}, \sigma_{yy}, \sigma_{xy}$. If the boundary conditions of a problem are such that surface tractions are all known, then the problem can be solved in terms of stresses, with no need to mention displacements unless they are desired. Even in a mixed boundary-value problem in which part of the boundary has prescribed displacements, it still may be advantageous to solve for the stress state first. These practical considerations lead to the method of Airy stress function.†

Airy's method is based on the observation that the left hand side of Eqs. (1) and (2) appears as the divergence of a vector. In hydrodynamics we are familiar with the fact that the conservation of mass, expressed in the equation of continuity

$$\frac{\partial u}{\partial x} + \frac{\partial v}{\partial y} = 0, \tag{11}$$

where $u, v$ are components of the velocity vector, can be derived from an arbitrary stream function $\psi(x, y)$:

$$u = \frac{\partial \psi}{\partial y}, \qquad v = -\frac{\partial \psi}{\partial x}. \tag{12}$$

In other words, if $u, v$ are derived from an arbitrary $\psi(x, y)$ according to (12), then Eq. (11) is satisfied identically.

Let us use the same technique for Eqs. (1) and (2). These equations can be put into the form of (11) if we assume that the body forces can be derived from a potential $V$, so that

$$X = -\frac{\partial V}{\partial x}, \qquad Y = -\frac{\partial V}{\partial y}. \tag{13}$$

A substitution of (13) into (1) and (2) results in

$$\frac{\partial}{\partial x}(\sigma_{xx} - V) + \frac{\partial \sigma_{xy}}{\partial y} = 0, \qquad \frac{\partial \sigma_{xy}}{\partial x} + \frac{\partial}{\partial y}(\sigma_{yy} - V) = 0. \tag{14}$$

Now, as in (11), these equations are identically satisfied if we introduce two stream functions $\Psi$ and $\chi$ in such a way that

$$\begin{aligned} \sigma_{xx} - V &= \frac{\partial \Psi}{\partial y}, & \sigma_{xy} &= -\frac{\partial \Psi}{\partial x}, \\ \sigma_{xy} &= -\frac{\partial \chi}{\partial y}, & \sigma_{yy} - V &= \frac{\partial \chi}{\partial x}. \end{aligned} \tag{15}$$

In other words, a substitution of (15) into (14) reduces (14) into an identity in $\Psi$ and $\chi$. Now, Eqs. (15) can be combined if we let

$$\chi = \frac{\partial \Phi}{\partial x}, \qquad \Psi = \frac{\partial \Phi}{\partial y}; \tag{16}$$

i.e.,

$$\blacktriangle \quad \sigma_{xx} - V = \frac{\partial^2 \Phi}{\partial y^2}, \qquad \sigma_{xy} = -\frac{\partial^2 \Phi}{\partial x\, \partial y}, \qquad \sigma_{yy} - V = \frac{\partial^2 \Phi}{\partial x^2}. \tag{17}$$

† For problems in which displacements are prescribed over the entire boundary, the displacement potential or other devices of the preceding chapter should be tried first.

It is readily verified that if $\sigma_{xx}$, $\sigma_{xy}$, $\sigma_{yy}$ are derived from an arbitrary function $\Phi(x, y)$ according to (17), then Eqs. (14) are identically satisfied. The function $\Phi(x, y)$ is called the *Airy stress function*, in deference to its inventor, the famous astronomer.

An arbitrary function $\Phi(x, y)$ generates stresses that satisfy the equations of equilibrium, but $\Phi$ is not entirely arbitrary: it is required to generate only those stress fields that satisfy the conditions of compatibility. Since the compatibility condition is given by (9) or (10), a substitution gives the requirement that, in the plane stress case,

$$(18) \quad \blacktriangle \qquad \frac{\partial^4 \Phi}{\partial x^4} + 2\frac{\partial^4 \Phi}{\partial x^2\, \partial y^2} + \frac{\partial^4 \Phi}{\partial y^4} = -(1-\nu)\left(\frac{\partial^2 V}{\partial x^2} + \frac{\partial^2 V}{\partial y^2}\right),$$

and that, in the plane strain case,

$$(19) \quad \blacktriangle \qquad \frac{\partial^4 \Phi}{\partial x^4} + 2\frac{\partial^4 \Phi}{\partial x^2\, \partial y^2} + \frac{\partial^4 \Phi}{\partial y^4} = -\frac{(1-2\nu)}{(1-\nu)}\left(\frac{\partial^2 V}{\partial x^2} + \frac{\partial^2 V}{\partial y^2}\right).$$

If the body forces vanish, then, in both plane stress and plane strain, $\Phi$ is governed by the equation

$$(20) \quad \blacktriangle \qquad \frac{\partial^4 \Phi}{\partial x^4} + 2\frac{\partial^4 \Phi}{\partial x^2\, \partial y^2} + \frac{\partial^4 \Phi}{\partial y^4} = 0.$$

A regular solution of Eq. (20) is called a *biharmonic function*. Solution of plane elasticity problems by biharmonic functions will be discussed in the following sections.

How about the other five compatibility conditions in Eqs. (6) left alone so far? In the case of plane strain it is clear that they are identically satisfied. In the case of plane stress, however, they cannot be satisfied in general if we assume $\sigma_{xx}$, $\sigma_{yy}$, $\sigma_{xy}$ to be independent of $z$. For, under such an assumption these compatibility conditions imply that

$$(21) \qquad \frac{\partial^2 e_{zz}}{\partial x^2} = \frac{\partial^2 e_{zz}}{\partial y^2} = \frac{\partial^2 e_{zz}}{\partial x\, \partial y} = 0.$$

Hence, $e_{zz}$, and hence $\sigma_{xx} + \sigma_{yy}$, must be a linear function of $x$ and $y$ $[e_{zz} = -\nu(\sigma_{xx} + \sigma_{yy})/E]$, which is the exception rather than the rule in the solution of plane stress problems. Hence, in general, the assumption that plane stress state is two-dimensional, so that $\sigma_{xx}$, $\sigma_{yy}$, $\sigma_{xy}$ are functions of $x, y$ only, cannot be true; and the solutions obtained under this assumption cannot be exact. However, as we have discussed previously (Sec. 9.1), they are close approximations for thin plates.

The stress-function method can be extended to three dimensions. The crucial observation is simply that the equation of equilibrium represents a vector divergence of the stress tensor. We are familiar with the stream function in hydrodynamics. In three-dimensions we need a triple of stream functions. Similarly, a generalization of Airy's procedure to equations of equilibrium in three dimensions requires a *tensor* of stress functions. Finzi[8.1] (1934) showed that a general solution to the equations

$$\sigma_{ij,j} = 0, \qquad \sigma_{ij} = \sigma_{ji} \tag{22}$$

is

$$\sigma_{ij} = e_{imr}e_{jns}\,\phi_{rs,mn}, \tag{23}$$

where $\phi_{rs}$ stands for the components of a symmetric second-order tensor of stress functions, while $e_{imr}$ is the usual alternator (Sec. 2.1). Specialization by taking $\phi_{rs} = 0$ ($r \neq s$), yields Maxwell's stress functions, while taking $\phi_{rr} = 0$ (no sum), yields Morera's stress function, see Sec. 10.9, Eqs. (10.9:19), et seq. If all elements $\phi_{rs}$, except $\phi_{33}$, are assumed to vanish, then (23) degenerates into Airy's solution of the two-dimensional equilibrium equations. An elegant proof of Finzi's result that is applicable to $n$-dimensional Euclidean space was given by Dorn and Schild.[8.1]

Finzi (1934) obtained further a beautiful extension to the equations of *motion* of a continuum, with an arbitrary density field, by introducing a fourth dimension. Finzi's arbitrary tensor yields by differentiation a motion and a stress field that satisfy the equation of motion.

For a curved space (nonEuclidean), Truesdell[16.1] obtained related results by methods of calculus of variations. The curved space problem arises naturally in the intrinsic theory of thin shells or membranes, for the two-dimensional surface is, to a two-dimensional observer who is not allowed to leave that surface, a nonEuclidean space (imbedded, of course, in a three-dimensional Euclidean space).

*Example 1*

The following polynomials of second and third degree are obviously biharmonic.

$$\Phi_2 = a_2x^2 + b_2xy + c_2y^2,$$
$$\Phi_3 = a_3x^3 + b_3x^2y + c_3xy^2 + d_3y^3.$$

By adjusting the constants $a_2$, $a_3$, etc., many problems in which the stresses are linearly distributed on rectangular boundaries can be solved. Examples of such problems are shown in Fig. 9.2:2.

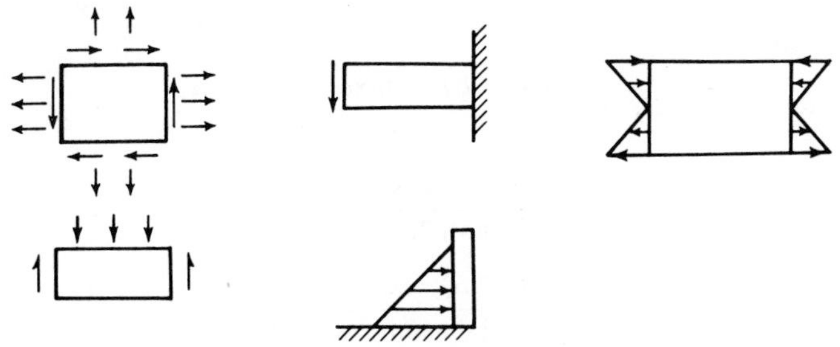

**Fig. 9.2:2.** Examples of problems solvable by simple polynomials.

*Example 2*

Consider a rectangular beam (Fig. 9.2:3) supported at the ends and subjected to the surface tractions

$$\text{on } y = +c \quad \sigma_{xy} = 0, \quad \sigma_{yy} = -B \sin \alpha x,$$
$$\text{on } y = -c \quad \sigma_{xy} = 0, \quad \sigma_{yy} = -A \sin \alpha x.$$

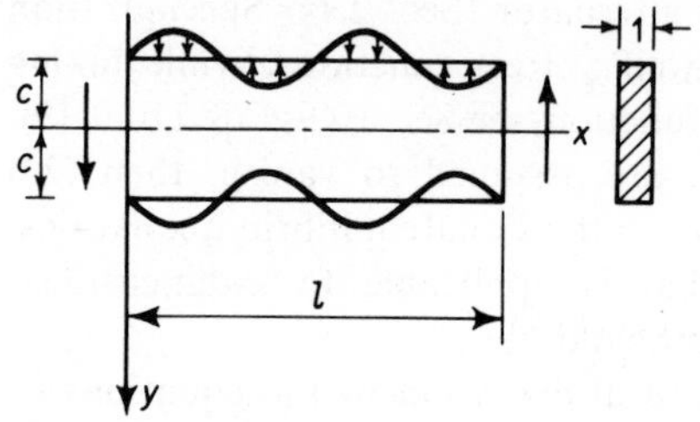

**Fig. 9.2:3.** A beam subjected to a sinusoidally distributed loading.

Other edge conditions are unspecified at the beginning, and the body forces are absent.

*Solution.* Let

$$\Phi = \sin \alpha x f(y),$$

where $f(y)$ is a function of $y$ only. A substitution into Eq. (20) yields

$$\alpha^4 f(y) - 2\alpha^2 f''(y) + f^{\text{iv}}(y) = 0.$$

The general solution is

$$f(y) = C_1 \cosh \alpha y + C_2 \sinh \alpha y + C_3 y \cosh \alpha y + C_4 y \sinh \alpha y.$$

Hence,

$$\Phi = \sin \alpha x\,(C_1 \cosh \alpha y + C_2 \sinh \alpha y + C_3 y \cosh \alpha y + C_4 y \sinh \alpha y)$$

$$\sigma_{xx} = \sin \alpha x\,[C_1 \alpha^2 \cosh \alpha y + C_2 \alpha^2 \sinh \alpha y + C_3(y\alpha^2 \cosh \alpha y + 2\alpha \sinh \alpha y) + C_4(y\alpha^2 \sinh \alpha y + 2\alpha \cosh \alpha y)]$$

$$\sigma_{yy} = -\alpha^2 \sin \alpha x\,[C_1 \cosh \alpha y + C_2 \sinh \alpha y + \ldots],$$

$$\sigma_{xy} = \alpha \cos \alpha x\,[C_1 \alpha \sinh \alpha y + C_2 \alpha \cosh \alpha y + C_3(\cosh \alpha y + y\alpha \sinh \alpha y) + C_4(\sinh \alpha y + y\alpha \cosh \alpha y)].$$

On application of the boundary conditions $\sigma_{xy} = 0$ on $y = \pm c$, we can express $C_3$ and $C_4$ in terms of $C_1$ and $C_2$. The other boundary conditions then yield the constants

$$C_1 = \frac{(A + B)}{\alpha^2} \frac{\sinh \alpha c + \alpha c \cosh \alpha c}{\sinh 2\alpha c + 2\alpha c},$$

$$C_2 = -\frac{(A - B)}{\alpha^2} \frac{\cosh \alpha c + \alpha c \sinh \alpha c}{\sinh 2\alpha c - 2\alpha c},$$

$$C_3 = \frac{A - B}{\alpha^2} \frac{\alpha \cosh \alpha c}{\sinh 2\alpha c - 2\alpha c},$$

$$C_4 = -\frac{(A + B)}{\alpha^2} \frac{\alpha \sinh \alpha c.}{\sinh 2\alpha c + 2\alpha c}.$$

The details can be found in Timoshenko and Goodier,[1.2] p. 48.

## 9.3. AIRY STRESS FUNCTION IN POLAR COORDINATES

For two-dimensional problems with circular boundaries, polar coordinates can be used to advantage. Let $\xi_r, \xi_\theta, \xi_z$ denote physical components of the displacement, and let $\epsilon_{rr}, \epsilon_{r\theta}, \ldots, \sigma_{rr}, \sigma_{r\theta}, \ldots$, etc., be the physical components of the strain and stress, respectively. The general equations in cylindrical polar coordinates are given in Sec. 4.12. For *plane stress* problems, we assume that

$$\sigma_{zz} = \sigma_{zr} = \sigma_{z\theta} = 0. \tag{1}$$

For *plane strain* problems, we assume that

$$\xi_z = 0, \tag{2}$$

and that all derivatives with respect to $z$ vanish. In both cases, the strain components are defined as follows:

$$\begin{aligned} \epsilon_{rr} &= \frac{\partial \xi_r}{\partial r}, \qquad \epsilon_{\theta\theta} = \frac{1}{r}\frac{\partial \xi_\theta}{\partial \theta} + \frac{\xi_r}{r}, \\ \epsilon_{r\theta} &= \frac{1}{2}\left(\frac{1}{r}\frac{\partial \xi_r}{\partial \theta} + \frac{\partial \xi_\theta}{\partial r} - \frac{\xi_\theta}{r}\right). \end{aligned} \tag{3}$$

Furthermore, in plane stress,

$$\epsilon_{rr} = \frac{1}{E}(\sigma_{rr} - \nu\sigma_{\theta\theta}), \qquad \epsilon_{\theta\theta} = \frac{1}{E}(\sigma_{\theta\theta} - \nu\sigma_{rr}), \qquad \epsilon_{r\theta} = \frac{1+\nu}{E}\sigma_{r\theta}, \tag{4}$$

and in plane strain,

$$\begin{aligned} \epsilon_{rr} &= \frac{1}{E}[(1-\nu^2)\sigma_{rr} - \nu(1+\nu)\sigma_{\theta\theta}], \\ \epsilon_{\theta\theta} &= \frac{1}{E}[(1-\nu^2)\sigma_{\theta\theta} - \nu(1+\nu)\sigma_{rr}], \\ \epsilon_{r\theta} &= \frac{1+\nu}{E}\sigma_{r\theta}, \end{aligned} \tag{5}$$

and the equations of equilibrium become

$$\begin{aligned} &\frac{1}{r}\frac{\partial(r\sigma_{rr})}{\partial r} + \frac{1}{r}\frac{\partial \sigma_{r\theta}}{\partial \theta} - \frac{\sigma_{\theta\theta}}{r} + F_r = 0, \\ &\frac{1}{r^2}\frac{\partial(r^2\sigma_{\theta r})}{\partial r} + \frac{1}{r}\frac{\partial \sigma_{\theta\theta}}{\partial \theta} + F_\theta = 0. \end{aligned} \tag{6}$$

If the body forces $F_r$, $F_\theta$ are zero, Eqs. (6) are satisfied identically if the stresses are derived from a function $\Phi(r, \theta)$:

$$\sigma_{rr} = \frac{1}{r}\frac{\partial \Phi}{\partial r} + \frac{1}{r^2}\frac{\partial^2 \Phi}{\partial \theta^2},$$

(7) ▲
$$\sigma_{\theta\theta} = \frac{\partial^2 \Phi}{\partial r^2},$$

$$\sigma_{r\theta} = -\frac{\partial}{\partial r}\left(\frac{1}{r}\frac{\partial \Phi}{\partial \theta}\right),$$

as can be verified by direct substitution. The function $\Phi(r, \theta)$ is the *Airy stress function.*

The compatibility conditions (9.2:9) and (9.2:10) are, if body forces are zero,

(8)
$$\left(\frac{\partial^2}{\partial x^2} + \frac{\partial^2}{\partial y^2}\right)(\sigma_{xx} + \sigma_{yy}) = 0.$$

The sum $\sigma_{xx} + \sigma_{yy}$ is an invariant with respect to rotation of coordinates. Hence,

(9)
$$\sigma_{xx} + \sigma_{yy} = \sigma_{rr} + \sigma_{\theta\theta}.$$

The Laplace operator is transformed as

(10)
$$\frac{\partial^2}{\partial x^2} + \frac{\partial^2}{\partial y^2} = \frac{\partial^2}{\partial r^2} + \frac{1}{r}\frac{\partial}{\partial r} + \frac{1}{r^2}\frac{\partial^2}{\partial \theta^2}.$$

Hence, on substituting (7) into (9) and using (10), Eq. (8) is transformed into

(11) ▲
$$\left(\frac{\partial^2}{\partial r^2} + \frac{1}{r}\frac{\partial}{\partial r} + \frac{1}{r^2}\frac{\partial^2}{\partial \theta^2}\right)\left(\frac{\partial^2 \Phi}{\partial r^2} + \frac{1}{r}\frac{\partial \Phi}{\partial r} + \frac{1}{r^2}\frac{\partial^2 \Phi}{\partial \theta^2}\right) = 0.$$

This is the compatibility equation to be satisfied by the Airy stress function $\Phi(r, \theta)$. If one can find *a* solution of (11) that also satisfies the boundary conditions, then the problem is solved; since by Kirchhoff's uniqueness theorem, such a solution is unique.

***Axially Symmetric Problems.*** If $\Phi$ is a function of $r$ alone and is independent of $\theta$, all derivatives with respect to $\theta$ vanish and Eq. (11) becomes

(12)
$$\frac{d^4\Phi}{dr^4} + \frac{2}{r}\frac{d^3\Phi}{dr^3} - \frac{1}{r^2}\frac{d^2\Phi}{dr^2} + \frac{1}{r^3}\frac{d\Phi}{dr} = 0.$$

This is the homogeneous differential equation which can be reduced to a linear differential equation with constant coefficients by introducing a new variable $t$ such that $r = e^t$. The general solution is

(13)
$$\Phi = A \log r + Br^2 \log r + Cr^2 + D,$$

which corresponds to

$$\sigma_{rr} = \frac{A}{r^2} + B(1 + 2\log r) + 2C,$$

$$\sigma_{\theta\theta} = -\frac{A}{r^2} + B(3 + 2\log r) + 2C, \tag{14}$$

$$\sigma_{r\theta} = 0.$$

The solutions of all problems of symmetrical stress distribution and no body forces can be obtained from this.

The displacement components corresponding to stresses given by (14) may be obtained as follows. Consider the *plane stress* case. Substituting (14) into the first of the Eqs. (4) and (3), we obtain

$$\frac{\partial \xi_r}{\partial r} = \frac{1}{E}\left[\frac{(1+\nu)A}{r^2} + 2(1-\nu)B\log r + (1-3\nu)B + 2(1-\nu)C\right],$$

from which, by integration,

$$\xi_r = \frac{1}{E}\left[-\frac{(1+\nu)A}{r} + 2(1-\nu)Br\log r - B(1+\nu)r + 2C(1-\nu)r\right] + f(\theta), \tag{15}$$

where $f(\theta)$ is an arbitrary function of $\theta$ only. From (14) and the second of the Eqs. (4) and (3), we obtain

$$\frac{\partial \xi_\theta}{\partial \theta} = \frac{4Br}{E} - f(\theta).$$

Hence, by integration,

$$\xi_\theta = \frac{4Br\theta}{E} - \int_0^\theta f(\theta)\,d\theta + f_1(r), \tag{16}$$

where $f_1(r)$ is a function of $r$ only. Finally, from the last of Eqs. (14), (4) and (3), we find, since $\sigma_{r\theta} = \epsilon_{r\theta} = 0$,

$$\frac{1}{r}\frac{df(\theta)}{d\theta} + \frac{df_1(r)}{dr} + \frac{1}{r}\int_0^\theta f(\theta)\,d\theta - \frac{1}{r}f_1(r) = 0.$$

Multiplying throughout by $r$, we find that the first and the third term are functions of $\theta$ only and the other two terms are functions of $r$ only. Hence, the only possibility for the last equation to be satisfied is

$$\frac{df(\theta)}{d\theta} + \int_0^\theta f(\theta)\,d\theta = \alpha, \qquad r\frac{df_1(r)}{dr} - f_1(r) = -\alpha,$$

where $\alpha$ is an arbitrary constant. The solutions are

$$f(\theta) = \alpha\sin\theta + \gamma\cos\theta, \qquad f_1(r) = \beta r + \alpha,$$

where $\alpha$, $\beta$, $\gamma$ are arbitrary constants. Substituting back into (15) and (16), we obtain, for the *plane stress case*,

$$\xi_r = \frac{1}{E}\left[-\frac{(1+\nu)A}{r} + 2(1-\nu)Br\log r - B(1+\nu)r + 2C(1-\nu)r\right] + \alpha\sin\theta + \gamma\cos\theta, \tag{17}$$

$$\xi_\theta = \frac{4Br\theta}{E} + \alpha\cos\theta - \gamma\sin\theta + \beta r.$$

The arbitrary constants $A, B, C, \alpha, \beta, \gamma$ are to be determined from the boundary conditions of each special problem. The corresponding expressions for the *plane strain case* are

$$\xi_r = \frac{1}{E}\left[-\frac{(1+\nu)A}{r} - B(1+\nu)r + 2B(1-\nu-2\nu^2)r\log r + 2C(1-\nu-2\nu^2)r\right] + \alpha\sin\theta + \gamma\cos\theta, \tag{18}$$

$$\xi_\theta = \frac{4Br\theta}{E}(1-\nu^2) + \alpha\cos\theta - \gamma\sin\theta + \beta r.$$

*Example 1. Uniform Pressure Acting on a Solid Cylinder*

The boundary conditions are, at $r = a$,

$$\sigma_{rr} = \sigma_{\theta\theta} = -p_0, \qquad \sigma_{r\theta} = 0.$$

There should be no singularity in the solid. A glance at Eq. (14) shows that the problem can be solved by taking

$$A = B = 0, \qquad C = -\frac{p_0}{2}.$$

*Example 2. Circular Cylindrical Tube Subjected to Uniform Internal and External Pressure*

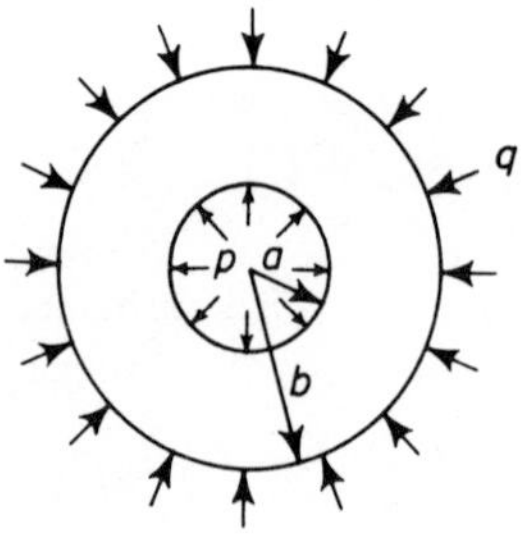


**Fig 9.3:1.** Circular tube.

For this problem (Fig. 9.3:1) the boundary conditions are

$$\sigma_{rr} = -p \quad \text{at} \quad r = a, \qquad \sigma_{rr} = -q \quad \text{at} \quad r = b. \tag{19}$$

The form of $\sigma_{rr}$ in Eq. (14) suggests that the boundary conditions can be satisfied by fixing two of the constants, say $A$ and $C$. From (19) alone, however, there is no reason to reject the term involving $B$; but inspection of the displacements given by Eq. (18) shows that if $B \neq 0$ the circumferential displacement $\xi_\theta$ will have a nonvanishing term

$$\xi_\theta = \frac{4Br\theta,}{E} \tag{20}$$

which is zero when $\theta = 0$ but becomes $\xi_\theta = 8\pi Br/E$ when one traces a circuit around the axis of symmetry and returns to the same point after turning around an angle $2\pi$. Thus the displacement given by (20) is not *single-valued.* Such a *multi-valued* expression for a displacement is physically impossible in a full cylindrical tube. Hence, $B = 0$.

It is simple to derive the constants $A$ and $C$ in the expression

$$\sigma_{rr} = \frac{A}{r^2} + 2C$$

such that (19) is satisfied. Thus we obtain Lamé's formulas for the stresses,

$$\sigma_{rr} = -p\frac{(b^2/r^2) - 1}{(b^2/a^2) - 1} - q\frac{1 - (a^2/r^2)}{1 - (a^2/b^2)}, \tag{21}$$
$$\sigma_{\theta\theta} = p\frac{(b^2/r^2) + 1}{(b^2/a^2) - 1} - q\frac{1 + (a^2/r^2)}{1 - (a^2/b^2)}.$$

It is interesting to note that $\sigma_{rr} + \sigma_{\theta\theta}$ is constant throughout the cylinder. If $q = 0$ and $p > 0$, $\sigma_{rr}$ is always a compressive stress and $\sigma_{\theta\theta}$ is always a tensile stress, the maximum value of which occurs at the inner radius and is always numerically greater than the internal pressure $p$.

*Example 3. Pure Bending of a Curved Bar*

The coefficients $A$, $B$, $C$ may be chosen to satisfy the conditions $\sigma_{rr} = 0$ for $r = a$ and $r = b$, and

$$\int_a^b \sigma_{\theta\theta}\, dr = 0,$$
$$\int_a^b \sigma_{\theta\theta} r\, dr = -M, \quad \text{and} \quad \sigma_{r\theta} = 0$$

at the boundary $\theta$ = const. (see Fig. 9.3:2). See Timoshenko and Goodier,[1.2] p. 61; Sechler,[1.2] p. 143.

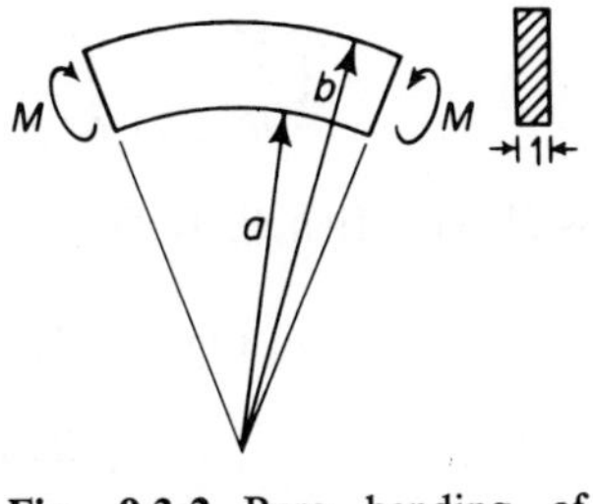


**Fig. 9.3:2.** Pure bending of a curved bar.

*Example 4. Initial Stress in a Welded Ring*

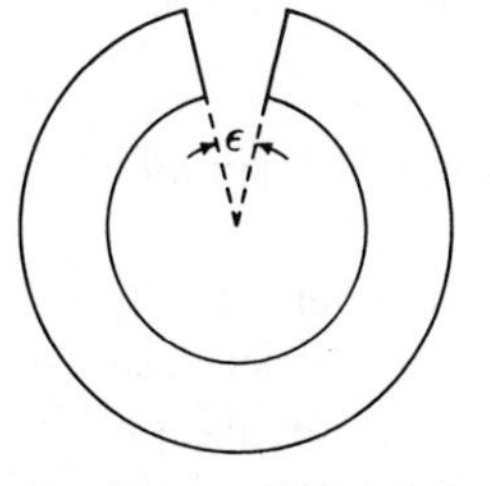


**Fig. 9.3:3.** Welding of a ring.

If the ends of an opening in a ring, as shown in Fig. 9.3:3, are joined together and welded, the initial stresses can be obtained from the expressions in Eq. (14) by taking

$$B = \frac{\epsilon E}{8\pi}. \tag{22}$$

See Eq. (17) for circumferential displacements. The constants $\alpha$, $\beta$, $\gamma$ are all zero.

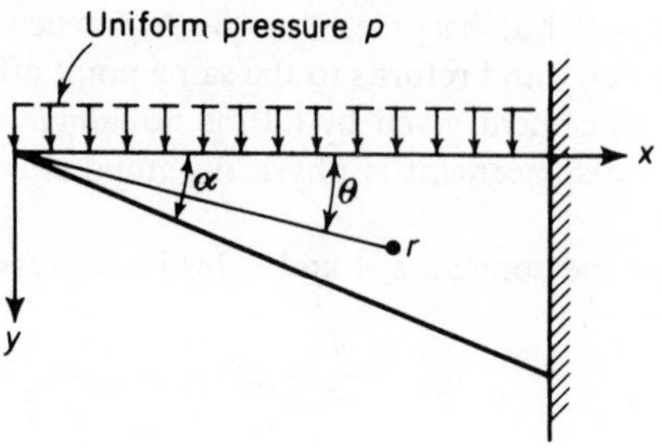


P9.1

**Problem 9.1.** Determine the constant $C$ in the stress function

$$\Phi = C[r^2(\alpha - \theta) + r \sin\theta \cos\theta - r^2 \cos^2\theta \tan\alpha]$$

required to satisfy the conditions on the upper and lower edges of the triangular plate shown in Fig. P9.1. Determine the components of displacement for points on the upper edge.

## 9.4. GENERAL CASE

By direct substitution, it can be verified that a solution of the equation

$$\left(\frac{\partial^2}{\partial r^2} + \frac{1}{r}\frac{\partial}{\partial r} + \frac{1}{r^2}\frac{\partial^2}{\partial \theta^2}\right)\left(\frac{\partial^2 \Phi}{\partial r^2} + \frac{1}{r}\frac{\partial \Phi}{\partial r} + \frac{1}{r^2}\frac{\partial^2 \Phi}{\partial \theta^2}\right) = 0$$

is (J. H. Michell, 1899)

$$\begin{aligned}\Phi = a_0 \log r &+ b_0 r^2 + c_0 r^2 \log r + d_0 r^2 \theta + a_0' \theta \\ &+ \frac{a_1''}{2} r\theta \sin\theta + (a_1 r + b_1 r^3 + a_1' r^{-1} + b_1' r \log r)\cos\theta \\ &+ \frac{c_1''}{2} r\theta \cos\theta + (c_1 r + d_1 r^3 + c_1' r^{-1} + d_1' r \log r)\sin\theta \\ &+ \sum_{n=2}^{\infty}(a_n r^n + b_n r^{n+2} + a_n' r^{-n} + b_n' r^{-n+2})\cos n\theta \\ &+ \sum_{n=2}^{\infty}(c_n r^n + d_n r^{n+2} + c_n' r^{-n} + d_n' r^{-n+2})\sin n\theta.\end{aligned} \tag{1}$$

By adjusting the coefficients $a_0, b_0, a_1, \ldots$, etc., a number of important problems can be solved. The general form of (1) as a Fourier series in $\theta$ and power series in $r$ provides a powerful means for solving problems involving circular and radial boundaries. The individual terms of the series provide beautiful solutions to several important engineering problems.

Many examples are given in Timoshenko and Goodier[1.2], pp. 73–130 and Sechler[1.2], pp. 149–171.

*Example 1. Bending of a Curved Bar By a Force at the End*

Consider a bar of a narrow rectangular cross section and with a circular axis, as shown in Fig. 9.4:1, loaded by a force $P$ in the radial direction. The bending moment at any cross section is proportional to $\sin\theta$. Since the elementary beam theory suggests that the normal stress $\sigma_{\theta\theta}$ is proportional to the bending moment, it is reasonable to try a solution for which $\sigma_{\theta\theta}$, and hence $\Phi$, is proportional to $\sin\theta$. Such a solution can be obtained from the term involving $\sin\theta$ in Eq. (1). It can be verified that the solution is

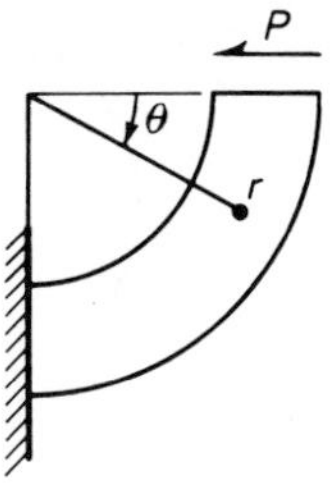


**Fig. 9.4:1.** Bending of a curved bar.

$$\Phi = \left(d_1 r^3 + \frac{c_1'}{r} + d_1' r \log r\right)\sin\theta, \tag{2}$$

with

$$d_1 = \frac{P}{2N}, \qquad c_1' = -\frac{Pa^2b^2}{2N}, \qquad d_1' = -\frac{P}{N}(a^2 + b^2),$$

where $N = a^2 - b^2 + (a^2 + b^2)\log(b/a)$. The stresses are

$$\begin{aligned}
\sigma_{rr} &= \frac{P}{N}\left(r + \frac{a^2b^2}{r^3} - \frac{a^2 + b^2}{r}\right)\sin\theta,\\
\sigma_{\theta\theta} &= \frac{P}{N}\left(3r - \frac{a^2b^2}{r^3} - \frac{a^2 + b^2}{r}\right)\sin\theta,\\
\sigma_{r\theta} &= -\frac{P}{N}\left(r + \frac{a^2b^2}{r^3} - \frac{a^2 + b^2}{r}\right)\cos\theta.
\end{aligned} \tag{3}$$

If the boundary stress distribution were exactly as prescribed by the equations above, namely,

$$\begin{aligned}
&\sigma_{rr} = \sigma_{r\theta} = 0 \quad \text{for} \quad r = a \quad \text{and} \quad r = b,\\
&\sigma_{\theta\theta} = 0, \quad \sigma_{r\theta} = -\frac{P}{N}\left[r + \frac{a^2b^2}{r^3} - \frac{1}{r}(a^2 + b^2)\right] \quad \text{for} \quad \theta = 0,\\
&\sigma_{r\theta} = 0, \quad \sigma_{\theta\theta} = \frac{P}{N}\left[3r - \frac{a^2b^2}{r^3} - (a^2 + b^2)\frac{1}{r}\right] \quad \text{for} \quad \theta = \frac{\pi}{2},
\end{aligned} \tag{4}$$

then an exact solution is obtained.

A detailed examination of the exact solution shows that a commonly used engineering approximation in the elementary beam theory, that plane cross sections of a beam remain plane during bending, gives very satisfactory results.

*Example 2. Concentrated Force at a Point on the Edge of a Semi-Infinite Plate*

Consider a concentrated vertical load $P$ acting on a horizontal straight boundary $AB$ of an infinitely large plate (Fig. 9.4:2). The distribution of the load along the thickness of the plate is uniform. The plate thickness is assumed to be unity, so that $P$ is the load per unit thickness.

This is the static counterpart of Lamb's problem (Secs. 8.15, 9.7). Its solution was obtained by Flamant (1892) from Boussinesq's (1885) three-dimensional solution (Sec. 8.10). In contrast to the dynamic or the three-dimensional cases, the solution to the present case is given simply by the Airy stress function

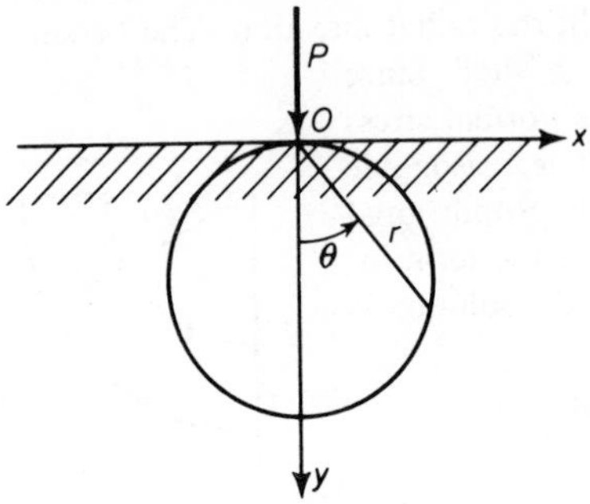


**Fig. 9.4:2.** Boussinesq-Flamant problem.

$$\Phi = -\frac{P}{\pi} r\theta \sin\theta, \tag{5}$$

which gives

$$\sigma_{rr} = -\frac{2P}{\pi}\frac{\cos\theta}{r}, \tag{6}$$

$$\sigma_{\theta\theta} = 0, \qquad \sigma_{r\theta} = 0.$$

From (6) it is seen that an element at a distance $r$ from the point of application of the load is subjected to a simple compression in the radial direction. Equation (6) shows also that the locus of points where the radial stress is a constant $-\sigma_{rr}$ is a circle $r = (-2P/\pi\sigma_{rr}) \cos\theta$, which is tangent to the $x$-axis, as shown in Fig. 9.4:2.

Using the relations (9.3:3) and (9.3:4) and Eq. (6), we can determine the displacement field $\xi_r$, $\xi_\theta$. If the constraint is such that the points on the $y$-axis have no lateral displacement ($\xi_\theta = 0$ when $\theta = 0$), then it can be shown that the elastic displacements are

$$\begin{aligned} \xi_r &= -\frac{2P}{\pi E}\cos\theta \log r - \frac{(1-\nu)P}{\pi E}\theta \sin\theta + B\cos\theta, \\ \xi_\theta &= \frac{2\nu P}{\pi E}\sin\theta + \frac{2P}{\pi E}\log r \sin\theta - \frac{(1-\nu)P}{\pi E}\theta\cos\theta + \frac{(1-\nu)P}{\pi E}\sin\theta - B\sin\theta, \end{aligned} \tag{7}$$

The constant $B$ can be fixed by fixing a point, say, $\xi_r = 0$ at $\theta = 0$, $r = a$. But the characteristic logarithmic singularity at $\infty$ cannot be removed. This is a peculiarity of the two-dimensional problem. The corresponding three-dimensional or dynamic cases do not have such logarithmically infinite displacements at infinity.

## PROBLEMS

**9.2.** Verify Eqs. (6) and (7) and show that Eq. (5) yields the exact solution of the problem posed in Example 2.

**9.3.** Find the stresses in a semi-infinite plate ($-\infty < x < \infty, 0 \le y < \infty$) due to a shear load of intensity $\tau \cos \alpha x$ acting on the edge $y = 0$, where $\tau$ and $\alpha$ are given constants. *Hint:* Consider $\Phi = (Ae^{\alpha y} + Be^{-\alpha y} + Cye^{\alpha y} + Dye^{-\alpha y}) \sin \alpha x$.

**9.4.** Show that the function $(M_t/2\pi)\theta$, where $M_t$ is a constant, is a stress function. Consider some boundary value problems which may be solved by such a stress function and give a physical meaning to the constant $M_t$.

**9.5.** Consider a two-dimensional wedge of perfectly elastic material as shown in Fig. P9.5. If one side of the wedge ($\theta = 0$) is loaded by a normal pressure distribution $p(r) = Pr^m$, where $P$ and $m$ are constants, while the other side ($\theta = \alpha$) is stress-free, show that the problem can be solved by an Airy stress function $\Phi(r, \theta)$ expressed in the following form. If $m \neq 0$ ($m$ may be $>0$ or $<0$),

$$(1)\quad \Phi = r^{m+2}[a \cos(m+2)\theta + b \sin(m+2)\theta + c \cos m\theta + d \sin m\theta].$$

If $m = 0$,

$$(2)\quad \Phi = Kr^2[-\tan\alpha \cos^2\theta + \tfrac{1}{2}\sin 2\theta + \alpha - \theta].$$

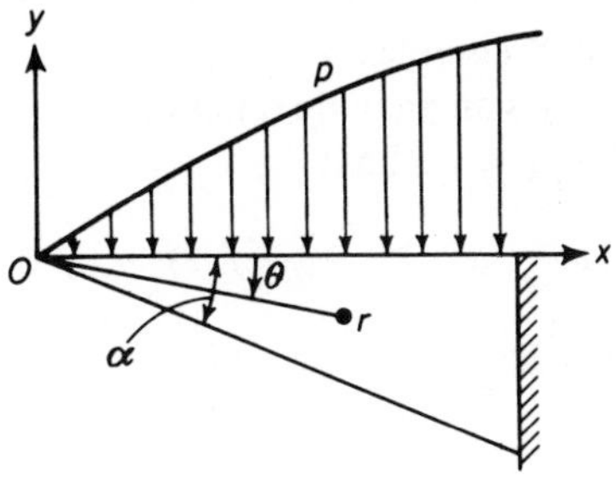


P9.5

Determine the constants $a$, $b$, $c$, $d$, $K$. Discuss the *boundedness* (i.e., whether they are zero, finite, or infinite) of the stresses $\sigma_{rr}$, $\sigma_{\theta\theta}$, $\sigma_{r\theta}$; the slope $\partial v/\partial r$, and the second derivative $\partial^2 v/\partial r^2$, in the neighborhood of the tip of the wedge, ($r \to 0$). The symbol $v$ denotes the component of displacement in the direction of increasing $\theta$.

When the wedge angle $\alpha$ is small, $v$ is approximately equal to the vertical displacement. This problem is of interest to the question of curling up of the sharp leading edge of a supersonic wing.

*Ref.* Fung, *J. Aeronautical Sciences*, **20,** 9 (1953).

**9.6.** A semi-infinite plate ($-\infty < x < \infty, 0 \leq y < \infty$) is subjected to the edge conditions

$$\sigma_{yy} = \frac{q_0}{2} - \frac{4q_0}{\pi^2}\left(\cos\frac{\pi x}{L} + \frac{1}{3^2}\cos\frac{3\pi x}{L} + \frac{1}{5^2}\cos\frac{5\pi x}{L} + \ldots\right),$$

$$\sigma_{xy} = 0,$$

on the edge $y = 0$ (Fig. P9.6). Find the Airy stress function that solves the problem.

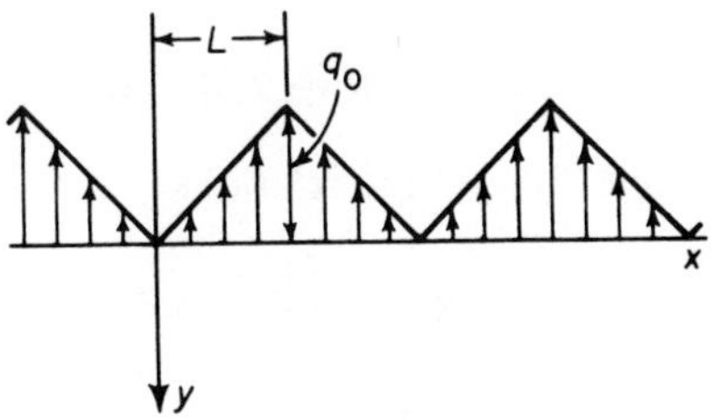


P9.6

Obtain expression for $\sigma_{xx}$ and $\sigma_{yy}$. Sketch $\sigma_{yy}$ as a function of $x$ on the line $y = d$, where $d > L$.

## 9.5. REPRESENTATION OF TWO-DIMENSIONAL BIHARMONIC FUNCTIONS BY ANALYTIC FUNCTIONS OF A COMPLEX VARIABLE

It is well-known that the real and imaginary part of any analytic function of a complex variable $z = x + iy$ is harmonic, where $i = \sqrt{-1}$. Thus, if $f(z)$ is an analytic function of $z$, and $u_1$ and $v_1$ are the real and imaginary parts of $f(z)$, we have

(1) $$f(z) = u_1 + iv_1,$$

(2) $$\frac{df}{dz} = \frac{\partial u_1}{\partial x} + i\frac{\partial v_1}{\partial x} = -i\frac{\partial u_1}{\partial y} + \frac{\partial v_1}{\partial y},$$

(3) $$\frac{\partial u_1}{\partial x} = \frac{\partial v_1}{\partial y}, \qquad \frac{\partial u_1}{\partial y} = -\frac{\partial v_1}{\partial x},$$

(4) $$\nabla^2 u_1 = 0, \qquad \nabla^2 v_1 = 0, \quad \text{where} \quad \nabla^2 = \frac{\partial^2}{\partial x^2} + \frac{\partial^2}{\partial y^2}.$$

Let $\phi(x, y)$ be a biharmonic function

$$\nabla^2(\nabla^2\phi) = 0.$$

Then $\nabla^2\phi$ is a harmonic function; let it be denoted by $P(x, y)$,

(5) $$\nabla^2\phi = P(x, y).$$

By means of the Cauchy-Riemann differential Eqs. (3), a conjugate harmonic function $Q(x, y)$ can be determined so that $P + iQ$ is an analytic function of $z$. Its integral is again analytic; let its real and imaginary parts be denoted by $R$ and $I$, respectively,

(6) $$\psi(z) = \tfrac{1}{4}\int_{z_0}^{z}(P + iQ)\,dz = R + iI.$$

We shall verify that

(7) $$\nabla^2[\phi - (xR + yI)] = 0.$$

This is readily seen from (5), and the relations

(8) $$\nabla^2(xR) = x\,\nabla^2 R + 2\frac{\partial R}{\partial x} = 2\,\text{Rl}\,\frac{d\psi}{dz} = \frac{P}{2},$$

(9) $$\nabla^2(yI) = 2\frac{\partial I}{\partial y} = 2\,\text{Rl}\,\frac{d\psi}{dz} = \frac{P}{2}.$$

Hence,

(10) $$\phi = xR + yI + u_1,$$

where $u_1$ is a harmonic function. The last formula can be written as the real part of an analytic function in the form

(11) $$\phi = \text{Rl}\,[\bar{z}\psi(z) + \chi(z)],$$

where $\bar{z} = x - iy$ is the complex conjugate of $z$, and $\chi(z)$ is an analytic function whose real part is $u_1$.

The formula (11) is due to the French mathematician Goursat. It forms the starting point of a very powerful method of solution for two-dimensional elasticity problems.

**Problem 9.7.**

(a) Show that $x - iy$ is not an analytic function of $z = x + iy$.

(b) Determine the real functions of $x$ and $y$ which are real and imaginary parts of the complex functions $z^n$ and $\tanh z$.

(c) Determine the real functions of $r$ and $\theta$ which are the real and imaginary parts of the complex function $z \log z$.

(d) $z = x + iy$, $\bar{z} = x - iy$, $a = \alpha + i\beta$, $\bar{a} = \alpha - i\beta$, where $x$, $y$, $\alpha$, $\beta$ are real numbers. Express the real and imaginary parts of the following functions explicitly in terms of $x$, $y$, $\alpha$, $\beta$ to get acquainted with the notations $f(z)$, $\bar{f}(\bar{z})$, $f(\bar{z})$:

1. $f(z) = az$, $\quad \bar{f}(\bar{z}) = \bar{a}\bar{z}$, $\quad f(\bar{z}) = a\bar{z}$.
2. $f(z) = e^{iaz}$, $\quad \bar{f}(\bar{z}) = e^{-i\bar{a}\bar{z}}$, $\quad f(\bar{z}) = e^{ia\bar{z}}$.
3. $f(z) = az^n$, $\quad \bar{f}(\bar{z}) = \bar{a}\bar{z}^n$, $\quad f(\bar{z}) = a\bar{z}^n$.

Show that the complex conjugate of $f(z)$ is $\bar{f}(\bar{z})$ in these examples.

(e) Show that the complex conjugate of $f(z) = \sum_{n=0}^{\infty} a_n z^n$ is

$$\bar{f}(\bar{z}) = \sum_{n=0}^{\infty} \bar{a}_n \bar{z}^n.$$

(f) Show that the derivative of $\bar{f}(\bar{z})$ with respect to $\bar{z}$ is equal to the complex conjugate of $df(z)/dz$.

(g) Find a function $v(x, y)$ of two real variables $x$, $y$, such that $\log(x^2 + y^2) + iv(x, y)$ is an analytic function of a complex variable $x + iy$. Use the Cauchy-Riemann differential equations.

## 9.6. KOLOSOFF-MUSKHELISHVILI METHOD

The representation of biharmonic functions by analytic functions leads to a general method of solving problems in plane stress and plane strain. We consider a region—simply or multiply connected—on the $x, y$-plane bounded by a number of contours (Fig. 9.6:1). The interior of the region is considered to represent a disk of unit thickness. Surface tractions are applied on the boundaries of this region. Body forces will be assumed to be zero to simplify the general solution. If body forces actually exist they can be represented by a particular solution so that the general problem is reduced to one without body force.

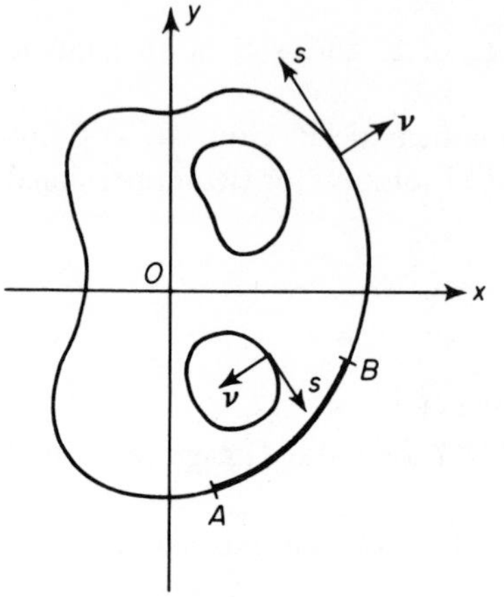


**Fig. 9.6:1.** Notations.

On introduction of the Airy stress function $\Phi(x, y)$ (Sec. 9.2), the plane stress and plane strain problems are reduced to the solution of a biharmonic equation under suitable boundary conditions. According to the results of Sec. 9.5, the Airy stress function $\Phi(x, y)$, defined by Eq. (9.2:17) as a function that yields a system of stresses compatible in a linear elastic body,

(1) ▲ $$\sigma_{xx} = \frac{\partial^2\Phi}{\partial y^2}, \qquad \sigma_{yy} = \frac{\partial^2\Phi}{\partial x^2}, \qquad \sigma_{xy} = -\frac{\partial^2\Phi}{\partial x\, \partial y},$$

can be represented in the form

(2) ▲ $$\Phi(x, y) = \text{Rl}\,[\bar{z}\psi(z) + \chi(z)];$$

i.e.,

(2a) ▲ $$2\Phi(x, y) = \bar{z}\psi(z) + z\bar{\psi}(\bar{z}) + \chi(z) + \bar{\chi}(\bar{z}),$$

where $\psi(z)$, $\chi(z)$ are two analytic functions of a complex variable $z = x + iy$. The problem in plane elasticity is to determine the functions $\psi(z)$ and $\chi(z)$ so that the boundary conditions are satisfied. For this purpose, we must express all the boundary conditions in terms of $\psi(z)$, $\chi(z)$ and their derivatives. This is done by computing first the stresses $\sigma_{xx}$, $\sigma_{xy}$, $\sigma_{yy}$ from $\Phi(x, y)$; then the strains $e_{xx}$, $e_{xy}$, $e_{yy}$ through the Hooke's law; then integrating to obtain expressions for the displacements $u, v$; and finally, representing the boundary conditions which specify surface tractions or displacements in terms of $\psi(z)$, $\chi(z)$. The lengthy but straightforward calculations will be outlined below. The results are

(3) ▲ $$\sigma_{xx} + \sigma_{yy} = 2[\psi'(z) + \bar{\psi}'(\bar{z})],$$

(4) ▲ $$\sigma_{yy} - \sigma_{xx} + 2i\sigma_{xy} = 2[\bar{z}\psi''(z) + \chi''(z)],$$

(5) ▲ $$2G(u + iv) = \kappa\psi(z) - z\bar{\psi}'(\bar{z}) - \bar{\chi}'(\bar{z}),$$

where a prime denotes differentiation with respect to $z$, thus $\psi'(z) = d\psi/dz$, and

$$\kappa = 3 - 4\nu \quad \text{for plane strain,} \tag{6}$$

$$\kappa = \frac{3 - \nu}{1 + \nu} \quad \text{for plane stress;} \tag{7}$$

$\nu$ is Poisson's ratio. In addition, we have the following expressions for the resultant forces $F_x$, $F_y$ and the moment $M$ about the origin of the surface tractions acting on an arc $AB$ of the boundary from a point $A$ to a point $B$:

$$\blacktriangle \quad F_x + iF_y = \int_A^B \left(\overset{\nu}{T}_x + i\overset{\nu}{T}_y\right) ds = -i\left[\psi(z) + z\bar{\psi}'(\bar{z}) + \bar{\chi}'(\bar{z})\right]_A^B, \tag{8}$$

$$\blacktriangle \quad M = \text{Rl}\left[\chi(z) - \bar{z}\bar{\chi}'(\bar{z}) - z\bar{z}\bar{\psi}'(\bar{z})\right]_A^B. \tag{9}$$

The physical quantities $\sigma_{xx}$, $\sigma_{yy}$, $\sigma_{xy}$, $u$, $v$, etc., are of course real-valued. If the complex-valued functions on the right-hand side of Eqs. (3)–(8) are known, a separation into real and imaginary parts determines all the stresses and displacements.

The derivation of these formulas is as follows. From (1), we have

$$\sigma_{xx} + i\sigma_{xy} = \frac{\partial^2\Phi}{\partial y^2} - i\frac{\partial^2\Phi}{\partial x\,\partial y} = -i\frac{\partial}{\partial y}\left(\frac{\partial\Phi}{\partial x} + i\frac{\partial\Phi}{\partial y}\right),$$

$$\sigma_{yy} - i\sigma_{xy} = \frac{\partial^2\Phi}{\partial x^2} + i\frac{\partial^2\Phi}{\partial x\,\partial y} = \frac{\partial}{\partial x}\left(\frac{\partial\Phi}{\partial x} + i\frac{\partial\Phi}{\partial y}\right).$$

From (2a) it is easily verified that, since

$$\frac{\partial\Phi}{\partial x} = \frac{\partial\Phi}{\partial z}\frac{\partial z}{\partial x} + \frac{\partial\Phi}{\partial \bar{z}}\frac{\partial \bar{z}}{\partial x}, \qquad \frac{\partial\Phi}{\partial y} = \frac{\partial\Phi}{\partial z}\frac{\partial z}{\partial y} + \frac{\partial\Phi}{\partial \bar{z}}\frac{\partial \bar{z}}{\partial y}, \tag{10}$$

we have

$$\frac{\partial\Phi}{\partial x} + i\frac{\partial\Phi}{\partial y} = \psi(z) + z\bar{\psi}'(\bar{z}) + \bar{\chi}'(\bar{z}). \tag{11}$$

A further differentiation and recombination and taking complex conjugates gives Eqs. (3) and (4).

To derive (5), we have to consider plane stress and plane strain separately. Consider the plane stress case. Hooke's law states

$$e_{xx} = \frac{\partial u}{\partial x} = \frac{1}{E}(\sigma_{xx} - \nu\sigma_{yy}) = \frac{1}{E}\left(\frac{\partial^2\Phi}{\partial y^2} - \nu\frac{\partial^2\Phi}{\partial x^2}\right), \tag{12}$$

$$e_{yy} = \frac{\partial v}{\partial y} = \frac{1}{E}(\sigma_{yy} - \nu\sigma_{xx}) = \frac{1}{E}\left(\frac{\partial^2\Phi}{\partial x^2} - \nu\frac{\partial^2\Phi}{\partial y^2}\right), \tag{13}$$

$$e_{xy} = \frac{1}{2}\left(\frac{\partial u}{\partial y} + \frac{\partial v}{\partial x}\right) = \frac{1}{2G}\sigma_{xy} = -\frac{1}{2G}\frac{\partial^2\Phi}{\partial x\,\partial y}. \tag{14}$$

But, from Eqs. (1) and (3), we have $\nabla^2\Phi = \sigma_{xx} + \sigma_{yy} = 4 \text{ Rl } \psi'$. Hence, $\partial^2\Phi/\partial y^2 = 4 \text{ Rl } \psi' - \partial^2\Phi/\partial x^2$. A substitution into (12) gives

$$\frac{\partial u}{\partial x} = \frac{1}{E}\left[4 \text{ Rl } \psi' - (1 + \nu)\frac{\partial^2\Phi}{\partial x^2}\right].$$

An integration with respect to $x$ gives

$$u = \frac{1}{E}\left[4 \text{ Rl } \psi - (1 + \nu)\frac{\partial\Phi}{\partial x}\right] + f(y), \tag{15}$$

where $f(y)$ is an arbitrary function of $y$. Similarly, from (13), we obtain

$$v = \frac{1}{E}\left[4 \text{ Im } \psi - (1 + \nu)\frac{\partial\Phi}{\partial y}\right] + g(x), \tag{16}$$

where $g(x)$ is an arbitrary function of $x$. Substituting (15), (16) into (14), and noticing that $\partial(\text{Rl } \psi)/\partial y = -\partial(\text{Im } \psi)/\partial x$, and $E/G = 2(1 + \nu)$, we obtain

$$f'(y) + g'(x) = 0.$$

Hence, $f(y) = \alpha y + \beta$, $g(x) = -\alpha x + \gamma$, where $\alpha, \beta, \gamma$ are constants. The forms of $f$ and $g$ indicate that they represent a rigid displacement and can

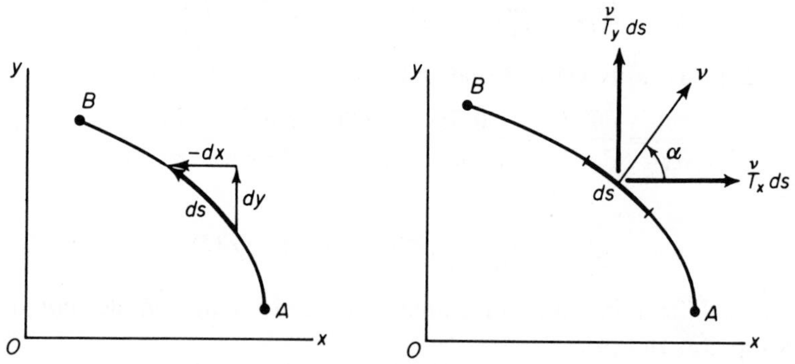


**Fig. 9.6:2.** Notations.

thus be disregarded in the analysis of deformation. If we set $f = g = 0$ in (15), (16), using Eq. (11), and combine properly, we obtain Eqs. (5) and (7). The plane strain case is similar, and (6) can be obtained without further analysis by applying a rule derived in Sec. 9.1, namely, replacing $\nu$ in (7) by an apparent value $\nu/(1 - \nu)$.

It remains to prove formulas (8) and (9). Refer to Fig. 9.6:2. Let $AB$ be a continuous arc, $s$ be the arc length, $ds$ be a vector tangent to the arc, $\boldsymbol{\nu}$ be the unit normal vector, and $\overset{\nu}{T}_x$, $\overset{\nu}{T}_y$ be the surface tractions acting on a surface perpendicular to $\boldsymbol{\nu}$. We note that

$$\cos\alpha = \frac{dy}{ds}, \qquad \sin\alpha = -\frac{dx}{ds}, \tag{17}$$

$$\overset{\nu}{T}_x = \sigma_{xx}\cos\alpha + \sigma_{xy}\sin\alpha, \qquad \overset{\nu}{T}_y = \sigma_{yy}\sin\alpha + \sigma_{xy}\cos\alpha. \tag{18}$$

Hence by Eq. (1),

$$\overset{\nu}{T}_x = \frac{\partial^2\Phi}{\partial y^2}\frac{dy}{ds} + \frac{\partial^2\Phi}{\partial x\,\partial y}\frac{dx}{ds} = \frac{d}{ds}\left(\frac{\partial\Phi}{\partial y}\right), \qquad \overset{\nu}{T}_y = -\frac{d}{ds}\left(\frac{\partial\Phi}{\partial x}\right). \tag{19}$$

Therefore, the resultant of forces acting on the arc $AB$ has the following components:

$$F_x = \int_A^B \overset{\nu}{T}_x\,ds = \frac{\partial\Phi}{\partial y}\bigg|_A^B, \qquad F_y = \int_A^B \overset{\nu}{T}_y\,ds = -\frac{\partial\Phi}{\partial x}\bigg|_A^B. \tag{20}$$

The moment about the origin $O$ of the forces acting on the arc $AB$ is:

$$\begin{aligned} M &= \int_A^B x\overset{\nu}{T}_x\,ds - \int_A^B y\overset{\nu}{T}_y\,ds = -\int_A^B \left[x\,d\left(\frac{\partial\Phi}{\partial x}\right) + y\,d\left(\frac{\partial\Phi}{\partial y}\right)\right] \\ &= \int\left(\frac{\partial\Phi}{\partial x}\frac{dx}{ds} + \frac{\partial\Phi}{\partial y}\frac{dy}{ds}\right)ds - \left(x\frac{\partial\Phi}{\partial x} + y\frac{\partial\Phi}{\partial y}\right)\bigg|_A^B \\ &= \Phi\bigg|_A^B - \left(x\frac{\partial\Phi}{\partial x} + y\frac{\partial\Phi}{\partial y}\right)\bigg|_A^B. \end{aligned} \tag{21}$$

A simple substitution of (20), (21), (11), and (2) into (8) and (9) verifies the desired results.

Equations (8) and (9), applied to a closed contour, show that if $\psi(z)$ and $\chi(z)$ are single-valued, the resultant forces and moment of the surface traction acting on the contour vanish, since the functions in brackets return to their initial values when the circuit is completed. If the resultant of tractions acting on a closed boundary does not vanish, the value of the functions in brackets must not return to their initial value when the circuit is completed. The function $\log z = \log r + i\theta$ does not return to its original value on completing a circuit around the origin, since $\theta$ increases by $2\pi$. Thus, if $\psi(z) = C\log z$, or $\chi(z) = Dz\log z$, where $C$ and $D$ are complex constants, Eq. (8) will yield a nonzero value of $F_x + iFy$. Similarly, $\chi(z) = D\log z$ will yield a nonzero value of $M$ according to Eq. (9) if $D$ is imaginary, but a zero value of $M$ if $D$ is real.

The examples below will show that a number of problems in plane stress and plane strain can be solved by simple stress functions. By reducing the problem into the determination of two analytic functions of a complex variable, Muskhelishvili and his school have succeeded in devising several general methods of approach. The principal tools for these general approaches are the conformal transformation and Cauchy integral equations. The details are well recorded in Muskhelishvili's books which an interested reader must consult.

## PROBLEMS

**9.8.** Discuss the conformal mapping specified by $\zeta = \dfrac{1}{2}\left(z + \dfrac{1}{z}\right)$. Obtain the inverse transformation. This transformation is the basis of airplane wing theory.

**9.9.** Consider a multiply connected region in the $x, y$-plane. Express the condition for the single-valuedness of displacements in terms of the stress components.

**9.10.** If $\partial\Phi/\partial x + i\partial\Phi/\partial y = \psi(z) + z\bar{\psi}'(\bar{z}) + \bar{\chi}'(\bar{z})$ is to be single-valued, how arbitrary are the functions $\psi(z)$, $\chi(z)$?

**9.11.** Consider a doubly connected region $R$ bounded by two concentric circles. If the stress is single-valued in $R$, how arbitrary are the functions $\psi(z)$, $\chi(z)$ in the Airy stress function? If both stresses and displacements are single-valued in $R$, how arbitrary are $\psi(z)$, $\chi(z)$? [Muskhelishvili,[1.2] p. 116–128]

**9.12.** Show that, if all stress and displacement components $\sigma_{ij}$ and $u_i$ vanish on an arc $AB$, then the stresses and displacements vanish identically in the entire region $R$ containing the arc $AB$. [Muskhelishvili,[1.2] p. 132]

**9.13.** Show that the transformation $z = e^{\zeta}$ defines a transformation from rectangular coordinates to polar coordinates.

**9.14.** A ring ($a < r < b$) is subjected to uniformly distributed shearing stress in the circumferential direction on the inner and outer surfaces as in Fig. P9.14. The couple of the shearing forces over the circumference is of magnitude $M$. Determine

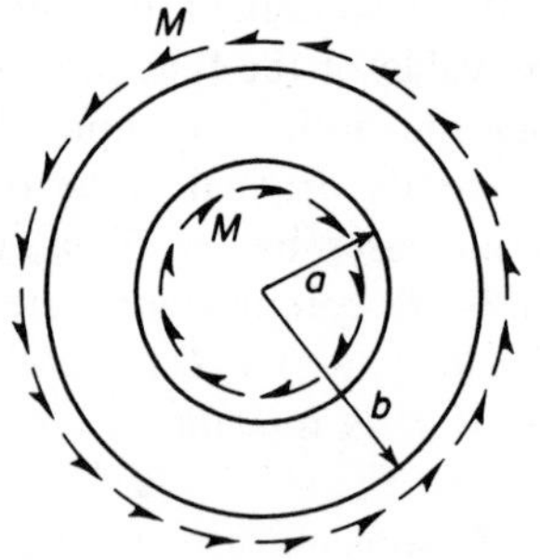


**P9.14.** Twisting of a disk.

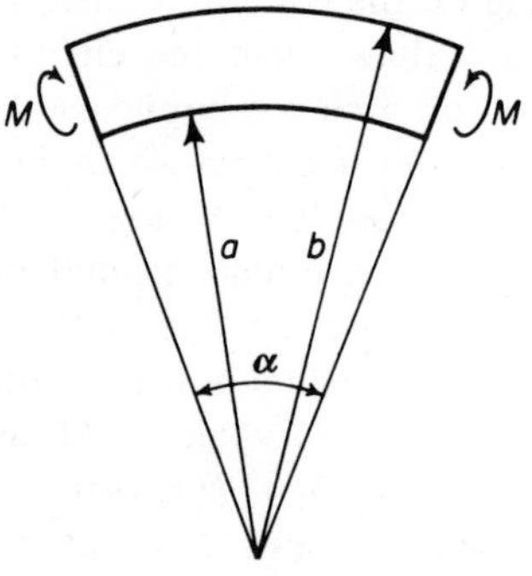


**P9.15.** Pure bending of a curved bar.

the stress and displacement components by means of complex potentials of the form

$$\psi(z) = 0, \qquad \chi(z) = A \log z,$$

where $A$ is a constant which can be complex-valued.

**9.15.** A curved bar is subjected to a bending moment $M$ at each end. The bar is defined by circular arcs of radius $a$ and $b$ and radial lines with an opening angle $\alpha$, ($\alpha < 2\pi$), as in Fig. P9.15. Show that the problem can be solved by complex potentials of the form

$$\psi(z) = Az \log z + Bz,$$
$$\chi(z) = C \log z.$$

Determine the constants $A$, $B$, and $C$.

**9.16.** Consider a rectangular plate of sides $2l$ and $2L$, and small thickness $2h$. Various loads are applied on the edges, Fig. P9.16. There is no body force. Determine the stresses and displacements in the plate with the suggested stress functions $\psi(z)$ and $\chi(z)$. Express the constants $a$, $b$ in terms of the external loads.

(a) All-round tension, Fig. P9.16(a):

$$\psi(z) = az, \qquad \chi(z) = 0.$$

(b) Uniaxial tension at an angle $\alpha$ to the $x$-axis, Fig. P9.16(b):

$$\psi(z) = az, \qquad \chi'(z) = -2az\, e^{-2i\alpha}.$$

(c) Pure bending, Fig. P9.16(c):

$$\psi(z) = aiz^2, \qquad \chi(z) = -\tfrac{1}{3}aiz^3.$$

(d) Bending with shear, Fig. P9.16(d):

$$\psi(z) = 2aiz^3, \qquad \chi'(z) = -2ai(z^3 + 6zb^2).$$

*Note:* In comparison with the bending of a beam in the three-dimensional case discussed in Sec. 7.7, the example in (d) shows how much simpler is the two-dimensional problem.

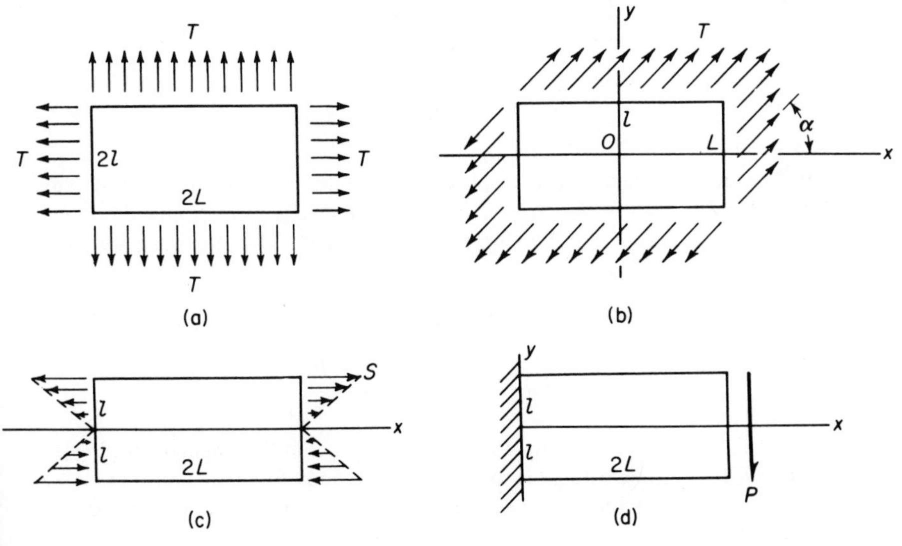


**P9.16**

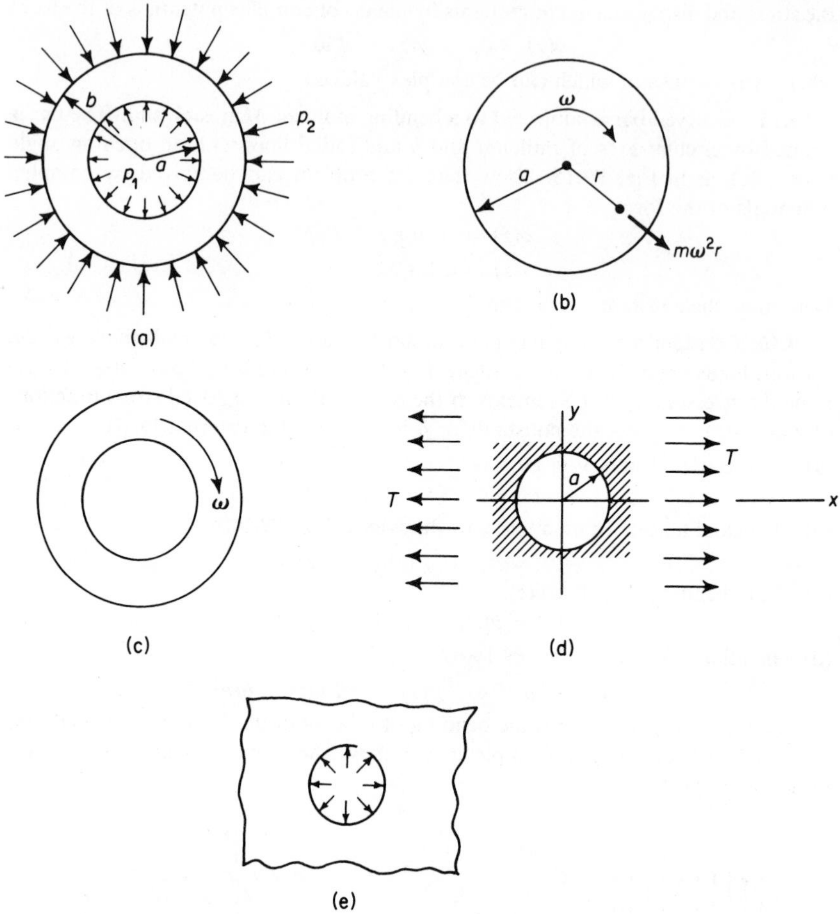


**P9.17**

**9.17.** Consider plane-stress or plane-strain states, parallel to the $x$, $y$-plane, of elastic bodies occupying regions bounded by circles. Determine the stresses and displacements for the following physical problems with the suggested stress functions (Fig. P9.17):

(a) Cylinder under internal pressure $p_1$ and external pressure $p_2$ [Fig. P9.17(a)]

$$\psi(z) = Az, \qquad \chi'(z) = \frac{B}{z}.$$

(b) A disk of radius $a$ rotating about its axis at a constant angular speed $\omega$ [Fig. P9.17(b)]: The centrifugal force constitutes a body force per unit mass $= \omega^2 r$, which has a potential $V = -\frac{1}{2}\omega^2 r^2$. Find a particular integral to Eq. (9.2:18) and (9.2:19).

(c) Rotating hollow disk [Fig. P9.17(c)]. Solve by combining (a) and (b).

(d) Large plate under tension and containing an unstressed circular hole [Fig. P9.17(d)]:

$$\psi(z) = \frac{1}{4} Tz + \frac{A}{z},$$

$$\chi'(z) = -\frac{1}{2} Tz + \frac{B}{z} + \frac{C}{z^3}.$$

Show that $A = \frac{1}{2}a^2$, $B = -\frac{1}{2}Ta^2$, $C = \frac{1}{2}Ta^4$.

(e) Large plate containing a circular hole under uniform pressure [Fig. P9.17(e)]:

$$\psi(z) = 0, \qquad \chi'(z) = \frac{A}{z}.$$

## 9.7. STEADY-STATE RESPONSE TO MOVING LOADS

We shall now again turn our attention to dynamics. Limiting ourselves to plane stress or plane strain, we shall deal with two-dimensional wave equations. The general equations of Sec. 9.1 are applicable; but the Airy stress function is no longer useful because the inertia forces now occupy the center of the stage.

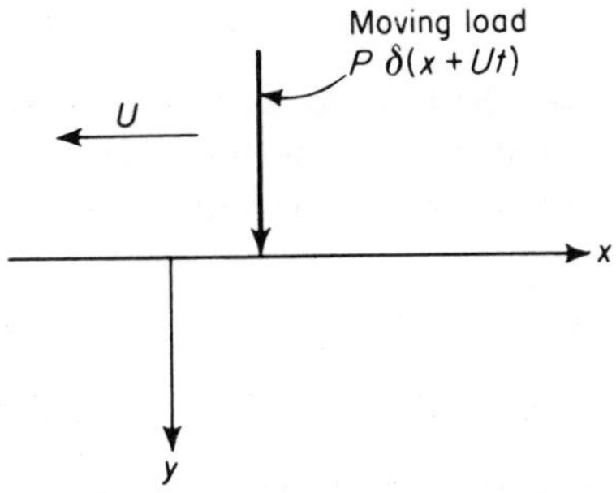

**Fig. 9.7:1.** A moving load over an elastic half-space.

Let us consider the following problem. A line load moves with a constant speed $U$ over an elastic half-space as illustrated in Fig. 9.7:1. We shall assume that a *plane strain* state prevails. At the beginning of applying the load, elastic waves are generated as discussed in Secs. 8.15 and 8.16. However, we shall assume that the load has been applied and moving for an infinitely long time so that a steady state prevails in the neighborhood of the loading, as seen by an observer moving with the load. We shall determine this steady-state response and shall not be concerned any further with the propagation of the initial disturbances.

Let $x, y, z$ be rectangular Cartesian coordinates fixed in the medium which occupies the half-space $y \geqslant 0$. The medium is assumed to be homogeneous, isotropic, and obeying Hooke's law. In a state of plane strain the elastic displacements $u, v, (w = 0)$ are derivable from the displacement potentials $\phi(x, y)$ and $\psi(x, y)$ so that

$$(1) \qquad u = \frac{\partial \phi}{\partial x} - \frac{\partial \psi}{\partial y}, \qquad v = \frac{\partial \phi}{\partial y} + \frac{\partial \psi}{\partial x}.$$

A solution of Navier's Eq. (9.1:8) is obtained if $\phi$ and $\psi$ satisfy the wave equations

$$\frac{\partial^2 \phi}{\partial x^2} + \frac{\partial^2 \phi}{\partial y^2} = \frac{1}{c_L^2}\frac{\partial^2 \phi}{\partial t^2}, \qquad \frac{\partial^2 \psi}{\partial x^2} + \frac{\partial^2 \psi}{\partial y^2} = \frac{1}{c_T^2}\frac{\partial^2 \psi}{\partial t^2}, \tag{2}$$

where

$$c_L = \sqrt{\frac{\lambda + 2G}{\rho}}, \qquad c_T = \sqrt{\frac{G}{\rho}}$$

are the dilatational and shear wave speeds, respectively (see Sec. 8.2). The stress components are, in plane strain,

$$\begin{gathered} \sigma_{xx} = (\lambda + 2G)\frac{\partial u}{\partial x} + \lambda\frac{\partial v}{\partial y}, \qquad \sigma_{yy} = (\lambda + 2G)\frac{\partial v}{\partial y} + \lambda\frac{\partial u}{\partial x}, \\ \sigma_{xy} = G\left(\frac{\partial u}{\partial y} + \frac{\partial v}{\partial x}\right). \end{gathered} \tag{3}$$

Now, in the problem stated above, the load moves in the negative $x$-axis direction at a constant speed $U$. An observer moving with the load at the same speed would see the load as stationary. Thus, if we introduce a Galilean transformation

$$x' = x + Ut, \qquad y' = y, \qquad t' = t, \tag{4}$$

then the boundary conditions would be independent of $t'$. In other words, $x$ and $t$ enter the boundary conditions only in the combination $x + Ut$. For a concentrated load the boundary conditions are, in the moving coordinates,

$$\sigma_{yy} = -P\delta(x'), \qquad \sigma_{xy} = 0, \qquad \text{at } y' = 0, \tag{5}$$

where $\delta(x')$ is the Dirac-delta function, which is zero everywhere except at $x' = 0$ where it tends to infinity in such a way that $\int_{-\infty}^{\infty} \delta(x')\, dx' = 1$.

Now, if the response in the elastic half-space is in a steady-state, $\phi$ and $\psi$ also shall be independent of $t'$ as seen by an observer moving with the load. In other words, $x$ and $t$ enter $\phi$ and $\psi$ only in the combination $x + Ut$. Under this assumption, Eqs. (2) are simplified into

$$\frac{\partial^2 \phi}{\partial x'^2} + \frac{\partial^2 \phi}{\partial y'^2} = \frac{U^2}{c_L^2}\frac{\partial^2 \phi}{\partial x'^2}, \qquad \frac{\partial^2 \psi}{\partial x'^2} + \frac{\partial^2 \psi}{\partial y'^2} = \frac{U^2}{c_T^2}\frac{\partial^2 \psi}{\partial x'^2}. \tag{6}$$

On introducing the Mach numbers

$$M_1 = \frac{U}{c_L}, \qquad M_2 = \frac{U}{c_T} \tag{7}$$

and the parameters

$$\beta_1 = \sqrt{1 - M_1^2}, \qquad \beta_2 = \sqrt{1 - M_2^2}, \qquad \text{if } M_1 < 1,\ M_2 < 1, \tag{8}$$

and

(9) $$\bar{\beta}_1 = \sqrt{M_1^2 - 1}, \qquad \bar{\beta}_2 = \sqrt{M_2^2 - 1}, \qquad \text{if } M_1 > 1,\ M_2 > 1,$$

we obtain the elliptic equations

(10) $$\beta_1^2 \frac{\partial^2 \phi}{\partial x'^2} + \frac{\partial^2 \phi}{\partial y'^2} = 0, \qquad \text{if } M_1 < 1,$$
$$\beta_2^2 \frac{\partial^2 \psi}{\partial x'^2} + \frac{\partial^2 \psi}{\partial y'^2} = 0, \qquad \text{if } M_2 < 1,$$

and the hyperbolic equations

(11) $$\bar{\beta}_1^2 \frac{\partial^2 \phi}{\partial x'^2} - \frac{\partial^2 \phi}{\partial y'^2} = 0, \qquad \text{if } M_1 > 1,$$
$$\bar{\beta}_2^2 \frac{\partial^2 \psi}{\partial x'^2} - \frac{\partial^2 \psi}{\partial y'^2} = 0, \qquad \text{if } M_2 > 1.$$

Expressions of the stress components may be reduced, by means of (3), (6), and (7), into

(12) $$\frac{\sigma_{xx}}{G} = (M_2^2 - 2M_1^2 + 2)\frac{\partial^2 \phi}{\partial x'^2} - 2\frac{\partial^2 \psi}{\partial x'\,\partial y'},$$
$$\frac{\sigma_{yy}}{G} = (M_2^2 - 2)\frac{\partial^2 \phi}{\partial x'^2} + 2\frac{\partial^2 \psi}{\partial x'\,\partial y'},$$
$$\frac{\sigma_{xy}}{G} = 2\frac{\partial^2 \phi}{\partial x'\,\partial y'} - (M_2^2 - 2)\frac{\partial^2 \psi}{\partial x'^2}.$$

The boundary conditions (5) become, at $y' = 0$,

(13) $$(M_2^2 - 2)\frac{\partial^2 \phi}{\partial x'^2} + 2\frac{\partial^2 \psi}{\partial x'\,\partial y'} = -\frac{P}{G}\,\boldsymbol{\delta}(x'),$$
$$2\frac{\partial^2 \phi}{\partial x'\,\partial y'} - (M_2^2 - 2)\frac{\partial^2 \psi}{\partial x'\,\partial y'} = 0,$$

which can be integrated once to give, at $y' = 0$,

(14) $$(M_2^2 - 2)\frac{\partial \phi}{\partial x'} + 2\frac{\partial \psi}{\partial y'} = -\frac{P}{G}\,\mathbf{1}(x'),$$
$$2\frac{\partial \phi}{\partial y'} - (M_2^2 - 2)\frac{\partial \psi}{\partial x'} = 0.$$

The solution of our problem is given by the functions $\phi$ and $\psi$, which satisfy the differential eqs. (10) or (11), the boundary conditions on the free surface (14), and the appropriate radiation and finiteness conditions at infinity (Sec. 8.15).

The simplicity of the problem is now readily seen. Equations (10) are potential equations which are satisfied by the real parts of arbitrary analytic functions $f$, $g$ of complex variables $x' + i\beta_1 y'$, $x' + i\beta_2 y'$, respectively,

$$\phi = \text{Rl}\, f(x' + i\beta_1 y'), \qquad \psi = \text{Rl}\, g(x' + i\beta_2 y'). \tag{15}$$

Equations (11) are wave equations which are satisfied by arbitrary functions of arguments $x' \pm \bar{\beta}_1 y'$, $x' \pm \bar{\beta}_2 y'$, respectively. The sign in front of $y'$ must be chosen according to the radiation condition. If we assume that the disturbance is initiated solely by the pressure pulse on the surface $y' = 0$, and the pulse moves in the negative $x$-direction, then the negative sign must be chosen and

$$\phi = f(x' - \bar{\beta}_1 y'), \qquad \psi = g(x' - \bar{\beta}_2 y'). \tag{16}$$

Our problem is reduced to finding the arbitrary functions $f$ and $g$ according to the boundary conditions (14).

The nature of the solution depends on the Mach numbers $M_1, M_2$. Three cases can be distinguished:

(a) $M_2 > M_1 > 1$, (b) $M_2 > 1 > M_1$, (c) $1 > M_2 > M_1$.

Since $c_L > c_T$, so that $M_2 > M_1$, the three cases above exhaust all possibilities. We shall call these supersonic, transonic, and subsonic cases, respectively.

***Supersonic Case,*** $M_1 > 1$, $M_2 > 1$. Straightforward substitution of (16) into (14) gives the following equations, which must hold when $y' = 0$:

$$\begin{aligned} (M_2^2 - 2)f' - 2\bar{\beta}_2 g' &= -\frac{P}{G}\mathbf{1}(x'), \\ -2\bar{\beta}_1 f' - (M_2^2 - 2)g' &= 0, \end{aligned} \tag{17}$$

where $f', g'$ denote the derivatives of $f(x' - \bar{\beta}_1 y')$ and $g(x' - \bar{\beta}_2 y')$ with respect to the arguments $x' - \bar{\beta}_1 y'$, $x' - \bar{\beta}_2 y'$, respectively. Hence, at $y' = 0$,

$$f' = -\frac{M_2^2 - 2}{2\bar{\beta}_1} g', \qquad g' = 2\bar{\beta}_1 \frac{P}{G\Delta}\mathbf{1}(x'), \tag{18}$$

where

$$\Delta = (2 - M_2^2)^2 + 4\bar{\beta}_1\bar{\beta}_2. \tag{18a}$$

This gives the solution

$$\begin{aligned} g'(x' - \bar{\beta}_2 y') &= 2\bar{\beta}_1 \frac{P}{G\Delta}\mathbf{1}(x' - \bar{\beta}_2 y'), \\ f'(x' - \bar{\beta}_1 y') &= (2 - M_2^2)\frac{P}{G\Delta}\mathbf{1}(x' - \bar{\beta}_1 y'). \end{aligned} \tag{19}$$

Hence, from Eqs. (16), (1), and (12), we have

$$u = \frac{P}{G\Delta}[(2 - M_2^2)\mathbf{1}(x' - \bar{\beta}_1 y) + 2\bar{\beta}_1\bar{\beta}_2\mathbf{1}(x' - \bar{\beta}_2 y')],$$

$$v = \frac{P}{G\Delta}[-\bar{\beta}_1(2 - M_2^2)\mathbf{1}(x' - \bar{\beta}_1 y') + 2\bar{\beta}_1\mathbf{1}(x' - \bar{\beta}_2 y')],$$

(20) $$\sigma_{xx} = \frac{P}{\Delta}\left[(2 - M_2^2)\left(2 + \frac{\lambda}{G}M_1^2\right)\boldsymbol{\delta}(x' - \bar{\beta}_1 y') + 4\bar{\beta}_1\bar{\beta}_2\boldsymbol{\delta}(x' - \bar{\beta}_2 y')\right],$$

$$\sigma_{yy} = \frac{P}{\Delta}\left[(2 - M_2^2)\left(2\bar{\beta}_1^2 + \frac{\lambda}{G}M_1^2\right)\boldsymbol{\delta}(x' - \bar{\beta}_1 y') - 4\bar{\beta}_1\bar{\beta}_2\boldsymbol{\delta}(x' - \bar{\beta}_2 y')\right],$$

$$\sigma_{xy} = -\frac{P}{\Delta}2\bar{\beta}_1(2 - M_2^2)[\boldsymbol{\delta}(x' - \bar{\beta}_1 y') - \boldsymbol{\delta}(x' - \bar{\beta}_2 y')].$$

These equations show that the disturbances are marked by two Mach waves

(21) $x' - \bar{\beta}_1 y' = 0, \qquad x' - \bar{\beta}_2 y' = 0.$

The medium is undisturbed in front of these Mach waves, as shown in Fig. 9.7:2.

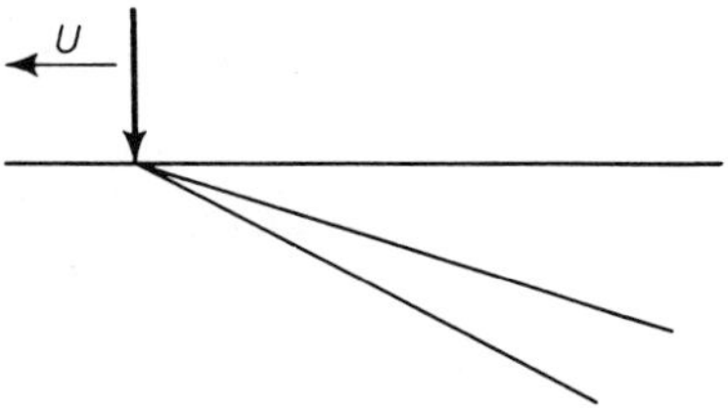


**Fig. 9.7:2.** Wave fronts induced by a load traveling at supersonic speed.

***Subsonic Case,*** $M_1 < 1$, $M_2 < 1$. In this case $\phi$ and $\psi$ are given by (15), which, on substituting into (14), reduces the boundary conditions at $y' = 0$ into

(22a) $$(2 - M_2^2)\,\mathrm{Rl}\,f' + 2\beta_2\,\mathrm{Im}\,g' = \frac{P}{G}\mathbf{1}(x'),$$

(22b) $$-2\beta_1\,\mathrm{Im}\,f' + (2 - M_2^2)\,\mathrm{Rl}\,g' = 0,$$

where $f'$, $g'$ denote the derivatives of $f(x' + i\beta_1 y')$, $g(x' + i\beta_2 y')$ with respect to the complex variables $x' + i\beta_1 y'$ and $x' + i\beta_2 y'$, respectively. These boundary conditions can be used to characterize the singularities of the analytic functions $f$ and $g$. Equation (22b) can be satisfied by

(23) $$g' = -i\frac{2\beta_1}{2 - M_2^2}f'.$$

Then Eq. (22a) shows that $\mathrm{Rl}\,f'$ has a jump across the load as the argument of $x' + i\beta_1 y'$ goes from $\pi$ to zero. This implies a logarithmic singularity in $f'$ and $g'$, and the solution can be found to be

(24)
$$f'(x' + i\beta_1 y') = \frac{K_1 P}{G}\left[\frac{i}{\pi}\log(x' + i\beta_1 y') + 1\right],$$

$$g'(x' + i\beta_2 y') = \frac{K_2 P}{G}\left[\frac{1}{\pi}\log(x' + i\beta_2 y') - i\right],$$

where

$$K_1 = \frac{2 - M_2^2}{(2 - M_2^2)^2 - 4\beta_1\beta_2}, \qquad K_2 = \frac{2\beta_1}{(2 - M_2^2)^2 - 4\beta_1\beta_2}. \tag{25}$$

From this solution the displacements and stresses are found to be

$$\begin{aligned} u &= \frac{K_1 P}{G}\left(1 - \frac{\theta_1}{\pi}\right) - \frac{\beta_2 K_2 P}{G}\left(1 - \frac{\theta_2}{\pi}\right), \\ v &= \frac{P}{G\pi}(K_2 \log r_2 - \beta_1 K_1 \log r_1), \end{aligned} \tag{26}$$

where

$$(x' + i\beta_1 y') = r_1 e^{i\theta_1}, \qquad (x' + i\beta_2 y') = r_2 e^{i\theta_2}, \qquad 0 \leqslant \theta_1, \theta_2 \leqslant \pi. \tag{27}$$

and

$$\begin{aligned} -\pi\sigma_{yy} &= K_1 P(2 - M_2^2)\frac{\sin\theta_1}{r_1} - 2K_2\beta_2 P\frac{\sin\theta_2}{r_2}, \\ -\pi\sigma_{xy} &= 2K_1\beta_1 P\left[\frac{\cos\theta_1}{r_1} - \frac{\cos\theta_2}{r_2}\right], \\ -\pi\sigma_{xx} &= -K_1 P(M_2^2 - 2M_1^2 + 2)\frac{\sin\theta_1}{r_1} + 2K_2\beta_2 P\frac{\sin\theta_2}{r_2}. \end{aligned} \tag{28}$$

Note that there is a singularity under the point of application of the load, where the vertical displacement is logarithmically singular. This is caused by the idealized concept of concentrated load. For any loads distributed over an area, this singularity is integrable. However, the vertical displacement also tends to infinity logarithmically as $r_1 \to \infty$, $r_2 \to \infty$. This unbounded displacement at infinity is not caused by load concentration. It is a general feature of the steady-state solutions of two-dimensional problems in an unbounded region. We have encountered it in the two-dimensional Boussinesq problem—a static *line* load over an elastic half-space (Sec. 9.3). It can be shown generally (cf. Muskhelishvili,[1.2] Sec. 36) that in two dimensions if static loads with a nonvanishing resultant act on a semi-infinite elastic body, the displacement at infinity is logarithmically divergent. On the other hand, the solution of the three-dimensional Boussinesq problem—a static *point* load over an elastic half-space (Sec. 8.9)—does not present such a difficulty; nor is the solution of the dynamic line load problem singular at infinity (Secs. 8.16, 8.17). Therefore, the feature of an unbounded displacement at infinity must be regarded as a mathematical consequence of idealizing a real physical problem into a static two-dimensional problem.

The horizontal displacement given by (26) is not antisymmetric with respect to $x = 0$. This again serves to emphasize that the displacements are not uniquely determined in this problem and that a constant displacement could be added to (26) to adjust the symmetry. However, the stresses are uniquely determined.

A boundary condition at infinity for this case states that the integrated stresses should balance the load; this is automatically satisfied.

The denominator of $K_1$ and $K_2$ in Eq. (25) vanishes when

$$(2 - M_2^2)^2 - 4\beta_1\beta_2 = 0. \tag{29}$$

If the definitions (7), (8) are substituted into this equation, we see that it is satisfied if the load moves at exactly the speed of Rayleigh surface waves. Hence if the load moves steadily at the Rayleigh wave speed, the responses will be infinitely large.

***Transonic Case,*** $M_1 < 1$, $M_2 > 1$. In this case the load is moving at a speed slower than the longitudinal wave speed but faster than the shear wave speed. According to Eqs. (15) and (16), the solution may be taken in the form

$$\phi = \mathrm{Rl}\, f(x' + i\beta_1 y'), \qquad \psi = g(x' - \bar\beta_2 y'). \tag{30}$$

Then boundary conditions at the surface $y' = 0$, Eqs. (14) become

$$\begin{aligned} (2 - M_2^2)\,\mathrm{Rl}\, f'(x') + 2\bar\beta_2 g'(x') &= \frac{P}{G}\,\mathbf{1}(x'), \\ -2\beta_1\,\mathrm{Im}\, f'(x') + (2 - M_2^2) g'(x') &= 0, \end{aligned} \tag{31}$$

where $f', g'$ denote derivatives of $f$ and $g$ with respect to their respective arguments. On eliminating $g'$ from Eqs. (31), we obtain

$$(2 - M_2^2)^2\,\mathrm{Rl}\, f' + 4\beta_1\bar\beta_2\,\mathrm{Im}\, f' = \frac{P}{G}(2 - M_2^2)\mathbf{1}(x'). \tag{32}$$

Hence, there is a jump in both $\mathrm{Rl}\, f'$ and $\mathrm{Im}\, f'$ on crossing the load at the surface. This again implies a logarithmic singularity. The solution is

$$f'(x' + i\beta_1 y') = \frac{(K_3 + iK_4)P}{\pi G}[\log(x' + i\beta_1 y') - i\pi], \tag{33}$$

where

$$K_3 = \frac{-4\beta_1\bar\beta_2(2 - M_2^2)}{(2 - M_2^2)^4 + 16\beta_1^2\bar\beta_2^2}, \qquad K_4 = \frac{(2 - M_2^2)^3}{(2 - M_2^2)^4 + 16\beta_1^2\bar\beta_2^2}. \tag{34}$$

From (29) and (31) we obtain

$$g'(x' - \bar\beta_2 y') = \frac{2\beta_1 P}{\pi(2 - \bar\beta_2^2)G}[K_4 \log|x' - \bar\beta_2 y'| + K_3\pi\mathbf{1}(x' - \bar\beta_2 y')]. \tag{35}$$

It is evident now that the solution becomes singular both under the load and along the characteristics $x' - \bar\beta_2 y' = 0$.

The problem discussed above was first proposed by Lamb[7.4] (1904), see Sec. 8.15. The solution given above is due to Cole and Huth[9.2] (1956). Sneddon[9.2] (1951) gave the solution for the subsonic case, and also extended the problem to include a moving load tangential to the free surface. In his book *Fourier Transforms* (1951), Sneddon[9.2] also considered moving point loads by the method of integral transforms. The problem has attracted much attention because of its engineering significance in the question of ground motion under pressure waves generated by nuclear explosions.

## 9.8. ALTERNATE METHOD OF SOLUTION

The Lamb's problem discussed in the previous section can be solved by other methods. We shall discuss here an approach which is useful in solving many problems in mathematical physics.

Let us consider first the *supersonic case* $M_1 > 1$, $M_2 > 1$. In this case, we must solve Eqs. (9.7:11), which admit the elementary solutions

$$\phi(x', y') = A\, e^{i\lambda x'}\, e^{-i\lambda \beta_1 y'},$$
$$\psi(x', y') = B\, e^{i\lambda x'}\, e^{-i\lambda \bar{\beta}_2 y'},$$

where $\lambda$, $A$, $B$ are arbitrary constants. These solutions satisfy the radiation condition at infinity—only backward running waves are admitted. The other possible solutions, of the form $\exp\,[i\lambda(x' + \bar{\beta}_1 y)]$, are rejected on the basis of the radiation condition because they represent disturbances which originate at infinity and converge toward the load.

Since $\lambda$ is arbitrary and the system is linear, we may let the constants $A$ and $B$ depend on $\lambda$ and integrate over $\lambda$. Hence, we assume a general solution in the form

(1) $$\phi(x', y') = \frac{1}{2\pi}\int_{-\infty}^{\infty} A(\lambda)\, e^{-i\lambda\beta_1 y'}\, e^{i\lambda x'}\, d\lambda,$$

(2) $$\psi(x', y') = \frac{1}{2\pi}\int_{-\infty}^{\infty} B(\lambda)\, e^{-i\lambda\beta_2 y'}\, e^{i\lambda x'}\, d\lambda.$$

The boundary conditions (9.7:14) then give

(3) $$(M_2^2 - 2)i\lambda A(\lambda) - 2B(\lambda)i\lambda\bar{\beta}_2 = \mathscr{P}(\lambda),$$

(4) $$-2i\lambda\bar{\beta}_1 A(\lambda) - (M_2^2 - 2)i\lambda B(\lambda) = 0,$$

where $\mathscr{P}(\lambda)$ is the Fourier transform of $-\dfrac{1}{G}\displaystyle\int_0^{x'} p(x')\, dx'$, and $p(x') = p(x + Ut)$ is the pressure distribution over the surface $y = 0$. Hence,

(5) $$A(\lambda) = \frac{(M_2^2 - 2)}{2\bar{\beta}_1}\,\frac{1}{\kappa}\,\frac{\mathscr{P}(\lambda)}{i\lambda},$$

(6) $$B(\lambda) = -\frac{\mathscr{P}(\lambda)}{\kappa\lambda i},$$

where

(7) $$\kappa = \frac{(M_2^2 - 2)^2}{2\bar{\beta}_1} + 2\bar{\beta}_2.$$

Substituting back into (1), we obtain

(8) $$\phi(x', y') = \frac{M_2^2 - 2}{4\pi\bar{\beta}_1\kappa}\int_{-\infty}^{\infty} \frac{\mathscr{P}(\lambda)}{i\lambda}\, e^{i\lambda(x'-\bar{\beta}_1 y')}\, d\lambda.$$

Hence,

(9) $$\begin{aligned}\frac{\partial\phi}{\partial x'} &= \frac{M_2^2 - 2}{4\pi\bar{\beta}_1\kappa}\int_{-\infty}^{\infty} \mathscr{P}(\lambda)\, e^{i\lambda(x'-\bar{\beta}_1 y')}\, d\lambda \\ &= \frac{2 - M_2^2}{(M_2^2 - 2)^2 + 4\bar{\beta}_1\bar{\beta}_2}\,\frac{1}{G}\int_0^{x'-\bar{\beta}_1 y'} p(x')\, dx'.\end{aligned}$$

In particular, if the load $p(x')$ is a concentrated force $P$, then

(10) $$\frac{\partial\phi}{\partial x'} = \frac{2 - M_2^2}{(M_2^2 - 2)^2 + 4\bar{\beta}_1\bar{\beta}_2}\,\frac{P}{G}\,\mathbf{1}(x' - \bar{\beta}_1 y').$$

Similarly,

(11) $$\frac{\partial\psi}{\partial x'} = \frac{2\bar{\beta}_1}{(M_2^2 - 2)^2 + 4\bar{\beta}_1\bar{\beta}_2}\,\frac{P}{G}\,\mathbf{1}(x' - \bar{\beta}_2 y').$$

These results are in agreement with those derived in the previous section.

Next, let us consider the *subsonic case*, $M_1 < 1$, $M_2 < 1$. In this case the differential eq. (9.7:10) admits the formal solution

(12) $$\phi = \frac{1}{2\pi}\int_{-\infty}^{\infty} A(\lambda)\, e^{i\lambda x'}\, e^{-\beta_1 y'|\lambda|}\, d\lambda,$$

(13) $$\psi = \frac{1}{2\pi}\int_{-\infty}^{\infty} B(\lambda)\, e^{i\lambda x'}\, e^{-\beta_2 y'|\lambda|}\, d\lambda.$$

The boundary conditions (9.7:14) become

(14) $$i\lambda(M_2^2 - 2)A(\lambda) - 2\beta_2\,|\lambda|\, B(\lambda) = \mathscr{P}(\lambda),$$

(15) $$-2\beta_1\,|\lambda|\, A(\lambda) - (M_2^2 - 2)i\lambda B(\lambda) = 0.$$

Hence,

(16) $$A(\lambda) = -\frac{(M_2^2 - 2)i}{(M_2^2 - 2)^2 - 4\beta_1\beta_2}\,\frac{\mathscr{P}(\lambda)}{\lambda},$$

(17) $$B(\lambda) = \frac{2\beta_1}{(M_2^2 - 2)^2 - 4\beta_1\beta_2}\,\frac{\mathscr{P}(\lambda)}{|\lambda|}.$$

If the loading is a concentrated line load, $\mathscr{P}(\lambda)$ is the Fourier transform of a step-function

(18) $$-\frac{P}{G}\,\mathbf{1}(x') = -\frac{P}{G}\,\frac{1}{2\pi}\lim_{\epsilon=0}\int_{-\infty}^{\infty} \frac{1}{\epsilon + i\lambda}\, e^{i\lambda x'}\, d\lambda.$$

Hence,

(19) $$\frac{\partial\phi}{\partial x'} = -\frac{1}{2\pi}\,\frac{P}{G}\,\frac{(M_2^2 - 2)}{(M_2^2 - 2)^2 - 4\beta_1\beta_2}\lim_{\epsilon=0}\int_{-\infty}^{\infty} \frac{1}{\epsilon + i\lambda}\, e^{i\lambda x'}\, e^{-\beta_1 y'|\lambda|}\, d\lambda.$$

The last integral is

$$2\int_0^\infty \frac{\sin \lambda x'}{\lambda} e^{-\beta_1 y' \lambda}\, d\lambda = \pi - 2\tan^{-1}\frac{\beta_1 y'}{x'} = 2\tan^{-1}\frac{x'}{\beta_1 y'}. \tag{20}$$

[See Sneddon, *Fourier Transform*,[9.2] p. 272, or Erdelyi, *Tables of Integral Transforms*, McGraw-Hill, Vol. 1, p. 72, 2.4(2).] Hence,

$$\frac{\partial \phi}{\partial x'} = -\frac{1}{G}\frac{(M_2^2 - 2)}{(M_2^2 - 2)^2 - 4\beta_1\beta_2}\left(\frac{1}{2} - \frac{1}{\pi}\tan^{-1}\frac{\beta_1 y'}{x'}\right). \tag{21}$$

Similarly,

$$\frac{\partial \psi}{\partial y'} = -\frac{1}{G}\frac{2\beta_1\beta_2}{(M_2^2 - 2)^2 - 4\beta_1\beta_2}\left(\frac{1}{2} - \frac{1}{\pi}\tan^{-1}\frac{\beta_2 y'}{x'}\right). \tag{22}$$

Hence,

$$u = P\frac{K_1}{G}\left(\frac{1}{2} - \frac{\theta_1}{\pi}\right) - P\frac{\beta_2 K_2}{G}\left(\frac{1}{2} - \frac{\theta_2}{\pi}\right), \tag{23}$$

which differs from Huth and Cole's solution, Eq. (9.7:24), only in a constant, which reveals the nonunique character of the solution with respect to displacement.

Finally, consider the *transonic case*, $M_1 < 1$, $M_2 > 1$. In this case the solution to the applicable differential equations may be posed in the form

$$\phi = \frac{1}{2\pi}\int_{-\infty}^{\infty} A(\lambda)\, e^{i\lambda x'}\, e^{-\beta_1 y' |\lambda|}\, d\lambda, \tag{24}$$

$$\psi = \frac{1}{2\pi}\int_{-\infty}^{\infty} B(\lambda)\, e^{-i\lambda\beta_2 y'}\, e^{i\lambda x'}\, d\lambda. \tag{25}$$

The boundary conditions can now be written as

$$(M_2^2 - 2)A - 2\beta_2 B = \frac{\mathscr{P}(\lambda)}{i\lambda}, \tag{26}$$

$$-2\beta_1 |\lambda|\, A - (M_2^2 - 2)i\lambda B = 0. \tag{27}$$

Hence,

$$A = -\frac{(M_2^2 - 2)i\lambda}{2\beta_1 |\lambda|} B, \tag{28}$$

$$B = -\frac{\mathscr{P}(\lambda)}{i\lambda}\left[2\beta_2 + \frac{(M_2^2 - 2)^2}{2\beta_1}\, i\,\frac{\lambda}{|\lambda|}\right]^{-1}, \tag{29}$$

which again lead to the desired solution.

## PROBLEMS

**9.18.** A unit concentrated load acts normal to the free edge of a semi-infinite elastic plate, and moves at constant speed $U$ over the free edge. Assuming a plane stress state, find the steady-state responses in the plate.

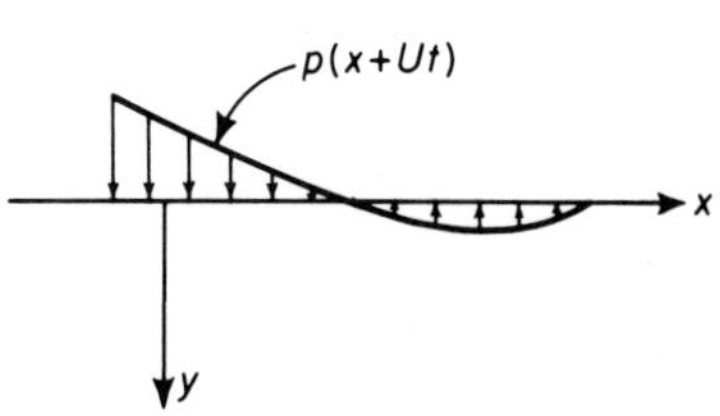


P9.19

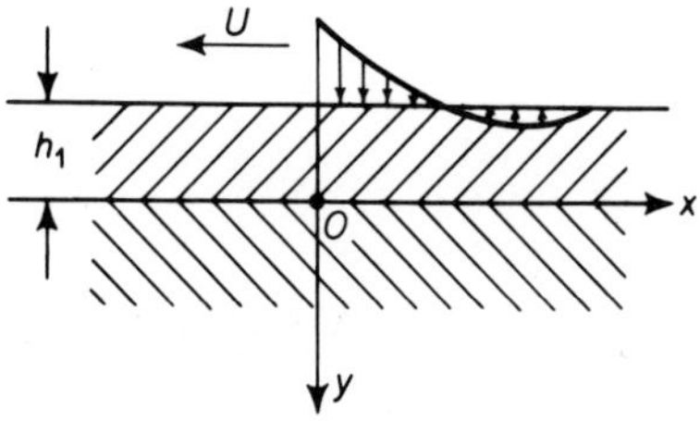


P9.21

**9.19.** Generalize the solutions obtained above to the case in which the traveling load is a distributed loading $p(x + Ut)$, as in Fig. P9.19.

**9.20.** For engineering purposes it is of interest to know the acceleration imparted to the elastic medium due to the traveling pressure loading. Derive expressions for the vertical and horizontal accelerations for the problems considered in Secs. 9.7 and 9.8.

**9.21.** Consider a half-space composed of a homogeneous layer of fluid on top of an elastic medium of infinite extent (Fig. P9.21). Let a pressure pulse $p(x + Ut)$ travel to the left at constant speed $U$. Find the response in the fluid and solid media.

**9.22.** Let the top layer in Prob. 9.21 be a linear elastic solid as well as the medium below. Let the two media be "welded" together at the interface. Let $\alpha_1, \beta_1$ be the longitudinal and shear wave speeds of the medium in the top layer, and let $\alpha_2, \beta_2$ be the corresponding speeds in the medium below. Now we have five characteristic speeds: $U$, $\alpha_1$, $\alpha_2$, $\beta_1$, $\beta_2$, with which four Mach numbers can be defined; namely,

$$M_1^{(1)} = \frac{U}{\alpha_1}, \qquad M_2^{(1)} = \frac{U}{\beta_1}, \qquad M_1^{(2)} = \frac{U}{\alpha_2}, \qquad M_2^{(2)} = \frac{U}{\beta_2}.$$

The nature of the response of the layered medium to the traveling pressure pulse will differ according to whether the $M$'s are greater than 1 or not. We have to distinguish the following cases.

(a) super-super: $M_2^{(1)} > M_1^{(1)} > 1, \quad M_2^{(2)} > M_1^{(2)} > 1.$

(b) super-tran: $M_2^{(1)} > M_1^{(1)} > 1, \quad M_2^{(2)} > 1 > M_1^{(2)}.$

(c) super-sub: $M_2^{(1)} > M_1^{(1)} > 1, \quad 1 > M_2^{(2)} > M_1^{(2)}.$

And similarly, tran-super, tran-tran, tran-sub, sub-super, sub-tran, and sub-sub (altogether 9 cases). The case (a) is the simplest to solve. Obtain the solution for the displacements and stresses in the elastic media for a traveling line load.

**9.23.** Consider a cylinder of radius $a$ rotating at a constant speed $\omega$ about its axis and subjected to a concentrated line load $P$ located at a fixed position in space, Fig. P9.23. Therefore the load $P$ travels at a constant speed relative to the cylinder surface. Assume a plane strain state. Discuss qualitatively the nature of the stresses and displacements in the cylinder.

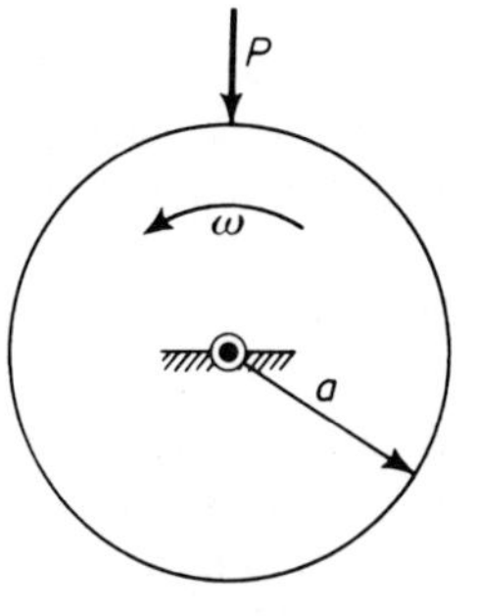


P9.23

# 10

# VARIATIONAL CALCULUS, ENERGY THEOREMS, SAINT-VENANT'S PRINCIPLE

There are at least three important reasons for taking up the calculus of variations in the study of continuum mechanics.

1. Because basic minimum principles exist, which are among the most beautiful of theoretical physics.
2. The field equations (ordinary or partial differential equations) and the associated boundary conditions of many problems can be derived from variational principles. In formulating an approximate theory, the shortest and clearest derivation is usually obtained through variational calculus.
3. The "direct" method of solution of variational problems is one of the most powerful tools for obtaining numerical results in practical problems of engineering importance.

In this chapter we shall discuss several variational principles and their applications.

A brief introduction to the calculus of variations is furnished below. Those readers who are familiar with the mathematical techniques of the calculus of variations may skip over the first six sections.

## 10.1. MINIMIZATION OF FUNCTIONALS

The calculus of variations is concerned with the minimization of functionals. If $u(x)$ is a function of $x$, defined for $x$ in the interval $(a, b)$, and if $I$ is a quantity defined by the integral

$$I = \int_a^b [u(x)]^2 \, dx,$$

then the value of $I$ depends on the function $u(x)$ as a whole. We may indicate this dependence by writing $I[u(x)]$. Such a quantity $I$ is said to be a *functional* of $u(x)$. Physical examples of functionals are the total kinetic energy of a flow field and the strain energy of an elastic body.

The basic problem of the calculus of variations may be illustrated by the following example. Let us consider a functional $J[u]$ defined by the integral

$$J[u] = \int_a^b F(x, u, u')\, dx. \tag{1}$$

We shall give our attention to *all* functions $u(x)$ which are continuous and differentiable, with continuous derivatives $u'(x)$ and $u''(x)$ in the interval $a \leqslant x \leqslant b$, and satisfying the boundary conditions

$$u(a) = u_0, \qquad u(b) = u_1, \tag{2}$$

where $u_0$ and $u_1$ are given numbers. We assume that the function $F(x, u, u')$ in Eq. (1) is continuous and differentiable with respect to $x$, and all such $u$, and $u'$, up to all second-order partial derivatives, which are themselves continuous.

Among all functions $u(x)$ satisfying these continuity conditions and boundary values, we try to find a special one $u(x) = y(x)$, with the property that $J[u]$ attains a minimum when $u(x) = y(x)$, with respect to a sufficiently small neighborhood of $y(x)$. The neighborhood $(h)$ of $y(x)$ is defined as follows. If $h$ is a positive quantity, a function $u(x)$ is said to lie in the neighborhood $(h)$ of $y(x)$ if the inequality

$$|y(x) - u(x)| < h \tag{3}$$

holds for all $x$ in $(a, b)$. The situation is illustrated in Fig. 10.1:1.

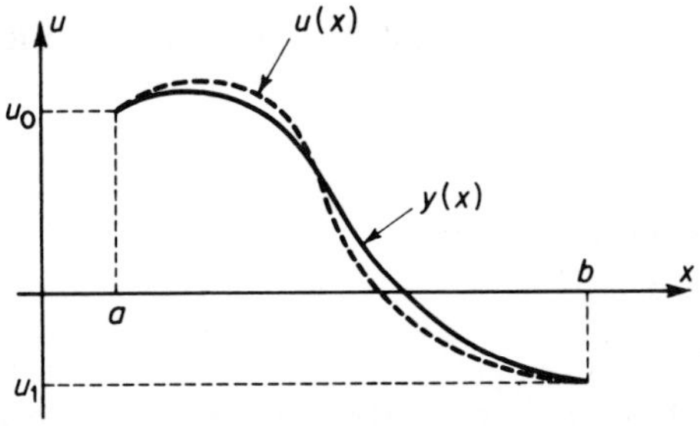


**Fig. 10.1:1.** Functions $u(x)$ and $y(x)$.

Let us assume that the problem posed above has a solution, which will be designated by $y(x)$; i.e., there exists a function $y(x)$ such that the inequality

$$J[y] \leqslant J[u] \tag{4}$$

holds for all functions $u(x)$ in a sufficiently small neighborhood $(h)$ of $y(x)$. Let us exploit the necessary consequences of this assumption.

Let $\eta(x)$ be an arbitrary function with the properties that $\eta(x)$ and its derivatives $\eta'(x)$, $\eta''(x)$ are continuous in the interval $a \leqslant x \leqslant b$, and that

$$\eta(a) = \eta(b) = 0. \tag{5}$$

Then the function

$$u(x) = y(x) + \epsilon\eta(x) \tag{6}$$

satisfies all the continuity conditions and boundary values specified at the beginning of this section. In fact, any function $u(x)$ satisfying these conditions can be represented with some function $\eta(x)$ in this manner. For a sufficiently small $\delta > 0$, this function $u(x)$ belongs, for all $\epsilon$ with $|\epsilon| < \delta$, to a prescribed neighborhood $(h)$ of $y(x)$. Now we introduce the function

$$\phi(\epsilon) = J[y + \epsilon\eta] = \int_a^b F[x, y(x) + \epsilon\eta(x), y'(x) + \epsilon\eta'(x)]\, dx. \tag{7}$$

Since $y(x)$ is assumed to be known, $\phi(\epsilon)$ is a function of $\epsilon$ for any specific $\eta(x)$. According to (4), the inequality

$$\phi(0) \leqslant \phi(\epsilon) \tag{8}$$

must hold for all $\epsilon$ with $|\epsilon| < \delta$. In other words, $\phi(\epsilon)$ attains a minimum at $\epsilon = 0$. The function $\phi(\epsilon)$ is differentiable with respect to $\epsilon$. Therefore, the necessary condition for $\phi(\epsilon)$ to attain a minimum at $\epsilon = 0$ must follow,

$$\phi'(0) = 0, \tag{9}$$

where a prime indicates a differentiation with respect to $\epsilon$. Now, a differentiation under the sign of integration yields

(10)

$$\phi'(\epsilon) = \int_a^b [F_u(x, y + \epsilon\eta, y' + \epsilon\eta')\eta(x) + F_{u'}(x, y + \epsilon\eta, y' + \epsilon\eta')\eta'(x)]\, dx.$$

where $F_u$, $F_{u'}$ indicates $\partial F/\partial u$, $\partial F/\partial u'$, respectively. Integrating the last integral by parts, we obtain

(11)

$$\phi'(\epsilon) = \int_a^b \left[ F_u(x, y + \epsilon\eta, y' + \epsilon\eta') - \frac{d}{dx} F_{u'}(x, y + \epsilon\eta, y' + \epsilon\eta') \right] \eta(x)\, dx + F_{u'}(x, y + \epsilon\eta, y' + \epsilon\eta')\eta(x) \Big|_a^b$$

The last term vanishes according to (5). $\phi'(\epsilon)$ must vanish at $\epsilon = 0$, at which $u$ equals $y$. Hence, we obtain the equation

$$0 = \phi'(0) = \int_a^b \left[ F_y(x, y, y') - \frac{d}{dx} F_{y'}(x, y, y') \right] \eta(x)\, dx, \tag{12}$$

which must be valid for an arbitrary function $\eta(x)$.

Equation (12) leads immediately to Euler's differential equation, by virtue of the following "fundamental lemma of the calculus of variations:"

LEMMA. *Let $\psi(x)$ be a continuous function in $a \leqslant x \leqslant b$. If the relation*

$$\int_a^b \psi(x)\eta(x)\, dx = 0 \tag{13}$$

*holds for all functions $\eta(x)$ which vanish at $x = a$ and $b$ and are continuous together with their first $2n$ derivatives, where $n$ is a positive integer, then $\psi(x) \equiv 0$.*

*Proof.* This lemma is easily proved indirectly. We shall show first that $\psi(x) \equiv 0$ in the open interval $a < x < b$. Let us suppose that this statement is not true, that $\psi(x)$ is different from zero, say positive, at $x = \xi$, where $\xi$ lies in the open interval. Then, according to the continuity of $\psi(x)$, there must exist an interval $\xi - \delta \leqslant x \leqslant \xi + \delta$ (with $a < \xi - \delta$, $\xi + \delta < b$, $\delta > 0$), in which $\psi(x)$ is positive. Now we take the function

$$\eta(x) = \begin{cases} (x - \xi + \delta)^{4n}(x - \xi - \delta)^{4n} \text{ in } \xi - \delta \leqslant x \leqslant \xi + \delta, \\ 0 \text{ elsewhere,} \end{cases} \tag{14}$$

where $n$ is a positive integer. (See Fig. 10.1:2.) This function $\eta(x)$ satisfies the continuity and boundary conditions specified above. But the choice of $\eta(x)$ as defined by (14) will make $\int_a^b \psi(x)\eta(x)\,dx > 0$, in contradiction to the hypothesis. Thus, the hypothesis is untenable, and $\psi(x) \equiv 0$ in $a < x < b$. According to the continuity of $\psi(x)$, we get $\psi(x) = 0$ also in $a \leqslant x \leqslant b$. Q.E.D. We remark that the lemma can be extended to hold equally well for multiple integrals.

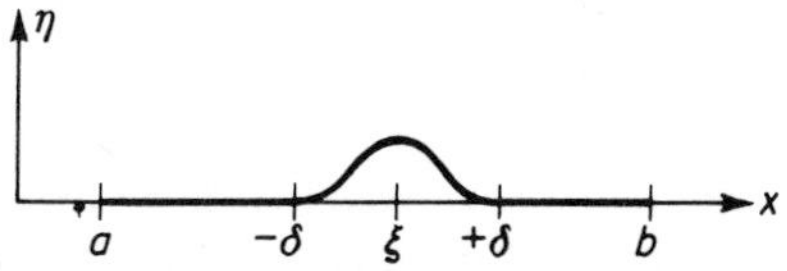

**Fig. 10.1:2.** The function $\eta(x)$.

From Eq. (12), it follows immediately from the lemma that the factor in front of $\eta(x)$ must vanish:

$$F_y(x, y, y') - \frac{d}{dx} F_{y'}(x, y, y') = 0, \qquad a \leqslant x \leqslant b. \tag{15a ▲}$$

This is a differential equation that $y(x)$ must satisfy and is known as Euler's equation. Written out *in extenso*, we have

$$\frac{d^2y}{dx^2}\frac{\partial^2 F}{\partial y'\,\partial y'}(x, y, y') + \frac{dy}{dx}\frac{\partial^2 F}{\partial y'\,\partial y}(x, y, y') + \frac{\partial^2 F}{\partial y'\,\partial x}(x, y, y') - \frac{\partial F}{\partial y}(x, y, y') = 0. \tag{15b ▲}$$

Should the problem be changed to finding the necessary condition for $J[u]$ to attain a maximum with respect to a sufficiently small neighborhood $(h)$ of $y(x)$, the same result would be obtained. Hence, the result: *The validity of Euler's differential Eq.* (15) *is a necessary condition for a function $y(x)$ to furnish an extremum of the functional $J[u]$ with respect to a sufficiently small neighborhood $(h)$ of $y(x)$.*

The satisfaction of Euler's equation is a necessary condition for $J[u]$ to attain an extremum; but it is not a sufficient condition. The question of sufficiency is rather involved; an interested reader must refer to treatises on calculus of variations such as those listed in Bib. 10.1, on p. 494.†

Now a point of notation. It is customary to call $\epsilon\eta(x)$ the *variation* of $u(x)$ and write

$$\epsilon\eta(x) = \delta u(x). \tag{16}$$

It is also customary to define the first variation of the functional $J[u]$ as

$$\delta J = \epsilon\, \phi'(\epsilon). \tag{17}$$

On multiplying both sides of Eq. (11) by $\epsilon$, it is evident that it can be written as

$$\delta J = \int_a^b \left[ F_u(x, u, u') - \frac{d}{dx} F_{u'}(x, u, u') \right] \delta u(x)\, dx + F_{u'}(x, u, u')\, \delta u(x) \Big|_a^b . \tag{18}$$

This is analogous to the notation of the differential calculus, in which the expression $df = \epsilon f'(x)$ for an arbitrary small parameter $\epsilon$ is called the "differential of the function $f(x)$." It is obvious that $\delta J$ depends on the function $u(x)$ and its variation $\delta u(x)$. Thus, a necessary condition for $J[u]$ to attain an extremum when $u(x) = y(x)$ is the vanishing of the first variation $\delta J$ for all variations $\delta u$ with $\delta u(a) = \delta u(b) = 0$.

*Example 1.*

$$I[u] = \int_a^b (1 + u'^2)\, dx = \min, \qquad u(a) = 0, \qquad u(b) = 1.$$

The Euler equation is

$$-\frac{d}{dx} F_{y'} = \frac{d}{dx} 2y' = 2y'' = 0.$$

Hence $y(x)$ is a straight line passing through the points $(a, 0)$ and $(b, 1)$.

*Example 2.*

Which curve minimizes the following functional?

$$I[u] = \int_0^{\pi/2} [(u')^2 - u^2]\, dx, \qquad u(0) = 0, \qquad u(\tfrac{1}{2}\pi) = 1.$$

*Ans.* $y = \sin x$.

† Any consideration of the sufficient conditions requires the concept of the second variations and the examination of its positive or negative definiteness. Several conditions are known; they are similar to, but more complex than, the corresponding conditions for maxima or minima of ordinary functions. It is useful to remember that many subtleties exist in the calculus of variations. A physicist or an engineer rarely worries about the mathematical details. His minimum principles are established on physical grounds and the existence of a solution is usually taken for granted. Mathematically, however, many examples can be constructed to show that a functional may not have a maximum or a minimum, or that a solution of the Euler equation may not minimize the functional. An engineer should be aware of these possibilities.

*Example 3. Minimum surface of revolution*

Find $y(x)$ that minimizes the functional

$$I[u] = 2\pi \int_a^b u\sqrt{1 + u'^2}\, dx, \qquad u(a) = A, \qquad u(b) = B.$$

*Ans.* The Euler equation can be integrated to give

$$y\sqrt{1 + y'^2} - \frac{yy'^2}{\sqrt{1 + y'^2}} = c_1.$$

Let $y' = \sinh t$, then $y = c_1 \cosh t$ and

$$dx = \frac{dy}{y'} = \frac{c_1 \sinh t\, dt}{\sinh t} = c_1\, dt, \qquad x = c_1 t + c_2.$$

Hence, the minimum surface is obtained by revolving a curve with the following parametric equations about the $x$-axis,

$$x = c_1 t + c_2, \qquad y = c_1 \cosh t, \quad \text{or} \quad y = c_1 \cosh \frac{x - c_2}{c_1},$$

which is a family of catenaries. The constants $c_1$, $c_2$ can be determined from the end points.

## 10.2. FUNCTIONAL INVOLVING HIGHER DERIVATIVES OF THE DEPENDENT VARIABLE

In an analogous manner one can treat the variational problem connected with the functional

$$J[u] = \int_a^b F(x, u, u^{(1)}, \ldots, u^{(n)})\, dx \tag{1}$$

involving $u(x)$ and its successive derivatives $u^{(1)}(x), u^{(2)}(x), \ldots, u^{(n)}(x)$, for $a \leqslant x \leqslant b$. To state the problem concisely we denote by $D$ the set of all real functions $u(x)$ with the following properties:

(2a) $\quad u(x), \ldots, u^{(2n)}(x)$ continuous in $a \leqslant x \leqslant b$,

(2b) $\quad u(a) = \alpha_0, \quad u^{(\nu)}(a) = \alpha_\nu, \quad \nu = 1, \ldots, n-1,$

(2c) $\quad u(b) = \beta_0, \quad u^{(\nu)}(b) = \beta_\nu, \quad \nu = 1, \ldots, n-1,$

where $\alpha_0$, $\beta_0$, $\alpha_\nu$ and $\beta_\nu$ are given numbers. A function $u(x)$ which possesses these continuity properties and boundary values is said to be in the set $D$. We assume that the function $F(x, u, u^{(1)}, \ldots, u^{(n)})$ has continuous partial derivatives up to the order $n + 1$ with respect to the $n + 2$ arguments $x, u, u^{(1)}, \ldots, u^{(n)}$ for $u$ in the set $D$. We seek a function $u(x) = y(x)$ in $D$ so that $J[u]$ is a minimum (or a maximum) for $u(x) = y(x)$, with respect to a sufficiently small neighborhood $(h)$ of $y(x)$.

Let $\eta(x)$ be an arbitrary function with the properties

$$\begin{aligned}&\eta(x),\ \eta^{(1)}(x),\ \ldots,\ \eta^{(2n)}(x) \text{ continuous in } a \leqslant x \leqslant b,\\ &\eta(a) = \ldots = \eta^{(n-1)}(a) = \eta(b) = \ldots = \eta^{(n-1)}(b) = 0.\end{aligned} \tag{3}$$

Then the function $u(x) = y(x) + \epsilon\eta(x)$ belongs to the set $D$ for all $\epsilon$. For sufficiently small $\epsilon$'s this function belongs to a prescribed neighborhood $(h)$ of $y(x)$. Once again, we introduce the function

$$\phi(\epsilon) = J[y + \epsilon\eta] = \int_a^b F(x, y + \epsilon\eta, y^{(1)} + \epsilon\eta^{(1)}, \ldots, y^{(n)} + \epsilon\eta^{(n)})\, dx. \tag{4}$$

The assumption that $y(x)$ minimizes $J[u]$ leads to the necessary condition $\phi'(0) = 0$. So we obtain by differentiating under the sign of integration, integrating by parts, and using Eqs. (3), the result

$$\begin{aligned}0 = \phi'(0) &= \int_a^b \left\{\sum_{\nu=0}^{n}\left[\frac{\partial}{\partial y^{(\nu)}} F(x, y, y^{(1)}, \ldots, y^{(n)})\right]\eta^{(\nu)}(x)\right\} dx\\ &= \int_a^b \left\{\sum_{\nu=0}^{n}(-1)^{\nu}\frac{d^{\nu}}{dx^{\nu}}\left[\frac{\partial}{\partial y^{(\nu)}} F(x, y, y^{(1)}, \ldots, y^{(n)})\right]\right\}\eta(x)\, dx,\end{aligned} \tag{5}$$

which holds for an arbitrary function $\eta(x)$ satisfying (3). According to the lemma proved in Sec. 10.1, we obtain the Euler equation

$$\blacktriangle \qquad \sum_{\nu=0}^{n}(-1)^{\nu}\frac{d^{\nu}}{dx^{\nu}}\left[\frac{\partial}{\partial y^{(\nu)}} F(x, y, y^{(1)}, \ldots, y^{(n)})\right] = 0, \tag{6}$$

which is a necessary condition for the minimizing (or maximizing) function $y(x)$.

The notations $\delta u$, $\delta J$, etc., introduced in Sec. 10.1, can be extended to this case by obvious changes.

*Example*

Find the extremal of $J[u] = \int_0^1 (1 + u''^2)\, dx$ satisfying the boundary conditions

$$u(0) = 0, \qquad u'(0) = 1, \qquad u(1) = 1, \qquad u'(1) = 1.$$

*Ans.* $y = x$.

## 10.3. SEVERAL UNKNOWN FUNCTIONS

The method used in the previous sections can be extended to more complicated functionals. For example, let

$$J[u_1, u_2, \ldots, u_m] = \int_a^b F(x, u_1, \ldots, u_m; u_1', \ldots, u_m')\, dx \tag{1}$$

be a functional depending on $m$ functions $u_1(x), \ldots, u_m(x)$. We assume that the functions $F(x, u_1, \ldots, u_m')$, $u_1(x), \ldots, u_m(x)$ are twice differentiable, and that the boundary values of $u_1, \ldots, u_m$ are given at $x = a$ and $b$. We seek a special set of functions $u_\mu(x) = y_\mu(x)$, $\mu = 1, \ldots, m$, in order that $J[u_1, \ldots, u_m]$ attains a minimum (or a maximum) when $u_\mu(x) = y_\mu(x)$, with respect to a sufficiently small neighborhood of the $y_\mu(x)$; i.e., for all $u_\mu(x)$ satisfying the relation

$$|y_\mu(x) - u_\mu(x)| < h_\mu, \qquad h_\mu > 0, \qquad \mu = 1, 2, \ldots, m.$$

Again it is easy to obtain the necessary conditions. Let the set of functions $y_1(x), \ldots, y_m(x)$ be a solution of the variational problem. Let $\eta_1(x), \ldots, \eta_m(x)$ be an arbitrary set of functions with the properties

$$\text{(2)} \qquad \begin{aligned} &\eta_\mu(x), \eta_\mu'(x), \eta_\mu''(x) \text{ continuous in } a \leqslant x \leqslant b, \\ &\eta_\mu(a) = \eta_\mu(b) = 0, \qquad \mu = 1, \ldots, m. \end{aligned}$$

Then consider the function

$$\text{(3)} \qquad \begin{aligned} \phi(\epsilon_1, \ldots, \epsilon_m) &= J[y_1 + \epsilon_1\eta_1, \ldots, y_m + \epsilon_m\eta_m] \\ &= \int_a^b F(x, y_1 + \epsilon_1\eta_1, \ldots, y_m + \epsilon_m\eta_m; y_1' + \epsilon_1\eta_1', \ldots, y_m' + \epsilon_m\eta_m')\, dx. \end{aligned}$$

Since $y_1(x), \ldots, y_m(x)$ minimize (or maximize) $J[y_1 + \epsilon\eta_1, \ldots, y_m + \epsilon_m\eta_m]$, the following inequality must hold for sufficiently small $\epsilon_1, \ldots, \epsilon_m$:

$$\text{(4)} \qquad \phi(\epsilon_1, \ldots, \epsilon_m) \geqslant \phi(0, \ldots, 0), \qquad [\text{or } \phi(\epsilon_1, \ldots, \epsilon_m) \leqslant \phi(0, \ldots, 0)].$$

The corresponding necessary conditions are

$$\text{(5)} \qquad \left(\frac{\partial \phi}{\partial \epsilon_\mu}\right) = 0, \quad \text{for} \quad \epsilon_1 = \epsilon_2 = \ldots = \epsilon_m = 0, \quad \mu = 1, \ldots, m,$$

which lead to

$$\text{(6)} \qquad \begin{aligned} 0 = \int_a^b [&F_{y_\mu}(x, y_1, \ldots, y_m, y_1', \ldots, y_m')\eta_\mu(x) \\ &+ F_{y'_\mu}(x, y_1, \ldots, y_m, y_1', \ldots, y_m')\eta_\mu'(x)]\, dx, \\ &\mu = 1, \ldots, m. \end{aligned}$$

Using integration by parts and the conditions (2), we obtain

$$\text{(7)} \qquad \begin{aligned} 0 = \int_a^b [&F_{y_\mu}(x, y_1, \ldots, y_m, y_1', \ldots, y_m') \\ &- \frac{d}{dx} F_{y'_\mu}(x, y_1, \ldots, y_m, y_1', \ldots, y_m')]\eta_\mu(x)\, dx \\ &+ F_{y'_\mu}(x, y_1, \ldots, y_m, y_1', \ldots, y_m')\eta_\mu(x)\Big|_a^b, \\ &\mu = 1, \ldots, m. \end{aligned}$$

Equation (7) must hold for any set of functions $\eta(x)$ satisfying (2). By the lemma of Sec. 10.1 we have the following Euler equations which must be satisfied by $y_1(x), \ldots, y_m(x)$ minimizing (or maximizing) the functional (1):

(8) ▲ $$F_{y_\mu}(x, y_1, \ldots, y_m; y_1', \ldots, y_m') - \frac{d}{dx} F_{y'_\mu}(x, y_1, \ldots, y_m; y_1', \ldots, y_m') = 0, \qquad \mu = 1, \ldots, m.$$

As a generalization of the variational notations given in Sec. 10.1 the expression

(12) $$\delta J = \sum_{\mu=1}^{m} \epsilon_\mu \left(\frac{\partial \phi}{\partial \epsilon_\mu}\right)$$

is called the "first variation of the functional."

Variational problems for functionals involving higher derivatives of $u_1, \ldots, u_m$ can be treated in the same way.

*Example*

Find the extremals of the functional

$$J[y, z] = \int_0^{\pi/2} (y'^2 + z'^2 + 2yz)\, dx,$$

$$y(0) = 0, \qquad y\left(\frac{\pi}{2}\right) = 1, \qquad z(0) = 0, \qquad z\left(\frac{\pi}{2}\right) = -1.$$

The Euler equations are

$$y'' - z = 0, \qquad z'' - y = 0.$$

Eliminating $z$, we have $y^{\text{iv}} - y = 0$. Hence,

$$y = c_1 e^x + c_2 e^{-x} + c_3 \cos x + c_4 \sin x,$$
$$z = y'' = c_1 e^x + c_2 e^{-x} - c_3 \cos x - c_4 \sin x.$$

From the boundary conditions we obtain the solution

$$y = \sin x, \qquad z = -\sin x.$$

## 10.4. SEVERAL INDEPENDENT VARIABLES

Consider the functional

(1) $$J[u] = \iint_G F(x, y, u, u_x, u_y)\, dx\, dy,$$

where $G$ is a finite, closed domain of the $x, y$-plane with a boundary curve $C$ which has a piecewise continuously turning tangent. The function $F(x, y, u, u_x, u_y)$ is assumed to be twice continuously differentiable with respect to its five arguments. Let $D$ be the set of all functions $u(x, y)$ with the following properties:

(2) $$D: \begin{cases} \text{(a)} & u(x, y),\quad u_x(x, y),\quad u_y(x, y),\quad u_{xx}(x, y), \\ & u_{xy}(x, y),\quad u_{yy}(x, y) \text{ continuous in } G, \\ \text{(b)} & u(x, y) \text{ prescribed on } C. \end{cases}$$

We now seek a special function $u(x, y) = v(x, y)$ in the set $D$, which minimizes (or maximizes) the functional $J[u]$.

Let $v(x, y)$ be a solution of this variational problem. Let $\eta(x, y)$ denote an arbitrary function with the properties

(3) $$\begin{cases} \text{(a)} & \eta(x, y) \text{ has continuous derivatives up to the second order in } G, \\ \text{(b)} & \eta(x, y) = 0 \text{ for } (x, y) \text{ on the boundary } C. \end{cases}$$

A consideration of the function

(4) $$\phi(\epsilon) = J[v + \epsilon\eta],$$

which attains an extremum when $\epsilon = 0$, again leads to the necessary condition

(5) $$\frac{d\phi}{d\epsilon}(0) = 0,$$

i.e., explicitly,

(6) $$0 = \iint_G \{F_v(x, y, v, v_x, v_y)\eta + F_{v_x}(x, y, v, v_x, v_y)\eta_x + F_{v_y}(x, y, v, v_x, v_y)\eta_y\}\, dx\, dy.$$

The last two terms can be simplified by Gauss' theorem after rewriting (6) as†

$$0 = \iint_G \left\{F_v \cdot \eta + \frac{\hat{\partial}}{\partial x}(F_{v_x} \cdot \eta) - \eta \frac{\hat{\partial}}{\partial x} F_{v_x} + \frac{\hat{\partial}}{\partial y}(F_{v_y} \cdot \eta) - \eta \frac{\hat{\partial}}{\partial y} F_{v_y}\right\} dx\, dy.$$

An application of Gauss' theorem to the sum of the second and fourth terms in the integrand gives

(7) $$0 = \iint_G \left\{F_v - \frac{\hat{\partial}}{\partial x} F_{v_x} - \frac{\hat{\partial}}{\partial y} F_{v_y}\right\}\eta(x, y)\, dx\, dy + \int_C \{F_{v_x} \cdot n_1(s) + F_{v_y} \cdot n_2(s)\}\eta\, ds$$

Here $n_1(s)$ and $n_2(s)$ are the components of the unit outer-normal vector $\mathbf{n}(s)$ of $C$.

† The symbol $(\hat{\partial}/\partial x)F_{v_x}$ means that $F_{v_x}$ should be considered as a function of $x$ and $y$, e.g.,

$$\frac{\hat{\partial}}{\partial x} F_{v_x} = \frac{\partial}{\partial x} F_{v_x} + \frac{\partial F_{v_x}}{\partial v}\frac{\partial v}{\partial x} + \frac{\partial F_{v_x}}{\partial v_x}\frac{\partial v_x}{\partial x} + \frac{\partial F_{v_x}}{\partial v_y}\frac{\partial v_y}{\partial x}.$$

The line integral in (7) vanishes according to (3). The function $\eta(x, y)$ in the surface integral is arbitrary. The generalized lemma of Sec. 10.1 then leads to the Euler equations

$$F_v - \frac{\partial}{\partial x} F_{v_x} - \frac{\partial}{\partial y} F_{v_y} = 0;$$

i.e.,

(9) ▲ $$\frac{\partial F}{\partial v} - \frac{\partial^2 F}{\partial v_x\, \partial x} - \frac{\partial^2 F}{\partial v_y\, \partial y} - \frac{\partial^2 F}{\partial v_x\, \partial v}\frac{\partial v}{\partial x} - \frac{\partial^2 F}{\partial v_y\, \partial v}\frac{\partial v}{\partial y} - \frac{\partial^2 F}{\partial v_x^2}\frac{\partial^2 v}{\partial x^2} - 2\frac{\partial^2 F}{\partial v_x\, \partial v_y}\frac{\partial^2 v}{\partial x\, \partial y} - \frac{\partial^2 F}{\partial v_y^2}\frac{\partial^2 v}{\partial y^2} = 0.$$

Equation (9) is a necessary condition for a function $v(x, y)$ in the set $D$ minimizing (or maximizing) the functional (1).

In the same way one may treat variational problems connected with functionals which involve more than two independent variables, higher derivatives, and several unknown functions.

*Example 1*

$$J[u] = \iint_D (u_x^2 + u_y^2)\, dx\, dy,$$

with a boundary condition that $u$ is equal to an assigned function $f(x, y)$ on the boundary $C$ of the domain $D$. The Euler equation is the Laplace equation $v_{xx} + v_{yy} = 0$.

*Example 2*

$$J[u] = \iint_D (u_{xx}^2 + u_{yy}^2 + 2u_{xy})^2\, dx\, dy = \text{min}.$$

Then $v$ must satisfy the biharmonic equation

$$\frac{\partial^4 v}{\partial x^4} + 2\frac{\partial^4 v}{\partial x^2 \partial y^2} + \frac{\partial^4 v}{\partial y^4} = 0.$$

## 10.5. SUBSIDIARY CONDITIONS—LAGRANGIAN MULTIPLIERS

In many problems, we are interested in the extremum of a function or functional under certain subsidiary conditions. As an elementary example, let us consider a function $f(x, y)$ defined for all $(x, y)$ in a certain domain $G$. Suppose that we are interested in the extremum of $f(x, y)$, not for all points in $G$, but only for those points in $G$ which satisfy the relation

(1) $$\phi(x, y) = 0.$$

Thus, if the domain $G$ and the curve $\phi(x, y) = 0$ are as shown in Fig. 10.5:1, then we are interested in finding the extremum of $f(x, y)$ among all points

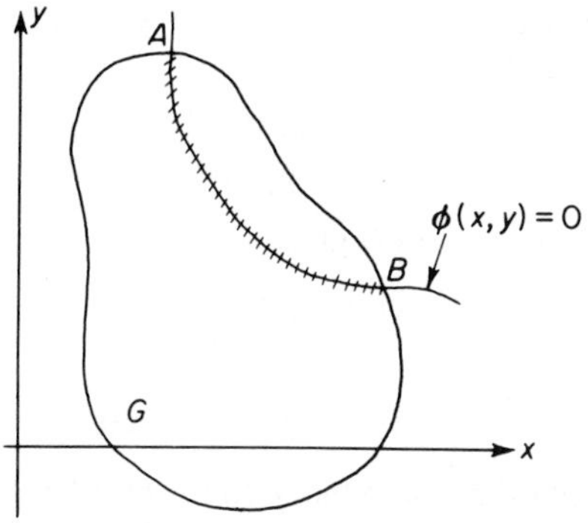


**Fig. 10.5:1.** Illustrating a subsidiary condition.

that lie on the segment of the curve $AB$. For conciseness of expression, we shall call such a subdomain—the segment $AB$—$G_\phi$.

Let us find the necessary conditions for $f(x, y)$ to attain an extreme value at a point $(\bar{x}, \bar{y})$ in $G_\phi$, with respect to all points of a sufficiently small neighborhood of $(\bar{x}, \bar{y})$ belonging to $G_\phi$.

Let us assume that the function $\phi(x, y)$ has continuous partial derivatives with respect to $x$ and $y$, and that at the point $(\bar{x}, \bar{y})$ not both derivatives are zero; say,

(2) $$\phi_y(\bar{x}, \bar{y}) \neq 0.$$

Then, according to a fundamental theorem on implicit functions, there exists a neighborhood of $\bar{x}$, say $\bar{x} - \delta \leqslant x \leqslant \bar{x} + \delta$ $(\delta > 0)$, where the equation $\phi(x, y) = 0$ can be solved uniquely in the form

(3) $$y = g(x).$$

The function $g(x)$ so defined is single-valued and differentiable, and $\phi[x, g(x)] = 0$ is an identity in $x$. Hence

(4) $$0 = \phi_x[x, g(x)] + \phi_y[x, g(x)]\frac{dg(x)}{dx}.$$

Now, let us consider the problem of the extremum value of $f(x, y)$ in $G_\phi$. According to (3), in a sufficiently small neighborhood of $(\bar{x}, \bar{y})$, $y$ is an implicit function of $x$, and $f(x, y)$ becomes

(5) $$\mathscr{F}(x) = f[x, g(x)].$$

If $\mathscr{F}(x)$ attains an extreme value at $\bar{x}$, then the first derivative of $\mathscr{F}(x)$ vanishes at $\bar{x}$; i.e.,

(6) $$0 = \frac{d\mathscr{F}}{dx}(\bar{x}) = f_x(\bar{x}, \bar{y}) + f_y(\bar{x}, \bar{y})\frac{dg}{dx}(\bar{x}).$$

But the derivative $dg/dx$ can be eliminated between (6) and (4). The result, together with (1), constitutes the necessary condition for an extremum of $f(x, y)$ in $G_\phi$.

The formalism will be more elegant by introducing a number $\lambda$, called *Lagrange's multiplier*, defined by

$$\lambda = -\frac{f_y(\bar{x}, \bar{y})}{\phi_y(\bar{x}, \bar{y})}. \tag{7}$$

A combination of (4), (6), and (7) gives

$$f_x(\bar{x}, \bar{y}) + \lambda\phi_x(\bar{x}, \bar{y}) = 0, \tag{8}$$

while (7) may be written as

$$f_y(\bar{x}, \bar{y}) + \lambda\phi_y(\bar{x}, \bar{y}) = 0. \tag{9}$$

Equations (1), (8), and (9) are necessary conditions for the function $f(x, y)$ to attain an extreme at a point $(\bar{x}, \bar{y})$, where $\phi_x^2 + \phi_y^2 > 0$. These conditions constitute three equations for the three "unknowns" $x$, $y$, and $\lambda$.

These results can be summarized in the following manner. We introduce a new function

$$F(x, y; \lambda) = f(x, y) + \lambda\phi(x, y). \tag{10}$$

If the function $f(x, y)$ has an extreme value at the point $(\bar{x}, \bar{y})$ with respect to $G_\phi$, and if

$$[\phi_x(\bar{x}, \bar{y})]^2 + [\phi_y(\bar{x}, \bar{y})]^2 > 0, \tag{11}$$

then there exists a certain number $\bar{\lambda}$ so that the three partial derivatives of $F(x, y; \lambda)$ with respect to $x$, $y$, and $\lambda$ are zero at $(\bar{x}, \bar{y}, \bar{\lambda})$:

$$\begin{aligned} \frac{\partial F}{\partial x}(\bar{x}, \bar{y}, \bar{\lambda}) &= f_x(\bar{x}, \bar{y}) + \bar{\lambda}\phi_x(\bar{x}, \bar{y}) = 0, \\ \frac{\partial F}{\partial y}(\bar{x}, \bar{y}, \bar{\lambda}) &= f_y(\bar{x}, \bar{y}) + \bar{\lambda}\phi_y(\bar{x}, \bar{y}) = 0, \\ \frac{\partial F}{\partial \lambda}(\bar{x}, \bar{y}, \bar{\lambda}) &= \phi(\bar{x}, \bar{y}) = 0. \end{aligned} \tag{12}$$

If there exist points $(x', y')$ in $G$ with $\phi(x', y') = 0$ and $\phi_x(x', y') = \phi_y(x', y') = 0$, additional considerations are necessary.

In the formulation (10) and (12), the theorem can be generalized to the case of $n$ variables $x_1, \ldots, x_n$ and several subsidiary conditions $0 = \phi_1(x_1, \ldots, x_n) = \cdots = \phi_m(x_1, \ldots, x_n)$, $m < n$.

The application of the Lagrange multiplier method to the minimization of functionals follows a similar reasoning.

As an example, let us consider the classical problem of geodesics: to find the line of minimal length lying on a given surface $\phi(x, y, z) = 0$ and joining

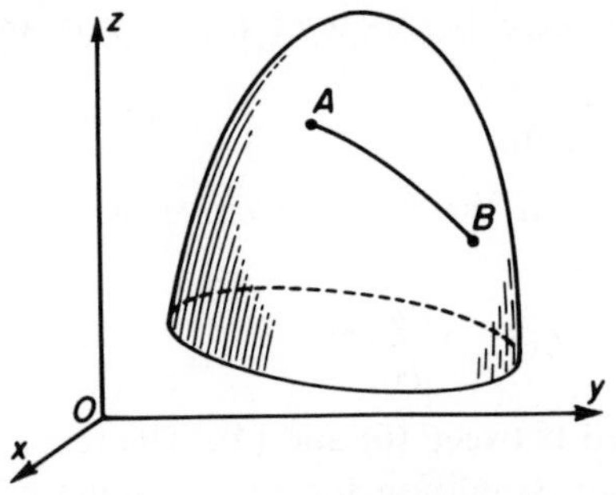

**Fig. 10.5:2.** A problem of geodesics.

two given points on this surface (Fig. 10.5:2). Here we must minimize the functional

(13) $$l = \int_{x_0}^{x_1} \sqrt{1 + y'^2 + z'^2}\, dx,$$

with $y(x)$, $z(x)$ satisfying the condition $\phi(x, y, z) = 0$. This problem was solved in 1697 by Johann Bernoulli, but a general method of solution was given by L. Euler and J. Lagrange.

We consider a new functional

(14) $$I^*[y, z, \lambda] = \int_{x_0}^{x_1} [\sqrt{1 + y'^2 + z'^2} + \lambda(x)\phi(x. y, z)]\, dx$$

and use the method of Sec. 10.3 to determine the functions $y(x)$, $z(x)$, and $\lambda(x)$ that minimizes $I^*$. The necessary conditions are

(15) $$\lambda \frac{\partial \phi}{\partial y} - \frac{d}{dx} \frac{y'}{\sqrt{1 + y'^2 + z'^2}} = 0,$$

(16) $$\lambda \frac{\partial \phi}{\partial z} - \frac{d}{dx} \frac{z'}{\sqrt{1 + y'^2 + z'^2}} = 0,$$

(17) $$\phi(x, y, z) = 0.$$

This system of equations determines the functions $y(x)$, $z(x)$, and $\lambda(x)$.

## 10.6. NATURAL BOUNDARY CONDITIONS

In previous sections we considered variational problems in which the admissible functions have prescribed values on the boundary. We shall now consider problems in which no boundary values are prescribed for the admissible functions. Such problems lead to the *natural boundary conditions.*

Consider again the functional (10.1:1) but now omit the boundary condition (10.1:2). Following the arguments of Sec. 10.1, we obtain the necessary condition (10.1:11),

(1) $$0 = \int_a^b \left\{F_y - \frac{d}{dx} F_{y'}\right\}\eta(x)\, dx + F_{y'} \cdot \eta(x) \Big|_a^b .$$

This equation must hold for all arbitrary functions $\eta(x)$ and, in particular, for arbitrary functions $\eta(x)$ with $\eta(a) = \eta(b) = 0$. This leads at once to the Euler equation (10.1:15)

(2) $$F_y(x, y, y') - \frac{d}{dx} F_{y'}(x, y, y') = 0, \qquad \text{in } a \leqslant x \leqslant b.$$

In contrast to Sec. 10.1, however, the last term $F_{y'} \cdot \eta(x) \,|_a^b$, does not vanish by prescription. Hence, by Eqs. (1) and (2), we must have

(3) $$(F_{y'} \cdot \eta)_{x=b} - (F_{y'} \cdot \eta)_{x=a} = 0$$

for all functions $\eta(x)$. But now $\eta(a)$ and $\eta(b)$ are arbitrary. Taking two functions $\eta_1(x)$ and $\eta_2(x)$ with

$$\eta_1(a) = 1, \qquad \eta_1(b) = 0; \qquad \eta_2(a) = 0, \qquad \eta_2(b) = 1; \tag{4}$$

we get, from (3),

$$F_{y'}[a, y(a), y'(a)] = 0, \qquad F_{y'}[b, y(b), y'(b)] = 0. \tag{5}$$

The conditions (5) are called the *natural boundary conditions* of our problem. They are the boundary conditions which must be satisfied by the function $y(x)$ if the functional $J[u]$ reaches an extremum at $u(x) = y(x)$, provided that $y(a)$ and $y(b)$ are entirely arbitrary.

Thus, if the first variation of a functional $J[u]$ vanishes at $u(x) = y(x)$, and if the boundary values of $u(x)$ at $x = a$ and $x = b$ are arbitrary, then $y(x)$ must satisfy not only the Euler equation but also the natural boundary conditions which, in general, involve the derivatives of $y(x)$. In contrast to the *natural* boundary conditions, the conditions $u(a) = \alpha$, $u(b) = \beta$, which specify the boundary values of $u(x)$ at $x = a$ and $b$, are called *rigid boundary conditions*.

The concept of natural boundary conditions is also important in a more general type of variational problem in which boundary values occur explicitly in the functionals. It can be generalized also to functionals involving several dependent and independent variables and higher derivatives of the dependent variables.

The idea of deriving natural boundary conditions for a physical problem is of great importance and will be illustrated again and again below. See, for example, Sec. 10.8, 11.2.

## 10.7. THEOREM OF MINIMUM POTENTIAL ENERGY UNDER SMALL VARIATIONS OF DISPLACEMENTS

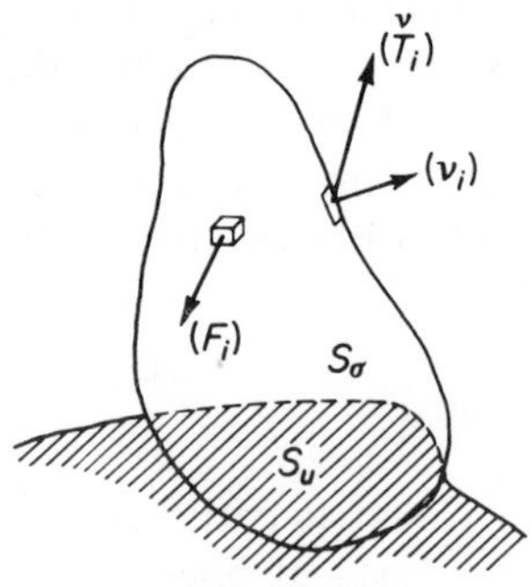

**Fig. 10.7:1.** Notations.

Let a body be *in static equilibrium* under the action of specified body and surface forces. (Fig. 10.7:1). The boundary surface $S$ shall be assumed to consist of two parts, $S_\sigma$ and $S_u$, with the following boundary conditions.

Over $S_\sigma$: The surface traction $\overset{\nu}{T}_i$ is prescribed.
Over $S_u$: The displacement $u_i$ is prescribed.

We assume that there exists a system of displacements $u_1$, $u_2$, $u_3$ satisfying the Navier's equations of equilibrium and the given boundary conditions. Let us consider a class of arbitrary displacements $u_i + \delta u_i$ consistent with the constraints imposed on the body. Thus $\delta u_i$ must vanish over $S_u$, but it is arbitrary over $S_\sigma$. We further restrict $\delta u_i$ to be triply differentiable and to be of such an order of magnitude that the material remains elastic. Such arbitrary displacements $\delta u_i$ are called *virtual displacements*.

Let us assume that static equilibrium prevails and compute the "virtual work" done by the body force $F_i$ per unit volume and the surface force $\overset{\nu}{T}{}_i^*$ per unit area:

$$\int_V F_i \, \delta u_i \, dv + \int_S \overset{\nu}{T}{}_i^* \, \delta u_i \, dS.$$

On substituting $\overset{\nu}{T}{}_i^* = \sigma_{ij}\nu_j$ and transforming according to Gauss' theorem, we have

$$\text{(1)} \qquad \begin{aligned} \int_S \overset{\nu}{T}{}_i^* \, \delta u_i \, dS &= \int_S \sigma_{ij} \, \delta u_i \nu_j \, dS \\ &= \int_V (\sigma_{ij} \, \delta u_i)_{,j} \, dv \\ &= \int_V \sigma_{ij,j} \, \delta u_i \, dv + \int_V \sigma_{ij} \delta u_{i,j} \, dv. \end{aligned}$$

According to the equation of equilibrium the first integral on the right-hand side is equal to $-\int F_i \, \delta u_i \, dv$. On account of the symmetry of $\sigma_{ij}$, the second integral may be written as

$$\int_V \sigma_{ij}[\tfrac{1}{2}(\delta u_{i,j} + \delta u_{j,i})] \, dv = \int_V \sigma_{ij} \, \delta e_{ij} \, dv.$$

Therefore, (1) becomes

$$\text{(2)} \ \blacktriangle \qquad \int_V F_i \, \delta u_i \, dv + \int_{S_\sigma} \overset{\nu}{T}{}_i^* \, \delta u_i \, dS = \int_V \sigma_{ij} \, \delta e_{ij} \, dv.$$

This equation expresses the *principle of virtual work*. The surface integral needs only be integrated over $S_\sigma$, since $\delta u_i$ vanishes over the surface $S_u$ where boundary displacements are given.

If the *strain-energy function* $W(e_{11}, e_{12}, \ldots)$ exists, so that $\sigma_{ij} = \partial W / \partial e_{ij}$, (Sec. 6.1), then it can be introduced into the right-hand side of Eq. (2). Since

$$\text{(3)} \qquad \int_V \sigma_{ij} \, \delta e_{ij} \, dv = \int_V \frac{\partial W}{\partial e_{ij}} \, \delta e_{ij} \, dv = \delta \int_V W \, dv,$$

the principle of virtual work can be stated as

$$\text{(4)} \ \blacktriangle \qquad \delta \int_V W \, dv - \int_V F_i \, \delta u_i \, dv - \int_{S_\sigma} \overset{\nu}{T}{}_i^* \, \delta u_i \, dS = 0.$$

Further simplification is possible if the body force $F_i$ and the surface tractions $\overset{\nu}{T}{}_i^*$ are *conservative* so that

$$\text{(5)} \qquad F_i = -\frac{\partial G}{\partial u_i}, \qquad \overset{\nu}{T}{}_i^* = -\frac{\partial g}{\partial u_i}.$$

The functions $G(u_1, u_2, u_3)$ and $g(u_1, u_2, u_3)$ are called the potential of $F_i$ and $\overset{\nu}{T}_i^*$, respectively. In this case,

$$(6) \qquad -\int_V F_i\, \delta u_i\, dv - \int_{S_\sigma} \overset{\nu}{T}_i^*\, \delta u_i\, dS = \delta \int_V G\, dv + \delta \int_S g\, dS.$$

Then Eq. (4) may be written as

$$(7) \ \blacktriangle \qquad \delta \mathscr{V} = 0,$$

where

$$(8) \ \blacktriangle \qquad \mathscr{V} \equiv \int_V (W + G)\, dv + \int_S g\, dS$$

The function $\mathscr{V}$ is called the *potential energy of the system.* This equation states that the potential energy has a stationary value in a class of admissible variations $\delta u_i$ of the displacements $u_i$ in the equilibrium state. Formulated in another way, it states that, *of all displacements satisfying the given boundary conditions, those which satisfy the equations of equilibrium are distinguished by a stationary (extreme) value of the potential energy.* For a rigid-body, $W$ vanishes and the familiar form is recognized. We emphasize that the linearity of the stress-strain relationship has not been invoked in the above derivation, so that *this principle is valid for nonlinear, as well as linear, stress-strain law, as long as the body remains elastic.*

That this stationary value is a miminum in the neighborhood of the natural, unstrained state follows the assumption that the strain energy function is positive definite in such a neighborhood (see Secs. 12.4, 12.5). This can be shown by comparing the potential energy $\mathscr{V}$ of the actual displacements $u_i$ with the energy $\mathscr{V}'$ of another system of displacements $u_i + \delta u_i$ satisfying the condition $\delta u_i = 0$ over $S_u$. We have

$$(9) \qquad \mathscr{V}' - \mathscr{V} = \int [W(e_{11} + \delta e_{11}, \dots,) - W(e_{11}, \dots,)]\, dv - \int F_i\, \delta u_i\, dv - \int \overset{\nu}{T}_i^*\, \delta u_i\, dS.$$

Expanding $W(e_{11} + \delta e_{11}, \dots,)$ into a power series, we have

$$(10) \qquad W(e_{11} + \delta e_{11}, \dots,) = W(e_{11}, \dots,) + \frac{\partial W}{\partial e_{ij}} \delta e_{ij} + \frac{1}{2} \frac{\partial^2 W}{\partial e_{ij}\, \partial e_{kl}} \delta e_{ij}\, \delta e_{kl} + \dots.$$

A substitution into (9) yields, up to the second order in $\delta e_{ij}$,

$$(11) \qquad \mathscr{V}' - \mathscr{V} = \int_V \frac{\partial W}{\partial e_{ij}} \delta e_{ij}\, dv - \int_V F_i\, \delta u_i\, dv - \int_S \overset{\nu}{T}_i^*\, \delta u_i\, dS + \int_V \frac{1}{2} \frac{\partial^2 W}{\partial e_{ij}\, \partial e_{kl}} \delta e_{ij}\, \delta e_{kl}\, dv.$$

The sum of the terms in the first line on the right-hand side vanishes on account of (4). The sum in the second line is positive for sufficiently small values of strain $\delta e_{ij}$ and can be seen as follows. Let us set $e_{ij} = 0$ in Eq. (10). The constant term is immaterial. The linear term must vanish because $\partial W/\partial e_{ij} = \sigma_{ij}$, which must vanish as $e_{ij} \to 0$. Hence, up to the second order,

$$\frac{1}{2}\frac{\partial^2 W}{\partial e_{ij}\,\partial e_{kl}}\,\delta e_{ij}\,\delta e_{kl} = W(\delta e_{ij}). \tag{12}$$

Therefore, (11) becomes

$$\mathscr{V}' - \mathscr{V} = \int W(\delta e_{ij})\,dv. \tag{13}$$

If $W(\delta e_{ij})$ is positive definite, then the last line in (11) is positive, and

$$\mathscr{V}' - \mathscr{V} \geqslant 0 \tag{14}$$

and that $\mathscr{V}$ is a minimum is proved. Accordingly, our principle is called the *principle of minimum potential energy*. The equality sign holds only if all $\delta e_{ij}$ vanish, i.e., if the virtual displacements consist of a virtual rigid-body motion. If there were three or more points of the body fixed in space, such a rigid-body motion would be excluded, and $\mathscr{V}$ is a strong minimum; otherwise it is a weak minimum.

To recapitulate, we remark again that the variational principle (2) is generally valid; (4) is established whenever the strain energy function $W(e_{11}, e_{12}, \ldots,)$ exists; and (7) is established when the potential energy $\mathscr{V}$ can be meaningfully defined, but the fact that $\mathscr{V}$ is a minimum for "actual" displacements is established only in the neighborhood of the stable natural state, where $W$ is positive definite.

Conversely, we may show that the variational principle gives the equations of elasticity. In fact, starting from Eq. (7) and varying $u_i$, we have

$$\delta\mathscr{V} = \int_V \frac{\partial W}{\partial e_{ij}}\,\delta e_{ij}\,dv - \int_V F_i\,\delta u_i\,dv - \int_S \overset{\nu}{T}{}_i^*\,\delta u_i\,dS = 0. \tag{15}$$

But

$$\int \frac{\partial W}{\partial e_{ij}}\,\delta e_{ij}\,dv = \frac{1}{2}\int \sigma_{ij}(\delta u_{i,j} + \delta u_{j,i})\,dv = -\int \sigma_{ij,j}\,\delta u_i\,dv + \int \sigma_{ij}\nu_j\,\delta u_i\,dS.$$

Hence,

$$\int_V (\sigma_{ij,j} + F_i)\,\delta u_i\,dv + \int_S (\sigma_{ij}\nu_j - \overset{\nu}{T}{}_i^*)\,\delta u_i\,dS = 0, \tag{16}$$

which can be satisfied for arbitrary $\delta u_i$ if

$$\sigma_{ij,j} + F_i = 0 \qquad \text{in } V \tag{17}$$

and either

$$\delta u_i = 0 \qquad \text{on } S_u \text{ (rigid boundary condition),} \tag{18}$$

or

$$\overset{\nu}{T}{}_i^* = \sigma_{ij}\nu_j \qquad \text{on } S_\sigma \text{ (natural boundary condition).} \tag{19}$$

The one-to-one correspondence between the differential equations of equilibrium and the variational equation is thus demonstrated; for we have first derived (7) from the equation of equilibrium and then have shown that, conversely, Eqs. (17)–(19) necessarily follow (7).

The commonly encountered external force systems in elasticity are conservative systems in which the body force $F_i$ and surface tractions $\overset{\nu}{T}_i$ are independent of the elastic deformation of the body. In this case, $\mathscr{V}$ is more commonly written as

$$\mathscr{V} \equiv \int_V W\,dv - \int_V F_i u_i\,dv - \int_{S_\sigma} \overset{\nu}{T}{}_i^* u_i\,dS. \tag{20}$$

A branch of mechanics in which the external forces are in general non-conservative is the theory of aeroelasticity. In aeroelasticity one is concerned with the interaction of aerodynamic forces and elastic deformation. The aerodynamic forces depend on the flow and the deformation of the entire body, not just the local deformation; thus in general it cannot be derived from a potential function. The principle of virtual work, in the form of Eq. (4), is still applicable to aeroelasticity.

**Problem 10.1.** The minimum potential energy principle states that elastic equilibrium is equivalent to the condition $J = \min$, where

(a) $$J = \int_V [W(e_{ij}) + F_i u_i]\,dV + \int_{S_\sigma} \overset{\nu}{T}_i u_i\,dS, \qquad \text{(varying } u_i\text{);}$$

(b) $$e_{ij} = \tfrac{1}{2}(u_{i,j} + u_{j,i}).$$

By the method of Lagrange multipliers, the subsidiary condition (b) can be incorporated into the functional $J$, and we are led to consider the variational equation

(c) $$\delta J' = 0, \qquad \text{(varying } e_{ij}, u_i \text{ independently),}$$

where

(d) $$J' = J + \int_V \lambda_{ij}[e_{ij} - \tfrac{1}{2}(u_{i,j} + u_{j,i})]\,dV.$$

Since the quantity in [ ] is symmetric in $i, j$, we may restrict the Lagrange multipliers $\lambda_{ij}$ to be symmetric, $\lambda_{ij} = \lambda_{ji}$, so that only six independent multipliers are needed. Derive the Euler equations for (c) and show that $\lambda_{ij}$ should be interpreted as the stress tensor.

*Note:* Once Lagrange's multipliers are employed, the phrase "minimum conditions" used in the principle of minimum potential energy has to be replaced by "stationary conditions."

## 10.8. EXAMPLE OF APPLICATION: STATIC LOADING ON A BEAM—NATURAL AND RIGID END CONDITIONS

As an illustration of the application of the minimum potential energy principle in formulating approximate theories of elasticity, let us consider the approximate theory of bending of a slender beam under static loading. Let the beam be perfectly straight and lying along the $x$-axis before the application of external loading, which consists of a distributed lateral load $p(x)$ per unit length, a bending moment $M_0$, and a shearing force $Q_0$ at the end $x = 0$, and a moment $M_l$ and a shear $Q_l$ at the end $x = l$ (Fig. 10.8:1). We assume that the principal axes of inertia of every cross section of the beam lie in two mutually orthogonal principal planes and that the loading $p$, $M$, $Q$ are

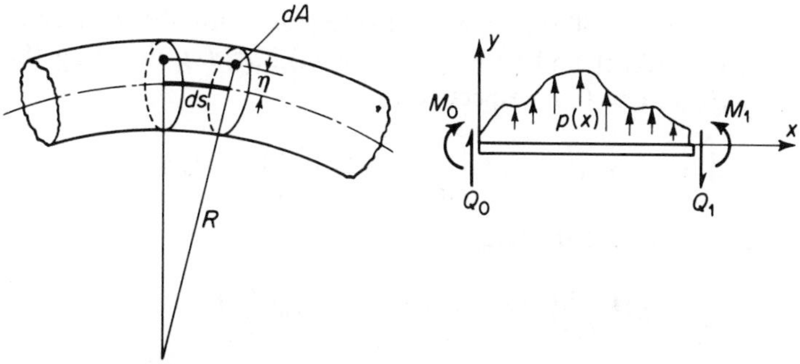


**Fig. 10.8:1.** Applications to a simple beam.

applied in one of the principal planes. In accordance with the approximate beam theory, we assume that every plane cross section of the beam remains plane during bending.

Let $ds$ be the arc length along the neutral axis. Under these assumptions, we see that when the neutral axis of the beam is bent from the initial straight line into a curve with radius of curvature $R$, the length of a filament, initially $ds$ and parallel to the neutral axis, is altered by the bending in the ratio $1:(1 + \eta/R)$, where $\eta$ is the distance between the filament and the neutral axis. The strain is $\eta/R$, and the force acting on the filament is $E\,\eta dA/R$, where $dA$ is the cross section of the filament. The resultant moment of these forces about the neutral axis is

$$\int_A \eta \cdot E\frac{\eta}{R}\,dA = \frac{E}{R}\int_A \eta^2\,dA = \frac{EI}{R},$$

where $I$ is the moment of inertia of the area of the beam cross section. The angle through which the cross sections rotate relative to each other is

$ds/R$. Hence, the work to bend a segment $ds$ of the beam to a curvature $1/R$ is

$$\frac{1}{2} EI \frac{ds}{R^2},$$

where the factor $\frac{1}{2}$ is added since the mean work is half the value of the product of the final moment and angle of rotation. Hence, integrating throughout the beam, we have the strain energy

$$U = \frac{1}{2}\int_0^l EI \frac{ds}{R^2}. \tag{1}$$

If we now assume that the deflection of the beam is *infinitesimal* so that if $y$ denotes the lateral deflection, the curvature $1/R$ is approximated by $d^2y/dx^2$, and $ds$ is approximated by $dx$, then

$$U = \frac{1}{2}\int_0^l EI\left(\frac{d^2y}{dx^2}\right)^2 dx. \tag{1a}$$

The potential energy of the external loading is, with the sign convention specified in Fig. 10.8:1 and with $y_0$, $y_1$, $(dy/dx)_0$,$(dy/dx)_l$ denoting the value of $y$ and $dy/dx$ at $x = 0$, $l$, respectively,

$$-\int p(x)y(x)\,dx + M_0\left(\frac{dy}{dx}\right)_0 - M_l\left(\frac{dy}{dx}\right)_l - Q_0y_0 + Q_ly_l. \tag{2}$$

Hence, the total potential energy is

$$\mathscr{V} = \frac{1}{2}\int_0^l EI\left(\frac{d^2y}{dx^2}\right)^2 dx - \int_0^l py\,dx + M_0\left(\frac{dy}{dx}\right)_0 - M_l\left(\frac{dy}{dx}\right)_l - Q_0y_0 + Q_ly_l. \tag{3}$$

At equilibrium, the variation of $\mathscr{V}$ with respect to the virtual displacement $\delta y$ must vanish. Hence,

$$\delta\mathscr{V} = \int_0^l EI\frac{d^2y}{dx^2}\,\delta\left(\frac{d^2y}{dx^2}\right) dx - \int_0^l p\,\delta y\,dx + M_0\,\delta\left(\frac{dy}{dx}\right)_0 - M_l\,\delta\left(\frac{dy}{dx}\right)_l - Q_0\,\delta y_0 + Q_l\,\delta y_l = 0.$$

Integrating the first term by parts twice and collecting terms, we obtain

$$\begin{aligned}\delta\mathscr{V} = &\int_0^l \left[\frac{d^2}{dx^2}\left(EI\frac{d^2y}{dx^2}\right) - p\right]\delta y\,dx \\ &+ \left[EI\left(\frac{d^2y}{dx^2}\right)_l - M_l\right]\delta\left(\frac{dy}{dx}\right)_l - \left[EI\left(\frac{d^2y}{dx^2}\right)_0 - M_0\right]\delta\left(\frac{dy}{dx}\right)_0 \\ &- \left[\frac{d}{dx}EI\left(\frac{d^2y}{dx^2}\right)_l - Q_l\right]\delta y_l + \left[\frac{d}{dx}EI\left(\frac{d^2y}{dx^2}\right)_0 - Q_0\right]dy_0 \\ = 0 &\end{aligned} \tag{4}$$

Since $\delta y$ is arbitrary in the interval $(0, l)$, we obtain the differential equation of the beam

$$\frac{d^2}{dx^2}\left(EI\frac{d^2y}{dx^2}\right) - p = 0, \qquad 0 \leqslant x \leqslant l. \tag{5}$$

In order that the remaining terms in (4) may vanish, it is sufficient to have the end conditions

(6a) Either $EI\left(\frac{d^2y}{dx^2}\right)_l - M_l = 0$ or $\delta\left(\frac{dy}{dx}\right)_l = 0.$

(6b) Either $EI\left(\frac{d^2y}{dx^2}\right)_0 - M_0 = 0$ or $\delta\left(\frac{dy}{dx}\right)_0 = 0.$

(6c) Either $\frac{d}{dx}\left(EI\frac{d^2y}{dx^2}\right)_l - Q_l = 0$ or $\delta y_l = 0.$

(6d) Either $\frac{d}{dx}\left(EI\frac{d^2y}{dx^2}\right)_0 - Q_0$ or $\delta y_0 = 0.$

If the deflection $y_0$ is prescribed at the end $x = 0$, then $\delta y_0 = 0$. If the slope $(dy/dx)_0$ is prescribed at the end $x = 0$, then $\delta(dy/dx)_0 = 0$. These are called "rigid" boundary conditions. On the other hand, if the value of $y_0$ is unspecified and perfectly free, then $\delta y_0$ is arbitrary and we must have

$$\frac{d}{dx}\left(EI\frac{d^2y}{dx^2}\right)_0 - Q_0 = 0 \tag{7}$$

as an end condition for a free end; otherwise $\delta\mathscr{V}$ cannot vanish for arbitrary variations $\delta y_0$. Equation (7) is called a "natural" boundary condition. Similarly, all the left-hand equations in (6a)–(6d) are natural boundary conditions, and all the right-hand equations are rigid boundary conditions.

The distinction between natural and rigid boundary conditions assumes great importance in the application of the direct methods of solution of variational problems; the assumed functions in the direct methods must satisfy the rigid boundary conditions. See Sec. 11.8.

It is worthwhile to consider the following question. The reader must be familiar with the fact that in the engineering beam theory, the end conditions often considered are:

(8a) *Clamped end:* $y = 0, \quad \frac{dy}{dx} = 0,$

(8b) *Free end:* $EI\frac{d^2y}{dx^2} = 0, \quad \frac{d}{dx}\left(EI\frac{d^2y}{dx^2}\right) = 0,$

(8c) *Simply supported end:* $y = 0, \quad EI\frac{d^2y}{dx^2} = 0,$

where $y(x)$ is the deflection function of the beam. May we ask why are the other two combinations, namely

$$y = 0, \qquad \frac{d}{dx}\left(EI\frac{d^2y}{dx^2}\right) = 0, \tag{9a}$$

$$\frac{dy}{dx} = 0, \qquad EI\frac{d^2y}{dx^2} = 0, \tag{9b}$$

never considered?

An acceptable answer is perhaps that the boundary conditions (9a) and (9b) cannot be realized easily in the laboratory. But a more satisfying answer is that they are not proper sets of boundary conditions. If the conditions (9a) or (9b) were imposed, then, according to (4), it can not at all be assured that the equation $\delta\mathscr{V} = 0$ will be satisfied. Thus, a basic physical law might be violated. These boundary conditions are, therefore, inadmissible.

From the point of view of the differential Eq. (5), one may feel that the end conditions (9a) or (9b) are legitimate. Nevertheless, they are ruled out by the minimum potential energy principle on physical grounds. In fact, in the theory of differential equations the Eq. (5) and the end conditions (8) are known to form a so-called *self-adjoint* differential system, whereas (5) and (9) would form a *nonself-adjoint* differential system. Very great difference in mathematical character exists between these two catagories. For example, a free vibration problem of a nonself-adjoint system may not have an eigenvector, or it may have complex eigenvalues or complex eigenvectors.

There are other conceivable admissible boundary conditions, such as to require

$$\frac{dy}{dx} = 0, \qquad \frac{d}{dx}\left(EI\frac{d^2y}{dx^2}\right) = 0, \qquad \text{at } x = 0. \tag{10}$$

Such an end, with zero slope and zero shear, cannot be easily established in the laboratory. Similarly, it is conceivable that one may require that at the end $x = 0$ the following ratios hold:

$$\delta\left(\frac{\partial y}{\partial x}\right) : \delta y = c, \quad \text{a constant,}$$
$$\left[\frac{d}{dx}\left(EI\frac{d^2y}{dx^2}\right) - Q_0\right] : \left[EI\frac{d^2y}{dx^2} - M_0\right] = c, \quad \text{the same constant.} \tag{11}$$

This pair of conditions are also admissible, but are unlikely to be encountered in practice.

## 10.9. THE COMPLEMENTARY ENERGY THEOREM UNDER SMALL VARIATIONS OF STRESSES

In contrast to the previous sections let us now consider the variation of stresses in order to investigate whether the "actual" stresses satisfy a minimum principle. We pose the problem as in Sec. 10.7 with a body held in equilibrium under the body force per unit volume $F_i$ and surface tractions per unit area $\overset{\nu}{T}{}_i^*$ over the boundary $S_\sigma$, whereas over the boundary $S_u$ the displacements are prescribed. Let $\sigma_{ij}$ be the "actual" stress field which satisfies the equations of equilibrium and boundary conditions

$$\sigma_{ij,j} + F_i = 0 \qquad \text{in } V,$$
(1)
$$\sigma_{ij}\nu_j = \overset{\nu}{T}{}_i^* \qquad \text{on } S_\sigma.$$

Let us now consider a system of variations of stresses which also satisfy the equations of equilibrium and the stress boundary conditions

$$(\delta\sigma_{ij})_{,j} + \delta F_i = 0 \qquad \text{in } V,$$
(2)
$$(\delta\sigma_{ij})\nu_j = \delta\overset{\nu}{T}_i \qquad \text{on } S_\sigma,$$
$$\delta\sigma_{ij} \text{ are arbitrary on } S_u.$$

In contrast to the previous sections, we shall now consider the *complementary virtual work*,

$$\int_V u_i\,\delta F_i\,dv + \int_S u_i\,\delta\overset{\nu}{T}_i\,dS,$$

which, by virtue of (2) and through integration by parts,

$$= -\int_V u_i(\delta\sigma_{ij})_{,j}\,dv + \int_S u_i(\delta\sigma_{ij})\nu_j\,dS$$

$$= \int_V (\delta\sigma_{ij})u_{i,j}\,dv - \int_S u_i\nu_j(\delta\sigma_{ij})\,dS + \int_S u_i(\delta\sigma_{ij})\nu_j\,dS$$

$$= \tfrac{1}{2}\int_V (\delta\sigma_{ij})(u_{i,j} + u_{j,i})\,dv$$

$$= \int_V e_{ij}\,\delta\sigma_{ij}\,dv.$$

Hence,

(3) ▲
$$\int_V e_{ij}\,\delta\sigma_{ij}\,dv = \int_V u_i\,\delta F_i\,dv + \int_S u_i\,\delta\overset{\nu}{T}_i\,dS.$$

This equation may be called the *principle of virtual complementary work.* Now, if we introduce *the complementary strain energy* $W_c$,† which is a function of the stress components $\sigma_{11}, \sigma_{12}, \ldots$, and which has the property that,

(4)
$$\frac{\partial W_c}{\partial \sigma_{ij}} = e_{ij}$$

† The Gibbs' thermodynamic potential (Sec. 12.3) per unit volume, $\rho\Phi$, is equal to the negative of the complementary strain energy function. If the stress-strain law were linear, then $W_c(\sigma_{ij})$ and $W(e_{ij})$ are equal: $-\rho\Phi = W_c = W$ (linear stress-strain law).

then the complementary virtual work may be written as

$$\text{(5)} \qquad \int_V u_i\, \delta F_i\, dv + \int_S u_i\, \delta \overset{\nu}{T}_i\, dS = \int_V \frac{\partial W_c}{\partial \sigma_{ij}}\, \delta\sigma_{ij}\, dv = \delta \int_V W_c\, dv.$$

Since the volume is fixed and $u_i$ are not varied, the result above can be written as

$$\text{(6)} \ \blacktriangle \qquad \delta \mathscr{V}^* = 0,$$

where $\mathscr{V}^*$, as a function of the stresses $\sigma_{11}, \sigma_{12}, \ldots$, the surface traction $\overset{\nu}{T}_i$ and the body force per unit volume $F_i$, is defined as the *complementary energy*

$$\text{(7)} \ \blacktriangle \qquad \mathscr{V}^*(\sigma_{11}, \ldots, F_i) \equiv \int_V W_c\, dv - \int_V u_i F_i\, dv - \int_S u_i \overset{\nu}{T}_i\, dS.$$

In practice, we would like to compare stress fields which all satisfy the equations of equilibrium, but not necessarily the conditions of compatibility. In other words, we would have $\delta F_i = 0$ in $V$ and $\delta \overset{\nu}{T}_i = 0$ on $S_\sigma$. In this case $\delta\sigma_{ij}$ and, hence, $\delta \overset{\nu}{T}_i$ are arbitrary only on that portion of the boundary where displacements are prescribed, $S_u$. Therefore, only a surface integral over $S_u$ is left in the left-hand side of (5) and we have

$$\text{(8)} \ \blacktriangle \qquad \mathscr{V}^*(\sigma_{11}, \ldots, \sigma_{33}) \equiv \int_V W_c\, dv - \int_{S_u} u_i \overset{\nu}{T}_i\, dS.$$

Therefore, we have the

THEOREM. *Of all stress tensor fields $\sigma_{ij}$ that satisfy the equation of equilibrium and boundary conditions where stresses are prescribed, the "actual" one is distinguished by a stationary (extreme) value of the complementary energy $\mathscr{V}(\sigma_{11}, \ldots, \sigma_{33})$ as given by* (8).

In this formulation, the linearity of the stress-strain relationship is *not* required, only the existence of the complementary strain energy function is assumed. However, if the stress-strain law were *linear*, and the material is isotropic and obeying Hooke's law, then

$$\text{(9)} \qquad W_c(\sigma_{11}, \ldots, \sigma_{33}) = -\frac{\nu}{2E}(\sigma_{\alpha\alpha})^2 + \frac{1+\nu}{2E}\sigma_{ij}\sigma_{ij}.$$

We must remark that the variational Eqs. (10.7:2) and Eq. (3) of the present section are applicable even if the body is *not elastic*, for which the energy functional cannot be defined. These variational equations are used in the analysis of inelastic bodies.

Before we proceed further, it may be well worthwhile to consider the concept of complementary work and complementary strain energy. Consider a simple, perfectly elastic bar subjected to a tensile load. Let the relationship between the load $P$ and the elongation of the bar $u$ be given by a unique curve as shown in Fig. 10.9:1. Then the work $W$ is the area between the displacement axis and the curve, while the complementary work $W_c$ is that included between the force axis and the curve. Thus, the two areas complement each other in the rectangular area (force) · (displacement), which would be the work if the force were acting with its full intensity from the beginning of the

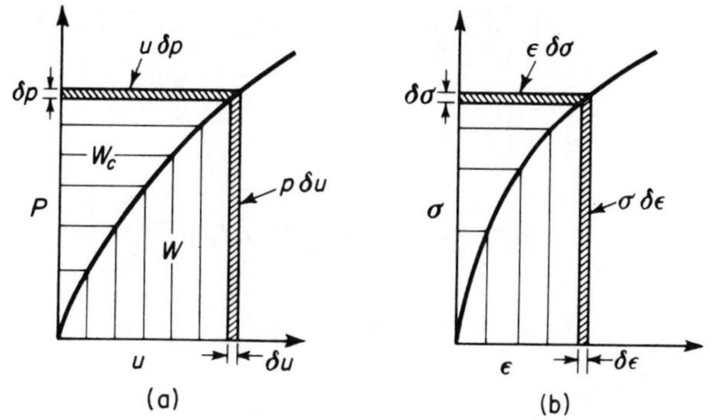


**Fig. 10.9:1.** Complementary work and strain energy.

displacement. Naturally, $W$ and $W_c$ are equal if the material follows Hooke's law.

The principle of minimum potential energy was formulated by Willard Gibbs; and many beautiful applications were shown by Lord Rayleigh. The complementary energy concept was introduced by F. Z. Engesser; its applications were developed by H. M. Westergaard. In the hands of Kirchhoff, the minimum potential energy theorem becomes the foundation of the approximate theories of plates and shells. Recently, Argyris has made the complementary energy theorem the starting point for practical methods of analysis of complex elastic structures using modern digital computers. See Bibliography 10.2, p. 494.

Let us now return to the complementary energy theorem. We shall show that *in the neighborhood of the natural state, the extreme value of the complementary energy* $\mathscr{V}$ *is actually a minimum.* By a natural state is meant a state of stable thermodynamic existence (see Sec. 12.4). In the neighborhood of a natural state, the thermodynamic potential per unit volume $\rho\Phi$ can be approximated by a homogeneous quadratic form of the stresses. For an isotropic material, $-\rho\Phi$ is given by Eq. (9), in which $W_c$ is positive definite.

The proof that $\mathscr{V}^*$ is a minimum is analogous to that in the previous section, and it can be sketched as follows. Comparing $\mathscr{V}^*(\sigma_{ij} + \delta\sigma_{ij})$ with $\mathscr{V}^*(\sigma_{ij})$, we have

$$(10) \quad \mathscr{V}^*(\sigma_{11} + \delta\sigma_{11}, \ldots,) - \mathscr{V}^*(\sigma_{11}, \ldots,)$$

$$= \int_V \left[ W_c(\sigma_{11} + \delta\sigma_{11}, \ldots,) - W_c(\sigma_{11}, \ldots,) \right] dv$$

$$- \int_{S_u} [\overset{\nu}{T}_i(\sigma_{11} + \delta\sigma_{11}, \ldots,) - \overset{\nu}{T}_i(\sigma_{11}, \ldots,)] u_i \, dS$$

$$= \int_V \left[ \left( \frac{1+\nu}{E} \sigma_{ij} - \frac{\nu}{E} \theta \, \delta_{ij} \right) \delta\sigma_{ij} + W_c(\delta\sigma_{11}, \ldots,) \right] dv$$

$$- \int_{S_u} (\delta\sigma_{ij}) \nu_j u_i \, dS$$

$$= \int_V e_{ij} \, \delta\sigma_{ij} \, dv - \int_{S_u} (\delta\sigma_{ij}) \nu_j u_i \, dS + \int W_c(\delta\sigma_{11}, \ldots,) \, dv$$

$$= \delta\mathscr{V}^* + \int_V W_c(\delta\sigma_{11}, \delta\sigma_{12}, \ldots,) \, dv,$$

where

$$(11) \quad W_c(\delta\sigma_{11}, \ldots,) = \frac{1+\nu}{2E} (\delta\sigma_{ij})(\delta\sigma_{ij}) - \frac{\nu}{2E} (\delta\sigma_{ii})^2 \geqslant 0.$$

$W_c(\delta\sigma_{11}, \ldots,)$ is positive definite for infinitesimal variations $\delta\sigma_{ij}$. Hence, when $\delta\mathscr{V}^* = 0$,

$$(12) \quad \mathscr{V}^*(\sigma_{11} + \delta\sigma_{11}, \ldots,) - \mathscr{V}^*(\sigma_{11}, \ldots,) \geqslant 0,$$

and that $\mathscr{V}^*(\sigma_{11}, \ldots,)$ is a minimum is proved.

The converse theorem reads:

*Let $\mathscr{V}^*$ be the complementary energy defined by Eq. (8). If the stress tensor field $\sigma_{ij}$ is such that $\delta\mathscr{V}^* = 0$ for all variations of stresses $\delta\sigma_{ij}$ which satisfy the equations of equilibrium in the body and on the boundary where surface tractions are prescribed, then $\sigma_{ij}$ also satisfies the equations of compatibility*. In other words, the conditions of compatibility are the Euler equation for the variational equation $\delta\mathscr{V}^* = 0$.

The proof was given by Richard V. Southwell[1.2] (1936) through the application of Maxwell and Morera stress functions. We begin with the variational equation

$$(13) \quad \delta\mathscr{V}^* = \int_V e_{ij} \, \delta\sigma_{ij} \, dv - \int_{S_u} u_i \, \delta\overset{\nu}{T}_i \, dS = 0.$$

The variations $\delta\sigma_{ij}$ are subjected to the restrictions

$$(\delta\sigma_{ij})_{,j} = 0 \qquad \text{in } V, \tag{14}$$

$$(\delta\sigma_{ij})\nu_j = 0 \qquad \text{on } S_\sigma. \tag{15}$$

To accommodate the restrictions (14) into (13), we make use of the celebrated result that the equations of equilibrium (14) are satisfied formally, as can be easily verified, by taking

$$\begin{aligned}
\delta\sigma_{11} &= \phi_{22,33} + \phi_{33,22} - 2\phi_{23,23}, \\
\delta\sigma_{22} &= \phi_{33,11} + \phi_{11,33} - 2\phi_{31,31}, \\
\delta\sigma_{33} &= \phi_{11,22} + \phi_{22,11} - 2\phi_{12,12}, \\
\delta\sigma_{23} &= \phi_{31,12} + \phi_{12,13} - \phi_{11,23} - \phi_{23,11}, \\
\delta\sigma_{31} &= \phi_{12,23} + \phi_{23,21} - \phi_{22,31} - \phi_{31,22}, \\
\delta\sigma_{12} &= \phi_{23,31} + \phi_{31,32} - \phi_{33,12} - \phi_{12,33},
\end{aligned} \tag{16}$$

where $\phi_{ij} = \phi_{ji}$ are arbitrary stress functions. On setting $\phi_{12} = \phi_{23} = \phi_{31} = 0$, we obtain the solutions proposed by James Clerk Maxwell. On taking $\phi_{11} = \phi_{22} = \phi_{33} = 0$, we obtain the solutions proposed by G. Morera.

Let us use the Maxwell system of arbitrary stress functions for the variations $\delta\sigma_{ij}$. Equation (13) may be written as

$$\begin{aligned}
\delta\mathscr{V}^* = &\int_V [e_{11}(\phi_{22,33} + \phi_{33,22}) + e_{22}(\phi_{33,11} + \phi_{11,33}) \\
&+ e_{33}(\phi_{11,22} + \phi_{22,11}) - 2e_{23}\phi_{11,23} - 2e_{31}\phi_{22,31} - 2e_{12}\phi_{33,12}]\, dv \\
&- \int_{S_u} u_i\, \delta\overset{\nu}{T}_i\, dS \\
= &\ 0.
\end{aligned} \tag{17}$$

Integrating by parts twice, we obtain

$$\begin{aligned}
\delta\mathscr{V}^* = &\int_V [(e_{22,33} + e_{33,22} - 2e_{23,23})\phi_{11} + (e_{33,11} + e_{11,33} - 2e_{31,31})\phi_{22} \\
&+ (e_{11,22} + e_{22,11} - 2e_{12,12})\phi_{33}]\, dv + \text{a surface integral} \\
= &\ 0.
\end{aligned} \tag{18}$$

Inasmuch as the stress functions $\phi_{11}, \phi_{22}, \phi_{33}$ are arbitrary in the volume $V$, the Euler's equations are

$$e_{22,33} + e_{33,22} - 2e_{23,23} = 0, \tag{19}$$

etc., which are Saint-Venant's compatibility equations (see Sec. 4.6). The treatment of the surface integral is cumbersome, but it says only that over

$S_u$ the values of $u_i$ are prescribed and the stresses are arbitrary; the derivation concerns a certain relationship between $\phi_{ij}$ and their derivatives to be satisfied on the boundary.

Similarly, the use of Morera system of arbitrary stress functions leads to the other set of Saint-Venant's compatibility equations

$$e_{11,23} = -e_{23,11} + e_{31,21} + e_{12,31}, \tag{20}$$

etc. [Eq. (4.6:4)].

If we start with $\delta\mathscr{V}^*$ in the form

$$\delta\mathscr{V}^* = 0 = \int_V \left(\frac{1+\nu}{E}\sigma_{ij} - \frac{\nu}{E}\sigma_{\alpha\alpha}\delta_{ij}\right)\delta\sigma_{ij}\,dv + \int_{S_u} u_i\,\delta\overset{\nu}{T}_i\,dS \tag{21}$$

and introduce the stress functions, the Beltrami-Michell compatibility equations

$$\nabla^2\sigma_{ij} + \frac{1}{1+\nu}\theta_{,ij} + \frac{\nu}{1-\nu}\delta_{ij}F_{k,k} + F_{i,j} + F_{j,i} = 0 \qquad \text{in } V \tag{22}$$

can be obtained directly.

In concluding this section, we remark once more that when we consider the variation of the stress field of a body in equilibrium, the principle of virtual complementary work has broad applicability. The introduction of the complementary energy functional $\mathscr{V}^*$, however, limits the principle to elastic bodies. That $\mathscr{V}^*$ actually is a minimum with respect to all admissible variations of stress field is established only if the complementary strain energy function is positive definite.

There are many fascinating applications of the minimum complementary energy principle. In Sec. 10.11 below we shall consider its application in proving Saint-Venant's principle in Zanaboni's formulation.

**Problem 10.2.** The principle of virtual complementary work states that

$$\int_V e_{ij}\delta\sigma_{ij}\,dv - \int_{S_u} u_i^*\delta\overset{\nu}{T}_i\,dS = 0 \tag{23}$$

under the restrictions that

$$\delta\sigma_{ij,j} = 0 \quad \text{in } V, \tag{24}$$

$$\delta\sigma_{ij}\nu_j = 0 \quad \text{on } S_\sigma. \tag{25}$$

$$u_i = u_i^* \quad \text{prescribed on } S_u, \text{ but } \delta\overset{\nu}{T}_i = \delta\sigma_{ij}\nu_j \text{ arbitrary on } S_u. \tag{26}$$

Using Lagrange multipliers, we may restate this principle as

$$\int_V e_{ij}\delta\sigma_{ij}\,dv - \int_{S_u} u_i^*\delta\overset{\nu}{T}_i\,dS + \int_V \lambda_i\delta\sigma_{ij,j}\,dv - \int_{S_\sigma} \mu_i\delta\sigma_{ij}\nu_j\,dS = 0 \tag{27}$$

The six equations (24), (25), $i = 1, 2, 3$, require six Lagrange multipliers $\lambda_i, \mu_i$ which are functions of $(x_1, x_2, x_3)$. Show that the Euler equations for (27) yield the physical interpretation

(28) $$\lambda_i = u_i, \qquad \mu_i = u_i.$$

## 10.10. REISSNER'S PRINCIPLE

Consider the functional $J(e_{ij}, u_k, \sigma_{ij})$, where $\sigma_{ij} = \sigma_{ji}$:

(1) $$J = \int_V [W(e_{ij}) - F_i u_i]\, dv - \int_V \sigma_{ij}[e_{ij} - \tfrac{1}{2}(u_{i,j} + u_{j,i})]\, dv - \int_{S_\sigma} \overset{\nu}{T}_i^* u_i\, dS - \int_{S_u} \sigma_{ij}\nu_j(u_i - u_i^*)\, dS, \qquad (S = S_\sigma + S_u).$$

Let us seek the necessary conditions for $J$ to be stationary. On setting the first variation of $J$ to zero, permitting the fifteen functions $e_{ij}$, $u_i$, and $\sigma_{ij}$, $(i = 1, 2, 3)$ to vary over the domain $V$, and $u_i$ to vary over $S_\sigma$, and $\sigma_{ij}$ to vary over $S_u$, while $F_i$, $\overset{\nu}{T}_i^*$, and $u_i^*$ are prescribed, we obtain

(2) $$0 = \delta J = \int_V \left\{ \frac{\partial W}{\partial e_{ij}} \delta e_{ij} - F_i\, \delta u_i - \delta\sigma_{ij}[e_{ij} - \tfrac{1}{2}(u_{i,j} + u_{j,i})] - \sigma_{ij}[\delta e_{ij} - \tfrac{1}{2}(\delta u_{i,j} + \delta u_{j,i})] \right\} dv - \int_{S_\sigma} \overset{\nu}{T}_i^*\, \delta u_i\, dS - \int_{S_u} [\delta\sigma_{ij}\nu_j(u_i - u_i^*) + \sigma_{ij}\nu_j\, \delta u_i]\, dS.$$

Integrating by parts those terms involving $\delta u_{i,j}$, we obtain the Euler equations:

(3) $$\frac{\partial W}{\partial e_{ij}} = \sigma_{ij} \quad \text{in } V,$$

(4) $$\sigma_{ij,j} + F_i = 0 \quad \text{in } V,$$

(5) $$e_{ij} = \tfrac{1}{2}(u_{i,j} + u_{j,i}) \quad \text{in } V,$$

(6) $$\sigma_{ij}\nu_j = \overset{\nu}{T}_i^* \quad \text{on } S_\sigma,$$

(7) $$u_i = u_i^*, \qquad \delta u_i = 0 \quad \text{on } S_u.$$

Clearly these are the basic equations of linear elasticity (Section 7.1).

The Eq. (2), $\delta J = 0$, where $J$ is defined in (1), is a statement of Reissner's principle as applied to linear elasticity.

The motivation for $J$ is clear. The functional

$$J_1 = \int_V \{W(e_{ij}) - F_i u_i\}\, dv - \int_{S_\sigma} \overset{\nu}{T}_i^* u_i\, dS$$

represents the sum of the strain energy and the potential energy of the external loads. Now if we wish to find a stationary value of $J_1$ by varying $e_{ij}$ and $u_i$, subjected to the subsidiary conditions (5), (6), and (7), we may do so by introducing Lagrange multipliers. Obviously $\sigma_{ij}$ plays the role of the Lagrangian multiplier in (1).

Eric Reissner,[10.2] announced this principle in 1950 and extended it to finite elastic deformations in 1953. In Reissner's original paper, the complementary energy functional was used instead of the potential energy. In the infinitesimal displacement theory, the complementary energy is defined as

$$W_c(\sigma_{ij}) = \sigma_{ij}e_{ij} - W(e_{ij}). \tag{8}$$

$W_c(\sigma_{ij})$ is a function of the stress components if $e_{ij}$ on the right-hand side are expressed in terms of $\sigma_{kl}$ by means of Hooke's law. $W_c(\sigma_{ij})$ has the property that

$$\frac{\partial W_c}{\partial \sigma_{ij}} = e_{ij}. \tag{9}$$

On substituting $W = \sigma_{ij}e_{ij} - W_c$ from Eq. (8) into Eq. (1), we have Reissner's functional $J_R$:

$$J_R = \int_V \{-W_c(\sigma_{ij}) - F_i u_i + \tfrac{1}{2}\sigma_{ij}(u_{i,j} + u_{j,i})\}\, dv - \int_{S_\sigma} \overset{\nu}{T_i}{}^{*} u_i \, dS - \int_{S_u} \sigma_{ij}\nu_j(u_i - u_i^*)\, dS \tag{10}$$

and Reissner's principle that the elastic equilibrium is distinguished by $\delta J_R = 0$, when $\sigma_{ij}$ and $u_i$ $(i, j = 1, 2, 3)$ are varied independently, provided that the tensor $\sigma_{ij}$ is symmetric.

If we vary $J_R$ with respect to $\sigma_{ij}$ only, $(\delta u_i \equiv 0)$, keeping the stress conditions (4) and (6) satisfied, then the complementary energy theorem is obtained.

## 10.11. SAINT-VENANT'S PRINCIPLE

In 1855, Barre de Saint-Venant enunciated the "principle of the elastic equivalence of statically equipollent systems of loads." According to this principle, the strains that are produced in a body by the application, to a small part of its surface, of a system of forces statically equivalent to zero force and zero couple, are of negligible magnitude at distances which are large compared with the linear dimensions of the part. When this principle is applied to the problem of torsion of a long shaft due to a couple applied at its ends, it states that the shear stress distribution in the shaft at a distance from the ends large compared with the cross-sectional dimension of the shaft will be practically independent of the exact distribution of the surface tractions of which the couple is the resultant. Such a principle is nearly always applied, consciously or unconsciously, when we try to simplify or idealize a problem in mathematical physics. It is used, for example, in devising a simple tension test for a material, when we clamp the ends of a test specimen in the jaws of a testing machine and assume that the action on the central part of the bar is nearly the same as if the forces were uniformly applied at the ends.

The justification of the principle is largely empirical and, as such, its interpretation is not entirely clear.

One possible way to formulate Saint-Venant's principle with mathematical precision is to state the principle in certain sense of average as follows (Zanaboni,[10.3] 1937, Locatelli,[10.3] 1940, 1941). Consider a body as shown in Fig. 10.11:1. A system of forces $P$ in static equilibrium (with zero resultant force and zero resultant couple) is applied to a region of the body enclosed in a small sphere $B$. Otherwise the body is free. Let $S'$ and $S''$ be

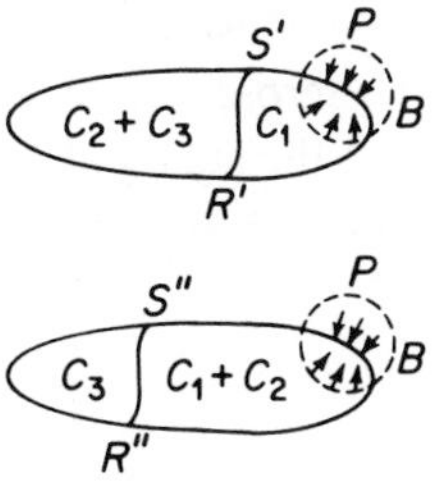


Fig. 10.11:1

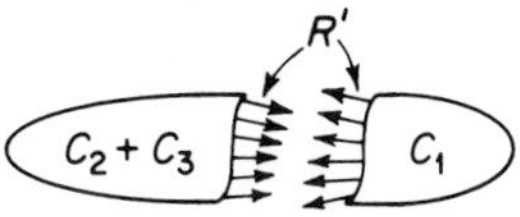


Fig. 10.11:2

two arbitrary nonintersecting cross sections, both outside of $B$, with $S''$ farther away from $B$ than $S'$. Due to the system of loads $P$, stresses are induced in the body. If we know these stresses, we can calculate the tractions acting on the surfaces $S'$ and $S''$. Let the body be considered as severed into two parts at $S'$, and let the system of surface tractions acting on the surface $S'$ be denoted by $R'$ which is surely a system of forces in equilibrium (see Fig. 10.11:2). Then a convenient measure of the magnitude of the tractions $R'$ is the total strain energy that would be induced in the two parts should they be loaded by $R'$ alone. Let this strain energy be denoted by $U_{R'}$. We have,

$$U_{R'} = \int_V W(\sigma_{ij}^{(R)})\, dv,$$

where $W$ is the strain energy density function, and the stresses $\sigma_{ij}^{(R)}$ correspond to the loading system $R'$. Similarly, let the magnitude of the stresses at the section $S''$ be measured by the strain energy $U_{R''}$, which would have been induced by the tractions $R''$ acting on the surface $S''$ over the two parts of the body severed at $S''$. Both $U_{R'}$ and $U_{R''}$ are positive quantities and they vanish only if $R'$, $R''$ vanish identically. Now we shall formulate Saint-Venant's principle in the following form (Zanaboni,[10.3] 1937).

*Let $S'$ and $S''$ be two nonintersecting sections both outside a sphere $B$. If the section $S''$ lies at a greater distance than the section $S'$ from the sphere $B$ in which a system of self-equilibrating forces $P$ acts on the body, then*

$$U_{R''} < U_{R'}. \tag{1}$$

In this form, the diminishing influence of the self-equilibrating system of loading $P$ as the distance from $B$ increases is expressed by the functional $U_R$, which is a special measure of the stresses induced at any section outside $B$. The reason for the choice of $U_R$ as a measure is its positive definiteness character and the simplicity with which the theorem can be proved. Further sharpening of the principle will be discussed later.

In order to prove the theorem (1), we first derive an auxiliary principle. Let a self-equilibrating system of forces $P$ be applied to a limited region $B$ at the surface of an otherwise free elastic body $C_1$, (Fig. 10.11:3). Let $U_1$ be the strain energy produced by $P$ in $C_1$. Let us now consider an enlarged body $C_1 + C_2$ by affixing to $C_1$ an additional body $C_2$ across a surface $S$ which does not intersect the region $B$. When $P$ is applied to the enlarged body $C_1 + C_2$, the strain energy induced is denoted by $U_{1+2}$. Then the lemma states that

$$U_{1+2} < U_1. \tag{2}$$

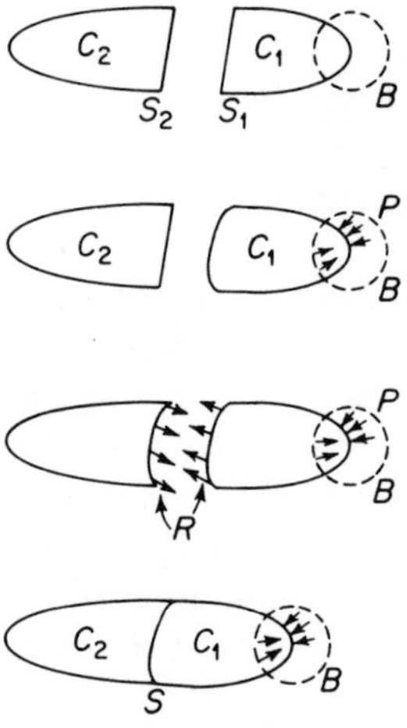


**Fig. 10.11:3**

*Proof of the Lemma.* To compute $U_{1+2}$, we imagine that the stresses in the enlarged body $C_1 + C_2$ are built up by the following steps (see Fig. 10.11:3). First the load $P$ is applied to $C_1$. The face $S_1$ of $C_1$ is deformed. Next a system of surface tractions $R$ is applied to $C_1$ and $C_2$ on the surfaces of separation, $S_1$ and $S_2$. $R$ will be so chosen that the deformed surfaces $S_1$ and $S_2$ fit each other exactly, so that displacements of material points in $C_1$ and $C_2$ are continuous as well as the stresses. Now $C_1$ and $C_2$ can be brought together and welded, and $S$ becomes merely an interface. The result is the same if $C_1$ and $C_2$ were joined in the unloaded state and the combined body $C_1 + C_2$ is loaded by $P$.

The strain energy $U_{1+2}$ is the sum of the work done by the forces in the above stages. In the first stage, the work done by $P$ is $U_1$. In the second stage, the work done by $R$ on $C_2$ is $U^*_{R2}$; the work done by $R$ on $C_1$ consists of two parts, $U^*_{R1}$ if $C_1$ were free, and $U^*_{PR}$, the work done by the system of loads $P$ due to the deformation caused by $R$. Hence,

$$U_{1+2} = U_1 + \overset{*}{U}_{R1} + \overset{*}{U}_{R2} + \overset{*}{U}_{PR}. \tag{3}$$

Now the system of forces $R$ represents the internal normal and shear stresses acting on the interface $S$ of the body $C_1 + C_2$. It is therefore determined by the minimum complementary energy theorem. Consider a special variation of stresses in which all the actual forces $R$ are varied in the ratio $1:(1 + \epsilon)$, where $\epsilon$ may be positive or negative. The work $U^*_{R1}$ will be changed to $(1 + \epsilon)^2 U^*_{R1}$, because the load and deformation will both be changed by a factor $(1 + \epsilon)$. Similarly, $U^*_{R2}$ is changed to $(1 + \epsilon)^2 U^*_{R2}$. But $U^*_{PR}$ is only changed to $(1 + \epsilon)U^*_{PR}$, because the load $P$ is fixed, while the deformation is varied by a factor $(1 + \epsilon)$. Hence, $U_{1+2}$ is changed to

$$U'_{1+2} = U_1 + (1 + \epsilon)^2 U^*_{R1} + (1 + \epsilon)^2 U^*_{R2} + (1 + \epsilon) U^*_{PR}. \tag{4}$$

The difference between (4) and (3) is

$$\Delta U_{1+2} = \epsilon(2U^*_{R1} + 2U^*_{R2} + U^*_{PR}) + \epsilon^2(U^*_{R1} + U^*_{R2}).$$

For $U_{1+2}$ to be a minimum, $\Delta U_{1+2}$ must be positive regardless of the sign of $\epsilon$. This is satisfied if

$$2U^*_{R1} + 2U^*_{R2} + U^*_{PR} = 0. \tag{5}$$

On substituting (5) into (3), we obtain

$$U_{1+2} = U_1 - (U^*_{R1} + U^*_{R2}). \tag{6}$$

Since $U^*_{R1}$ and $U^*_{R2}$ are positive definite, we see that lemma (2) is proved.

Now we shall prove Saint-Venant's principle embodied in Eq. (1). Consider an elastic body consisting of three parts $C_1 + C_2 + C_3$ loaded by $P$ in $B$, as shown in Fig. 10.11:1. Let this body be regarded first as a result of adjoining $C_2 + C_3$ to $C_1$ with an interface force system $R'$, and then as a result of adjoining $C_3$ to $C_1 + C_2$ with an interface force $R''$. We have, by repeated use of (6),

$$\begin{aligned} U_{1+(2+3)} &= U_1 - (U^*_{R'1} + U^*_{R'(2+3)}), \\ U_{(1+2)+3} &= U_{1+2} - (U^*_{R''(1+2)} + U^*_{R''3}) \\ &= U_1 - (U^*_{R1} + U^*_{R2}) - (U^*_{R''(1+2)} + U^*_{R''3}). \end{aligned}$$

Equating these expressions, we obtain

$$U^*_{R'1} + U^*_{R'(2+3)} = U^*_{R1} + U^*_{R2} + U^*_{R''(1+2)} + U^*_{R''3},$$

or, since $U^*_{R1}$ and $U^*_{R2}$ are essentially positive quantities,

$$U^*_{R'1} + U^*_{R'(2+3)} > U^*_{R''(1+2)} + U^*_{R''3}. \tag{7}$$

This is Eq. (1), on writing $U_{R'}$ for $U^*_{R'1} + U^*_{R'(2+3)}$, etc. Hence, the principle is proved.

## 10.12. SAINT-VENANT'S PRINCIPLE—BOUSSINESQ-VON MISES-STERNBERG FORMULATION

The Saint-Venant principle, as enunciated in terms of the strain energy functional, does not yield any detailed information about individual stress components at any specific point in an elastic body. However, such information is clearly desired. To sharpen the principle, it may be stated as follows (von Mises, 1945). "If the forces acting upon a body are restriced to several small parts of the surface, each included in a sphere of radius $\epsilon$, then the strains and stresses produced in the interior of the body at a finite distance from all those parts are smaller in order of magnitude when the forces for each single part are in equilibrium than when they are not."

The classical demonstration of this principle is due to Boussinesq (1885), who considered an infinite body filling the half-space $z \geqslant 0$ and subjected to several concentrated forces, each of magnitude $F$, normal to the boundary $z = 0$. If these normal forces are applied to points in a small circle $B$ with diameter $\epsilon$, Boussinesq proved that the largest stress component at a point $P$ which lies at a distance $R$ from $B$ is

(1) of order $F/R^2$ if the resultant of the forces is of order $F$,
(2) of order $(\epsilon/R)(F/R^2)$ if the resultant of the forces is zero,
(3) of order $(\epsilon/R)^2(F/R^2)$ if both the resultant force and the resultant moment vanish.

(Cf. Secs. 8.8, 8.10.)

These relative orders of magnitude were believed to have general validity until von Mises (1945) showed that a modification is necessary.

Consider the half-space $z \geqslant 0$ again. Let forces of magnitude $F$ *tangent* to the boundary $z = 0$ be applied to points in a small circle $B$ of diameter $\epsilon$. Making use of the well-known Cerruti solution (Sec. 8.8), von Mises obtained

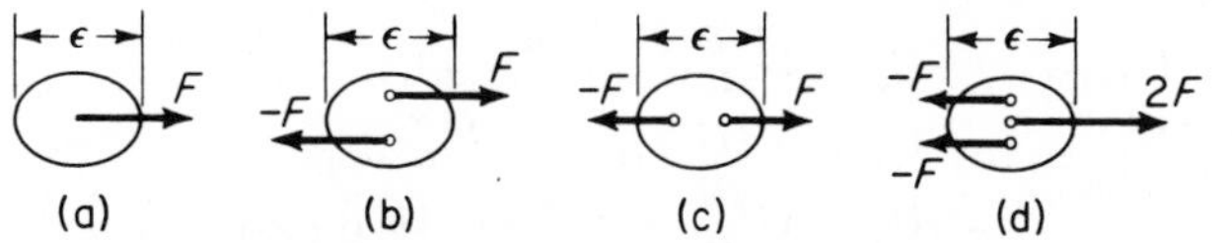


**Fig. 10.12:1.** von Mises' examples in which forces tangential to the surface of an elastic half-space are applied in a small area.

the following results for the four cases illustrated in Fig. 10.12:1. The order of magnitude of the largest stress component at a point $P$ which lies at a distance $R$ from $B$ is

(1) of order $\sigma_0 = F/R^2$ in case (a),
(2) of order $(\epsilon/R)\sigma_0$ in case (b),
(3) of order $(\epsilon/R)\sigma_0$ in case (c),
(4) of order $(\epsilon/R)^2\sigma_0$ in case (d).

The noteworthy case is (c), which is drastically different from what one would expect from an indiscriminating generalization of Boussinesq's result, for in this case the forces are in static equilibrium, with zero moment about any axes, but all one could expect is a stress magnitude of order no greater than $(\epsilon/R)\sigma_0$, not $(\epsilon/R)^2\sigma_0$. von Mises found that in this case the order of magnitude of the largest stress component is reduced to $(\epsilon/R)^2\sigma_0$ if and only if the external forces acting upon a small part of the surface are such as to remain in equilibrium when all the forces are turned through an arbitrary angle. (Such a case is called *astatic equilibrium.*)

von Mises examined next the stresses in a finite circular disk due to loads acting on the circumference, and a similar conclusion was reached. (See Fig. 10.12:2.)

These examples show that Saint-Venant's principle, as stated in the traditional form at the beginning of this section, does not hold true.

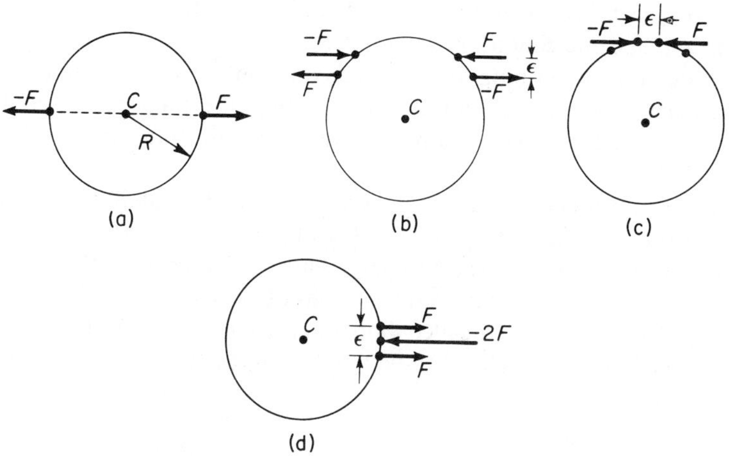


**Fig. 10.12:2.** von Mises' examples of a circular disk subjected to loads on the circumference.

Accordingly, von Mises[10.3] proposed, and later Sternberg[10.3] proved (1954), the following mathematical statement of the Saint-Venant principle:

Let a body be acted on by surface tractions which are finite and are distributed over a number of regions all no greater than a sphere of diameter $\epsilon$. Consider an interior point $x$ whose distance to any of these loading areas in no less than a characteristic length which will be taken as unity. $\epsilon$ is nondimensionalized with respect to this characteristic length. Then, as $\epsilon \to 0$, the order of magnitude of the strain components at $x$ is as follows:

$$e(x, \epsilon) = O(\epsilon^p)$$

where

(a) If the tractions have nonvanishing vector sums in at least one area, then in general, $\rho \geqslant 2$. (Note that the surface traction is assumed to be finite, so the resultant force $\to 0$ as $\epsilon^2$, since the area on which the surface tractions act $\to 0$ as $\epsilon^2$.)

(b) If the resultant of the surface tractions in every loading area vanishes, then $\rho \geqslant 3$.

(c) If, in addition, the resultant moment of the surface tractions in every loading area also vanishes, then still we can be assured only of $\rho \geqslant 3$.

(d) $\rho \geqslant 4$ in case of astatic equilibrium in every loading area, which may be described by the 12 scalar conditions

$$\int_{S(\epsilon)} T_i^* \, dS = 0, \qquad \int_{S(\epsilon)} T_i^* x_j \, dS = 0, \qquad (i, j = 1, 2, 3)$$

for each loading area $S(\epsilon)$, where $T_i^*$ represents the specified surface traction over $S(\epsilon)$.

If the tractions applied to $S(\epsilon)$ are parallel to each other and not tangential to the surface, then if they are in equilibrium they are also in astatic equilibrium, and the condition $\rho \geqslant 4$ prevails.

If, instead of prescribing finite surface traction, we consider finite forces being applied which remain finite as $\epsilon \to 0$, then the exponent $\rho$ should be replaced by 0, 1, 2 in the cases named above, as was illustrated in Figs. 10.12:1 and 10.12:2.

The theorem enunciated above does not preclude the validity of a stronger Saint-Venant principle for special classes of bodies, such as thin plates or shells or long rods. With respect to perturbations that occur at the edges of a thin plate or thin shell, a significant result was obtained by K. O. Friedrichs[10.3] (1950) in the form of a so-called boundary-layer theory. With respect to lateral loads on shells, Naghdi[10.3] (1960) obtained results similar to von Mises-Sternberg's.

## 10.13. PRACTICAL APPLICATIONS OF SAINT-VENANT'S PRINCIPLE

It is well-known that Saint-Venant's principle has its analogy in hydrodynamics and that these features are associated with the elliptic nature of the partial differential equations. If the differential equations were hyperbolic and two-dimensional, local disturbances may be propagated far along the characteristics without attenuation. Then the concept of the Saint-Venant's principle will not apply. For example, in the problem of the response of an elastic half-space to a line load traveling at supersonic speeds over the free surface, as discussed in Sec. 9.7, the governing equations (9.7:11) are hyperbolic, and we know that any fine structure of the surface pressure distribution is propagated all the way to infinity.

On the other hand, one feels intuitively that the validity of Saint-Venant's principle is not limited to linear elastic solid or infinitesimal displacements. One expects it to apply in the case of rubber for finite strain or to steel even when yielding occurs. Although no precise proof is available, Goodier[10.3] (1937) has argued on the basis of energy as follows:

Let a solid body be loaded in a small area whose linear dimensions are of order $\epsilon$, with tractions which combine to give zero resultant force and couple. Such a system of tractions imparts energy to the solid through the *relative* displacements of the points in the small loaded area, because no work is done by the tractions in any translation or rotation of the area as a rigid body. Let the tractions be of order $p$. Let the slope of the stress strain curve of the material be of order $E$. (The stress-strain relationship does not have to be linear.) Let one element of the loaded area be regarded as fixed in position and orientation. Then since the strain must be of order $p/E$, the displacements of points within the area are of order $(p\epsilon/E)$. The work done by the traction acting on an element $dS$ is of order $(p\epsilon p \cdot dS/E)$. The order of magnitude of total work is, therefore, $p^2\epsilon^3/E$. Since a stress of order $p$ implies a strain energy of order $p^2/E$ *per unit volume*, the region in which the stress is of order $p$ must have a volume comparable with $\epsilon^3$. Hence, the influence of tractions cannot be appreciable at a distance from the loaded area which are large compared with $\epsilon$.

Goodier's argument can be extended to bodies subjected to limited plastic deformation. In fact, the argument provides an insight to practical judgement of how local self-equilibrating tractions should influence the strain and stress in the interior of a body.

An engineer needs to know not only the order of magnitude comparison such as stated in von Mises-Sternberg theorem; he needs to know also how numerically trustworthy Saint-Venant's principle is to his particular problem. Hoff[10.3] (1945) has considered several interesting examples, two of which are given in Figs. 10.13:1 and 10.13:2 and will be explained below.

In the first example, the torsion of beams with different cross sections is considered. One end of the beam is clamped, where cross sectional warping is prevented. The other end is free, where a torque is applied by means of shear stresses distributed according to the requirements of the theory of pure torsion. The difference between the prescribed end conditions at the clamped end from those assumed in Saint-Venant's torsion theory (Sec. 7.5) may be stated in terms of a system of self-equilibrating tractions that act at the clamped end. Timoshenko has given approximate solutions to these problems. Hoff's example refers to beams with dimensions as shown in the figure and subjected to a torque of 10 in. lb. Due to the restrictions of warping, normal stresses are introduced in the bar in addition to the shearing stresses of Saint-Venant's torsion. For the rectangular beam, the maximum normal stress at the fixed end is equal to 157 lb/sq in. For the other two thin-walled channel sections, the maximum normal stresses at the fixed end are 1230 and 10,900 lb/sq in., respectively, for the thicker and thinner sections. In Fig.

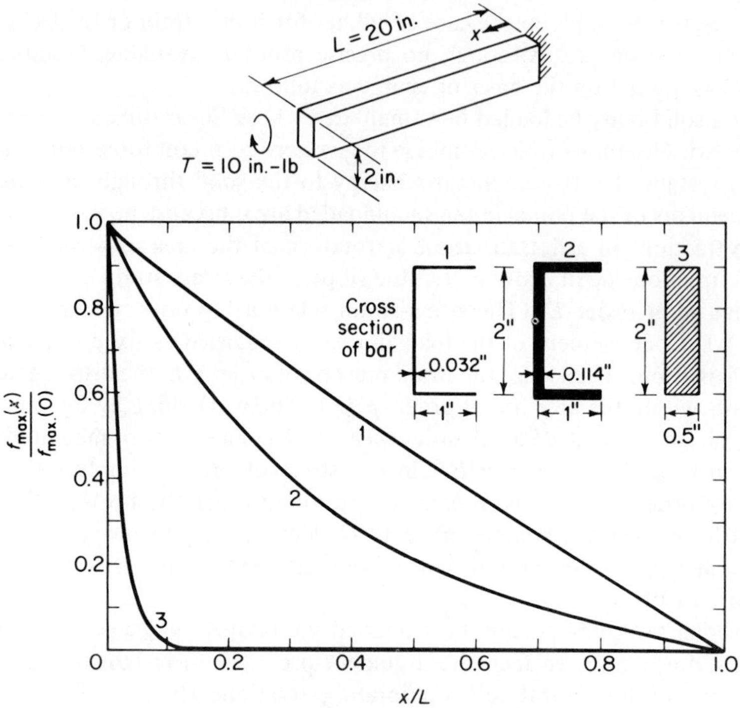


**Fig. 10.13:1.** Hoff's illustration of St-Venant's principle.

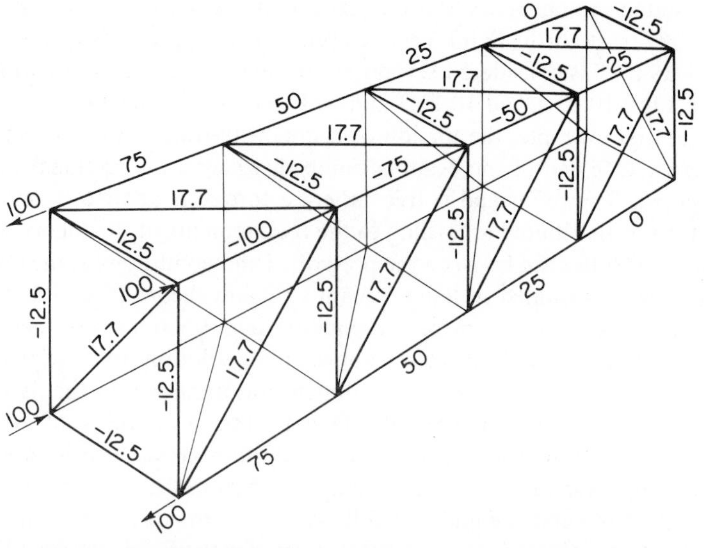


**Fig. 10.13:2.** Hoff's example illustrating the slow decay of self-equilibrating forces in a space framework.

10.13:1 curves are shown for the ratio of maximum normal stress $f_{\max}(x)$ in any section $x$ divided by the maximum normal stress $f_{\max}(0)$ in the fixed-end section, plotted against the ratio distance $x$ of the section from the fixed end of the bar, divided by the total length $L$ of the bar. Inspection of the curves reveals that, while in the case of the solid rectangular section the normal stress caused by the restriction to warping at the fixed end is highly localized, it has appreciable values over the entire length of the channel section bars. Consequently, reliance on Saint-Venant's principle in the calculation of stresses caused by torsion is entirely justified with the bar of rectangular section. In contrast, stresses in the thin-walled section bars depend largely upon the end conditions.

Hoff's second example refers to pin-jointed space frameworks. In one case a statically determinate framework is considered, to one end of which a set of four self-equilibrating concentrated loads are applied, as shown in Fig. 10.13:2. The figures written on the elements of the framework represent the forces acting in the bars measured in the same units as the applied loads. Negative sign indicates compression. It can be seen that the effect of the forces at one end of the structure is still noticeable at the other end.

Hoff's examples show that Saint-Venant's principle works only if there is a possibility for it to work; in other words, only if there exist paths for the internal forces to follow in order to balance one another within a short distance of the region at which a group of self-equilibrating external forces is applied. This point of view is in agreement with Goodier's reasoning.

Hoff's examples are not in conflict with the von Mises-Sternberg theorem, for the latter merely asserts a certain order of magnitude comparison for the stress and strain as the size of the region of self-equilibrating loading shrinks to zero, and does not state how the stresses are propagated. On the other hand, although Goodier's reasoning does not provide a definitive theorem, it is very suggestive in pointing out the basic reason for Saint-Venant's principle and can be used in estimating the practical efficiency of the principle.

## PROBLEMS

**10.3.** Derive the differential equation and permissible boundary conditions for a membrane stretched over a simply connected regular region $A$ by minimizing the functional $I$ with respect to the displacement $w$; $p(x, y)$ being a given function

$$I = \tfrac{1}{2}\iint_A (w_x^2 + w_y^2)\, dx\, dy - \iint_A pw\, dx\, dy.$$

**10.4.** Obtain Euler's equation for the functional

$$I' = \tfrac{1}{2}\iint_A [w_x^2 + w_y^2 + (\nabla^2 w + p)^2 - 2pw]\, dx\, dy.$$

*Ans.* $\nabla^4 w - \nabla^2 w + \nabla^2 p - p = 0.$

*Note:* Compare $I'$ with $I$ of Prob. 10.3. Since $\nabla^2 w + p = 0$ for a membrane, the solution of Prob. 10.3 also satisfies the present problem. But the integrand of $I'$ contains higher derivatives of $w$ and the equation $\delta I' = 0$ is a sharper condition than the equation $\delta I = 0$. These examples illustrate Courant's method of sharpening a variational problem and accelerating the convergence of an approximating sequence in the direct method of solution of variational problems.

**10.5.** Bateman's principle in fluid mechanics states that in a flow of an ideal, nonviscous fluid (compressible or incompressible) in the absence of external body forces, the "pressure integral"

$$-\iiint p\, dx\, dy\, dz,$$

which is the integral of pressure over the entire fluid domain, is an extremum. Consider the special case of a steady, irrotational flow of an incompressible fluid, for which the Bernoulli's equation gives

$$p = \text{const.} - \frac{\rho V^2}{2} = \text{const.} - \frac{\rho}{2}\left[\left(\frac{\partial \phi}{\partial x}\right)^2 + \left(\frac{\partial \phi}{\partial y}\right)^2\right],$$

where $\phi$ is the velocity potential. Derive the field equation governing $\phi$ and the corresponding natural boundary conditions according to Bateman's principle. [H. Bateman, *Proc. Roy. Soc. London*, **A, 125** (1929) 598–618.]

**10.6.** Let

$$T = \sum_{\mu,\nu=1}^{n} P_{\mu\nu}(t, q_1, \ldots, q_m)\, \dot{q}_\mu(t)\, \dot{q}_\nu(t),$$

$$U = U(t, q_1, \ldots, q_m),$$

and

$$L = T - U,$$

where $\dot{q}_\mu(t) = dq_\mu/dt$, and $q_\mu(t_0) = a_\mu$, $q_\mu(t_1) = b_\mu$, $(\mu = 1, \ldots, m)$; $a_\mu$, $b_\mu$ are given numbers. Derive the Euler differential equations for the variational problem connected with the functional

$$J[q_1, \ldots, q_m] = \int_{t_0}^{t_1} L\, dt.$$

Prove that

$$\frac{d(T + U)}{dt} = 0,$$

if $\partial U/\partial t = 0$, $\partial P_{\mu\nu}/\partial t = 0$, $\mu, \nu = 1, \ldots, m$. *Hint:* Use the Euler's relation for homogeneous functions.

**10.7.** Find the curves in the $x$, $y$-plane such that

$$\int_a^b \sqrt{2E - n^2 y^2}\, ds, \qquad ds^2 = dx^2 + dy^2$$

is stationary, where $E$ and $n$ are constants and the integral is taken between fixed end points.

**10.8.** Let $D$ be the set of all functions $u(x)$ with the following properties:

$$D: \begin{cases} (1) & u(x) = a_1 \sin \pi x + a_2 \sin 2\pi x, \\ (2) & a_1, a_2 \text{ are real, arbitrary numbers,} \\ (3) & a_1^2 + a_2^2 > 0. \end{cases}$$

Consider the functional

$$J[u] = \frac{\int_0^1 [u'(x)]^2 \, dx}{\int_0^1 [u(x)]^2 \, dx}$$

for all functions $u(x)$ in $D$.

*Questions:*

(a) Give necessary conditions for a function $u^*(x)$ in $D$ so that $u^*(x)$ furnishes an extremum of the functional $J$.

(b) Show that there exist exactly two functions

$$u_1(x) = a_{11} \sin \pi x + a_{12} \sin 2\pi x,$$
$$u_2(x) = a_{21} \sin \pi x + a_{22} \sin 2\pi x,$$

notwithstanding a constant factor, which satisfy these necessary conditions mentioned in (a).

(c) Show that for one of the functions mentioned in (b), say, $u_1(x)$, the inequality

$$J[u_1] \leqslant J[u]$$

holds for all functions $u(x)$ in $D$. Then show that for the other solution, $u_2(x)$, the inequality

$$J[u] \leqslant J[u_2]$$

holds for all functions $u(x)$ in $D$.

(d) Does there exist any relation between $J[u_1]$, $J[u_2]$ and the eigenvalues of the eigenvalue problem

$$-u''(x) = \lambda u(x), \qquad 0 \leqslant x \leqslant 1,$$

with

$$u(0) = u(1) = 0.$$

**10.9.** Clapeyron's theorem states that, if a linear elastic body is in equilibrium under a given system of body forces $F_i$ and surface forces $T_i$, then the strain energy of deformation is equal to one-half the work that would be done by the external forces (of the equilibrium state) acting through the displacements $u_i$ from the unstressed state to the state of equilibrium; i.e.,

$$\int_V F_i u_i \, dv + \int_S \overset{\nu}{T}_i u_i \, dS = 2 \int_V W \, dv.$$

Demonstrate Clapeyron's Theorem for:

(a) A simply supported beam under its own weight [Fig. P10.9(a)].

(b) A rod under its own weight [Fig. P10.9(b)],

$$u(x) = \frac{g\rho L^2}{E}\left[\frac{1}{2}\left(\frac{x}{L}\right)^2 - \frac{x}{L}\right].$$

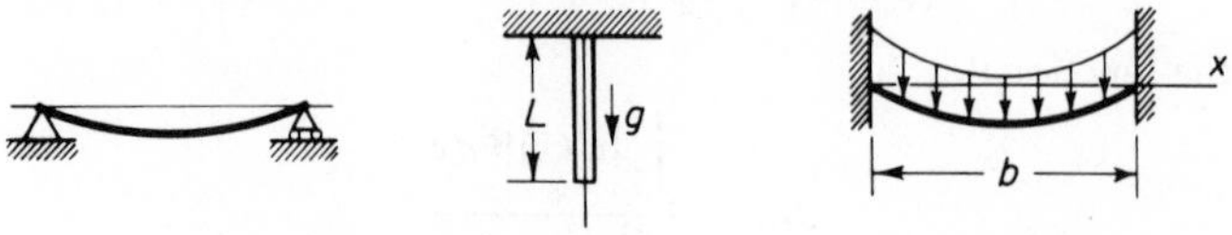


(a) (b) (c)

**P10.9.**

(c) A strip of membrane of infinite length and width $b$ [Fig. P10.9(c)], under a constant pressure $p_0$, for which

$$w = w_0(1 - x^2), \qquad \frac{w_0}{h} = \sqrt[3]{\frac{3(1 - \nu^2)}{64}\frac{p_0 b^4}{Eh^4}}.$$

**10.10.** *The "poker chip" problem* (*Max Williams*). To obtain a nearly triaxial tension test environment for polymer materials, a "poker chip" test specimen (a short circular cylinder) is glued between two circular cylinders (Fig. P10.10). When the cylinder is subjected to simple tension, the center of the poker chip is subjected to a triaxial tension stress field. Let the elastic constants of the media be $E_1$, $G_1$, $\nu_1$ and $E_2$, $G_2$, $\nu_2$, as indicated in Fig. P10.10, with $E_1 \ll E_2$, $\nu_1 \cong 0.5$. Assume cylindrical symmetry, and obtain approximate expressions of the stress field by the following methods.

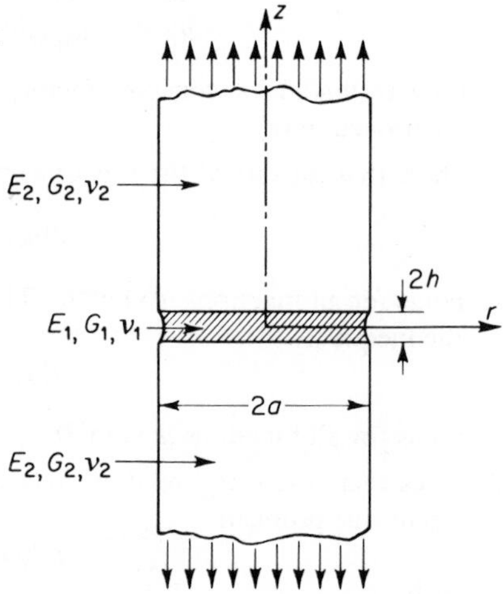


**P10.10** Poker chip specimen.

(a) Use the complementary energy theorem and a stress field satisfying stress boundary conditions and the equations of equilibrium.

(b) Use the potential energy theorem and assumed displacements satisfying displacement boundary conditions.

(c) Experimental results suggest that the following displacement field is reasonable:

$$w = w_0\frac{z}{h}, \qquad -h \leqslant z \leqslant h,$$

$$u = \left(1 - \frac{z^2}{h^2}\right)g(r),$$

where $g(r)$ is as yet an unknown function of $r$. Obtain the governing equation for $g(r)$ and its appropriate solution for this problem by attempting to satisfy the following equilibrium equations in some average sense with respect to the $z$-direction:

$$\frac{\partial \sigma_{rr}}{\partial r} + \frac{\partial \sigma_{rz}}{\partial z} + \frac{\sigma_{rr} - \sigma_{\theta\theta}}{r} = 0,$$

$$\frac{\partial \sigma_{rz}}{\partial r} + \frac{\partial \sigma_{zz}}{\partial z} + \frac{\sigma_{rz}}{r} = 0.$$

**10.11.** Consider a pin-ended column of length $L$ and bending stiffness $EI$ subjected to an end thrust $P$. $A$ spring is attached at the middle of the column as shown in Fig. P10.11. When the column is straight the spring tension is zero. If the column deflects by an amount $\Delta$, the spring exerts a force $R$ on the column:

$$R = K\Delta - \alpha\Delta^3, \qquad (K > 0,\ \alpha > 0).$$

Derive the equation of equilibrium of the system. It is permissible to use the Euler-Bernoulli approximation for a beam, for which the strain energy per unit length is

$$\tfrac{1}{2}EI \cdot (\text{curvature})^2$$

and

$$\text{bending moment} = EI \cdot (\text{curvature}).$$

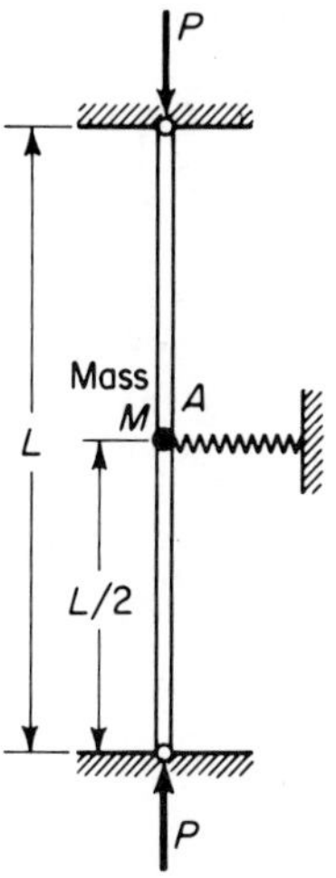

P10.11

**10.12.** For the column of Prob. 10.11 under the axial load $P$, is the solution unique? Under what situation is the solution nonunique? What are the possible solutions when the uniqueness is lost?

**10.13.** A linear elastic beam of bending stiffness $EI$ is supported at four equidistant points $ABCD$, Fig. P10.13. The supports at $A$ and $D$ are pin-ended. At $B$ and $C$, the beam rests on two identical nonlinear pillars. The characteristic of the pillars may be described as "hardening" and is expressed by the equation

$$K\Delta = R - \beta R^3 \qquad (K > 0,\ \beta > 0),$$

where $R$ is the reaction of the pillar, $\Delta$ is the downward deflection of the beam at the point of attachment of the beam to the pillar, $K$ and $\beta$ are constants.

A load $P$ is applied to the beam at the mid-span point. Find the reactions at the supports $B$ and $C$. One of the two minimum principles (of potential energy and of complementary energy) is easier to apply to this problem. Solve the problem by a variational method with the appropriate minimum principle.

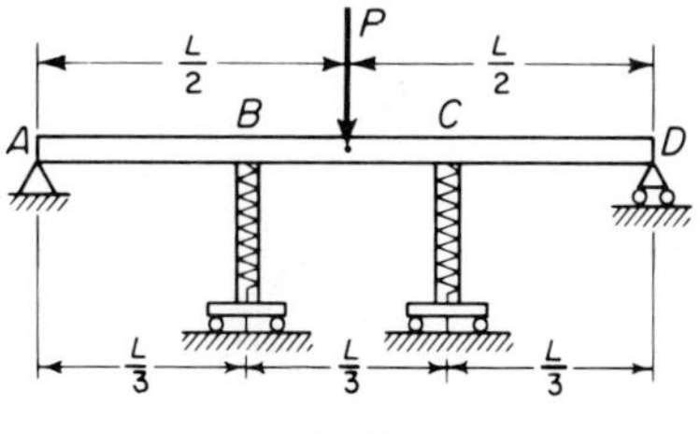

P10.13

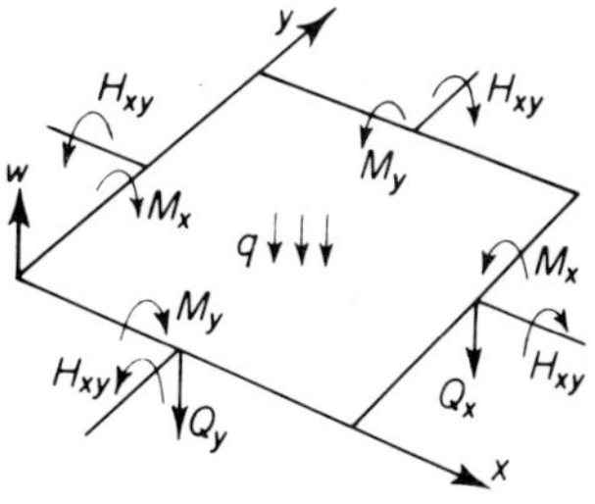

P10.14

**10.14.** Consider a square plate loaded in the manner shown in Fig. P10.14. Derive Euler's equation and the natural boundary conditions for $V$ to be a minimum when $w(x, y)$ is varied.

$$V = U + A,$$

$$U = \frac{D}{2}\int_0^1\int_0^1 [(w_{xx} + w_{yy})^2 - 2(1 - \nu)(w_{xx}w_{yy} - w_{xy}^2)]\,dx\,dy,$$

$$\begin{aligned} A = \int_0^1\int_0^1 qw\,dx\,dy - \int_0^1 \left(M_x\frac{\partial w}{\partial x}\right)\Bigg|_{x=0}^{x=1} dy + \int_0^1 (Q_x w)\Bigg|_{x=0}^{x=1} dy \\ + \int_0^1 \left(H_{xy}\frac{\partial w}{\partial x}\right)\Bigg|_{x=0}^{x=1} dy - \int_0^1 \left(M_y\frac{\partial w}{\partial y}\right)\Bigg|_{y=0}^{y=1} dx \\ + \int_0^1 (Q_y w)\Bigg|_{y=0}^{y=1} dx + \int_0^1 \left(H_{xy}\frac{\partial w}{\partial y}\right)\Bigg|_{y=0}^{y=1} dx. \end{aligned}$$

# 11

# HAMILTON'S PRINCIPLE, WAVE PROPAGATION, APPLICATIONS OF GENERALIZED COORDINATES

In dynamics, the counterpart of the minimum potential energy theorem is Hamilton's principle. In this chapter we shall discuss this important principle and its applications to vibrations and wave propagations in beams.

Toward the end of the chapter a brief discussion of the so-called *direct methods* of solving variational problems is given. The basic idea is to apply the concept of generalized coordinates to obtain approximate solutions for a continuous system by reducing it to one with a finite number of degrees of freedom. Several important methods—those of Euler, Rayleigh-Ritz-Galerkin, and Kantrovich—will be outlined. The question of convergence, however, must be referred to mathematical treatises. See Bibliography 11.3 on p. 498.

## 11.1. HAMILTON'S PRINCIPLE

For an oscillating body, with displacements $u_i$ so small that the acceleration is given by $\partial^2 u_i/\partial t^2$ in Eulerian coordinates (Sec. 5.2) the equation of small motion is

$$\sigma_{ij,j} + F_i = \rho \frac{\partial^2 u_i}{\partial t^2}, \tag{1}$$

where $\rho$ is the density of the material and $F_i$ is the body force per unit volume. Let us again consider virtual displacements $\delta u_i$ as specified in Sec. 10.7, but instead of a body in static equilibrium we now consider a vibrating body. The variations $\delta u_i$ must vanish over the boundary surface $S_u$, where values of displacements are prescribed; but are arbitrary, triply differentiable over the domain $V$; and are arbitrary also over the rest of the boundary surface $S_\sigma$, where surface tractions are prescribed.

The virtual work done by the body and surface forces is, as before,

$$\int_V F_i\,\delta u_i\,dv + \int_S \overset{\nu}{T}_i\,\delta u_i\,dS.$$

The last integral can be transformed on introducing $\overset{\nu}{T}_i = \sigma_{ij}\nu_j$ and using Gauss' theorem, as in Sec. 10.7:

$$\int_S \overset{\nu}{T}_i\,\delta u_i\,dS = \int_S \sigma_{ij}\nu_j\,\delta u_i\,dS = \int_V (\sigma_{ij}\,\delta u_i)_{,j}\,dv$$
$$= \int_V \sigma_{ij,j}\,\delta u_i\,dv + \int_V \sigma_{ij}\,\delta u_{i,j}\,dv$$

By Eq. (1), and by the symmetry of $\sigma_{ij}$, the right hand side is equal to

$$\int_V \left(\rho\frac{\partial^2 u_i}{\partial t^2} - F_i\right)\delta u_i\,dv + \int_V \sigma_{ij}\,\delta e_{ij}\,dv.$$

Therefore, we obtain the *variational equation of motion*

$$(2)\quad \blacktriangle \qquad \int_V \sigma_{ij}\,\delta e_{ij}\,dv = \int_V \left(F_i - \rho\frac{\partial^2 u_i}{\partial t^2}\right)\delta u_i\,dv + \int_S \overset{\nu}{T}_i\,\delta u_i\,dS$$

As before, this general equation can be stated more concisely if we introduce various levels of restrictions. Thus, if the body is perfectly elastic and a strain energy function $W$ exists, then the variational equation of motion can be written as

$$(3)\quad \blacktriangle \qquad \delta\int_V W\,dv = \int_V \left(F_i - \rho\frac{\partial^2 u_i}{\partial t^2}\right)\delta u_i\,dv + \int_S \overset{\nu}{T}_i\,\delta u_i\,dS.$$

The variations $\delta u_i$ are assumed to vanish over the part of the boundary $S_u$ where surface displacements are prescribed. Hence, the limit for the surface integral can be replaced by $S_\sigma$. If the variations $\delta u_i$ were identified with the actual displacements $(\partial u_i/\partial t)\,dt$, then the result above states that, in an arbitrary time interval, the sum of the energy of deformation and the kinetic energy increases by an amount that is equal to the work done by the external forces during the same time interval.

If the virtual displacements $\delta u_i$ are regarded as functions of time and space, not to be identified with the actual displacements, and the variational equation of motion (3) is integrated with respect to time between two arbitrary instants $t_0$ and $t_1$, an important variational principle for the moving body can be derived. We have

$$(4)\qquad \int_{t_0}^{t_1}\int_V \delta W\,dv\,dt = \int_{t_0}^{t_1} dt\int_V F_i\,\delta u_i\,dv + \int_{t_0}^{t_1} dt\int_{S\sigma} \overset{\nu}{T}_i\,\delta u_i\,dS$$
$$- \int_{t_0}^{t_1} dt\int_V \rho\frac{\partial^2 u_i}{\partial t^2}\,\delta u_i\,dv.$$

Calling the last term $J$, inverting the order of integration, and integrating by parts, we obtain

(5) $$J = \int_V \rho \frac{\partial u_i}{\partial t} \delta u_i \, dv \Big|_{t_0}^{t_1} - \int_V dv \int_{t_0}^{t_1} \frac{\partial u_i}{\partial t}\left(\rho \frac{\partial \, \delta u_i}{\partial t} + \frac{\partial \rho}{\partial t} \delta u_i\right) dt.$$

The $\partial\rho/\partial t$ term can be ignored because $\partial\rho/\partial t = -\rho(\partial \dot{u}_i/\partial x_i)$ according to the equation of continuity, and thus the term $\delta u_i \partial\rho/\partial t$ is an order of magnitude smaller than other terms in this equation. Let us now impose the restriction that at the time $t_0$ and $t_1$, the variations $\delta u_i$ are zero at all points of the body; i.e.,

(6) $$\delta u_i(t_0) = \delta u_i(t_1) = 0 \qquad \text{in } V.$$

Then

(7) $$\begin{aligned} J &= -\int_{t_0}^{t_1}\int_V \rho \frac{\partial u_i}{\partial t}\frac{\partial \, \delta u_i}{\partial t} \, dv \, dt \\ &= -\int_{t_0}^{t_1}\int_V \rho \frac{\partial u_i}{\partial t} \, \delta \frac{\partial u_i}{\partial t} \, dv \, dt = -\int_{t_0}^{t_1} \delta \int_V \frac{1}{2} \rho \frac{\partial u_i}{\partial t}\frac{\partial u_i}{\partial t} \, dv \, dt \\ &= -\int_{t_0}^{t_1} \delta K \, dt, \end{aligned}$$

where

(8) $$K = \frac{1}{2}\int \rho \frac{\partial u_i}{\partial t}\frac{\partial u_i}{\partial t} \, dv$$

is the kinetic energy of the moving body. Therefore, under the assumption (6), Eq. (4) becomes,

(9) ▲ $$\int_{t_0}^{t_1} \delta(U - K) \, dt = \int_{t_0}^{t_1}\int_V F_i \, \delta u_i \, dv \, dt + \int_{t_0}^{t_1}\int_{S_\sigma} \overset{\nu}{T}_i \, \delta u_i \, dS \, dt;$$

where $U$ represents the total strain energy of the body,

$$U = \int_V W \, dv.$$

If the external forces acting on the body are such that the sum of the integrals on the right-hand side of (9) represents the variation of a single function—*the potential energy of the loading*, $-A$,

(10) $$\int_V F_i \, \delta u_i \, dv + \int_{S_\sigma} \overset{\nu}{T}_i \, \delta u_i \, dS = -\delta A,$$

then (9) can be written as

(11) ▲ $$\delta \int_{t_0}^{t_1} (U - K + A) \, dt = 0.$$

The term

$$L \equiv U - K + A \tag{12}$$

(or sometimes $-L$) is called the *Lagrangian function* and the equation (11) represents *Hamilton's principle*, which states that:

*The time integral of the Lagrangian function over a time interval $t_0$ to $t_1$ is an extremum for the "actual" motion with respect to all admissable virtual displacements which vanish, first, at instants of time $t_0$ and $t_1$ at all points of the body, and, second, over $S_u$, where the displacements are prescribed, throughout the entire time interval.*

To formulate this principle in another way, let us call $u_i(x_1, x_2, x_3; t)$ a dynamic path. Then Hamilton's principle states that *among all dynamic paths that satisfy the boundary conditions over $S_u$ at all times and that start and end with the actual values at two arbitrary instants of time $t_0$ and $t_1$ at every point of the body, the "actual" dynamic path is distinguished by making the Lagrangian function an extremum.*

In rigid body dynamics the term $U$ drops out, and we obtain Hamilton's principle in the familiar form. The symbol $A$ replaces the usual symbol $V$ in books on dynamics because we have used $V$ for something else.

Note that the potential energy $-A$ of the external loads exists and is a linear function of the displacements if the loads are independent of the elastic displacements, as is commonly the case. In aeroelastic problems, however, the aerodynamic loading is sensitive to the small surface displacements $u_i$; moreover, it depends on the time history of the displacements and cannot be derived from a potential. Hence, in aeroelasticity we are generally forced to use the variational form (9) of Hamilton's principle.

In some applications of the direct method of calculation, it is even desirable to liberalize the variations $\delta u_i$ at the instants $t_0$ and $t_1$ and use Hamilton's principle in the variational form (4) which cannot be expressed elegantly as the minimum of a well-defined functional. On the other hand, such a formulation will be accessible to the direct methods of solution. On introducing (5), (7), and (10), we may rewrite Eq. (4) in the following form:

$$\int_{t_0}^{t_1} \delta(U - K + A)\, dt = \int_{t_0}^{t_1} \int_V F_i\, \delta u_i\, dv\, dt + \int_{t_0}^{t_1} \int_S \overset{\nu}{T}_i\, \delta u_i\, dS - \int_V \rho \frac{\partial u_i}{\partial t}\, \delta u_i\, dv \Big|_{t_0}^{t_1}. \tag{13}$$

Here $U$ is the total strain energy, $K$ is the total kinetic energy, $A$ is the potential energy for the conservative external forces, $F_i$ and $\overset{\nu}{T}_i$ are, respectively, those external body and surface forces that are not included in $A$, and $\delta u_i$ are the virtual displacements.

**Problem 11.1.** Prove the converse theorem that, for a conservative system, the variational Eq. (11) leads to the equation of motion

$$\rho \frac{\partial^2 u_i}{\partial t^2} = F_i + \frac{\partial}{\partial x_j} \frac{\partial W}{\partial e_{ij}}$$

and the boundary conditions

$$\text{either} \quad \delta u_i = 0 \quad \text{or} \quad \frac{\partial W}{\partial e_{ij}} \nu_j = \overset{\nu}{T}_i.$$

## 11.2. EXAMPLE OF APPLICATION—EQUATION OF VIBRATION OF A BEAM

As an example of the application of Hamilton's principle in the formulation of approximate theories in elasticity, let us consider the free, lateral vibration of a straight simple beam. We assume that the beam possesses principal planes and that the vibration takes place in one of the principal planes, and let $y$ denote the small deflection of the neutral axis of the beam from its initial, straight configuration. In Sec. 10.8 it is shown that the strain energy of the beam is, for small deflections,

$$U = \frac{1}{2} \int_0^l EI \left( \frac{\partial^2 y}{\partial x^2} \right)^2 dx, \tag{1}$$

where $E$ is the Young's modulus of the beam material, $I$ is the cross-sectional moment of interia, and $l$ is the length of the beam.

The kinetic energy of the beam is derived partly from the translation, parallel to $y$, of the elements composing it, and partly from the rotation of the same elements about an axis perpendicular to the neutral axis and the plane of vibration. The former part is

$$\frac{1}{2} \int_0^l m \left( \frac{\partial y}{\partial t} \right)^2 dx,$$

where $m$ is the mass per unit length of the beam. The latter part is, for each element $dx$, the product of moment of inertia times one-half of the square of the angular velocity. Let $I_\rho$ denote the mass moment of inertia about the neutral axis per unit length of the beam. The angular velocity being $\partial^2 y/\partial t\, \partial x$, the kinetic energy of the beam is

$$K = \frac{1}{2} \int_0^l m \left( \frac{\partial y}{\partial t} \right)^2 dx + \frac{1}{2} \int_0^l I_\rho \left( \frac{\partial^2 y}{\partial x\, \partial t} \right)^2 dx. \tag{2}$$

If the beam is loaded by a distributed lateral load of intensity $p(x, t)$ per unit length and moment and shear $M$ and $Q$, respectively, at the ends as shown in Fig. 11.2:1, then the potential energy of the external loading is

$$A = -\int_0^l p(x, t) y(x)\, dx - M_l \left( \frac{\partial y}{\partial x} \right)_l + M_0 \left( \frac{\partial y}{\partial x} \right)_0 + Q_l y_l - Q_0 y_0. \tag{3}$$

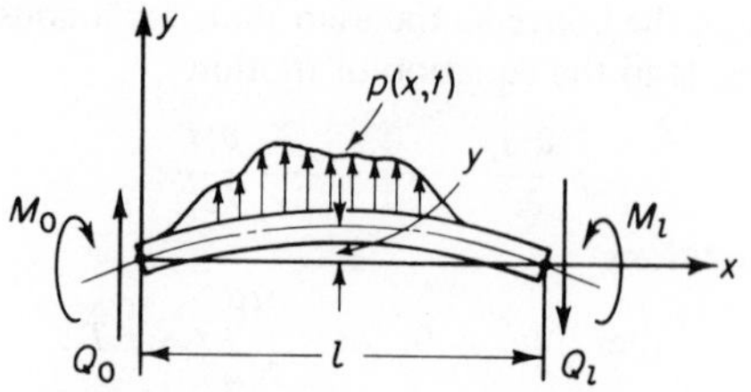


**Fig. 11.2:1.** Applications to a beam.

The equation of motion is given by Hamilton's principle:

$$\delta \int_{t_0}^{t_1} (U - K + A)\, dt = 0; \tag{4}$$

i.e.,

$$\delta \int_{t_0}^{t_1} \left\{ \int_0^l \left[ \frac{1}{2} EI \left( \frac{\partial^2 y}{\partial x^2} \right)^2 - \frac{1}{2} m \left( \frac{\partial y}{\partial t} \right)^2 - \frac{1}{2} I_\rho \left( \frac{\partial^2 y}{\partial x\, \partial t} \right)^2 - py \right] dx \right.$$
$$\left. - M_l \left( \frac{\partial y}{\partial x} \right)_l + M_0 \left( \frac{\partial y}{\partial x} \right)_0 + Q_l y_l - Q_0 y_0 \right\} dt = 0. \tag{5}$$

Following the usual procedure of the calculus of variations, noting that the virtual displacement must be so specified that $\delta y \equiv 0$ at $t_0$ and $t_1$, and, hence, $\partial(\delta y)/\partial x = \delta(\partial y/\partial x) \equiv 0$ at $t_0$ and $t_1$, we obtain

$$\int_{t_0}^{t_1} \left[ \int_0^l \left( EI \frac{\partial^2 y}{\partial x^2} \frac{\partial^2\, \delta y}{\partial x^2} - m \frac{\partial y}{\partial t} \frac{\partial\, \delta y}{\partial t} - I_\rho \frac{\partial^2 y}{\partial x\, \partial t} \frac{\partial^2\, \delta y}{\partial x\, \partial t} - p\, \delta y \right] dx \right.$$
$$\left. - M_l\, \delta \left( \frac{\partial y}{\partial x} \right)_l + M_0\, \delta \left( \frac{\partial y}{\partial x} \right)_0 + Q_l\, \delta y_l - Q_0\, \delta y_0 \right] dt = 0.$$

Integrating by parts, we obtain

$$\int_{t_0}^{t_1} \int_0^l \left[ \frac{\partial^2}{\partial x^2} \left( EI \frac{\partial^2 y}{\partial x^2} \right) + m \frac{\partial^2 y}{\partial t^2} - \frac{\partial^2}{\partial x\, \partial t} \left( I_\rho \frac{\partial^2 y}{\partial x\, \partial t} \right) - p(x, t) \right] \delta y\, dx\, dt$$
$$- \int_{t_0}^{t} \left[ EI \frac{\partial^2 y}{\partial x^2} - M \right] \delta \left( \frac{\partial y}{\partial x} \right) \Bigg|_0^l \Bigg]\, dt$$
$$- \int_{t_0}^{t} \left[ \frac{\partial}{\partial x} \left( EI \frac{\partial^2 y}{\partial x^2} \right) - \frac{\partial}{\partial t} \left( I_\rho \frac{\partial^2 y}{\partial x\, \partial t} \right) - Q \right] \delta y \Bigg|_0^l dt = 0. \tag{6}$$

Hence, the Euler equation of motion is

$$\frac{\partial^2}{\partial x^2} \left( EI \frac{\partial^2 y}{\partial x^2} \right) + m \frac{\partial^2 y}{\partial t^2} - \frac{\partial^2}{\partial x\, \partial t} \left( I_\rho \frac{\partial^2 y}{\partial x\, \partial t} \right) = p(x, t), \tag{7}$$

and a proper set of boundary conditions at each end is

$$\text{either} \quad EI \frac{\partial^2 y}{\partial x^2} = M \quad \text{or} \quad \delta \left( \frac{\partial y}{\partial x} \right) = 0 \tag{8a}$$

and

(8b) either $$\frac{\partial}{\partial x}\left(EI\frac{\partial^2 y}{\partial x^2}\right) - \frac{\partial}{\partial t}\left(I_\rho \frac{\partial^2 y}{\partial x\,\partial t}\right) = Q \quad \text{or} \quad \delta y = 0.$$

*These are equations governing the motion of a beam including the effect of the rotary inertia, due to Lord Rayleigh, and known as* ***Rayleigh's equations.*** If the rotary inertia is neglected and if the beam were uniform, then the governing equation is simplified into:

(9) $$\frac{\partial^2 y}{\partial t^2} + c_0^2 R^2 \frac{\partial^4 y}{\partial x^4} = \frac{1}{EI}\, p,$$

where

(10) $$c_0^2 = \frac{E}{\rho}, \qquad R^2 = \frac{I}{A}.$$

The constant $c_0$ has the dimension of speed and can be identified as the phase velocity of longitudinal waves in a uniform bar.† $R$ is the radius of gyration of the cross section. $A$ is the cross-sectional area, so that $m = \rho A$.

In the special case of a uniform beam of infinite length free from lateral loading, $p = 0$, Eq. (9) becomes

(11) $$\frac{\partial^2 y}{\partial t^2} + c_0^2 R^2 \frac{\partial^4 y}{\partial x^4} = 0.$$

It admits a solution in the form

(12) $$y = a \sin \frac{2\pi}{\lambda}(x - ct),$$

which represents a progressive wave of phase velocity $c$ and wave length $\lambda$. On substituting (12) into (11), we obtain the relation

(13) $$c = \pm c_0 R \frac{2\pi}{\lambda},$$

which states that the phase velocity depends on the wave length and that it tends to infinity for very short wave lengths. Somewhat disconcerting is the fact that, according to Eq. (13), the group velocity (see Sec. 11.3) also tends to infinity as the wave length tends to zero. Since group velocity is the velocity at which energy is transmitted, this result is physically unreasonable. If Eq. (13) were correct, then the effect of a suddenly applied concentrated load will be felt at once everywhere in the beam, as the Fourier representation for a concentrated load contains harmonic components with infinitesimal wave length, and hence infinite wave speed. Thus, Eq. (11) cannot be very accurate in describing the effect of impact loads on a beam.

† See Prob. 11.2, p. 325.

This difficulty of infinite wave speed is removed by the inclusion of the rotary inertia. However, the speed versus wave length relationship obtained from Rayleigh's Eq. (7) for a uniform beam of circular cross section with radius $a$, as is shown in Fig. 11.2:2, still deviates appreciably from Pochhammer and Chree's results, which were derived from the exact three-dimensional linear elasticity theory. A much better approximation is obtained by including the shear deflection of the beam, as was first shown by Timoshenko.

To incorporate the shear deformation, we note that the slope of the deflection curve depends not only on the rotation of cross sections of the beam but also on the shear. Let $\psi$ denote the slope of the deflection curve when the shearing force is neglected and $\beta$ the angle of shear at the neutral axis in the same cross section. Then the total slope is

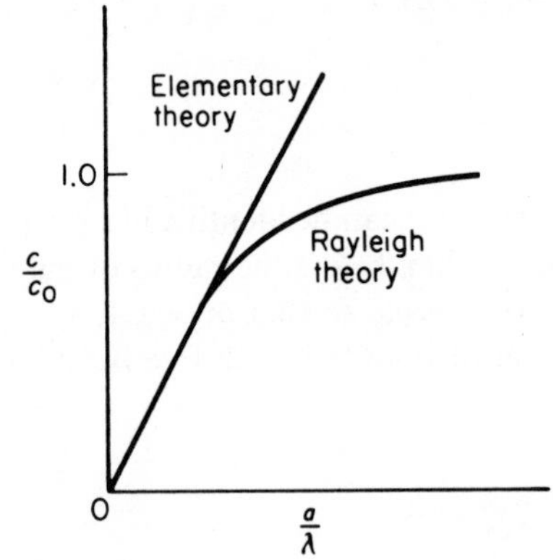


**Fig. 11.2:2.** Phase velocity curves for flexural elastic waves in a circular cylinder of radius $a$.

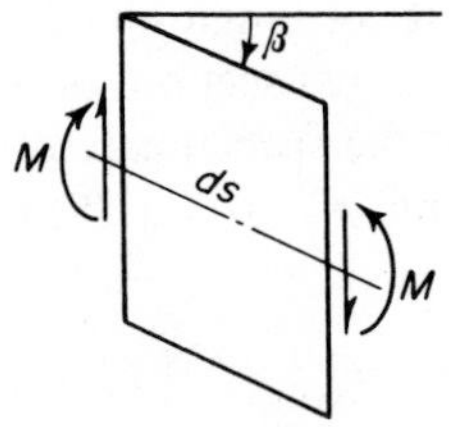


**Fig. 11.2:3.** A Timoshenko beam element.

$$\frac{\partial y}{\partial x} = \psi + \beta. \tag{14}$$

The strain energy due to bending, Eq. (1), must be replaced by

$$\frac{1}{2}\int_0^l EI\left(\frac{\partial \psi}{\partial x}\right)^2 dx, \tag{15}$$

because the internal bending moment does no work when shear deformation takes place (see Fig. 11.2:3). The strain energy due to shearing strain $\beta$ must be a quadratic function of $\beta$ if linear elasticity is assumed. We shall write

$$\frac{1}{2}\int_0^l k\beta^2\, dx = \frac{1}{2}\int_0^l k\left(\frac{\partial y}{\partial x} - \psi\right)^2 dx \tag{16}$$

for the strain energy for shear. The kinetic energy is

$$K = \frac{1}{2}\int_0^l m\left(\frac{\partial y}{\partial t}\right)^2 dx + \frac{1}{2}\int_0^l I_\rho\left(\frac{\partial \psi}{\partial t}\right)^2 dx, \tag{17}$$

because the translational velocity is $\partial y/\partial t$, but the angular velocity is $\partial \psi/\partial t$. Hence, Hamilton's principle states that

$$\delta \int_{t_0}^{t_1}\int_0^l \frac{1}{2}\left[EI\left(\frac{\partial \psi}{\partial x}\right)^2 + k\left(\frac{\partial y}{\partial x} - \psi\right)^2 - m\left(\frac{\partial y}{\partial t}\right)^2 - I_\rho\left(\frac{\partial \psi}{\partial t}\right)^2\right] dx\, dt + \delta A = 0, \tag{18}$$

where $A$ is given by (3) except that $\partial y/\partial x$ at the ends is to be replaced by $\psi$. The virtual displacements now consist of $\delta y$ and $\delta \psi$, which must vanish at $t_0$ and $t_1$ and also where displacements are prescribed. On carrying out the calculations, the following two Euler equations are obtained:

$$\frac{\partial}{\partial x}\left(EI\frac{\partial \psi}{\partial x}\right) + k\left(\frac{\partial y}{\partial x} - \psi\right) - I_\rho\frac{\partial^2 \psi}{\partial t^2} = 0, \tag{19a}$$

$$m\frac{\partial^2 y}{\partial t^2} - \frac{\partial}{\partial x}\left[k\left(\frac{\partial y}{\partial x} - \psi\right)\right] - p = 0. \tag{19b}$$

The appropriate boundary conditions are, at each end of the beam,

$$\text{Either} \quad EI\frac{\partial \psi}{\partial x} = M \quad \text{or} \quad \delta\psi = 0, \tag{20a}$$

and

$$\text{either} \quad k\left(\frac{\partial y}{\partial x} - \psi\right) = Q \quad \text{or} \quad \delta y = 0. \tag{20b}$$

These are the differential equation and boundary conditions of the so-called *Timoshenko beam theory.*

For a uniform beam, $EI$, $k$, $m$, etc., are constants, and the function $\psi$ can be eliminated from the equations above to obtain the well-known *Timoshenko equation for lateral vibration of prismatic beams,*

$$EI\frac{\partial^4 y}{\partial x^4} + m\frac{\partial^2 y}{\partial t^2} - \left(I_\rho + \frac{EIm}{k}\right)\frac{\partial^4 y}{\partial x^2\,\partial t^2} + I_\rho\frac{m}{k}\frac{\partial^4 y}{\partial t^4} = p + \frac{I_\rho}{k}\frac{\partial^2 p}{\partial t^2} - \frac{EI}{k}\frac{\partial^2 p}{\partial x^2}. \tag{21}$$

So far we have not discussed the constants $m$, $I_\rho$, and $k$. For a beam of uniform material, $m = \rho A$, $I_\rho = \rho A R^2$, where $\rho$ is the mass density of the beam material, $A$ is the cross-sectional area, and $R$ is the radius of gyration of the cross section about an axis perpendicular to the plane of motion and through the neutral axis. But $k$ depends on the distribution of shearing stress in the beam cross section. Timoshenko writes

$$k = k'AG, \tag{22}$$

where $G$ is the shear modulus of elasticity and $k'$ is a numerical factor depending on the shape of the cross section, and ascertains that according to the elementary beam theory, $k' = \frac{2}{3}$ for a rectangular cross section. The use of such a value of $k$ is, however, a subject of controversy in the literature. Mindlin[11.1] suggests that the value of $k$ can be so selected that the solution of Eq. (21) be made to agree with certain solution of the exact three-dimensional equations of Pochhammer (1876) and Chree (1889) (see Love,[1.2] *Elasticity*, 4th ed., pp. 287–92). Indeed, $I_\rho$, which arises in the assumption of plane sections remain plane in bending, may also be regarded, when such an assumption is relaxed, as an empirical factor to be determined by comparison with exact solutions.

For a uniform beam free from lateral loadings, Eq. (21) can be written as

$$\frac{\partial^4 y}{\partial x^4} - \left(\frac{1}{c_0^2} + \frac{1}{c_Q^2}\right)\frac{\partial^4 y}{\partial x^2\,\partial t^2} + \frac{1}{c_0^2 c_Q^2}\frac{\partial^4 y}{\partial t^4} + \frac{1}{c_0^2 R^2}\frac{\partial^2 y}{\partial t^2} = 0, \tag{23}$$

where

$$c_0^2 = \frac{E}{\rho}, \qquad c_Q^2 = \frac{k'G}{\rho}, \qquad R^2 = \frac{I}{A}. \tag{24}$$

If the beam is of infinite length, a solution of the form (12) may be substituted into (23), and we see that the wave speed $c$ must satisfy the equation

$$1 - \left(\frac{c^2}{c_0^2} + \frac{c^2}{c_Q^2}\right) + \frac{c^4}{c_0^2 c_Q^2} - \frac{c^2}{c_0^2 R^2}\left(\frac{\lambda}{2\pi}\right)^2 = 0. \tag{25}$$

The solution of this equation for $c/c_0$ versus $\lambda$ yields two branches, corresponding to two "modes" of motion (two different shear-to-bending deflection ratios for the same wavelength). They are plotted in Fig. 11.2:4 for the special case of a beam of circular cross section with radius $a$. The results of the exact solution of Pochhammer and Chree for Poisson's ratio $\nu = 0.29$ are also plotted there for comparison. It is seen that the Timoshenko theory agrees reasonably well with the exact theory in the first mode, but wide discrepancy occurs in the second mode. The approximate theory gives no information about higher modes: an infinite number of which exist in the exact theory.

The equations derived above are, of course, appropriate for the determination of the free-vibration modes and frequencies of a beam. The effects of rotary inertia and shear are unimportant if the wavelength of the vibration mode is large compared with the cross-sectional dimensions of the beam; but these effects become more and more important with a decrease of wavelength, i.e., with an increase in the frequency of vibration. In the example of a uniform beam with rectangular cross section and simply supported at both ends, with $E = \frac{8}{3}G$ and $k' = \frac{2}{3}$, we find that the shear deflection and rotary inertia reduce the natural frequencies. If the wavelength

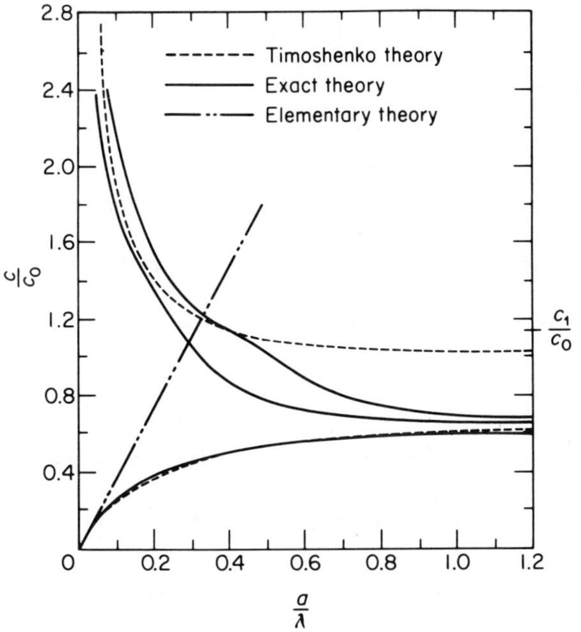


**Fig. 11.2:4.** Phase velocity curves for flexural elastic waves in a solid circular cylinder of radius $a$. (From Abramson,[11.1] *J. Acoust. Soc. Am.*, 1957.)

is ten times larger than the depth of the beam, the correction on the frequency due to rotary inertia alone is about 0.4 per cent, and the correction due to rotary inertia and shear together will be about 2 per cent.

The Timoshenko beam theory has attracted much attention in recent years. For a survey of literature, see Abramson, Plass, and Ripperger.[11.1]

## PROBLEMS

**11.2.** Consider the free longitudinal vibration of a rod of uniform cross section and length $L$, as shown in Fig. P11.2. Let us assume that plane cross sections remain plane, that only axial stresses are present, being uniformly distributed over the cross section, and that radial displacements are negligible (i.e., the displacements consist of only one nonvanishing component $u$ in the $x$-direction). Derive

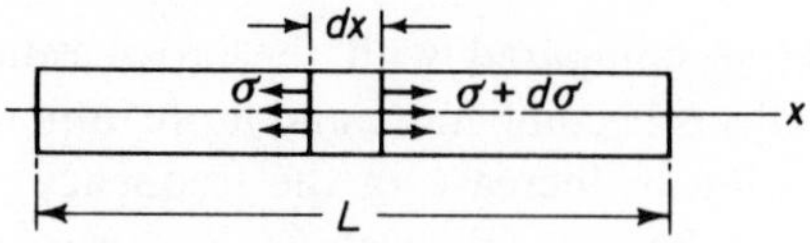


**P11.2.** Longitudinal vibration of a rod.

expressions for the potential and kinetic energy and show that the equation of motion is

$$\frac{\partial^2 u}{\partial x^2} - \frac{1}{c_0^2}\frac{\partial^2 u}{\partial t^2} = 0, \qquad c_0^2 = \frac{E}{\rho}. \tag{26}$$

Show that the general solution is of the form

$$u = f(x - c_0 t) + F(x + c_0 t), \tag{27}$$

where $f$ and $F$ are two arbitrary functions.

**11.3.** Consider the same problem as above, but now incorporate approximately the transverse inertia associated with the lateral expansion or contraction connected with axial compression and extension, respectively. Let the (Love's) assumption be made that the displacement in the radial direction $v$ is proportional to the radial coordinate $r$, measured from a centroidal axis, and to the axial strain $\partial u/\partial x$; i.e.,

$$v = -\nu r \frac{\partial u}{\partial x}, \tag{28}$$

where $\nu$ is Poisson's ratio. Derive expressions of the kinetic and potential energy and obtain the equation of motion according to Hamilton's principle,

$$\rho\left[\frac{\partial^2 u}{\partial t^2} - (\nu R)^2 \frac{\partial^4 u}{\partial x^2\,\partial t^2}\right] - E\frac{\partial^2 u}{\partial x^2} = 0, \tag{29}$$

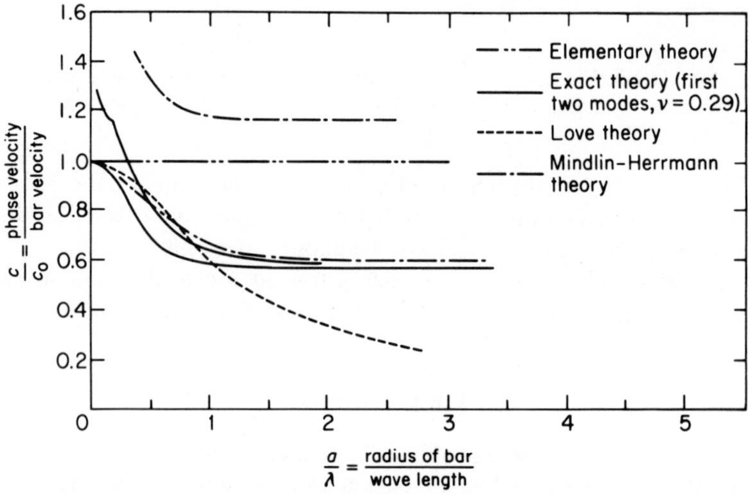


**P11.3.** Phase velocity curves for longitudinal elastic waves in a solid circular cylinder of radius $a$. (After Abramson et al., *Adv. Applied Mech.*, *5*, 1958.)

where $R$ is the polar radius of gyration of the cross section. The natural boundary condition at the end $x = 0$, if that end is subject to a stress $\sigma_0(t)$, is

(30)
$$\rho v^2 R^2 \frac{\partial^3 u}{\partial x\, \partial t^2} + E \frac{\partial u}{\partial x} = \sigma_0(t) \qquad \text{at } x = 0.$$

*Note.* It is important to note that, according to the last equation, the familiar proportionality between axial stress $\sigma$ and axial strain $\partial u/\partial x$ does not exist in this theory.

Comparison of the dispersion curves obtained from the elementary theory (Prob. 11.2), the Love theory, the Pochhammer-Chree "exact" theory, and another approximate theory due to Mindlin and Herrmann,[11.1] are shown in Fig. P11.3 The last-mentioned theory accounts for the strain energy associated with the transverse displacement $v$, of which the most important contribution comes from the shearing strain caused by the lateral expansion of the cross section near a wave front.

**11.4.** The method of derivation of the various forms of equations of motion of beams as presented above has the advantage of being straightforward, but it does not convey the physical concepts as clearly as in an elementary derivation. Hence, rederive the basic equations by considering the forces that act on an element of length $dx$, as shown in Fig. P11.2 and Fig. P11.4. Obtain the following equations, and then derive the wave equations by proper reductions.

*Longitudinal waves, elementary theory* (Fig. P11.2):

(31)
$$\frac{\partial \sigma_{xx}}{\partial x} = \rho \frac{\partial^2 u}{\partial t^2} \quad \text{(equation of motion)},$$

(32)
$$\frac{\partial^2 u}{\partial x\, \partial t} = \frac{\partial \epsilon_{xx}}{\partial t} \quad \text{(equation of strain)},$$

(33)
$$\sigma_{xx} = E\epsilon_{xx} \quad \text{(equation of material behavior)},$$

where $\sigma_{xx}$ = axial stress, $\epsilon_{xx}$ = axial strain, $\partial u/\partial t$ = axial particle velocity, $x$ = axial coordinate, $t$ = time, $E$ = modulus of elasticity, and $\rho$ = mass density.

*Flexural waves, Timoshenko theory* (Fig. P11.4, p. 328):

(34)
$$\frac{\partial M}{\partial x} - Q = \rho I \frac{\partial \omega}{\partial t} \quad \text{(rotational)}$$

(35)
$$\frac{\partial Q}{\partial x} = \rho A \frac{\partial v}{\partial t} \quad \text{(transverse)}$$

(equations of motion),

(36)
$$\frac{\partial K}{\partial t} = \frac{\partial \omega}{\partial x} \quad \text{(bending)}$$

(37)
$$\frac{\partial \beta}{\partial t} = \frac{\partial v}{\partial x} + \omega \quad \text{(shear)}$$

(38)
$$M = EIK \quad \text{(bending)}$$

(39)
$$Q = A_s \beta \quad \text{(shear)}$$

(equations of material behavior),

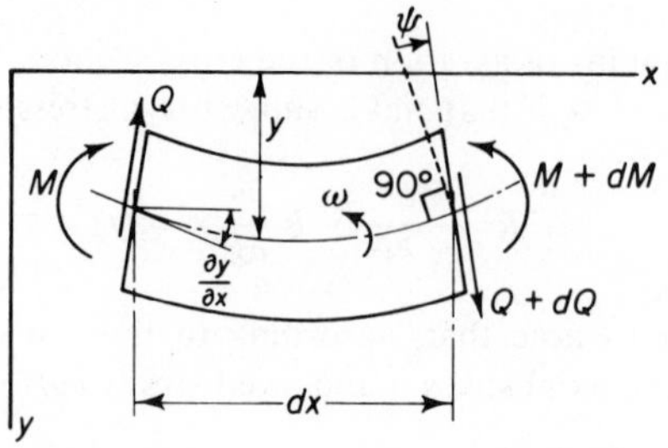


**P11.4.** Element of a beam in bending.

where $M$ = moment, $Q$ = shear force, $K$ = axial rate of change of section angle = $-\partial\psi/\partial x$, $\beta$ = shear strain = $\partial y/\partial x - \psi$, $\omega$ = angular velocity of section = $-\partial\psi/\partial t$, $v$ = transverse velocity = $\partial y/\partial t$, $I$ = section moment of inertia, $A$ = section area, and $A_s$ = area parameter defined by $\iint \gamma(z)\, dA = \beta A_s$ where $\gamma(z)$ is the shear strain at a point $z$ in the cross section.

## 11.3. GROUP VELOCITY

Since we have been concerned in the preceding sections about wave propagations in beams, it seems appropriate to make a digression to explain the concept of *group velocity* as distinguished from the *phase velocity*. We have seen that for certain equations a solution of the following form exists:

$$u = a \sin(\mu x - \nu t). \tag{1}$$

If $x$ is increased by $2\pi/\mu$, or $t$ by $2\pi/\nu$, the sine takes the same value as before, so that $\lambda = 2\pi/\mu$ is the wavelength and $T = 2\pi/\nu$ is the period of oscillation. If $\mu x - \nu t$ = constant, i.e. $x$ = const. + $\nu t/\mu$, the argument of the sine function remains constant in time; which means that the whole waveform is displaced towards the right with a velocity $c = \nu/\mu$. The quantity $c$ is called the phase velocity, in terms of which Eq. (1) may be exhibited as

$$u = a \sin \frac{2\pi}{\lambda}(x - ct). \tag{2}$$

If the phase velocity $c$ depends on the wavelength $\lambda$, the wave is said to exhibit *dispersion*. Our examples in the previous section show that dispersion exists in both longitudinal and flexural waves in rods and beams.

What happens when two sine waves of the same amplitude but slightly different wavelengths and frequencies are superposed? Let these two waves be characterized by two sets of slightly different values $\mu$, $\nu$ and $\mu'$, $\nu'$. The resultant of the superposed waves is

$$u + u' = A[\sin(\mu x - \nu t) + \sin(\mu' x - \nu' t)].$$

Using the well-known formula

$$\sin \alpha + \sin \beta = 2 \sin \tfrac{1}{2}(\alpha + \beta) \cos \tfrac{1}{2}(\alpha - \beta),$$

we have

(3)
$$u + u' = 2A \sin [\tfrac{1}{2}(\mu + \mu')x - \tfrac{1}{2}(\nu + \nu')t] \cos [\tfrac{1}{2}(\mu - \mu')x - \tfrac{1}{2}(\nu - \nu')t].$$

This expression represents the well-known phenomenon of "beats." The sine factor represents a wave whose wave number and frequency are equal to the mean of $\mu$, $\mu'$ and $\nu$, $\nu'$, respectively. The cosine factor, which varies very slowly when $\mu - \mu'$, $\nu - \nu'$ are small, may be regarded as a varying amplitude,

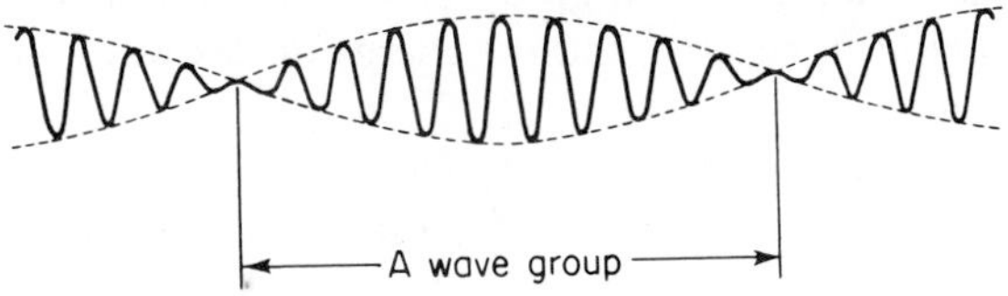


**Fig. 11.3:1.** An illustration of a wave group.

as shown in Fig. 11.3:1. The "wave group" ends wherever the cosine becomes zero. The velocity of advance of these points is called the *group velocity*; its value $U$ is equal to $(\nu - \nu')/(\mu - \mu')$. For long groups (or slow beats), the group velocity may be written with sufficient accuracy as

(4)
$$U = \frac{d\nu}{d\mu}.$$

In terms of the wavelength $\lambda$ $(= 2\pi/\mu)$, we have

(5)
$$U = \frac{d(\mu c)}{d\mu} = c - \lambda \frac{dc}{d\lambda},$$

where $c$ is the phase velocity.

From the fact that no energy can travel past the nodes, one can infer that the rate of transfer of energy is identical with the group velocity. This fact is capable of rigorous proof for single trains of waves.

The most familiar examples of propagation of wave groups are perhaps the water waves. It has often been noticed that when an isolated group of waves, of sensibly the same length, advancing over relatively deep water, the velocity of the group as a whole is less than that of the individual waves composing it. If attention is fixed on a particular wave, it is seen to advance through the group, gradually dying out as it approaches the front, while its former place in the group is occupied in succession by other waves which have come forward from the rear. Another familiar example is the wave train set up by ships. The explanation as presented above seems to have been first given by Stokes (1876). Other derivations and interpretations of the group velocity concept can be found in Lamb, *Hydrodynamics*,[11.1] Secs. 236, 237.

If a concentrated lateral load is suddenly and impulsively applied on an infinitely long beam, the disturbance is propagated out along the beam by flexural waves. The initial loading may be regarded as composed of an infinite number of sine-wave components of all wavelengths but of the same amplitude, with proper phase relationship, so that they reinforce each other in the limited region where the force is applied, but cancel each other everywhere outside the region of load application. As time increases, these sine waves propagate with their own phase velocities, and the pattern of interference changes with time. Thus, at time $t$ and at a point which is at a distance $x$ from the initial loading, only a group of waves in the neighborhood of a specific wavelength can be seen, and the energy of this wave group is propagated at the group velocity $U$.

From the dispersion curves of Fig. 11.2:4 for the phase velocity of flexural elastic waves in a circular cylinder of radius $a$, the group wave velocity

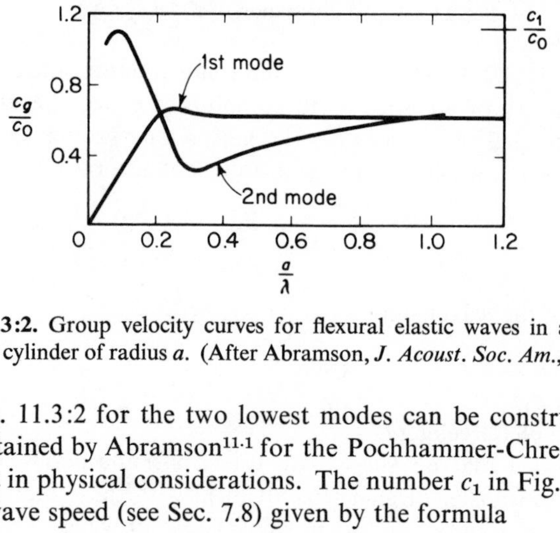


**Fig. 11.3:2.** Group velocity curves for flexural elastic waves in a solid circular cylinder of radius $a$. (After Abramson, *J. Acoust. Soc. Am.*, 1957.)

curves of Fig. 11.3:2 for the two lowest modes can be constructed. These curves are obtained by Abramson[11.1] for the Pochhammer-Chree theory and are important in physical considerations. The number $c_1$ in Fig. 11.3:2 is the dilatational wave speed (see Sec. 7.8) given by the formula

$$c_1 = \sqrt{\frac{(1-\nu)E}{(1+\nu)(1-2\nu)\rho}}\,.$$

The number $c_o$ is $\sqrt{(E/\rho)}$. The group velocity is denoted by $c_g$. The ratio $c_1/c_0$ depends on the Poisson's ratio $\nu$:

$$\nu = 0.25, \qquad \frac{c_1}{c_0} = 1.095,$$

$$\nu = 0.30, \qquad \frac{c_1}{c_0} = 1.16.$$

Note that the phase velocities in the higher branches exceed the dilatational wave velcoity (at long wavelengths), but the group velocities do not. Similar group velocity curves can be constructed for longitudinal waves from Fig. P11.3. Note that the greatest possible velocity of energy transmission is the dilatational velocity, for both flexural and longitudinal waves.

From Fig. 11.3:2 for a circular cylinder of radius $a$, it can be seen that for the first branch the group velocity reaches a maximum when $a/\lambda$ is between 0.25 and 0.30, where $\lambda$ is the wavelength. Hence, when a pulse is propagated along a beam in accordance with the first branch, Fourier components of wavelengths about 0.25 or 0.30 times the radius will be found at the head of the pulse. A detailed study of the transmission of a pulse along the beam can be found in R. M. Davies' paper.[11.2]

We must remark that elastic substances are not inherently dispersive, but the dispersion effect appears in waves traveling through rods and beams because of wave reflections from the boundaries. Therefore, the dispersion in the case of an elastic medium is simply an interface phenomenon and not a physical property of the material. The phenomenon is analogous to that obtained in electrical wave guides and can be discussed in similar terms. In this light, we can answer the question of the existence of phase velocities greater than the dilatational velocity, although no disturbance can be propagated with a velocity greater than that. The point is that energy is propagated at the group velocity and that group velocities greater than the dilatational velocity will not occur. The phase velocities are merely velocities of propagation in the axial direction of the loci of constant phase in specific mode patterns and their values are thus not limited by physical considerations.

## 11.4. HOPKINSON'S EXPERIMENTS

John Hopkinson published in 1872 an account of an interesting experiment, the explanation of which will help us understand the nature of elastic wave propagation and its significance in engineering. Hopkinson measured the strength of steel wires when they were suddenly stretched by a falling weight. A ball-shaped weight pierced by a hole was threaded on the wire and was dropped from a known height so that it struck a clamp attached to the bottom of the wire (Fig. 11.4:1). For a given weight we expect the existence of a critical height beyond which the falling weight would break the wire. Using different weights dropped from different heights, however, Hopkinson obtained the remarkable result that the minimum height from which a weight had to be dropped to break the wire was nearly *independent* of the size of the weight!

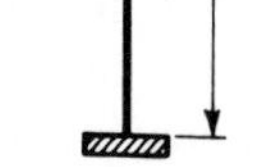

**Fig. 11.4:1.** A sketch of Hopkinson's experiment.

Now, when different weights are dropped from a given height, the velocity reached at the end is independent of the size of the weight. Hopkinson's result suggests that in breaking the wire it is the velocity of the loading end that counts. Following this lead, Hopkinson explains his result on the basis of elastic wave propagation (ignoring the effects of plasticity). Assuming that the stress state in the wire is approximately one-dimensional, we see from Eqs. (11.2:27), or (11.2:31)–(11.2:33), that the stress in the wire is proportional to the particle velocity:

$$\sigma_{xx} = \rho c_0 \frac{\partial u}{\partial t}, \tag{1}$$

where $c_0 = \sqrt{E/\rho}$ is the speed of sound of longitudinal waves in the wire. For steel, $c_0$ is about 16,000 ft/sec. The largest particle velocity in the wire, however, is not reached at the instant of blow, but is reached at the top after the elastic waves propagate up and down the wire several times. When this largest particle velocity induces a stress equal to the ultimate stress of the wire, the wire breaks.

When the weight hits the clamp in Hopkinson's test, the end of the wire acquires a particle velocity $V_0$ equal to that of the weight and clamp. A steep-fronted tension wave is generated and propagated up the wire. In the meantime, the weight and the clamp are slowed down exponentially by the tensile load imposed by the wire:

$$M\frac{dV}{dt} = A\sigma_{xx} = A\rho c_0 V, \tag{2}$$

where $M$ is the mass of the weight and clamp, $V$ is its velocity, and $A$ is the cross-sectional area of the wire. The elastic wave, on reaching the fixed end at the top, is reflected as a tension wave of twice the intensity of the incident wave. The reflected wave, going through the tail of the incident pulse, is reflected again at the lower end as a compression wave, and so on. If the stress at the first reflection (equal to $2\rho c_0 V_0$), is sufficient to break the wire, then fracture will be expected to take place near the top. In a systematic test of gradually increasing the height of drop $h$ (i.e., gradually increasing $V_0$), this breaking at the first reflection does not happen. In J. Hopkinson's experiments, the head of the stress waves was able to travel the length of wire several times before the weight $M$ was decelerated sufficiently, and the stress wave pattern in the wire was very complicated. Bertram Hopkinson (1905), in repeating his father's experiment, used smaller weights so that the rate of decay was rapid. Nevertheless, as shown by G. I. Taylor[11.2] (1946), the maximum tensile stress in B. Hopkinson's experiment did not occur at the first reflection, when the stress was $2\rho c_0 V_0$, but at the third reflection, i.e., the second reflection at the top of the wire, when the tensile stress reached $2.15\rho c_0 V_0$.

To work out the details, we have to determine the function $f(x - c_0 t)$ and $F(x + c_0 t)$ of Eq. (11.2:27) to satisfy the initial conditions $u = \partial u/\partial t = 0$ at $t = 0$, and the boundary conditions $u = 0$ at the top; $\partial u/\partial t = V$, the velocity of the weight and clamp, at the bottom. The velocity $V$ is determined by Eq. (2). The full solution is given in Taylor's paper.[11.2]

Incidentally, Hopkinson found that the tensile strength of steel wires under rapid loading is much greater than that under a static load.

## 11.5. GENERALIZED COORDINATES

We shall now turn our attention to the actual construction of the minimizing sequence in solving a variational problem. Our aim is to obtain efficient approximate solutions. For this purpose, we shall discuss first the idea of generalized coordinates and the definitions of the "best approximations."

Let us consider a function $u(x)$ of a real variable $x$, defined over an interval $(a, b)$. If $u(x)$ can be represented by a power series

(1) $$u(x) = a_0 + a_1 x + a_2 x^2 + \dots + a_n x^n + \dots$$

or by a Fourier series

(2) $$u(x) = \sum_{n=1}^{\infty} b_n \sin nx$$

or, more generally, by a series of the form

(3) $$u(x) = \sum_{n=1}^{\infty} c_n \phi_n(x),$$

where $\{\phi_n(x)\}$ is a given set of functions, then $u(x)$ can be specified by the sets of coefficients $\{a_n\}$, $\{b_n\}$, or $\{c_n\}$. These coefficients may be regarded as the generalized coordinates of $u(x)$ referred to the bases $\{x^n\}$, $\{\sin nx\}$, $\{\phi_n\}$, respectively. In (3), each term $\phi_n(x)$ represents a degree-of-freedom, and the coefficient $c_n$ specifies the extent to which $\phi_n(x)$ participates in the function $u(x)$. For example, if $u(x)$ represents a disturbed state of an elastic string which is anchored at the points $x = 0$ and $x = \pi$, then Eq. (3) states that $u(x)$ can be obtained by a superposition of successive normal modes $\phi_n = \sin nx$, and that the $n$th mode participates with an amplitude $c_n$.

This terminology is conventional when $c_n$ is a function of time $t$. For example, in the dynamics of a vibrating string the displacement $u(x, t)$ may be written as

(4) $$u(x, t) = \sum_{n=1}^{\infty} q_n(t) \sin nx.$$

The quantity $q_n(t)$ is known as the generalized coordinate with respect to the mode sin $nx$. The possibility of generalizing this terminology to cases in which $\mathbf{x}$ and $\mathbf{u}$ represent vectors in vector spaces of dimensions $m$ and $k$, respectively, is obvious.

The fundamental idea of approximate methods of solution is to reduce a continuum problem (with infinite number of degrees-of-freedom) into one of a finite number of degrees-of-freedom. The reduced problem is solved by usual methods. Then a limiting process is used to extend the restricted solution to the solution of the original problem. In the following discussion we shall describe the ideas about the first step—reducing a given problem to one of finite degrees-of-freedom. The second step—the proof of the validity of the limiting process—is the main mathematical problem. Examples of the mathematical theories which deal with the second step are the theories of Taylor's series, Fourier series, eigenfunction expansions, etc. The theories are beyond the scope of our discussion, and the reader is referred to the Bibliography 11.3.

## 11.6. APPROXIMATE REPRESENTATION OF FUNCTIONS

Let a function $f(x_1, x_2, \ldots, x_n)$ of $n$ variables $x_1, \ldots, x_n$ be defined in a domain $\mathscr{R}$ $(x_1, x_2, \ldots, x_n)$; or, in brief notation, $f(x_j)$ in $\mathscr{R}$ $(x_j)$. To represent $f(x_j)$ "as well as possible" by a linear combination of given functions $\phi_n(x_j)$, $n = 1, 2, \ldots, N$, say, $Q_N(x_j)$, where

$$Q_N(x_j) = \sum_{n=1}^{N} a_n \phi_n(x_j), \tag{1}$$

the question arises as to the meaning of the term "as well as possible." As soon as the "best approximation" is defined, a definite procedure for calculating $a_n$ can be devised.

The usual definitions of the "best approximation" may be classified in two classes.† In the first class the concept of *norm* is introduced. For any function $g(x_j)$ in the domain $\mathscr{R}(x_j)$, a real positive number, called the *norm of* $g(x_j)$, and denoted by $\|g(x_j)\|$, is defined, with the stipulation that $\|g(x_j)\| = 0$ only for $g(x_j) \equiv 0$ in $\mathscr{R}$ $(x_j)$. Then the best approximation is taken to mean that

$$\left\|f(x_j) - \sum_{n=1}^{N} a_n \phi_n(x_j)\right\| = \min. \tag{2}$$

In the second class, a set of linear homogeneous expressions $L_k$, $k = 1, 2, \ldots, N$, defined for functions in the domain $\mathscr{R}(x_j)$, is selected, and the best approximation is taken to mean that

$$L_k(\varepsilon_N) = 0, \qquad k = 1, 2, \ldots, N, \tag{3}$$

† We follow Collatz[11.3] (1955) in the subsequent exposition.

where $\varepsilon_N$ is the "error"

$$\varepsilon_N \equiv \varepsilon_N(x_j) \equiv f(x_j) - \sum_{n=1}^{N} a_n \phi_n(x_j). \tag{4}$$

Examples of the first class are

(A) *Absolute error method:* In this method the norm for any function $g(x_j)$ in $\mathscr{R}(x_j)$ is defined to be the maximum absolute value of $g(x_j)$ in $\mathscr{R}(x_j)$:

$$\|g(x_j)\| = \max_{x \text{ in } \mathscr{R}} |g(x_j)|. \tag{5}$$

Hence, the best approximation is defined by requiring that

$$\|\epsilon_N\| = \max_{x \text{ in } \mathscr{R}} |\epsilon_N(x_j)| = \min. \tag{6}$$

This method was used in the early nineteenth century, but is not in favor now because it is difficult to apply.

(B) *Least squares method:* In this method we take a positive integrable function $W(x_j)$ and define the norm

$$\|g(x_j)\| = \sqrt{\int_{\mathscr{R}} W g^2 \, dv}, \tag{7}$$

where $W(x_j)$ is called the *weighting function.* The volume integral, which may be defined in Riemann or Lebesgue sense, is assumed to exist, $dv$ being a volume element. In this method, the constants $a_n$ are to be determined from the equations

$$\frac{1}{2}\frac{\partial \|\epsilon_N\|^2}{\partial a_k} = \int_{\mathscr{R}} \epsilon_N W \phi_k \, dv = 0, \qquad k = 1, 2, \ldots, N. \tag{8}$$

Examples of the second class are

(C) *Collocation method:* We take $N$ points $P_k$ in the domain $\mathscr{R}$ $(x_j)$ and let

$$L_k(\epsilon_N) = \epsilon_N(P_k). \tag{9}$$

Hence Eq. (3) implies that the constants $a_n$ are so chosen that $f(x_j)$ is represented exactly at $N$ points. A substituting leads to the equations

$$\sum a_N \phi_N(P_k) = f(P_k), \qquad k = 1, 2, \ldots, N. \tag{10}$$

If the determinant $\Delta$ of coefficients of $a_n$ does not vanish,

$$\Delta = \begin{vmatrix} \phi_1(P_1) & \cdots & \phi_1(P_N) \\ \cdot & \cdots & \cdot \\ \cdot & \cdots & \cdot \\ \cdot & \cdots & \cdot \\ \phi_N(P_1) & \cdots & \phi_N(P_N) \end{vmatrix} \neq 0,$$

the constants $a_n$ can be uniquely determined.

(D) *Orthogonality method:* We choose a set of $N$ linearly independent functions $g_k(x_j)$ and let

$$L_k(\epsilon_N) = \int_{\mathscr{R}} \epsilon_N g_k \, dv, \qquad k = 1, 2, \ldots, N. \tag{11}$$

If we choose $g_k = W\phi_k$, then we get the conditions

$$\int_{\mathscr{R}} \epsilon_N W \phi_k \, dv = 0 \tag{12}$$

identical with the least squares method.

(E) *Subregion method:* As a generalization of the collocation method, we subdivide the region $\mathscr{R}(x_j)$ into $N$ subregions $\mathscr{R}_1, \mathscr{R}_2, \ldots, \mathscr{R}_N$ and define

$$L_k(\epsilon) = \int_{\mathscr{R}_k} \epsilon \, dv, \qquad k = 1, \ldots, N. \tag{13}$$

then follow the steps of the collocation method.

## 11.7. APPROXIMATE SOLUTION OF DIFFERENTIAL EQUATIONS

If we try to approximate the solution $u(x_j)$ of a differential equation defined over a domain $\mathscr{R}(x_1, \ldots, x_N)$ with certain boundary conditions by an expression

$$Q_N(x_j) = \sum_{n=1}^{N} a_n \phi_n(x_j), \tag{1}$$

we may take one of the following approaches: (A) each $\phi_N(x_j)$ satisfies the boundary conditions, (B) each $\phi_N(x_j)$ satisfies the differential equation. In either case an error function can be defined, and it is required to be "as small as possible."

For example, consider the case (A). Let the differential equation be

$$\mathscr{L}\{u(x_j)\} = 0. \tag{2}$$

A substitution of (1) into (2) gives the error

$$\epsilon_N(x_j) \equiv \mathscr{L}\{\sum a_n \phi_n(x_j)\}, \tag{3}$$

which must be minimized. The definition of the "best approximation" can be varied, but all methods of the previous section can be used here.

A possible generalization of (1) to the form

$$Q_N(x_j) = w(x_1, \ldots, x_N; a_1, \ldots, a_N),$$

which depends on the $N$ parameters $a_n$, seems evident, although a mathematical theory of such a generalization is by no means simple.

## 11.8. DIRECT METHODS OF VARIATIONAL CALCULUS

Ideas similar to those presented in the preceding sections can be applied to variational problems. Let $\mathscr{D}$ be the class of admissible functions $u(x_j)$ of $n$ variables $x_1, x_2, \ldots, x_n$, or $x_j$ for short, defined over a domain $\mathscr{R}(x_j)$ and satisfying all the "rigid" boundary conditions (cf. Sec. 10.8). Let $y(x_j)$, belonging to $\mathscr{D}$, minimize the functional $I[u]$. The following approximate representations of $y(x_j)$ are suggested:

***Method of Finite Differences.*** We choose a set of points $P_k$, $k = 1, 2, \ldots, N$, in $\mathscr{R}$ and let

(1) $$y(P_k) = y_k, \qquad P_k \text{ in } \mathscr{R},\ k = 1, 2, \ldots, N.$$

The derivatives of $y(x_j)$ are then replaced by expressions involving successive finite differences of $y_1, y_2, \ldots, y_N$, according to the calculus of finite differences, and the integrations are replaced by a finite summation. Then the functional $I[y]$ becomes a function $\Psi(y_1, y_2, \ldots, y_N)$. We choose the values $y_1, \ldots, y_N$ so that $\Psi(y_j)$ has an extremum; i.e.,

(2) $$\frac{\partial \Psi}{\partial y_k} = 0, \qquad k = 1, \ldots, N.$$

This method was used by Euler.

***Rayleigh-Ritz Method.*** We set

(3) $$y_N(x_j) = \sum_{n=1}^{N} a_n \phi_n(x_j),$$

where $\phi_n(x_j)$ are known functions which are so chosen that $y_N(x_j)$ are admissible, i.e., they belong to the class $\mathscr{D}$. When (3) is substituted into $I[u]$, the latter becomes a function $\Psi(a_1, a_2, \ldots, a_N)$ of the coefficients $a_j$. These coefficients $a_1, \ldots, a_N$ are then so chosen that $\Psi(a_j)$ has an extremum; i.e.,

(4) $$\frac{\partial \Psi}{\partial a_j} = 0, \qquad j = 1, 2, \ldots, N.$$

***Kantrovich's Method.*** We choose a set of coordinate functions

$$\phi_1(x_1, x_2, \ldots, x_n),\ \phi_2(x_1, x_2, \ldots, x_n), \ldots, \phi_N(x_1, x_2, \ldots, x_n)$$

and try an approximate solution in the form

(5) $$y_N = \sum_{k=1}^{N} a_k(x_1)\phi_k(x_1, x_2, \ldots, x_n),$$

where the coefficients $a_k(x_1)$ are no longer constants, but are unknown functions of one of the independent variables. The functional $I[u]$ is reduced to a new functional $\Psi[a_1(x_1), a_2(x_1), \ldots, a_N(x_1)]$, which depends on $N$ functions of one independent variable,

$$a_1(x_1),\ a_2(x_1), \ldots, a_N(x_1). \tag{6}$$

The functions (6) must be chosen to minimize the functional $\Psi$ (Sec. 14.3).

It should be recognized that Kantrovich's method is a generalization of a method commonly used in the classical treatment of small oscillations of an elastic body, in which the displacements are represented in the form

$$u(x, t) = \sum_{k=1}^{N} q_k(t)\phi_k(x), \tag{7}$$

where $\phi_k(x)$, depending solely on the spatial coordinates, are the bases of the generalized coordinates. The Euler equations of the energy integral, according to Hamilton's principle, are the Lagrangian equations of motion.

***Galerkin's Method.*** Galerkin's idea of minimization of errors by orthogonalizing with respect to a set of given functions is best illustrated by an example. Consider the simple beam discussed in Sec. 10.8. The minimization of the functional $V$, Eq. (10.8:3), leads to the variational equation

$$\begin{aligned} 0 = \delta V = & \int_0^l \left[\frac{d^2}{dx^2}\left(EI\,\frac{d^2y}{dx^2}\right) - p\right] \delta y\, dx \\ & + \left[EI\,\frac{d^2y}{dx^2} - M\right] \delta\left(\frac{dy}{dx}\right)\Big|_0^l - \left[\frac{d}{dx}\left(EI\,\frac{d^2y}{dx^2}\right) - Q\right] \delta y\,\Big|_0^l . \end{aligned} \tag{8}$$

Here $EI$, and $p$ are known functions of $x$; $M$, $Q$ are unspecified if an end is "rigid" [where $\delta y = \delta(dy/dx) = 0$]; or are given numbers if an end is "natural" [where $\delta y$ and $\delta(dy/dx)$ are arbitrary].

Let us consider a clamped-clamped beam (both ends "rigid"). If we assume

$$y_N = \sum_{k=1}^{N} a_k \phi_k(x), \tag{9}$$

where

$$\phi_k = \frac{d\phi_k}{dx} = 0 \quad \text{at } x = 0 \text{ and } x = l, \qquad k = 1, \ldots, N, \tag{10}$$

then, in general, Eq. (8) cannot be satisfied. The quantity

$$\frac{d^2}{dx^2}\left(EI\,\frac{d^2y_N}{dx^2}\right) - p = \epsilon_N(x) \tag{11}$$

may be called an *error*. Now we demand that the coefficients $a_k$ be so chosen that Eq. (8) be satisfied when $\delta y$ is identified with any of the functions $\phi_1(x), \phi_2(x), \ldots, \phi_N(x)$:

$$\int_0^l \epsilon_N(x)\phi_k(x)\, dx = 0, \qquad k = 1, \ldots, N. \tag{12}$$

In other words, we demand that the error be orthogonal to $\phi_k(x)$.

If some of the end conditions are "natural," we still represent the approximate solution in the form (9) and require $y_N(x)$ to satisfy only the "rigid" boundary conditions wherever they apply. In this case, Eq. (8) leads to the following in place of (12).

$$\int_0^l \epsilon_N(x)\phi_k(x\ dx + \left[EI\frac{d^2y_N}{dx^2} - M\right]\frac{d\phi_k}{dx}\bigg|_0^l - \left[\frac{d}{dx}\left(EI\frac{d^2y_N}{dx^2}\right) - Q\right]\phi_k\bigg|_0^l = 0, \qquad k = 1, 2, \ldots, N. \tag{13}$$

Equations (12) or (13) are sets of $N$ linear equations to solve for the $N$ unknowns constants $a_1, \ldots, a_N$.

***Trefftz' Method.*** Consider again Eq. (8). In the Trefftz method, the approximate solution (9) is so chosen that $\phi_k(x)$ satisfies the Euler equation

$$\frac{d^2}{dx^2}\left(EI\frac{d^2\phi_k}{dx^2}\right) - p = 0$$

but not necessarily the boundary conditions. We now select the coefficients $a_k$ in such a way that (8) is satisfied if the variations $\delta y$ were limited to the set of functions $\phi_k$:

$$\left(EI\frac{d^2y_N}{dx^2} - M\right)\frac{d\phi_k}{dx}\bigg|_0^l - \left[\frac{d}{dx}\left(EI\frac{d^2y_N}{dx^2}\right) - Q\right]\phi_k\bigg|_0^l = 0.$$

## PROBLEMS

**11.5.** St. Venant's problem of torsion of a shaft with a cross section occupying a region $R$ on the $x$, $y$ plane bounded by a curve (or curves) $C$, is discussed in Section 7.5. Show that the St. Venant's theory is equivalent to finding a function $\psi(x, y)$ (Prandtl's stress function, p. 167) which minimizes the functional

$$I = \iint_R \left[\left(\frac{\partial\psi}{\partial x}\right)^2 + \left(\frac{\partial\psi}{\partial y}\right)^2 - 4G\alpha\psi\right]dx\,dy,$$

under the boundary condition that $d\psi/ds = 0$ on $C$. If the region $R$ is simply connected, we may take $\psi = 0$ on $C$. If $R$ is multiply connected, we take $\psi$ equal to a different constant on each boundary. The torque is then given by Eq. (7.5:24).

Consider a shaft of rectangular cross section. Obtain approximate solutions to the torsion problem by two different direct methods outlined in Section 11.8. Compare the values of the torsional rigidity of the shaft obtained in each method. (Ref., Sokolnikoff[1.2], 2nd ed., Ch. 7, pp. 400, 416, 423, 430, 437, 442.)

**11.6.** Consider the pin-ended column described in Prob. 10.11, p. 313. Assume that the column and the spring are massless but that a lumped mass $M$ is situated on the column at the point $A$. At the instant of time $t = 0$, the column is plucked so that the deflection at $A$ is $\Delta_0 \neq 0$. By Hamilton's principle, or otherwise, find the equation of motion of the mass. Integrate the equation of motion to determine the motion of the mass in the case in which the lateral spring is not there ($K = 0, \alpha = 0$).

**11.7.** Let the load $P$ of Probs. 10.11 and 11.6 be an oscillatory load,

$$P = P_0 + P_1 \cos \omega t.$$

Could the column become unstable (i.e., with the amplitude of motion unbounded as $t \to \infty$)? In particular, if $P_0 = P_{cr}$, the buckling load, is there a chance that by properly selecting $P_1$ and $\omega$, the column remains stable? Discuss the case of no lateral spring ($K = 0$, $\alpha = 0$) in greater detail.

**11.8.** Consider longitudinal wave propagation in a slender rod which in various segments is made of different materials. It is desired to transmit the wave through the rod with as little distortion as possible. How should the material constants be matched? (This is a practical problem which often occurs in instrumentation, such as piezoelectric sensing, etc.) *Hint:* Consider transmission of harmonic waves without reflection at the interfaces between segments. If quantities pertaining to the $n$th segment are indicated by a subscript $n$, show that we must have $\rho_n c_n = \text{const.}$ for all $n$ or, in another form, $E_n \rho_n = \text{const.}$

**11.9.** Consider a beam on an elastic foundation. A traveling load moves at a constant speed over the beam. Determine the steady-state elastic responses of the beam.

**11.10.** Consider the longitudinal vibrations of a slender elastic rod of variable cross-sectional area $A(x)$. Determine the speed of propagation of longitudinal waves, assuming that the rod is so slender that the variation of the elastic displacement over the cross section is negligible. If one end of the beam is subjected to a harmonic longitudinal oscillation with the boundary condition $(u)_{x=0} = ae^{i\omega t}$, whereas the other end is free, determine the beam response for the following cases:

(1) $$A(x) = \text{const.},$$

(2) $$A(x) = A_0 x^2, \qquad (0 < a \leqslant x \leqslant b),$$

(3) $$A(x) = A_0 e^{2x/L}, \qquad (0 \leqslant x \leqslant L).$$

# 12

# ELASTICITY AND THERMODYNAMICS

In this chapter we shall consider what restrictions classical thermodynamics imposes on the theory of solids and what information concerning solid continua can be deduced from thermodynamics. We shall discuss also the question of existence and positive definiteness of the strain energy function which is of central importance in the theory of elasticity.

## 12.1. THE LAWS OF THERMODYNAMICS

A brief summary of the basic structure of the classical theory of thermodynamics is given in this section. We call a particular collection of matter which is being studied a *system*. Only *closed* systems, i.e., systems which do not exchange matter with their surroundings, are considered here. Occasionally, a further restriction is made that no interactions between the system and its surroundings occur; the system is then said to be *isolated*.

Consider a given system. When all the information required for a complete characterization of the system for a purpose at hand is available, it will be said that the *state* of the system is known. For example, for a certain homogeneous elastic body at rest, a complete description of its thermodynamic state requires a specification of its material content, i.e., the quantity of each chemical substance contained; its geometry in the natural or unstrained state; its deviation from the natural state, or strain field; its stress field; and, if some physical properties depend on whether the body is hot or cold, one extra independent quantity which fixes the degree of hotness or coldness. These quantities are called *state variables*. If a certain state variable can be expressed as a single-valued function of a set of other state variables, then the functional relationship is said to be an *equation of state*, and the variable so described is called a *state function*. The selection of a particular set of independent state variables is important in each problem, but the choice is to a certain extent arbitrary.

If, for a given system, the values of the state variables are independent of time, the system is said to be in *thermodynamic equilibrium*. If the state variables vary with time, then the system is said to undergo a *process*. The number of state variables required to describe a process may be larger than that required to describe the system at thermodynamic equilibrium. For example, in describing the flow of a fluid we may need to know the viscosity.

The boundary or *wall* separating two systems is said to be *insulating* if it has the following property: If any system in complete internal equilibrium is completely surrounded by an insulating wall, then no change can be produced in the system by an external agency except by movement of the wall or by long-range forces such as gravitation. A system surrounded by an insulating wall is said to be *thermally insulated*, and any process taking place in the system is called *adiabatic*.

A system is said to be *homogeneous* if the state variables does not depend on space coordinates. The classical thermodynamics is concerned with the conditions of equilibrium within a homogeneous system or in a heterogeneous system which is composed of homogeneous parts ("phases"). A generalization to nonhomogeneous systems requires certain additional hypotheses which will be considered in Sec. 13.1. *In this chapter we shall restrict our attention to homogeneous systems.*

The first step in the formulation of thermodynamics is to introduce the concept of temperature. It is postulated that *if two systems are each in thermal equilibrium with a third system, they are in thermal equilibrium with each other.* From this it can be shown to follow that the condition of thermal equilibrium between several systems is the equality of a certain single-valued function of the thermodynamic states of the systems, which may be called the temperature $\mathcal{T}$: any one of the systems being used as a "thermometer" reading the temperature $\mathcal{T}$ on a suitable scale. The temperature whose existence is thus postulated is measured on a scale which is determined by the arbitrary choice of the thermometer and is called the *empirical temperature.*

The *first law of thermodynamics* can be formulated as follows. *If a thermally insulated system can be taken from a state I to a state II by alternative paths, the work done on the system has the same value for every such (adiabatic) path.* From this we can deduce that there exists a single-valued function of the state of a system, called its *energy*, such that for any adiabatic process the increase of the energy is equal to the work done on the system. Thus,

$$\Delta \text{ energy} = \text{work done} \qquad \text{(adiabatic process).} \tag{1}$$

It is to be noticed that for this definition of energy it is necessary and sufficient that it be possible by an adiabatic process to change the system either from state I to state II or from state II to state I.

We now define the heat $Q$ absorbed by a system as the increase in energy of the system, less the work done on the system. Thus,

$$Q = \Delta \text{ energy} - \text{work done} \qquad \text{(all processes),} \tag{2}$$

or

$$\Delta \text{ energy} = Q + \text{work done.} \tag{3}$$

If this is regarded as a statement of the conservation of energy and is compared with (1), we observe that the energy of a system can be increased either by work done on it or by absorption of heat.

It is customary to identify several types of energy which make up the total: the kinetic energy $K$, the gravitational energy $G$, and the internal energy $E$. Thus, the equation

$$\text{energy} = K + G + E \tag{4}$$

may be regarded as the definition of the internal energy.

We now formulate the *second law of thermodynamics for a homogeneous system* as follows.

*There exist two single-valued functions of state, $T$, called the absolute temperature, and $S$, called the entropy, with the following properties:*

I. *$T$ is a positive number which is a function of the empirical temperature $\mathscr{T}$ only.*

II. *The entropy of a system is equal to the sum of entropies of its parts.*

III. *The entropy of a system can change in two distinct ways, namely, by interaction with the surroundings and by changes taking place inside the system. Symbolically, we may write*

$$dS = d_eS + d_iS, \tag{5}$$

*where $dS$ denotes the increase of entropy of the system, $d_eS$ denotes the part of this increase due to interaction with the surroundings, and $d_iS$ denotes the part of this increase due to changes taking place inside the system. Then, if $dQ$ denotes the heat absorbed by the system from its surroundings, we have*

$$d_eS = \frac{dQ}{T}. \tag{6}$$

*The change $d_iS$ is never negative. If $d_iS$ is zero the process is said to be reversible. If $d_iS$ is positive the process is said to be irreversible:*

$$\begin{aligned} d_iS &> 0 \qquad \textit{(irreversible process)}, \\ d_iS &= 0 \qquad \textit{(reversible processes)}. \end{aligned} \tag{7}$$

*The remaining case, $d_iS < 0$, never occurs in nature.*

The absolute temperature $T$ and the entropy $S$ are two fundamental quantities. We will not attempt to define them in terms of other quantities regarded as simpler. They are defined merely by their properties expressed in the second law. Lord Kelvin has shown how it is possible to calibrate any thermometer into absolute temperature scale. (See Prob. 12.3.) The scale of the absolute temperature is fixed by defining the temperature at equilibrium between liquid water and ice at a pressure of 1 atm at 273.16°K, i.e., 273.16 degrees on Kelvin's scale.

These postulates form the basis of the classical thermodynamics. The justification of these postulates is the empirical fact that all conclusions derived from these assumptions are without exception in agreement with the experimentally observed behavior of systems in nature at the macroscopic scale. The form in which these principles are enunciated above is essentially that used by Max Born[12.1] (1921) and I. Prigogine[13.1] (1947).

It is important to familiarize ourselves with the concept that entropy is an attribute to a material body, just as its mass or its electric charges are. A pound of oxygen has a definite amount of entropy, which can be changed by changing the temperature and specific volume of the gas. A pound of steel at a given temperature and in a given state of strain has a definite amount of entropy. It is instructive to consider Problems 12.5 and 12.6 to obtain some intuitive feeling about entropy.

## PROBLEMS

**12.1.** Calculate the work performed by 10 g of hydrogen expanding isothermally at 20°C from 1 to 0.5 atm of pressure. *Note:* The equation of state for a system composed of $m$ grams of a gas whose molecular weight is $M$ is given approximately by $pv = (m/M)RT$, where $R$ is a universal constant (same for all gases); $R = 8.314 \times 10^7$ ergs/degree, or 1.986 cal/degrees.)

**12.2.** Calculate the energy variation of a system which performs $8 \times 10^8$ ergs of work and absorbs 60 cal of heat. (*Note:* 1 cal $= 4.185 \times 10^7$ ergs.)

**12.3.** To calibrate an empirical temperature scale against the absolute temperature scale, let us consider a heat engine. A working substance is assumed which undergoes a cycle of operations involving changes of temperature, volume, and pressure and which is eventually brought back to its original condition. In a Carnot cycle, an ideal gas is first brought to contact with a large heat source which is maintained at a constant empirical temperature $\mathscr{T}_1$. The gas at the equilibrium temperature $\mathscr{T}_1$ is allowed to expand (isothermally) and absorbs an amount of heat $Q_1$ from the heat source. Then it is removed and insulated from the heat source and is allowed to expand adiabatically to perform work, lowering its empirical temperature to a value $\mathscr{T}_2$. Next it is put in contact with a heat sink of temperature $\mathscr{T}_2$ and is compressed isothermally to release an amount of heat $Q_2$. Finally, it is recompressed adiabatically back to its original state. If every process in the Carnot cycle is reversible, the engine is said to be reversible, otherwise it is irreversible.

From the second law of thermodynamics prove that:

(a) When the temperatures of the source and sink are given, a reversible engine is the most efficient type of engine, i.e., it yields the maximum amount of mechanical work for a given abstraction of heat from the source. *Note:* Efficiency $= (Q_1 - Q_2)/Q_1$.

(b) The efficiency of a reversible engine is independent of the construction or material of the engine and depends only on the temperatures of the source and sink.

(c) In a reversible Carnot engine, we must have

$$\frac{Q_2}{Q_1} = f(\mathscr{T}_1, \mathscr{T}_2),$$

where $f(\mathscr{T}_1, \mathscr{T}_2)$ is a universal function of the two temperatures $\mathscr{T}_1$ and $\mathscr{T}_2$. Show that the function $f(\mathscr{T}_1, \mathscr{T}_2)$ has the property

$$f(\mathscr{T}_1, \mathscr{T}_2) = \frac{f(\mathscr{T}_0, \mathscr{T}_2)}{f(\mathscr{T}_0, \mathscr{T}_1)},$$

where $\mathscr{T}_0$, $\mathscr{T}_1$, and $\mathscr{T}_2$ are three arbitrary temperatures. If $\mathscr{T}_0$ is kept constant, we may consider $f(\mathscr{T}_0, \mathscr{T})$ as being a function of the temperature $\mathscr{T}$ only. Hence, we may write

$$Kf(\mathscr{T}_0, \mathscr{T}) = \theta(\mathscr{T}),$$

where $K$ is a constant. Thus,

$$\frac{Q_1}{\theta(\mathscr{T}_1)} = \frac{Q_2}{\theta(\mathscr{T}_2)}.$$

Lord Kelvin perceived (1848) that the numbers $\theta(\mathscr{T}_1)$, $\theta(\mathscr{T}_2)$, may be taken to define an absolute temperature scale. Identify this with the $T$ introduced above.

(d) It is suggested that the reader review the theory of the Joule-Thomson experiment, which is used in practice to calibrate a gas thermometer to absolute temperature scale.

**12.4.** By considering an ideal gas as the working fluid of a Carnot cycle, show that the gas thermometer reads exactly the absolute temperature.

**12.5.** Derive the following formula for the entropy of one mole of an ideal gas.

$$\mathscr{S} = C_v \log T + R \log V + a,$$

where $C_v$ is the specific heat at constant volume, $R$ is the universal gas constant, and $a$ is a constant of integration.

**12.6.** Prove the following analogy. The work that one can derive from a waterfall is proportional to the product of the height of the waterfall and the quantity of the water flow. The work that can be derived from a heat engine performing a Carnot cycle is proportional to the product of the temperature difference between the heat source and the heat sink, and the quantity of entropy flow.

## 12.2. THE ENERGY EQUATION

We shall now express the equation of balance of energy, (12.1:3) and (12.1:4), in a form more convenient for continuum mechanics. As an exception in the present chapter, this section is not limited to homogeneous systems.

For a body of particles in a configuration occupying a region $V$ bounded by a surface $S$, the *kinetic energy* is

$$K = \int_V \tfrac{1}{2}\rho v_i v_i \, dv, \tag{1}$$

where $v_i$ are the components of the velocity vector of a particle occupying an element of volume $dv$, and $\rho$ is the density of the material. The *internal energy* is written in the form

$$E = \int_V \rho \mathscr{E} \, dv, \tag{2}$$

where $\mathscr{E}$ is the *internal energy per unit mass.*† The heat input into the body must be imparted through the boundary. A new vector, the *heat flux* $\mathbf{h}$, with components $h_1, h_2, h_3$, is defined as follows. Let $dS$ be a surface element in the body, with unit outer normal $\nu_i$. Then the rate at which heat is transmitted across the surface $dS$ in the direction of $\nu_i$ is assumed to be representable as $h_i \nu_i \, dS$. To be specific, we insist on defining the heat flux in the case of a moving medium that the surface element $dS$ be composed of the same particles. The rate of heat input is therefore

$$\dot{Q} = -\int_S h_i \nu_i \, dS = -\int_V h_{j,j} \, dv \tag{3}$$

The rate at which work is done on the body by the body force per unit volume $F_i$ in $V$ and surface traction $\overset{\nu}{T}_i$ in $S$ is the *power*

$$\begin{aligned} P &= \int F_i v_i \, dv + \int \overset{\nu}{T}_i v_i \, dS \\ &= \int F_i v_i \, dv + \int \sigma_{ij} \nu_j v_i \, dS \\ &= \int F_i v_i \, dv + \int (\sigma_{ij} v_i)_{,j} \, dv. \end{aligned} \tag{4}$$

The first law states that

$$\dot{K} + \dot{E} = \dot{Q} + P, \tag{5}$$

where the dot denotes the material derivative $D/Dt$.

Formula for computing the material derivative of an integral has been given in Eq. (5.3:4). A little calculation yields

$$\begin{aligned} \frac{1}{2}\rho \frac{Dv^2}{Dt} + \frac{v^2}{2}\frac{D\rho}{Dt} + \frac{v^2}{2}\rho \operatorname{div} \mathbf{v} + \rho \frac{D\mathscr{E}}{Dt} + \mathscr{E}\frac{D\rho}{Dt} + \mathscr{E}\rho \operatorname{div} \mathbf{v} \\ = -h_{j,j} + F_i v_i + \sigma_{ij,j} v_i + \sigma_{ij} v_{i,j}. \end{aligned} \tag{6}$$

† It is interesting to verify that if one chooses to use the internal energy *per unit volume* $\mathscr{E}_v$ instead of $\mathscr{E}$ per unit mass, so that $E = \int \mathscr{E}_v \, dv$ then the energy equation, corresponding to Eq. (7) infra, will be somewhat more complicated.

On substituting the equation of continuity and the equations of motion

$$\frac{D\rho}{Dt} + \rho \operatorname{div} \mathbf{v} = 0, \qquad \rho \frac{Dv_i}{Dt} = F_i + \sigma_{ij,j},$$

respectively, into Eq. (6), we obtain

(7) ▲
$$\rho \frac{D\mathscr{E}}{Dt} = -\frac{\partial h_j}{\partial x_j} + \sigma_{ij} V_{ij}$$

where

(8)
$$V_{ij} = \frac{1}{2}\left(\frac{\partial v_i}{\partial x_j} + \frac{\partial v_j}{\partial x_i}\right)$$

is the symmetric part of the tensor $v_{i,j}$, called the *rate of deformation tensor.* The antisymmetric part of $v_{i,j}$ contributes nothing to the sum $\sigma_{ij}v_{i,j}$ since $\sigma_{ij}$ is symmetric.

In classical thermodynamics we are concerned with the small neighborhood of thermodynamic equilibrium. It suffices in the present chapter to consider infinitesimal strain, which is imposed very slowly. In this case we may write Eq. (7) in a form commonly used in thermodynamics books:

(9)
$$\rho\, d\mathscr{E} = dQ + \sigma_{ij}\, de_{ij}$$

Our Eq. (7) gives precise meaning to this Eq. (9). If there is no internal entropy production in the process, so that $d_i\mathscr{S} = 0$, then the second law gives $dQ = T\rho\, d\mathscr{S}$, where $\mathscr{S}$ denotes the *specific entropy*, or the *entropy per unit mass*. Hence,

(10) ▲
$$d\mathscr{E} = T\, d\mathscr{S} + \frac{1}{\rho}\sigma_{ij}\, de_{ij}$$

**Problem 12.7.** In the case of a fluid under hydrostatic pressure, so that

$$\sigma_{11} = \sigma_{22} = \sigma_{33} = -p, \qquad \sigma_{12} = \sigma_{23} = \sigma_{31} = 0,$$

show that

(11)
$$d\mathscr{E} = T\, d\mathscr{S} - p\, d\mathscr{V},$$

where $\mathscr{V} = 1/\rho$ is the specific volume, i.e., the volume per unit mass of the fluid.

## 12.3. THE STRAIN ENERGY FUNCTION

In the theory of elasticity, it is often desired to define a function $W(e_{11}, e_{12}, \ldots)$ of the strain components $e_{ij}$, with the property that†

(1)
$$\frac{\partial W}{\partial e_{ij}} = \sigma_{ij} \qquad (i, j = 1, 2, 3).$$

† We limit our attention to infinitesimal deformations in this chapter. Then $e_{ij}$ is the infinitesimal strain tensor. For finite deformations, see Chap. 16, especially Secs. 16.6 and 16.7.

Such a function $W$ is called the *strain energy function.* We shall now identify $W$ with the internal energy in an isentropic process and the free energy in an isothermal process.

By definition, the state of stress in an elastic body is a unique function of the strain, and vice versa. Hence, it is sufficient to choose one of these two tensors, $e_{ij}$ and $\sigma_{ij}$, as an independent state variable. Now, Eq. (12.2:10),

$$d\mathscr{E} = \frac{1}{\rho}\,\sigma_{ij}\,de_{ij} + T\,d\mathscr{S}, \tag{2a}$$

shows that $\mathscr{E}$ is a state function, $\mathscr{E}(e_{ij}, \mathscr{S})$, of the strain $e_{ij}$ and the entropy per unit mass $\mathscr{S}$. By ordinary rules of differentiation, we have

$$d\mathscr{E} = \left(\frac{\partial \mathscr{E}}{\partial \mathscr{S}}\right)_{e_{ij}} d\mathscr{S} + \left(\frac{\partial \mathscr{E}}{\partial e_{ij}}\right)_{\mathscr{S}} de_{ij}. \tag{2b}$$

On comparing (2a) and (2b), we see that

$$\rho\left(\frac{\partial \mathscr{E}}{\partial e_{ij}}\right)_{\mathscr{S}} = \sigma_{ij}, \qquad \left(\frac{\partial \mathscr{E}}{\partial \mathscr{S}}\right)_{e_{ij}} = T. \tag{3}$$

Since in infinitesimal deformations the density $\rho$ remains constant, Eq. (3) states that in a reversible adiabatic process there exists a scalar function $\rho\mathscr{E}$ whose partial derivative with respect to a strain component gives the corresponding stress component.†

On the other hand, if the process is isothermal ($T =$ const.), we introduce the Helmholtz' free energy function per unit mass (also called *Gibbs' work function*) $\mathscr{F}$:

$$\mathscr{F} \equiv \mathscr{E} - T\mathscr{S} \tag{4}$$

Then, from Eqs. (4) and (2a), we obtain

$$d\mathscr{F} = -\mathscr{S}\,dT + \frac{1}{\rho}\,\sigma_{ij}\,de_{ij}, \tag{5}$$

whence

$$\rho\left(\frac{\partial \mathscr{F}}{\partial e_{ij}}\right)_{T} = \sigma_{ij}, \qquad \left(\frac{\partial \mathscr{F}}{\partial T}\right)_{e_{ij}} = -\mathscr{S}. \tag{6}$$

Hence, in an isothermal process there also exists a scalar function, $\rho\mathscr{F}$, whose partial derivative with respect to a strain component gives the corresponding stress component in an elastic body.

Comparing Eq. (1) with (3) or (6), we see that $W$ can be identified with $\rho\mathscr{E}$ or $\rho\mathscr{F}$, depending on whether an isentropic or an isothermal condition is being considered.

† In the corresponding equation (16.7:1) for finite deformation, the density $\rho_0$ is the uniform density at the natural state of the material.

A counterpart to Eq. (6), namely,

$$\rho\left(\frac{\partial \Phi}{\partial \sigma_{ij}}\right)_T = -e_{ij}, \tag{7}$$

is obtained from the *Gibbs thermodynamic potential* $\Phi$, which is defined as

$$\Phi = \mathscr{E} - T\mathscr{S} - \frac{1}{\rho}\sigma_{ij}e_{ij} = \mathscr{F} - \frac{1}{\rho}\sigma_{ij}e_{ij}. \tag{8}$$

For, on differentiating (8) and using (5), we have

$$d\Phi = -\mathscr{S}\,dT - \frac{1}{\rho}e_{ij}\,d\sigma_{ij}, \tag{9}$$

from which (7) follows. Thus, $-\rho\Phi$ may be identified with the complementary energy function $W_c$, which has the property that

$$\frac{\partial W_c}{\partial \sigma_{ij}} = e_{ij}. \tag{10}$$

We have identified the strain energy function for two well-known thermodynamic processes—adiabatic and isothermal. For these two processes, no explicit display of temperature in the function $W$ is necessary. In other words, the stress-strain relation can be written without reference to temperature. In other thermomechanical processes the situation is not so simple; explicit display of temperature is required. In general, $W$ is a function of $e_{ij}$ and $T$. A simple example is discussed in Sec. 12.7.

## 12.4. THE CONDITIONS OF THERMODYNAMIC EQUILIBRIUM

The second law of thermodynamics connects the theory with nature in the statement that the case $d_i\mathscr{S} < 0$ never occurs in nature. It states a direction of motion of events in the world. According to the proverb, "Time is an arrow;" the second law endows the arrow to time.

Thus, the second law indicates a trend in nature. What is the end point of this trend? It is thermodynamic equilibrium. In Sec. 12.1 we have defined thermodynamic equilibrium as a situation in which no state variable varies with time. For a system to be in thermodynamic equilibrium no change in boundary conditions is permissible and no spontaneous process which is consistent with the boundary conditions will occur in it.

To derive the necessary and sufficient conditions for thermodynamic equilibrium, we compare the system in equilibrium with those neighboring systems whose state variables are slightly different from those of the equilibrium state, but all subjected to the same boundary conditions. This comparison is to be made in the sense of variational calculus. We shall write

$\delta T$, $\delta\mathscr{S}$, etc., to denote the first-order infinitesimals, and $\Delta T$, $\Delta\mathscr{S}$, etc., to denote the variations including the first-, second-, and higher-order infinitesimal terms. Thus, if we consider a homogeneous system $A$ and a neighboring state $B$, we have the following variables:

| *System A* | *System B* |
| --- | --- |
| *In equilibrium* | *Neighboring to A* |
| $\mathscr{E}, e_{ij}, \sigma_{ij}, \mathscr{S}, T$ | $\mathscr{E} + \Delta\mathscr{E}, e_{ij} + \Delta e_{ij}, \sigma_{ij} + \Delta\sigma_{ij}$ |
| | $\mathscr{S} + \Delta\mathscr{S}, \quad T + \Delta T$ |

We shall require that the first law be satisfied so that all variations must be subjected to the restriction

$$(1) \qquad \Delta\mathscr{E} = \Delta Q + \frac{1}{\rho}(\sigma_{ij} + \Delta\sigma_{ij})\,\Delta e_{ij}, \qquad \delta\mathscr{E} = \delta Q + \frac{1}{\rho}\,\sigma_{ij}\,\delta e_{ij}.$$

Let us consider the following particular case. We require the system $B$ to have the same energy as $A$ and that the boundaries are rigid so that no work can be done; thus $\Delta\mathscr{E} = 0$, $\Delta e_{ij} = 0$. It follows from (1) that $\Delta Q = 0$ too. Now system $B$ has the entropy $\mathscr{S} + \Delta\mathscr{S}$. If any spontaneous changes can occur in $B$ at all, it would occur in the direction of increasing entropy. Therefore, if $\Delta\mathscr{S} < 0$, we must say that the second law would permit system $B$ to change itself into the state of $A$. If $\Delta\mathscr{S} > 0$, no such change is permitted. The second law does not say that if $\Delta\mathscr{S} < 0$ the change must occur. But if *we make an additional assumption*—as an addition to the second law—*that which is permitted by the second law will actually occur in nature*, then $\Delta\mathscr{S} < 0$ necessarily implies that $B$ will change itself into $A$. On the other hand, if $\Delta\mathscr{S} < 0$ the second law prohibits $A$ from changing into $B$. Therefore, *if energy variation is prohibited the necessary and sufficient condition for the system A to be in the state of thermodynamic equilibrium is*

$$(2) \ \blacktriangle \qquad (\Delta\mathscr{S})_{\mathscr{E}} < 0.$$

In other words, the entropy $\mathscr{S}$ of the system $A$ is the maximum with respect to all the neighboring states which have the same energy. Hence, the variational equation

$$(3) \ \blacktriangle \qquad \mathscr{S} = \max. \qquad (\mathscr{E} = \text{const.}, \ e_{ij} = \text{const.})$$

for thermodynamic equilibrium. If we restrict ourselves to the first-order infinitesimals, the condition for thermodynamic equilibrium may be written

$$(4) \ \blacktriangle \qquad (\delta\mathscr{S})_{\mathscr{E}} \leqslant 0.$$

This is the celebrated condition of equilibrium of Willard Gibbs. Gibbs gave an alternative statement, which is more directly applicable to continuum mechanics: *For the equilibrium of any isolated system it is necessary and sufficient that in all possible variations in the state of the system which do not alter its entropy, the variation of its energy shall either vanish or be positive:*

$$(5) \quad \blacktriangle \qquad (\Delta\mathscr{E})_{\mathscr{S}} > 0, \qquad (\delta\mathscr{E})_{\mathscr{S}} \geqslant 0.$$

In other words, *the internal energy shall be a minimum with respect to all neighboring states which have the same entropy.*

Gibbs proves the equivalence of conditions (2) and (5) from the consideration that it is always possible to increase both the energy and the entropy of a system, or to decrease both together, viz., by imparting heat to any part of the system or by taking it away. Now, if the condition (2) is not satisfied, there must be some neighboring systems for which

$$\Delta\mathscr{S} > 0 \quad \text{and} \quad \Delta\mathscr{E} = 0.$$

Then, by diminishing both the energy and the entropy of the system in this varied state, we shall obtain a state for which

$$\Delta\mathscr{S} = 0 \quad \text{and} \quad \Delta\mathscr{E} < 0;$$

therefore, (5) is not satisfied. Conversely, if condition (5) is not satisfied, there must be a variation in the state of the system for which

$$\Delta\mathscr{E} < 0 \quad \text{and} \quad \Delta\mathscr{S} = 0.$$

Hence, there must also be one for which

$$\Delta\mathscr{E} = 0 \quad \text{and} \quad \Delta\mathscr{S} > 0.$$

Therefore, condition (2) is not satisfied. Taken together, the equivalence of (2) and (5) is established.

The derivation of Gibbs' conditions for thermodynamic equilibrium requires a little more than the first and second laws of thermodynamics. The conditions also guarantee more than equilibrium in the ordinary sense—they guarantee *stable* equilibrium, stable in the sense that a neighboring disturbed state will actually tend to return to the equilibrium state.

## 12.5. THE POSITIVE DEFINITENESS OF THE STRAIN ENERGY FUNCTION

A piece of steel can exist for a long time without any spontaneous changes taking place in it. Thus, it is in a stable thermodynamic equilibrium condition, or said to be in a *natural state*. When we strain a piece of steel with an application of forces, the strained body is in a state disturbed from the natural one. If the metal remains elastic, the strained body will return to its natural state upon removal of the applied forces. This tendency to return to the natural state endows a property to a strain energy function which we shall now investigate.

Since the natural state of a solid body is a stable equilibrium state, it follows from Eq. (12.4:5) that at constant entropy the strain energy must be a minimum. Consider now a strained solid. Let the internal energy of a strained body be $\mathscr{E}$ and that of an unstrained body be $\mathscr{E}_0$. According to Gibbs' theorem, therefore, the stability of the unstrained state implies that the difference $\Delta\mathscr{E} = \mathscr{E} - \mathscr{E}_0$ is positive, and it vanishes only in the trivial case of the natural state itself. Therefore, in isentropic processes

$$\Delta\mathscr{E} = \mathscr{E} - \mathscr{E}_0 \geqslant 0 \tag{1}$$

in the neighborhood of the natural state; the equality sign holds only in the natural state. In other words, $\mathscr{E} - \mathscr{E}_0$ is positive definite.

In the case of an isothermal process, it can be deduced from the Helmholtz free energy function that $\mathscr{F}$ attains its minimum $\mathscr{F}_0$ at thermodynamic equilibrium. A similar consideration as above shows that $\Delta\mathscr{F} = \mathscr{F} - \mathscr{F}_0$ is positive definite in the neighborhood of the natural state in the isothermal condition.

In either the adiabatic or the isothermal condition, the strain energy function $W$ can be identified as $\rho(\mathscr{E} - \mathscr{E}_0)$ or $\rho(\mathscr{F} - \mathscr{F}_0)$ respectively. Therefore, since $\rho > 0$, the strain energy function $W$ is positive definite in the neighborhood of the natural state.

From the positive definiteness of the strain energy function, the following important theorems are deduced.

1. The uniqueness of the solution in elastostatics and dynamics. (Sec. 7.4.)
2. The minimum potential energy theorem. (Sec. 10.7.)
3. The minimum complementary energy theorem. (Sec. 10.9.)
4. Saint-Venant's principle (in a certain sense). (Sec. 10.11.)

## 12.6. THERMODYNAMIC RESTRICTIONS ON THE STRESS-STRAIN LAW OF AN ISOTROPIC ELASTIC MATERIAL

We shall consider those restrictions imposed by thermodynamics on the stress-strain law proposed in Sec. 6.2, which is supposed to hold in isentropic or isothermal conditions.

By definition of the strain energy function, $dW = \sigma_{ij}\, de_{ij}$. If we substitute the stress-strain law Eq. (6.2:7) into this equation, we have

$$\begin{aligned} dW &= (\lambda e\, \delta_{ij} + 2Ge_{ij})\, de_{ij} \\ &= \lambda e\, de + 2Ge_{ij}\, de_{ij}, \end{aligned} \qquad e = e_{\alpha\alpha}. \tag{1}$$

The elastic constants $\lambda$ and $G$ differ slightly in adiabatic and isothermal conditions (Sec. 14.3), but this difference is of no concern here. On integrating Eq. (1) from the natural to the strained state, we have

$$W = \int_0^e \lambda e \, de + \int_0^{e_{ij}} 2Ge_{ij} \, de_{ij}; \tag{2}$$

i.e.,

$$W = \frac{\lambda e^2}{2} + Ge_{ij}e_{ij}. \tag{3}$$

This function is required to be positive definite from thermodynamic considerations. Since all terms appear as squares, an obvious set of sufficient conditions are

$$\lambda > 0, \qquad G > 0. \tag{4}$$

But this condition is not necessary because the term $e_{ij}e_{ij}$ is not independent of $e$. Therefore, let us introduce the strain deviation $e'_{ij} = e_{ij} - \frac{1}{3}e\,\delta_{ij}$, whose first invariant vanishes, and rewrite (3) in the form

$$W = \tfrac{1}{2}(\lambda + \tfrac{2}{3}G)e^2 + G(e_{ij} - \tfrac{1}{3}e\,\delta_{ij})(e_{ij} - \tfrac{1}{3}e\,\delta_{ij}), \tag{5}$$

On recognizing that $\lambda + \frac{2}{3}G$ is the bulk modulus $K$ and $\frac{1}{2}(e_{ij} - \frac{1}{3}e\,\delta_{ij})^2$ is the second invariant of the strain deviation $e'_{ij}$,

$$J_2 = \tfrac{1}{2}e'_{ij}e'_{ij} \tag{6}$$

we see that Eq. (5) may be written as

$$W = \tfrac{1}{2}Ke^2 + 2GJ_2. \tag{7}$$

The two groups of square terms are now independent of each other. Hence, the necessary and sufficient condition for $W$ to be positive definite is

$$K = \lambda + \tfrac{2}{3}G > 0, \qquad G > 0. \tag{8}$$

Since $K$ and $G$ are related to Young's modulus $E$ and Poisson's ratio $\nu$ by

$$K = \frac{E}{3(1 - 2\nu)}, \qquad G = \frac{E}{2(1 + \nu)}, \tag{9}$$

an equivalent set of conditions are

$$E > 0, \qquad -1 < \nu < \tfrac{1}{2}. \tag{10}$$

In reality, materials with negative values of $\nu$ are unknown. Common steels have $\nu$ about 0.25, and aluminum alloys have $\nu$ about 0.32; but the value of $\nu$ for beryllium is near zero (0.01–0.06) and for lead is near 0.45.

## 12.7. GENERALIZED HOOKE'S LAW, INCLUDING THE EFFECT OF THERMAL EXPANSION

We shall now consider the stress-strain law in a general thermal environment, not limited to isothermal or isentropic conditions. Let us assume that the strain energy function $W$ can be expanded in a power series

(1) $$2W = C_0 + C_{ij}e_{ij} + C'_{ijkl}e_{ij}e_{kl} + \dots,$$

where the coefficients $C_0$, $C_{ij}$, etc., are functions of the entropy $\mathscr{S}$ and the temperature $T$. Since $W = 0$ when $e_{ij} = 0$ by definition, the constant $C_0$ must vanish. As to $C_{ij}$, we note that since

$$\sigma_{ij} = \frac{\partial W}{\partial e_{ij}} = \frac{C_{ij}}{2} + 0(e_{ij}),$$

$C_{ij}$ means twice the value of the stress component $\sigma_{ij}$ at zero strain. Such stresses at zero strain can be brought about by a linear expansion if there are temperature changes from a standard state. The simplest assumption is to let $C_{ij}$ be linearly proportional to the temperature change:

$$C_{ij} = -2\beta_{ij}(T - T_0),$$

where $\beta_{ij}$ are numerical constants. Hence, we have, when $e_{ij}$ are small and terms higher than the second order are neglected,

(2) ▲ $$W = -\beta_{ij}(T - T_0)e_{ij} + \tfrac{1}{2}C'_{ijkl}e_{ij}e_{kl}.$$

The stress components are, therefore,

(3) $$\sigma_{ij} = \frac{\partial W}{\partial e_{ij}} = \frac{1}{2}(C'_{ijkl} + C'_{klij})e_{kl} - \beta_{ij}(T - T_0).$$

This is the general *Duhamel-Neumann form of Hooke's law*. If the quadratic form in Eq. (2) is symmetrized in advance, we can write (3) in the form

(4) ▲ $$\sigma_{ij} = C_{ijkl}e_{kl} - \beta_{ij}(T - T_0),$$

where

(5) $$C_{ijkl} = C_{klij}$$

is a tensor of rank 4. Since $e_{ij}$ and $\sigma_{ij}$ are symmetric tensors, we also have

(6) $$C_{ijkl} = C_{jikl}, \qquad C_{ijkl} = C_{ijlk}, \qquad \beta_{ij} = \beta_{ji}.$$

Thus, for an anisotropic medium there are 21 independent elastic constants $C_{ijkl}$ and six independent constants $\beta_{ij}$. This number is reduced to two $C$'s and one $\beta$ if the medium is isotropic. The relationship for an isotropic material is

(7) ▲ $$\sigma_{ij} = \lambda e\, \delta_{ij} + 2Ge_{ij} - \beta(T - T_0)\, \delta_{ij},$$

or

(8) ▲ $$e_{ij} = -\frac{\nu}{E} s\, \delta_{ij} + \frac{1+\nu}{E} \sigma_{ij} + \alpha(T - T_0)\, \delta_{ij},$$

where

(9) $$e = e_{11} + e_{22} + e_{33}, \qquad s = \sigma_{11} + \sigma_{22} + \sigma_{33},$$

(10) $$\lambda = \frac{E\nu}{(1+\nu)(1-2\nu)}, \qquad \beta = \frac{E\alpha}{1-2\nu}.$$

The corresponding strain energy function is

(11) ▲ $$\begin{aligned} W &= Ge_{ij}e_{ij} + \frac{\lambda}{2} e^2 - \beta(T - T_0)e \\ &= G\left[ e_{ij}e_{ij} + \frac{\nu}{1-2\nu} e^2 - \frac{2(1+\nu)}{1-2\nu} \alpha(T - T_0)e \right]. \\ &= \frac{1}{4G}\left[ \sigma_{ij}\sigma_{ij} - \frac{\nu}{1+\nu} s^2 \right] + \alpha(T - T_0)s. \end{aligned}$$

Here $\alpha$ is the linear coefficient of thermal expansion, $T_0$ the temperature of a reference state of the body, $T - T_0$ the rise of temperature above the reference state.

The strain energy function must be a relative minimum at the natural state (see Sec. 12.4), which implies that, when $T = T_0$, the quadratic form

(12) $$W = \tfrac{1}{2} C_{ijkl} e_{ij} e_{kl}$$

must be positive definite.

## PROBLEMS

**12.8.** State the conditions which must be satisfied by the constants $C_{ijkl}$ in order that the quadratic form (12) be positive definite.

**12.9.** Let $T = T_0$, and derive Clapeyron's formula for a linear elastic material: $W = \frac{1}{2}\sigma_{ij}e_{ij}$.

**12.10.** In a conventional structural analysis, honeycomb sandwich material is often treated as a homogeneous orthotropic material, provided that the dimensions of the structure are much larger than the dimensions of the individual honeycomb cell. Suggest a stress-strain law for an orthotropic material. Determine explicitly the limitations imposed on the elastic constants that appear in the stress-strain law by the first and second laws of thermodynamics.

## 12.8. THERMODYNAMIC FUNCTIONS FOR ISOTROPIC HOOKEAN MATERIALS

It will be interesting to record the thermodynamic functions for an isotropic elastic material. Restricted to infinitesimal strains, the derivation of the formulas is quite simple. Consider a perfectly elastic, isotropic body which obeys the Duhamel-Neumann generalization of Hooke's law, Eq. (12.7:7). In a reversible, infinitesimal deformation, we have

(1) $$d\mathscr{F} = -\mathscr{S}\,dT + \frac{1}{\rho}\,\sigma_{ij}\,de_{ij},$$

where $T, \mathscr{E}, \mathscr{S}$, and $\mathscr{F} = \mathscr{E} - T\mathscr{S}$ are, respectively, the absolute temperature, internal energy, entropy, and Helmholtz free energy *per unit mass*. If we substitute $\sigma_{ij}$ from Eq. (12.7:7) into (1), we obtain

(2) $$\rho\,d\mathscr{F} = -\rho\mathscr{S}\,dT + \lambda e\,de - \beta(T - T_0)\,de + 2Ge_{ij}\,de_{ij}, \qquad e = e_{\alpha\alpha},$$

which can be integrated with respect to the strain to yield

(3) $$\rho\mathscr{F} = \frac{\lambda e^2}{2} - \beta(T - T_0)e + Ge_{ij}e_{ij} + C_1(T),$$

where $C_1(T)$ is a function of $T$. To evaluate $C_1$, we first note that Eq. (1) implies that $(\partial\mathscr{F}/\partial T)_{e_{ij}} = -\mathscr{S}$, [see Eq. (12.3:6)]. Hence, from (3),

(4) $$\rho\mathscr{S} = -\frac{e^2}{2}\frac{\partial\lambda}{\partial T} - e_{ij}e_{ij}\frac{\partial G}{\partial T} + e\frac{\partial}{\partial T}[\beta(T - T_0)] - \frac{dC_1}{dT}.$$

Now, the function $C_1$ can be expressed in terms of the heat capacity per unit mass at constant strain $C_v$. For, on the one hand, since $\mathscr{S}$ is a function of $T$ and $e_{ij}$, we have that, if $de_{ij} = 0$,

(5) $$dQ = T\rho\,d\mathscr{S} = T\rho\left(\frac{\partial\mathscr{S}}{\partial T}\right)_{e_{ij}} dT.$$

On the other hand, we have, by definition, that at zero strain

(6) $$dQ = \rho C_v\,dT.$$

Hence

(7) $$C_v = T\left(\frac{\partial\mathscr{S}}{\partial T}\right)_{e_{ij}},$$

which relates $C_v$ with $C_1$ when $\mathscr{S}$ is substituted from (4). Simple relationships are obtained if $C_v$ is evaluated at zero strain. Then

(8) $$(C_v)_{e_{ij}=0} = T\left(\frac{\partial\mathscr{S}}{\partial T}\right)_{e_{ij}=0} = -\frac{T}{\rho}\frac{d^2C_1}{dT^2},$$

or, if we assume that $\mathscr{S}$ and $\mathscr{F}$ vanish for zero strain at $T = T_0$,

$$C_1 = -\int_{T_0}^{T} dT \int_{T_0}^{T} \frac{\rho}{T} (C_v)_{e_{ij}=0}\, dT. \tag{9}$$

Now, the internal energy $\mathscr{E}$ per unit volume can be obtained. By Eqs. (3), (4) and (12.6:3)–(12.6:9), we have

$$\begin{aligned} \rho\mathscr{E} &= \rho(\mathscr{F} + T\mathscr{S}) \\ &= \frac{1}{2} e^2 \left[K - T\frac{\partial K}{\partial T}\right] + 2J_2\left(G - T\frac{\partial G}{\partial T}\right) \\ &\quad + e\left[\beta T_0 + T(T - T_0)\frac{\partial \beta}{\partial T}\right] + C_1 - T\frac{dC_1}{dT}. \end{aligned} \tag{10}$$

If $\lambda$, $G$, $\beta$, and $C_v$ are independent of temperature, Eqs. (9) and (10) are simplified to

$$C_1 = -C_v T_0\left(\frac{T}{T_0} \log \frac{T}{T_0} - \frac{T}{T_0} + 1\right), \tag{11}$$

$$\rho\mathscr{E} = \tfrac{1}{2}Ke^2 + 2GJ_2 + \beta T_0 e + (C_v)_{e_{ij}=0}(T - T_0), \tag{12}$$

where $J_2$ is the second strain invariant:

$$J_2 = \tfrac{1}{2}(e_{ij} - \tfrac{1}{3}e\delta_{ij})(e_{ij} - \tfrac{1}{3}e\delta_{ij}). \tag{13}$$

**Problem 12.11.** Starting with the equation

$$d\Phi = -\mathscr{S}\, dT - \frac{1}{\rho} e_{ij}\, d\sigma_{ij}, \tag{14}$$

where $\Phi$ is Gibbs' thermodynamic potential per unit volume, show that

$$\rho\Phi = \frac{\nu}{2E} s^2 - \frac{1 + \nu}{2E} \sigma_{ij}\sigma_{ij} - \alpha(T - T_0)s + C_2(T), \tag{15}$$

$$C_2 = \rho \int_{T_0}^{T} dT \int_{T_0}^{T} (C_p)_{\sigma_{ij}=0} \frac{1}{T}\, dT, \tag{16}$$

where $s = \sigma_{ii}$ and $C_p$ is the heat capacity per unit volume at constant stresses.

## 12.9. EQUATIONS CONNECTING THERMAL AND MECHANICAL PROPERTIES OF A SOLID

It is possible to derive relations which connect the specific heats, the moduli of elasticity, the latent heats of change of strain or stress at constant temperature, etc. These results were given by Lord Kelvin and by Gibbs.

The specific heat at constant strain and the various latent heats of change of strain are defined by the following equation.

(1) $$dQ = C_v\, dT + \ell_{e_{ij}}\, de_{ij},$$

where $dQ$ is the heat required to change the temperature of a unit mass by $dT$ and the strain by $de_{ij}$, $C_v$ is the specific heat per unit mass measured in the state of constant strain, and $\ell_{e_{ij}}$ are the six latent heats per unit mass due to a change in a component of strain (in each case the temperature and the remaining five strain components are unchanged). Now, under reversible conditions, the change of entropy is

(2) $$d\mathscr{S} = \frac{dQ}{T} = \frac{C_v}{T}\, dT + \frac{\ell_{e_{ij}}}{T}\, de_{ij}.$$

Hence,

(3) $$\ell_{e_{ij}} = T\frac{\partial \mathscr{S}}{\partial e_{ij}}, \qquad C_v = T\frac{\partial \mathscr{S}}{\partial T}.$$

But from Eq. (12.3:6) we have, by further differentiations,

$$\frac{\partial \mathscr{S}}{\partial e_{ij}} = -\frac{\partial^2 \mathscr{F}}{\partial T\, \partial e_{ij}} = -\frac{1}{\rho}\frac{\partial \sigma_{ij}}{\partial T}, \qquad \frac{\partial^2 \mathscr{S}}{\partial e_{ij}\, \partial T} = -\frac{1}{\rho}\frac{\partial^2 \sigma_{ij}}{\partial T^2}.$$

It follows that

(4) $$\ell_{e_{ij}} = -\frac{T}{\rho}\frac{\partial \sigma_{ij}}{\partial T}, \qquad \frac{\partial C_v}{\partial e_{ij}} = -\frac{T}{\rho}\frac{\partial^2 \sigma_{ij}}{\partial T^2}.$$

In a similar manner we may choose the temperature and the six components of stress as thermodynamic variables and define the specific heat per unit mass at constant stress and the latent heats per unit mass due to change of stress by an equation analogous to (2),

(5) $$d\mathscr{S} = \frac{C_p}{T}\, dT + \frac{1}{T} L_{\sigma_{ij}}\, d\sigma_{ij}$$

The specific heat $C_p(T, \sigma)$ is more practically obtainable under usual conditions of measurement where the external forces on the solid are unchanged. The latent heats $L_{\sigma_{11}}$, etc., are defined each at a constant temperature and with five of the stress-components unchanged.

A similar analysis with Gibbs' potential $\Phi$ leads to the results

(6) $$L_{\sigma_{ij}} = \frac{T}{\rho}\frac{\partial e_{ij}}{\partial T}, \qquad \frac{\partial C_p}{\partial \sigma_{ij}} = \frac{T}{\rho}\frac{\partial^2 e_{ij}}{\partial T^2}.$$

The difference between the two specific heats can be expressed as follows. Let an infinitesimal change take place at constant stress. The change of entropy can be expressed in two ways. By Eq. (2) it is

$$d\mathscr{S} = \frac{C_v}{T}\, dT + \frac{1}{T}\ell_{e_{ij}}\frac{\partial e_{ij}}{\partial T}\, dT;$$

by Eq. (5) it is

$$d\mathscr{S} = \frac{C_p}{T}\, dT.$$

Equating these two expressions, we obtain

(7) $$C_p = C_v + \ell_{e_{ij}} \frac{\partial e_{ij}}{\partial T}.$$

By Eq. (4), this becomes

(8) $$C_p - C_v = -\frac{T}{\rho} \frac{\partial e_{ij}(T, \sigma)}{\partial T} \frac{\partial \sigma_{ij}(T, e)}{\partial T}.$$

**Problem 12.12.** Prove the following reciprocal relations.

$$\frac{\partial \ell_{e_{ij}}}{\partial e_{kl}} = \frac{\partial \ell_{e_{kl}}}{\partial e_{ij}}, \qquad \frac{\partial \sigma_{ij}}{\partial e_{kl}} = \frac{\partial \sigma_{kl}}{\partial e_{ij}},$$

$$\frac{\partial L_{\sigma_{ij}}}{\partial \sigma_{kl}} = \frac{\partial L_{\sigma_{kl}}}{\partial \sigma_{ij}}, \qquad \frac{\partial e_{ij}}{\partial \sigma_{kl}} = \frac{\partial e_{kl}}{\partial \sigma_{ij}}.$$

# 13

# IRREVERSIBLE THERMODYNAMICS AND VISCOELASTICITY

Problems involving heat conduction, viscosity, and plasticity belong to the realm of irreversible thermodynamics. In this chapter we shall outline the methods of irreversible thermodynamics, limiting ourselves to linear processes. Biot's treatment of relaxation modes and hidden variables will be presented and applied to the stress-strain relationship of viscoelastic materials.

## 13.1. BASIC ASSUMPTIONS

The classical thermodynamics of Chap. 12 deals essentially with equilibrium conditions of a uniform system (or a heterogeneous system with uniform phases). As far as irreversible processes are concerned, it contains little more than a statement about the trend toward thermodynamic equilibrium. To bring the irreversibility to a sharper focus, we would like to be able to write down definite "equations of evolution" (or equations of "motion"), which describe in precise terms the way an irreversible process evolves. To make the theory useful to continuum mechanics, we must extend the formulation to include nonuniform systems. These two items are the main objectives of the theory of irreversible thermodynamics. Since these objectives are beyond the scope of classical thermodynamics, some new hypotheses must be introduced, whose justification can only be sought by comparing any theoretical deductions with experiments.

The first assumption is that the entropy is a function of state in irreversible processes as well as in reversible processes. In Chap. 12 we considered reversible processes and showed that the value of the entropy of a system can be computed when the values of the state variables are known (except for an integration constant which is, in general, of no importance.) (Near zero absolute temperature, the additive constant of entropy does become important, but in that case its value can be fixed according to Nerst's theorem—the third law of thermodynamics.) We now assume that this same function of state defines entropy even if the system is not in equilibrium. The motivation for this assumption is analogous to our attitude toward the mass distribution of any mechanical system: once the mass of a particle in a continuum is determined, we assume that it remains the same no matter how fast the particle moves.

Just as the constancy of mass must be subjected to relativistic restrictions, the validity of the assumption of entropy as a state function must be restricted to relatively mild processes. Prigogine has investigated this question by comparing our assumption with results of statistical mechanics for some particular models, such as the Chapman-Enskog theory of nonuniform gases. He shows that the domain of validity of this assumption extends throughout the domain of validity of linear phenomenological laws (Fourier's law of heat conduction, Fick's law of mass diffusion, etc.). In the case of chemical reactions, he shows that the reaction rates must be sufficiently slow so as not to perturb Maxwell-Boltzmann equilibrium distribution of velocities of each component to an appreciable extent. This means, for example, that the temperature changes over the length of one mean free path must be much smaller than the absolute temperature itself; and the like for other properties.

The second assumption consists of an extension of the second law of thermodynamics locally to every portion of a continuum, whether it is uniform or nonuniform, and may be explained as follows. We notice first that entropy is an extensive quantity, so it must be subjected to a conservation law: for a given set of particles occupying a domain $V$ the total change of entropy must be equal to the total amount of entropy transferred to these particles through the boundary, plus the entropy produced inside this domain. Let the boundary of this domain be denoted by $B$. Then we may write

$$\text{(1)} \qquad \frac{D}{Dt}\int_V \rho\mathscr{S}\, dV = -\int_B \dot{\boldsymbol{\phi}}\cdot d\mathbf{A} + \frac{D}{Dt}\int_V \rho({}_i\mathscr{S})\, dV,$$

where $\mathscr{S}$ is the *specific entropy* per unit mass, $({}_i\mathscr{S})$ is the *entropy source* (internal entropy production per unit mass), $\dot{\boldsymbol{\phi}}$ is the *entropy flow vector* on the boundary, $d\mathbf{A}$ is a surface element whose vector direction coincides with the outer normal, so that $\dot{\boldsymbol{\phi}} \cdot d\mathbf{A}$ is a scalar product representing the *outflow*, which is equal to the normal component of $\dot{\boldsymbol{\phi}}$ times the area $dA$. The material derivative $D/Dt$ is taken with respect to a given set of particles. On transforming the second term into a volume integral and using Eq. (5.3:4) to reduce the material derivative of an integral, and realizing that the domain $V$ is arbitrary, we obtain

$$\rho\frac{D\mathscr{S}}{Dt} + \mathscr{S}\left(\frac{D\rho}{Dt} + \rho \operatorname{div} \mathbf{v}\right) = -\operatorname{div}\dot{\boldsymbol{\phi}} + \rho\frac{D_i\mathscr{S}}{Dt} + ({}_i\mathscr{S})\left(\frac{D\rho}{Dt} + \rho \operatorname{div} \mathbf{v}\right)$$

where $\mathbf{v}$ is the particle velocity vector of the flow field. The sum in the parentheses vanishes by the equation of continuity, Eq. (5.4:4). Hence, we have

$$\text{(2)} \ \blacktriangle \qquad \rho\frac{D\mathscr{S}}{Dt} = -\operatorname{div}\dot{\boldsymbol{\phi}} + \rho\frac{D_i\mathscr{S}}{Dt}.$$

We can now state the second hypothesis (an extension of the second law of thermodynamics) as follows.

(3) ▲ $$\dot{\boldsymbol{\phi}} = \frac{\mathbf{h}}{T}, \qquad \mathbf{h} = \text{heat flux vector},$$

(4) ▲ $$\frac{D_i\mathcal{S}}{Dt} = 0, \qquad \text{reversible processes},$$

(5) ▲ $$\frac{D_i\mathcal{S}}{Dt} > 0, \qquad \text{irreversible processes},$$

(6) $$\frac{D_i\mathcal{S}}{Dt} < 0, \qquad \text{never occurs in nature.}$$

In Eq. (3) $\mathbf{h}$ is the heat flux vector representing the heat flow per unit area and $T$ is the local absolute temperature. In a moving medium, $\mathbf{h}$ is defined convectively, i.e., with respect to elements of surfaces that are composed always of the same sets of particles.

When Eqs. (3)–(5) are compared with Eqs. (12.1:6) and (12.1:7), we see that the entropy production is now required to be nonnegative everywhere in the system. Such a formulation may be called a *local formulation of the second law* in contrast to the *global formulation* of the classical thermodynamics.

Corresponding to $\dot{\boldsymbol{\phi}}$ we have a vector field $\boldsymbol{\phi}$ which is the *entropy displacement field* introduced by Biot[13.1] (1956). If we require that

(7) $$\boldsymbol{\phi} = 0 \quad \text{at} \quad t = t_0,$$

then the points of the body at time $t > t_0$ are assigned an entropy displacement vector $\boldsymbol{\phi}$, just as they are assigned a material displacement vector $\mathbf{u}$. The two vector fields $\mathbf{u}$ and $\boldsymbol{\phi}$ connect the state of the body at time $t > t_0$ with the natural state at $t_0$. The material displacement field defines the reversible change of mass distribution by $-\operatorname{div}(\rho\mathbf{u})$; the entropy displacement field $\boldsymbol{\phi}$ defines a reversible change of entropy by $-\operatorname{div}\boldsymbol{\phi}$. The material displacement field can be analyzed through the displacement gradient tensor $u_{i,j}$; the entropy displacement field can be analyzed through the second order tensor $\phi_{i,j}$.

If we introduce a notation analogous to that of Eq. (12.1:5), we may write the rate of reversible change of entropy as

(8) $$\rho\frac{D_e\mathcal{S}}{Dt} = -\operatorname{div}\dot{\boldsymbol{\phi}}.$$

Further development of irreversible thermodynamics requires a detailed description of the entropy production $D_i\mathcal{S}/Dt$. To this end we shall first consider a simple example in the following section.

## 13.2. ONE-DIMENSIONAL HEAT CONDUCTION

Consider the heat transfer in a slender solid bar with continuous temperature gradient in the direction of the lengthwise axis of the bar, $x$; Fig. 13.2:1. The temperature is assumed to be uniform in each cross section of the bar, and the walls (except the ends) are thermally insulated. It will be assumed that the heat flow takes place in the direction of the temperature gradient, so that the problem is mathematically one-dimensional. Furthermore, we assume that the bar is free of stresses and that the cross section is of unit area.

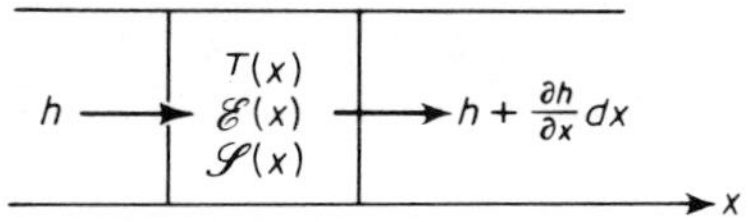

**Fig. 13.2:1.** Heat conduction.

Let $T(x)$, $\mathscr{E}(x)$, $\mathscr{S}(x)$ denote, respectively, the temperature, the energy per unit mass, and the entropy per unit mass of the solid. Let $h$ denote the heat flux per unit area per unit time in the $x$-direction; i.e., if $Q$ represents the heat transported to the right across a unit cross-sectional area, then

$$h = \frac{\partial Q}{\partial t}. \tag{1}$$

Now consider the changes occurring in a small element of length $dx$ in a small time interval $dt$. The net increment of heat in this element is evidently

$$dQ = h\,dt - \left(h + \frac{\partial h}{\partial x}\,dx\right)dt = -\frac{\partial h}{\partial x}\,dx\,dt.$$

The increase in energy in this element is $(\rho\, D\mathscr{E}/Dt\, dt)\, dx$. The stresses being zero, the first law states that the quantities above must be equal. Cancelling the factor $dx\,dt$ we obtain

$$\rho\frac{D\mathscr{E}}{Dt} = -\frac{\partial h}{\partial x}. \tag{2}$$

The change of entropy in this element is caused by the addition of heat $dQ$ alone. Hence,

$$\rho\frac{D\mathscr{S}}{Dt}\,dx\,dt = \frac{dQ}{T} = -\frac{1}{T}\frac{\partial h}{\partial x}\,dx\,dt. \tag{3}$$

Hence,

$$\rho\frac{D\mathscr{S}}{Dt} = -\frac{1}{T}\frac{\partial h}{\partial x} = -\frac{\partial}{\partial x}\left(\frac{h}{T}\right) - h\frac{1}{T^2}\frac{\partial T}{\partial x}. \tag{4}$$

This equation is of the form of Eq. (13.1:2). The entropy flow $\dot{\boldsymbol{\phi}}$ may be identified with $h/T$. The entropy production per unit time per unit volume is

$$\rho\frac{D_i\mathscr{S}}{Dt} = -\frac{h}{T^2}\frac{\partial T}{\partial x}. \tag{5}$$

The postulate that $D_i\mathscr{S}/Dt$ must be positive implies that $h$ has an oppositive sign to that of $\partial T/\partial x$; i.e., the heat flows in a direction against the temperature gradient.

**Problem 13.1.** Generalize the results above to the three-dimensional case: The entropy production per unit volume is

$$(6) \quad \blacktriangle \qquad \rho\frac{D_i\mathscr{S}}{Dt} = -\left(\frac{h_x}{T^2}\frac{\partial T}{\partial x} + \frac{h_y}{T^2}\frac{\partial T}{\partial y} + \frac{h_z}{T^2}\frac{\partial T}{\partial z}\right) = -\frac{\mathbf{h}}{T}\cdot\frac{\text{grad } T}{T},$$

where $h_x$, $h_y$, $h_z$ are the three components of the heat flux vector $\mathbf{h}$, i.e., $h_x$ is the heat flux per unit area across a surface element normal to the $x$-axis, etc.

## 13.3. PHENOMENOLOGICAL RELATIONS—ONSAGER PRINCIPLE

In the preceding section it was shown that in heat conduction the entropy production can be written as the product of a "generalized force" $-(\text{grad } T)/T$ and the corresponding flux $\mathbf{h}/T$. This is true also for other types of irreversible processes such as diffusion, chemical reactions. In chemical reactions the "forces" and "fluxes" are often termed affinities and reaction rates. In general, then, we may write, for an irreversible process,

$$(1) \qquad \frac{D_i\mathscr{S}}{Dt} = \sum_k J_k X_k > 0,$$

where $J_k$ denotes the $k$th flux and $X_k$ the corresponding generalized force. For example, in heat conduction

$$(2) \qquad J_k = \frac{h_k}{T}, \qquad X_k = -\frac{1}{T}\frac{\partial T}{\partial x_k}, \qquad k = 1, 2, 3.$$

At thermodynamic equilibrium all processes stop and we have simultaneously for all irreversible processes

$$(3) \qquad J_k = 0, \qquad X_k = 0, \qquad \text{at equilibrium.}$$

It is quite natural to assume, at least in the neighborhood of an equilibrium condition, that the relations between generalized forces and generalized fluxes are linear. Fourier's law for heat conduction is such an example. Linear laws of this kind are called *phenomenological relations*. For such phenomenological relations, an important symmetry is known among the coefficients that appear in these relations. This symmetry is embodied in the so-called *Onsager principle*.

Let us consider heat conduction again. It has been known for a long time that in anisotropic crystals a symmetry exists in the heat conduction coefficient matrix, which could not be explained by crystallographic symmetry properties. It was found that when we write

$$
\begin{aligned}
h_x &= L_{11}\frac{\partial T}{\partial x} + L_{12}\frac{\partial T}{\partial y} + L_{13}\frac{\partial T}{\partial z}, \\
h_y &= L_{21}\frac{\partial T}{\partial x} + L_{22}\frac{\partial T}{\partial y} + L_{23}\frac{\partial T}{\partial z}, \qquad (4) \\
h_z &= L_{31}\frac{\partial T}{\partial x} + L_{32}\frac{\partial T}{\partial y} + L_{33}\frac{\partial T}{\partial z},
\end{aligned}
$$

the matrix $L_{kl}$ is always symmetric:

$$L_{kl} = L_{lk}, \qquad l, k = 1, 2, 3, \qquad (5)$$

for arbitrary orientations of the coordinate axes, irrespective of the crystallographic axes. Considerable effort has been spent on accurate experiments to detect any deviation from relation (5). The most ingenious experiments are those of Soret[13.1] and Voigt.[13.1] The symmetry relation is always verified with great accuracy.

A similar symmetry property occurs in other phenomenological relations describing interferences between several simultaneous irreversible processes. For example, consider the case of one-dimensional thermodiffusion for which two phenomenological relations may be written as

$$
\begin{aligned}
J_1 &= L_{11}X_1 + L_{12}X_2, \\
J_2 &= L_{21}X_1 + L_{22}X_2,
\end{aligned} \qquad (6)
$$

where $J_1$ represents heat flux, $J_2$ represents the mass flux of a particular component in a mixture, $X_1$ represents the temperature gradient, and $X_2$ represents the concentration gradient of that particular component. Then $L_{11}$ is the heat conductivity, $L_{22}$ is the diffusion coefficient, and $L_{12}$ and $L_{21}$ are coefficients describing the *interference* of the two irreversible processes of heat conduction and diffusion. The coefficient $L_{21}$ is connected with the appearance of a concentration gradient when a temperature gradient is imposed, called the *Soret effect*. The coefficient $L_{12}$ is connected with the inverse phenomenon, the *Dufour effect*, which describes the appearance of temperature difference when a concentration gradient exists. The first equation in (6) is a modification of Fourier's law of heat conduction to include the Dufour effect. The second equation in (6) is a modification of the Fick's law of diffusion to include the Soret effect. In this case also, we have

$$L_{12} = L_{21}.$$

Other linear phenomenological laws verified by experiments include Ohm's law between electrical current and potential gradient, Newton's law between shearing force and velocity gradient, the chemical reaction law between reaction rate and chemical potentials, etc. The reciprocal phenomena of thermoelectricity arising from the interference of heat conduction and electrical conduction include the *Peltier effect* (discovered in 1834), which refers to the evolution or absorption of heat at junctions of different metals resulting from the flow of an electric current, the *Seebeck effect* (discovered in 1822), which relates to the electromotive force developed in a circuit made up of different conducting elements, when not all of the junctions are at the same temperature, and the *Thomson effect* (discovered by Lord Kelvin in 1856), which refers to the reversible heat absorption which occurs when an electric current flows in a homogeneous conductor in which there is a temperature gradient.

The simultaneous action of several irreversible processes may cause interferences as typified by the examples named above. As a general form of phenomenological laws, let $J_k (k = 1, 2, \dots, m)$ represent the generalized *fluxes* (heat flow, electric current, chemical reaction rate, etc.), and let $X_k (k = 1, 2, \dots, m)$ represent the generalized *forces* (temperature gradient, electric potential gradient, chemical affinity, etc.). Then a linear phenomenological law is

$$J_k = \sum_{l=1}^{m} L_{kl} X_l, \qquad k = 1, 2, \dots, m. \tag{7}$$

To this general form, Onsager principle now states a symmetry property of the coefficients $L_{kl}$ as follows.

*When a proper choice of the fluxes $J_k$ and the forces $X_l$ is made, so that the entropy production per unit time may be written as*

$$\frac{D_i \mathscr{S}}{Dt} = \sum_{k=1}^{m} J_k X_k, \tag{8}$$

*and if the flux and forces are related by linear phenomenological relations,*

$$J_k = \sum_{k=1}^{m} L_{kl} X_l,$$

*then the matrix of the coefficients $L_{kl}$ is symmetric; i.e.,*

$$L_{kl} = L_{lk}, \qquad k, l = 1, 2, \dots, m. \tag{9}$$

The identities (9) are called the *Onsager reciprocal relations.*

On substituting (9) into (8), we obtain a quadratic form

$$\frac{D_i \mathscr{S}}{Dt} = \sum_{k=1}^{m} \sum_{l=1}^{m} L_{kl} X_l X_k > 0. \tag{10}$$

This quadratic form has to be positive for all positive or negative values of the variables $X_k$ except when $X_1 = X_2 = \ldots = 0$, in which case the entropy production vanishes. The necessary and sufficient conditions which must be satisfied by the coefficients $L_{kl}$ in order that the quadratic form (10) be positive definite are well known (cf. Probs. 1.16, 1.17).

Onsager derived (1931) this principle from statistical mechanics considerations under the assumption of microscopic reversibility; i.e., the symmetry of all mechanical equations of motion of individual particles with respect to time. It furnishes an explanation of the observed relation (5) of the heat conduction for anisotropic crystals. The statistical derivation may be found in Onsager's papers, or in Prigogine[13.1] and De Groot's[13.1] books.

Some further remarks on the coupling of irreversible processes are given in Notes 13.1, p. 479.

## 13.4. BASIC EQUATIONS OF THERMOMECHANICS

Thermomechanical systems are subjected to the same general conservation laws with regard to mass and momentum as were discussed in Chap. 5. However, the law of conservation of energy contains both mechanical and thermal energy. The change of thermal energy is related to the change of entropy. Thus, a complete description of the evolution of a system requires a knowledge of the entropy production. Phenomenological relations, together with the Onsager principle, provide sufficient details about the entropy production. These laws, taken together, determine the evolution of the system.

To demonstrate the procedure let us consider a solid body occupying a regular domain $V$ with boundary $B$. A set of rectangular Cartesian coordinates will be used for reference; the instantaneous position of a particle will be denoted by $(x_1, x_2, x_3)$. Equations will be written with respect to the instantaneous configuration of the body, i.e., in Eulerian coordinates.

The conservation of mass is expressed by the equation of continuity (5.4:3)

(1) $$\frac{\partial \rho}{\partial t} + \frac{\partial \rho v_i}{\partial x_i} = 0.$$

The conservation of momentum is expressed by the Eulerian equation of motion (5.5:7), and Cauchy's formula (3.3:2)

(2) $$\rho \frac{Dv_i}{Dt} = \sigma_{ij,j} + \rho F_i \qquad \text{in } V,$$

(2a) $$\overset{\nu}{T}_i = \sigma_{ij}\nu_j \qquad \text{on } B,$$

(2b) $$\sigma_{ij} = \sigma_{ji} \qquad \text{in } V + B.$$

The conservation of energy is given by Eq. (12.2:7)

$$\rho \frac{D\mathscr{E}}{Dt} = \sigma_{ij} v_{i,j} - h_{i,i}. \tag{3}$$

In these equations, $\rho$ is the mass density, $(v_1, v_2, v_3)$ is the velocity vector of a particle, $D/Dt$ is the material derivative

$$\frac{D}{Dt} = \frac{\partial}{\partial t} + v_j \frac{\partial}{\partial x_j}, \tag{4}$$

$F_i$ is the body force per unit mass, $\overset{\nu}{T}_i$ is the surface traction per unit area, $\mathscr{E}$ is the internal energy per unit mass, $\sigma_{ij}$ is the stress tensor, $h_i$ is the heat flux vector, and a comma followed by $i$ indicates partial differentiation with respect to $x_i$. The indices $i$, $j$ range over 1, 2, 3; and the summation convention is used.

Next we need the entropy balance equation. According to the first assumption named in Sec. 13.1, we shall assume, *for a solid body*, that the specific entropy $\mathscr{S}$ per unit mass is a function of the internal energy $\mathscr{E}$ and the strain tensor $e_{ij}$,

$$\mathscr{S} = \mathscr{S}(\mathscr{E}, e_{ij}). \tag{5}$$

This is also expressed by the fact that, in equilibrium, the total differential of $\mathscr{S}$ is given by Gibbs' relation (12.2:10)

$$\rho T\, d\mathscr{S} = \rho\, d\mathscr{E} - \sigma_{ij}\, de_{ij}. \tag{6}$$

Since our hypothesis asserts that $\mathscr{S}$ is related to $\mathscr{E}$ and $e_{ij}$ in the same way even if the system is not in equilibrium, it follows that along the path of motion

$$\rho T \frac{D\mathscr{S}}{Dt} = \rho \frac{D\mathscr{E}}{Dt} - \sigma_{ij} V_{ij}. \tag{7}$$

where $V_{ij} = \frac{1}{2}(v_{i,j} + v_{j,i})$ is the symmetric part of $v_{i,j}$, called the *rate of deformation tensor*. On substituting Eq. (3) into (7), we obtain

$$\rho T \frac{D\mathscr{S}}{Dt} = -h_{i,i}, \tag{8}$$

or

$$\rho \frac{D\mathscr{S}}{Dt} = -\frac{h_{i,i}}{T} = -\left(\frac{h_i}{T}\right)_{,i} - h_i \frac{T_{,i}}{T^2}. \tag{9}$$

The first term on the right-hand side is the divergence of the entropy flow, the second term is the entropy production which must be positive. This result is in agreement with what was discussed in Sec. 13.2.

The entropy production $-h_i T_{,i}/T^2$ is a product of the flux $h_i/T$, and the "force," $-T_{,i}/T$. A phenomenological law relates the flux to the force in the form

$$h_i = -k_{ij} T_{,i} \tag{10}$$

where

$$k_{ij} = k_{ji}. \tag{11}$$

This is all we can get from the general considerations of mass, momentum, energy, and entropy. Equations (1) through (11), taken together, still do not define the strain field uniquely. To complete the formulation, a constitutive law must be added, which relates the stress tensor $\sigma_{ij}$ to the strain tensor $e_{ij}$. It can be verified that with the addition of such a constitutive law, a sufficient number of differential equations are obtained for which boundary value problems can be formulated.

As another illustration let us consider a body of *fluid* instead of a solid. A fluid is distinguished from a solid by the fact that it cannot sustain shear stress without motion. In an equilibrium condition, the internal energy must depend only on the volumetric change. Hence, instead of (5), we now make the assumption

$$\mathscr{S} = \mathscr{S}(\mathscr{E}, e), \tag{12}$$

where

$$e = e_{ii} = u_{i,i} \tag{13}$$

is the first invariant of the strain tensor. Correspondingly, the stress tensor is separated into a pressure and a stress deviator:

$$\sigma_{ij} = -p\delta_{ij} + \sigma'_{ij}, \tag{14}$$

$$-p = \frac{\sigma_{ii}}{3}. \tag{15}$$

The Gibbs' relation in thermostatics (12.2:11) is now generalized to read, according to Eq. (12) and our basic assumptions listed in Sec. 13.1,

$$\rho T \frac{D\mathscr{S}}{Dt} = \rho \frac{D\mathscr{E}}{Dt} + p \frac{De}{Dt}. \tag{16}$$

A substitution of Eq. (3) yields

$$\rho T \frac{D\mathscr{S}}{Dt} = -h_{i,i} + \sigma'_{ij} v_{i,j} - p v_{i,i} + p v_{i,i}. \tag{17}$$

Hence,

$$\rho \frac{D\mathscr{S}}{Dt} = -\left(\frac{h_i}{T}\right)_{,i} - h_i \frac{T_{,i}}{T^2} + \sigma'_{ij} v_{i,j}. \tag{18}$$

The first term on the right-hand side of (18) is the divergence of the entropy flow. The last two terms have the significance of entropy production,

$$\rho \frac{D_i\mathscr{S}}{Dt} = -h_i \frac{T_{,i}}{T^2} + \sigma'_{ij}v_{i,j}. \tag{19}$$

Now $v_{i,j}$ can be split into a symmetric part $V_{ij}$ and an antisymmetric part $\omega_{ij}$, called the *rate of deformation* tensor and the *vorticity* tensor respectively,

$$v_{i,j} = V_{ij} + \omega_{ij}, \tag{20}$$

$$V_{ij} = V_{ji}, \qquad \omega_{ij} = -\omega_{ji}, \tag{21}$$

$$V_{ij} = \tfrac{1}{2}(v_{i,j} + v_{j,i}), \qquad \omega_{ij} = \tfrac{1}{2}(v_{i,j} - v_{j,i}) \tag{22}$$

The contraction of the symmetric stress tensor $\sigma'_{ij}$ with the antisymmetric tensor $\omega_{ij}$ vanishes. Hence, $\sigma'_{ij}v_{i,j} = \sigma'_{ij}V_{ij}$ and (19) becomes

$$\rho \frac{D_i\mathscr{S}}{Dt} = -h_i \frac{1}{T^2}\frac{\partial T}{\partial x_i} + \sigma'_{ij}V_{ij} \tag{23}$$

The linear phenomenological laws assume the form

$$\frac{h_i}{T} = -C^{(1)}_{ij}\frac{1}{T}\frac{\partial T}{\partial x_j} + C^{(2)}_{ijk}V_{jk}, \tag{24}$$

$$\sigma'_{ij} = C^{(3)}_{ijk}\frac{1}{T}\frac{\partial T}{\partial x_k} + C^{(4)}_{ijkl}V_{kl}, \tag{25}$$

where, according to Onsager's principle, the symmetry relation

$$C^{(2)}_{ijk} = C^{(3)}_{jki} \tag{26}$$

prevails. Equation (25) is a law for viscous flow. The coupling terms involving $C^{(2)}_{ijk}$, $C^{(3)}_{ijk}$ express the possible interference between viscous flow and heat conduction.†

In the case of a fluid, the Eqs. (1)–(3) and (12)–(26) provide the right number of field equations for the determination of the flow field.

## 13.5. EQUATIONS OF EVOLUTION FOR A LINEAR HEREDITARY MATERIAL

The remainder of this chapter will be devoted to the question of stress-strain law for a linear hereditary material, following a treatment first given by Maurice A. Biot[13.1] (1954). We have in mind such materials as

† It follows further from Curie's principle that $C^{(3)}_{ijk} = C^{(2)}_{ijk} = 0$. See p. 480.

polycrystalline metals, high polymers, etc. When such a material is subjected to a variable strain, many things happen inside, which, however, are not explicitly observed in formulating the stress-strain law. For example, when a polycrystalline metal is uniformly strained in a macroscopic sense, the individual anisotropic crystals are strained differently, and thermal currents that circulate among the crystals are generated. The interstitial atoms move in the crystals or among the crystals. These processes can be accounted for explicitly. However, often we are not interested in them. Our interest may be limited to the extent of their interference with the deformation. They are the "hidden variables" in the stress-strain relationship.

We follow Biot to formulate a problem as follows. Consider a system I with $n$ degrees of freedom defined by $n$ state variables $q_1, q_2, \ldots, q_n$. These $q$'s may represent strain components, local temperature, piezoelectric charges, concentrations such as induced by chemical or solubility processes, etc. It is assumed that the system is under the action of *generalized external forces* denoted by $Q_j$, the units and senses of which are such that for each $j = 1, 2, \ldots n$, $Q_j\, dq_j$ ($j$ not summed) represents the energy furnished to the system when $q_j$ is changed by an amount $dq_j$. These forces may be externally applied forces, electromotive forces, or may result from deviations of the Gibbs and chemical potential from the equilibrium state. Inertia forces, if any, are considered as external forces according to D'Alembert's principle.

Let the system I be adjoined to a system II which is a large reservoir at constant temperature $T_0$, and *let the total system I + II be insulated.* The variables $q_j$ will be defined as the departure from the state of thermodynamic equilibrium; i.e., $q_j = 0$ at the condition of equilibrium at which the temperature is uniform and equal to $T_0$.

Let the system I be given an initial disturbance from equilibrium. The forces $Q_j$ as functions of time are given. If all coordinates $q_i(t)$ are determined for $t \geqslant 0$, we say that the *evolution* of the system is known. The time history of $q_j(t)$ is said to be a *trajectory*. The differential equations that describe the evolution of $q_j(t)$ from their initial values are called the *equations of evolution*.

A little reflection will show that if we consider the body represented by the system I to be macroscopically homogeneous and subjected to macroscopically homogeneous stresses, we may let $q_1, q_2, \ldots, q_6$ represent the six independent strain components $e_{11}, e_{12} = e_{21}, \ldots, e_{33}$, and $Q_i$ the corresponding stress components. The relations between $Q_1, \ldots, Q_6$ and $q_1, \ldots, q_6$ are influenced by the other $q$'s and $Q$'s. However, if all other $Q$'s vanish and if we can eliminate all $q$'s other than the first six, then the stress-strain relationship of the material is obtained. It will be seen that the influence of the hidden variables is revealed in the hereditary character of the material. In the present section we shall derive the equations of evolution. The solution of free evolution following an initial disturbance will be discussed in Sec. 13.6, and the forced motion and the elimination of hidden variables will be treated in Secs. 13.7–13.8. Biot's results about viscoelastic material will be presented in Sec. 13.9.

We shall now consider what happens in a small time interval. Let the change of the coordinate $q_j$ in a time interval $dt$ be denoted by $dq_j$. According to the first law of thermodynamics, the heat, $\mathscr{Q}^{(\mathrm{I})}$, required by the system I to bring about a change of state $dq_j$ is the difference between the change in internal energy $\mathscr{E}_{\mathrm{I}}$ and the work done. Without loss of generality, we shall write the following equations under the assumption that the system I is of unit mass.

$$(1) \qquad d\mathscr{Q}^{(\mathrm{I})} = d\mathscr{E}_{\mathrm{I}} - \sum_{j=1}^{n} Q_j \, dq_j.$$

Since the total system I + II is insulated, we have

$$d\mathscr{Q}^{(\mathrm{I})} + d\mathscr{Q}^{(\mathrm{II})} = 0,$$

where $d\mathscr{Q}^{(\mathrm{II})}$ denotes the heat received by system II. According to our basic assumption stated in Sec. 13.1, Gibbs' relation for change of entropy holds. Hence, the change of entropy of system II is

$$d\mathscr{S}_{\mathrm{II}} = \frac{d\mathscr{Q}^{(\mathrm{II})}}{T_0} = -\frac{d\mathscr{Q}^{(\mathrm{I})}}{T_0} = -\frac{d\mathscr{E}_{\mathrm{I}}}{T_0} + \sum_{j=1}^{n} \frac{Q_j}{T_0} dq_j.$$

The change of entropy of the isolated total system I + II is

$$(2) \qquad d_i\mathscr{S}_{\mathrm{I+II}} = d\mathscr{S}_{\mathrm{I}} + d\mathscr{S}_{\mathrm{II}} = d\mathscr{S}_{\mathrm{I}} - \frac{d\mathscr{E}_{\mathrm{I}}}{T_0} + \sum_{j=1}^{n} \frac{Q_j}{T_0} dq_j.$$

This is the entropy production in the time interval $dt$. Since $\mathscr{S}_{\mathrm{I}}$ and $\mathscr{E}_{\mathrm{I}}$ are functions of state, the total differentials of $d\mathscr{S}_{\mathrm{I}}$ and $d\mathscr{E}_{\mathrm{I}}$ are

$$d\mathscr{S}_{\mathrm{I}} = \sum_j \frac{\partial \mathscr{S}_{\mathrm{I}}}{\partial q_j} dq_j, \qquad d\mathscr{E}_{\mathrm{I}} = \sum_j \frac{\partial \mathscr{E}_{\mathrm{I}}}{\partial q_j} dq_j.$$

Hence,

$$(3) \qquad d_i\mathscr{S}_{\mathrm{I+II}} = \sum_{j=1}^{n} \left( \frac{\partial \mathscr{S}_{\mathrm{I}}}{\partial q_j} - \frac{1}{T_0} \frac{\partial \mathscr{E}_{\mathrm{I}}}{\partial q_j} + \frac{Q_j}{T_0} \right) dq_j.$$

According to the second law of thermodynamics, $d_i\mathscr{S}_{\mathrm{I+II}}$ is nonnegative. If we write $dq_j/dt$ for the generalized flux $J_j$ of the previous section, we have the following general expression for the entropy production:

$$(4) \ \blacktriangle \qquad d_i\mathscr{S}_{\mathrm{I+II}} = \sum_{j=1}^{n} X_j \, dq_j,$$

where $X_j$ are the generalized force conjugate to $q_j$ in the entropy production. To distinguish $X_j$ from the "generalized external forces" $Q_i$, we shall call $X_j$ the *generalized entropy-production forces*. We remark that $X_j$ may be identically zero for some index $j$. Hence, the number of significant terms in the sum on the right-hand side of Eq. (3) may be less than $n$.

Identifying the general expression (4) with (3), we see that in our problem

(5) ▲ $$X_j = \frac{\partial}{\partial q_j}\left(\mathscr{S}_{\text{I}} - \frac{1}{T_0}\mathscr{E}_{\text{I}}\right) + \frac{Q_j}{T_0}.$$

Now, if the phenomenological laws connecting $X_j$ and $\dot{q}_j$ are linear so that

(6) ▲ $$X_j = \frac{1}{T_0}\sum_{k=1}^{n} b_{jk}\dot{q}_k,$$

then Eq. (5) becomes

(7) ▲ $$\sum_k b_{jk}\dot{q}_k - \frac{\partial}{\partial q_j}(T_0\mathscr{S}_{\text{I}} - \mathscr{E}_{\text{I}}) = Q_j, \qquad j = 1, 2, \ldots, n,$$

which is the equation of evolution.

Onsager's principle assures that the coefficients $b_{ij}$ in Eq. (6) are symmetric:

(8) ▲ $$b_{ij} = b_{ji}.$$

Let us introduce the quadratic form

(9) ▲ $$\mathscr{D} = \frac{1}{2}\sum_{i,j} b_{ij}\dot{q}_i\dot{q}_j.$$

Then Eq. (6) may be written as

(10) ▲ $$X_i = \frac{1}{T_0}\frac{\partial\mathscr{D}}{\partial\dot{q}_i}.$$

The quadratic form $\mathscr{D}$ is nonnegative because it is proportional to the entropy production, $d_i\mathscr{S}_{\text{I+II}}/dt$, and because $T_0$ is positive by definition, as can be seen from the following equation:

(11) ▲ $$\frac{d_i\mathscr{S}_{\text{I+II}}}{dt} = \sum_i X_i\dot{q}_i = \sum_i \frac{1}{T_0}\frac{\partial\mathscr{D}}{\partial\dot{q}_i}\dot{q}_i = \frac{2\mathscr{D}}{T_0} \geqslant 0$$

In addition, let us define

(12) ▲ $$\mathscr{V} = \mathscr{E}_{\text{I}} - T_0\mathscr{S}_{\text{I}}.$$

Then the equation of evolution (7) can be written in the familiar Lagrangian form

(13) ▲ $$\frac{\partial\mathscr{D}}{\partial\dot{q}_i} + \frac{\partial\mathscr{V}}{\partial q_i} = Q_i.$$

It is easy to see that in the neighborhood of an equilibrium state (in which $q_i = 0$, $\dot{q}_i = 0$, $Q_i = 0$, $X_i = 0$), $\mathscr{V}$ is a quadratic form in $q_i$. For, on expanding $\mathscr{V}$ into a power series in $q_i$, the constant term has no significance and can be removed, whereas the linear terms must all vanish, because otherwise Eq. (13) cannot be satisfied at the equilibrium condition. Hence, if sufficiently small values of $q_i$ are considered, a quadratic form is obtained if higher powers of $q_i$ are neglected:

(14) ▲ $$\mathscr{V} = \tfrac{1}{2} \sum_{i,j} a_{ij} q_i q_j.$$

Since only the sum is of interest, we may assume, without loss of generality, that this quadratic form is symmetric,

(15) ▲ $$a_{ij} = a_{ji}.$$

Hence, Eq. (7) can be put in the form

(16) ▲ $$\sum_j b_{ij} \dot{q}_j + \sum_j a_{ij} q_j = Q_i.$$

Equation (16) assumes a form which occurs frequently in the theory of vibration of discrete masses. Evidently, $\mathscr{V}$ plays the role of a potential energy and $\mathscr{D}$ that of the dissipation function which occur in the usual vibration theory.

## 13.6. RELAXATION MODES

Let us now consider the solutions of the equation of evolution

(1) $$\sum_{j=1}^{n} a_{ij} q_j + \sum_{j=1}^{n} b_{ij} \dot{q}_j = Q_i, \qquad i = 1, 2, \ldots, n,$$

where the coefficients $a_{ij}$ and $b_{ij}$ are real-valued and symmetric,

$$a_{ij} = a_{ji}, \qquad b_{ij} = b_{ji}.$$

We shall ignore the inertia forces for the moment, so that $Q_i$ do not involve the acceleration $\ddot{q}_i$. The nature of the solution depends very much on the nature of the quadratic form

(2) $$\mathscr{D} = \tfrac{1}{2} \sum_{i=1}^{n} \sum_{j=1}^{n} b_{ij} \dot{q}_i \dot{q}_j,$$

which has been shown above to be proportional to the entropy production $d_i\mathscr{S}_{\mathrm{I+II}}/dt$ and is thus nonnegative. However, $\mathscr{D}$ can be identically zero if the process is reversible. In any case, a coefficient $b_{ij}$ $(i \neq j)$ will be zero if there is no interference between the processes $\dot{q}_i$ and $\dot{q}_j$. For a certain $i$, the diagonal coefficient $b_{ii}$ will be zero if $\dot{q}_i$ do not participate in the irreversible entropy production (i.e., if $q_i$ is a reversible variable), in which case all coefficients $b_{i1}, \ldots, b_{in}$ vanish. Hence, in general, the quadratic form $\mathscr{D}$ is not positive definite. However, it can be reduced into a reduced positive definite form if all the reversible variables are eliminated (unless $\mathscr{D} = 0$, in which case there is no problem). Let us assume that this reduction has been done, so that the variables $\dot{q}_1, \dot{q}_2, \ldots, \dot{q}_m$ $(m \leqslant n)$ really participate in the irreversible entropy production. Then we can assert that the quadratic form

(3) $$\mathscr{D} = \frac{1}{2}\sum_{i=1}^{m}\sum_{j=1}^{m} b_{ij}\dot{q}_i\dot{q}_j$$

is positive definite in the $m$ variables $\dot{q}_i$ ($i = 1, \ldots, m$).

Equations (1) can then be separated into a group of $m$ equations involving $\dot{q}_1, \ldots, \dot{q}_m$, and another group of $(n - m)$ equations which does not contain time derivatives. The latter group of equations can be solved for $q_{m+1}, \ldots, q_n$, which can be substituted in turn into the first $m$ equations. In this way we obtain a sharper restatement of the equations of evolution:

(4) $$\sum_{j=1}^{m} a'_{ij}q_j + \sum_{j=1}^{m} b_{ij}\dot{q}_j = Q'_i, \qquad i = 1, 2, \ldots, m,$$

where the matrix $[b_{ij}]$ is real-valued, symmetric, and positive definite. It is easy to see that the new coefficients $a'_{ij}$ are again real and symmetric.

Let us now consider the solutions of Eqs. (4) in the homogeneous case in which $Q'_i = 0$. The equations

(5) $$\sum_{j=1}^{m} a'_{ij}q_j + \sum_{j=1}^{m} b_{ij}\dot{q}_j = 0, \qquad i = 1, \ldots, m,$$

admit a solution of the form

(6) $$q_j = \psi_j e^{-\lambda t}, \qquad j = 1, 2, \ldots, m,$$

where $\psi_j$ are constants. On substituting (6) into (5), we have

(7) $$\sum_{j=1}^{m}(a'_{ij} - \lambda b_{ij})\psi_j = 0, \qquad i = 1, \ldots, m.$$

To find a set of nontrivial solutions $\psi_j$ poses an eigenvalue problem. The eigenvalues satisfy the determinantal equation

(8) $$\det |a_{ij} - \lambda b_{ij}| = 0.$$

Since the square matrices $[a'_{ij}]$, $[b_{ij}]$ are real, symmetric, and $[b_{ij}]$ is positive definite, all eigenvalues $\lambda_1, \ldots, \lambda_m$ are real-valued. Whether they are all positive or not depends on the nature of the quadratic form

(9) $$\mathscr{V} = \frac{1}{2}\sum_{i=1}^{m}\sum_{j=1}^{m} a'_{ij}q_iq_j.$$

The irreversible process is considered as a disturbed motion about an equilibrium state. If the equilibrium state is *stable*, then $\mathscr{V}$ is positive definite and all roots $\lambda_1, \ldots, \lambda_m$ are positive. In that case, every disturbance tends to zero with increasing time [see Eq. (6)]. If the equilibrium is *unstable*, then $\mathscr{V}$ is indefinite and some of the roots $\lambda$ will be negative. In that case, the disturbed motion will increase exponentially with time. If the equilibrium is neutrally stable with respect to some coordinates, then $\mathscr{V}$ is positive semi-definite; some of the roots $\lambda$ may be zero, while the others are positive.

It is known that corresponding to the $m$ eigenvalues $\lambda_1, \ldots, \lambda_m$ (some of them may be zero or may be multiple roots), there exist $m$ eigenvectors $\{\psi_j^{(1)}\}, \ldots, \{\psi_j^{(m)}\}$ which are linearly independent and orthonormal:

$$\sum_{i=1}^{m}\sum_{j=1}^{m} b_{ij}\psi_i^{(k)}\psi_j^{(l)} = \begin{cases} 0 & \text{if } k \neq l, \\ 1 & \text{if } k = l, \end{cases} \tag{10}$$
$$\sum_{i=1}^{m}\sum_{j=1}^{m} a'_{ij}\psi_i^{(k)}\psi_j^{(l)} = \begin{cases} 0 & \text{if } k \neq l, \\ \lambda_k & \text{if } k = l. \end{cases}$$

The general solution of Eqs. (5) is

$$q_j = \sum_{k=1}^{m} \psi_j^{(k)}\, e^{-\lambda_k t}, \qquad j = 1, 2, \ldots, m. \tag{11}$$

Each solution corresponding to a $\lambda_k$ $(k = 1, \ldots, m)$,

$$q_j^{(k)} = \psi_j^{(k)}\, e^{-\lambda_k t}, \qquad j = 1, 2, \ldots, m, \tag{12}$$

is called a *relaxation mode*.

Thus, for an irreversible process with $m$ fluxes $\dot{q}_1, \ldots, \dot{q}_m$ participating, there exist $m$ linearly independent orthonormal relaxation modes.

## 13.7. NORMAL COORDINATES

Now we shall consider the nonhomogeneous case, so that we may find the evolution of an irreversible process under a given set of forcing functions. As in the theory of mechanical vibrations, an introduction of the relaxation modes as the basis of generalized coordinates will decouple the equations of motion and thus lead to simple solutions.

We introduce the linear transformation from the state variables $\{q_i\}$ to the normal coordinates $\{\xi_i\}$ by the equations

$$q_i = \sum_{k=1}^{m} \psi_i^{(k)} \xi_k, \qquad i = 1, \ldots, m, \tag{1}$$

where $\{\psi_i^{(k)}\}$ is the modal column of the $k$th relaxation mode. Let the relaxation modes be normalized as in Eq. (13.6:10). Then the potential and dissipation functions assume the form

$$\mathscr{V} = \tfrac{1}{2} \sum_{k=1}^{m} \lambda_k \xi_k^2, \tag{2}$$
$$\mathscr{D} = \tfrac{1}{2} \sum_{k=1}^{m} \dot{\xi}_k^2.$$

The Lagrangian equations of motion, Eq. (13.6:4), now become decoupled,

(3) $$\frac{\partial \mathscr{V}}{\partial \xi_k} + \frac{\partial \mathscr{D}}{\partial \dot{\xi}_k} = \Xi_k,$$

where $\Xi_k$ is the generalized force corresponding to $\xi_k$,

(4) $$\Xi_k = \sum_{j=1}^{m} \psi_j^{(k)} Q'_j.$$

From (2) and (3) we have

(5) $$\dot{\xi}_k + \lambda_k \xi_k = \Xi_k, \qquad k = 1, 2, \ldots, m.$$

The Laplace transform of (5) is, for zero initial condition ($\xi_k = 0$ when $t = 0$),

(6) $$(s + \lambda_k)\bar{\xi}_k = \bar{\Xi}_k, \qquad k = 1, 2, \ldots, m,$$

where

(7) $$\bar{\xi}_k = \int_0^\infty e^{-st} \xi_k(t)\, dt, \qquad \bar{\Xi}_k = \int_0^\infty \Xi_k(t)\, e^{-st}\, dt.$$

The solution of (6) as algebraic equations in $s$ is

(8) $$\bar{\xi}_k = \frac{\bar{\Xi}_k}{s + \lambda_k}, \qquad k = 1, 2, \ldots, m.$$

Substituting back into Eq. (1), we obtain the Laplace transform of $q_i(t)$,

(9) $$\bar{q}_i = \int_0^\infty e^{-st} q_i(t)\, dt = \sum_{k=1}^{m} \psi_i^{(k)} \frac{\bar{\Xi}_k}{s + \lambda_k}$$
$$= \sum_{k=1}^{m} \sum_{j=1}^{m} \frac{\psi_i^{(k)} \psi_j^{(k)} \bar{Q}'_j}{s + \lambda_k}.$$

Equation (9) provides the solution to the response problem in Laplace transformation language. On transforming back to the physical plane, the time history of the evolution of an irreversible process in response to external forcing functions is obtained:

(10) ▲ $$q_i(t) = \sum_{j=1}^{m} \sum_{k=1}^{m} \psi_i^{(k)} \psi_j^{(k)} \int_0^t e^{-\lambda_k \tau} Q'_j(t - \tau)\, d\tau,$$

or

(10a) $$q_i(t) = \sum_{j=1}^{m} \sum_{k=1}^{m} \psi_i^{(k)} \psi_j^{(k)} \int_0^t e^{-\lambda_k (t-\tau)} Q'_j(\tau)\, d\tau.$$

This gives the complete solution if $m = n$. The case $m < n$ [cf. discussion in Sec. 13.6 following Eq. (13.6:2)] is treated in the exercise below (Prob. 13.2).

It is seen from Eq. (17) below, that, in general, if some coordinates are reversible, $q_i$ are related to $Q_i$ by equations of the form

$$(11) \quad ▲ \qquad \bar{q}_i(s) = \sum_{j=1}^{n} \left( \sum_{k=1}^{m} \frac{A_{ij}^{(k)}}{s + \lambda_k} + A_{ij} \right) \bar{Q}_j(s),$$

$$(12) \quad ▲ \qquad q_i(t) = \sum_{j=1}^{n} \left[ \sum_{k=1}^{m} A_{ij}^{(k)} \int_0^t e^{-\lambda_k(t-\tau)} Q_j(\tau)\, d\tau + A_{ij} Q_j(t) \right],$$

where

$$(13) \quad ▲ \qquad A_{ij}^{(k)} = A_{ji}^{(k)}, \qquad A_{ij} = A_{ji}.$$

**Problem 13.2.** Write Eq. (13.6:1) in the form of partitioned matrices

$$(14) \qquad \left[\begin{array}{c|c} \mathbf{A} & \mathbf{B} \\ \hline \mathbf{C} & \mathbf{H} \end{array}\right] \left[\begin{array}{c} \mathbf{q}^{(1)} \\ \hline \mathbf{q}^{(2)} \end{array}\right] + \left[\begin{array}{c|c} \mathbf{F} & 0 \\ \hline 0 & 0 \end{array}\right] \left[\begin{array}{c} \dot{\mathbf{q}}^{(1)} \\ \hline 0 \end{array}\right] = \left[\begin{array}{c} \mathbf{Q}^{(1)} \\ \hline \mathbf{Q}^{(2)} \end{array}\right],$$

where

$$\mathbf{q}^{(1)} = \begin{bmatrix} q_1 \\ \cdot \\ \cdot \\ \cdot \\ q_m \end{bmatrix}, \qquad \mathbf{Q}^{(1)} = \begin{bmatrix} Q_1 \\ \cdot \\ \cdot \\ \cdot \\ Q_m \end{bmatrix},$$

$$\mathbf{q}^{(2)} = \begin{bmatrix} q_{m+1} \\ \cdot \\ \cdot \\ \cdot \\ q_n \end{bmatrix}, \qquad \mathbf{Q}^{(2)} = \begin{bmatrix} Q_{m+1} \\ \cdot \\ \cdot \\ \cdot \\ Q_n \end{bmatrix},$$

$$\mathbf{A} = \begin{bmatrix} a_{11} & a_{12} & \cdots \\ \cdot & \cdot & \\ \cdot & \cdot & \\ \cdot & \cdot & \\ \cdot & \cdot & a_{mm} \end{bmatrix}, \qquad \mathbf{F} = \begin{bmatrix} b_{11} & b_{12} & \cdots \\ \cdot & \cdot & \\ \cdot & \cdot & \\ \cdot & \cdot & \\ \cdot & \cdot & b_{mm} \end{bmatrix},$$

$$\mathbf{B} = \begin{bmatrix} a_{1,m+1} & \cdots & a_{1n} \\ \cdot & & \cdot \\ \cdot & & \cdot \\ \cdot & & \cdot \\ a_{m1} & \cdots & a_{mn} \end{bmatrix}, \qquad \mathbf{C} = \begin{bmatrix} a_{m+1,1} & \cdots & a_{m+1,m} \\ \cdot & & \cdot \\ \cdot & & \cdot \\ \cdot & & \cdot \\ a_{n1} & \cdots & a_{nm} \end{bmatrix} = \mathbf{B}^T,$$

$$\mathbf{H} = \begin{bmatrix} a_{m+1,m+1} & \cdots & a_{m+1,n} \\ \cdot & & \cdot \\ \cdot & & \cdot \\ \cdot & & \cdot \\ a_{n,m+1} & \cdots & a_{nn} \end{bmatrix}.$$

Assume that $\det|\mathbf{H}| \neq 0$ so that $\mathbf{H}^{-1}$ exists and that at $t = 0$, $\mathbf{q}^{(1)} = 0$, $\mathbf{q}^{(2)} = 0$. Show that if Eq. (13.6:4) is written as

$$\mathbf{A}' \cdot \mathbf{q}^{(1)} + \mathbf{F} \cdot \dot{\mathbf{q}}^{(1)} = \mathbf{Q}', \tag{15}$$

then

$$\begin{aligned} \mathbf{A}' &= \mathbf{A} - \mathbf{B}\mathbf{H}^{-1}\mathbf{C}, \\ \mathbf{Q}' &= \mathbf{Q}^{(1)} - \mathbf{B}\mathbf{H}^{-1}\mathbf{Q}^{(2)}. \end{aligned} \tag{16}$$

Let $\mathbf{\Psi}$ be a square matrix formed by the columns of relaxation modes, $\mathbf{\Psi}^T$ be the transpose of $\mathbf{\Psi}$, and $\mathbf{\Lambda}$ be a diagonal matrix of the relaxation spectrum:

$$\mathbf{\Psi} = \begin{bmatrix} \psi^{(1)} & \psi^{(2)} & \cdots & \psi^{(m)} \\ \cdot & \cdot & & \cdot \\ \cdot & \cdot & & \cdot \\ \cdot & \cdot & & \cdot \\ \cdot & \cdot & & \cdot \end{bmatrix}, \qquad \mathbf{\Lambda} = \begin{bmatrix} \lambda_1 & 0 & \cdots & 0 \\ 0 & \lambda_2 & \cdots & 0 \\ \cdot & \cdot & & \cdot \\ \cdot & \cdot & & \cdot \\ \cdot & \cdot & & \cdot \\ 0 & 0 & \cdots & \lambda_m \end{bmatrix}.$$

Then the Laplace transforms of $\mathbf{q}^{(1)}$, $\mathbf{q}^{(2)}$, $\mathbf{Q}^{(1)}$, $\mathbf{Q}^{(2)}$ are related by

$$\begin{aligned} \bar{\mathbf{q}}^{(1)}(s) &= \mathbf{\Psi}(\mathbf{\Lambda} + s\mathbf{I})^{-1}\mathbf{\Psi}^T(\bar{\mathbf{Q}}^{(1)} - \mathbf{B}\mathbf{H}^{-1}\bar{\mathbf{Q}}^{(2)}), \\ \bar{\mathbf{q}}^{(2)}(s) &= \mathbf{H}^{-1}\bar{\mathbf{Q}}^{(2)} - \mathbf{H}^{-1}\mathbf{C}\mathbf{\Psi}(\mathbf{\Lambda} + s\mathbf{I})^{-1}\mathbf{\Psi}^T(\bar{\mathbf{Q}}^{(1)} - \mathbf{B}\mathbf{H}^{-1}\bar{\mathbf{Q}}^{(2)}), \end{aligned} \tag{17}$$

where

$$(\mathbf{\Lambda} + s\mathbf{I})^{-1} = \begin{bmatrix} \dfrac{1}{s + \lambda_1} & 0 & \cdots & 0 \\ \cdot & \ddots & & \cdot \\ \cdot & & \ddots & \cdot \\ \cdot & & & \cdot \\ 0 & \cdots & 0 & \dfrac{1}{s + \lambda_m} \end{bmatrix}.$$

## 13.8. HIDDEN VARIABLES AND THE FORCE-DISPLACEMENT RELATIONSHIP

In many physical problems a great many variables are "hidden," that is, we do not observe them. For example, we may have a system with $n$ variables in which only $k$ internal forces $Q_1, \ldots, Q_k$, corresponding to the displacements $q_1, \ldots, q_k$, are examined, while the remaining coordinates constitute an "internal system." An example of this occurs in the stress-strain relationship of a polycrystalline metal with stresses $Q_1 = \sigma_{11}$, $Q_2 = \sigma_{12} = \sigma_{21}, \ldots,$ $Q_6 = \sigma_{33}$, strains $q_1 = e_{11}$, $q_2 = e_{12} = e_{21}, \ldots, q_6 = e_{33}$; whereas the intercrystalline thermal currents, movement of interstitial atoms, twining, etc., are represented by $q_7, q_8, \ldots, q_n$, for which no corresponding external forces are applied. We wish to know the influence of these hidden variables on the stress-strain relationship.

In the general case, some of the hidden variables $q_{k+1}, q_{k+2}, \ldots, q_n$ may not participate in the entropy production: they may be reversible. These reversible hidden variables are related to the other variables by linear algebraic equations, and can be eliminated easily. To simplify our expressions we shall assume that this elimination has been done so that every flux $\dot{q}_{k+1}, \dot{q}_{k+2}, \ldots, \dot{q}_n$ participates in the entropy production.

We now divide all the variables into two groups:

$$\mathbf{q}^{(1)} = \begin{bmatrix} q_1 \\ q_2 \\ \cdot \\ \cdot \\ \cdot \\ q_k \end{bmatrix}, \qquad \mathbf{q}^{(2)} = \begin{bmatrix} q_{k+1} \\ q_{k+2} \\ \cdot \\ \cdot \\ \cdot \\ q_n \end{bmatrix}, \tag{1}$$

where $\mathbf{q}^{(1)}$ is the vector to be observed, the elements of which correspond to the generalized forces $Q_1, \ldots, Q_k$ written as a vector

$$\mathbf{Q} = \begin{bmatrix} Q_1 \\ Q_2 \\ \cdot \\ \cdot \\ \cdot \\ Q_k \end{bmatrix}. \tag{2}$$

We may now write the equations of evolution in the form of partitioned matrices:

$$\left[\begin{array}{c:c} \mathbf{A}_{11} & \mathbf{A}_{12} \\ \hdashline \mathbf{A}_{12}^T & \mathbf{A}_{22} \end{array}\right] \left[\begin{array}{c} \mathbf{q}^{(1)} \\ \hdashline \mathbf{q}^{(2)} \end{array}\right] + \left[\begin{array}{c:c} \mathbf{B}_{11} & \mathbf{B}_{12} \\ \hdashline \mathbf{B}_{12}^T & \mathbf{B}_{22} \end{array}\right] \left[\begin{array}{c} \dot{\mathbf{q}}^{(1)} \\ \hdashline \dot{\mathbf{q}}^{(2)} \end{array}\right] = \left[\begin{array}{c} \mathbf{Q} \\ \hdashline 0 \end{array}\right], \tag{3}$$

where a superscript $T$ indicates the transpose of the matrix. The submatrices $\mathbf{A}_{11}$, $\mathbf{A}_{22}$, $\mathbf{B}_{11}$, $\mathbf{B}_{22}$ are symmetric. The matrix $\mathbf{B}_{22}$ is positive definite because we have assumed that every element of $\dot{\mathbf{q}}^{(2)}$ participates in the entropy production.

Our problem is to eliminate $\mathbf{q}^{(2)}$ and relate $\mathbf{Q}$ to $\mathbf{q}^{(1)}$ directly. Let us first consider the subsystem

$$\mathbf{A}_{22}\mathbf{q}^{(2)} + \mathbf{B}_{22}\dot{\mathbf{q}}^{(2)} = 0. \tag{4}$$

Since $\mathbf{B}_{22}$ is a positive definite square matrix of order $n - k$, there exists $n - k$ linearly independent orthonormal relaxation modes $\boldsymbol{\psi}^{(1)}, \boldsymbol{\psi}^{(2)}, \ldots, \boldsymbol{\psi}^{(n-k)}$, corresponding to the real-valued eigenvalues $\lambda_1, \lambda_2, \ldots, \lambda_{n-k}$. Let $\boldsymbol{\Psi}$ be a square matrix composed of the columns of relaxation modes,

$$\boldsymbol{\Psi} = \begin{bmatrix} \boldsymbol{\psi}^{(1)} & \boldsymbol{\psi}^{(2)} & \cdots & \boldsymbol{\psi}^{(n-k)} \\ \cdot & \cdot & & \cdot \\ \cdot & \cdot & & \cdot \\ \cdot & \cdot & & \cdot \\ \cdot & \cdot & & \cdot \end{bmatrix}, \tag{5}$$

and let the relaxation modes be so normalized that

$$\Psi^T \mathbf{B}_{22} \Psi = \mathbf{I}, \qquad \Psi^T \mathbf{A}_{22} \Psi = \mathbf{\Lambda}, \tag{6}$$

where $\mathbf{I}$ is a unit matrix and $\mathbf{\Lambda}$ is diagonal:

$$\mathbf{I} = \begin{bmatrix} 1 & \cdots & 0 \\ \vdots & \ddots & \vdots \\ 0 & \cdots & 1 \end{bmatrix}, \qquad \mathbf{\Lambda} = \begin{bmatrix} \lambda_1 & \cdots & 0 \\ \vdots & \ddots & \vdots \\ 0 & \cdots & \lambda_{n-k} \end{bmatrix}. \tag{7}$$

By introducing the normal coordinates $\boldsymbol{\xi}$ defined by the relation

$$\mathbf{q}^{(2)} = \Psi \boldsymbol{\xi}, \tag{8}$$

we are ready to simplify Eq. (3). Let us introduce Laplace transformation with respect to time for the $\mathbf{q}$'s and $\mathbf{Q}$'s:

$$\bar{q}_i = \int_0^\infty e^{-st} q_i(t)\, dt, \tag{9}$$

etc., and write $\bar{\mathbf{q}}^{(1)}$ for the column matrix whose elements are $\bar{q}_1, \bar{q}_2, \ldots, \bar{q}_k$, etc. For our purpose it is sufficient to consider zero initial condition, $\mathbf{q}^{(1)} = 0$, $\mathbf{q}^{(2)} = 0$, when $t = 0$. Then, since the Laplace transformation of $\dot{q}_i$ is equal to $s\bar{q}_i$, we can reduce the differential Eq. (3) into the algebraic equation

$$\left\{ \begin{bmatrix} \mathbf{A}_{11} & \mathbf{A}_{12} \\ \mathbf{A}_{12}^T & \mathbf{A}_{22} \end{bmatrix} + s \begin{bmatrix} \mathbf{B}_{11} & \mathbf{B}_{12} \\ \mathbf{B}_{12}^T & \mathbf{B}_{22} \end{bmatrix} \right\} \begin{Bmatrix} \bar{\mathbf{q}}^{(1)} \\ \bar{\mathbf{q}}^{(2)} \end{Bmatrix} = \begin{Bmatrix} \bar{\mathbf{Q}} \\ 0 \end{Bmatrix}. \tag{10}$$

Finally, let us define an $n \times n$ matrix $\mathbf{\Phi}$

$$\mathbf{\Phi} = \begin{bmatrix} \mathbf{I} & 0 \\ 0 & \Psi \end{bmatrix}. \tag{11}$$

Then it is easy to verify, on account of Eqs. (6) and (7), that

$$\mathbf{\Phi} \begin{bmatrix} \bar{\mathbf{q}}^{(1)} \\ \boldsymbol{\xi} \end{bmatrix} = \begin{bmatrix} \bar{\mathbf{q}}^{(1)} \\ \bar{\mathbf{q}}^{(2)} \end{bmatrix}, \tag{12}$$

$$\mathbf{\Phi}^T \begin{pmatrix} \mathbf{A}_{11} & \mathbf{A}_{12} \\ \mathbf{A}_{12}^T & \mathbf{A}_{22} \end{pmatrix} \mathbf{\Phi} = \begin{pmatrix} \mathbf{A}_{11} & \mathbf{A}_{12}\Psi \\ \Psi^T \mathbf{A}_{12}^T & \mathbf{\Lambda} \end{pmatrix}, \tag{13}$$

$$\mathbf{\Phi}^T \begin{pmatrix} \mathbf{B}_{11} & \mathbf{B}_{12} \\ \mathbf{B}_{12}^T & \mathbf{B}_{22} \end{pmatrix} \mathbf{\Phi} = \begin{bmatrix} \mathbf{B}_{11} & \mathbf{B}_{12}\Psi \\ \Psi^T \mathbf{B}_{12}^T & \mathbf{I} \end{bmatrix}. \tag{14}$$

Premultiply (10) by $\mathbf{\Phi}^T$, substituting (12), and using (13), (14), we obtain

$$\begin{bmatrix} \mathbf{A}_{11} + s\mathbf{B}_{11} & \mathbf{A}_{12}\Psi + s\mathbf{B}_{12}\Psi \\ \Psi^T \mathbf{A}_{12}^T + s\Psi^T \mathbf{B}_{12}^T & \mathbf{\Lambda} + s\mathbf{I} \end{bmatrix} \begin{bmatrix} \bar{\mathbf{q}}^{(1)} \\ \boldsymbol{\xi} \end{bmatrix} = \begin{bmatrix} \bar{\mathbf{Q}} \\ 0 \end{bmatrix}. \tag{15}$$

Expanding by rows, we see that this is equivalent to two equations:

$$(\mathbf{A}_{11} + s\mathbf{B}_{11})\bar{\mathbf{q}}^{(1)} + (\mathbf{A}_{12} + s\mathbf{B}_{12})\boldsymbol{\Psi}\bar{\boldsymbol{\xi}} = \bar{\mathbf{Q}}. \tag{16}$$

$$\boldsymbol{\Psi}^T(\mathbf{A}_{12}^T + s\mathbf{B}_{12}^T)\bar{\mathbf{q}}^{(1)} + (\boldsymbol{\Lambda} + s\mathbf{I})\bar{\boldsymbol{\xi}} = 0. \tag{17}$$

From (17), we have

$$\bar{\boldsymbol{\xi}} = -(\boldsymbol{\Lambda} + s\mathbf{I})^{-1}\boldsymbol{\Psi}^T(\mathbf{A}_{12}^T + s\mathbf{B}_{12}^T)\bar{\mathbf{q}}^{(1)}. \tag{18}$$

A substitution back into (16) then gives the final result:

$$\bar{\mathbf{Q}} = [\mathbf{A}_{11} + s\mathbf{B}_{11} - (\mathbf{A}_{12} + s\mathbf{B}_{12})\boldsymbol{\Psi}(\boldsymbol{\Lambda} + s\mathbf{I})^{-1}\boldsymbol{\Psi}^T(\mathbf{A}_{12}^T + s\mathbf{B}_{12}^T)]\bar{\mathbf{q}}^{(1)}. \tag{19}$$

It is clear that the matrix in the brackets is symmetric. The inverse matrix $(\boldsymbol{\Lambda} + s\mathbf{I})^{-1}$ is simply

$$(\boldsymbol{\Lambda} + s\mathbf{I})^{-1} = \begin{Bmatrix} \frac{1}{s+\lambda_1} & 0 & \cdots & 0 \\ 0 & \frac{1}{s+\lambda_2} & \cdots & 0 \\ \cdot & \cdot & & \\ \cdot & \cdot & & \\ \cdot & \cdot & & \\ 0 & 0 & \cdots & \frac{1}{s+\lambda_{n-k}} \end{Bmatrix} \tag{20}$$

The general term in (19) may be written in the form

$$\blacktriangle \quad \bar{Q}_i = \sum_{j=1}^{k} \left( c_{ij} + sc'_{ij} + \sum_{\alpha=1}^{n-k} \frac{c_{ij}^{(\alpha)}}{s + \lambda_\alpha} \right) \bar{q}_j^{(1)}, \tag{21}$$

where

$$\blacktriangle \quad c_{ij} = c_{ji}, \qquad c'_{ij} = c'_{ji}, \qquad c_{ij}^{(\alpha)} = c_{ji}^{(\alpha)}. \tag{22}$$

By taking the inverse transform, we obtain the general form of the generalized force-generalized displacement relationship,

$$\blacktriangle \quad Q_i(t) = \sum_{j=1}^{k} \left[ c_{ij} q_j(t) + c'_{ij} \dot{q}_j(t) + \sum_{\alpha=1}^{n-k} c_{ij}^{(\alpha)} \int_0^t e^{-\lambda_\alpha (t-\tau)} q_j(\tau)\, d\tau \right]. \tag{23}$$

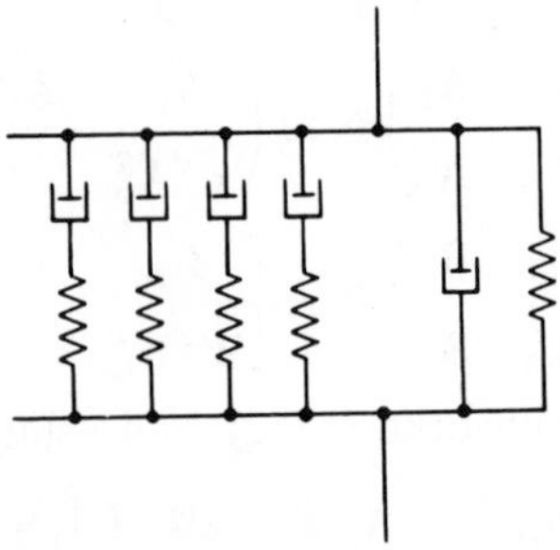

**Fig. 13.8:1.** Model suggested by Eq. (13.8:23).

If this equation is interpreted in terms of mechanical models, we see that any relaxation phenomena may be represented by a spring, a dashpot, and a sum of a great many elements made up of a Maxwell type material (Fig. 13.8:1).

## 13.9. ANISOTROPIC LINEAR VISCOELASTIC MATERIALS

Let the six independent components of the stress tensor $\sigma_{ij}$ play the role of the generalized forces $Q_1, \ldots, Q_6$, and let the six independent components of the strain tensor $e_{ij}$ play the role of the generalized coordinates $q_1, \ldots, q_6$. According to the results of Sec. 13.8, the effect of relaxation modes renders the stress-strain relationship into the form (with summation convention used)

(1) ▲ $$\bar{\sigma}_{ij} = \bar{C}_{ij}^{kl}(s)\bar{e}_{kl},$$

with

(2) ▲ $$\bar{C}_{ij}^{kl}(s) = \sum_{\alpha=1}^{n} \frac{D_{ij}^{kl(\alpha)}}{s + \lambda_\alpha} + D_{ij}^{kl} + sD_{ij}^{\prime kl}.$$

The restrictions imposed by thermodynamics are

(3) ▲ $$\bar{C}_{ij}^{kl} = \bar{C}_{ji}^{kl} = \bar{C}_{ij}^{lk}, \qquad \bar{C}_{ij}^{kl} = \bar{C}_{kl}^{ij}.$$

Strictly speaking, the last two Laplace transform terms in (2) have no inverse in the time domain. But the inverse of $\bar{C}_{ij}^{kl}/s^2$ exists. Hence, if $s^2\bar{e}_{kl}$ has a Laplace transform, the inverse of $\bar{\sigma}_{ij}$ given by Eq. (1) can still be defined. However, as the delta function and higher order delta functions are often used in mathematical physics as generalized functions, we can formally write

(4) $$C_{ij}^{kl}(t) = \sum_{\alpha} D_{ij}^{kl(\alpha)} e^{-\lambda_\alpha t}\mathbf{1}(t) + D_{ij}^{kl}\,\boldsymbol{\delta}(t) + D_{ij}^{\prime kl}\,\boldsymbol{\delta}_1(t),$$

where $\mathbf{1}(t)$, $\boldsymbol{\delta}(t)$, $\boldsymbol{\delta}_1(t)$ are, respectively, the unit-step function, the delta function, and the first-order delta function. The inverse of (1) then gives, formally,

(5) ▲ $$\sigma_{ij}(t) = \sum_{\alpha} \int_0^t e^{-\lambda_\alpha(t-\tau)} D_{ij}^{kl(\alpha)} e_{kl}(\tau)\, d\tau + D_{ij}^{kl} e_{kl}(t) + D_{ij}^{\prime kl} \frac{\partial e_{kl}}{\partial t}.$$

A further generalization is possible if the hidden variables are so numerous so that the summation over $\alpha$ is replaced by an integration over $\lambda_\alpha$. See M. A. Biot[13.1] (1956).

On the other hand, in analogy to the Neumann-Duhamel generalization of Hooke's law in thermoelasticity, we may wish to include the deviation of local temperature from the uniform reference temperature, $T - T_0$, as an observable variable in the stress-strain relationship. This can be done by adding another coordinate $q_7$ to be the local temperature deviation, while the corresponding force $Q_7$ is taken to mean the local value of entropy per unit volume above that at the reference state.

## PROBLEMS

**13.3.** Specialize the results of Sec. 13.9 to the case of an isotropic material. (See M. A. Biot,[13.1] 1954.)

**13.4.** From Eq. (13.7:9) or (13.7:11), derive the form of the coefficients $A_{ij}^{kl}$ that relates strain to stress:

$$\bar{e}_{ij} = \bar{A}_{ij}^{kl}(s)\bar{\sigma}_{kl}.$$

**13.5.** If, for purely mathematical generality, we write the most general linear relation between the Laplace transforms of stress and strain as

$$\left(\sum_{\alpha} a_{\mu\nu}^{ij(\alpha)} s^{\alpha}\right)\bar{\sigma}_{\mu\nu} = \left(\sum_{\alpha} b_{\mu\nu}^{ij(\alpha)} s^{\alpha}\right)\bar{e}_{\mu\nu},$$

where $s$ is the Laplace transformation variable, and

$$\bar{\sigma}_{\mu\nu}(s) = \int_0^{\infty} \sigma_{\mu\nu}(t)\, e^{-st}\, dt,$$

etc., enumerate the ways that this general expression may not satisfy thermodynamic requirements. (See Biot,[13.1] 1958.)

# 14

# THERMOELASTICITY

Basic equations of thermoelasticity will be discussed in this chapter, and their applications will be illustrated in some typical problems.

## 14.1. BASIC EQUATIONS

We shall consider a solid body subjected to external forces and heating. We assume that the material is linear elastic, and that it is stress-free at a uniform temperature $T_0$ when all external forces are removed. The stress-free state will be referred to as the *reference state*, and the temperature $T_0$ as the *reference temperature*. A system of rectangular Cartesian coordinates $x_i$ will be chosen. The displacement $u_i$ of every particle in the instantaneous state from its position in the reference state will be assumed to be small, so that the infinitesimal strain components are

(1) $$e_{ij} = \tfrac{1}{2}(u_{i,j} + u_{j,i}), \qquad i, j = 1, 2, 3.$$

The instantaneous absolute temperature will be denoted by $T$, and the difference $T - T_0$ by $\theta$:

(2) $$\theta = T - T_0.$$

Under these conditions the basic equations of thermoelasticity are

*The Constitutive Equation.* Duhamel-Neumann law. (12.7:4.)

(3) $$\sigma_{ij} = C_{ijkl}e_{kl} - \beta_{ij}(T - T_0).$$

*Conservation of Mass.* Continuity equation.

(4) $$\frac{\partial \rho}{\partial t} + \frac{\partial \rho v_i}{\partial x_i} = 0, \qquad v_i = \frac{\partial u_i}{\partial t}.$$

*Conservation of Momentum.* Newton's law.

(5) $$\rho \dot{v}_i = \frac{\partial \sigma_{ij}}{\partial x_j} + X_i.$$

*Conservation of Energy.* (12.2:7).

(6) $$\dot{\mathscr{E}} = T\dot{\mathscr{S}} + \frac{1}{\rho}\sigma_{ij}V_{ij}, \qquad V_{ij} = \frac{1}{2}\left(\frac{\partial v_i}{\partial x_j} + \frac{\partial v_j}{\partial x_i}\right).$$

*Rate of Change of Entropy.* (13.1:2) and (13.2:6).

$$\rho\dot{\mathscr{S}} = -\frac{1}{T}\frac{\partial h_i}{\partial x_i} = -\frac{\partial}{\partial x_i}\left(\frac{h_i}{T}\right) - \frac{h_i}{T^2}\frac{\partial T}{\partial x_i}. \tag{7}$$

*Heat Conduction.* Fourier's law. (13.3:4), (13.4:10).

$$h_i = -k_{ij}\frac{\partial T}{\partial x_j}. \tag{8}$$

*Definition of Specific Heat.* If $\dot{e}_{ij} = 0$ $(i, j = 1, 2, 3)$, then

$$-\frac{\partial h_i}{\partial x_i} = \rho C_v \dot{T}. \tag{9}$$

In these equations, all the indices range over 1, 2, 3; $\rho$ is the mass density, $\sigma_{ij}$ are components of the stress tensor, $C_{ijkl}$ are the elastic moduli; $\beta_{ij}$ are the thermal moduli; $v_i$ are components of the velocity vector; $X_i$ are components of the body force per unit volume; $\mathscr{E}$ is the internal energy per unit mass; $\mathscr{S}$ is the entropy per unit mass; $h_i$ are the components of the heat flux vector; $k_{ij}$ are the heat conduction coefficients; and $C_v$ the heat capacity per unit mass at constant volume of the solid. A dot above a variable denotes the material derivative of that variable:

$$(\dot{\ }) = \frac{D(\ )}{Dt} = \frac{\partial(\ )}{\partial t} + v_j\frac{\partial(\ )}{\partial x_j}.$$

which, under the approximation of small velocities, is the same as the partial derivative with respect to time. In the following, we assume

$$(\dot{\ }) = \frac{\partial(\ )}{\partial t}.$$

A comma indicates partial differentiation, thus $\theta_{,i}$ means $\partial\theta/\partial x_i$. Summation convention for repeated indices is used. Equation (3) is discussed in Sec. 12.7; Eqs. (4) and (5) are discussed in Sec. 5.4; Eq. (6) is discussed in Sec. 12.2; Eq. (7) is discussed in Secs. 13.1 and 13.2, where $h_i/T$ is identified as the rate of change of entropy displacement and $-(h_i/T^2)\,\partial T/\partial x_i$ as the entropy production. We have assumed in Eq. (7) that there is no heat source in the material; otherwise, to the right-hand side of Eq. (7) we should add a term representing the strength of the heat source (heat generation per unit time per unit volume). Equation (8) is discussed in Sec. 13.3. The coefficients $C_{ijkl}$, $\beta_{ij}$, and $k_{ij}$ have the following symmetry properties as discussed in Secs. 12.7 and 13.3.

$$C_{ijkl} = C_{klij} = C_{ijlk} = C_{jikl}, \tag{10}$$

$$\beta_{ij} = \beta_{ji}, \tag{11}$$

$$k_{ij} = k_{ji}. \tag{12}$$

The energy equation may be put in a more convenient form as follows. On introducing the free energy $\mathscr{F}$

$$\mathscr{F} = \mathscr{E} - T\mathscr{S},$$

we obtain, according to Eq. (12.3:6),

$$\rho\left(\frac{\partial \mathscr{F}}{\partial e_{ij}}\right)_T = \sigma_{ij}, \qquad \left(\frac{\partial \mathscr{F}}{\partial T}\right)_{e_{ij}} = -\mathscr{S}. \tag{13}$$

With (13), Eq. (7) can be written as

$$\begin{aligned} -\frac{1}{T}\frac{\partial h_i}{\partial x_i} = \rho\dot{\mathscr{S}} &= \rho\frac{\partial \mathscr{S}}{\partial e_{ij}}\dot{e}_{ij} + \rho\frac{\partial \mathscr{S}}{\partial T}\dot{T} \\ &= -\rho\frac{\partial^2 \mathscr{F}}{\partial e_{ij}\,\partial T}\dot{e}_{ij} - \rho\frac{\partial^2 \mathscr{F}}{\partial T^2}\dot{T}. \end{aligned} \tag{14}$$

A multiplication by $T$ and comparison with (9) when $\dot{e}_{ij} = 0$ yields

$$C_v = -T\frac{\partial^2 \mathscr{F}}{\partial T^2}. \tag{15}$$

Also, from (13) and (3), it is seen that

$$\rho\frac{\partial^2 \mathscr{F}}{\partial e_{ij}\,\partial T} = \frac{\partial \sigma_{ij}}{\partial T} = -\beta_{ij}. \tag{16}$$

Equation (14) therefore takes the form

$$-\frac{\partial h_i}{\partial x_i} = \rho C_v\dot{T} + T\beta_{ij}\dot{e}_{ij}. \tag{17 ▲}$$

Finally, on substituting Eq. (8) into (17), we obtain

$$\frac{\partial}{\partial x_i}\left(k_{ij}\frac{\partial T}{\partial x_j}\right) = \rho C_v\frac{\partial T}{\partial t} + T\beta_{ij}\frac{\partial e_{ij}}{\partial t}. \tag{18 ▲}$$

If the material is isotropic, these equations are simplified as follows.

$$\sigma_{ij} = \lambda e_{\mu\mu}\,\delta_{ij} + 2Ge_{ij} - \beta\,\delta_{ij}\theta, \tag{19}$$

$$\beta_{ij} = \beta\,\delta_{ij}, \qquad \beta = \frac{\alpha E}{1 - 2\nu} = (3\lambda + 2G)\alpha, \tag{20}$$

$$k_{ij} = k\,\delta_{ij}. \tag{21}$$

The constant $\alpha$ is the thermal coefficient of linear expansion and $k$ is the heat conductivity.

A problem in thermoelasticity is formulated when appropriate boundary conditions and initial conditions are specified. Several simple examples will be discussed below. A proof for the existence and uniqueness of solution under suitable continuity conditions can be constructed in a way analogous to that discussed in Sec. 7.4; an explicit proof in the case of isotropic materials can be found in Boley and Weiner[14.1], pp. 38–40.

## 14.2. THERMAL EFFECTS DUE TO A CHANGE OF STRAIN; KELVIN'S FORMULA

It is a familiar fact that an adiabatic expansion of a gas is accompanied by a drop in its temperature. Similarly, a solid body changes its temperature when the state of strain of the body is altered adiabatically. For a material like steel, there will be a fall of temperature when the body is strained to expand adiabatically.

Equation (14.1:18) gives the relationship between the rate of change of the temperature and the strain with the heat conduction:

$$\frac{\partial}{\partial x_i}\left[k_{ij}\frac{\partial T}{\partial x_j}\right] = \rho C_v \frac{\partial T}{\partial t} + T\beta_{ij}\frac{\partial e_{ij}}{\partial t}.$$

If heat conduction is prevented, the left-hand side should vanish, and we obtain at once

$$(1) \qquad \frac{\partial T}{\partial t} = -\frac{T}{\rho C_v}\beta_{ij}\frac{\partial e_{ij}}{\partial t},$$

which is Kelvin's formula for the change of temperature of an insulated body due to a uniform strain.

## 14.3. RATIO OF ADIABATIC TO ISOTHERMAL ELASTIC MODULI

If an elastic modulus is measured on a sample that is completely insulated, or if the change in strain takes place so rapidly that the heat does not have time to escape, the measured value may be called the *adiabatic modulus*. On the other hand, it is called the *isothermal* modulus if it is measured in such a way that the temperature is kept uniform and constant throughout the process.

The coefficients $C_{ijkl}$ in the Duhamel-Neumann law (14.1:3) is the isothermal modulus of elasticity at $\theta = 0$:

$$(1) \qquad \sigma_{ij} = C_{ijkl}e_{kl} - \beta_{ij}\theta, \qquad \theta = T - T_0.$$

If a deformation is adiabatic, and no heat conduction takes place, then Eq. (14.1:17) gives

$$(2) \qquad \rho C_v\dot{\theta} + T_0\beta_{ij}\dot{e}_{ij} = 0,$$

or

$$(3) \qquad \rho C_v\theta + T_0\beta_{kl}e_{kl} = \text{const.}$$

The constant of integration is zero if $e_{ij} = 0$ when $\theta = 0$. On solving (3) for $\theta$ and substituting into (1), we obtain in an adiabatic process

$$(4) \qquad \sigma_{ij} = \left(C_{ijkl} + \frac{T_0}{\rho C_v}\beta_{ij}\beta_{kl}\right)e_{kl} = C'_{ijkl}e_{kl}.$$

Thus, the adiabatic modulus of elasticity $C'_{ijkl}$ is related to the isothermal modulus $C_{ijkl}$ by the equation

$$C'_{ijkl} = C_{ijkl} + \frac{T_0}{\rho C_v} \beta_{ij}\beta_{kl} \tag{5}$$

## PROBLEMS

**14.1.** Derive the relationship of the temperature and the elastic constant $(\partial p/\partial V)_T$ for an ideal gas, and compare qualitatively the temperature change in an adiabatic expansion of a gas with the temperature change of a similar process in a solid.

**14.2.** Show that the Young's modulus of steel determined by an adiabatic process such as the propagation of elastic waves or the longitudinal vibration of a rod is higher than that determined by a slow static test in which the strains are maintained constant for a sufficiently long time for the temperature to become uniform.

**14.3.** For steel at $T = 274.7°$ K, or nearly 1.6° C, $\rho = 7.0$, $\alpha = 1.23 \times 10^{-5}$, $C_v = 0.102$ gm-cal/(gm) (°C), show that $dT = -0.125°$ C if a wire is subjected adiabatically to a sudden increment of tensile stress of $1.09 \times 10^9$ dyne/sq cm. Over a century ago, Joule [*Phil. Trans. Roy. Soc. London*, **149** (1859), p. 91] gave an observed value $dT = -0.1620°$ C for the above case by experiments on cylindrical bars.

**14.4.** Consider simple tension and derive the following relationship between the adiabatic and isothermal Young's modulus, $E'$ and $E$, respectively:

$$\frac{1}{E'} = \frac{1}{E} - \frac{\alpha^2 T_0}{J C_v \rho},$$

which shows that the adiabatic modulus $E'$ is always greater than the isothermal modulus $E$. In the equation above $\alpha$ is the linear coefficient of expansion, $T_0$ is the equilibrium temperature, $J$ is the mechanical equivalent of heat, $C_v$ is the specific heat per unit mass at constant strain, $\rho$ is the density of the material.

**14.5.** Show that the theoretical value of the ratio $E'/E$ is of order 1.003 for steel and copper at room temperature.

## 14.4. UNCOUPLED, QUASI-STATIC THERMOELASTIC THEORY

The basic equations given in Sec. 14.1 combine the theory of elasticity with heat conduction under transient conditions. Boundary-value problems involving these equations are rather difficult to solve. Fortunately, in most engineering applications it is possible to omit the mechanical coupling term in the energy equation (14.1:6) or (14.1:18) and the inertia term in the equation of motion (14.1:5) without significant error. When these simplifying assumptions are introduced, the theory is referred to as an *uncoupled, quasi-static theory*; it degenerates into heat conduction and thermoelasticity as two separate problems.

A plausible argument for the smallness of the thermoelastic coupling is as follows. We have seen in Sec. 14.2 that the change of temperature of an elastic body due to adiabatic straining is, in general, very small. If this interaction between strain and temperature is ignored, then the only effects of elasticity on the temperature distribution are effects of change in dimensions of the body under investigation. The change in dimension of a body is of the order of the product of the linear dimension of the body $L$, the temperature rise $T$, and the coefficient of thermal expansion $\alpha$. If $L = 1$ in., $\theta = 1000°$ F, $\alpha = 10^{-5}$ per °F, the change in dimension is $10^{-2}$ in., which is negligible in problems of heat conduction.

Again, if the temperature rise from 0 to 1000° F were achieved in a time interval of 0.1 sec, then the acceleration is of the order $10^{-2} \div (0.1)^2 = 1$ in./sec². The change of stress due to this acceleration may be estimated from the equation of equilibrium

$$\Delta\sigma_{xx} \cong \Delta x \rho \frac{d^2u}{dt^2} .$$

If the specific gravity of the material is ten and the material is 1 in. thick, we have

$$\Delta\sigma_{xx} = \frac{1}{12} \times 10 \times 62.4 \times \frac{1}{32.2} \times \frac{1}{12} \times \frac{1}{144} \cong 0.001 \text{ lb/sq in.}$$

This stress is negligible in most structural problems in which the magnitude of the stresses concerned are of the order of the yielding stress or ultimate stress of the material.

Some examples of the coupled theory are presented in Chap. 2 of Boley and Weiner's book.[14.1] It is pointed out that the thermomechanical coupling is important in the problem of internal friction of metals.

## 14.5. TEMPERATURE DISTRIBUTION

We shall denote the temperature above a reference temperature by $\theta$, the space coordinates by $(x, y, z)$, and time by $t$. In an uncoupled theory for an isotropic material, Eq. (14 1:18) becomes

$$\text{(1)} \qquad \frac{\partial}{\partial x}\left(k\frac{\partial\theta}{\partial x}\right) + \frac{\partial}{\partial y}\left(k\frac{\partial\theta}{\partial y}\right) + \frac{\partial}{\partial z}\left(k\frac{\partial\theta}{\partial z}\right) = \rho C_v \frac{\partial\theta}{\partial t} .$$

In the following discussion we shall assume $k$ and $C_v$ to be constants. Then,

$$\text{(2)} \qquad k\left[\frac{\partial^2\theta}{\partial x^2} + \frac{\partial^2\theta}{\partial y^2} + \frac{\partial^2\theta}{\partial z^2}\right] = \rho C_v \frac{\partial\theta}{\partial t} .$$

In cylindrical coordinates $(x, r, \psi)$, we have

$$k\left[\frac{\partial^2\theta}{\partial x^2} + \frac{\partial^2\theta}{\partial r^2} + \frac{1}{r}\frac{\partial\theta}{\partial r} + \frac{1}{r^2}\frac{\partial^2\theta}{\partial\psi^2}\right] = \rho C_v \frac{\partial\theta}{\partial t}. \tag{3}$$

In spherical polar coordinates $(r, \psi, \varphi)$, we have

$$k\left[\frac{1}{r^2}\frac{\partial}{\partial r}\left(r^2\frac{\partial\theta}{\partial r}\right) + \frac{1}{r^2 \sin\psi}\frac{\partial}{\partial\psi}\left(\sin\psi\frac{\partial\theta}{\partial\psi}\right) + \frac{1}{r^2\sin^2\psi}\frac{\partial^2\theta}{\partial\varphi^2}\right] = \rho C_v\frac{\partial\theta}{\partial t}. \tag{4}$$

For the special case of steady heat flow, $\partial\theta/\partial t$ vanishes, then

$$\nabla^2\theta = 0. \tag{5}$$

Numerous examples of solution of these equations can be found in books on the conduction of heat in solids, such as Carslaw and Jaeger, McAdams, etc. (See Bibliographies 14.1, 14.2.)

*Example. Steady Temperature Distribution in a Disk Cooled by Air*

If there is a heat source in the material, of *Q Btu* per unit volume per unit time, the equation of heat conduction is modified into

$$k\,\nabla^2\theta + Q = \rho C_v\frac{\partial\theta}{\partial t}. \tag{6}$$

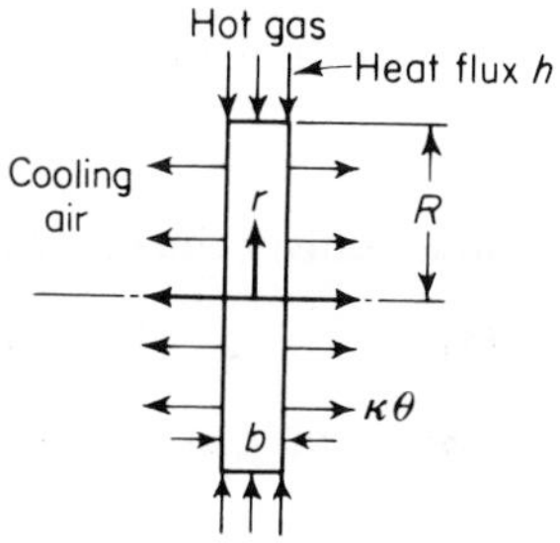


**Fig. 14.5:1.** A turbine disk.

An atomic reactor with a radiation source may present problems of this kind. A turbine disk may also be described approximately by an equation of this type. Consider a circular disk as shown in Fig. 14.5:1. The heat flux from the rim represents the heat from the turbine blades operating in hot combustion gas. The heat loss on the disk surface represents the effect of cooling air. Let $\theta(r)$ be the average temperature (above the cooling air) across the disk at any radius $r$. The heat loss per unit area on each face is $-\kappa\theta$. The total heat loss per unit volume is $Q = -2\kappa\theta/b$. Hence, for the axially symmetric temperature distribution, the heat conduction equation at the steady state is

$$\frac{d^2\theta}{dr^2} + \frac{1}{r}\frac{d\theta}{dr} - \frac{2\kappa}{kb}\theta = 0. \tag{7}$$

This is the differential equation for the modified Bessel functions. The appropriate solution is

$$\theta = AI_0\left(\sqrt{\frac{2\kappa}{kb}}\,r\right), \tag{8}$$

where $I_0(z)$ is the modified Bessel function of the first kind and zeroth order.

The constant $A$ is determined by the boundary condition at the rim $r = R$, where a heat flux $h$ enters the disk from the hot gas:

$$k\frac{d\theta}{dr} = h \qquad \text{at } r = R. \tag{9}$$

Therefore,

$$kA\sqrt{\frac{2\kappa}{kb}}\,I_0'\left(\sqrt{\frac{2\kappa}{kb}}\,R\right) = h, \tag{10}$$

where $I_0'$ is the derivative of $I_0$ and is equal to $I_1$. Thus, finally,

$$\theta = \frac{h}{\sqrt{2\kappa k/b}}\,\frac{I_0(\sqrt{2\kappa/kb}\,r)}{I_1(\sqrt{2\kappa/kb}\,R)}. \tag{11}$$

## 14.6. THERMAL STRESSES

In the uncoupled, quasi-static theory, the stress and strain fields are computed for each instantaneous temperature distribution $\theta(x_1, x_2, x_3)$ according to Eqs. (14.1:1) to (14.1:5), with appropriate boundary conditions. For an isotropic material, we have

$$\sigma_{ij} = \lambda e_{\mu\mu}\,\delta_{ij} + 2Ge_{ij} - \beta\,\delta_{ij}\theta, \tag{1}$$

$$\sigma_{ij,j} + X_i = 0. \tag{2}$$

The strain field

$$e_{ij} = \tfrac{1}{2}(u_{i,j} + u_{j,i}) \tag{3}$$

must satisfy the compatibility conditions

$$e_{ij,kl} + e_{kl,ij} - e_{ik,jl} - e_{jl,ik} = 0. \tag{4}$$

As a conjugate to (1), we have

$$e_{ij} = \frac{1+\nu}{E}\,\sigma_{ij} - \frac{\nu}{E}\,\sigma_{\mu\mu}\,\delta_{ij} + \alpha\theta\,\delta_{ij}, \qquad \beta = \frac{E\alpha}{1-2\nu}. \tag{5}$$

Let us first observe that if $\sigma_{ij} = 0$, then

$$e_{ij} = \alpha\theta\,\delta_{ij} \tag{6}$$

and Eq. (4) is reduced to

$$\theta_{,kl}\,\delta_{ij} + \theta_{,ij}\,\delta_{kl} - \theta_{,jl}\,\delta_{ik} - \theta_{,ik}\,\delta_{jl} = 0, \tag{7}$$

which is satisfied if

$$\theta_{,ij} = 0, \qquad i, j = 1, 2, 3. \tag{8}$$

The solution is

$$\theta = a_0 + a_1x_1 + a_2x_2 + a_3x_3, \tag{9}$$

with arbitrary coefficients $a_i$. Hence, if $\theta$ is a linear function of spatial coordinates $x_1, x_2, x_3$, and if the displacements on the boundary are unrestrained, then it is possible to satisfy the compatibility condition without calling stresses into play. Generally, for an arbitrary temperature field, the strain field corresponding to thermal expansion alone, Eq. (6), will not be compatible, in which case thermal stresses must be called into action.

If we substitute (1) and (3) into (2), we obtain the generalized Navier's equation

$$Gu_{i,\mu\mu} + (\lambda + G)u_{\mu,\mu i} + X_i - \beta\theta_{,i} = 0. \tag{10}$$

This is particularly convenient to use if the boundary condition is specified in terms of displacements:

$$u_i = f_i(x_1, x_2, x_3) \qquad \text{on the boundary.} \tag{11}$$

On the other hand, if tractions are specified on the boundary, i.e., if

$$\overset{\nu}{T}_i = \sigma_{ij}\nu_j = g_i(x_1, x_2, x_3) \qquad \text{on the boundary,} \tag{12}$$

where $\nu_j$ is the outer normal vector to the boundary surface, and $g_i$'s are specified, we must have

$$\nu_j[\lambda u_{\mu,\mu}\,\delta_{ij} + Gu_{i,j} + Gu_{j,i} - \beta\,\delta_{ij}\theta] = g_i \quad \text{on the boundary.} \tag{13}$$

By comparing Eqs. (10)–(13) with the corresponding equations in linear elasticity (Secs. 7.1 and 7.3), we see that the effect of the temperature change $\theta$ is equivalent to replacing the body force $X_i$ in Navier's equations by $X_i - \beta\theta_{,i}$ and to substituting for the surface tractions $g_i$ by $g_i + \beta\nu_i\theta$. Thus the displacements $u_1, u_2, u_3$ produced by a temperature change $\theta$ are the same as those produced by the body forces $-\beta\theta_{,i}$ and the normal tractions $\beta\theta$ acting on the surface of a body of the same shape but throughout which the temperature is uniform. These facts can be stated in a theorem.†

THEOREM: DUHAMEL-NEUMANN ANALOGY. *Consider two bodies of exactly the same shape but with conditions prescribed as shown in Fig.* 14.6:1. Then

$$u_i^{(\mathrm{I})}(x_1, x_2, x_3; t) = u_i^{(\mathrm{II})}(x_1, x_2, x_3; t),$$
$$\sigma_{ij}^{(\mathrm{I})} = \sigma_{ij}^{(\mathrm{II})} - \beta\theta^{(\mathrm{I})}\,\delta_{ij}.$$

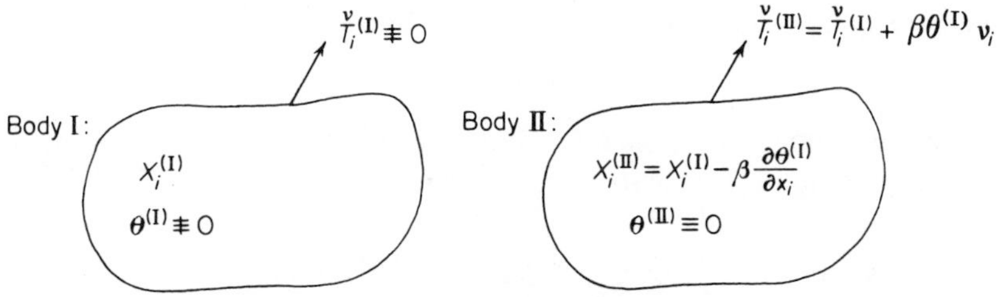


**Fig. 14.6:1.** Duhamel-Neumann analogy.

† The Duhamel-Neumann theorem was stated in this form by A. J. A. Morgan.

**Problem 14.6.** *Thermal Stresses in Plates with Clamped Edges.* Consider a flat plate of arbitrary shape clamped at all edges. Let the midplane of the plate in the unloaded position be the $x, y$-plane and let the temperature be uniform in the plane of the plate but variable throughout the thickness, i.e., $\theta = \theta(z)$. Show that under these conditions the plate will remain flat, i.e., no bending will be introduced except possibly near the edges. Determine the thermal stresses in the plate. This example shows the extreme importance of the edge constraints on the thermal stress problem.

## 14.7. PARTICULAR INTEGRAL. GOODIER'S METHOD

By introducing suitable particular integrals, the problem of thermal stresses can be reduced to the solution of the homogeneous Navier's equations. Let us assume a solution of Eq. (14.6:10) in the form

$$u_i = \frac{\partial \phi}{\partial x_i} \tag{1}$$

where $\phi$ is a *displacement potential.* Let us assume also that the body forces are conservative so that

$$X_i = -\frac{\partial P}{\partial x_i}, \tag{2}$$

then the generalized Navier's Eq. (14.6:10) may be written as

$$(\lambda + 2G)\phi_{,i\mu\mu} = P_{,i} + \beta\theta_{,i}. \tag{3}$$

An integration yields

$$\phi_{,\mu\mu} = \frac{1}{\lambda + 2G}(P + \beta\theta). \tag{4}$$

The constant of integration may be absorbed in the potential $P$. Since Eq. (4) is linear its solution may be written

$$\phi = \phi^{(p)} + \phi^{(c)}, \tag{5}$$

where $\phi^{(p)}$ is a particular integral satisfying Eq. (4), and $\phi^{(c)}$ is the complementary solution satisfying the equation

$$\nabla^2\phi^{(c)} = 0. \tag{6}$$

A particular integral can be taken in the form of the gravitational potential due to a distribution of matter of density $(P + \beta\theta)/(\lambda + 2G)$, i.e.,

$$\phi^{(p)}(x_i) = \frac{1}{4\pi(\lambda + 2G)}\iiint \frac{1}{r}\,[P(x_i') + \beta\theta(x_i')]\,dx_1'\,dx_2'\,dx_3', \tag{7}$$

where

$$r = [(x_1 - x_1')^2 + (x_2 - x_2')^2 + (x_3 - x_3')^2]^{1/2}. \tag{8}$$

Once $\phi^{(p)}$ is determined, the boundary conditions for the complementary function $\phi^{(c)}$ can be derived.

There are problems which cannot be solved by the method of scalar potential. However, the particular solution $\phi^{(p)}$ can still be used. From $\phi^{(p)}$, the displacements $u_i^{(p)}$ are computed according to Eq. (1). Then we put

$$u_i = u_i^{(c)} + u_i^{(p)}. \tag{9}$$

The equations governing $u_i^{(c)}$ are the homogeneous equations

$$(\lambda + G)\frac{\partial e^{(c)}}{\partial x_i} + G\,\nabla^2 u_i^{(c)} = 0, \qquad e^{(c)} = u_{\alpha,\alpha}^{(c)}, \qquad i = 1, 2, 3. \tag{10}$$

The boundary conditions for $u_i^{(c)}$ must be derived, of course, from the original boundary conditions by subtracting the contributions of the particular integral on the boundary.

**Problem 14.7.** Consider the special case in which $X_i = 0$. Show that, in view of the heat conduction equation

$$k\,\nabla^2\theta = \rho C_v \frac{\partial \theta}{\partial t},$$

a particular solution $\phi^{(p)}$ of Eq. (4) is

$$\phi^{(p)}(x_1, x_2, x_3; t) = \frac{\alpha k}{\rho C_v}\frac{1+\nu}{1-\nu}\int_t^\infty \theta(x_1, x_2, x_3; \xi)\,d\xi, \tag{11}$$

provided that $\theta \to 0$ as $t \to \infty$.

## 14.8. PLANE STRAIN

A long cylindrical body is said to be in the state of plane strain parallel to the $x, y$-plane if the displacement component $w$ vanishes and the components $u$ and $v$ are functions of $x, y$, but not of $z$. The temperature distribution must be independent of $z$ also. Let the body occupy a domain $V(x, y)$ with boundary $B$ in the $x, y$-plane (Fig. 14.8:1). Then, in $V + B$,

$$u = u(x, y), \qquad v = v(x, y), \qquad w = 0, \qquad \theta = \theta(x, y). \tag{1}$$

It is easily verified that

$$e_{xz} = e_{yz} = e_{zz} = 0, \tag{2}$$

$$\sigma_{xz} = \sigma_{yz} = 0, \tag{3}$$

$$\begin{aligned}\sigma_{zz} &= \lambda(e_{xx} + e_{yy}) - \beta\theta \\ &= \nu(\sigma_{xx} + \sigma_{yy}) - \alpha E\theta.\end{aligned} \tag{4}$$

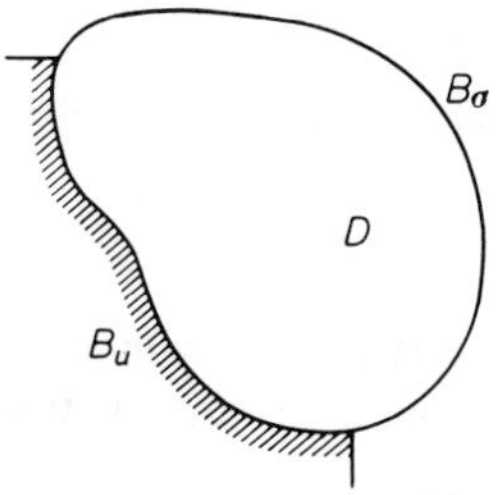

**Fig. 14.8:1.** Notations.

We may derive Duhamel's analogy explicitly in this two-dimensional case. For a body in $V + B$, subjected to a temperature rise $\theta(x, y)$, but with zero body force and zero surface traction over $B_\sigma$, and assigned displacements $u_i$ on $B_u$, where $B = B_\sigma + B_u$, the stress field is the same as that given by a superposition of a hydrostatic pressure

(5) $$\sigma_{xx} = \sigma_{yy} = -\beta\theta, \qquad \sigma_{xy} = 0,$$

with that given by another problem in which the same body in $V + B$ is kept at the uniform reference temperature, but subjected to a body force $X = -\beta(\partial\theta/\partial x)$; $Y = -\beta(\partial\theta/\partial y)$, and a tension $\beta\theta$ over $B_\sigma$.

Let us see how the last-mentioned problem can be solved. The equation of equilibrium is

(6) $$\frac{\partial\sigma_{xx}}{\partial x} + \frac{\partial\sigma_{xy}}{\partial y} - \beta\frac{\partial\theta}{\partial x} = 0, \qquad \frac{\partial\sigma_{xy}}{\partial x} + \frac{\partial\sigma_{yy}}{\partial y} - \beta\frac{\partial\theta}{\partial y} = 0,$$

and the compatibility condition is

(7) $$\frac{\partial^2 e_{xx}}{\partial y^2} + \frac{\partial^2 e_{yy}}{\partial x^2} = 2\frac{\partial^2 e_{xy}}{\partial x\,\partial y},$$

which can be reduced to

(8) $$\nabla^2(\sigma_{xx} + \sigma_{yy}) = \frac{\beta}{1 - \nu}\nabla^2\theta.$$

This equation can be obtained by substituting $-\beta\,\partial\theta/\partial x$ and $-\beta\,\partial\theta/\partial y$ for $X$ and $Y$ in Eq. (9.2:10). If we introduce the Airy stress function $\Phi$ so that

(9) $$\sigma_{xx} = \frac{\partial^2\Phi}{\partial y^2} + \beta\theta, \qquad \sigma_{yy} = \frac{\partial^2\Phi}{\partial x^2} + \beta\theta, \qquad \sigma_{xy} = -\frac{\partial^2\Phi}{\partial x\,\partial y},$$

then (6) is identically satisfied and (7) becomes

(10) $$\nabla^4\Phi = -\frac{\alpha E}{1 - \nu}\nabla^2\theta, \qquad \nabla^2 = \frac{\partial^2}{\partial x^2} + \frac{\partial^2}{\partial y^2},$$

where $\alpha = (1 - 2\nu)\beta/E$ is the linear coefficient of expansion. But

(11) $$k\,\nabla^2\theta = \rho C_v\frac{\partial\theta}{\partial t}.$$

Hence, finally,

(12) $$\nabla^4\Phi = -\frac{\alpha E\rho C_v}{k(1 - \nu)}\frac{\partial\theta}{\partial t}.$$

The details are left to the reader.

*Example*

If a long cylinder with axis $z$ is subjected to *steady heat flow* with resulting temperature distribution $\theta(x, y)$ independent of $t$, then $\nabla^2\theta = 0$ and (12) implies that

(13) $$\nabla^4\Phi = 0.$$

Let the boundary surface of the cylinder be unrestrained. According to Duhamel analogy, the boundary condition is a tensile traction $\beta\theta$. Hence, according to Eq. (9), the boundary conditions can be satisfied by taking

$$\frac{\partial^2\Phi}{\partial y^2} = 0, \qquad \frac{\partial^2\Phi}{\partial x^2} = 0, \qquad \frac{\partial^2\Phi}{\partial x\, \partial y} = 0 \qquad \text{on the boundary.} \tag{14}$$

A solution is evidently $\Phi \equiv 0$. For this solution, a superposition of (5) and (9) yields

$$\sigma_{xx} = \sigma_{yy} = \sigma_{xy} = 0, \tag{15}$$

whereas, from Eq. (4),

$$\sigma_{zz} = -\alpha E\theta. \tag{16}$$

If the region $V(x, y)$ occupied by the cylinder is simply connected, and if plane strain condition can be assumed, we can quote the uniqueness theorem to say that this is *the* solution. Hence, we conclude that *if a cylinder is simply connected and unrestrained on the surface, then under steady two-dimensional heat flow and plane strain condition, there will be no thermal stress in any surface element parallel to the cylinder axis.*

## PROBLEMS

**14.8.** *Plane Stress.* A body is said to be in a state of plane stress parallel to the $x_1, x_2$-plane if

$$\sigma_{13} = \sigma_{23} = \sigma_{33} = 0$$

so that

$$e_{33} = -\frac{\nu}{1-\nu}\frac{\partial u_i}{\partial x_i} + \frac{1+\nu}{1-\nu}\alpha\theta, \qquad e_{13} = e_{23} = 0,$$

where $i = 1, 2$. Prove the following Duhamel analogy (Fig. 14.8:2):

*Problem I: Heating* *Problem II: Nonheating*

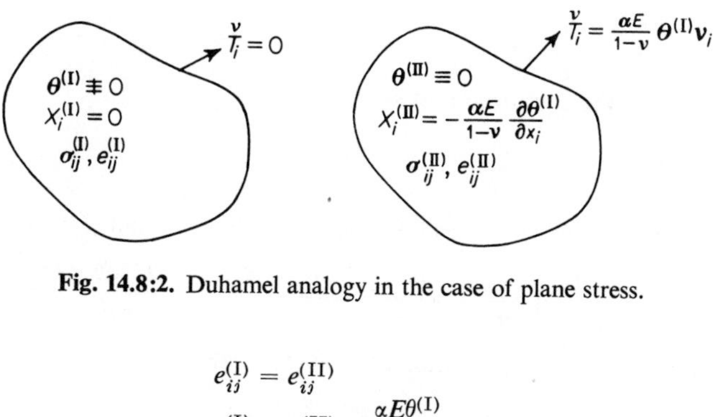


**Fig. 14.8:2.** Duhamel analogy in the case of plane stress.

Then

$$e_{ij}^{(I)} = e_{ij}^{(II)}$$

$$\sigma_{ij}^{(I)} = \sigma_{ij}^{(II)} - \frac{\alpha E\theta^{(I)}}{1-\nu}\sigma_{ij}.$$

It is seen that the role of the parameter $\beta = \alpha E/(1 - 2\nu)$ is now played by the parameter $\alpha E/(1 - \nu)$. If Prob. II in plane stress is posed in terms of stresses and Airy stress functions so that

$$\sigma_{11} = \frac{\partial^2 \Phi}{\partial x_2^2} + \frac{\alpha E \theta}{1 - \nu}, \qquad \sigma_{22} = \frac{\partial^2 \Phi}{\partial x_1^2} + \frac{\alpha E \theta}{1 - \nu}, \qquad \sigma_{12} = -\frac{\partial^2 \Phi}{\partial x_1 \, \partial x_2},$$

then

$$\nabla^4 \Phi = -\alpha E \, \nabla^2 \theta, \qquad \nabla^2 = \frac{\partial^2}{\partial x_1^2} + \frac{\partial^2}{\partial x_2^2}.$$

**14.9.** Consider a thin plate parallel to the $x, y$-plane subjected to a temperature distribution $T(x, y)$ which is independent of $z$. The Duhamel analogy can be applied either in the three-dimensional form or in the plane stress form with the vanishing of $\sigma_{zz}$, $\sigma_{zx}$, $\sigma_{zy}$ introduced explicitly at the outset. Demonstrate the similarity and differences of these two analogies.

## 14.9. AN EXAMPLE—STRESSES IN A TURBINE DISK

Let the cooled turbine disk which is treated in Sec. 14.5 (see Fig. 14.5:1) be rotating at an angular velocity $\omega$. The body force per unit volume is then radial and equal to $\omega^2 \rho r$, where $\rho$ is the density of the disk material. If the thickness of the disk is small compared with its radius $R$, the axial stress is negligible. The disk is then in a state of plane stress.

On account of the fact that all stresses are considered to be functions of $r$ only and the boundary condition $\tau_{rz} = 0$ at $r = R$, the shearing stress $\tau_{rz}$ vanishes and only the tension strains $e_{rr}$, $e_{\theta\theta}$ and tension stress $\sigma_{rr}$, $\sigma_{\theta\theta}$ are to be determined. The elastic displacement has only the radial component $u$, which is a function of $r$. Then

$$e_{rr} = \frac{du}{dr}, \qquad e_{\theta\theta} = \frac{u}{r}, \tag{1}$$

$$\frac{du}{dr} = \frac{1}{E}(\sigma_{rr} - \nu \sigma_{\theta\theta}) + \alpha \theta, \tag{2}$$

$$\frac{u}{r} = \frac{1}{E}(\sigma_{\theta\theta} - \nu \sigma_{rr}) + \alpha \theta. \tag{3}$$

The equation for the equilibrium of stresses and body forces is

$$\frac{d\sigma_{rr}}{dr} + \frac{\sigma_{rr} - \sigma_{\theta\theta}}{r} + \omega^2 \rho r = 0. \tag{4}$$

To solve for stresses directly, we eliminate $u$ from Eqs. (2) and (3). Thus,

$$\frac{1}{1 + \nu}\left(\frac{d\sigma_{\theta\theta}}{dr} - \nu \frac{d\sigma_{rr}}{dr}\right) = \frac{\sigma_{rr} - \sigma_{\theta\theta}}{r} - \frac{\alpha E}{1 + \nu} \frac{d\theta}{dr} \tag{5}$$

An elimination of $(\sigma_{rr} - \sigma_{\theta\theta})/r$ from (4) and (5) yields

$$\frac{d\sigma_{rr}}{dr} + \frac{d\sigma_{\theta\theta}}{dr} + \alpha E \frac{d\theta}{dr} + (1 + \nu)\rho\omega^2 r = 0. \tag{6}$$

But (4) gives

$$\sigma_{\theta\theta} = \frac{d}{dr}(r\sigma_{rr}) + \omega^2\rho r^2. \tag{7}$$

Therefore, from (6) and (7), the equation for $\sigma_{rr}$ is

$$\frac{1}{r^2}\frac{d}{dr}\left(r^3\frac{d\sigma_{rr}}{dr}\right) + \alpha E\frac{d\theta}{dr} + (3 + \nu)\omega^2\rho r = 0. \tag{8}$$

On integrating twice, we obtain

$$\sigma_{rr} = -\frac{1}{2}\frac{C_1}{r^2} + C_2 - \alpha E\int_0^r \frac{1}{\eta^3}\,d\eta\left[\int_0^\eta \xi^2\frac{d\theta(\xi)}{d\xi}\,d\xi\right] - \frac{3+\nu}{8}\,\omega^2\rho r^2. \tag{9}$$

The integral involving $\theta$ may be simplified by interchanging the order of integration. Thus

$$I \equiv \int_0^r \xi^2\frac{d\theta(\xi)}{d\xi}\,d\xi\int_\xi^r\frac{1}{\eta^3}\,d\eta = \int_0^r \xi^2\frac{d\theta(\xi)}{d\xi}\left(\frac{1}{2\xi^2} - \frac{1}{2r^2}\right)d\xi$$
$$= -\frac{\theta(0)}{2} + \frac{1}{r^2}\int_0^r \xi\theta(\xi)\,d\xi.$$

Hence, (9) may be written

$$\sigma_{rr} = -\frac{1}{2}\frac{C_1}{r^2} + C_2 + \frac{\alpha E\theta(0)}{2} - \frac{\alpha E}{r^2}\int_0^r \xi\theta(\xi)\,d\xi - \frac{3+\nu}{8}\,\omega^2\rho r^2. \tag{10}$$

The boundary conditions are that $\sigma_{rr} = 0$ at $r = R$ and $\sigma_{rr}$ remain finite at $r = 0$. Hence,

$$\sigma_{rr} = \alpha E\left[\frac{1}{R^2}\int_0^R \xi\theta(\xi)\,d\xi - \frac{1}{r^2}\int_0^r \xi\theta(\xi)\,d\xi\right] + \frac{3+\nu}{8}\,\omega^2\rho(R^2 - r^2). \tag{11}$$

Equation (7) then gives

$$\sigma_{\theta\theta} = \alpha E\left[\frac{1}{R^2}\int_0^R \xi\theta(\xi)\,d\xi + \frac{1}{r^2}\int_0^r \xi\theta(\xi)\,d\xi - \theta(r)\right] + \omega^2\rho\,\frac{(3+\nu)R^2 - (1+3\nu)r^2}{8}. \tag{12}$$

It is interesting to note that the first integral in (11) is one-half the average temperature throughout the whole disk. The second integral is one-half the average temperature in the inner portion of the disk (between 0 and $r$). Therefore, the thermal stress in the radial direction, i.e., the part of $\sigma_{rr}$ caused by $\theta(r)$, is proportional to the difference between the average temperature throughout the whole disk and the average temperature for the inside portion between 0 and $r$. The circumferential thermal stress differs from that in the radial direction by

$$\sigma_{\theta\theta} - \sigma_{rr} = \alpha E[\theta_{\text{avg}\,(0\text{ to }r)} - \theta(r)] + \frac{(1-\nu)\rho r^2\omega^2}{4}. \tag{13}$$

For the particular problem of the turbine disk treated in Sec. 14.5, the temperature is given by Eq. (14.5:11). Since

$$\int_0^z zI_0(z)\,dz = zI_1(z),$$

we have

$$\frac{1}{r^2}\int_0^r \xi\theta(\xi)\,d\xi = \frac{h}{2\kappa}\frac{b}{r}\frac{I_1(\sqrt{2\kappa/kb}\;r)}{I_1(\sqrt{2\kappa/kb}\;R)}$$

and

$$\frac{1}{R^2}\int_0^R \xi\theta(\xi)\,d\xi = \frac{h}{2\kappa}\frac{b}{R}.$$

With these relations, the thermal stresses can be easily calculated.

## 14.10. VARIATIONAL PRINCIPLE FOR UNCOUPLED THERMOELASTICITY

In uncoupled thermoelasticity theory, the temperature field is determined by heat conduction and the influence of the latent heat due to change of strain is ignored. Variational principles analogous to those of Chap. 10 can be derived. In fact, the only difference between the linear elasticity of Chaps. 7–9, which is valid in isothermal or isentropic conditions, and the thermoelasticity of the present chapter lies in a difference in the stress-strain law. Furthermore, in linear thermoelasticity concerning infinitesimal displacement, even this difference in stress-strain is hidden in the following expressions which apply to both elasticity and thermoelasticity:

(1) $$\frac{\partial W}{\partial e_{ij}} = \sigma_{ij}, \qquad \frac{\partial W_c}{\partial \sigma_{ij}} = e_{ij}.$$

Here $W(e_{ij}, T)$ and $W_c(\sigma_{ij}, T)$ are the strain energy function and the complementary strain energy function, respectively. $W(e_{ij}, T)$ is expressed in strain components; $W_c(\sigma_{ij}, T)$ is expressed in stress components; in linear theory they are equal: $W = W_c$. For isothermal or isentropic elasticity (Chap. 7) $W$ is given by Eq. (12.6:3). For general thermoelasticity (Chap. 14), $W$ is given by Eq. (12.7:2) or (12.7:11).

With these remarks we may derive the following theorems.

Let $\mathscr{U}$ be the total strain energy of a body which occupies a region $V$ with boundary $B$,

(2) $$\mathscr{U} = \int_V W(e_{ij}, T)\,dv,$$

where $W(e_{ij}, T)$ is given by Eq. (12.7:2).

Let the body be subjected to a body force $X_i$ per unit volume in $V$, and surface tractions $\overset{\nu}{T}_i$ on $B_\sigma$ ($B_\sigma$ in $B$). Both $X_i$ and $\overset{\nu}{T}_i$ are assumed to be functions of time and space not influenced by the small elastic displacements. Let $\mathscr{K}$ be the kinetic energy of the body and $\mathscr{A}$ be the potential of the external loading:

$$\mathscr{K} = \frac{1}{2}\int_V \frac{\partial u_i}{\partial t}\frac{\partial u_i}{\partial t}\,\rho\, dv \tag{3}$$

$$\mathscr{A} = -\int_V X_i u_i\, dv - \int_{B_\sigma} \overset{\nu}{T}_i u_i\, dS. \tag{4}$$

Then *the Hamilton's principle states that of all displacements $u_i$ that satisfy the boundary condition that*

$$u_i = g_i(x_1, x_2, x_3) \quad \text{over } B_u = B - B_\sigma, \tag{5}$$

*the one that also satisfies the equations of motion minimizes the integral*

$$\int_{t_0}^{t_1} (\mathscr{U} - \mathscr{K} + \mathscr{A})\, dt = \text{minimum}, \tag{6}$$

*where $t_0$, $t_1$ are two arbitrary instants, and where the admissible variations $\delta u_i$ are triply differentiable and satisfy the conditions*

$$\delta u_i = 0 \quad \text{on } B_u \text{ in } t_0 \leqslant t \leqslant t_1, \tag{7}$$

$$\delta u_i = 0 \quad \text{at } t = t_0 \text{ and } t = t_1 \text{ in } V + B. \tag{8}$$

If we consider static equilibrium, then *the principle of minimum potential energy for thermoelasticity states that of all displacements that are continuous and triply differentiable in a body $V + B$, and satisfying the specified boundary displacements $u_i$ over the surface $B_u$, the one that also satisfies the condition of equilibrium minimizes the potential*

$$\int_V W(e_{ij}, T)\, dv - \int_V X_i u_i\, dv - \int_{B_\sigma} \overset{\nu}{T}_i u_i\, dS = \text{minimum}. \tag{9}$$

Similarly, on varying the stresses $\sigma_{ij}$, *the complementary energy theorem in thermoelasticity states that the stress system $\sigma_{ij}$ that satisfies the equations of elasticity minimizes the complementary energy with respect to variations of $\sigma_{ij}$,*

$$\int_V W_c(\sigma_{ij}, T)\, dv - \int X_i u_i\, dv - \int_{B_u} \overset{\nu}{T}_i u_i\, dS = \text{minimum}. \tag{10}$$

$W_c = -\rho\Phi$ is given by Eq. (12.8:15). The variations $\delta\sigma_{ij}$ are subjected to the conditions $(\delta\sigma_{ij})_{,j} = 0$ in $V$ and $\delta\sigma_{ij}\nu_j = 0$ over $B_\sigma$, where surface tractions are specified.

**Problem 14.10.** Derive Euler equations for the functionals in Eqs. (6), (9), and (10) and compare the results with the basic equations of Sec. 14.1.

## 14.11. VARIATIONAL PRINCIPLE FOR HEAT CONDUCTION

The basic equations for heat conduction, Eqs. (14.1:8) and (14.1:9), are

$$(1) \qquad h_i = -k_{ij}\frac{\partial \theta}{\partial x_j}, \qquad i = 1, 2, 3,$$

$$(2) \qquad -\frac{\partial h_i}{\partial x_i} = \rho C_v \frac{\partial \theta}{\partial t},$$

where $C_v$ is the specific heat per unit mass at constant volume, $\rho$ is the density, $k_{ij}$ are the coefficients of heat conduction, $\theta = T - T_0$ is the difference of local temperature $T$ from a uniform reference temperature $T_0$, and $\mathbf{h} = (h_1, h_2, h_3)$ is the heat flux vector per unit area. A problem in heat conduction for a body occupying a volume $V$ with boundary $B$ is to find a continuously differentiable function $\theta(x_1 x_{,2}, x_3; t)$ which satisfies Eqs. (1) and (2), the boundary conditions

$$(3) \qquad \theta = \theta_0(x, t) \qquad \text{on } B_1,$$

$$(4) \qquad h_n = \kappa(x, t)[\theta_0(x, t) - \theta(x, t)] \qquad \text{on } B - B_1,$$

and the initial condition at $t = 0$

$$(5) \qquad \theta = \theta_0(x, 0) \qquad \text{in } V \text{ and } B,$$

where $h_n$ is the component of the heat flux vector $\mathbf{h}$ in the direction of the *outer normal* to the boundary, i.e., $h_n = h_i \nu_i$ and $\theta_0$ is an assigned temperature. Equation (4) represents Newton's law of surface heat transfer; $\theta_0(x, t)$ is the temperature just outside the boundary layer, and $\kappa$ is the heat transfer coefficient which usually depends on the space coordinate $\mathbf{x}$.

We shall now prove *Biot's variational principle which states that if $C_v$ is a constant, then Eqs.* (1)–(4) *are the necessary conditions for the following variational equation to hold for an arbitrary variation of the vector $H_i$*

$$(6) \qquad \delta \int_V \frac{\rho C_v \theta^2}{2}\, dv + \int_V \lambda_{ij} \dot{H}_j\, \delta H_i\, dv + \int_{B_1} \theta_0\, \delta H_i \nu_i\, dS = 0,$$

*under the stipulations that*

$$(7) \qquad \rho C_v \theta = -H_{i,i}$$

*and that $\delta H_i = 0$ over the boundary $B - B_1$, where heat flux is so specified that $\partial H_i/\partial t = h_i$ satisfies Eq.* (4). *The matrix $(\lambda_{ij})$, called the thermal resistivity matrix, is the inverse of the conductivity matrix $(k_{ij})$:*

$$(8) \qquad (\lambda_{ij}) = (k_{ij})^{-1}.$$

*Proof.* First, we have

$$\delta \int_V \rho C_v \frac{\theta^2}{2}\, dv = \delta \int_V \frac{1}{2\rho C_v} H_{i,i} H_{j,j}\, dv \qquad \text{(by substitution)}$$

$$= \int_V \frac{1}{\rho C_v} H_{i,i}\, \delta H_{j,j}\, dv \qquad (i, j \text{ being dummy indices})$$

$$= \int_V \left[ \left( \frac{1}{\rho C_v} H_{i,i}\, \delta H_j \right)_{,j} - \left( \frac{1}{\rho C_v} H_{i,i} \right)_{,j} \delta H_j \right] dv$$

$$= \int_B \frac{1}{\rho C_v} H_{i,i}\, \delta H_j \nu_j\, dS - \int_V \left( \frac{1}{\rho C_v} H_{i,i} \right)_{,j} \delta H_j\, dv$$

(by Gauss' theorem)

$$= -\int_B \theta\, \delta H_j \nu_j\, dS + \int_V \theta_{,j}\, \delta H_j\, dv.$$

Hence, Eq. (6) is

$$0 = \int_V (\theta_{,j} + \lambda_{ij} \dot{H}_j)\, \delta H_i\, dv - \int_{B_1} (\theta - \theta_0)\, \delta H_i \nu_i\, dS - \int_{B-B_1} \theta\, \delta H_i \nu_i\, dS.$$

To satisfy this equation by arbitrary $\delta H_i$ in $V$ and on $B_1$, the necessary conditions are

$$\theta_{,j} + \lambda_{ij} \dot{H}_j = 0, \quad \text{or} \quad \dot{H}_j = -k_{ij} \theta_{,j} \qquad \text{in } V,$$

$$\theta = \theta_0 \qquad \text{on } B_1,$$

$$\delta H_j \nu_j = 0 \qquad \text{on } B - B_1.$$

The first equation, together with (7), is exactly Eq. (1). The second equation is the same as (3), the last equation is in accordance with the stipulation stated. Q.E.D.

Applications of this variational principle will be shown later in Secs. 14.13. A few remarks will be made here. First, the vector $H_i$ is related to the heat flux $h_i$:

$$h_i = \dot{H}_i. \tag{9}$$

If we impose the condition that

$$H_i = 0 \quad \text{when} \quad \theta = 0, \tag{10}$$

then Eqs. (2) and (7) are consistent since $C_v$ is a constant. Comparing (9) with Eq. (13.1:3) it is seen that $H_i$ is proportional to the entropy displacement introduced by Biot.

Second, the special way in which the second term in Eq. (6) is posed should be noted. Biot denotes this term by $\delta D$ and calls it *the variation of a dissipation function.* The justification of this terminology becomes clear when generalized coordinates are introduced. Let $q_1, \ldots, q_n$ be the generalized coordinates so that

$$H_j = H_j(q_1, q_2, \ldots, q_n; x_1, x_2, x_3). \tag{11}$$

Then

$$\dot{H}_j = \frac{\partial H_j}{\partial q_k} \dot{q}_k.$$

Hence,

$$\frac{\partial \dot{H}_j}{\partial \dot{q}_k} = \frac{\partial H_j}{\partial q_k}. \tag{12}$$

Therefore,

$$\begin{aligned}\int_V \lambda_{ij} \dot{H}_j \, \delta H_i \, dv &= \int_V \lambda_{ij} \dot{H}_j \frac{\partial H_i}{\partial q_k} \, \delta q_k \, dv \\ &= \delta q_k \int_V \lambda_{ij} \dot{H}_j \frac{\partial \dot{H}_i}{\partial \dot{q}_k} \, dv \\ &= \delta q_k \frac{\partial}{\partial \dot{q}_k} \int_V \frac{1}{2} \lambda_{ij} \dot{H}_j \dot{H}_i \, dv.\end{aligned}$$

It is then clear that if we define a dissipation function

$$D = \tfrac{1}{2} \int_V \lambda_{ij} \dot{H}_j \dot{H}_i \, dv, \tag{13}$$

then

$$\int_V \lambda_{ij} \dot{H}_j \, \delta H_i \, dv = \frac{\partial D}{\partial \dot{q}_k} \, \delta q_k. \tag{14}$$

The application of generalized coordinates will be illustrated in Sec. 14.13.

It may appear unnatural to introduce the vector field $H_i$ rather than vary the temperature $\theta$ directly. A little reflection will show, however, that the thermal evolution is determined by the heat flow. The three components of the heat flow vector are capable of independent variations. Hence, as a proper choice of variables, we use the components of the heat flow vector. Equivalently, the three components of the temperature gradient may be used also, but then the analysis will be purely formal; the functionals do not have as simple an interpretation as those encountered above.

## 14.12. COUPLED THERMOELASTICITY

A variational equation corresponding to the basic equations of Sec. 14.1 is given by Biot under the assumptions that the elastic constants $C_{ijkl}$, the thermal stress constants $\beta_{ij}$, the specific heat per unit mass at constant strain $C_v$, and the heat conductivity matrix $(k_{ij})$ or its inverse $(\lambda_{ij}) = (k_{ij})^{-1}$, are independent of the temperature and time, and that the temperature change

$\theta$ is small compared with $T_0$, the absolute temperature of the reference state. Under these assumptions, Eq. (14.1:17) can be integrated to give

$$(1) \qquad -H_{i,i} = \rho C_v \theta + T_0 \beta_{ij} e_{ij},$$

where

$$\frac{\partial H_i}{\partial t} \cong h_i,$$

$$H_i = 0 \quad \text{when} \quad \theta = e_{ij} = 0.$$

The vector $H_i$ so defined is proportional to the entropy displacement. Biot's principle now states that *the thermoelastic equations are necessary conditions for the following variational equation to hold for arbitrary $\delta u_i$ and $\delta H_i$, except that $\delta u_i = 0$ over the portion of the boundary where the displacement $u_i$ is prescribed, and $\delta H_i = 0$ where the heat flow is prescribed:*

$$(2) \qquad \delta \mathscr{V} + \delta \mathscr{D} = \int_V (X_i - \rho \ddot{u}_i)\, \delta u_i \, dv + \int_B \left( \overset{\nu}{T}_i \, \delta u_i + \theta_0 \frac{\delta H_i}{T_0} \nu_i \right) dS,$$

where

$$(3) \qquad \mathscr{V} = \int_V \left( \frac{1}{2} C_{ijkl} e_{ij} e_{kl} + \frac{1}{2} \frac{\rho C_v \theta^2}{T_0} \right) dv,$$

$$(4) \qquad \delta \mathscr{D} = \int_V \frac{1}{T_0} \lambda_{ij} \dot{H}_j \, \delta H_i \, dv,$$

$\theta_0$ is the boundary value of $\theta$, $\nu_i$ are the direction cosines of the outward normal to the surface $B$ (see Sec. 14.11), and $\mathscr{V}$ is the *thermoelastic potential* of Biot. On solving (1) for $\theta$, substituting $\theta$ and $e_{ij} = \frac{1}{2}(u_{i,j} + u_{j,i})$ into (3), and varying $u_i$ and $H_i$, the statement above can be verified readily. We may also use the method of Lagrangian multipliers as in Sec. 14.10. The details are left to the reader.

Note that if Eq. (2) is integrated with respect to time between an arbitrary interval $(t_0, t_1)$, under the stipulation that $\delta u_i = 0$ at $t = t_0, t_1$, the term involving the inertia force $-\rho \ddot{u}_i$ can be expressed as $-\delta \mathscr{K}$, where $\mathscr{K}$ is the kinetic energy. Then a formula similar to the Hamilton's principle, now including a dissipation function, is obtained.

Biot[14.5] derived the expressions above for $\mathscr{V}$ and $\mathscr{D}$ from a reasoning which has been discussed in Sec. 13.5. For small deviations from equilibrium, the equation of evolution (13.5:13) is a necessary condition for the variational equation

$$(5) \qquad \left( \frac{\partial \mathscr{V}}{\partial q_i} + \frac{\partial \mathscr{D}}{\partial \dot{q}_i} - Q_i \right) \delta q_i = 0$$

to hold for arbitrary variations $\delta q_i$. It was shown in Sec. 13.5 that $\mathscr{V}$ is the sum of $\mathscr{E} - T_0 \mathscr{S}$ over the entire system, and $2\mathscr{D}/T_0$ is equal to the rate of entropy production. Although no heat input on the boundary was considered in Sec. 13.5, the generalization of $Q_i$ to include heat conduction on the boundary is justified by the theorem just proved above.

Let us now evaluate $\mathscr{V} = \int \rho(\mathscr{E} - T_0\mathscr{S})\, dv$ when a temperature increment $\theta(x_1, x_2, x_3; t)$ is imposed on the body without changing any other state variables. The absolute temperature distribution is $T_0 + \theta$. To change the temperature by $d\theta$ at a particular point requires an amount of heat $\rho C_v\, d\theta$ per unit mass, where $C_v$ is the specific heat per unit mass at constant strain. Hence, the change of internal energy per unit volume, $\mathscr{E}$, is $\rho C_v\, d\theta$, and the change of entropy per unit volume is $\rho C_v\, d\theta/(T_0 + \theta)$. Thus, the imposition of a temperature increment $\theta$ changes $\mathscr{V}$ by

(6)
$$\mathscr{V}_c = \int_V dv\left(\int_0^\theta \rho C_v\, d\theta - T_0\int_0^\theta \frac{\rho C_v\, d\theta}{T_0 + \theta}\right)$$
$$= \int_V dv \int_0^\theta \frac{\rho C_v \theta}{T_0 + \theta}\, d\theta.$$

If $\theta \ll T_0$, then

(7)
$$\mathscr{V}_c \doteq \frac{1}{2}\int_V \frac{\rho C_v \theta^2}{T_0}\, dv.$$

The total value of $\mathscr{V}$ is obtained by adding to $\mathscr{V}_c$ the value of $\mathscr{V}$ at the constant temperature $T_0$. The latter is the classical Helmholtz free energy at $T_0$. The term $\mathscr{V}_c$ is familiar to power engineering: it represents the heat that may be transformed into useful work. Indeed, $\mathscr{V}_c$ is an integral of the heat $\rho C_v\, d\theta$ multiplied by the Carnot efficiency $\theta/(T_0 + \theta) \doteq \theta/T_0$, integrated first with respect to $\theta$ from 0 to $\theta$, then over the region $V$.

The special treatment of the dissipation function $\mathscr{D}$ in the variational equations should be noted. The variational invariant $\delta\mathscr{D}$ was defined directly. When the equations of motion are written in the Lagrangian form, Eq. (5), $\mathscr{D}$ can be identified as a functional which is equal to $T_0/2$ times the rate of entropy production of the entire system [see Eq. (13.5:11), assuming $\theta \ll T_0$]. From Eq. (13.2:6), the entropy production per unit volume is $-h_j T_{,j}/T_0^2$, where $h_j$ is the heat flux and $T_{,j}$ is the temperature gradient. On introducing Fourier's law of heat conduction in the form $T_{,j} = -\lambda_{jl} h_l$, the rate of entropy production per unit volume is $\lambda_{jl} h_j h_l / T_0^2$. Hence,

(8)
$$\mathscr{D} = \frac{1}{2T_0}\int_V \lambda_{jl} h_j h_l\, dv.$$

In Sec. 14.11, for the boundary condition over $B - B_1$, where the wall temperature is unspecified, it is required that the heat flow be specified exactly. For a variational approach, this boundary condition can be relaxed by including in the dissipation function a term corresponding to entropy production at the boundary. Let there be a heat flux per unit area $h_j\nu_j$ leaving the body at a boundary, where $\nu_j$ is the direction cosine of the outer normal to the surface. Let $T_a$ be the temperature *outside* the heat transfer layer at the wall and $T$ be that of the body inside. (For example, $T_a$ represents the adiabatic wall temperature in the boundary-layer flow in aerodynamic heating.) If $T_a \neq T$, the entropy production per unit wall area is $-h_j\nu_j(T^{-1} - T_a^{-1})$. A phenomenological law may be posed as

$$h_n = h_j\nu_j = -K\left(\frac{1}{T} - \frac{1}{T_a}\right). \tag{9}$$

Therefore, the contribution of heat transfer at the boundary to the dissipation function is

$$\mathscr{D}_B = \frac{T_0}{2}\int_B K\left(\frac{1}{T} - \frac{1}{T_a}\right)^2 dS = \frac{T_0}{2}\int_B \frac{1}{K}\, h_n^2\, dS. \tag{10}$$

If $\mathscr{D}_B$ is added to $\mathscr{D}$ from (8), we have

$$\mathscr{D} = \frac{1}{2T_0}\int_V \lambda_{jl}h_jh_l\, dv + \frac{T_0}{2}\int_B \frac{1}{K}\, h_n^2\, dS, \tag{11}$$

which must be used if the heat flow at the wall is to be considered as a natural boundary condition. When the expression (11) is used, the variation $\delta H_i$ can be regarded as arbitrary over the entire boundary.

## 14.13. LAGRANGIAN EQUATIONS FOR HEAT CONDUCTION AND THERMOELASTICITY

When $n$ generalized coordinates $q_1, q_2, \ldots, q_n$ are introduced to represent the displacement vector and heat flow vector by expressions

$$\begin{aligned} u_i &= u_i(q_1, q_2, \ldots, q_n; x_1, x_2, x_3), & i &= 1, 2, 3, \\ H_i &= H_i(q_1, q_2, \ldots, q_n; x_1, x_2, x_3), & i &= 1, 2, 3. \end{aligned} \tag{1}$$

The variational principles of the preceding section may be written as

$$\frac{\partial \mathscr{V}}{\partial q_j} + \frac{\partial \mathscr{D}}{\partial \dot{q}_j} = Q_j, \qquad j = 1, 2, \ldots, n, \tag{2}$$

where $\mathscr{V}$ and $\mathscr{D}$ are given by Eqs. (14.12:3), (14.12:8) respectively, and

$$Q_j = \int_V (X_i - \rho\ddot{u}_i)\frac{\partial u_i}{\partial q_j}\, dv + \int_B \left(\overset{\nu}{T}_i \frac{\partial u_i}{\partial q_j} + \frac{\theta_0}{T_0}\frac{\partial H_n}{\partial q_j}\right) dS. \tag{3}$$

The applications of these Lagrangian equations are well-known for the stress problems (Chaps. 10, 11). In what follows we shall give illustration to heat conduction problems.

*Example 1. Prescribed Wall Temperature*

Consider a semi-infinite homogeneous, isotropic solid, with constant parameters $k$ and $C_v$, initially at a uniform temperature $\theta = 0$ (Fig. 14.13:1). The boundary at $x = 0$ is heated to a variable temperature $\theta_0(t)$. On ignoring the inertia force and thermoelastic coupling, it is desired to find the temperature distribution in the half-space.

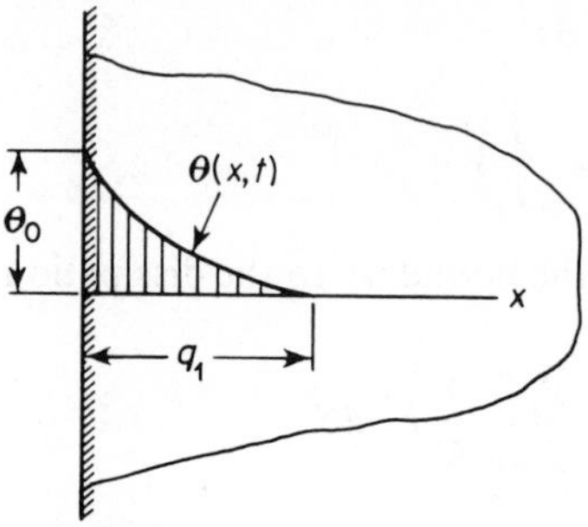


**Fig. 14.13:1.** Penetration of heat in one-dimensional flow.

We propose to solve this problem approximately. Following Biot[14.5] (1957), we assume the temperature distribution to be parabolic and represented by

$$\text{(4)} \qquad \begin{aligned} \theta(x, t) &= \theta_0(t)\left[1 - \frac{x}{q_1(t)}\right]^2, && x < q_1, \\ \theta(x, t) &= 0, && x > q_1. \end{aligned}$$

The function $q_1(t)$ can be interpreted physically as a "penetration depth," and will be taken as our generalized coordinate (see Fig. 14.13:1).

The heat flow vector $H_i$ is defined by Eqs. (14.11:7), (14.11:9), and (14.11:10).

$$\text{(5)} \qquad \dot{H}_i = h_i, \qquad -H_{i,i} = \rho C_v \theta.$$

In the one-dimensional problem posed here, we may assume that only one component $H_1$ is different from zero and that $H_1$ is a function of $x$ alone. Hence, from (5),

$$\text{(6)} \qquad \frac{\partial H_1}{\partial x} = -\rho C_v \theta.$$

For $x > q_1$, thermal equilibrium is undisturbed so that $H_1 = 0$. Hence, Eqs. (6) and (4) can be integrated to give

$$\text{(7)} \qquad H_1(x, t) = \rho C_v \int_x^{q_1(t)} \theta(\xi, t)\, d\xi = \rho C_v \theta_0 \left(\frac{q_1}{3} - x + \frac{x^2}{q_1} - \frac{x^3}{3q_1^2}\right).$$

To evaluate $\mathscr{V}$ and $\mathscr{D}$, it is sufficient to consider a semi-infinite cylinder of unit cross section with axis parallel to $x$. Hence,

$$\text{(8)} \qquad \mathscr{V} = \frac{\rho C_v}{2T_0} \int_0^{q_1} \theta^2\, dx = \frac{\rho C_v \theta_0^2}{10 T_0} q_1,$$

$$\text{(9)} \qquad \mathscr{D} = \frac{1}{2kT_0} \int_0^{q_1} \left(\frac{\partial H_1}{\partial t}\right)^2 dx,$$

which, from (7),

$$= \frac{\rho^2 C_v^2}{2kT_0} \int_0^{q_1} \left[\dot{\theta}_0 \left(\frac{q_1}{3} - x + \frac{x^2}{q_1} - \frac{x^3}{3q_1^2}\right) + \theta_0 \dot{q}_1 \left(\frac{1}{3} - \frac{x^2}{q_1^2} + \frac{2x^3}{3q_1^3}\right)\right]^2 dx$$

$$= \frac{\rho^2 C_v^2}{2kT_0} q_1 \left(\frac{13}{315} \dot{q}_1^2 \theta_0^2 + \frac{1}{21} q_1 \theta_0 \dot{q}_1 \dot{\theta}_0 + \frac{1}{63} \dot{\theta}_0^2 q_1^2\right).$$

The generalized force $Q_1$ is, according to Eq. (3),

$$Q_1(t) = \frac{\theta_0}{T_0}\left(\frac{\partial H_1}{\partial q_1}\right)_{x=0} = \frac{\rho C_v \theta_0^2}{3T_0}. \tag{10}$$

Hence, the Lagrangian equation for heat conduction (2) is

$$\frac{\rho C_v \theta_0^2}{10T_0} + \frac{\rho^2 C_v^2}{2kT_0} q_1 \left(\frac{26}{315}\dot{q}_1 \theta_0^2 + \frac{1}{21} q_1 \theta_0 \dot{\theta}_0\right) = \frac{\rho C_v \theta_0^2}{3T_0}. \tag{11}$$

Let

$$z = q_1^2. \tag{12}$$

Then Eq. (11) reduces to

$$\dot{z} + \frac{15}{13}\frac{\dot{\theta}_0}{\theta_0} z = \frac{147}{13}\frac{k}{\rho C_v}. \tag{13}$$

This is a standard differential equation with an integration factor

$$e^{\int (15/13)(\dot{\theta}_0/\theta_0)dt} = [\theta_0(t)]^{15/13};$$

i.e.

$$\frac{d}{dt}[z\theta_0{}^{15/13}] = \frac{147}{13}\frac{k}{\rho C_v}\theta_0^{15/13}.$$

Hence, on noting the initial condition $q_1 = z = 0$ when $t = 0$, we obtain the solution

$$z = \frac{147}{13}\frac{k}{\rho C_v}[\theta_0(t)]^{-15/13}\int_0^t [\theta_0(\tau)]^{15/13}\, d\tau. \tag{14}$$

If it is assumed that the wall temperature follows a power law,

$$\theta_0(t) = \alpha t^n, \qquad n \geqslant 0, \tag{15}$$

we find

$$q_1^2 = z = \frac{147}{13}\frac{k}{\rho C_v}\frac{t}{\frac{15}{13}n + 1}. \tag{16}$$

This shows that the penetration depth $q_1$ varies with $\sqrt{t}$. It is independent of $\alpha$ and depends on the exponent $n$ only through a constant factor $[\frac{15}{13}n + 1]^{-1/2}$. The case where $\theta_0 = $ const. corresponds to $n = 0$ and yields the result

$$q_1 = 3.36\sqrt{\frac{kt}{\rho C_v}}. \tag{17}$$

Comparison of this simple solution, with only one generalized coordinate, with the exact solution of the problem was made by Biot[14.5] (1957) who shows that the approximation (4) is valid if the temperature increases or decreases monotonically. Biot also points out that if this is not the case, one may split up the time interval into segments for which $\theta_0$ varies monotonically, and then apply (13) to each segment using the principle of superposition. Use can also be made of the power law solution (16) by dividing the time history of the temperature into segments, each of which may be approximated by a power law, or by an additive combination of such terms, including the constant value.

*Example 2. Prescribed Heat Flux*

The semi-infinite solid described in Example 1 is heated at the wall $x = 0$ with a uniform, constant heat flux $h_x = F$ at the wall (see Fig. 14.13:2) (Lardner,[14.5] 1963). The problem is assumed to be one-dimensional with only one nonvanishing component of the vector $\mathbf{H}$; namely, $H_1$, which will be written as $H$. Then

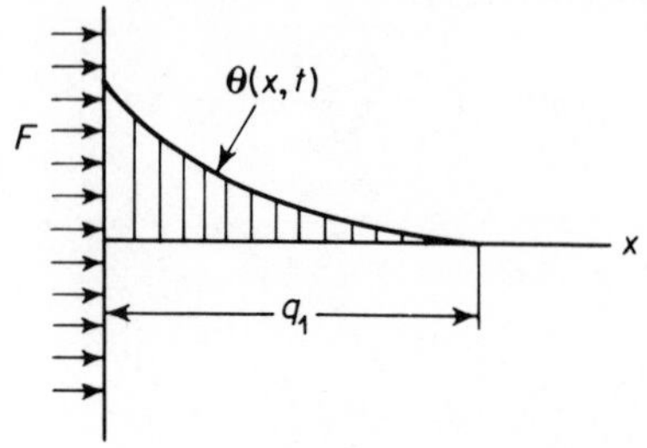


**Fig. 14.13:2.** Heating of a half-space with prescribed heat flux.

$$\frac{\partial H}{\partial x} = -\rho C_v \theta \tag{18}$$

Again, we assume a parabolic temperature distribution

$$\theta(x, t) = A(t)\left[1 - \frac{x}{q_1(t)}\right]^2, \tag{19}$$

with $q_1$ representing a "penetration depth." The function $A(t)$ must be so chosen to satisfy the boundary condition at $x = 0$. An integration of (18) gives

$$H = \frac{A\rho C_v q_1}{3}\left(1 - \frac{x}{q_1}\right)^3, \tag{20}$$

whence

$$h_x = \dot{H} = \dot{A}\frac{\rho C_v}{3} q_1\left(1 - \frac{x}{q_1}\right)^3 + \frac{A}{3}\rho C_v \dot{q}_1\left(1 - \frac{x}{q_1}\right)^3 + \frac{xA\rho C_v \dot{q}_1}{q_1}\left(1 - \frac{x}{q_1}\right)^2. \tag{21}$$

At $x = 0$, the boundary condition $h_x = F$ requires that

$$(\dot{A}q_1 + \dot{q}_1 A)\frac{\rho C_v}{3} = F,$$

or

$$A = \frac{3Ft}{\rho C_v q_1}. \tag{22}$$

On substituting this expression for $A$ into (19) and (20) and then computing the thermal potential $\mathscr{V}$ and dissipation function $\mathscr{D}$ according to Eqs. (14.12:3) and (14.12:11), we obtain

$$\frac{\partial \mathscr{V}}{\partial q_1} = -\frac{9F^2t^2}{10\rho C_v q_1^2},$$

$$\frac{\partial \mathscr{D}}{\partial \dot{q}_1} = \frac{F^2 t}{k}\left(\frac{3\dot{q}_1 t}{35 q_1} + \frac{3}{42}\right),$$

$$Q_1 = \frac{9}{10}t.$$

The governing differential equation (2) becomes

$$\frac{k}{\rho C_v}\left[\frac{3}{35}q_1\dot{q}_1 t + \frac{3}{42}q_1^2\right] = \frac{9}{10}t. \tag{23}$$

The solution for the penetration depth is

(24) $$q_1 = 2.81\left(\frac{k}{\rho C_v} t\right)^{1/2}.$$

The surface temperature is

(25) $$\theta_0 = \frac{3Ft}{\rho C_v q_1} = 1.065\,\frac{F}{k}\left(\frac{kt}{\rho C_v}\right)^{1/2}.$$

The exact solution is known to be

(26) $$\theta_0 = 1.128\,\frac{F}{k}\left(\frac{kt}{\rho C_v}\right)^{1/2}.$$

## PROBLEMS

**14.11.** Consider a slab of thickness $b$, heated on the surface $x = 0$. Initial condition at $t = 0$ is $\theta = 0$ throughout the plate. The boundary condition at the surface $x = 0$ is $\theta = \theta_0(t)$ for $t > 0$ (same as in Example 1), and that at $x = b$ is $h_x = \partial\theta/\partial x = 0$. At certain time $t_0$, the penetration depth $q_1$ will be equal to $b$. For time $t > t_0$, the temperature profile will be as shown in Fig. P14.11. Obtain an approximate solution by taking $q_3$ as the generalized coordinate.

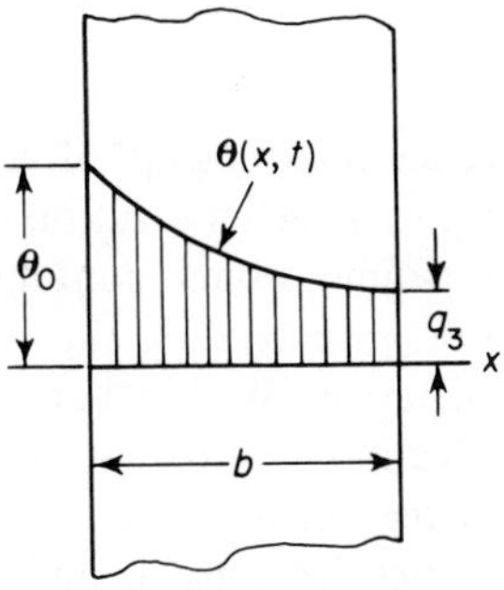


P14.11

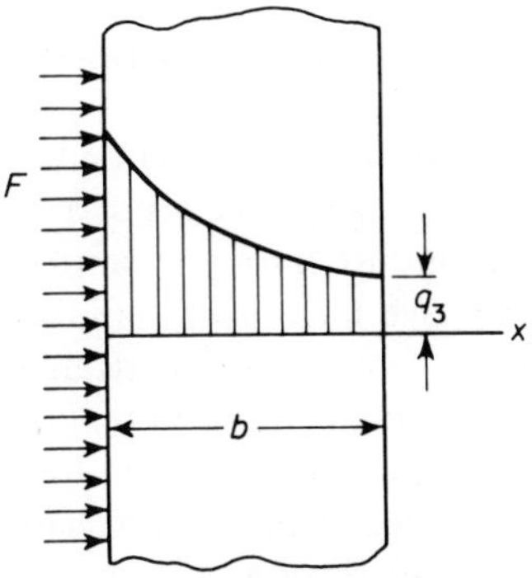


P14.12

**14.12.** Consider the same problem as in Prob. 14.11 except with the boundary conditions shown in Fig. P14.12: $x = 0$, is $h_x = F$, a const. (same as in Example 2).

**14.13.** A long, circular, cylindrical, solid-fuel rocket, upon curing, releases its heat of polymerization throughout its mass at a steady rate, say, $Q_0$. The external surfaces may be assumed held at constant temperature $T_0$. Assuming that the cylinder is so long that over the body of the cylinder the heat flow is essentially two-dimensional, determine approximately the temperature distribution in the cross section of the cylinder.

Consider other cylindrical rockets whose cross-sectional shapes are (1) an isosceles right triangle, (2) an equilateral triangle, (3) a square. Let all of these rockets be made of the same material and have the same cross-sectional area. Compare the steady-state temperatures at the centroids of these rockets.

# 15

# VISCOELASTICITY

In this chapter we shall generalize the ideas discussed in Sec. 1.6 to deal with a three-dimensional continuum. It is popular to call a material that obeys a linear hereditary law *viscoelastic*, although the logic of this terminology is by no means certain. The origin of the term lies in the simple models such as those of Maxwell, Voigt, or the standard linear solid, which are built of springs and dashpots. The springs are elastic, the dashpots are viscous; hence the name. Such an etymology is entirely different from that of viscoplasticity, hyperelasticity, hypoelasticity, elastic-ideally plastic materials, etc. However, the terminology of identifying a material obeying a linear hereditary law as viscoelastic is popular and well-accepted.

## 15.1. VISCOELASTIC MATERIAL

In Sec. 1.6 we described the load-deflection relationship of a linear viscoelastic bar by means of a convolution integral. A generalization of such a relationship into one applicable to a three-dimensional continuum is desired. We have had a similar occasion before: in extending the Hooke's law (1.1:1) into the stress-strain relationship of Chap. 6. We see that a proper generalization consists in replacing the load-deflection relation by a tensorial stress-strain relation. From this remark, we proceed as follows.

Let a rectangular Cartesian frame of reference be chosen, and let the position vector of a point be denoted by $(x_1, x_2, x_3)$. A function of position $f(x_1, x_2, x_3)$ will be written as $f(x)$ for short. Let $\sigma_{ij}$ and $e_{ij}$ be the stress and strain tensors defined at every point $x$ of a body and in the time interval $(-\infty < t < \infty)$. The strain field $e_{ij}(x, t)$ and the displacement field $u_i(x, t)$, as well as the velocity field $v_i(x, t)$, will be assumed to be infinitesimal, and

(1) $$e_{ij} = \tfrac{1}{2}(u_{i,j} + u_{j,i}),$$

where a comma indicates a partial differentiation. Under the assumption of infinitesimal strain and velocity, the partial derivative with respect to time $\partial e_{ij}/\partial t$ is equal to the material derivative $\dot{e}_{ij}$ within the first order. Now we define a *linear viscoelastic* material to be one for which $\sigma_{ij}(x, t)$ is related to $e_{ij}(x, t)$ by a convolution integral as follows.

(2) ▲ $$\sigma_{ij}(x, t) = \int_{-\infty}^{t} G_{ijkl}(x, t - \tau) \frac{\partial e_{kl}}{\partial \tau}(x, \tau)\, d\tau,$$

where $G_{ijkl}$ is a tensor field of order 4 and is called the *tensorial relaxation function* of the material. Equation (2) is called the stress-strain law of the *relaxation type.* Its inverse,

$$(3)\ \blacktriangle \qquad e_{ij}(x, t) = \int_{-\infty}^{t} J_{ijkl}(x, t - \tau) \frac{\partial \sigma_{k,l}}{\partial \tau}(x, \tau)\, d\tau,$$

if it exists, is called the stress-strain law of the *creep type.* The fourth-order tensor $J_{ijkl}$ is called the *tensorial creep function.* It can be shown (Sternberg and Gurtin[15.1] (1962), Theorem 3.3), that the inverse (3) of (2) exists if $G_{ijkl}(x, t)$ is twice differentiable and if the initial value of $G_{ijkl}(x, t)$ at $t = 0$ is not zero.

The lower limits of integration in Eqs. (2) and (3) are taken as $-\infty$, which is to mean that the integration is to be taken before the very beginning of

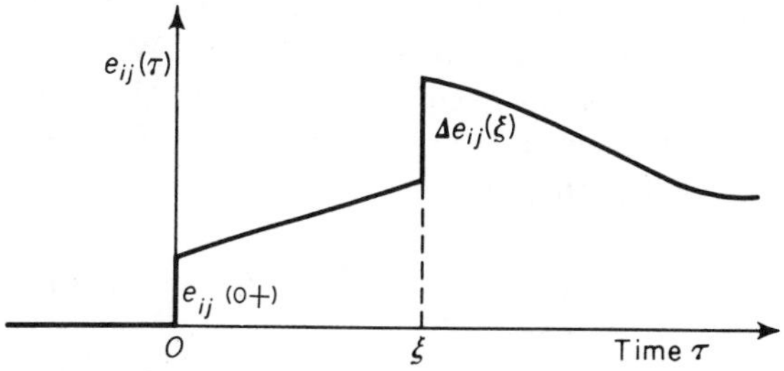


**Fig. 15.1:1.** Illustration for a loading history with jumps.

motion. If the motion starts at time $t = 0$, and $\sigma_{ij} = e_{ij} = 0$ for $t < 0$, Eq. (2) reduces to

$$(4) \qquad \sigma_{ij}(x, t) = e_{kl}(x, 0+)G_{ijkl}(x, t) + \int_{0}^{t} G_{ijkl}(x, t - \tau) \frac{\partial e_{kl}}{\partial \tau}(x, \tau)\, d\tau,$$

where $e_{ij}(x, 0+)$ is the limiting value of $e_{ij}(x, t)$ when $t \to 0$ from the positive side. The first term in Eq. (4) gives the effect of initial disturbance; it arises from the jump of $e_{ij}(x, t)$ at $t = 0$. In fact, it was tacitly assumed that $e_{ij}(x, t)$ is continuous and differentiable when Eq. (2) was written. Any discontinuity of $e_{ij}(x, t)$ in the form of a jump will contribute a term similar to the first term in (4). For example, if $e_{ij}(x, t)$ has another jump $\Delta e_{ij}(x, \xi)$ at $t = \xi$, as shown in Fig. 15.1:1, while $G_{ijkl}$ and $\partial e_{ij}/\partial t$ are continuous elsewhere, then we have

$$(5) \qquad \sigma_{ij}(x, t) = e_{kl}(x, 0+)G_{ijkl}(x, t) + \Delta e_{kl}(x, \xi)G_{ijkl}(x, t - \xi)\mathbf{1}(t - \xi) + \int_{0}^{t} G_{ijkl}(x, t - \tau) \frac{\partial e_{kl}}{\partial \tau}(x, \tau)\, d\tau,$$

where $\mathbf{1}(t)$ is the unit-step function. In such cases the following forms, which are equivalent to (4) when $\partial e_{ij}/\partial t$, $\partial G_{ijkl}/\partial t$ exist and are continuous in $0 \leqslant t < \infty$, may be used to advantage:

$$\tag{6} \begin{aligned} \sigma_{ij}(x, t) &= G_{ijkl}(x, 0)e_{kl}(x, t) + \int_0^t e_{kl}(x, t-\tau)\frac{\partial G_{ijkl}}{\partial \tau}(x, \tau)\, d\tau \\ &= \frac{\partial}{\partial t}\int_0^t e_{kl}(x, t-\tau)G_{ijkl}(x, \tau)\, d\tau. \end{aligned}$$

These constitutive equations are appropriate for *isothermal conditions*. For the influence of temperature, see references listed in Bib. 15.2, p. 505.

A viscoelastic material is defined by a specific relaxation function or creep function. In the following discussions we shall regard $G_{ijkl}$ or $J_{ijkl}$ as experimentally determined functions. Restrictions imposed on these functions by thermodynamics, as well as their origin in hidden coordinates, are discussed in Chap. 13.

In the treatment of viscoelasticity, we would have to write many convolution integrals. A shorthand is therefore desired. We shall introduce the notation of *composition products* commonly used in abstract algebra. Let $\phi$ and $\psi$ be functions defined on the intervals $0 \leqslant t < \infty$ and $-\infty < t < \infty$, respectively, and let the integral

$$\tag{7} I(t) = \int_0^t \phi(t-\tau)\frac{d\psi}{d\tau}(\tau)\, d\tau + \phi(t)\psi(0)$$

exist for all $t$ in $(0, \infty)$. Then the function $I(t)$ is called the *convolution of* $\phi$ and $\psi$ and is denoted by a 'composition' product

$$\tag{8} I(t) = \phi * d\psi.$$

The integral in (7) may be understood in Riemannian or Stieltjes' sense, the latter being more general and better suited to our purpose. Sternberg and Gurtin[15.1] developed the theory of viscoelasticity on the basis of Stieltjes' convolution.

The following properties of the convolution of $\phi$ with $\psi$ and $\theta$ (also defined over $-\infty < t < \infty$) can be verified.

$$\tag{9} \phi * d\psi = \psi * d\phi \qquad \text{(commutativity)},$$

$$\tag{10} \phi * d(\psi * d\theta) = (\phi * d\psi) * d\theta = \phi * d\psi * d\theta \qquad \text{(associativity)},$$

$$\tag{11} \phi * d(\psi + \theta) = \phi * d\psi + \phi * d\theta \qquad \text{(distributivity)},$$

$$\tag{12} \phi * d\psi = 0 \ \text{ implies } \ \phi \equiv 0 \ \text{ or } \ \psi \equiv 0 \ \text{(Titchmarsh theorem)}.$$

With these notations, Eqs. (2) and (3) may be written as

(2a) ▲ $$\sigma_{ij} = G_{ijkl} * de_{kl} = e_{kl} * dG_{ijkl},$$

(3a) ▲ $$e_{ij} = J_{ijkl} * d\sigma_{kl} = \sigma_{kl} * dJ_{ijkl}.$$

The symmetry of the stress and strain tensors either requires or permits that

(13) $$G_{ijkl} = G_{jikl} = G_{ijlk},$$

(14) $$J_{ijkl} = J_{jikl} = J_{ijlk}.$$

Furthermore, it is natural to require that the action of a loading at a time $t_0$ will produce a response only for $t \geqslant t_0$. Hence, we must have

(15) $$G_{ijkl} = 0, \qquad J_{ijkl} = 0, \qquad \text{for } -\infty < t < 0.$$

This requirement is sometimes called *the axiom of nonretroactivity.*

If $G_{ijkl}$ is invariant with respect to rotation of Cartesian coordinates, the material is said to be *isotropic*. It can be shown that a material is isotropic if and only if $G_{ijkl}$ is an isotropic tensor. A fourth-order isotropic tensor with the symmetry properties (13) can be written as

(16) $$G_{ijkl} = \frac{G_2 - G_1}{3}\,\delta_{ij}\,\delta_{kl} + \frac{G_1}{2}(\delta_{ik}\,\delta_{jl} + \delta_{il}\,\delta_{jk}),$$

where $G_1$, $G_2$ are scalar functions satisfying (15). If (16) is substituted into (2a), we obtain the stress-strain law which can be put in the form

(17) ▲ $$\sigma'_{ij} = e'_{ij} * dG_1 = G_1 * de'_{ij}, \qquad \sigma_{kk} = e_{kk} * dG_2 = G_2 * de_{kk},$$

where $\sigma'_{ij}$ and $e'_{ij}$ are the components of the stress and strain deviators:

(18) $$\sigma'_{ij} = \sigma_{ij} - \tfrac{1}{3}\delta_{ij}\sigma_{kk}, \qquad e'_{ij} = e_{ij} - \tfrac{1}{3}\delta_{ij}e_{kk}.$$

*The functions $G_1$ and $G_2$ are referred to as the relaxation functions in shear and in isotropic compression*, respectively.

The corresponding stress-strain laws of the creep type for an isotropic material is

(19) ▲ $$e'_{ij} = \sigma'_{ij} * dJ_1 = J_1 * d\sigma'_{ij}, \qquad e_{kk} = \sigma_{kk} * dJ_2 = J_2 * d\sigma_{kk},$$

where $J_1$, $J_2$ *are called the creep functions in shear and in isotropic compression, respectively.*

If $G_{ijkl}$, $J_{ijkl}$ or $G_1$, $G_2$, $J_1$, $J_2$ are step functions in time, then the stress-strain laws (2), (3), (17), and (19) reduce to those of linear elastic solids.

Although Eqs. (2), (3), (17), and (19) are tensorial equations, for fixed indices $i$ and $j$ they are of the same form as those discussed in Sec. 1.6. Therefore, what was deduced in Secs. 1.6 and 1.7 applies equally well here. In particular, we see that the Fourier transforms of $G$ and $J$ can be interpreted as the responses to appropriate harmonic forcing functions. Within a certain range of frequencies, measurements of harmonic responses of many materials are feasible. The bulk of our information about the viscoelastic behavior of high polymers is presented in the form of frequency responses.

## 15.2. STRESS-STRAIN RELATIONS IN DIFFERENTIAL EQUATION FORM

In Secs. 13.8 and 13.9 it was shown that when the relaxation function consists of a finite discrete spectrum, the stress-strain relation may be put in the form of a differential equation. Simple examples are the Maxwell, Voigt, or the standard linear models discussed in Sec. 1.6. A more general expression may be given as follows. Let $D$ denote the time-derivative operator defined by

$$Df = \frac{\partial f(t)}{\partial t}, \qquad D^2 f = \frac{\partial^2 f}{\partial t^2}, \quad \text{etc.,} \tag{1}$$

where $f$ is a function of time. Let us consider the polynomials

$$\begin{aligned} P_1(D) &= \sum_{k=0}^{n_1} a_k D^k, & Q_1(D) &= \sum_{k=0}^{m_1} b_k D^k, \\ P_2(D) &= \sum_{k=0}^{n_2} c_k D^k, & Q_2(D) &= \sum_{k=0}^{m_2} d_k D^k, \end{aligned} \tag{2}$$

where $a_k$, $b_k$, $c_k$, $d_k$ are real-valued functions of the spatial coordinates $x_1, x_2, x_3$. We assume that the leading coefficients $a_{n_1}, b_{m_1}, c_{n_2}, d_{m_2}$ are different from zero, so that $P_1$, $P_2$, $Q_1$, $Q_2$ are polynomials of degree $n_1$, $n_2$, $m_1$, $m_2$, respectively. Then the stress-strain relations

$$\blacktriangle \qquad P_1(D)\sigma'_{ij} = Q_1(D)e'_{ij}, \qquad P_2(D)\sigma_{kk} = Q_2(D)e_{kk} \tag{3}$$

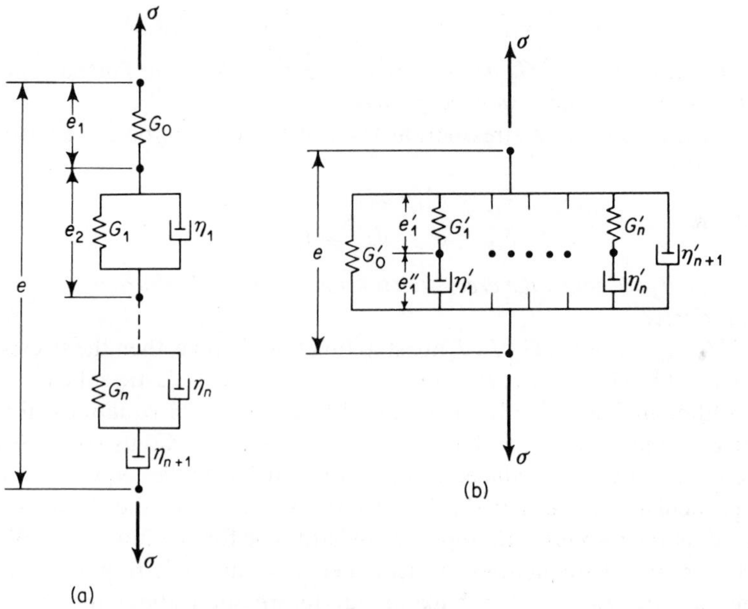


**Fig. 15.2:1.** Generalized Kelvin and Maxwell models.

specify an isotropic linear viscoelastic material. Here $\sigma_{ij}$, $e_{ij}$, $\sigma'_{ij}$, $e'_{ij}$ [see Eq. (15.1:17)] shall be understood to be functions of $x_1$, $x_2$, $x_3$ and $t$. In Eq. (3) et seq., the dependence on $x$ is not explicitly shown, but shall be understood.

If we consider the load-deflection relationship of a network of springs and dashpots such as those pictured in Fig. 15.2:1, it can be shown that the load $\sigma$ and the deflection $e$ are related by an equation of the form of (3) (see Prob. 15.1). For this reason, a viscoelastic body is often represented by a mechanical model. Naturally, not all conceivable polynomials (2) may represent physically realizable systems. Some restrictions are imposed by thermodynamic considerations, as discussed in Sec. 13.9 and Prob. 13.5.

If the stress-strain law for a given material can be expressed in both the integral form and the differential form, then the relationship between the relaxation (or creep) function and the differential operators can be determined. The result is particularly simple if we assume that

(4) $$\sigma_{ij}(t) = e_{ij}(t) = 0 \qquad \text{for } t < 0,$$

and that the Laplace transformation (with respect to time) of all the functions concerned exist. Let the Laplace transformation of a function $f(t)$ be indicated by a bar, thus

$$\bar{f}(s) = \int_0^\infty e^{-st} f(t)\, dt, \qquad \text{Rl } s > s_0,$$

where $s_0$ is the abscissa of convergence of the Laplace integral.

The Laplace transformations of Eqs. (15.1:17) and (15.1:19) are

(5) $$\bar{\sigma}'_{ij}(s) = s\bar{G}_1(s)\bar{e}'_{ij}(s), \qquad \bar{\sigma}_{kk}(s) = s\bar{G}_2(s)\bar{e}_{kk}(s),$$

(6) $$\bar{e}'_{ij}(s) = s\bar{J}_1(s)\bar{\sigma}'_{ij}(s), \qquad \bar{e}_{kk}(s) = s\bar{J}_2(s)\bar{\sigma}_{kk}(s).$$

The Laplace transformations of Eqs. (3) are

(7a) $$\bar{P}_1(s)\bar{\sigma}'_{ij}(s) - \sum_{k=1}^{n_1} a_k\left[s^{k-1}\sigma_{ij}(0) + s^{k-2}\frac{\partial\sigma'_{ij}}{\partial t}(0) + \ldots + \frac{\partial^{k-1}\sigma'_{ij}}{\partial t^{k-1}}(0)\right]$$
$$= \bar{Q}_1(s)\bar{e}'_{ij}(s) - \sum_{k=1}^{m_1} b_k\left[s^{k-1}e'_{ij}(0) + s^{k-2}\frac{\partial e'_{ij}}{\partial t}(0) + \ldots + \frac{\partial^{k-1}e'_{ij}}{\partial t^{k-1}}(0)\right],$$

(7b) $$\bar{P}_2(s)\bar{\sigma}_{jj}(s) - \sum_{k=1}^{n_2} c_k\left[s^{k-1}\sigma_{jj}(0) + \ldots + \frac{\partial^{k-1}\sigma_{jj}}{\partial t^{k-1}}(0)\right]$$
$$= \bar{Q}_2(s)\bar{e}_{jj}(s) - \sum_{k=1}^{m_2} d_k\left[s^{k-1}e_{jj}(0) + \ldots + \frac{\partial^{k-1}e_{jj}}{\partial t^{k-1}}(0)\right],$$

where

(8) $$\bar{P}_1(s) = \sum_{k=0}^{n_1} a_k s^k, \qquad \bar{Q}_1(s) = \sum_{k=0}^{m_1} b_k s^k,$$
$$\bar{P}_2(s) = \sum_{k=0}^{n_2} c_k s^k, \qquad \bar{Q}_2(s) = \sum_{k=0}^{m_2} d_k s^k,$$

and $\sigma'_{ij}(0)$, $(\partial\sigma'_{ij}/\partial t)(0)$, etc., are the initial values of $\sigma'_{ij}$, $\partial\sigma'_{ij}/\partial t$, etc., i.e., the value of $\sigma'_{ij}(x, t)$ and its time derivatives as $t \to 0$ from the positive side.

Whether Eqs. (7) should be identified with the relaxation law (5) or with the creep law (6) depends on whether $n_1 \geqslant m_1$, $n_2 \geqslant m_2$ or not. If $n_1 \geqslant m_1$, then (7a) may be written as *

$$\text{(9)} \blacktriangle \qquad \bar{\sigma}'_{ij}(s) = \frac{\bar{Q}_1(s)}{\bar{P}_1(s)} \bar{e}'_{ij}(s),$$

*provided that the following initial condition holds:*

$$\text{(10)} \qquad \sum_{k=1}^{n_1} a_k \left[ s^{k-1}\sigma'_{ij}(0) + \ldots + \frac{\partial^{k-1}\sigma'_{ij}}{\partial t^{k-1}}(0) \right] - \sum_{k=1}^{m_1} b_k \left[ s^{k-1} e'_{ij}(0) + \ldots + \frac{\partial^{k-1} e'_{ij}}{\partial t^{k-1}}(0) \right] \equiv 0.$$

In this case we may identify (9) with the relaxation law (5), with

$$\text{(11)} \blacktriangle \qquad \bar{G}_1(s) = \frac{\bar{Q}_1(s)}{s\bar{P}_1(s)}.$$

The initial condition (10) is an identity as a polynomial in $s$; every coefficient of the polynomial must vanish. Thus, if $m_1 = n_1$,

$$\text{(12)} \blacktriangle \qquad \begin{aligned} &a_{n_1}\sigma'_{ij}(0) = b_{n_1} e'_{ij}(0), \\ &\ldots\ldots\ldots\ldots\ldots\ldots\ldots\ldots\ldots\ldots\ldots\ldots, \\ &a_{n_1}\frac{\partial^{n_1-1}\sigma'_{ij}}{\partial t^{n_1-1}}(0) + a_{n_1-1}\frac{\partial^{n_1-2}\sigma'_{ij}}{\partial t^{n_1-2}}(0) + \ldots + a_1\sigma'_{ij}(0) \\ &\qquad = b_{n_1}\frac{\partial^{n_1-1} e'_{ij}}{\partial t^{n_1-1}}(0) + \ldots + b_1 e'_{ij}(0). \end{aligned}$$

If $m_1 < n_1$, then those coefficients $b_k$ in Eqs. (12) with subscript $k > m_1$ must be replaced by zero.

In the alternative case $n_1 \leqslant m_1$, Eq. (7a) may be written as *

$$\text{(13)} \qquad \bar{e}'_{ij}(s) = \frac{\bar{P}_1(s)}{\bar{Q}_1(s)} \bar{\sigma}'_{ij}(s),$$

provided that the initial condition (10) holds. We may identify Eq. (13) with the creep law (6), with

$$\text{(14)} \qquad \bar{J}_1(s) = \frac{\bar{P}_1(s)}{s\bar{Q}_1(s)}.$$

If $n_1 = m_1$, then the material may be represented by stress-strain laws of both the relaxation and the creep type.

An analogous situation exists for the laws governing the mean stress and mean strain, $\sigma_{kk}$ and $e_{kk}$.

* Note that a polynomial of $s$ has no continuous inverse function. The inverse of $s$ is $\delta(t)$; those of $s^2$, $s^3$ are higher order singularities.

The initial conditions (12) are interesting and important. They represent the proper initial conditions that must be imposed when Eqs. (3) are regarded as differential equations to solve for $\sigma'_{ij}$, with $e'_{ij}$ regarded as given forcing functions; or vice versa.

What is the physical significance of the initial conditions (12)? We recognized their necessity through an identification of Eq. (15.1:17), which is valid for $t$ in $(-\infty, \infty)$, with Eq. (3), which applies for $t$ in $(0, \infty)$. For the purpose of identification we have assumed $e'_{ij} = \sigma'_{ij} = 0$ for $t$ in $(-\infty, 0)$. If

(15) $$e'_{ij} = \frac{\partial e'_{ij}}{\partial t} = \cdots = \frac{\partial^{n_1-1} e'_{ij}}{\partial t^{n_1-1}} = 0 \qquad \text{at } t = 0,$$

then Eqs. (12) imply

(16) $$\sigma'_{ij} = \frac{\partial \sigma'_{ij}}{\partial t} = \cdots = \frac{\partial^{n_1-1} \sigma'_{ij}}{\partial t^{n_1-1}} = 0 \qquad \text{at } t = 0.$$

Thus the transition is smooth, as expected. The surprising aspect of Eqs. (12) arises only when $e'_{ij}(t)$ is continuous in $(0, \infty)$ but the limiting values $e'_{ij}(0+)$, $(\partial e'_{ij}/\partial t)(0+)$, etc., are nonzero as $t \to 0$ from the positive side. In this case there is a jump in the value of $e'_{ij}$ or its derivatives in the neighborhood of $t = 0$. The initial conditions (12) must stem from the fact that the differential operator in Eq. (3) is expected to hold, in a certain sense, during the jump.

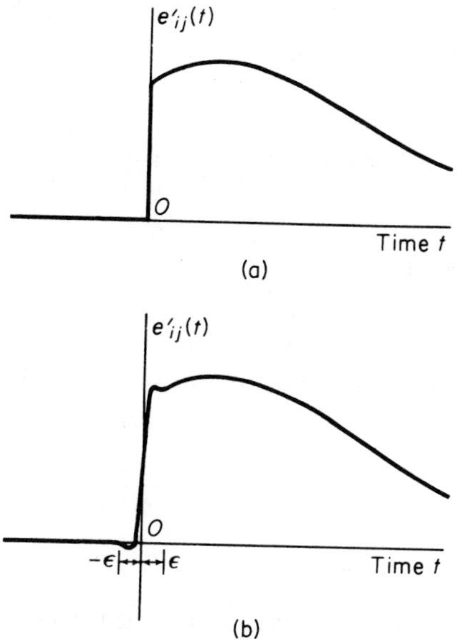


**Fig. 15.2:2.** Smoothing of a jump.

To see that this is indeed the case, let us smooth off the jump in an interval $(-\epsilon, \epsilon)$ (see Fig. 15.2:2). Let $e'_{ij}(t, \epsilon)$ be a smooth function (bounded, continuous, and $n_1$ times differentiable) in the range $(-\epsilon, \epsilon)$, which has the property that $e'_{ij}(t, \epsilon)$ and its derivatives vanish at the left end, $t = -\epsilon$; whereas assume the values $e'_{ij}(0+)$, $(\partial e'_{ij}/\partial t)(0+)$, etc., at the right end, $t = +\epsilon$. Such a function serves to describe the jump if we let $\epsilon \to 0$. Let $\sigma'_{ij}(t, \epsilon)$ be the corresponding smoothing function describing the jump in the stress deviation. We now assume that $e'_{ij}(t, \epsilon)$ and $\sigma'_{ij}(t, \epsilon)$ are connected by the viscoelasticity law (3); i.e.,

(17) $$\sum_{k=0}^{N} a_k \frac{\partial^k \sigma'_{ij}(t, \epsilon)}{\partial t^k} = \sum_{k=0}^{N} b_k \frac{\partial^k e'_{ij}(t, \epsilon)}{\partial t^k},$$

where $N$ is equal to the larger of $m_1$ and $n_1$, with some of the coefficients $a_k$, $b_k$ set equal to zero to eliminate any terms which may be absent from (3). An integration of (17) with respect to $t$ gives

(18)
$$\sum_{k=1}^{N} a_k \frac{\partial^{k-1}\sigma'_{ij}(t, \epsilon)}{\partial t^{k-1}} + a_0 \int_{-\epsilon}^{t} \sigma'_{ij}(\tau, \epsilon)\, d\tau = \sum_{k=1}^{N} b_k \frac{\partial^{k-1}e'_{ij}(t, \epsilon)}{\partial t^{k-1}} + b_0 \int_{-\epsilon}^{t} e'_{ij}(\tau, \epsilon)\, d\tau.$$

Now let $t = \epsilon$, and take the limit $\epsilon \to 0$. Since $\sigma'_{ij}(t, \epsilon)$, $e'_{ij}(t, \epsilon)$ are bounded in $(-\epsilon, \epsilon)$, the integrals in (18) are of order $\epsilon$ and vanish as $\epsilon \to 0$. Therefore, at $t = \epsilon$ and in the limit $\epsilon \to 0$, we obtain the necessary initial condition

$$\sum_{k=1}^{N} a_k \frac{\partial^{k-1}\sigma'_{ij}}{\partial t^{k-1}} (0+) = \sum_{k=1}^{N} b_k \frac{\partial^{k-1}e'_{ij}}{\partial t^{k-1}} (0+),$$

which is one of the equations in (12). Integrating (18) again from $-\epsilon$ to $\epsilon$ and repeating the arguments as above, we obtain the second initial condition

$$\sum_{k=2}^{N} a_k \frac{\partial^{k-2}\sigma'_{ij}}{\partial t^{k-2}} (0+) = \sum_{k=2}^{N} b_k \frac{\partial^{k-2}e'_{ij}}{\partial t^{k-2}} (0+).$$

All Eqs. (12) can be obtained in this way by repeated applications of the same process.

**Problem 15.1.** Let $\sigma$ denote loads and $e$ denote displacements. Show that, for the generalized Kelvin model shown in Fig. 15.2:1(a),

$$G_0 e_1 = \sigma, \qquad G_1 e_2 + \eta_1 D e_2 = \sigma, \quad \ldots, \quad e = e_1 + e_2 + \ldots + e_n,$$

so that

$$\left(\frac{1}{G_0} + \frac{1}{G_1 + \eta_1 D} + \ldots + \frac{1}{G_n + \eta_n D} + \frac{1}{\eta_{n+1} D}\right)\sigma = e.$$

For the generalized Maxwell model shown in Fig. 15.2:1(b), we have

$$G'_1 e'_1 = \sigma_1, \qquad \eta'_1 D e''_1 = \sigma_1,$$

$$e = e'_1 + e''_1 = \left(\frac{1}{G'_1} + \frac{1}{\eta'_1 D}\right)\sigma_1,$$

$$\sigma = \sigma_0 + \sigma_1 + \ldots + \sigma_{n+1} = \left(G'_0 + \frac{G_1\eta_1 D}{G_1 + \eta_1 D} + \ldots + \frac{G_n \eta_n D}{G_n + \eta_n D} + \eta_{n+1} D\right) e.$$

Reduce these relations to the form of Eq. (3).

**Problem 15.2.** Let a special case of Eq. (3) be specified by

$$P_1 \sigma_{ij} = a_2 \ddot{\sigma}_{ij} + a_1 \dot{\sigma}_{ij} + a_0 \sigma_{ij},$$
$$Q_1 e_{ij} = b_2 \ddot{e}_{ij} + b_1 \dot{e}_{ij} + b_0 e_{ij},$$

where a dot denotes a differentiation with respect to time. Deduce proper initial conditions. Interpret the usual statement that "the initial response of a viscoelastic body is purely elastic."

## 15.3. BOUNDARY-VALUE PROBLEMS AND INTEGRAL TRANSFORMATIONS

The motion of a viscoelastic body is governed by the laws of conservation of mass and momentum, the stress-strain relations, and the boundary conditions and initial conditions. Let $u_i$, $e_{ij}$, $\sigma_{ij}$, and $X_i$ denote the Cartesian components of the displacement, strain, stress, and body force per unit volume, respectively, and $\rho$ the mass density. Let us restrict our consideration to homogeneous and isotropic viscoelastic solids, and infinitesimal displacements. Under these limitations the field equations are:

(1) *Definition of strain:*

$$e_{ij} = \tfrac{1}{2}(u_{i,j} + u_{j,i}),$$

(2) *Equation of continuity:*

$$\frac{\partial \rho}{\partial t} + \frac{\partial}{\partial x_i}\left(\rho \frac{\partial u_i}{\partial t}\right) = 0,$$

(3) *Equations of motion:*

$$\sigma_{ij,j} + X_i = \rho \frac{\partial^2 u_i}{\partial t^2}, \qquad \sigma_{ij} = \sigma_{ji}.$$

The stress-strain relation may assume any of the following forms (see Sec. 15.2):

(4a) *Relaxation law:*

$$\sigma'_{ij} = e'_{ij} * dG_1, \qquad \sigma_{kk} = e_{kk} * dG_2,$$

(4b) *Creep law:*

$$e'_{ij} = \sigma'_{ij} * dJ_1, \qquad e_{kk} = \sigma_{kk} * dJ_2,$$

(4c) *Differential operator law:*

$$P_1(D)\sigma'_{ij} = Q_1(D)e'_{ij}, \qquad P_2(D)\sigma_{kk} = Q_2(D)e_{kk},$$

where $e'_{ij}$, $\sigma'_{ij}$ are the stress and strain deviators defined by

(5) $$\sigma'_{ij} = \sigma_{ij} - \tfrac{1}{3}\delta_{ij}\sigma_{kk}, \qquad e'_{ij} = e_{ij} - \tfrac{1}{3}\delta_{ij}e_{kk},$$

$G_1$, $G_2$ are the relaxation functions, $J_1$, $J_2$ are the creep functions, and $P_1$, $P_2$, $Q_1$, $Q_2$ are polynomials of the time-derivative operator $D$ as discussed in Sec. 15.2. In view of the assumed homogeneity of the material, these functions and operators are independent of position.

If the body is originally undisturbed, then the initial conditions are

(6) $$u_i = e_{ij} = \sigma_{ij} = 0 \qquad \text{for } -\infty < t < 0.$$

If the differential operator law (4c) is used and there is a jump condition at $t = 0$, then the initial conditions may assume the form of specific assigned values of $e'_{ij}(0+)$, $\frac{\partial e'_{ij}}{\partial t}(0+), \ldots, \frac{\partial^n e'_{ij}}{\partial t^n}(0+)$, and of $\sigma'_{ij}(0+)$, $\frac{\partial \sigma'_{ij}}{\partial t}(0+), \ldots, \frac{\partial^n \sigma'_{ij}}{\partial t^n}(0+)$, which must be connected by the necessary conditions [Eq. (15.2:12)]:

$$\sum_{k=r}^{n} a_k \frac{\partial^{k-r}\sigma'_{ij}}{\partial t^{k-r}}(0+) = \sum_{k=r}^{n} b_k \frac{\partial^{k-r} e'_{ij}}{\partial t^{k-r}}(0+), \quad r = 1, 2, \ldots, n. \tag{7}$$

where $n$ is the larger of the degrees of $P_1(D)$ and $Q_1(D)$. Similar statements hold for initial values of mean stress and mean strain.

The boundary conditions may take the form of either assigned traction over a surface $S_\sigma$ with an outward-pointing unit normal vector $\nu_i$,

$$\overset{\nu}{T}_i = \sigma_{ij}\nu_j = f_i \qquad \text{over } S_\sigma, \tag{8}$$

or of assigned displacement over a surface $S_u$,

$$u_i = g_i \qquad \text{over } S_u, \tag{9}$$

where $f_i$, $g_i$ are prescribed functions of position and time, and $S_\sigma + S_u = S$, the total surface of the body.

The problems in the theory of linear viscoelasticity usually consist of determining $u_i$, $e_{ij}$, and $\sigma_{ij}$ for prescribed $X_i$, $f_i$, $g_i$ and initial conditions. Except for the stress-strain law, these same equations occur in the theory of linear elasticity.

In the theory of linear elasticity we have applied the Laplace transformation to solve dynamic problems. The transformation translates an original problem involving derivatives and convolution integrals with respect to the time $t$ into an algebraic problem with respect to a parameter $s$. After the algebraic problem is solved, the solution is translated back to the time domain. The last step, of course, may not be easy.

It appears natural to apply the Laplace transformation also to problems in linear viscoelasticity. In fact, the transformed problem, involving the parameter $s$, can be put into the same form as that in the theory of linear elasticity. If the latter problem can be solved, the viscoelasticity solution can be obtained by an inverse transformation from $s$ back to the time domain. The last step, again, may not be easy.

The Fourier transform may be used in place of the Laplace transform. The selection of the appropriate transform depends on the nature of the functions involved. A function which has a Laplace transform may not necessarily have a Fourier transform, and vice versa. For example, the unit-step function $\mathbf{1}(t)$, or a power $t^n$, has no Fourier transform in the interval $(0, \infty)$.

The identification of a problem in linear elasticity with one in viscoelasticity in the transformed plane, is called the *correspondence principle.* Applications of the correspondence principle will become apparent by examining some examples in Secs. 15.4 and 15.5.

**Problem 15.3.** Assuming that the functions $u_i$, $G_1$, $G_2$ are continuous and can be differentiated as many times as may be desired, show that the equation of motion may be written (in the form of Navier's equation) as

$$(u_i)_{,jj} * dG_1 + (u_{j,j})_{,i} * dK + 2X_i = 0, \qquad i = 1, 2, 3,$$

where $3K = G_1 + 2G_2$.

(Gurtin and Sternberg (1962)[15.1])

**Problem 15.4.** From the strain equations of compatibility

$$e_{ij,kk} + e_{kk,ij} - e_{ik,jk} - e_{jk,ik} = 0,$$

the stress-strain relations

$$e_{ij} = \sigma_{ij} * dJ_1 + \tfrac{1}{3}\delta_{ij}\sigma_{kk} * d(J_2 - J_1),$$

and the equation of equilibrium

$$\sigma_{ik,kj} = -X_{i,j},$$

deduce the stress equations of compatibility

$$(\sigma_{ij})_{,kk} * dJ_1 + \sigma_{kk,ij} * d\Lambda = \Theta_{ij},$$

where

$$\Theta_{ij} = \delta_{ij}X_{k,k} * d\Omega - (X_{i,j} + X_{j,i}) * dJ_1,$$

$$3\Lambda = 2J_1 + J_2,$$

$$9\Omega = J_1 * d(J_2 - J_1) * d(J_1 + 2J_2)^{-1}.$$

(Gurtin and Sternberg (1962)[15.1])

**Problem 15.5.** Let $\mathbf{u}(u_i)$, $\boldsymbol{\epsilon}(e_{ij})$, $\boldsymbol{\sigma}(\sigma_{ij})$ satisfy the equations of viscoelasticity, and $G_1 \not\equiv 0$, $2G_1 + G_2 \not\equiv 0$. Show that, when the body force $X_i$ is absent,

$$\nabla^2(\nabla \cdot \mathbf{u}) = 0, \qquad \nabla^2(\nabla \times \mathbf{u}) = 0, \qquad \nabla^2 e_{kk} = 0, \qquad \nabla^2 \sigma_{kk} = 0,$$

$$\nabla^4 \mathbf{u} = 0, \qquad \nabla^4 \boldsymbol{\epsilon} = 0, \qquad \nabla^4 \boldsymbol{\sigma} = 0.$$

(Gurtin and Sternberg (1962)[15.1])

## 15.4. WAVES IN AN INFINITE MEDIUM

Let us consider a viscoelastic medium of infinite extent and look for solutions of Eqs. (15.3:1)–(15.3:6), in which all dependent variables, including body forces, vary sinusoidally with time. We shall use the complex representation for sinusoidal oscillations. Since $\cos \omega t$ is equal to the real part of $e^{i\omega t}$, a real-valued function $f(x, t)$ that varies like $\cos \omega t$ at every point $x$ may be represented as $\mathrm{Rl}\,[F(x)\, e^{i\omega t}]$, where $F(x)$ is real. A real-valued function $f(x, t)$ that varies like $\cos(\omega t + \phi)$ may be represented as $\text{Rl}\,[F(x)\,e^{i(\omega t+\phi)}]$ or as $\text{Rl}\,[\bar{F}(x)\,e^{i\omega t}]$, where $\bar{F}(x)$ now stands for a complex number $F(x)\,e^{i\phi}$. It is elementary to show that

$$\text{Rl}\,\bar{F}(x) = F(x)\cos\phi, \qquad \text{Im}\,\bar{F}(x) = F(x)\sin\phi,$$

$$\begin{aligned}\text{Rl}\,[\bar{F}(x)\,e^{i\omega t}] &= \text{Rl}\,\bar{F}\,\text{Rl}\,e^{i\omega t} - \text{Im}\,\bar{F}\,\text{Im}\,e^{i\omega t}\\ &= F\cos\phi\cos\omega t - F\sin\phi\sin\omega t\\ &= F\cos(\omega t + \phi).\end{aligned}$$

Thus it is clear that the real part of $\bar{F}(x)$ is *in phase* with $\cos\omega t$ and that the imaginary part of $\bar{F}(x)$ *leads* $\cos\omega t$ by a phase angle $\phi = \pi/2$. In general, $\phi$ is a function of $x$. In this way a sinusoidal oscillation of a solid medium whose amplitude and phase angle vary from point to point may be represented by a complex function $\bar{F}(x)$. A multiplication of $\bar{F}(x)$ by the imaginary number $i$ means an advance of phase angle by $\pi/2$; a multiplication by $-i$ means a lag by $\pi/2$.

With these interpretations of complex representation we let

(1) $$e_{ij} = \text{Rl}\,(\bar{e}_{ij}\,e^{i\omega t}), \qquad \sigma_{ij} = \text{Rl}\,(\bar{\sigma}_{ij}e^{i\omega t}), \quad \text{etc.},$$

where $\bar{e}_{ij}$, $\bar{\sigma}_{ij}$, etc., are complex-valued functions of the spatial coordinates only. The basic Eqs. (15.3:1)–(15.3:4) now read

(2) $$\bar{e}_{ij} = \tfrac{1}{2}(\bar{u}_{i,j} + \bar{u}_{j,i}),$$

(3) $$\bar{\sigma}_{ij,j} + \bar{X}_i + \rho\omega^2\bar{u}_i = 0,$$

(4) $$\bar{\sigma}'_{ij} = i\omega\bar{G}_1(\omega)\bar{e}'_{ij}, \qquad \bar{\sigma}_{kk} = i\omega\bar{G}_2(\omega)\bar{e}_{kk},$$

where $\bar{G}_1(\omega)$, $\bar{G}_2(\omega)$ are the deviatoric and the dilatational complex moduli, and are the Fourier transforms of the relaxation functions $G_1(t)$ and $G_2(t)$, respectively. If we write

(5) $$\bar{\lambda}(\omega) = \tfrac{1}{3}i\omega[\bar{G}_2(\omega) - \bar{G}_1(\omega)], \qquad \bar{G}(\omega) = \tfrac{1}{2}i\omega\bar{G}_1(\omega),$$

and call $\bar{\lambda}(\omega)$ and $\bar{G}(\omega)$ the complex Lamé constants for a viscoelastic material, we see that Eqs. (2)–(5) are identical with those governing linear elasticity theory (Sec. 7.1), except that the Lamé constants are replaced by complex moduli. Hence, without much ado, we can write down the Navier equation [cf. Eq. (7.1:9)]

(6) $$\bar{G}(\omega)\bar{u}_{i,jj} + [\bar{\lambda}(\omega) + \bar{G}(\omega)]\bar{u}_{j,ji} + \bar{X}_i + \rho\omega^2\bar{u}_i = 0$$

and the solutions, in case $\bar{X}_i = 0$,

(7) $$u_j = \text{Rl}\left\{An_j\exp\left[i\omega\left(t \pm \sqrt{\frac{\rho}{3\bar{\lambda}(\omega) + 2\bar{G}(\omega)}}\,n_k x_k\right)\right]\right\}$$

and

(8) $$u_j = \text{Rl}\left\{C_j\exp\left[i\omega\left(t \pm \sqrt{\frac{\rho}{\bar{G}(\omega)}}\,n_k x_k\right)\right]\right\},$$

where $A$ is a complex constant, $n_j$ is an arbitrary vector, and $C_j$ are components of a vector perpendicular to $n_j$, i.e., $C_j n_j = 0$.

Equation (7) represents a plane dilatational wave. Equation (8) represents a plane shear wave. The exponential factors in (7) and (8) are complex. A little reflection shows that the real parts of the factors in front of $n_k x_k$ are the inverse of wave velocities, $v_D$, $v_R$; whereas the imaginary parts are attenuation factors $\alpha_D$, $\alpha_R$.

(9) *Dilatational waves:*

$$v_D = \left\{ \mathrm{Rl}\left[\sqrt{\left(\frac{\rho}{3\bar{\lambda}(\omega) + 2\bar{G}(\omega)}\right)}\right]\right\}^{-1},$$

$$\alpha_D = -\omega\,\mathrm{Im}\left[\sqrt{\left(\frac{\rho}{3\bar{\lambda}(\omega) + 2\bar{G}(\omega)}\right)}\right],$$

(10) *Rotational waves:*

$$v_R = \left\{ \mathrm{Rl}\left[\sqrt{\left(\frac{\rho}{\bar{G}(\omega)}\right)}\right]\right\}^{-1},$$

$$\alpha_R = -\omega\,\mathrm{Im}\left[\sqrt{\left(\frac{\rho}{\bar{G}(\omega)}\right)}\right].$$

If the material is elastic, $\bar{\lambda}(\omega)$, $\bar{G}(\omega)$ become real numbers, independent of $\omega$, in which case the attenuation factors $\alpha_D$ and $\alpha_R$ vanish and the wave speeds $v_D$, $v_R$ are independent of frequency.

Any formal solution of Navier's equation in the classical theory of linear elasticity, of the form $f = \mathrm{Rl}\,[\bar{f}(x_1, x_2, x_3)\, e^{i\omega t}]$, offers a corresponding solution for a linear viscoelastic body, if the elastic moduli that occur in $\bar{f}(x_1, x_2, x_3)$ are replaced by the corresponding complex moduli of the material. If the boundary conditions for the elastic and viscoelastic problems are identical, then a solution in viscoelasticity can be obtained by this correspondence principle. Note, however, that the operation of separating a complex modulus into its real and imaginary parts is one that has no counterpart in the theory of linear elasticity. Hence, such an operation is excluded from the correspondence principle. For example, if we wish to find the maximum of $|f|$ with respect to $\omega$, we must separate the real and imaginary parts, and no direct analogy will be available.

**Problem 15.6.** Work out the details of the derivation of Eqs. (7) and (8) by reference to Sec. 7.8.

**Problem 15.7.** Find the speed and attenuation factor for Rayleigh surface waves propagating over a viscoelastic half-space (see Sec. 7.9). *Ref.* Bland, *Linear Viscoelasticity*[15.1] pp. 73–5.

**Problem 15.8.** Consider a uniform cantilever beam the clamped end of which is forced to oscillate at a constant amplitude $w_0$ (the clamping wall moves, while the other end is free). Fig. P15.8.

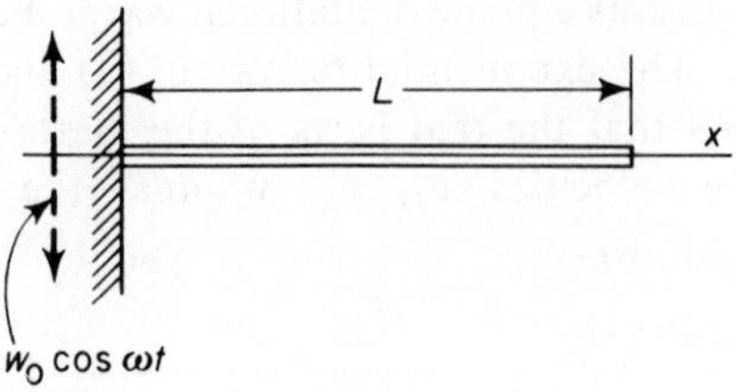


**P15.8**

For an elastic beam, the equation of motion of the beam is

$$EI\frac{\partial^4 w}{\partial x^4} + m\frac{\partial^2 w}{\partial t^2} = 0,$$

where $w$ is the beam deflection, $m$ is the mass per unit length, $E$ is the Young's modulus, and $I$ the moment of inertia of the cross section. The boundary conditions are at $x = 0$:

$$w = w_0 \cos \omega t, \qquad \frac{\partial w}{\partial x} = 0;$$

at $x = L$:

$$\frac{\partial^2 w}{\partial x^2} = \frac{\partial^3 w}{\partial x^3} = 0.$$

Determine a solution $w(x, t)$ in the form $\bar{w}(x)\, e^{i\omega t}$.

If the beam material is viscoelastic, determine (a) the corresponding complex bending rigidity $EI(\omega)$ from the stress-strain relationship of the material and (b) the solution $w(x, t)$.

The amplitude ratio and phase lag of the oscillations at the free and the clamped ends can be used for experimental determination of the complex modulus. *Ref.* Bland and Lee, *J. Appl. Phys.*, *26* (1955), p. 1497.

## 15.5. QUASI-STATIC PROBLEMS

If all functions of concern vanish for $t < 0$ and have Laplace transformations, the transforms of the basic Eqs. (15.3:1)–(15.3:9) assume exactly the same form as in the theory of linear elasticity. In quasi-static problems, the loading is assumed to be so slow that inertia forces may be neglected. In this case, the corresponding elastic problems are static.

Let the Laplace transform of a variable be denoted by a bar over it. The basic equations for quasi-static problems of a viscoelastic body occupying a region $V$ with boundary $B = B_\sigma + B_u$ are

(1) $$\bar{e}_{ij} = \tfrac{1}{2}(\bar{u}_{i,j} + \bar{u}_{j,i}) \qquad \text{in } V,$$

(2) $$\bar{\sigma}_{ij,j} + \bar{X}_i = 0 \qquad \text{in } V,$$

(3a) $$\bar{\sigma}'_{ij} = s\bar{G}_1(s)\bar{e}'_{ij}, \qquad \bar{\sigma}_{kk} = s\bar{G}_2(s)\bar{e}_{kk} \qquad \text{in } V,$$

or

(3b) $$\bar{e}'_{ij} = s\bar{J}_1(s)\bar{\sigma}'_{ij}, \qquad \bar{e}_{kk} = s\bar{J}_2(s)\bar{\sigma}_{kk} \qquad \text{in } V,$$

or

(3c) $$P_1(s)\bar{\sigma}'_{ij} = Q_1(s)\bar{e}'_{ij}, \qquad P_2(s)\bar{\sigma}_{kk} = Q_2(s)\bar{e}_{kk} \qquad \text{in } V,$$

(4) $$\bar{\sigma}_{ij}\nu_j = \bar{f}_i \qquad \text{on } B_\sigma,$$

(5) $$\bar{u}_i = \bar{g}_i \qquad \text{on } B_u.$$

The boundary surface enclosing $V$ is $B = B_\sigma + B_u$. The initial conditions (15.3:6) and (15.3:7) are assumed to be satisfied both in $V$ and on $B$. The material properties $\bar{G}_1(s)$, $\bar{G}_2(s)$; or $\bar{J}_1(s)$, $\bar{J}_2(s)$; or $P_1(s)$, $P_2(s)$, $Q_1(s)$, $Q_2(s)$; are given. The problem is to determine $\bar{u}_i$, $\bar{e}_{ij}$, $\bar{\sigma}_{ij}$ for assigned forcing functions $\bar{X}_i$ in $V$ and $\bar{f}_i$, $\bar{g}_i$ on the boundary. If the bars were removed, these would be the same equations that govern the static equilibrium of an elastic body of the same geometry. If the solution of the latter were known, the transformed solution would be obtained, and the solution of the original problem could be obtained by an inversion.

The counterpart of Young's modulus $E$, the shear modulus $G$, the bulk modulus $K$, Poisson's ratio $\nu$, and one of Lamé constants $\lambda$ (the other Lamé constant is $\mu = G$) can be derived from Eqs. (3a), (3b), and (3c), and are

(6) $$\bar{E}(s) = \frac{3Q_1(s)Q_2(s)}{Q_1(s)P_2(s) + 2P_1(s)Q_2(s)} = \frac{3s\bar{G}_1(s)\bar{G}_2(s)}{2\bar{G}_2(s) + \bar{G}_1(s)} = \frac{3}{s[2\bar{J}_1(s) + \bar{J}_2(s)]},$$

(7) $$\bar{G}(s) = \frac{1}{2}\frac{Q_1(s)}{P_1(s)} = \frac{1}{2}\,s\bar{G}_1(s) = \frac{1}{2}\frac{1}{s\bar{J}_1(s)},$$

(8) $$\bar{K}(s) = \frac{1}{3}\frac{Q_2(s)}{P_2(s)} = \frac{1}{3}\,s\bar{G}_2(s) = \frac{1}{3}\frac{1}{s\bar{J}_2(s)},$$

(9) $$\bar{\nu}(s) = \frac{P_1(s)Q_2(s) - Q_1(s)P_2(s)}{Q_1(s)P_2(s) + 2P_1(s)Q_2(s)} = \frac{\bar{G}_2(s) - \bar{G}_1(s)}{2\bar{G}_2(s) + \bar{G}_1(s)} = \frac{\bar{J}_1(s) - \bar{J}_2(s)}{2\bar{J}_1(s) + \bar{J}_2(s)},$$

(10) $$\bar{\lambda}(s) = \bar{K}(s) - \tfrac{2}{3}\bar{G}(s) = \tfrac{1}{3}s[\bar{G}_2(s) - \bar{G}_1(s)].$$

*Example* 1. *Boussinesq Problem*

Consider the problem of a concentrated load $Z$ on a half-space (Fig. 15.5:1). The elastic solution is given in Sec. 8.10. For example, the stress component $\sigma_{rr}$ is given by Eq. (8.10:7) as

(11) $$\sigma_{rr}(r, z) = \frac{Z}{2\pi}\left\{(1 - 2\nu)\left[\frac{1}{r^2} - \frac{z}{r^2}\frac{1}{(r^2 + z^2)^{1/2}}\right] - 3r^2z\frac{1}{(r^2 + z^2)^{5/2}}\right\}.$$

On applying the correspondence principle, the solution of $\bar{\sigma}_{rr}$ is

(12) $$\bar{\sigma}_{rr}(r, z;\ s) = \frac{\bar{Z}(s)}{2\pi}\left\{[1 - 2\,\bar{\nu}(s)]\left[\frac{1}{r^2} - \frac{z}{r^2}\frac{1}{(r^2 + z^2)^{1/2}}\right] - 3r^2z\frac{1}{(r^2 + z^2)^{5/2}}\right\}.$$

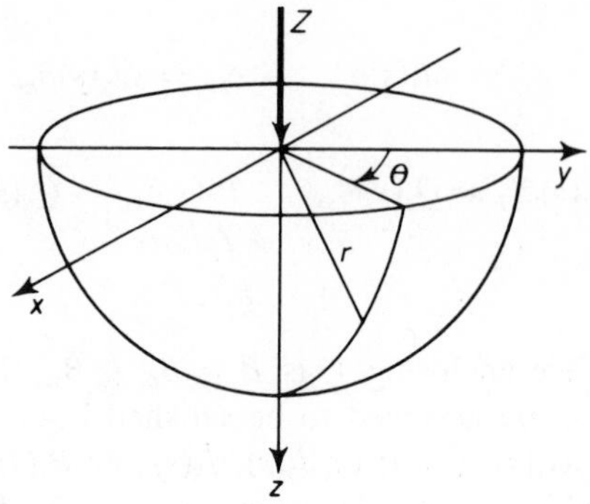


**Fig. 15.5:1.** Boussinesq problem over a viscoelastic material.

where $\bar{\nu}(s)$ is given by Eq. (9). An inverse Laplace transformation gives, therefore,

$$\sigma_{rr}(r, z;\ t) = \frac{1}{2\pi}\left\{\left[\frac{1}{r^2} - \frac{z}{r^2}\frac{1}{(r^2 + z^2)^{1/2}}\right]\int_0^t Z(t-\tau)\,\Phi(\tau)\,d\tau - 3r^2 z\,\frac{1}{(r^2+z^2)^{5/2}}\,Z(t)\right\}, \tag{13}$$

where

$$\Phi(t) = \int_{c-i\infty}^{c+i\infty} [1 - 2\bar{\nu}(s)]e^{st}\,ds \tag{14}$$

is the inverse Laplace transformation of $1 - 2\bar{\nu}(s)$, the constant $c$ being any number greater than the abscissa of convergence.

**Problem 15.9.** A concentrated load $Z(t) = Z_0\mathbf{1}(t)$ ($Z_0$ a constant, $\mathbf{1}(t)$ a unit-step function) acts normal to the free surface of a half-space of Voigt material, for which

$$\sigma'_{ij} = (2\eta D + 2G)e'_{ij}, \qquad \sigma_{jj} = 3Ke_{jj}.$$

Show that

$$1 - 2\bar{\nu}(s) = \frac{3(\eta s + G)}{3K + \eta s + G}$$

and

$$\sigma_{rr} = \frac{Z_0}{2\pi}\left\{\frac{3}{3K+G}\left[G + 3K\exp\left(-\frac{3K+G}{\eta}t\right)\right]\left[\frac{1}{r^2} - \frac{z}{r^2}\frac{1}{(r^2+z^2)^{1/2}}\right] - 3r^2 z\,\frac{1}{(r^2+z^2)^{5/2}}\right\}\mathbf{1}(t).$$

**Problem 15.10.** Consider a long, thick-walled circular cylinder subjected to a uniform internal pressure $p_i$ and uniform external pressure $p_0$. Fig. P15.10. Regarding this as a plane strain problem, show that the stress distribution in the cylinder is the same whether the cylinder material is linear elastic or linear viscoelastic. Derive the history of deformation $u_i(t)$ if the cylinder is viscoelastic.

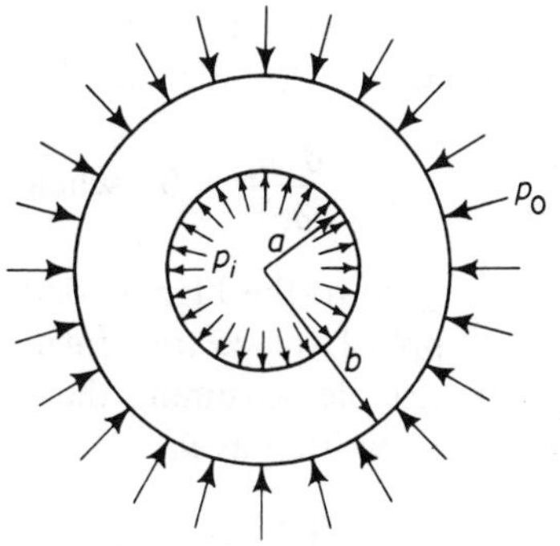


**P15.10**

*Note*: In the theory of elasticity, the biharmonic equation governing the Airy stress function in plane stress or plane strain problems is independent of the elastic constants if the body force vanishes. Note, however, that if the boundary conditions of a two-dimensional problem involve displacements, then the associated elastic problem will involve elastic moduli and the correspondence principle applies.

## 15.6. RECIPROCITY RELATIONS

In the beginning of Sec. 1.1 we derived the Maxwell and Betti-Rayleigh reciprocity relations. These relations can be generalized to an elastic or viscoelastic continuum and, with proper interpretations, they are valid not only for statics but also for dynamics.

The most general reciprocal theorem in dynamics was asserted by Horace Lamb[15.3] to be derivable from a remarkable formula established by Lagrange in the *Mécanique Analytique* (1809) by way of a prelude to Lagrange's theory of the variation of arbitrary constants. Lamb showed how the reciprocal theorems of von Helmholtz, in the theory of least action in acoustics and optics, and of Lord Rayleigh, in acoustics, can be derived from Lagrange's formula. Rayleigh[15.3] extended the reciprocal theorem to include the action of dissipative forces, and Lamb[15.3] showed the complete reciprocity relationships in a moving fluid with reversed-flow conditions.

In the theory of elasticity, a generalization of the reciprocity theorem to dynamic problems was given by Graffi,[15.3] and certain applications of Graffi's results to the problem of elastic wave propagation were pointed out by Di Maggio and Bleich.[15.3]

We shall derive the dynamic reciprocity relationship for a viscoelastic solid, which includes the linear elastic solid as a special case. We assume the stress-strain law to be given by one of the equations (15.3:4). The equations of motion and boundary conditions are given by Eqs. (15.3:3) and (15.3:8) or (15.3:9), respectively. Let us consider the problems in which the body force $X_i(x_1, x_2, x_3; t)$, the specified surface tractions $f_i(x_1, x_2, x_3; t)$ and the specified displacements $g_i(x_1, x_2, x_3; t)$ are given functions of space and time, which starts its action at $t > 0$, under the initial conditions

$$u_i = \frac{\partial u_i}{\partial t} = \ldots = \frac{\partial^N u_i}{\partial t^N} = 0 \quad \text{when } t \leqslant 0, \tag{1}$$

where $N$ is either unity or equal to the highest derivative that occurs in the stress-strain law (4c). We apply Laplace transformation with respect to the time $t$ to every dependent variable, assuming that the transforms exist. Let the Laplace transform of $u$ be written as $\bar{u}$,

$$\bar{u} = \int_0^\infty e^{-st} u(t)\, dt. \tag{2}$$

We have, due to the initial conditions indicated above,

$$\begin{aligned} \bar{\sigma}_{ij} &= \bar{\lambda}(s)\bar{u}_{k,k}\,\delta_{ij} + \bar{G}(s)(\bar{u}_{i,j} + \bar{u}_{j,i}), \\ s^2\rho\bar{u}_i &= \bar{X}_i + \bar{\sigma}_{ij,j} \quad \text{in } V, \\ \bar{\sigma}_{ij}\nu_j &= \bar{f}_i \quad \text{on } S_\sigma \\ \bar{u}_i &= \bar{g}_i \quad \text{on } S_u \end{aligned} \tag{3}$$

where $\bar{\lambda}(s)$ and $\bar{G}(s)$ are given by Eqs. (15.5:10) and (15.5:7), respectively.

Now consider two problems where the applied body force and the surface tractions and displacements are specified differently. Let the variables involved in these two problems be distinguished by superscripts in parentheses. Then,

(4a) In $V$: $\quad s^2\rho\bar{u}_i^{(1)} = \bar{X}_i^{(1)} + \bar{\sigma}_{ij,j}^{(1)}$ $\qquad$ (4b) $\quad s^2\rho\bar{u}_i^{(2)} = \bar{X}_i^{(2)} + \bar{\sigma}_{ij,j}^{(2)}$

(5a) On $S_\sigma$: $\quad \bar{\sigma}_{ij}^{(1)}\nu_j = \bar{f}_i^{(1)}$ $\qquad$ (5b) $\quad \bar{\sigma}_{ij}^{(2)}\nu_j = \bar{f}_i^{(2)}$

(6a) On $S_u$: $\quad \bar{u}_i^{(1)} = \bar{g}_i^{(1)}$ $\qquad$ (6b) $\quad \bar{u}_i^{(2)} = \bar{g}_i^{(2)}$

Multiplying Eq. (4a) by $\bar{u}_i^{(2)}$ and (4b) by $\bar{u}_i^{(1)}$, subtracting, and integrating over the region $V$, we obtain

$$\int_V \bar{X}_i^{(1)}\bar{u}_i^{(2)}\, dv + \int_V \bar{\sigma}_{ij,j}^{(1)}\bar{u}_i^{(2)}\, dv = \int_V \bar{X}_i^{(2)}\bar{u}_i^{(1)}\, dv + \int_V \bar{\sigma}_{ij,j}^{(2)}\bar{u}_i^{(1)}\, dv \tag{7}$$

Now

$$\begin{aligned} \int_V \bar{\sigma}_{ij,j}^{(1)}\bar{u}_i^{(2)}\, dv &= \int_V [(\bar{\lambda}\bar{u}_{k,k}^{(1)}\,\delta_{ij})_{,j}\bar{u}_i^{(2)} + (\bar{G}\bar{u}_{i,j}^{(1)} + \bar{G}\bar{u}_{j,i}^{(1)})_{,j}\bar{u}_i^{(2)}]\, dv \\ &= \int_S \bar{\lambda}\bar{u}_{k,k}^{(1)}\bar{u}_i^{(2)}\nu_i\, dS - \int_V \bar{\lambda}\bar{u}_{k,k}^{(1)}\bar{u}_{j,j}^{(2)}\, dv \\ &\quad + \int_S \bar{G}\bar{u}_{i,j}^{(1)}\bar{u}_i^{(2)}\nu_j\, dS + \int_S \bar{G}\bar{u}_{j,i}^{(1)}\bar{u}_i^{(2)}\nu_j\, dS \\ &\quad - \int_V \bar{G}\bar{u}_{i,j}^{(1)}\bar{u}_{i,j}^{(2)}\, dv - \int_V \bar{G}\bar{u}_{j,i}^{(1)}\bar{u}_{i,j}^{(2)}\, dv. \end{aligned} \tag{8}$$

A similar expression is obtained for the integral

$$\int_V \bar{\sigma}^{(2)}_{ij,j}\bar{u}^{(1)}_i\, dv. \tag{9}$$

When these expressions are substituted into Eq. (7), we see that a number of volume integrals cancel each other. The surface integrals contributed by the integrals (8) and (9) to (7) are

$$\begin{aligned} &\int_S \bar{\lambda}\bar{u}^{(1)}_{k,k}\bar{u}^{(2)}_i \nu_i\, dS + \int_S \bar{G}\bar{u}^{(1)}_{i,j}\bar{u}^{(2)}_i \nu_j\, dS + \int_S \bar{G}\bar{u}^{(1)}_{j,i}\bar{u}^{(2)}_i \nu_j\, dS \\ &\quad - \left(\int_S \bar{\lambda}\bar{u}^{(2)}_{k,k}\bar{u}^{(1)}_i \nu_i\, dS + \int_S \bar{G}\bar{u}^{(2)}_{i,j}\bar{u}^{(1)}_i \nu_j\, dS + \int_S \bar{G}\bar{u}^{(2)}_{j,i}\bar{u}^{(1)}_i \nu_j\, dS\right) \\ &\qquad = \int_S \bar{\sigma}^{(1)}_{ij}\bar{u}^{(2)}_i \nu_j\, dS - \int_S \bar{\sigma}^{(2)}_{ij}\bar{u}^{(1)}_i \nu_j\, dS. \end{aligned} \tag{10}$$

Recall that $S_\sigma$ is the part of $S$ over which surface tractions are specified, $S_u$ is that part of $S$ over which displacements are specified, $S = S_\sigma + S_u$. Now, substituting (8), (9), (10) and the boundary conditions (5a), (5b), (6a), (6b) into Eq. (7), we obtain finally:

$$\begin{aligned} &\int_V \bar{X}^{(1)}_i\bar{u}^{(2)}_i\, dv + \int_{S_\sigma} \bar{f}^{(1)}_i\bar{u}^{(2)}_i\, dS + \int_{S_u} \bar{\sigma}^{(1)}_{ij}\bar{g}^{(2)}_i \nu_j\, dS \\ &\qquad = \int_V \bar{X}^{(2)}_i\bar{u}^{(1)}_i\, dv + \int_{S_\sigma} \bar{f}^{(2)}_i\bar{u}^{(1)}_i\, dS + \int_{S_u} \bar{\sigma}^{(2)}_{ij}\bar{g}^{(1)}_i \nu_j\, dS. \end{aligned} \tag{11}$$

This is the *general reciprocal relation in the Laplace transformation form.* If we remove the bars and consider the variables to be in the real-time domain, then it becomes Betti's reciprocal relation in elastostatics.

Since the inverse transform of the product of two functions is the convolution of the inverses, we obtain:

$$\begin{aligned} &\int_V\int_0^t X^{(1)}_i(x, t-\tau)u^{(2)}_i(x,\tau)\, d\tau\, dv \\ &+ \int_{S_\sigma}\int_0^t f^{(1)}_i(x, t-\tau)u^{(2)}_i(x,\tau)\, d\tau\, dS + \int_{S_u}\int_0^t \sigma^{(1)}_{ij}(x, t-\tau)g^{(2)}_i(x,\tau)\nu_j\, d\tau\, dS \\ &= \int_V\int_0^t X^{(2)}_i(x, t-\tau)u^{(1)}_i(x,\tau)\, d\tau\, dv + \int_{S_\sigma}\int_0^t f^{(2)}_i(x, t-\tau)u^{(1)}_i(x,\tau)\, d\tau\, dS \\ &\qquad + \int_{S_u}\int_0^t \sigma^{(2)}_{ij}(x, t-\tau)g^{(1)}_i(x,\tau)\nu_j\, d\tau\, dS \end{aligned} \tag{12}$$

This is the *general reciprocal relation for elasto-kinetics.* Whether the material is viscoelastic or purely elastic makes no difference in the final result. Note that this result holds for variable density $\rho(x_1, x_2, x_3)$ and nonhomogeneous material properties.

***Generalization to Infinite Region.*** A generalization of the above result to an infinite or semi-infinite region is possible. Since with a finite wave speed there always exists a finite boundary surface, at any finite $t > 0$, which is yet uninfluenced by the loading initiated at $t = 0$. Let $S_u$ be such a surface. Then $g_i = 0$ on $S_u$, and the remainder of the equation holds without question.

*Examples of Applications*

(a) *Space-time-separable body forces, surface tractions and displacements.* If

$$\begin{aligned} X_i^{(1)} &= \Xi_i^{(1)}(x)h(t), & X_i^{(2)} &= \Xi_i^{(2)}(x)h(t), \\ f_i^{(1)} &= P_i^{(1)}(x)h(t), & f_i^{(2)} &= P_i^{(2)}(x)h(t), \\ g_i^{(1)} &= W_i^{(1)}(x)h(t), & g_i^{(2)} &= W_i^{(2)}(x)h(t), \end{aligned}$$

then Eq. (11) can be written, on cancelling $\bar{h}(s)$ from every term, as

$$\int_V \Xi_i^{(1)}\bar{u}_i^{(2)}\,dv + \int_{S\sigma} P_i^{(1)}\bar{u}_i^{(2)}\,dS + \int_{S_u} W_i^{(2)}\,\bar{\sigma}_{ij}^{(1)}\,\nu_j\,dS$$
$$= \int_V \Xi_i^{(2)}\bar{u}_i^{(1)}\,dv + \int_{S\sigma} P_i^{(2)}\bar{u}_i^{(1)}\,dS + \int_{S_u} W_i^{(1)}\,\bar{\sigma}_{ij}^{(2)}\nu_j\,dS.$$

The inverse transformation gives

$$\int_V \Xi_i^{(1)}u_i^{(2)}(x, t)\,dv + \int_{S\sigma} P_i^{(1)}u_i^{(2)}(x, t)\,dS + \int_{S_u} W_i^{(2)}\sigma_{ij}^{(2)}(x, t)\nu_j\,dS = \big]_1^2,$$

or

$$\int_V X_i^{(1)}(x, t)u_i^{(2)}(x, t)\,dv + \int_{S\sigma} f_i^{(1)}(x, t)u_i^{(2)}(x, t)\,dS + \int_{S_u} \sigma_{ij}^{(1)}(x, t)g_i^{(2)}(x, t)\nu_j\,dS = \big]_1^2$$

where $\big]_1^2$ indicates the same expression as on the left-hand side except that the superscripts (1) and (2) are interchanged. Graffi's well-known formula results if $g_i^{(1)} = g_i^{(2)} = 0$ on $S_u$.

(b) *Forces applied at different times.* If

$$\begin{aligned} X_i^{(1)} &= \Xi_i^{(1)}(x)h(t - T_1), & X_i^{(1)} &= \Xi_i^{(1)}(x)h(t - T_2) \\ f_i^{(1)} &= P_i^{(1)}(x)h(t - T_1), & f_i^{(2)} &= P_i^{(2)}(x)h(t - T_2), \\ g_i^{(1)} &= 0, & g_i^{(2)} &= 0, \end{aligned}$$

where $h(t) = 0$ for $t \leq 0$, then Eq. (12) becomes, on cancelling $\bar{h}(s)$,

$$\int_V e^{-sT_1}\rho\,\Xi_i^{(1)}\bar{u}_i^{(2)}\,dv + \int_{S\sigma} e^{-sT_1}P_i^{(1)}u_i^{(2)}\,dS = \big]_1^2.$$

The inverse transform gives

$$\int_V \Xi_i^{(1)}(x)u_i^{(2)}(x, t - T_1)\,dv + \int_{S\sigma} P_i^{(1)}(x)u_i^{(2)}(x, t - T_1)\,dS$$
$$= \int_V \Xi_i^{(2)}(x)u_i^{(1)}(x, t - T_2)\,dv + \int_{S\sigma} P_i^{(2)}(x)u_i^{(1)}(x, t - T_2)\,dS.$$

(c) *Concentrated forces.* If the loading consists of concentrated loads $F_i^{(1)}$ and $F_i^{(2)}$ acting at points $p_1$, $p_2$ respectively, we may consider $\Xi_i$ or $P_i$ of the cases (*a*), (*b*) above as suitable delta functions and obtain at once

$$F_i^{(1)}(p_1)u_i^{(2)}(p_1, t - T_1) = F_i^{(2)}(p_2)u_i^{(1)}(p_2, t - T_2),$$

or, if $T_1 = T_2$,

$$F_i^{(1)}(p_1)u_i^{(2)}(p_1, t) = F_i^{(2)}(p_2)u_i^{(1)}(p_2, t).$$

This is the extension of the conventional elastostatic Betti-Rayleigh reciprocal relation to kinetics.

(d) *Impulsive and traveling concentrated forces.* Let an impulsive concentrated force act at a point $p_1$,

$$X_i^{(1)} = F_i^{(1)}\,\delta(p_1)\,\delta(t),$$

where $\delta(t)$ is a unit-impulse or delta function, and let a concentrated force $F_i^{(2)}$ be applied at the origin at $t = 0$, and thereafter moved along the $x_1$ axis at uniform speed $U$:

$$X_i^{(2)} = F_i^{(2)}\delta\left(t - \frac{x_1}{U}\right)\,\delta(x_2)\,\delta(x_3).$$

No other surface loading or displacement is imposed. Then **Eq. (12)** gives

$$F_i^{(1)}u_i^{(2)}(p_1, t) = F_i^{(2)} \iiint \delta(x_2)\,\delta(x_3)\,dx_1\,dx_2\,dx_3 \int_0^t \delta\left(\tau - \frac{x_1}{U}\right) u_i^{(1)}(x_1, x_2, x_3, t - \tau)\,d\tau$$

and therefore,

$$F_i^{(1)}u_i^{(2)}(p_1, t) = F_i^{(2)} \int_{-\infty}^{x_1/U} u_i\left(x_1, 0, 0, t - \frac{x_1}{U}\right) dx_1$$

If $u_i^{(1)}\,(x_1, 0, 0, t - x_1/U)$ is known, then $u_i^{(2)}(p_1, t)$ can be found from the above equation.

(e) *Suddenly started line load over an elastic half-space.* Ang[9.2] considered the problem of suddenly started line load acting on the surface of a half-space. Now, according to the reciprocal theorem Ang's problem can be solved by one integration of the solution of Lamb's problem: the impulsive loading at one point (not traveling) inside a two-dimensional half-space. Only the surface displacement due to the point loading needs to be known.

**Problem 15.11.** The equation of transverse motion of a string stretched between two points is

$$c^2\frac{\partial^2 w}{\partial x^2} - \frac{\partial^2 w}{\partial t^2} = F(x, t), \qquad 0 \leqslant x \leqslant L.$$

Let $F^{(1)}(x, t)$ be a dynamic loading corresponding to a solution $w^{(1)}(x, t)$: $F^{(2)}(x, t)$ be a dynamic loading corresponding to a solution $w^{(2)}(x, t)$. Generalize the dynamic reciprocity relationship to the present problem by integrating over the length $0 \leqslant x \leqslant L$ as well as over the time $(0, t)$.

Next consider the specific case in which $F^{(2)}(x, t)$ is an impulsive concentrated load acting at $x$: $F^{(2)}(x, t) = \delta(x)\,\delta(t)$; and $F^{(1)}(x, t)$ is a traveling load $F^{(1)}(x, t) = \delta(x - Vt)$, where $V$ is a constant. Show that $w^{(1)}(x, t)$ can be derived by the reciprocity relation from the solution $w^{(2)}(x, t)$ corresponding to $F^{(2)}(x, t)$.

# 16

# FINITE DEFORMATION

In Chap. 4 we considered the analysis of deformation without being restricted to infinitesimal displacements, and we defined and distinguished Almansi's and Green's strain tensors. In subsequent chapters, however, we restricted ourselves to problems involving small displacements only. Now we shall remove this restriction and consider some basic questions involving finite displacements.

The generalization from the infinitesimal strain theory to finite deformation opens up a tremendous field. The nonlinear field theory is difficult and extensive. In this chapter we shall discuss only some of the most important concepts. Our objective is to include the material needed for a precise treatment of the theory of plates and shells, which is of great importance in modern technology. The application of the general concepts to the derivation of the fundamental large-deflection equations in the theory of plates will be illustrated.

## 16.1. STRAIN TENSORS

Consider the transformation

$$x_i = a_i + u_i, \tag{1}$$

which maps the location of a particle $P$ at a point $\{a_i\}$ in a three-dimensional space into a point $\{x_i\}$ in the same space. We shall assume that *both* $\{x_i\}$ *and* $\{a_i\}$ *are referred to the same set of rectangular Cartesian coordinates.* The vector $\{u_i\}$ is the displacement of the particle, the coordinates $\{a_i\}$ describe the location of the particle in the original position, and $\{x_i\}$, its displaced position.

Now let us consider another particle $Q$ with coordinates $\{a_i + da_i\}$ in a neighborhood of the point $P$ whose coordinates are $\{a_i\}$. The vector $\{da_i\}$ defines the line segment $PQ$ in the original configuration. After deformation, $\{da_i\}$ becomes $\{dx_i\}$, which, according to Eq. (1), has components

$$dx_i = \frac{\partial x_i}{\partial a_j}\, da_j = \left(\delta_{ij} + \frac{\partial u_i}{\partial a_j}\right) da_j. \tag{2}$$

When the displacement $\{u_i\}$ is regarded as a function of $a_1$, $a_2$, $a_3$, we define

$$\bar{\epsilon}_{ij} = \frac{1}{2}\left(\frac{\partial u_i}{\partial a_j} + \frac{\partial u_j}{\partial a_i}\right), \tag{3}$$

$$\bar{\omega}_{ij} = \frac{1}{2}\left(\frac{\partial u_j}{\partial a_i} - \frac{\partial u_i}{\partial a_j}\right). \tag{4}$$

When $\{u_i\}$ is regarded as function of $x_1$, $x_2$, $x_3$, we define

$$\epsilon_{ij} = \frac{1}{2}\left(\frac{\partial u_i}{\partial x_j} + \frac{\partial u_j}{\partial x_i}\right), \tag{5}$$

$$\omega_{ij} = \frac{1}{2}\left(\frac{\partial u_j}{\partial x_i} - \frac{\partial u_i}{\partial x_j}\right). \tag{6}$$

With these symbols, we may write

$$dx_i = \frac{\partial x_i}{\partial a_j}\, da_j = (\delta_{ij} + \bar{\epsilon}_{ij} - \bar{\omega}_{ij})\, da_j, \tag{7}$$

$$da_i = \frac{\partial a_i}{\partial x_j}\, dx_j = (\delta_{ij} - \epsilon_{ij} + \omega_{ij})\, dx_j, \tag{8}$$

or

$$du_i = dx_i - da_i = (\bar{\epsilon}_{ij} - \bar{\omega}_{ij})\, da_j = (\epsilon_{ij} - \omega_{ij})\, dx_i. \tag{9}$$

The Green and Almansi strain tensors $E_{ij}$ and $e_{ij}$, respectively, arise from the relations (see Sec. 4.2)

$$ds^2 - ds_0^2 = dx_i\, dx_i - da_i\, da_i = 2E_{ij}\, da_i\, da_j = 2e_{ij}\, dx_i\, dx_j \tag{10}$$

and are defined by

(11) ▲
$$E_{ij} = \frac{1}{2}\left[\frac{\partial x_k}{\partial a_i}\frac{\partial x_k}{\partial a_j} - \delta_{ij}\right] = \frac{1}{2}\left[\frac{\partial u_i}{\partial a_j} + \frac{\partial u_j}{\partial a_i} + \frac{\partial u_k}{\partial a_i}\frac{\partial u_k}{\partial a_j}\right] \quad \text{(Green)},$$

(12) ▲
$$e_{ij} = \frac{1}{2}\left[\delta_{ij} - \frac{\partial a_k}{\partial x_i}\frac{\partial a_k}{\partial x_j}\right] = \frac{1}{2}\left[\frac{\partial u_i}{\partial x_j} + \frac{\partial u_j}{\partial x_i} - \frac{\partial u_k}{\partial x_i}\frac{\partial u_k}{\partial x_j}\right] \quad \text{(Almansi)}.$$

A substitution of (5) yields

(13) ▲
$$E_{ij} = \bar{\epsilon}_{ij} + \tfrac{1}{2}(\bar{\epsilon}_{ki} - \bar{\omega}_{ki})(\bar{\epsilon}_{kj} - \bar{\omega}_{kj}),$$

(14) ▲
$$e_{ij} = \epsilon_{ij} - \tfrac{1}{2}(\epsilon_{ki} - \omega_{ki})(\epsilon_{kj} - \omega_{kj}).$$

In the case of infinitesimal displacements, we know the geometric meaning of Eqs. (3)–(6): $\epsilon_{ij}$, $\bar{\epsilon}_{ij}$ are strains and $\omega_{ij}$, $\bar{\omega}_{ij}$ are *rotations*. For infinitesimal displacements there is no need of distinguishing $\bar{\epsilon}_{ij}$, $\epsilon_{ij}$, $E_{ij}$, and

$e_{ij}$; they are the same. In finite displacements, however, they are all different. We see that the strain components $E_{ij}$ depend on both $\bar{\epsilon}_{ij}$ and $\bar{\omega}_{ij}$; and $e_{ij}$ depend on both $\epsilon_{ij}$ and $\omega_{ij}$. In finite displacements, the vanishing of $E_{ij}$ or $e_{ij}$ characterizes the absence of strain, whereas $\epsilon_{ij}$, $\bar{\epsilon}_{ij}$ no longer enjoy such good roles.

The geometric meaning of various components of $E_{ij}$ and $e_{ij}$ has been discussed in Chap. 4. In particular, we see that since $E_{ij}$ is symmetric, there exists at least one set of principal axes with respect to which the matrix $E_{ij}$ is diagonal. Therefore, there exists a triple of material directions at $P$ that are orthogonal both in the initial and the displaced state.

The necessity of specifying whether the stresses and strains are measured with respect to the original configuration (Lagrangian description) or with respect to the deformed configuration (Eulerian description) is characteristic of a finite strain analysis. When the original state is a *natural* state such as the homogeneous, stress-free state of an elastic body, the kinematic equations are simple, but the dynamic equations are complicated in the Lagrangian description; whereas the reverse is true for Eulerian description: the dynamic equations are simple but the kinematic equations are complicated. These remarks will become evident in the following sections.

## 16.2. LAGRANGE'S AND KIRCHHOFF'S STRESS TENSORS

When we defined stresses in Chap. 3, we considered a continuum in a strained state. The resulting equation of equilibrium is simply expressed in terms of a symmetric stress tensor $\sigma_{ij}$:

$$\frac{\partial \sigma_{ij}}{\partial x_j} + X_i = 0.$$

A stress tensor referred to the strained state is a natural physical concept. However, in the course of analysis we must relate stresses to strains. Hence, if strains were referred to the original position of particles in a continuum, it would be convenient to define stresses similarly with respect to the original configuration.

Consider an element of a strained solid as shown on the right-hand side of Fig. 16.2:1. Assume that in the original (undeformed) state this element has the configuration as shown on the left side of Fig. 16.2:1. A force vector $d\mathbf{T}$ acts on the surface $PQRS$. A corresponding force vector $d\mathbf{T}_0$ acts on the surface $P_0Q_0R_0S_0$. If we assign a rule of correspondence between $d\mathbf{T}$ and $d\mathbf{T}_0$ for every corresponding pair of surfaces, and define stress vectors in each case as the limiting ratios $d\mathbf{T}/dS$, $d\mathbf{T}_0/dS_0$, where $dS$ and $dS_0$ are the areas of $PQRS$, $P_0Q_0R_0S_0$, respectively, then by the method of Chap. 3 we can define stress tensors in both configurations. The assignment of a

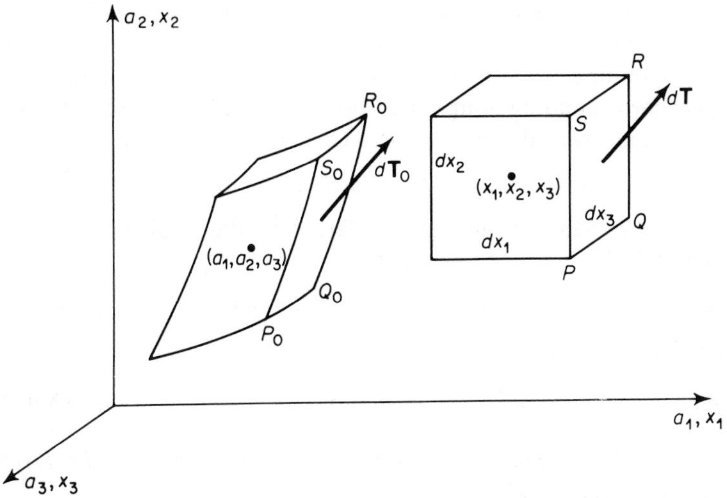


**Fig. 16.2:1.** The corresponding tractions in the original and deformed state of a body.

correspondence rule between $d\mathbf{T}$ and $d\mathbf{T}_0$ is arbitrary, but must be mathematically consistent.

The following alternative rules are known as Lagrangian and Kirchhoff rules, respectively (see Fig. 16.2:2):

(1) ▲ $\quad dT_{0_i}^{(L)} = dT_i \qquad (d\mathbf{T}_0^{(L)} = d\mathbf{T}),$

(2) ▲ $\quad dT_{0_i}^{(K)} = \dfrac{\partial a_i}{\partial x_j}\, dT_j$ $\left(\text{i.e., } d\mathbf{T}_0^{(K)} \text{ and } d\mathbf{T} \text{ are related by the same rule as the transformation } da_i = \dfrac{\partial a_i}{\partial x_j}\, dx_j\right).$

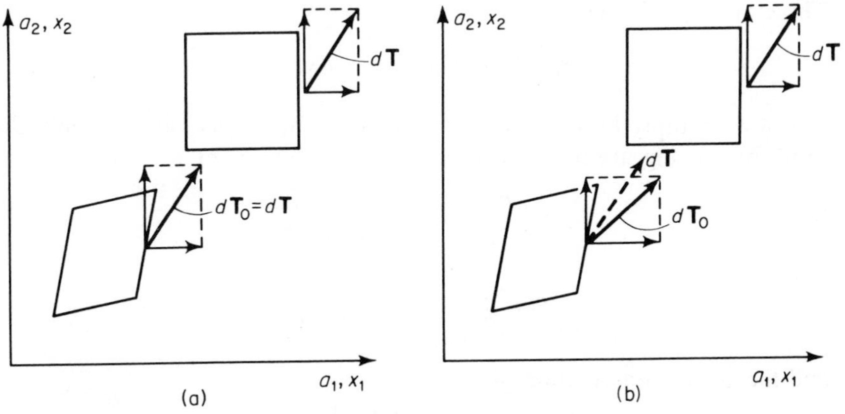


**Fig. 16.2:2.** The correspondence of force vectors in defining (a) Lagrange's and (b) Kirchhoff's stresses, illustrated in a two-dimensional case.

The vector $d\mathbf{T}$, with components $dT_i$, denotes the force acting on a surface element of area $dS$ with a unit outer normal $\boldsymbol{\nu}$; whereas $d\mathbf{T}_0$ denotes the corresponding force assigned to the corresponding original area $dS_0$, with the corresponding unit outer normal $\boldsymbol{\nu}_0$. If $\sigma_{ij}$ is the stress tensor referred to the strained state, we have Cauchy's relation

$$(3) \qquad dT_i = \sigma_{ji}\nu_j \, dS.$$

We now define stress components referred to the original state by a law similar to (3). If (1) is used, we write

$$(4) \ \blacktriangle \qquad dT_{0_i}^{(L)} = T_{ji}\nu_{0_j} \, dS_0 = dT_i \qquad \text{(Lagrange)}.$$

If (2) is used, we write

$$(5) \ \blacktriangle \qquad dT_{0_i}^{(K)} = S_{ji}\nu_{0_j} \, dS_0 = \frac{\partial a_i}{\partial x_\alpha} dT_\alpha \qquad \text{(Kirchhoff)}.$$

*$\sigma_{ij}$, $T_{ij}$, $S_{ij}$ will be called the Eulerian, the Lagrangian, and the Kirchhoff's stress tensors, respectively.*

In order to find the relationship between $\sigma_{ij}$, $T_{ij}$, and $S_{ij}$, we must find the relation between $\nu_j \, dS$ and $\nu_{0_j} \, dS_0$. Consider two material line elements $d\mathbf{x}(dx_1, dx_2, dx_3)$, and $\delta\mathbf{x}(\delta x_1, \delta x_2, \delta x_3)$, in the strained state, which correspond to $d\mathbf{a}(da_i)$ and $\delta\mathbf{a}(\delta a_i)$ in the original state. The area $dS$ of a parallelogram with $d\mathbf{x}$ and $\delta\mathbf{x}$ as sides is given by the magnitude of the vector product $d\mathbf{x} \times \delta\mathbf{x}$, which may be written with the help of the permutation tensor $e_{ijk}$ as

$$\nu_i \, dS = e_{ijk} \, dx_j \, \delta x_k.$$

The unit vector normal to the plane of $d\mathbf{x}$ and $\delta\mathbf{x}$ has the components $\nu_i$. Similarly, in the original state, the area $dS_0$ and the normal $\nu_{0_i}$ of the parallelogram formed by the vectors $da_i$ and $\delta a_i$ is given by the product

$$\nu_{0_i} \, dS_0 = e_{ijk} \, da_j \, \delta a_k.$$

Therefore, by (16.1:8),

$$\nu_{0_i} \, dS_0 = e_{ijk} \frac{\partial a_j}{\partial x_\alpha} \frac{\delta a_k}{\partial x_\beta} dx_\alpha \, \delta x_\beta.$$

We now multiply both sides of this equation by $\partial a_i / \partial x_\gamma$ and simplify the result by making use of the definition of the determinant

$$(6) \qquad \det \left| \frac{\partial a_l}{\partial x^m} \right| = e_{ijk} \frac{\partial a_i}{\partial x_1} \frac{\partial a_j}{\partial x_2} \frac{\partial a_k}{\partial x_3},$$

i.e.,

$$e_{ijk} \frac{\partial a_i}{\partial x_\gamma} \frac{\partial a_j}{\partial x_\alpha} \frac{\partial a_k}{\partial x_\beta} = e_{\gamma\alpha\beta} \det \left| \frac{\partial a_l}{\partial x_m} \right|,$$

and the law of conservation of mass

$$(7) \qquad \frac{\rho}{\rho_0} = \det \left| \frac{\partial a_l}{\partial x_m} \right|.$$

where $\rho$ is the density of the material in the deformed configuration, and $\rho_0$, that in the original configuration. We obtain

$$\frac{\partial a_i}{\partial x_\gamma} \nu_{0_i}\, dS_0 = \frac{\rho}{\rho_0} e_{\gamma\alpha\beta}\, dx_\alpha\, \delta x_\beta = \frac{\rho}{\rho_0} \nu_\gamma\, dS. \tag{8}$$

Hence, from (4), (3), and (8),

$$T_{ji}\nu_{0_j}\, dS_0 = \sigma_{ji}\nu_j\, dS = \sigma_{mi} \frac{\rho_0}{\rho} \frac{\partial a_j}{\partial x_m} \nu_{0_j}\, dS_0, \tag{9}$$

i.e.,

$$T_{ji} = \frac{\rho_0}{\rho} \frac{\partial a_j}{\partial x_m} \sigma_{mi}. \tag{10 ▲}$$

Similarly, from (5), (3), and (8),

$$S_{ji} = \frac{\rho_0}{\rho} \frac{\partial a_i}{\partial x_\alpha} \frac{\partial a_j}{\partial x_\beta} \sigma_{\beta\alpha}. \tag{11 ▲}$$

The Eulerian stress tensor $\sigma_{ij}$ is symmetric. From (10), we see that, in general, the Lagrangian stress tensor $T_{ji}$ is not symmetric. From (11), we see that the Kirchhoff's stress tensor $S_{ji}$ is symmetric. The Lagrangian tensor will be inconvenient to use in a stress-strain law in which the strain tensor is always symmetric. The Kirchhoff stress tensor is more suitable for this purpose.

From (11), we have

$$S_{ji} = \frac{\partial a_i}{\partial x_\alpha} T_{j\alpha}. \tag{12 ▲}$$

Hence, from the identities

$$\delta_{ij} = \frac{\partial a_i}{\partial x_p} \frac{\partial x_p}{\partial a_j}, \qquad \delta_{ij} = \frac{\partial x_i}{\partial a_p} \frac{\partial a_p}{\partial x_j},$$

we find at once the relations

$$\sigma_{ji} = \frac{\rho}{\rho_0} \frac{\partial x_i}{\partial a_p} T_{pj} = \frac{\rho}{\rho_0} \frac{\partial x_i}{\partial a_\alpha} \frac{\partial x_j}{\partial a_\beta} S_{\beta\alpha}, \tag{13 ▲}$$

$$T_{ij} = S_{ip} \frac{\partial x_j}{\partial a_p}. \tag{14 ▲}$$

Expressed in terms of the displacement components $u_1$, $u_2$, $u_3$, [Eq. (16.1:1)], Eqs. (11) and (13) may be written as

$$S_{ji} = \frac{\rho_0}{\rho}\left[\sigma_{ji} - \left(\delta_{j\beta} \frac{\partial u_i}{\partial x_\alpha} + \delta_{i\alpha} \frac{\partial u_j}{\partial x_\beta} - \frac{\partial u_i}{\partial x_\alpha} \frac{\partial u_j}{\partial x_\beta}\right)\sigma_{\beta\alpha}\right], \tag{15 ▲}$$

$$\sigma_{ji} = \frac{\rho}{\rho_0}\left[S_{ji} + \left(\delta_{j\beta} \frac{\partial u_i}{\partial a_\alpha} + \delta_{i\alpha} \frac{\partial u_j}{\partial a_\beta} + \frac{\partial u_i}{\partial a_\alpha} \frac{\partial u_j}{\partial a_\beta}\right) S_{\beta\alpha}\right]. \tag{16 ▲}$$

## 16.3. EQUATION OF MOTION IN LAGRANGIAN DESCRIPTION

Consider a body of a continuum occupying a region $V$ with a boundary surface $S$ in the deformed state, which correspond to a region $V_0$ with a boundary surface $S_0$ in the original (natural) state (Fig. 16.3:1). The body is subjected to external loads. In Eulerian description the external loads consist of a body force $\mathbf{F}$ (with components $F_1, F_2, F_3$) per unit mass, and a surface traction $\overset{\nu}{T}_i$ per unit area acting on a surface element $dS$ whose unit outer normal vector is $\nu_i$. In Lagrangian description, we shall write the body force per unit mass as $\mathbf{F}_0$ (with components $F_{0_1}, F_{0_2}, F_{0_3}$) which will be associated with the location $a_i$ of a particle of mass $\rho_0\, dv_0$. The original density $\rho_0$ corresponds to the density $\rho$ in the deformed state. We shall specify that

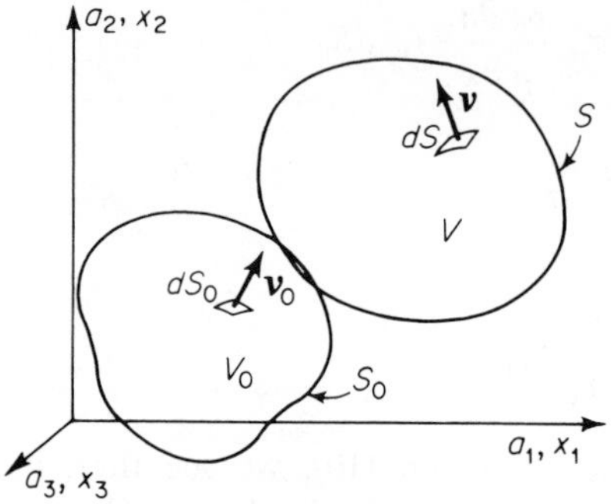

**Fig. 16.3:1.** Notations.

(1) $$\rho\, dv = \rho_0\, dv_0, \qquad F_{0_i} = F_i.$$

We shall consider the static equilibrium of the body under the body force and the surface traction. If the body is in motion, we shall consider the product of the mass and the negative of the acceleration as the inertia force and apply D'Alembert's principle to reduce an equation of motion to an equation of equilibrium. Hence $F_i$ and $F_{0_i}$ include the inertia force. The notations described above are shown in the accompanying table (see p. 441).

Now the resultant body force acting on the region $V(V_0)$ is

(2) $$\int_V F_i \rho\, dv = \int_{V_0} F_{0_i} \rho_0\, dv_0$$

The resultant of the surface tractions acting on $S(S_0)$ is

(3) $$\int_S \overset{\nu}{T}_i\, dS = \int_S \sigma_{ji}\nu_j\, dS$$

According to Eq. (16.2:4) this is equal to $\int_{S_0} T_{ji}\nu_{0_j}\, dS_0$, which can be transformed by Gauss' theorem into $\int_{V_0} \frac{\partial T_{ji}}{\partial a_j}\, dv_0$. Hence

(4) $$\int_S \overset{\nu}{T}_i\, dS = \int_{V_0} \frac{\partial T_{ji}}{\partial a_j}\, dv_0.$$

| | In Deformed Configuration | In Natural Configuration |
|---|---|---|
| Region | $V$ | $V_0$ |
| Boundary surface | $S$ | $S_0$ |
| Volume element | $dv = dx_1\, dx_2\, dx_3$ | $dv_0 = da_1\, da_2\, da_3$ |
| Surface element | $dS$ | $dS_0$ |
| Unit outer normal | $\boldsymbol{\nu}, (\nu_1, \nu_2, \nu_3)$ | $\boldsymbol{\nu}_0, (\nu_{0_1}, \nu_{0_2}, \nu_{0_3})$ |
| Particle coordinates | $x_1, x_2, x_3$ | $a_1, a_2, a_3$ |
| Body force per unit mass | $\mathbf{F}, (F_1, F_2, F_3)$ | $\mathbf{F}_0, (F_{0_1}, F_{0_2}, F_{0_3})$ |
| Surface traction | $\overset{\nu}{\mathbf{T}}, (\overset{\nu}{T}_1, \overset{\nu}{T}_2, \overset{\nu}{T}_3)$ | $\overset{\nu}{\mathbf{T}}_0, (\overset{\nu}{T}_{0_1}, \overset{\nu}{T}_{0_2}, \overset{\nu}{T}_{0_3})$ |
| Density | $\rho$ | $\rho_0$ |
| Stresses | $\sigma_{ij}$ (Eulerian) | $S_{ij}$ (Kirchhoff)<br>$T_{ij}$ (Lagrangian) |

The condition of equilibrium is obtained when the sum of the body forces and surface tractions vanishes. Hence, by (2) and (4), we have, at equilibrium,

$$\int_{V_0} \left( \rho_0 F_{0_i} + \frac{\partial T_{ji}}{\partial a_j} \right) dv_0 = 0. \tag{5}$$

Since this equation must be valid for an arbitrary region $V_0$, the integrand must vanish. Hence the equation of equilibrium

(6) ▲
$$\frac{\partial T_{ji}}{\partial a_j} + \rho_0 F_{0_i} = 0.$$

On substituting Eq. (16.2:14), we obtain the equation of equilibrium expressed in terms of the Kirchhoff stress tensor:

(7) ▲
$$\frac{\partial}{\partial a_j}\left( S_{jk} \frac{\partial x_i}{\partial a_k} \right) + \rho_0 F_{0_i} = 0,$$

i.e.,

(8) ▲
$$\frac{\partial}{\partial a_j}\left[ S_{jk}\left( \delta_{ik} + \frac{\partial u_i}{\partial a_k} \right) \right] + \rho_0 F_{0_i} = 0.$$

If we introduce displacements $u_i$ and tensors $\bar{\epsilon}_{ij}$ and $\bar{\omega}_{ij}$ as in Eq. (16.1:1), (16.1:3), and (16.1:4), we have

(9) ▲
$$\frac{\partial}{\partial a_j}[S_{jk}(\delta_{ik} + \bar{\epsilon}_{ik} - \bar{\omega}_{ik})] + \rho_0 F_{0_i} = 0.$$

This equation exhibits the geometric effect of finite deformation through the "strains" $\bar{\epsilon}_{ik}$ and "rotations" $\bar{\omega}_{ik}$.

## 16.4. RATE OF DEFORMATION

We shall discuss some of the applications of the concepts developed in previous sections, and we shall consider several classes of materials with particular constitutive equations. In some of these definitions the rate of deformation appears prominent. Hence, we shall digress into this topic.

In Sec. 5.2, we have discussed the velocity field of a continuum in both the material (Lagrangian) and the spatial (Eulerian) methods of description. In this section, we shall consider the relationship of velocities at neighboring points. Since we are concerned with the instantaneous motion, we choose the spatial description and write $v_i(x_1, x_2, x_3)$ for the velocity components of a typical particle $P$ as functions of the instantaneous position $x_1, x_2, x_3$ of this particle.

Consider two neighboring particles $P$ and $P'$ with instantaneous coordinates $x_i$ and $x_i + dx_i$, respectively. The difference in velocities at these two points is

$$dv_i = \frac{\partial v_i}{\partial x_j}\, dx_j, \tag{1}$$

where the partial derivatives $\partial v_i/\partial x_j$ are evaluated at the particle $P$. Now

$$\frac{\partial v_i}{\partial x_j} = \frac{1}{2}\left(\frac{\partial v_i}{\partial x_j} + \frac{\partial v_j}{\partial x_i}\right) - \frac{1}{2}\left(\frac{\partial v_j}{\partial x_i} - \frac{\partial v_i}{\partial x_j}\right). \tag{2}$$

Let us define the *rate-of-deformation tensor* $V_{ij}$ and the *spin tensor* $\Omega_{ij}$ as

$$V_{ij} \equiv \frac{1}{2}\left(\frac{\partial v_i}{\partial x_j} + \frac{\partial v_j}{\partial x_i}\right), \tag{3}$$

$$\Omega_{ij} \equiv \frac{1}{2}\left(\frac{\partial v_j}{\partial x_i} - \frac{\partial v_i}{\partial x_j}\right). \tag{4}$$

It is evident that $V_{ij}$ is symmetric and $\Omega_{ij}$ is antisymmetric; i.e.,

$$V_{ij} = V_{ji}, \qquad \Omega_{ij} = -\Omega_{ji}. \tag{5}$$

Hence, there exists a dual vector of $\Omega_{ij}$,

$$\Omega_k \equiv \tfrac{1}{2} e_{kij}\Omega_{ij}, \quad \text{i.e.,} \quad \mathbf{\Omega} = \tfrac{1}{2}\,\text{curl}\,\mathbf{V}, \tag{6}$$

which is called the *vorticity* vector. With these definitions, Eqs. (2) and (1) may be written.

$$\frac{\partial v_i}{\partial x_j} = V_{ij} - \Omega_{ij}, \tag{7}$$

$$dv_i = (V_{ij} - \Omega_{ij})\, dx_j. \tag{8}$$

Equations (3), (4), and (8) are similar in form to Eqs. (16.1:5), (16.1:6), and (16.1:9), respectively. Their interpretations are similar too. Indeed, if we multiply $v_i$ by an infinitesimal interval of time $dt$, the result is a small displacement $u_i = v_i\, dt$. On multiplying all the Eqs. (1)–(6) above by $dt$, we can interpret them as relating to an infinitesimal displacement field. Thus, from Eqs. (3), (4), and (6),

$$V_{ij}\,dt = \frac{1}{2}\left(\frac{\partial u_i}{\partial x_j} + \frac{\partial u_j}{\partial x_i}\right) = \epsilon_{ij},$$

$$\Omega_{ij}\,dt = \frac{1}{2}\left(\frac{\partial u_j}{\partial x_i} - \frac{\partial u_i}{\partial x_j}\right) = \omega_{ij},$$

$$\Omega_k\,dt = e_{kij}\omega_{ij} = \omega_k,$$

where $\epsilon_{ij}$, $\omega_{ij}$, $\omega_k$ are, respectively, the strain tensor, the rotation tensor, and the rotation vector of an infinitesimal displacement field. With the well-known interpretations of $\epsilon_{ij}$, $\omega_{ij}$, and $\omega_k$ (see Sec. 4.3 and 4.4), it is seen that $V_{ij}$, $\Omega_{ij}$, and $\Omega_k$ are, respectively, the rate of change with respect to time of the infinitesimal strain and rotations. Thus, the names assigned above, namely, the rate of deformation, the spin, and the vorticity, are justified.

## 16.5. VISCOUS FLUID

Before defining various kinds of elastic materials, let us consider the related topic of viscous fluids.

The mechanics of viscous fluids is as old as that of the linear elasticity. A *viscous fluid* is specified by the following constitutive equation:

(1) ▲ $$\sigma_{ij} = -p\delta_{ij} + \mathscr{D}_{ijkl}V_{kl}$$

where $\sigma_{ij}$ is the stress tensor, $V_{kl}$ is the rate-of-deformation tensor, and $\mathscr{D}_{ijkl}$ is a tensor of rank 4 defining the fluid, all referred to a rectangular Cartesian frame of reference in the Eulerian description, and $p$ is called the *static pressure*. The term $-p\delta_{ij}$ represents the state of stress possible in a fluid at rest (when $V_{kl} = 0$). The static pressure $p$ is assumed to depend on the density and temperature of the fluid according to an equation of state. For Newtonian fluids, we assume that the tensor $\mathscr{D}_{ijkl}$ may depend on the temperature, but not on the stress or the rate of deformation. By an argument analogous to that of Sec. 6.1, we can discuss the various symmetry properties of $\mathscr{D}_{ijkl}$ and count the number of independent constants that characterize $\mathscr{D}_{ijkl}$. In particular, if the fluid is *isotropic*, i.e., if the tensor $\mathscr{D}_{ijkl}$ has the same array of components in any right-handed system of rectangular Cartesian coordinates, then $\mathscr{D}_{ijkl}$ can be expressed in terms of two independent constants:

(2) $$\mathscr{D}_{ijkl} = \lambda\delta_{ij}\delta_{kl} + \mu(\delta_{ik}\delta_{jl} + \delta_{il}\delta_{jk}),$$

and

(3) ▲ $$\sigma_{ij} = -p\delta_{ij} + \lambda V_{kk}\delta_{ij} + 2\mu V_{ij}.$$

A contraction of (3) gives

(4) $$\sigma_{kk} = -3p + (3\lambda + 2\mu)V_{kk}.$$

If it is assumed that the mean normal stress $\frac{1}{3}\sigma_{kk}$ is *independent* of the rate of dilatation $V_{kk}$, then we must set

$$3\lambda + 2\mu = 0; \tag{5}$$

thus the constitutive equation becomes

$$\sigma_{ij} = -p\delta_{ij} + 2\mu V_{ij} - \tfrac{2}{3}\mu V_{kk}\delta_{ij}. \tag{6 ▲}$$

This formulation is due to George G. Stokes, *Trans. Cambridge Phil. Soc.*, *8* (1844–49), p. 287. A fluid that obeys (6) is called a *Stokes fluid.* The only characteristic constant $\mu$ is called the *coefficient of viscosity.*

If $\mu = 0$, we obtain the constitutive equation of the *perfect fluid:*

$$\sigma_{ij} = -p\delta_{ij}. \tag{7}$$

The presence of the static pressure term $p$ marks a fundamental difference between fluid mechanics and elasticity. To accommodate this new variable, it is often assumed that an *equation of state* exists which relates the pressure $p$, the density $\rho$, and the absolute temperature $T$,

$$f(p, \rho, T) = 0. \tag{8}$$

For example, for an *ideal gas*, we have

$$\frac{p}{\rho} = RT, \tag{9}$$

where $R$ is the gas constant.

Fluids obeying Eq. (1) or (3), whose viscosity effects are represented by terms that are linear in the components of the rate of deformation, are called *Newtonian fluids*. Fluids that behave otherwise are said to be *non-Newtonian*. For example, a fluid whose coefficient of viscosity depends on the basic invariants of $V_{ij}$ is non-Newtonian.

## 16.6. ELASTICITY, HYPERELASTICITY, AND HYPOELASTICITY

If one leaves the linear world of Hookean solids and Newtonian fluids, one encounters a world of much greater variety.

In the small-strain linear elasticity in isothermal or adiabatic conditions, the stress-strain relationship may be stated in three equivalent ways:

$$\sigma_{ij} = C_{ijkl}e_{kl}, \qquad C_{ijkl} = \text{constants}; \tag{1}$$

$$\sigma_{ij} = \frac{\partial W}{\partial e_{ij}}, \qquad W = \frac{1}{2} C_{ijkl}e_{ij}e_{kl}; \tag{2}$$

$$\frac{\partial}{\partial t}\sigma_{ij} = C_{ijkl}\frac{\partial e_{kl}}{\partial t} \tag{3}$$

The constants $C_{ijkl}$ may depend on temperature, but are independent of the stress or strain. In mathematically generalizing these statements into less restrictive ones that still retain the concept of "elasticity," one arrives at constitutive equations that are no longer equivalent. In fact, three different orders of generality are obtained, which are named by Truesdell the elasticity, the hyperelasticity, and the hypoelasticity. The definitions are as follows and will be explained in the sequel.

DEFINITION 1. A material is said to be *elastic*, if it possesses a homogeneous stress-free *natural state*, and if in an appropriately defined finite neighborhood of this state there exists a one-to-one correspondence between the Eulerian stress tensor $\sigma_{ij}$ and the Almansi strain tensor $e_{ij}$:

(4) ▲
$$\sigma_{ij} = \mathscr{F}(e_{ij}).$$

DEFINITION 2. A material is said to be *hyperelastic*, if it possesses a strain energy function per unit mass, $\mathscr{W}$, which is an analytic function of the strain tensor formed with respect to the homogeneous stress-free natural state, such that the rate of change of the strain energy per unit mass† is equal to the rate of doing work by the stresses. Thus, in Eulerian variables

(5a) ▲
$$\frac{D}{Dt}\mathscr{W} = \frac{1}{\rho}\sigma_{ij}V_{ij},$$

whereas in Lagrangian variables

(5b) ▲
$$\frac{D}{Dt}\mathscr{W} = \frac{1}{\rho_0}S_{ij}\frac{D}{Dt}E_{ij} = \frac{1}{\rho_0}S_{ij}\frac{\partial}{\partial t}E_{ij},$$

where $V_{ij}$ is the rate of deformation [see Eq. (16.4:3)], $\sigma_{ij}$ is the Eulerian stress tensor, $E_{ij}$ is the Green strain tensor, and $S_{ij}$ is the Kirchhoff's stress tensor; $\rho_0$ and $\rho$ are the density in the original and deformed states respectively.

DEFINITION 3. A material is said to be *hypoelastic* if the components of the stress rate are homogeneous linear function of the components of the rate of deformation, i.e., if

(6) ▲
$$\sigma_{ij}^{\nabla} = \mathscr{C}_{ijkl}V_{kl},$$

where

(7) ▲
$$\sigma_{ij}^{\nabla} = \frac{D}{Dt}\sigma_{ij} - \sigma_{ip}\Omega_{pj} - \sigma_{jp}\Omega_{pi}.$$

$V_{kl}$ and $\Omega_{ij}$ are defined in Eqs. (16.4:3) and (16.4:4), respectively. The tensor $\mathscr{C}_{ijkl}$ may depend on the stress tensor.

† For finite deformation it is natural to consider internal energy per unit mass. If we base our discussion on energy per unit volume, the resulting formulas will be more complicated.

The first definition is based on Cauchy's approach. It is understood that the natural state is a state of thermodynamic equilibrium, at which all components of stress and strain are zero throughout the body. This definition incorporates the idea that the stress and the strain are uniquely related and that the body returns to the natural state when all loads are removed.

Since the Eulerian stress tensor $\sigma_{ij}$ and the Almansi strain tensor $e_{ij}$ are related uniquely to the Kirchhoff stress tensor $S_{ij}$ and the Green's strain tensor $E_{ij}$, respectively, an elastic material may be described in Lagrangian variables as

$$(8) \quad ▲ \qquad S_{ij} = \mathscr{G}(E_{ij}) \qquad \text{(elastic material)},$$

where $\mathscr{G}$ is a single-valued function.

The simplest of such relations is the linear isotropic law

$$(9) \qquad \sigma_{ij} = \lambda e_{\alpha\alpha}\delta_{ij} + 2Ge_{ij},$$

$$(10) \qquad S_{ij} = \lambda' E_{\alpha\alpha}\delta_{ij} + 2G'E_{ij},$$

where $\lambda$, $G$, $\lambda'$, $G'$ are constants, and $e_{ij}$, $E_{ij}$ are no longer limited to be infinitesimal. In finite elasticity there exists a beautiful theory that determines and classifies the functions $\mathscr{F}(e_{ij})$, $\mathscr{G}(E_{ij})$ axiomatically on the basis of various invariant principles. The most important contributors to this theory are Rivlin, Reiner, Noll, Truesdell, etc. See Bibliography 16.1.

The second definition that defines hyperelasticity is based on Green's method. If we compare Eq. (5a) with Eq. (12.2:7), which describes the first law of thermodynamics, we see that in adiabatic conditions ($h_i = 0$), $\mathscr{W}$ can be interpreted as the internal energy per unit mass. Following the reasoning of Sec. 12.3 we see that in an isothermal condition $\mathscr{W}$ can be identified with the free energy per unit mass. In these cases the temperature $T$ does not have to appear explicitly in $\mathscr{W}$. On the other hand, a reference to Sec. 12.7 shows that $\mathscr{W}$ may exist in other thermal conditions, but then it would be necessary to exhibit the temperature $T$ in $\mathscr{W}$. Of course, when $T$ is variable it is necessary to define a reference state that is at a uniform temperature $T_0$ and is stress-free. The strain components are formed with respect to this reference state. The function $\mathscr{W}$ is assumed to be an analytic function of $T$ and the strain components.

The use of the strain energy function in variational principles will be discussed in Sec. 16.7. At present we shall consider only the definitions. To show that the statement of hyperelasticity is consistent, we must prove that the statements (5a) and (5b) are equivalent. To prove this equivalence we need a relationship between $V_{ij}$ and $DE_{ij}/Dt$, which can be derived as follows. Consider Eq. (16.1:10). Taking the material derivative gives

$$(11) \qquad \frac{D}{Dt}(ds^2 - ds_0^2) = \frac{D}{Dt}(dx_i dx_j \delta_{ij} - da_i\, da_j \delta_{ij}) = 2\left(\frac{DE_{ij}}{Dt}\right) da_i\, da_j$$

Now,

$$\frac{D}{Dt}x_i = v_i, \qquad \frac{D}{Dt}dx_i = dv_i = \frac{\partial v_i}{\partial x_l}dx_l$$

$$\frac{D}{Dt}a_i = 0, \qquad \frac{D}{Dt}da_i = 0.$$

Hence the left-hand side of (11) is

$$\begin{aligned} dv_i\,dx_j\delta_{ij} + dx_i\,dv_j\delta_{ij} &= \frac{\partial v_i}{\partial x_l}dx_l\,dx_i + \frac{dv_j}{\partial x_l}dx_j\,dx_l \\ &= 2V_{il}\,dx_i\,dx_l \\ &= 2V_{il}\frac{\partial x_i}{\partial a_k}\frac{\partial x_l}{\partial a_m}da_k\,da_m. \end{aligned}$$

A substitution back into (11), with several changes in the dummy indices, yields the desired relation:

$$(12) \quad \blacktriangle \qquad \frac{D}{Dt}E_{ij} = V_{pq}\frac{\partial x_p}{\partial a_i}\frac{\partial x_q}{\partial a_j}.$$

Equation (12) shows that a rigid-body motion, $V_{ij} = 0$, implies $DE_{ij}/Dt = 0$. We now have, according to Eq. (16.2:13) and Eq. (12) above,

$$\frac{1}{\rho}\sigma_{ij}V_{ij} = \frac{1}{\rho_0}\frac{\partial x_i}{\partial a_\alpha}\frac{\partial x_j}{\partial a_\beta}S_{\beta\alpha}V_{ij} = \frac{1}{\rho_0}S_{\beta\alpha}\frac{D}{Dt}E_{\beta\alpha}.$$

Thus the equivalence of (5a) and (5b) is demonstrated.

The simplest examples of strain energy functions are provided by those of linear elasticity, Eqs. (12.6:5) and (12.7:2). In finite deformation theory, certain restrictions on the form of the function $\mathscr{W}$ can be determined according to the type of crystal symmetry and various invariance axioms. See Green and Adkins,[1.2] and Noll and Truesdell.[1.2]

The third definition embodies the simplest rate theory, which was named hypoelasticity by Truesdell (1955). A crucial point in the third definition lies in the concept of stress rate. Now, the rate-of-deformation tensor $V_{ij}$ is a well-defined quantity; it vanishes when the body performs a rigid-body motion. But, for a stressed body performing rigid-body rotation, neither the time derivative $\partial\sigma_{ij}/\partial t$ nor the material derivative $D\sigma_{ij}/Dt$ vanishes identically. As an example, consider a bar subjected to simple tension and rotating about the $z$-axis. At one instant the bar is parallel to the $x$-axis so that $\sigma_{xx} \neq 0$, $\sigma_{yy} = 0$. At another instant, when the bar becomes parallel to the $y$-axis, we have $\sigma_{xx} = 0$, $\sigma_{yy} \neq 0$. Thus a rigid-body rotation changes the stress tensor, even though the *state* of stress in the bar remains unchanged. Thus neither $\partial\sigma_{ij}/\partial t$ nor $D\sigma_{ij}/Dt$ can serve as an appropriate stress rate measure to be related simply with the deformation rate $V_{ij}$. This is why a new rate derivative such as $\overset{\nabla}{\sigma}_{ij}$ is introduced. The need for such a new derivative was emphasized in recent years by Oldroyd (1950), Truesdell

(1953), and others, while Prager has pointed out that Jaumann (1911) defined a stress rate $\sigma_{ij}^{\nabla}$ as given by Eq. (7), which serves our purpose. The desired stress rate derivative must be invariant with respect to rigid-body rotations; but this requirement demands no unique answer. Truesdell (1953), Green (1956), Cotter and Rivlin (1955), and Naghdi (1961) introduced somewhat different definitions, while Oldroyd pointed out that the differences among the various definitions of stress rate are not essential as far as constitutive laws of the type Eq. (6) are concerned.

To see the meaning of Jaumann's definition, let us consider a field of flow with velocity components $v_i$ with respect to a rectangular Cartesian frame of reference $x_1, x_2, x_3$. For convenience of discussion we may take the origin of the coordinate system at a generic point $P$ in the flow field. Let $x_1', x_2', x_3'$ be another rectangular Cartesian frame of reference which rotates with the medium, with the same origin at $P$, at an angular velocity $\mathbf{\Omega}$, with components $\Omega_i$:

$$\Omega_1 = \frac{1}{2}\left(\frac{\partial v_2}{\partial x_3} - \frac{\partial v_3}{\partial x_2}\right), \quad \Omega_2 = \frac{1}{2}\left(\frac{\partial v_3}{\partial x_1} - \frac{\partial v_1}{\partial x_3}\right), \quad \Omega_3 = \frac{1}{2}\left(\frac{\partial v_1}{\partial x_2} - \frac{\partial v_2}{\partial x_1}\right) \tag{13}$$

Let $x_i'$ coincide with $x_i$ at the instant of time $t$, when the stress tensor at $P$ is $\sigma_{ij}'(t) = \sigma_{ij}(t)$. At a later instant of time $t + dt$ let the stress at the particle $P$ be denoted by $\sigma_{ij}'(t + dt)$ referred to the rotating axes $x_i'$. Then Jaumann defines the stress rate as

$$\sigma_{ij}^{\nabla} = \lim_{dt\to 0} \frac{1}{dt}\,[\sigma_{ij}'(t + dt) - \sigma_{ij}'(t)] \tag{14}$$

Now, the coordinates $x_i'$ and $x_i$ are related by

$$x_i' = x_i + e_{ijk}\Omega_j\, dt x_k = (\delta_{ik} + e_{ijk}\Omega_j\, dt)x_k \tag{15}$$

whereas the stress tensor at the particle $P$ at the instant $t + dt$ referred to the fixed coordinates $x_j$ is

$$\sigma_{ij}(t) + \frac{D\sigma_{ij}}{Dt}\,dt \tag{16}$$

Transforming the stress tensor (16) under the coordinate transformation (15) to the $x_j'$ axes, we obtain

$$\begin{aligned}\sigma_{ij}'(t + dt) &= (\delta_{ip} + e_{imp}\Omega_m\, dt)(\delta_{jq} + e_{jnq}\Omega_n\, dt)\left[\sigma_{pq}(t) + \frac{D\sigma_{pq}}{Dt}\,dt\right] \\ &= \sigma_{ij}(t) + \left\{\frac{D\sigma_{ij}}{Dt} + e_{imp}\Omega_m\sigma_{pj} + e_{jnq}\Omega_n\sigma_{iq}\right\} dt + O(dt^2)\end{aligned} \tag{17}$$

Hence, according to (14), we obtain

$$\sigma_{ij}^{\nabla} = \frac{D\sigma_{ij}}{Dt} + e_{imp}\Omega_m\sigma_{pj} + e_{jnq}\Omega_n\sigma_{iq} \tag{18}$$

This reduces to (7) when we recall the definition of vorticity:

$$\Omega_{jk} = e_{ijk}\Omega_i. \qquad \text{Q.E.D.}$$

Hypoelasticity, elasticity, and hyperelasticity are not equivalent. However, Noll[16.1] has shown that the definition of elastic materials characterizes a special case of hypoelastic behavior, and that an isotropic hyperelastic material is a particular elastic and hence hypoelastic material. If we restrict our attention to the neighborhood of the stress-free state, and linearize the constitutive equation (6), we obtain the linear elasticity law of Chap. 6. See Noll and Truesdell[1.2] for a complete summary of the literature.

Only experiments can indicate in which way a real material behaves—which constitutive equation describes it best. At present, a number of rubbers are identified as hyperelastic, but much research remains to be done.

## 16.7. THE STRAIN ENERGY FUNCTION AND VARIATIONAL PRINCIPLES

In this section we shall describe some salient properties of the strain energy function $\mathscr{W}$ that characterizes a hyperelastic solid. We shall assume that $\mathscr{W}$ as a function of the nine components of a symmetric strain tensor is written in a form which is symmetric in symmetric components ($E_{ij}$ and $E_{ji}$), and that in forming the derivative of $\mathscr{W}$ with respect to a typical tensor component ($E_{ij}$), symmetric components will be formally treated as independent variables. With this convention, we shall prove that the Kirchhoff's stress tensor can be expressed as

(1) ▲
$$S_{ij} = \frac{\partial(\rho_0 \mathscr{W})}{\partial E_{ij}}$$

where $\rho_0$ is the uniform mass density at the natural state, and $E_{ij}$ are the components of Green's strain tensor. On the other hand, if $\mathscr{W}$ is considered as a function of the components of the *deformation gradient tensor* $\partial x_i/\partial a_j = \delta_{ij} + \partial u_i/\partial a_j$ [see Eqs. (16.1:1), (16.1:2)],† then we shall show that the Lagrange's stress tensor can be expressed as

(2) ▲
$$T_{ji} = \frac{\partial(\rho_0 \mathscr{W})}{\partial(\partial x_i/\partial a_j)} = \frac{\partial(\rho_0 \mathscr{W})}{\partial(\partial u_i/\partial a_j)}.$$

With these relations we shall establish several variational principles: that of the potential energy, the complementary energy, and the Reissner principle, in analogy to those discussed in Chaps. 10 and 11, but extended to finite deformation.

† We use $x_i$ to describe the position of a particle in the deformed configuration of the body, and $a_i$ for that in the original, undeformed configuration. $x_i$ is also used as the geometric variable in Eulerian description of the deformed body, and $a_i$ for that in Lagrangian description of the undeformed body. $u_i$ is the displacement vector. $x_i, a_i, u_i$ are referred to the same rectangular Cartesian frame of reference.

To prove Eq. (1), we note that by the convention stated above we have

$$\frac{\partial \mathscr{W}}{\partial E_{ij}} = \frac{\partial \mathscr{W}}{\partial E_{ji}} \tag{3}$$

and

$$\frac{D}{Dt}\mathscr{W} = \frac{\partial \mathscr{W}}{\partial E_{ij}} \frac{DE_{ij}}{Dt} \tag{4}$$

A comparison of (4) with Eq. (16.6:5b) shows that

$$\left(\frac{1}{\rho_0} S_{ij} - \frac{\partial \mathscr{W}}{\partial E_{ij}}\right) \frac{DE_{ij}}{Dt} = 0. \tag{5}$$

Both factors of this equation are symmetric with respect to $i, j$. Since Eq. (5) is valid for arbitrary values of the rate of strain $DE_{ij}/Dt$, we must have

$$\frac{1}{\rho_0} S_{ij} - \frac{\partial \mathscr{W}}{\partial E_{ij}} = 0 \tag{6}$$

which is Eq. (1).

If all the components of rate-of-strain tensor $DE_{ij}/Dt$ are not allowed to vary independently, we cannot claim the validity of (6) on the basis of (5). For example, if we insist that the material is *incompressible*, then it follows from Eq. (5.4:4) that the divergence of velocity field vanishes: $\partial v_i/\partial x_i = V_{ii} = 0$. This can be expressed in terms of $DE_{ij}/Dt$ as follows. Multiply Eq. (16.6:12) by $(\partial a_i/\partial x_r)(\partial a_j/\partial x_s)$ and note that

$$\frac{\partial x_p}{\partial a_i}\frac{\partial a_i}{\partial x_r} = \delta_{pr}, \qquad \frac{\partial x_q}{\partial a_j}\frac{\partial a_j}{\partial x_s} = \delta_{qs} \tag{7}$$

we have

$$V_{rs} = \frac{DE_{ij}}{Dt}\frac{\partial a_i}{\partial x_r}\frac{\partial a_j}{\partial x_s}. \tag{8}$$

Hence the condition of incompressibility implies that

$$\frac{DE_{ij}}{Dt}\frac{\partial a_i}{\partial x_r}\frac{\partial a_j}{\partial x_r} = 0. \tag{9}$$

Writing

$$\frac{\partial a_i}{\partial x_r}\frac{\partial a_j}{\partial x_r} = B_{ij}, \tag{10}$$

we have

$$B_{ij}\frac{DE_{ij}}{Dt} = 0. \tag{11}$$

$B_{ij}$ is called *Finger's strain tensor*. Since $DE_{ij}/Dt$ are restricted by Eq. (11), we cannot say that the first factor in Eq. (5) must vanish, but only that it must be proportional to $B_{ij}$. Therefore, for an *incompressible hyperelastic material* we must have

(12) ▲
$$S_{ij} = \frac{\partial \rho_0 \mathscr{W}}{\partial E_{ij}} - pB_{ij}, \tag{12}$$

where $p$ is an arbitrary scalar which can be identified with pressure.

To derive Eq. (2), we return to the general compressible material and regard $\mathscr{W}$ as a function of the nine components of the tensor $\partial x_i/\partial a_j$. Let us write $x_{i,j}$ for $\partial x_i/\partial a_j$. Then

$$\frac{\partial \rho_0 \mathscr{W}}{\partial x_{i,j}} = \frac{\partial \rho_0 \mathscr{W}}{\partial E_{kl}} \frac{\partial E_{kl}}{\partial x_{i,j}} \tag{13}$$

From Eq. (16.1:11) we have

$$\frac{\partial E_{kl}}{\partial x_{i,j}} = \frac{1}{2}\left(\frac{\partial x_i}{\partial a_l}\,\delta_{kj} + \frac{\partial x_i}{\partial a_k}\,\delta_{lj}\right) \tag{14}$$

Hence

$$\begin{aligned} \frac{\partial \rho_0 \mathscr{W}}{\partial x_{i,j}} &= \frac{1}{2}\left(\frac{\partial \rho_0 \mathscr{W}}{\partial E_{jl}} \frac{\partial x_i}{\partial a_l} + \frac{\partial \rho_0 \mathscr{W}}{\partial E_{kj}} \frac{\partial x_i}{\partial a_k}\right) \\ &= \frac{\partial \rho_0 \mathscr{W}}{\partial E_{jk}} \frac{\partial x_i}{\partial a_k} = S_{jk} \frac{\partial x_i}{\partial a_k} = T_{ji}. \end{aligned} \tag{15}$$

The last equality is obtained in accordance with Eq. (16.2:14). Thus Eq. (2) is verified.

If one likes to consider $\mathscr{W}$ as a function of the derivatives of the displacements $u_{i,j} = \partial u_i/\partial a_j$, we note that $u_{i,j} = x_{i,j} - \delta_{ij}$, where $\delta_{ij}$ is the Kronecker delta. From Eq. (16.1:11) we obtain the following equation, which is equivalent to Eq. (14):

$$\frac{\partial E_{kl}}{\partial u_{i,j}} = \frac{1}{2}\left(\delta_{ki}\delta_{lj} + \delta_{li}\delta_{kj} + \frac{\partial u_i}{\partial a_l}\,\delta_{kj} + \frac{\partial u_i}{\partial a_k}\,\delta_{lj}\right). \tag{14a}$$

Hence, according to Eq. (13) we obtain

$$\frac{\partial \rho_0 \mathscr{W}}{\partial u_{i,j}} = \frac{\partial \rho_0 \mathscr{W}}{\partial E_{jk}}\left(\delta_{ik} + \frac{\partial u_i}{\partial a_k}\right) = S_{jk} \frac{\partial x_i}{\partial a_k} = T_{ji},$$

which is again Eq. (15) or (2).

With these results we shall now consider the variational principles. Consider the functional

$$\begin{aligned} \mathscr{V} = \int_{V_0} \rho_0 \mathscr{W}\, dv_0 - \int_{V_0} \rho_0 F_{0_i} u_i\, dv_0 \\ - \int_{S_{0u}} (u_i^* - u_i) \overset{\nu}{T}_{0_i}\, dS_0 - \int_{S_{0\sigma}} u_i \overset{\nu}{T}{}^*_{0_i}\, dS_0 \end{aligned} \tag{16 ▲}$$

for a body occupying a region $V$ bounded by a surface $S = S_u + S_\sigma$ in the deformed state, which corresponds to a region $V_0$ within a surface $S_0 = S_{0u} + S_{0\sigma}$ in the natural (undeformed) state. The body is subjected to a body-force per unit mass $F_{0i}$, a surface traction $\overset{\nu}{T}{}^*_{0i}$ over the portion of boundary surface $S_{0\sigma}$, and specified displacements $u_i^*$ over the portion of the boundary surface $S_{0u}$. The body force $F_{0i}\rho_0\,dv_0$ is defined for the natural state as one associated with the position of the particle of mass $\rho_0\,dv_0$ in the undeformed configuration, and in value identical with that acting on the particle (of mass $\rho\,dv = \rho_0\,dv_0$) in the deformed position. The surface traction $\overset{\nu}{T}{}_{0i}^*$ is defined in the position of the surface in the natural state; its magnitude is referred to the original area of the boundary $dS_0$, and acts in the same direction as the surface traction that acts on the surface element $dS$ in the deformed configuration. Thus

$$F_{0i}\rho_0\,dv_0 = F_i\rho\,dv, \qquad \overset{\nu}{T}_{0i}\,dS_0 = \overset{\nu}{T}_i\,dS \tag{17}$$

These are illustrated in Fig. 16.7:1.

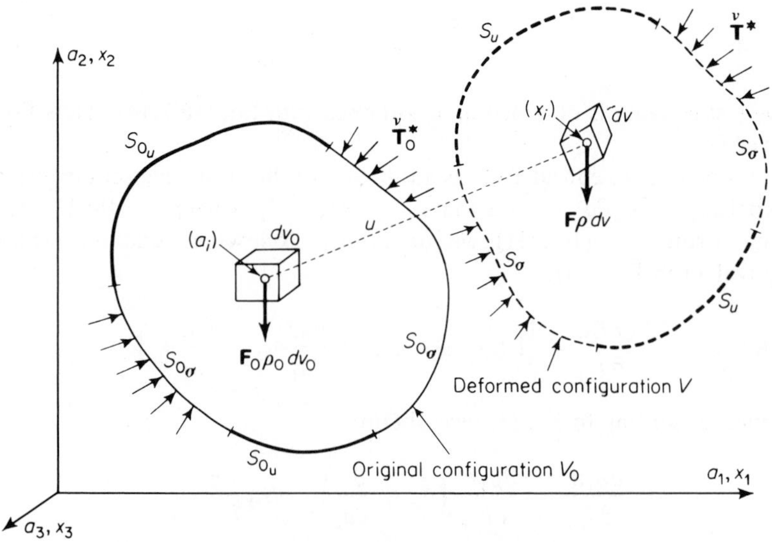


**Fig. 16.7:1.** Correspondence between original and deformed states.

As before, we describe the position of a particle before and after deformation by $a_i$ and $x_i$ respectively. In the functional (16), all quantities in the integrand are regarded as functions of $a_1$, $a_2$, $a_3$, and the integrations are referred to the undeformed configuration of the body.

Consider small variations of $u_i$, which are single-valued, continuous, and satisfy the boundary conditions over $S_{0u}$, so that $\delta u_i = 0$ on $S_{0u}$. Let $\mathscr{W}$ be regarded as a symmetric function of the nine derivatives $u_{i,j} \equiv \partial u_i/\partial a_j$. The variation of $\mathscr{V}$ with respect to the variation of $u_i$ is

$$\delta\mathscr{V} = \int_{V_0} \frac{\partial\rho_0\mathscr{W}}{\partial u_{i,j}}\,\delta u_{i,j}\,dv_0 - \int_{V_0} \rho_0 F_{0i}\delta u_i\,dv_0 - \int_{S_{0\sigma}} \overset{\nu}{T}{}_{0i}^*\delta u_i\,dS_0 \tag{18}$$

The first integral on the right-hand side can be transformed by Gauss theorem:

$$(19)\qquad \int_{V_0} \frac{\partial \rho_0 \mathscr{W}}{\partial u_{i,j}} \delta u_{i,j}\, dv_0 = \int_{V_0} \left( \frac{\partial \rho_0 \mathscr{W}}{\partial u_{i,j}} \delta u_i \right)_{,j} dv_0 - \int_{V_0} \left( \frac{\partial \rho_0 \mathscr{W}}{\partial u_{i,j}} \right)_{,j} \delta u_i\, dv_0$$
$$= \int_{S_0} \frac{\partial \rho_0 \mathscr{W}}{\partial u_{i,j}} \delta u_i \nu_j\, dS_0 - \int_{V_0} \left( \frac{\partial \rho_0 \mathscr{W}}{\partial u_{i,j}} \right)_{,j} \delta u_i\, dv_0.$$

Substituting into (13), we see that if $\delta \mathscr{V} = 0$ for arbitrary $\delta u_i$ in $V_0$ and over $S_{0\sigma}$, then the following Euler equation and boundary conditions must be satisfied:

$$(20)\qquad \frac{\partial}{\partial a_j} \left[ \frac{\partial \rho_0 \mathscr{W}}{\partial (u_{i,j})} \right] + \rho_0 F_{0i} = 0 \qquad \text{in } V_0,$$

$$(21)\qquad \overset{\nu}{T}{}^{*}_{0i} = \frac{\partial \rho_0 \mathscr{W}}{\partial (u_{i,j})} \nu_j \qquad \text{over } S_{0\sigma}.$$

By Eq. (2), we may write the results as

$$(22)\qquad \frac{\partial T_{ji}}{\partial a_j} + \rho_0 F_{0i} = 0 \qquad \text{in } V_0,$$

$$(23)\qquad T_{ji} \nu_j = \overset{\nu}{T}{}^{*}_{0i} \qquad \text{on } S_{0\sigma}.$$

These are precisely the equations of equilibrium and stress boundary conditions, Eqs. (16.3:1) and (16.2:4), respectively. Thus we conclude that the equation of equilibrium is obtained as the Euler equation of the variational equation $\delta \mathscr{V} = 0$.

If $\mathscr{W}$ is regarded as a symmetric function of $E_{ij}$, we have

$$(24)\qquad \delta \int_{V_0} \rho_0 \mathscr{W}(E_{ij})\, dv_0 = \int_{V_0} \frac{\partial \rho_0 \mathscr{W}}{\partial E_{kl}} \frac{\partial E_{kl}}{\partial (u_{i,j})} \delta u_{i,j}\, dv_0$$
$$= \frac{1}{2} \int_{V_0} S_{kl} \left[ \delta_{ki} \delta_{lj} + \delta_{kj} \delta_{li} + \frac{\partial u_i}{\partial a_l} \delta_{kj} + \frac{\partial u_i}{\partial a_k} \delta_{lj} \right] \delta u_{i,j}\, dv_0 \qquad \text{by Eq. (14a)},$$
$$= \int_{V_0} \left( S_{jl} \delta_{il} + S_{jl} \frac{\partial u_i}{\partial a_l} \right) \delta u_{i,j}\, dv_0 \qquad \text{since } S_{ij} = S_{ji},$$
$$= \int_{V_0} S_{jl} \frac{\partial x_i}{\partial a_l} \frac{\partial \delta u_i}{\partial a_j}\, dv_0 \qquad \text{by Eq. (16.1:2)},$$
$$= -\int_{V_0} \frac{\partial}{\partial a_j} \left( S_{jl} \frac{\partial x_i}{\partial a_l} \right) \delta u_i\, dv_0 + \int_{S_0} S_{jl} \frac{\partial x_i}{\partial a_l} \delta u_i \nu_j\, dS_0$$
$$\text{by Gauss' theorem.}$$

Hence, on replacing the first term in (18) with the above, we see that since the variations $\delta u_i$ are arbitrary but vanish over $S_{0u}$, the Euler equation and the boundary conditions are

$$\frac{\partial}{\partial a_j}\left(S_{jl}\frac{\partial x_i}{\partial a_l}\right) + \rho_0 F_{0_i} = 0 \qquad \text{in } V_0, \tag{25}$$

$$\nu_j S_{jl}\frac{\partial x_i}{\partial a_l} = \overset{\nu}{T}{}^*_{0i} \qquad \text{over } S_{0\sigma}. \tag{26}$$

These are again precisely in agreement with Eqs. (16.3:2) and (16.2:5).

Therefore the vanishing of $\delta\mathscr{V}$ for arbitrary small variation in $u_i$, which satisfies the specified boundary condition on displacements, implies the complete set of relations governing the finite deformation of a hyperelastic solid, and conversely the equations of finite elasticity imply $\delta\mathscr{V} = 0$. This is the potential energy principle extended to finite deformation of elastic bodies.

If the constitutive equation

$$S_{ij} = \frac{\partial \rho_0 \mathscr{W}}{\partial E_{ij}} \tag{27}$$

can be inverted; i.e., if $E_{ij}$ can be expressed as $E_{ij}(S_{11}, S_{12}, \ldots)$, then there exists a function $\mathscr{W}_c(S_{11}, S_{12}, \ldots)$ of the stress components, such that

$$\mathscr{W}_c = S_{ij}E_{ij} - \rho_0\mathscr{W} \tag{28}$$

and

$$E_{ij} = \frac{\partial \mathscr{W}_c}{\partial S_{ij}}. \tag{29}$$

In Eq. (28), $E_{ij}$ and $\mathscr{W}(E_{ij})$ are considered as functions of stresses after $E_{ij}$ is replaced by $E_{ij}(S_{11}, S_{12}, S_{22}, \ldots)$. Equation (28) defines the *complementary strain energy density* $\mathscr{W}_c$ by a Legendre (contact) transformation on $\mathscr{W}$. In this case we can substitute $\mathscr{W}$ from (28) into (16), and we are lead to consider Reissner's functional

$$\text{(30)} \ \blacktriangle \quad J_R = \int_{V_0} \{S_{kl}E_{kl} - \mathscr{W}_c(S_{ij}) - \rho_0 F_{0_i} u_i\}\, dv_0 - \int_{S_{0\sigma}} u_i \overset{\nu}{T}{}^*_{0_i}\, dS_0 - \int_{S_{0_u}} (u_i^* - u_i)\overset{\nu}{T}_{0_i}\, dS_0.$$

We shall consider the variation of $J_R$ when both $S_{ij}$ and $u_i$ are permitted to be varied. However, we assume that $u_i$ always satisfies the boundary conditions on displacements, and $S_{ij}$ always satisfy the boundary conditions on $S_{0\sigma}$. These restrictions can be stated as

$$\delta u_i = 0 \quad \text{on} \quad S_{0u}, \qquad \delta S_{ij}\nu_j = 0 \quad \text{on } S_{0\sigma}. \tag{31}$$

Now

$$(32)\qquad \delta J_R = \int_{V_0}\left[S_{kl}\frac{\partial E_{kl}}{\partial u_{i,j}}\,\delta u_{i,j} + E_{kl}\delta S_{kl} - \frac{\partial \mathscr{W}_c}{\partial S_{ij}}\,\delta S_{ij} - \rho_0 F_{0_i}\delta u_i\right] dv_0 - \int_{S_{0\sigma}} \delta u_i \overset{\nu}{T}{}^*_{0i}\, dS_0 + \int_{S_{0_u}}\left[\delta u_i \overset{\nu}{T}_{0_i} + (u_i^* - u_i)\delta\overset{\nu}{T}_{0_i}\right] dS_0.$$

The first term on the right-hand side can be reduced in the same way as in Eq. (24). The result is

$$(33)\qquad \int_{V_0} S_{kl}\frac{\partial E_{kl}}{\partial u_{i,j}}\,\delta u_{i,j}\, dv_0 = -\int_{V_0}\frac{\partial}{\partial a_j}\left(S_{jl}\frac{\partial x_i}{\partial a_l}\right)\delta u_i\, dv_0 + \int_{S_0} S_{jl}\frac{\partial x_i}{\partial a_l}\,\delta u_i \nu_j\, dS_0$$

On substituting back into Eq. (32) and reducing, we see that if $\delta J_R = 0$ for arbitrary $\delta S_{ij}$ and $\delta u_i$, subjected only to the restrictions (31), then the Euler equations are precisely Eqs. (25), (26), (29), and $u_i = u_i^*$ on $S_{0_u}$. These are the basic equations of elasticity under finite strain. Thus the vanishing of $\delta J_R$ for arbitrary $\delta u_i$ and $\delta S_{ij}$, subjected only to the restrictions (31), implies the field equations of finite elasticity. Conversely, Eqs. (25), (26), (29), and $u_i = u_i^*$ imply $\delta J_R = 0$. This is known as Reissner's theorem.

To generalize the complementary energy theorem of Sec. 10.9, consider variations of stresses which satisfy the equilibrium conditions (25), (26). Then, by expressing $E_{ij}$ according to (16.1:11), using Gauss' theorem and Eqs. (25), (26), we can reduce Reissner's functional $J_R$ to

$$(34)\ \blacktriangle\qquad J_c = -\int_{V_0} \overline{W}_c\, dv_0 + \int_{S_{0_u}} u_i^* \overset{\nu}{T}_{0_i}\, dS_0$$

$$(35)\ \blacktriangle\qquad \overline{W}_c = \tfrac{1}{2}S_{ij}u_{k,i}u_{k,j} + \mathscr{W}_c(S_{ij})$$

It can be shown (by differentiating $\overline{W}_c$ with respect to $u_{l,m}$ while holding $T_{ij}$ fixed) that $\overline{W}_c$ is a function of the Lagrangian stresses $T_{ij}$ only. Thus $J_c[T_{ij}]$ is a functional of $T_{ij}$. The complementary energy theorem of finite deformation states that $\delta J_c[T_{ij}] = 0$ for variations of $T_{ij}$ that satisfy the equations of equilibrium.

Finally, the reader may verify that an extension of the general equations (10.10:1) to (10.10:7) is

$$(36)\qquad \delta J = 0$$

under arbitrary simultaneous variations of $E_{ij}$, $u_i$, and $S_{ij}$, where

$$(37)\ \blacktriangle\quad J = \int_{V_0}\left\{\rho_0\mathscr{W}(E_{ij}) + S_{ij}\left[\frac{1}{2}\left(\frac{\partial u_i}{\partial a_j} + \frac{\partial u_j}{\partial a_i} + \frac{\partial u_k}{\partial a_i}\frac{\partial u_k}{\partial a_j}\right) - E_{ij}\right] - \rho_0 F_{0_i}u_i\right\} dv_0 - \int_{S_{0\sigma}} u_i \overset{\nu}{T}{}^*_{0i}\, dS_0 - \int_{S_{0_u}} (u_i - u_i^*)\overset{\nu}{T}_{0_i}\, dS_0$$

$$(38)\qquad S_{ij} = S_{ji}$$

and where $\overset{\nu}{T}{}^*_{0_i}$ and $u_i^*$ are prescribed over $S_{0_\sigma}$ and $S_{0_u}$ respectively. This may be regarded as the fundamental variational principle of finite elasticity because of its great generality. The practical applications of these principles are also similar to those discussed in Chaps. 10 and 11, and will not be discussed further. For the special case of incompressible material, see Levinson and Blatz.[10.2]

## 16.8. THEORY OF THIN PLATES—SMALL DEFLECTIONS

In the following three sections we shall consider the basic equations in the theory of plates. Section 16.8 serves as an introduction. Section 16.9 treats the main subject of large deflection of plates. From the standpoint of engineering applications, the theory of plates is currently one of the most important and one of the most interesting topics in the theory of elasticity.

A *plate* is a body bounded by two surfaces of small curvature, the distance between these surfaces, called the *thickness*, being small in comparison with the dimensions of the surface. The thickness of the plate will be denoted by $h$. A surface equidistant to the bounding surfaces is called the *middle surface*. When $h$ is constant, the plate is said to be of *uniform thickness*; when the middle surface is a plane in the underformed configuration, the plate is said to be *flat*. We shall consider only flat plates of uniform thickness and of homogeneous linear elastic material. Our purpose is to illustrate the application of the concepts developed in Secs. 16.1 to 16.3 to the derivation of basic equations for the theory of plates.

We shall develop the theory of plates in the light of their applications in engineering. A principal feature in straining a plate or a shell is the relative smallness of the tractions acting on surfaces parallel to the middle surface as compared with the maximum bending or stretching stresses in the body. For example, the aerodynamic pressure acting on the wings of an airplane in flight is of the order of 1 to 10 pounds per square inch, whereas the bending stress in the wing skin is likely to range from 10,000 to 200,000 pounds per square inch. In many other applications, plates and shells are used to transmit forces and moments acting on their edges, and no load acts on the faces at all. This practical situation is important in simplifying the theory.

When a plate is very thin, the smallness of tractions on the external faces implies the smallness of tractions on any surface parallel to the middle surface.

Let us consider a flat plate and choose a fixed right-handed rectangular Cartesian frame of reference with the $x, y$ plane coinciding with the middle surface of a plate in its initial, unloaded state, and the $z$-axis normal to it

(Fig. 16.8:1). Let the components of the displacement of a particle located initially at $(x, y, z)$ be denoted by $u_x$, $u_y$, $u_z$. As an introduction to the large-deflection analysis, let us consider in this section small deflections only. *We assume that $u_x$, $u_y$ $u_z$ are infinitesimal; in particular, $|u_i| \ll h$, the plate thickness.*

With the coordinates so chosen, our discussion above indicates that the stresses $\sigma_{zx}$, $\sigma_{zy}$, $\sigma_{zz}$ are small throughout the plate. Now the deformation pattern of the plate is determined if we assume that

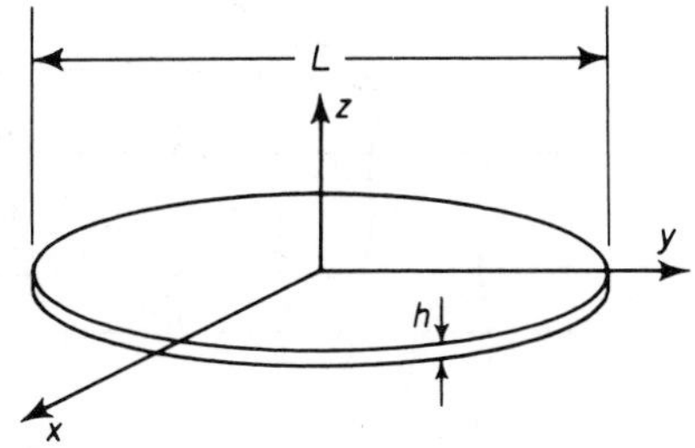


**Fig. 16.8:1.** A thin plate; coordinate system.

$$(1) \quad \sigma_{zx} = 0, \quad \sigma_{zy} = 0, \quad \sigma_{zz} = 0, \qquad -\frac{h}{2} \leqslant z \leqslant \frac{h}{2},$$

and that $\sigma_{xx}$, $\sigma_{yy}$ vary linearly throughout the thickness, i.e.,

$$(2) \qquad \sigma_{xx} = a_1(x, y) + b_1(x, y)z, \qquad \sigma_{yy} = a_2(x, y) + b_2(x, y)z.$$

By Hooke's law, Eqs. (1) imply the following statements concerning strain and displacements:

$$(3) \qquad e_{zx} = \frac{1}{2}\left(\frac{\partial u_x}{\partial z} + \frac{\partial u_z}{\partial x}\right) = 0, \qquad e_{zy} = \frac{1}{2}\left(\frac{\partial u_y}{\partial z} + \frac{\partial u_z}{\partial y}\right) = 0.$$

$$(4) \qquad e_{zz} = \frac{\partial u_z}{\partial z} = -\frac{\nu}{E}(\sigma_{xx} + \sigma_{yy}).$$

Equations (4) and (2) together yield

$$(5) \quad u_z = w(x, y) - \frac{\nu}{E}[a_1(x, y) + a_2(x, y)]z - \frac{\nu}{E}[b_1(x, y) + b_2(x, y)]\frac{z^2}{2}.$$

The function $w(x, y)$ represents the vertical deflection of the middle surface of the plate. In view of the thinness of the plate, the last two terms in (5) are in general small and will be neglected in comparison with $w(x, y)$.† Inserting $w(x, y)$ for $u_z$ in Eq. (3) and integrating, we obtain

$$(6) \qquad \begin{aligned} u_x &= u(x, y) - z\frac{\partial w(x, y)}{\partial x}, \\ u_y &= v(x, y) - z\frac{\partial w(x, y)}{\partial y}, \end{aligned}$$

† The consistency of this statement has to be verified *a posteriori*. Note, however, that $\sigma_{xx}$, $\sigma_{yy}$ must be bounded by the yielding stress, $\sigma_{Y.P.}$, of the material. Hence, according to (4), $e_{zz}$ is bounded by $2\nu\sigma_{Y.P.}/E$. Thus, the last two terms in (5) must be smaller than $2\nu(\sigma_{Y.P.}/E)h$. For most structural materials, $(\sigma_{Y.P.}/E)$ is of order $10^{-3}$. Hence the last two terms in (5) are negligible if $w/h \gg 10^{-3}$.

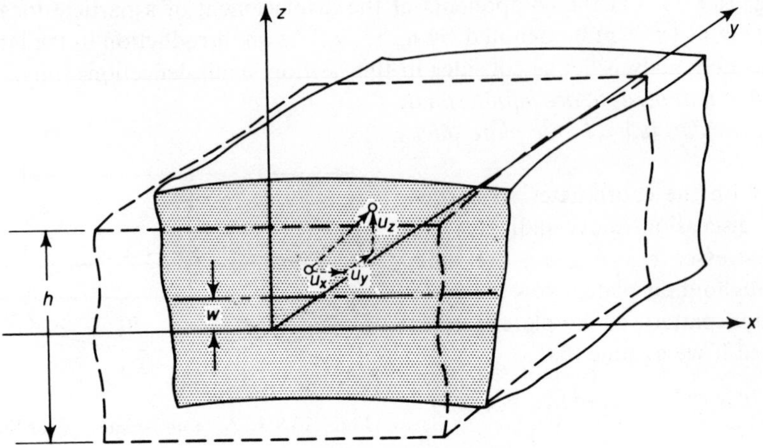


**Fig. 16.8:2.** Notations for components of displacement.

where $u(x, y)$, $v(x, y)$ are the tangential displacements of points lying on the middle surface (Fig. 16.8:2). From (6) we deduce the strain components:

$$e_{xx} \equiv \frac{\partial u_x}{\partial x} = \frac{\partial u}{\partial x} - z\,\frac{\partial^2 w}{\partial x^2},$$

$$(7) \qquad e_{yy} \equiv \frac{\partial u_y}{\partial y} = \frac{\partial v}{\partial y} - z\,\frac{\partial^2 w}{\partial y^2},$$

$$e_{xy} = \frac{1}{2}\left(\frac{\partial u_x}{\partial y} + \frac{\partial u_y}{\partial x}\right) = \frac{1}{2}\left(\frac{\partial u}{\partial y} + \frac{\partial v}{\partial x}\right) - z\,\frac{\partial^2 w}{\partial x\,\partial y}.$$

By means of Hooke's law for a homogeneous isotropic material, we find the stresses

$$\sigma_{xx} = \frac{E}{1-\nu^2}(e_{xx} + \nu e_{yy}) = \frac{E}{1-\nu^2}\left[\frac{\partial u}{\partial x} + \nu\,\frac{\partial v}{\partial y} - z\left(\frac{\partial^2 w}{\partial x^2} + \nu\,\frac{\partial^2 w}{\partial y^2}\right)\right],$$

$$(8) \qquad \sigma_{yy} = \frac{E}{1-\nu^2}(e_{yy} + \nu e_{xx}) = \frac{E}{1-\nu^2}\left[\frac{\partial v}{\partial y} + \nu\,\frac{\partial u}{\partial x} - z\left(\frac{\partial^2 w}{\partial y^2} + \nu\,\frac{\partial^2 w}{\partial x^2}\right)\right],$$

$$\sigma_{xy} = 2Ge_{xy} = 2G\left[\frac{1}{2}\left(\frac{\partial u}{\partial y} + \frac{\partial v}{\partial x}\right) - z\,\frac{\partial^2 w}{\partial x\,\partial y}\right].$$

In this way, the stress and deformation patterns in the plate are determined on the basis of the assumptions (1) and (2).

The approximate results embodied in Eqs. (6) are called *Kirchhoff's hypothesis: it states that every straight line in the plate that was originally perpendicular to the plate middle surface remains straight after the strain and perpendicular to the deflected middle surface.*

In most cases the *ad hoc* assumptions (1) and (2) can only be satisfied approximately. In general, they are not even consistent. Kirchhoff alluded to the theory of simple beams for justification of his hypothesis; but we know from Saint-Venant's beam theory (see Sec. 9.7) that the assumption "plane cross sections remain plane" is incorrect if the resultant shear does not vanish. For these reasons, many authors dislike these *ad hoc* assumptions, and have tried other formulations. The most successful of these new formulations reduces the equations of three-dimensional elastic continuum to those of a two-dimensional non-Euclidean continuum of the middle surface, describing the behavior of the plate in terms of the displacements (and their derivatives) of points on the middle surface. A systematic scheme of successive approximation has been developed, of which the "first-order" approximation coincides with those of the classical Kirchhoff theory.

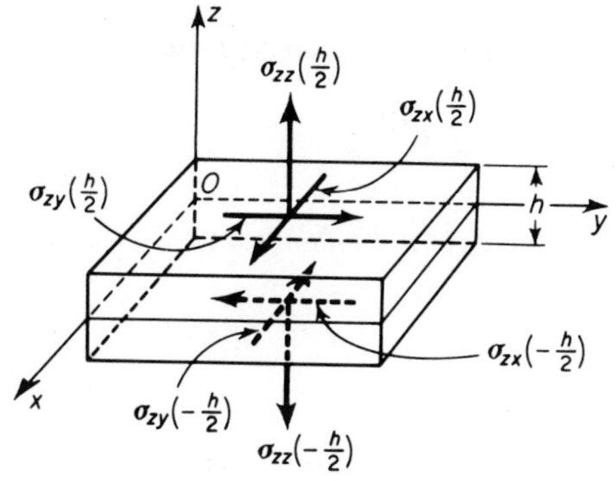


**Fig. 16.8:3.** External loads acting on the plate surfaces.

These criticisms provide impetus for further developments of plate theory. However, in retrospect, the truth of the matter is that the recognition of Kirchhoff's hypothesis (6) was the most important discovery in the theory of plates.

Let us now consider the equations of equilibrium. The three-dimensional equations of equilibrium are, of course,

$$\frac{\partial \sigma_{xx}}{\partial x} + \frac{\partial \sigma_{yx}}{\partial y} + \frac{\partial \sigma_{zx}}{\partial z} + X = 0, \tag{9}$$

$$\frac{\partial \sigma_{xy}}{\partial x} + \frac{\partial \sigma_{yy}}{\partial y} + \frac{\partial \sigma_{zy}}{\partial z} + Y = 0, \tag{10}$$

$$\frac{\partial \sigma_{xz}}{\partial x} + \frac{\partial \sigma_{yz}}{\partial y} + \frac{\partial \sigma_{zz}}{\partial z} + Z = 0, \tag{11}$$

where $X, Y, Z$ are the body forces per unit volume. Since the approximate stress distribution throughout the thickness is already known [Eq. (8)], we shall derive the approximate equations of equilibrium by integrating Eqs. (9)–(11) over the thickness. We assume that over the faces of the plate ($z = \pm h/2$) there act the normal stresses $\sigma_{zz}(\pm h/2)$ and the shear stresses $\sigma_{zx}(\pm h/2)$, $\sigma_{zy}(\pm h/2)$. Figure 16.8:3 shows the stress vector directions if the external loads $\sigma_{zx}(\pm h/2)$, $\sigma_{zy}(\pm h/2)$, $\sigma_{zz}(\pm h/2)$ are all positive. Now if we multiply Eqs. (9) by $dz$ and integrate from $-h/2$ to $h/2$, we have

$$\int_{-h/2}^{h/2} \frac{\partial}{\partial x} \sigma_{xx}\, dz + \int_{-h/2}^{h/2} \frac{\partial}{\partial y} \sigma_{yx}\, dz + \int_{-h/2}^{h/2} \frac{\partial}{\partial z} \sigma_{zx}\, dz + \int_{-h/2}^{h/2} X\, dz = 0. \tag{12}$$

The order of integration and differentiation can be interchanged in the first two terms. The third term contributes

$$\int_{-h/2}^{h/2} d\sigma_{zx} = \sigma_{zx}\Big|_{-h/2}^{h/2} = \sigma_{zx}\left(\frac{h}{2}\right) - \sigma_{zx}\left(-\frac{h}{2}\right).$$

Hence, defining the *stress resultants* (forces per unit length) $N_x$, $N_y$, $N_{xy}$,

$$\text{(13)}\quad N_x = \int_{-h/2}^{h/2} \sigma_{xx}\, dx, \quad N_{xy} = \int_{-h/2}^{h/2} \sigma_{xy}\, dz = N_{yx}, \quad N_y = \int_{-h/2}^{h/2} \sigma_{yy}\, dz,$$

and the *external loading tangential to the plate* (forces per unit area):

$$\text{(14)}\quad \begin{aligned} f_x &= \sigma_{zx}\left(\frac{h}{2}\right) - \sigma_{zx}\left(-\frac{h}{2}\right) + \int_{-h/2}^{h/2} X\, dz, \\ f_y &= \sigma_{zy}\left(\frac{h}{2}\right) - \sigma_{zy}\left(-\frac{h}{2}\right) + \int_{-h/2}^{h/2} Y\, dz; \end{aligned}$$

we can write (12) as

$$\text{(15)}\quad \frac{\partial N_x}{\partial x} + \frac{\partial N_{xy}}{\partial y} + f_x = 0.$$

Similarly, Eq. (10) can be integrated to

$$\text{(16)}\quad \frac{\partial N_{xy}}{\partial x} + \frac{\partial N_y}{\partial y} + f_y = 0.$$

Now if we multiply Eq. (9) by $z\, dz$ and integrate from $-h/2$ to $h/2$, we shall obtain the results

$$\text{(17)}\quad \begin{aligned} \frac{\partial M_x}{\partial x} + \frac{\partial M_{xy}}{\partial y} - Q_x + m_x &= 0, \\ \frac{\partial M_{yx}}{\partial x} + \frac{\partial M_y}{\partial y} - Q_y + m_y &= 0, \end{aligned}$$

where

$$\text{(18)}\quad M_x = \int_{-h/2}^{h/2} \sigma_{xx} z\, dz, \quad M_{xy} = \int_{-h/2}^{h/2} \sigma_{xy} z\, dz = M_{yx}, \quad M_y = \int_{-h/2}^{h/2} \sigma_{yy} z\, dz,$$

$$\text{(19)}\quad Q_x = \int_{-h/2}^{h/2} \sigma_{xz}\, dz, \qquad Q_y = \int_{-h/2}^{h/2} \sigma_{yz}\, dz,$$

$$\text{(20)}\quad \begin{aligned} m_x &= \frac{h}{2}\left[\sigma_{zx}\left(\frac{h}{2}\right) + \sigma_{zx}\left(-\frac{h}{2}\right)\right] + \int_{-h/2}^{h/2} zX\, dz, \\ m_y &= \frac{h}{2}\left[\sigma_{zy}\left(\frac{h}{2}\right) + \sigma_{zy}\left(-\frac{h}{2}\right)\right] + \int_{-h/2}^{h/2} zY\, dz, \end{aligned}$$

The quantities $M_x$, $M_y$, $M_{xy}$, of physical dimensions of moment per unit length, are called the *stress moments* in the plate. Specifically, $M_x$, $M_y$ are called the *bending moments* and $M_{xy}$ the *twisting moments*. $Q_x$, $Q_y$, of the dimensions force per unit length, are called the *transverse shear*; $m_x$, $m_y$, of physical dimensions of moments per unit area, are the *resultant external moment* per unit area about the middle surface. Clearly, the moment arm is $h/2$ for shear on the faces and $z$ in the plate. These stress resultants and stress moments are illustrated in Figs. 16.8:4 and 16.8:5.

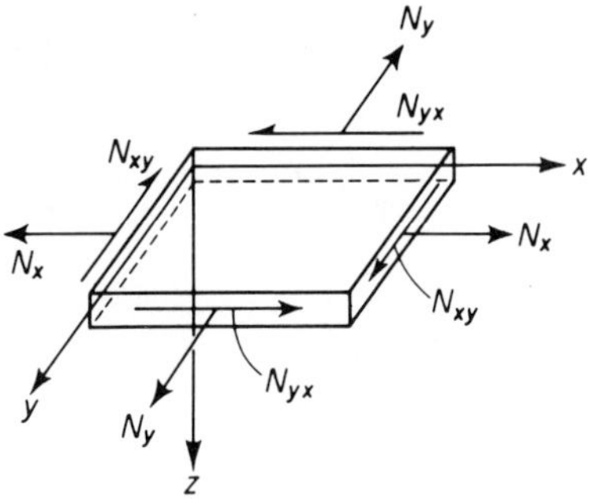

**Fig. 16.8:4.** Stress resultants.

If Eq. (11) is multiplied by $dz$ and integrated from $-h/2$ to $h/2$, we obtain

$$\frac{\partial Q_x}{\partial x} + \frac{\partial Q_y}{\partial y} + q = 0, \tag{21}$$

where $q$ is the external load per unit area normal to the middle surface,

$$q = \sigma_{zz}\left(\frac{h}{2}\right) - \sigma_{zz}\left(-\frac{h}{2}\right) + \int_{-h/2}^{h/2} Z\,dz. \tag{22}$$

Eliminating $Q_x$, $Q_y$ from (19) and (21), we have the equation of equilibrium in moments,

$$\frac{\partial^2 M_x}{\partial x^2} + 2\frac{\partial^2 M_{xy}}{\partial x\,\partial y} + \frac{\partial^2 M_y}{\partial y^2} = -\frac{\partial m_k}{\partial x} - \frac{\partial m_y}{\partial y} - q. \tag{23}$$

To complete the formulation, we substitute $\sigma_{xx}$, $\sigma_{yy}$, $\sigma_{xy}$ from (8) into (13) and (18), to obtain

$$(24)\quad \begin{aligned} N_x &= \frac{Eh}{1-\nu^2}\left(\frac{\partial u}{\partial x} + \nu\frac{\partial v}{\partial y}\right), \qquad N_y = \frac{Eh}{1-\nu^2}\left(\frac{\partial v}{\partial y} + \nu\frac{\partial u}{\partial x}\right), \\ N_{xy} &= N_{yx} = Gh\left(\frac{\partial u}{\partial y} + \frac{\partial v}{\partial k}\right) = \frac{1-\nu}{2}\,\frac{Eh}{1-\nu^2}\left(\frac{\partial u}{\partial y} + \frac{\partial v}{\partial x}\right), \end{aligned}$$

$$(25)\quad \begin{aligned} M_x &= -D\left(\frac{\partial^2 w}{\partial x^2} + \nu\frac{\partial^2 w}{\partial y^2}\right), \qquad M_y = -D\left(\frac{\partial^2 w}{\partial y^2} + \nu\frac{\partial^2 w}{\partial x^2}\right), \\ M_{xy} &= M_{yx} = -(1-\nu)D\frac{\partial^2 w}{\partial x\,\partial y}, \end{aligned}$$

where the quantity

$$D = \frac{Eh^3}{12(1-\nu^2)} \tag{26}$$

is called the *bending rigidity of the plate*.

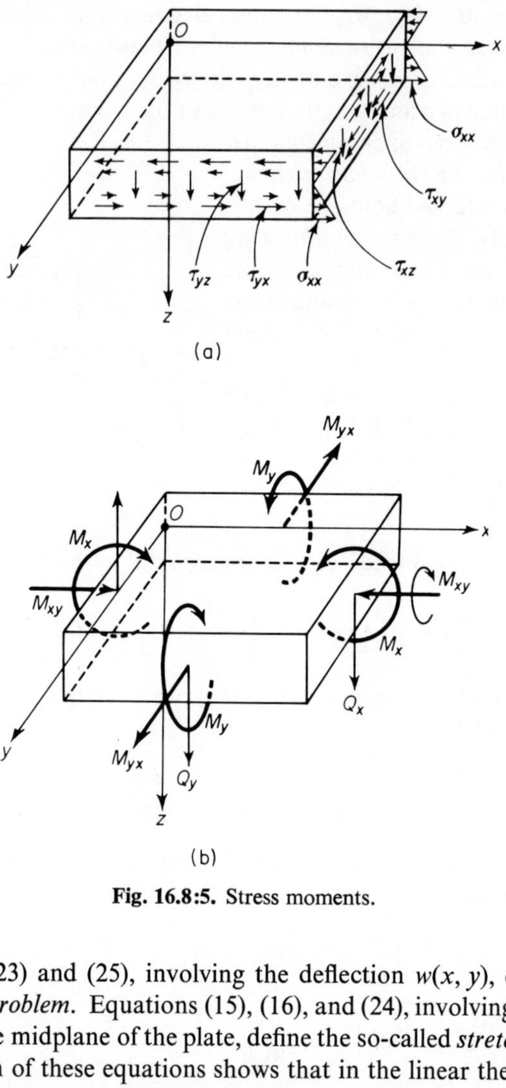


**Fig. 16.8:5.** Stress moments.

Equations (23) and (25), involving the deflection $w(x, y)$, define the so-called *bending problem*. Equations (15), (16), and (24), involving the displacements $u$, $v$ in the midplane of the plate, define the so-called *stretching problem*. An examination of these equations shows that in the linear theory of plates, under the assumption of infinitesimal displacements, the bending and the stretching of the plate are independent of each other. There is no coupling between these two responses. It will be seen later that for large deflections, the bending and stretching become coupled and introduce a nonlinearity which is the principal difficulty in the theory of plates.

A substitution of (25) into (23) yields the fundamental equation of the linear theory of bending of a plate of uniform thickness.

$$\frac{\partial^4 w}{\partial x^4} + 2\frac{\partial^4 w}{\partial x^2\,\partial y^2} + \frac{\partial^4 w}{\partial y^4} = \frac{1}{D}\left(q + \frac{\partial m_x}{\partial x} + \frac{\partial m_y}{\partial y}\right). \tag{27}$$

On the other hand, a substitution of Eqs. (24) into Eqs. (15) and (16) gives the basic equations for stretching of a plate of uniform thickness:

$$
(28)\qquad
\begin{aligned}
&\frac{\partial^2 u}{\partial x^2} + \frac{1-\nu}{2}\frac{\partial^2 u}{\partial y^2} + \frac{1+\nu}{2}\frac{\partial^2 v}{\partial x\,\partial y} + \frac{1-\nu^2}{Eh} f_x = 0,\\
&\frac{\partial^2 v}{\partial y^2} + \frac{1-\nu}{2}\frac{\partial^2 v}{\partial x^2} + \frac{1+\nu}{2}\frac{\partial^2 u}{\partial x\,\partial y} + \frac{1-\nu^2}{Eh} f_y = 0.
\end{aligned}
$$

Equation (27) shows that the problem of bending is reduced to the solution of biharmonic equations with appropriate boundary conditions. The mathematical problem is, therefore, the same as that of the Airy stress functions in plane elasticity. The mathematical problem of plate stretching is identical with the plane stress problem, and Airy stress functions can be introduced.

## 16.9. LARGE DEFLECTIONS OF PLATES

The theory of small deflections of plates was derived in the previous section under the assumption of infinitesimal displacements. Unfortunately, the results are valid literally only for very small displacements. When the deflections are as large as the thickness of the plate, the results become quite inaccurate. This is in sharp contrast with the theory of beams, for which the linear equation is valid as long as the slope of the deflection curve is everywhere small in comparison to unity.

A well-known theory of large deflections of plates is due to von Kármán. In this theory the following assumptions are made.

(H1) *The plate is thin. The thickness $h$ is much smaller than the typical plate dimension $L$, i.e. $h \ll L$.*

(H2) *The magnitude of the deflection $w$ is of the same order as the thickness of the plate, $h$, but small compared with the typical plate dimension $L$, i.e.,*

$$|w| = O(h), \qquad |w| \ll L.$$

(H3) *The slope is everywhere small, $|\partial w/\partial x| \ll 1$, $|\partial w/\partial y| \ll 1$.*

(H4) *The tangential displacements, $u$, $v$ are infinitesimal. In the strain-displacement relations only those nonlinear terms which depend on $\partial w/\partial x$, $\partial w/\partial y$ are to be retained. All other nonlinear terms are to be neglected.*

(H5) *All strain components are small. Hooke's law holds.*

(H6) *Kirchhoff's hypotheses hold; i.e., tractions on surfaces parallel to the middle surface are negligible, and strains vary linearly within the plate thickness.*

Thus von Kármán's theory differs from the linear theory only in retaining certain powers of the derivatives $\partial w/\partial x$, $\partial w/\partial y$ in the strain-displacement relationship.

We now have a basic small parameter $h$. The elastic displacement $w$ is assumed to be of the same order of magnitude as $h$. The term *large deflection* refers to the fact that $w$ is no longer small compared with $h$. Thus, as shown in Fig. 16.9:1, the deformed configuration differs considerably from the original one. We cannot imitate the procedure of Sec. 16.8 without further care. For example, if we retain the rectangular coordinates, with the $z$ plane fixed as the original middle plane of the plate in the undeformed position,

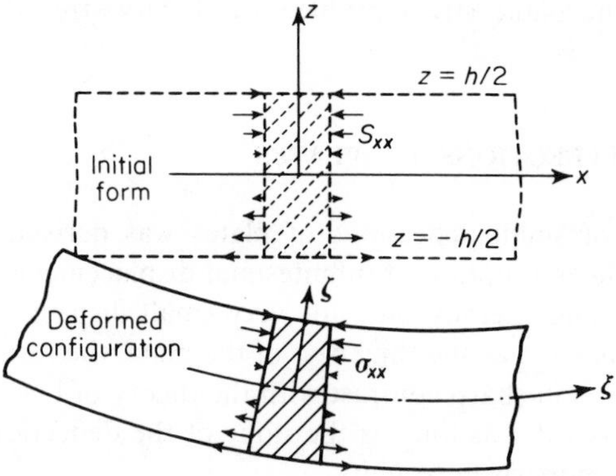


**Fig. 16.9:1.** Corresponding stresses in the initial and deformed configurations of a plate.

then the limits of integration across the plate thickness for a loaded plate can no longer be from $-h/2$ to $h/2$. In fact in a deformed plate the $z$ coordinates of the plate surfaces are now variable. On the other hand, if we wish to retain the convenience of fixed coordinates for the surfaces, we have two alternative courses. The first is to introduce "convective" coordinates which become curvilinear in the deformed state, i.e., a set of curvilinear coordinates imbedded in the plate so that the middle surface is always the surface $z = 0$, and the faces of the plate $z = \pm h/2$. This method is reasonable but complicated. The other is to adopt the Lagrangian description for the plate. In Lagrangian coordinates the plate surfaces are always designated as $z = \pm h/2$. The second alternative is by far the simplest approach, and will be pursued below.

In the following *we shall choose the Lagrangian description*. Let a fixed right-handed rectangular Cartesian frame of reference be used, with the $x$, $y$ plane coinciding with the middle surface of the plate in its initial, unloaded state, and the $z$-axis normal to it. Let the components of the displacement of a particle located initially at $(x, y, z)$ be denoted by $u_x$, $u_y$, $u_z$. In the Lagrangian description, the Green's strain tensor, referred to the initial configuration, is used, whose components are [see Eq. (16.1:11), identifying $a_1$, $a_2$, $a_3$ with $x, y, z$]

$$E_{xx} = \frac{\partial u_x}{\partial x} + \frac{1}{2}\left[\left(\frac{\partial u_x}{\partial x}\right)^2 + \left(\frac{\partial u_y}{\partial x}\right)^2 + \left(\frac{\partial u_z}{\partial x}\right)^2\right],$$

(1)
$$E_{yy} = \frac{\partial u_y}{\partial y} + \frac{1}{2}\left[\left(\frac{\partial u_x}{\partial y}\right)^2 + \left(\frac{\partial u_y}{\partial y}\right)^2 + \left(\frac{\partial u_z}{\partial y}\right)^2\right],$$

$$E_{xy} = \frac{1}{2}\left[\frac{\partial u_x}{\partial y} + \frac{\partial u_y}{\partial x} + \frac{\partial u_x}{\partial x}\frac{\partial u_x}{\partial y} + \frac{\partial u_y}{\partial x}\frac{\partial u_y}{\partial y} + \frac{\partial u_z}{\partial x}\frac{\partial u_z}{\partial y}\right].$$

Now, according to the *ad hoc* assumptions (H6), we have

(2)
$$u_x = u(x, y) - z\frac{\partial w}{\partial x}, \quad u_y = v(x, y) - z\frac{\partial w}{\partial y}, \quad u_z = w(x, y),$$

as was demonstrated in the previous section [Eqs. (16.8:5) and (16.8:6)]. Furthermore, according to (H4) the higher powers of the derivatives of $u$ and $v$ are negligible in comparison with their first powers, and since $|z| \leqslant h \ll L$, it can be argued that the only nonlinear terms to be retained in the strain-displacement relations (1) are the squares and products of $\partial w/\partial x$, $\partial w/\partial y$. Hence,

$$E_{xx} = \frac{\partial u}{\partial x} - z\frac{\partial^2 w}{\partial x^2} + \frac{1}{2}\left(\frac{\partial w}{\partial x}\right)^2,$$

(3)
$$E_{yy} = \frac{\partial v}{\partial y} - z\frac{\partial^2 w}{\partial y^2} + \frac{1}{2}\left(\frac{\partial w}{\partial y}\right)^2,$$

$$E_{xy} = \frac{1}{2}\left[\frac{\partial u}{\partial y} + \frac{\partial v}{\partial x} - 2z\frac{\partial^2 w}{\partial x\,\partial y} + \frac{\partial w}{\partial x}\frac{\partial w}{\partial y}\right].$$

Corresponding to Green's strain tensor, we have Kirchhoff's stress tensor $S_{ij}$, which can be used in the Lagrangian description. (See Sec. 16.2.) We assume, according to (H5), that $S_{ij}$ is related to $E_{ij}$ through Hooke's law. Hence, for an isotropic material,

(4)
$$S_{ij} = \lambda E_{kk}\delta_{ij} + 2GE_{ij}.$$

Identifying individual components, we obtain, from Eqs. (3) and (4),

$$S_{xx} = \frac{E}{1-\nu^2}(E_{xx} + \nu E_{yy})$$

(5)
$$= \frac{E}{1-\nu^2}\left[\frac{\partial u}{\partial x} + \nu\frac{\partial v}{\partial y} - z\left(\frac{\partial^2 w}{\partial x^2} + \nu\frac{\partial^2 w}{\partial y^2}\right) + \frac{1}{2}\left(\frac{\partial w}{\partial x}\right)^2 + \frac{\nu}{2}\left(\frac{\partial w}{\partial y}\right)^2\right],$$

$$S_{yy} = \frac{E}{1-\nu^2}\left[\frac{\partial v}{\partial y} + \nu\frac{\partial u}{\partial x} - z\left(\frac{\partial^2 w}{\partial y^2} + \nu\frac{\partial^2 w}{\partial x^2}\right) + \frac{1}{2}\left(\frac{\partial w}{\partial y}\right)^2 + \frac{\nu}{2}\left(\frac{\partial w}{\partial x}\right)^2\right],$$

$$S_{xy} = G\left[\frac{\partial u}{\partial y} + \frac{\partial v}{\partial x} - 2z\frac{\partial^2 w}{\partial x\,\partial y} + \frac{\partial w}{\partial x}\frac{\partial w}{\partial y}\right].$$

We now define the stress resultants

$$(6)\qquad N'_x = \int_{-h/2}^{h/2} S_{xx}\,dz, \qquad N'_{xy} = \int_{-h/2}^{h/2} S_{xy}\,dz, \qquad N'_y = \int_{-h/2}^{h/2} S_{yy}\,dz,$$

and the stress moments

$$(7)\qquad M'_x = \int_{-h/2}^{h/2} S_{xx}z\,dz, \qquad M'_{xy} = \int_{-h/2}^{h/2} S_{xy}z\,dz, \qquad M'_y = \int_{-h/2}^{h/2} S_{yy}z\,dz.$$

A prime is used over the $N$'s and $M$'s to indicate that they are based on Kirchhoff stresses and are referred to the initial, unloaded configuration of the plate. A substitution of (5) into (6) and (7) gives the following stress-deflection relations.

$$(8)\qquad \begin{aligned}
N'_x &= \frac{Eh}{1-\nu^2}\left[\frac{\partial u}{\partial x} + \nu\frac{\partial v}{\partial y} + \frac{1}{2}\left(\frac{\partial w}{\partial x}\right)^2 + \frac{\nu}{2}\left(\frac{\partial w}{\partial y}\right)^2\right],\\
N'_y &= \frac{Eh}{1-\nu^2}\left[\frac{\partial v}{\partial y} + \nu\frac{\partial u}{\partial x} + \frac{1}{2}\left(\frac{\partial w}{\partial y}\right)^2 + \frac{\nu}{2}\left(\frac{\partial w}{\partial x}\right)^2\right],\\
N'_{xy} &= \frac{1-\nu}{2}\,\frac{Eh}{1-\nu^2}\left[\frac{\partial u}{\partial y} + \frac{\partial v}{\partial x} + \frac{\partial w}{\partial x}\frac{\partial w}{\partial y}\right],\\
M'_x &= -D\left(\frac{\partial^2 w}{\partial x^2} + \nu\frac{\partial^2 w}{\partial y^2}\right),\\
M'_y &= -D\left(\frac{\partial^2 w}{\partial y^2} + \nu\frac{\partial^2 w}{\partial x^2}\right),\\
M'_{xy} &= -(1-\nu)D\frac{\partial^2 w}{\partial x\,\partial y}.
\end{aligned}$$

The stress tensor $S_{ij}$, the resultants $N'_x$, $N'_y$, $N'_{xy}$, and the moments $M'_x$, $M'_y$, $M'_{xy}$ are defined with respect to the original, unloaded configuration of the plate. However, since all derivatives $\partial u/\partial x$, $\partial u/\partial y$, $\partial v/\partial x$, $\partial v/\partial y$, $\partial w/\partial x$, $\partial w/\partial y$ are small in comparison with unity, the values of $S_{ij}$, $N'_x$, $N'_y$, $N'_{xy}$, and $M'_x$, $M'_y$, $M'_{xy}$ are, in the first order approximation, *equal*, respectively, to the Eulerian stress tensor $\sigma_{ij}$, the stress resultants $N_x$, $N_y$, $N_{xy}$, and the stress moments $M_x$, $M_y$, $M_{xy}$ defined in the deformed configuration of the plate. See Eq. (16.2:15), which relates the Eulerian stress tensor $\sigma_{ij}$ with the Kirchhoff stress tensor $S_{ij}$.

Let us now consider the equations of equilibrium. In Lagrangian description, the condition of equilibrium is given by Eq. (16.3:8):

$$\frac{\partial}{\partial a_j}\left[S_{jk}\left(\delta_{ik} + \frac{\partial u_i}{\partial a_k}\right)\right] + \bar{X}_i = 0.$$

where $\bar{X}_i$ are the components of the body force per unit volume referred to the original volume. Identifying $a_1$, $a_2$, $a_3$ with $x$, $y$, $z$, and identifying $x_1$, $x_2$, $x_3$ with $x + u_x$, $y + u_y$, $z + u_z$, we obtain, on substituting (2) and retaining only the first-order terms, the equations

$$\frac{\partial}{\partial x} S_{xx} + \frac{\partial}{\partial y} S_{xy} + \frac{\partial}{\partial z} S_{zx} + \bar{X} = 0, \tag{9}$$

$$\frac{\partial}{\partial x} S_{xy} + \frac{\partial}{\partial y} S_{yy} + \frac{\partial}{\partial z} S_{zy} + \bar{Y} = 0,$$

$$\frac{\partial}{\partial x}\left(S_{xx}\frac{\partial w}{\partial x} + S_{xy}\frac{\partial w}{\partial y} + S_{xz}\right) + \frac{\partial}{\partial y}\left(S_{yx}\frac{\partial w}{\partial x} + S_{yy}\frac{\partial w}{\partial y} + S_{yz}\right) + \frac{\partial}{\partial z}\left(S_{zx}\frac{\partial w}{\partial x} + S_{zy}\frac{\partial w}{\partial y} + S_{zz}\right) + \bar{Z} = 0. \tag{10}$$

The rest of the development of the large deflection theory parallels closely that of the classical linear theory. Hence, our presentation will be brief, and we refer to the preceding section for many details. Equations (9) and (10) correspond to Eqs. (16.8:9), (16.8:10) and (16.8:11). If equations (9) are multiplied successively by $dz$ and $z\,dz$ and integrated from $-h/2$ to $h/2$, we get, with the definitions (6) and (7),

$$\frac{\partial N'_x}{\partial x} + \frac{\partial N'_{xy}}{\partial y} + f'_x = 0, \qquad \frac{\partial N'_{xy}}{\partial x} + \frac{\partial N'_y}{\partial y} + f'_y = 0, \tag{11}$$

$$\frac{\partial M'_x}{\partial x} + \frac{\partial M'_{xy}}{\partial y} - Q'_x + m'_x = 0, \qquad \frac{\partial M'_{xy}}{\partial x} + \frac{\partial M'_y}{\partial y} - Q'_y + m'_y = 0. \tag{12}$$

These correspond with Eqs. (16.8:15)–(16.8:17): $f'_x, f'_y$, $Q'_x$, $Q'_y$, $m'_x$, and $m'_y$ are given by the same formulas as in Sec. 16.8, with $\sigma_{ij}$ replaced by $S_{ij}$. The primes are used to indicate that these forces and moments are referred to the undeformed plate.

Finally, an integration of Eq. (10) with respect to $z$ from $-h/2$ to $h/2$ gives

$$\frac{\partial}{\partial x}\left(Q'_x + N'_x\frac{\partial w}{\partial x} + N'_{xy}\frac{\partial w}{\partial y}\right) + \frac{\partial}{\partial y}\left(Q'_y + N'_{xy}\frac{\partial w}{\partial x} + N'_y\frac{\partial w}{\partial y}\right) + q' = 0, \tag{13}$$

where $q'$ is the lateral load per unit area (of the undeformed middle plane) acting on the plate,

$$q' = \left(S_{zz} + S_{zx}\frac{\partial w}{\partial x} + S_{zy}\frac{\partial w}{\partial y}\right)\Bigg|_{-h/2}^{h/2} + \int_{-h/2}^{h/2} \bar{Z}\,dz. \tag{14}$$

The first and the last terms on the right-hand side of (14) are the familiar vertical load. The second and third terms represent contributions to the lateral load due to shear acting on the surfaces that are rotated in the deformed position. Equation (13) is substantially different from the corresponding Eq. (16.8:21) of the small-deflection theory. By means of (11), Eq. (13) can be written as

(15)

$$\frac{\partial Q'_x}{\partial x} + \frac{\partial Q'_y}{\partial y} + q' + N'_x \frac{\partial^2 w}{\partial x^2} + 2N'_{xy} \frac{\partial^2 w}{\partial x\,\partial y} + N'_y \frac{\partial^2 w}{\partial y^2} - f'_x \frac{\partial w}{\partial x} - f'_y \frac{\partial w}{\partial y} = 0.$$

We can now eliminate $Q'_x$, $Q'_y$ between (12) and (15) to obtain

$$\text{(16)} \qquad \frac{\partial^2 M'_x}{\partial x^2} + 2\frac{\partial^2 M'_{xy}}{\partial x\,\partial y} + \frac{\partial^2 M'_y}{\partial y^2} = -q' - \frac{\partial m'_x}{\partial x} - \frac{\partial m'_y}{\partial y} - N'_x \frac{\partial^2 w}{\partial x^2} - 2N'_{xy} \frac{\partial^2 w}{\partial x\,\partial y} - N'_y \frac{\partial^2 w}{\partial y^2} + f'_x \frac{\partial w}{\partial x} + f'_y \frac{\partial w}{\partial y}.$$

Finally, a substitution of Eqs. (8) into the above gives, for a plate of uniform thickness,

$$\text{(17)} \qquad \frac{\partial^4 w}{\partial x^4} + 2\frac{\partial^4 w}{\partial x^2\,\partial y^2} + \frac{\partial^4 w}{\partial y^4} = \frac{1}{D}\left[q' + \frac{\partial m'_x}{\partial x} + \frac{\partial m'_y}{\partial y} + N'_x \frac{\partial^2 w}{\partial x^2} + 2N'_{xy} \frac{\partial^2 w}{\partial x\,\partial y} + N'_y \frac{\partial^2 w}{\partial y^2} - f'_x \frac{\partial w}{\partial x} - f'_y \frac{\partial w}{\partial y}\right].$$

The Eq. (11) governing the membrane stresses $N'_x$, $N'_{xy}$, $N'_y$ are identical with Eqs. (16.8:15) and (16.8:16) of the linear theory. A substitution of (8) into (11) yields immediately the basic equations

$$\frac{\partial}{\partial x}\left[\frac{\partial u}{\partial x} + \nu\frac{\partial v}{\partial y} + \frac{1}{2}\left(\frac{\partial w}{\partial x}\right)^2 + \frac{\nu}{2}\left(\frac{\partial w}{\partial y}\right)^2\right] + \frac{1-\nu}{2}\frac{\partial}{\partial y}\left[\frac{\partial u}{\partial y} + \frac{\partial v}{\partial x} + \frac{\partial w}{\partial x}\frac{\partial w}{\partial y}\right] + f'_x = 0,$$

(18)

$$\frac{\partial}{\partial y}\left[\frac{\partial v}{\partial y} + \nu\frac{\partial u}{\partial x} + \frac{1}{2}\left(\frac{\partial w}{\partial y}\right)^2 + \frac{\nu}{2}\left(\frac{\partial w}{\partial x}\right)^2\right] + \frac{1-\nu}{2}\frac{\partial}{\partial x}\left[\frac{\partial u}{\partial y} + \frac{\partial v}{\partial x} + \frac{\partial w}{\partial x}\frac{\partial w}{\partial y}\right] + f'_y = 0.$$

Alternatively, a stress function $F(x, y)$ can be introduced from which $N'_x$, $N'_y$ are derived that satisfy Eqs. (11) identically.

$$\text{(19)} \qquad N'_x = \frac{\partial^2 F}{\partial y^2} - P_1, \qquad N'_y = \frac{\partial^2 F}{\partial x^2} - P_2, \qquad N'_{xy} = -\frac{\partial^2 F}{\partial x\,\partial y},$$

where $P_1$, $P_2$ are the potentials of the tangential forces given by the indefinite integrals

$$P_1 = \int f'_x(x, y)\,dx, \qquad P_2 = \int f'_y(x, y)\,dy.$$

The compatibility condition can be derived by eliminating $u$, $v$ from Eqs. (8), and using (19). We obtain, for a plate of uniform thickness,

$$\text{(20)}\qquad \frac{\partial^4 F}{\partial x^4} + 2\frac{\partial^4 F}{\partial x^2\,\partial y^2} + \frac{\partial^4 F}{\partial y^4} = Eh\left[\left(\frac{\partial^2 w}{\partial x\,\partial y}\right)^2 - \frac{\partial^2 w}{\partial x^2}\frac{\partial^2 w}{\partial y^2}\right] + \frac{\partial^2}{\partial y^2}(P_1 - \nu P_2) + \frac{\partial^2}{\partial x^2}(P_2 - \nu P_1).$$

Equations (17) and (20) are the famous *von Kármán equations for large deflection of plates.*

In most problems the external loads $f_x'$, $f_y'$, $m_x'$, $m_y'$, $P_1$, $P_2$ are zero. We then have the well-known pair of equations†

$$\text{(21)}\qquad \nabla^4 w = \frac{1}{D}\left[q + \frac{\partial^2 F}{\partial y^2}\frac{\partial^2 w}{\partial x^2} - 2\frac{\partial^2 F}{\partial x\,\partial y}\frac{\partial^2 w}{\partial x\,\partial y} + \frac{\partial^2 F}{\partial x^2}\frac{\partial^2 w}{\partial y^2}\right],$$

$$\text{(22)}\qquad \nabla^4 F = Eh\left[\left(\frac{\partial^2 w}{\partial x\,\partial y}\right)^2 - \frac{\partial^2 w}{\partial x^2}\frac{\partial^2 w}{\partial y^2}\right],$$

where $\nabla^4$ is the two-dimensional biharmonic operator.

The most important features of Eqs. (21) and (22) are that they are coupled and nonlinear. The nonlinear terms are fairly easy to interpret. Equation (15) shows that the membrane stresses $N_x'$, $N_y'$, $N_{xy}'$ multiplied by the corresponding curvatures and twist, $\partial^2 w/\partial x^2$, $\partial^2 w/\partial y^2$, $\partial^2 w/\partial x\,\partial y$, respectively, are equivalent to a lateral force. The nonlinear terms in Eq. (17) or (21) represent such feedback of induced membrane stress on bending. On the other hand, the reader may recall from differential geometry that if a surface is represented by the equation $z = w(x, y)$, the total, or Gaussian, curvature of the surface is given by the quantity

$$\frac{\partial^2 w}{\partial x^2}\frac{\partial^2 w}{\partial y^2} - \left(\frac{\partial^2 w}{\partial x\,\partial y}\right)^2.$$

The Gaussian curvature of a surface is equal to the product of the principal curvatures at a point. The Gaussian curvature vanishes if the surface is developable. Hence, the right-hand side of (22) vanishes if the deflection surface is developable. If a flat plate is deformed into a nondevelopable surface, its middle surface will be stretched in some way and the right-hand side of Eq. (22) does not vanish. Thus, the nonlinear term on the right-hand side of (22) arises from the stretching of the middle surface of a flat plate due to bending into a nondevelopable surface.

When the nonlinear terms in Eqs. (18)–(22) are neglected, these equations reduce to the corresponding equations of the small-deflection theory.

† These equations were given without proof by von Kármán[16.3] in 1910. Most books give derivations of these equations without a clear indication as to whether Lagrangian or Eulerian descriptions were used. The explicit use of Lagrangian description introduces a degree of clarity not hitherto achieved.

There are a large number of papers on the theory of plates. Some are devoted to justifying the *ad hoc* assumptions (H1)–(H6), or to their removal or generalization; others are concerned with the solution of boundary-value problems. See Bibliographies 16.2–16.4. But we shall conclude our discussion here, merely pointing out that the mathematical structure of the theory of small deflection of plates is the same as the mathematical structure of the plane problems of linear elasticity, whereas the large-deflection equations are essentially nonlinear. The theory of large deflection of plates includes many interesting problems on the stability, responses, vibrations, subharmonic resonances, edge layers, etc., most of which are not yet entirely resolved, and are still subjected to intensive theoretical and experimental studies. The explicit introduction of both Eulerian and Lagrangian descriptions of stress and strain clarifies the foundation of the large-deflection theory of plates. It would be worthwhile to do the same for the theory of shells to clarify that subject too.

# EPILOGUE

On reaching this point in our study, the reader must have a feeling that the panorama of solid mechanics is somehow unreal. The examples selected and discussed in the book are idealized problems. When we think of any practical situation in the real world, we realize that none of the solutions is directly applicable. Much more complex situations would have to be tackled. The real material behaves in a complicated way. The real object has an irregular geometric shape. Nonhomogeneity and anisotropy prevail, nonlinearity reveals itself almost everywhere. We might ask: if the examples are too simple, are the underlying methods sufficiently general to deal with all practical situations?

Unfortunately, no sweeping answer can be given. It is true that with the application of modern computing machines, the power of some analytic methods can be extended beyond the dreams of their inventors. But many fundamental problems still are unsolved; and, beyond linearization, not only is our capability in analysis very limited, but in many instances we do not know even how to ask questions.

I do not wish to leave the impression that engineering problems can be solved by arbitrary idealizations. But by proper interpretation of idealized results we can often obtain adequate solutions to important problems. One may reflect on the fact that for generations bridges and buildings were designed on the basis of simple truss and beam theory. Products of our civilization, airplanes, ocean liners, even space ships, were designed mainly on the basis of the theory of elasticity. It seems that by a proper use of the known knowledge, amazing things can be achieved.

Central questions constantly asked by an engineer are: how can a problem be idealized so that it can be handled? and how can an idealization be evaluated as to its adequacy with respect to the real situation? There are no quick answers to these questions. By accumulating experiences, by refining analysis, by creating more general and more rigorous theories, we gradually learn to understand the real world. In a book like the present one, whose main aim is a survey of the principal methods in solid mechanics, much is left to the thoughtful contemplation of the reader.

Solid mechanics is going through a period of intensive development. On the one hand, the movement of rational mechanics has produced remarkable results. On the other hand, rapid developments in all kinds of engineering have created a tremendous literature in technical mechanics. To furnish a guide to literature in both of these fields, we give in the following (pp. 481–507) a selected list of references. The bases of selection are:

1. If an adequate review article is available from which a comprehensive bibliography can be found, the review article is quoted and most of the bibliography will be omitted.
2. In rational mechanics emphasis is laid on methods and ideas.
3. In technical mechanics emphasis is laid on quantitative information, both experimental and theoretical.

# NOTES

## CHAPTER 1

### 1.1. HISTORICAL REMARKS

The best-known constitutive equation for a solid is the Hooke's law of linear elasticity, which was discovered by Robert Hooke in 1660. This law furnishes the foundation for the mathematical theory of elasticity. By 1821, Louis M. H. Navier (1785–1836) had succeeded in formulating the general equations of the three-dimensional theory of elasticity. All questions of the small strain of elastic bodies were thus reduced to a matter of mathematical calculation. In the same year, 1821, Fresnel (1788–1827) announced his wave theory of light. The concept of transverse oscillations through an elastic medium attracted the attention of Cauchy and Poisson. Augustin L. Cauchy (1789–1857) developed the concept of stress and strain, and formulated the linear stress-strain relationship that is now called Hooke's law. Simeon D. Poisson (1781–1840) developed a molecular theory of elasticity and arrived at the same equation as Navier's. Both Navier and Poisson based their analysis on Newtonian conception of the constitution of bodies, and assumed certain laws of intermolecular forces. Cauchy's general reasoning, however, made no use of the hypothesis of material particles. In the ensuing years, with the contributions of George Green (1793–1841), George Stokes (1819–1903), Lord Kelvin (William Thomson, 1824–1907), and others, the mathematical theory was established. The fundamental questions of continuum mechanics have received renewed attention in recent years. Through the efforst of C. Truesdell, R. S. Rivlin, W. Noll, J. L. Erikson, A. E. Green, and others, theories of finite strain and nonlinear constitutive equations have been developed. New developments in continuum mechanics since late 1940's have been truly remarkable.

The great impact of solid mechanics on civilization, however, is felt through its application to technology. In this respect a fine tradition was established by early masters. Galileo Galilei (1564–1642) considered the question of strength of beams and columns. James (or Jacques) Bernoulli (1654–1705) introduced the simple beam theory. Leonhard Euler (1707–1783) gave the column formula. Charles Augustin Coulomb (1736–1806) considered the failure criterion. Joseph Louis Lagrange (1736–1813) gave the equation that governs the bending and vibration of plates. Navier and Poisson gave numerous applications of their general theory to special problems. The best example was set by Barre de Saint-Venant (1797–1886), whose general solution of the problems of torsion and bending of prismatical bars is of great importance in engineering. With characteristic consideration for those who would use his results he solved these problems with a completeness that included many numerical coefficients and graphical presentations.

It will not be possible to outline the history of our subject in a few pages. The reader is referred to the many excellent references listed in Bibliography 1.1, p. 481.

# CHAPTER 2

## 2.1. DERIVATION OF Eq. (2.15:4)

To establish Eq. (2.15:4), we differentiate the equation $g_{ij} = \mathbf{g}_i \cdot \mathbf{g}_j$ to obtain

$$\frac{\partial g_{ij}}{\partial \theta^k} = \frac{\partial \mathbf{g}_i}{\partial \theta^k} \cdot \mathbf{g}_j + \frac{\partial \mathbf{g}_j}{\partial \theta^k} \cdot \mathbf{g}_i.$$

On permuting the indices $i$, $j$, $k$, we can obtain the derivatives $\partial g_{ij}/\partial \theta^j$, $\partial g_{ik}/\partial \theta^i$. Furthermore, since $\mathbf{g}_i = \partial \mathbf{R}/\partial \theta^i$, we have

$$\frac{\partial \mathbf{g}_i}{\partial \theta^j} = \frac{\partial}{\partial \theta^j}\left(\frac{\partial \mathbf{R}}{\partial \theta^i}\right) = \frac{\partial}{\partial \theta^i}\left(\frac{\partial \mathbf{R}}{\partial \theta^j}\right) = \frac{\partial \mathbf{g}_j}{\partial \theta^i}.$$

Hence, by substitution, it is easy to show that

$$\frac{1}{2}\left(\frac{\partial g_{ik}}{\partial \theta^j} + \frac{\partial g_{jk}}{\partial \theta^i} - \frac{\partial g_{ij}}{\partial \theta^k}\right) = \frac{\partial \mathbf{g}_i}{\partial \theta^j} \cdot \mathbf{g}_k.$$

On multiplying the two sides of the equation by $g^{\alpha k}$ and summing over $k$, the left-hand side becomes $\Gamma^{\alpha}_{ij}$, according to the definition of Christoffel symbols. We then multiply the scalar quantity on both sides of the equation by the vectors $\mathbf{g}_\alpha$ and sum over $\alpha$ to obtain

(6)
$$\Gamma^{\alpha}_{ij}\, \mathbf{g}_\alpha = \left(\frac{\partial \mathbf{g}_i}{\partial \theta^j} \cdot \mathbf{g}_k\right) g^{\alpha k} \mathbf{g}_\alpha$$
$$= \left(\frac{\partial \mathbf{g}_i}{\partial \theta^j} \cdot \mathbf{g}_k\right) \mathbf{g}^k.$$

The right-hand side, the sum of the vectors $\mathbf{g}^k$ multiplied by the scalar quantities $\left(\frac{\partial \mathbf{g}_i}{\partial \theta^j} \cdot \mathbf{g}_k\right)$ is exactly equal to the vector $\partial \mathbf{g}_i/\partial \theta^j$ itself. To see this, we note that since the set of vectors $\mathbf{g}^1$, $\mathbf{g}^2$, $\mathbf{g}^3$ are linearly independent and form a basis of the vector space, $\partial \mathbf{g}_i/\partial \theta^j$ can be expressed as a linear combination

(7)
$$\frac{\partial \mathbf{g}_i}{\partial \theta^j} = \lambda_1 \mathbf{g}^1 + \lambda_2 \mathbf{g}^2 + \lambda_3 \mathbf{g}^3,$$

where $\lambda_1$, $\lambda_2$, $\lambda_3$ are scalars. If we multiply this equation by $\mathbf{g}_1$, we obtain

(8)
$$\lambda_1 = \frac{\partial \mathbf{g}_i}{\partial \theta^j} \cdot \mathbf{g}_1.$$

Similarly, $\lambda_2$, $\lambda_3$ can be evaluated. A comparison of (6), (8), and (7) thus shows the truth of Eq. (2.15:4):

$$\Gamma^{\alpha}_{ij} \mathbf{g}_\alpha = \frac{\partial \mathbf{g}_i}{\partial \theta^j}. \qquad \text{Q.E.D.}$$

## 2.2. ON THE DEFINITION OF PERMUTATION TENSORS

As defined in Sec. 2.6, the components of our permutation tensors $\epsilon_{ijk}$, $\epsilon^{ijk}$ do not have values 1, $-1$ or 0 in general coordinates. In some books, symbols $e_{ijk}$, $e^{ijk}$ are defined for general coordinates to have the same values 1, $-1$, or 0 as in rectangular coordinates. As is shown by Eqs. (2.6:6) and (2.6:7), it is clear that $e_{ijk}$ is a relative tensor of weight $-1$, and $e^{ijk}$ is a relative tensor of weight $+1$.

## 2.3. REMARKS ON TENSOR EQUATIONS. RELATIVE TENSORS

To apply the powerful procedure discussed in Sec. 2.13, one must make sure that all quantities involved are tensors. In particular, we must ascertain that all scalars are "absolute" constants. This remark is very important because in physics we also use quantities that transform like *relative tensors. A relative tensor of weight* $w$ is an object with components whose transformation law differs from the tensor transformation law by the appearance of the Jacobian to the $w$th power as a factor. Thus,

$$\bar{\phi}(\theta) = \left|\frac{\partial x_i}{\partial \theta_\alpha}\right|^w \phi(x), \tag{1}$$

$$\bar{\xi}^i(\theta) = \left|\frac{\partial x_i}{\partial \theta_\alpha}\right|^w \xi^\sigma(x)\frac{\partial \theta_i}{\partial x_\sigma} \tag{2}$$

are the transformation laws for a relative scalar field of weight $w$ and a relative contravariant vector field of weight $w$, respectively. If $w = 0$, we have the previous notion of a tensor field. Whether an object is a tensor or a relative tensor is often a matter of definition.

As an example, consider the total mass enclosed in a volume expressed in terms of density. Let $x_i$ be rectangular coordinates which are transformed into curvilinear coordinates $\theta_j$. We have

$$\begin{aligned} M &= \iiint_V \rho_0(x_1, x_2, x_3)\, dx_1, dx_2\, dx_3 \\ &= \iiint_V \rho_0[\theta(x)] \left|\frac{\partial x_i}{\partial x_j}\right| d\theta_1\, d\theta_2\, d\theta_3 \\ &= \iiint_V \bar{\rho}(\theta_1, \theta_2, \theta_3)\, d\theta_1\, d\theta_2\, d\theta_3. \end{aligned} \tag{3}$$

If $\bar{\rho}(\theta)$ in the last term is defined as the density distribution in the $\theta$-coordinates, then it is a relative scalar of weight one. On the other hand, $\rho_0[\theta(x)] = \rho_0(x)$ is an absolute scalar which defines the (physical) density of the medium.

As another example, consider the determinant $g$ of the Euclidean metric tensor $g_{ij}$ whose transformation law is

$$\bar{g}_{\alpha\beta}(\theta) = g_{ij}(x)\frac{\partial x_i}{\partial \theta_\alpha}\frac{\partial x_j}{\partial \theta_\beta} \tag{4}$$

Let $\bar{g} = |\bar{g}_{\alpha\beta}|$, the determinant of $\bar{g}_{\alpha\beta}$. By a double use of the formula for the product of two determinants when applied to (4), it is easy to prove that

$$\bar{g}(\theta) = \left|\frac{\partial x_i}{\partial \theta_\alpha}\right|^2 g(x) \tag{5}$$

which shows that $g$ is a relative scalar of weight two. It follows that $\sqrt{g}$ is a relative scalar of weight one. We note that if $x_i$ are rectangular coordinates, $g = 1$. Hence Eq. (5) shows that $\sqrt{g(\theta)} = |\partial x_i/\partial\theta_j|$, the Jacobian of the transformation. Thus the volume enclosed by a closed surface can be written as

$$\begin{aligned} V &= \iiint dx_1\, dx_2\, dx_3 \\ &= \iiint \left|\frac{\partial x_i}{\partial \theta_\alpha}\right| d\theta_1\, d\theta_2\, d\theta_3 \\ &= \iiint \sqrt{g(\theta)}\, d\theta_1\, d\theta_2\, d\theta_3 \end{aligned} \tag{6}$$

The last integral shows the importance of $\sqrt{g}$ in mechanics.

The method of tensor equations does not apply to relative tensors. Therefore it is important to properly define all quantities involved in an equation to be tensors.

# CHAPTER 6

## 6.1. DEFORMATION THEORIES OF PLASTICITY

In contrast to the *incremental* theories of plasticity discussed in Chap. 6, there is another group called the *deformation* theories of plasticity. The difference lies in the flow rule. Boundary-value problems in deformation theories are simpler than those in the incremental theories; hence the great popularity of the deformation theory among mathematicians.

A typical deformation theory is specified as follows (with the same notations as in Sec. 6.3 and 6.5):

(1) The mean stress and mean strain are linearly related:

$$\sigma_{\alpha\alpha} = 3Ke_{\alpha\alpha}, \qquad K = \text{bulk modulus.}$$

(2) The total strain deviators are composed of an elastic part and a plastic part,

$$e'_{ij} = e'^{(e)}_{ij} + e^{(p)}_{ij},$$

where

$$e'^{(e)}_{ij} = \frac{1}{2G}\sigma'_{ij}.$$

(3) Yielding is specified by a function $f(\sigma'_{ij}, e^{(p)}_{ij}, \kappa)$. The material remains elastic, i.e. $e^{(p)}_{ij} = 0$, when $f < 0$. When $f = 0$, and when the loading process

continues, the plastic strain $e_{ij}^{(p)}$ is a unique function of the stress deviation $\sigma'_{ij}$. A relation given in Prager's form is

$$e_{ij}^{(p)} = P\sigma'_{ij} + Qt_{ij},$$

where

$$t_{ij} = \sigma'_{ik}\sigma'_{kj} - \tfrac{2}{3}J_2\delta_{ij}$$

is the deviation of the square of the stress deviation. For an isotropic material $P$ and $Q$ are functions of the invariants $J_2$ and $J_3$ of the stress tensor. With the help of the Cayley-Hamilton theorem in the theory of matrices, it can be shown that Prager's form is the most general form under the assumption that the plastic strain can be expressed in powers of $\sigma'_{ij}$.

(4) If unloading occurs at any stage of plastic deformation, the material becomes elastic at once. In other words, as soon as $f(\sigma'_{ij}, e_{ij}^{(p)}, \kappa) < 0$, no change in $e_{ij}^{(p)}$ is possible.

(5) On reloading, the material remains elastic until the yield condition is reached, i.e., until $f(\sigma'_{ij}, e_{ij}^{(p)}, \kappa) = 0$; then the plastic strain follows the same rule (3).

Different choices of the yielding function $f(\sigma'_{ij}, e_{ij}^{(p)}, \kappa)$, the functions $P$ and $Q$, and the hardening parameter $\kappa$ lead to different theories of plasticity. But the characterizing feature of the deformation theory is the unique relationship between the plastic strain and the stresses in the plastic state. This uniqueness is what the incremental theory avoids.

The following comparison of the flow rules shows the differences:

Deformation theory:

$$e_{ij}^{(p)} = P\sigma'_{ij} + Qt_{ij}.$$

Incremental theory, ideal plastic material:

$$\dot{e}_{ij}^{(p)} = \lambda \frac{\partial f}{\partial \sigma'_{ij}}, \qquad f = f(\sigma'_{ij}), \qquad \lambda \text{ indefinite.}$$

Incremental theory, work-hardening material:

$$\dot{e}_{ij}^{(p)} = G \frac{\partial f}{\partial \sigma_{ij}} \frac{\partial f}{\partial \sigma_{kl}} \dot{\sigma}_{kl}, \qquad f = f(\sigma_{ij}, e_{ij}^{(p)}, \kappa).$$

Deformation theories were developed by Hencky and Nadai and their followers. Handelman, Lin, and Prager[6.3] discussed some conceptual difficulties of the deformation theory. See Hill[6.2], pp. 45–48.

In certain class of problems a deformation theory yields the same answer as the corresponding incremental theory. Such is the case if in a loading process the ratio of the stresss components

$$\sigma_{11} : \sigma_{22} : \sigma_{33} : \sigma_{12} : \sigma_{23} : \sigma_{31}$$

is held constant at all times. Many practical problems are concerned with such a loading. For these problems the incremental flow rule can be integrated to yield the deformation rule. More recently, Budiansky[6.3] and Kliushnikoff[6.3] have shown independently that the deformation rule can be derived from the incremental rule for a much wider class of loadings, provided that "corners" develop in the loading surface as plastic deformation proceeds.

# CHAPTER 7

## 7.1. ON UNIQUENESS OF SOLUTION AND STABILITY OF ELASTIC SYSTEMS

The uniqueness theorems of Kirchhoff and Neumann (Sec. 7.4) are the foundation for the method of potentials (Chaps. 8, 9). For, when uniqueness of solution is established, one needs to find only *a* solution of a given boundary-value problem: that solution is *the* solution.

But it is essential for our theory to be able to violate the uniqueness of solution in one way or another. We know that elastic columns can buckle, thin shells may collapse, airplane wings can flutter, machinery can become unstable in one sense or other. The word "stability" has many meanings; to define a stability problem we must define the sense of the word stability. But a large class of practical stability problems is connected with the loss of uniqueness of solution. Under certain circumstances two or more solutions may become possible; some of these may be dangerous from engineering point of view or undesirable for the function of the machinery; and the circumstances are said to cause instability.

The uniqueness theorem can be violated by violating any one of its assumptions. Referring to Kirchhoff and Neumann's theorem, we have two possibilities:

(a) Loss of positive-definiteness of the strain energy function, $W(e_{ij})$.
(b) Basic changes in the equations of equilibrium (or of motion).

The first possibility arises when the material becomes unstable, as yielding or flow occurs (cf. Sec. 12.5). It is relevant to the plastic buckling of columns, plates, and shells.

The second possibility may arise in a variety of forms. The most important are those due to

(1) finite deformation,
(2) nonconservative forces, and
(3) forces that are functionals of the deformation or history of deformation.

Most buckling problems can be understood only if we realize the basic changes in the equation of equilibrium introduced by finite deformation. For finite deformations the equation of equilibrium (or of motion) is given by Eq. (16.3:7) or (16.3:8). These equations are basically nonlinear because the strains and rotations depend on the stresses. The corresponding equations of equilibrium or motion for a plate are given in Sec. 16.9, e.g., Eqs. (16.9:21) and (16.9:22). The situation with a column is similar. The linearized versions of these equations retain the basic features of these large-deflection equations, so that these problems generally become eigenvalue problems or bifurcation problems.

Nonconservative forces generally depend on the deformation of the structure in a certain specific manner. For example, consider an axial load applied to the end of a cantilever column. If this load is fixed in direction, it is conservative. If this load is not fixed in direction, but may rotate in the process of buckling, then it is nonconservative. The case (3) listed above is a special but broad class of nonconservative forces. It occurs commonly if a solid body is placed in a flowing fluid. The aerodynamic or hydrodynamic pressure acting on the body depends on the deformation, or on the local velocity and acceleration of the body as a whole. If the wake and vorticity are important, the aerodynamic pressure will depend also on the history of deformation. This is commonly the case of an aircraft lifting surface. Under these loadings the terms $X_i$ and $\overset{\nu}{T}_i$ in Eqs. (7.4:3), (7.4:4), and (7.4:8), (7.4:10) are functions or functionals of $u_i(x_1, x_2, x_3, t)$.

In any of the cases mentioned above the basic equations differ from those assumed in the Kirchhoff and Neumann theorems. Loss of uniqueness does not necessarily follow, but it becomes a possibility and must be investigated.

# CHAPTER 13

## 13.1. FURTHER REMARKS ON THE COUPLING OF IRREVERSIBLE PROCESSES

Casimir [*Philips Res. Repts.*, *1* (1945), p. 185; *Revs. Mod. Phys.*, *17* (1945), p. 343] and Tellegen [*Philips Res. Repts.*, *3* (1948), pp. 81, 321; *4* (1949), pp. 31, 366] classify the irreversible processes into two classes according to whether

(1) $$J_k(t) = J_k(-t) \qquad J_k \text{ an even function of time,}$$

(2) $$J_k(t) = -J_k(-t) \qquad J_k \text{ an odd function of time.}$$

Heat flow, mass flow, and so on, defined by expressions such as $\delta Q/\delta t$, are odd functions of time. From the point of view of the kinetic theory of matter, $J$ is even or odd according to whether $J$ does or does not change sign as particle velocity is reversed. So far, we have considered cases in which generalized flows are so chosen that they are all odd. In the more general case, Onsager's principle should be generalized to state: If the irreversible processes $i$ and $k$ are both even or both odd, then

$$L_{ik} = L_{ki}$$

If one is even and the other is odd, then

$$L_{ik} = -L_{ki}$$

A further modification is necessary if there exist forces depending explicitly on the velocity, such as Lorentz forces, Coriolis forces, and so on. For example, in the presence of a magnetic field, $\mathbf{H}$, two odd or two even processes are related by:

$$L_{ik}(\mathbf{H}) = L_{ki}(-\mathbf{H})$$

The generalized theorem is sometimes called the *Onsager-Casimir theorem*.

For simultaneous action of several irreversible processes a general phenomenological law may be written in the form discussed in Sec. 13.3. The Onsager-Casimir theorem simplifies these laws. A further simplification exists in the form of certain symmetry requirements which may dictate whether certain interference coefficients should vanish or not. This is *Curie's symmetry principle* [P. Curie, *Oeuvres* (Paris: Société Français de Physique 1908)], which states that macroscopic causes always have fewer elements of symmetry than the effects they produce. An interference of the irreversible processes is possible only if this general principle is satisfied. For example, chemical affinity (a scalar) cannot produce a directed heat flow (a vector) and the interference coefficient between the heat flow and chemical affinity must vanish.

# BIBLIOGRAPHY

## 1.1. References for historical remarks.

Todhunter, I. and K. Pearson. A History of the Theory of Elasticity and of the Strength of Materials. Cambridge: University Press, Vol. 1, 1886; Vol. 2, 1893. Reprinted New York: Dover Publications, 1960. This monumental work demonstrates the amazing vigor with which the theory of elasticity was developed until the time of Lord Kelvin. Originated by the mathematician Todhunter (1820–1884, St. John's College, Cambridge), the book was completed after Todhunter's death by Karl Pearson (1857–1936). Pearson, who later became a famous statistician and biologist, writes in the preface: "The use of a work of this kind is twofold. It forms on the one hand the history of a peculiar phase of intellectual development, worth studying for the many side lights it throws on general human progress. On the other hand it serves as a guide to the investigator in what has been done, and what ought to be done. In this latter respect the individualism of modern science has not infrequently led to a great waste of power; the same bit of work has been repeated in different countries at different times, owing to the absence of such histories as Dr. Todhunter set himself to write. . . . As it is, the would-be-researcher either wastes time in learning the history of his subject, or else works away regardless of earlier investigators. The latter course has been singularly prevalent with even some first class British and French mathematicians" (vol. 1, p. xi).

Love, A. E. H. The Mathematical Theory of Elasticity. Cambridge: University Press, 1st ed., 1892, 1893; 4 th ed., 1927. Reprinted New York: Dover Publications, 1944. Note especially the "Historical Introduction," 1–31.

Encyklopädie der mathematischen Wissenschaften. Leipzig: Teubner. Vol. 4, sub. vol. 4, 1907–1914, "Elastizitäts-und Festigkeitslehre," 1–770.

Lorenz, H. Techniche Elastizitätslehre. Munich and Berlin: Oldenbourg, 1913. The last chapter, 644–83, is on the history.

Auerbach, F. and W. Hort (eds.). Handbuch der physikalischen und technischen Mechanik. Leipzig: Barth, Vol. 3, 1927, Vol. 4, sub. vols. 1 and 2, 1931.

Westergaard, H. M. Theory of Elasticity and Plasticity. Cambridge, Mass.: Harvard University Press, 1952. Chapter 2, "Historical Notes," contains a very interesting outline of developments until 1940.

Timoshenko, S. P. History of Strength of Materials. New York: McGraw-Hill, 1953 This book is beautifully written and illustrated.

## 1.2. Basic references, frequently referred to within the text, follow. The reader will benefit greatly by parallel reading of these references, which are classics in the field. Most of them contain extensive bibliography.

Biezeno, C. B. and R. Grammel. Engineering Dynamics. Vols. 1 and 2. (Transl. by M. L. Meyer from the 2nd German edition. Glasgow: Blackie, 1955, 1956.

Biot, M. A. Mechanics of Incremental Deformations (Theory of Elasticity and Viscoelasticity of Initially Stressed Solids and Fluids, Including Thermodynamic Foundations and Applications to Finite Strain). New York: Wiley, 1965.

Eringen, A. C. Nonlinear Theory of Continuous Media. New York: McGraw-Hill, 1962.

Green, A. E. and W. Zerna. Theoretical Elasticity. London: Oxford University Press, 1954. General tensor notations and convective coordinates are used. The first part concerns finite deformation; the second part is small perturbation about finite deformation; the third part is infinitesimal theory.

Green, A. E. and J. E. Adkins. Large Elastic Deformations and Nonlinear Continuum Mechanics. London: Oxford University Press, 1960. This is an advanced treatise. with many explicit solutions.

Love, A. E. H. A Treatise on the Mathematical Theory of Elasticity. Cambridge: University Press, 1st ed, 1892, 4th ed., 1927. Reprinted New York: Dover Publications, 1963. There is a great wealth of information presented in this classical treatise.

Muskhelishvili, N. I. Some Basic Problems of the Theory of Elasticity. (Translation of third Russian edition by J. R. M. Radok.) Groningen, Netherlands: Noordhoff, 1953.

Noll, W. and C. A. Truesdell. The Non-linear Field Theories of Mechanics. Vol. 3, part 3, of Encyclopedia of Physics, edited by Flügge. Berlin; Springer, 1964. This treatise presents the most advanced picture of continuum mechanics known today.

Prager, W. Introduction to Mechanics of Continua. Boston: Ginn and Co., 1961. An excellent book which treats both solids and fluids.

Rayleigh, J. W. S., *baron*, 1842–1919. The Theory of Sound, 1st ed. 1877, 2nd ed. revised. New York: Dover Publications, 1945.

Sechler, E. E. Elasticity in Engineering. New York: Wiley, 1952. An effective introduction to elasticity with many examples worked out in detail.

Sokolnikoff, I. S. Mathematical Theory of Elasticity, 2nd ed. New York: McGraw-Hill, 1956. Contains a thorough treatment of the theory of bending and torsion of beams, complex variable technique in two dimensional problems, and variational principles.

Southwell, R. V. Theory of Elasticity. Oxford: The Clarendon Press, 1936, 1941.

Timoshenko, S. and N. Goodier. Theory of Elasticity. New York: McGraw-Hill, 1st ed. 1934, 2nd ed. 1951.

Timoshenko, S and J. M. Gere. Theory of Elastic Stability. New York: McGraw-Hill, 1st ed. 1936, 2nd ed. 1961.

Timoshenko, S. and S. Woinowsky-Krieger. Theory of Plates and Shells. New York: McGraw-Hill, 1st ed. 1940, 2nd ed. 1959.

Timoshenko, S. Vibration Problems in Engineering, 2nd ed. New York: D. Van Nostrand, 1937.

Truesdell, C. A. and R. A. Toupin. The Classical Field Theories. Vol. 3, part 1, of Encyclopedia of Physics, edited by Flügge. Berlin: Springer, 1954. This book explains the development of basic field concepts through exhaustive historic and scientific detail.

von Karman, Theodore and M. A. Biot. Mathematical Methods in Engineering. New York: McGraw-Hill, 1940.

Westergaard, H. M. Theory of Elasticity and Plasticity. Cambridge Mass.: Harvard University Press, 1952.

**1.3.** The following introductory texts contain applications of the principles and methods discussed in Chapter 1. See also Sechler (Bib. 1.2), Southwell (Bib. 1.2).

Barton, M. V. Fundamentals of Aircraft Structures. Englewood Cliffs, N.J.: Prentice-Hall, Inc., 1948.

Cross, Hardy, and N. D. Morgan. Continuous Frames of Reinforced Concrete. New York: Wiley, 1954.

Hoff, N. J. The Analysis of Structures. New York: Wiley, 1956.

Niles, A. S. and J. S. Newell. Airplane Structures (2 vol.), 3rd ed. New York: Wiley, 1955.

Sutherland, H., and H. L. Bowman. Structural Theory, 4th ed. New York: Wiley, 1958.

Timoshenko, S. Strength of Materials, 3rd ed. New York: Van Nostrand, 1955.

Williams, D. An Introduction to the Theory of Aircraft Structures. London: Edward Arnold Publishers, 1960.

**1.4.** Additional references.

Argyris, J. H. and P. C. Dunne. Structural Principles and Data. New York: Pitman, 1952.

Azaroff, L. V. Introduction to Solids. New York: McGraw-Hill, 1960.

Coker, E. G. and L. N. G. Filon. Photoelasticity. Cambridge: University Press, 1931. A classic which presents mathematical solutions as well as experimental results.

Cosserat, E. and F. Cosserat. Theorie des corps deformables. Paris: Hermann, 1909. A remarkable general treatise containing full discussion of curvature and torque stresses.

Cox, H. L. The Buckling of Plates and Shells. New York: Pergamon Press, 1963.

Durelli, A. J., E. A. Phillips, and C. H. Tsao. Introduction to the Theoretical and Experimental Analysis of Stress and Strain. New York: McGraw-Hill, 1958.

Flügge, S. (ed). Handbuch der Physik, Vol. VI, Elasticity and Plasticity. Berlin: Springer, 1958. Contains the following chapters: Sneddon, I. N. and D. S. Berry: The classical theory of elasticity. Jessop, H. T.: Photoelasticity. Freudenthal, A. M. and H. Geiringer: The mathematical theories of the inelastic continuum. Reiner, M.: Rheology. Irwin, G. R.: Fracture. Freudenthal, A. M.: Fatigue. This comprehensive handbook will be an invaluable source of reference.

Flügge, W. (ed.). Handbook of Engineering Mechanics. New York: McGraw-Hill, 1962.

Föppl, L. Drang und Zwang, Vol. 3. Munich: Leibniz Verlag, 1947.

Fredrickson, A. G. Principles and Applications of Rheology. Englewood Cliffs, N.J.: Prentice-Hall, Inc., 1964.

Frocht, M. M. Photoelasticity (2 vol.). New York: Wiley, 1941, 1948.

Girkmann, K.: Flächentragwerke, 5th ed. Vienna: Springer, 1959.

Godfrey, D. E. R. Theoretical Elasticity and Plasticity for Engineers. London: Thames and Hudson, 1959.

Goodier, J. N. and N. J. Hoff (eds.). Structural Mechanics. Proc. 1st Sym. on Naval Structural Mechanics. New York: Pergamon Press, 1960.

Grioli, G. Mathematical Theory of Elastic Equilibrium (Recent Results). New York: Academic Press, 1962.

Hetenyi, M. Handbook of Experimental Stress Analysis. New York: Wiley, 1950.

Hodge, P. G. Plastic Analysis of Structures. New York: McGraw-Hill, 1959.

Jaeger, J. C. Elasticity, Fracture and Flow, with Engineering and Geological Applications. New York: Wiley, 1956.

Kauderer, H. Nichtlineare Mechanik. Berlin: Springer, 1958. Part I about elasticity, Part II on vibration.

Landau, L. D. and E. M. Lifshitz. Theory of Elasticity (translated from Russian by J. B. Sykes and W. H. Reid). London: Pergamon, 1959.

Langhaar, H. L. Energy Methods in Applied Mechanics. New York: Wiley, 1962.

Lekhnitskii, S. G. Theory of Elasticity of an Aniostropic Elastic Body. San Francisco: Holden-Day, 1963.

Lubahn, J. D. and R. P. Felgar. Plasticity and Creep of Metals. New York: Wiley, 1961.

Milne-Thomson, L. M. Antiplane Elastic Systems. New York: Academic Press, 1962. Presents author's method of solving static problems by complex variables.

Morse, P. M. Vibration and Sound, 2nd ed. New York: McGraw-Hill, 1948.

Murnaghan, Francis D. Finite Deformation of an Elastic Solid. New York: Wiley, 1951.

Nadai, A. Theory of Flow and Fracture of Solids (2 vol). New York: McGraw-Hill, 1963.

Novozhilov, V. V. Theory of Elasticity (translated from Russian by J. K. Lusher). New York: Pergamon Press, 1961. The first half deals with finite strain, the second half with infinitesimal theory. It does not make use of the tensor notation.

Novozhilov, V. V. Foundations of the Nonlinear Theory of Elasticity (English translation by F. Bagemihl, H. Komm, and W. Seidel). Rochester, New York: Graylock Press, 1953.

Olszak, W. (ed.). Non-Homogeneity in Elasticity and Plasticity. New York: Pergamon Press, 1959.

Pearson, C. E. Theoretical Elasticity. Cambridge: Harvard University Press, 1959. A brief outline of theoretical elasticity.

Pestel, E. C. and F. A. Leckie. Matrix Methods in Elasto Mechanics. New York: McGraw-Hill, 1961.

Prescott, J. Applied Elasticity. New York: Dover Publications, 1946.

Roark, R. J. Formulas for Stress and Strain, 3rd ed., pp. 174–179. New York: McGraw-Hill, 1954.

Shames, I. H. Mechanics of Deformable Solids. Englewood Cliffs, N.J.: Prentice-Hall, Inc., 1964.

Simmons, J. A., F. Hauser, and J. E. Dorn. Mathematical Theories of Plastic Deformation Under Impulsive Loading. Berkeley: Univ. of Calif. Press, 1962.

Sneddon, I. N. and R. Hill (eds.). Progress in Solid Mechanics, Vol. I, 1960, Vol. 2, 1961, Vol. 3, 1963, Vol. 4, 1963. Amsterdam: North-Holland Pub. Co., 1960.

Southwell, R. V. Relaxation Methods in Theoretical Physics, Vol. 2. New York: Oxford Univ. Press, 1956.

Theimer, O. F. Hilfstafeln zur Berechnung wandartiger Stahlbetonträger. Berlin: Ernst, 1956.

Thomas, T. Plastic Flow and Fracture in Solids. New York: Academic Press, 1961.

Trefftz, E. Mechanik der elastischen Körper, in H. Geiger, K. Scheel (eds.), Handbuch der Physik, Vol. 6. Berlin: Springer, 1928.

Volterra, Vito and Enrico. Sur les distorsions des corps e'lestiques. Paris: Gauthier-Villars, 1960.

Wang, C. T. Applied Elasticity. New York: McGraw-Hill, 1953.

Weber, C. and Günther, W. Torsionstheorie. Braunschweig: Verlag Friedr. Vieweg and Sohn, 1958.

Zener, C. Elasticity and Anelasticity of Metals. Chicago: Univ. of Chicago Press, 1948.

**2.1.** Tensor analysis has its historical origin in the works on differential geometry by Gauss, Riemann, Beltrami, Christoffel, and others. As a formal system it was first developed and published by Ricci and Levi-Civita.[2.1] A brief historical introduction can be found in Wills.[2.1] An exhaustive presentation is given by Coleman in an appendix to Truesdell and Toupin (Bib. 1.2).

Aris, R. Vectors, Tensors, and the Basic Equations of Fluid Mechanics. Englewood Cliffs, N.J.: Prentice-Hall, Inc., 1962.

Borg, S. F. Matrix-Tensor Methods in Continuum Mechanics. Princeton, N.J.: D. Van Nostrand Co., 1963.

Brillouin, L. Les Tenseurs en mécanique et en élasticité (Paris, 1938). New York: Dover Publications, 1946.

Jeffreys, H. Cartesian Tensors. Cambridge: University Press, 1931.

Landau, L. and E. Lifshitz. The Classical Theory of Fields (Translated from Russian by H. Hammermesh). Reading, Mass.: Addison-Wesley Press, 1951. Covers tensor analysis in enough detail and with exceptional physical judgement.

Levi-Civita, T. The Absolute Differential Calculus (translated from Italian by M. Long). London: Blackie, 1927.

Lichnerowicz, A. Éléments de Calcul tensoriel. Paris: Lib. Armand Colin., 1958.

McConnell, A. J. Applications of the Absolute Differential Calculus. London: Blackie 1946.

Michal, A. D. Matrix and Tensor Calculus, with Applications to Mechanics, Elasticity and Aeronautics. New York: Wiley, 1947.

Ricci, G. and T. Levi-Civita: Méthodes du calcul différentiel absolu et leurs applications. Math. Ann., *54* (1901) 125–201.

Sokolnikoff, I. S. Tensor Analysis, Theory and Applications. New York: Wiley, 1951, 1958.

Synge, J. L. and A. Schild. Tensor Calculus. Toronto: Univ. Press, 1949.

Wills, A. P. Vector Analysis with an Introduction to Tensor Analysis. Englewood Cliffs N.J.: Prentice-Hall, Inc., 1938.

**3.1.** Couple-stress. E. and F. Cosserat (Bib. 1.4) presents a thorough formulation of the couple-stress theory. Hellinger (Bib. 10.2) gives the variational principles including couple-stress.

Koiter, W. T. Couple-Stresses in the Theory of Elasticity. Kon. Nederl. Akad. v. Wetenschappen-Amsterdam, Proc. Ser. B, *67*, No. 1 (1964).

Kröner, E. On the physical reality of torque stress in continuum mechanics. Int. J. Eng. Sci., *1* (1963) 261–78.

Mindlin, R. D. and H. F. Tiersten. Effects of Couple-Stresses in Linear Elasticity. Arch. Rational Mech. and Analysis. *11* (1962) 415–48.

Sadowsky, M. A., Y. C. Hsu, and M. A. Hussain. Boundary layers in couple-stress elasticity and stiffening of thin layers in shear. Watervliet Arsenal, New York, Report RR-6320, 1963.

Voigt, W. Theoretische Studien über die Elasticitätsverhältnisse der Krystalle. Abh. Ges. Wiss. Göttingen, *34* (1887).

**4.1.** Our presentation of the compatibility conditions follows Sokolnikoff, (Bib. 1.2), p. 25. The original reference is

Cesaro, E. Sulle formole del Volterra, fondamentali nella teoria delle distorsioni elastiche. Rendiconto dell' Accademia della Scienze Fisiche e Matematiche (Società Reale di Napoli), (1906) 311–21.

**5.1.** More rigorous treatment of Gauss' theorem can be found in

Courant, R. Differential and Integral Calculus (transl. by E. J. McShane), New York: Interscience.

Kellogg, O. D. Foundations of Potential Theory. Berlin: J. Springer, 1929.

**6.1.** The determination of elastic constants is discussed in the following books.

Born, M. and K. Huang. Dynamical Theory of Crystal Lattices. Oxford: Clarendon Press, 1954.

Duffy, J. and R. D. Mindlin. Stress-Strain Relations of a Granular Medium. Jour. Appl. Mech., *24* (1957) 585–93.

Huntington, H. B. The Elastic Constants of Crystals. New York: Academic Press, 1958.

Koehler, J. S. Imperfections in Nearly Perfect Crystals. (See p. 197 for influence on elastic constants.) New York: Wiley, 1952.

Walsh, J. M. and R. H. Christian. Equation of state of metals from shock wave measurements. Phy. Rev., *97* (1955) 1544–56.

## 6.2. Books on mathematical theory of plasticity.

Freudenthal, A. M. and H. Geiringer. The Mathematical Theories of the Inelastic Continuum. Vol. VI of Encyclopedia of Physics, S. Flügge (ed.). Berlin: Springer-Verlag, 1958.

Goodier, J. N. and P. G. Hodge, Jr. Elasticity and Plasticity. New York: Wiley, 1958.

Hill, R. The Mathematical Theory of Plasticity. Oxford: Clarendon Press, 1950.

Hodge, P. G.: Plastic Analysis of Structures. New York: McGraw-Hill, 1959.

Hodge, P. G., Jr. Limit Analysis of Rotationally Symmetric Plates and Shells. Englewood Cliffs, N.J.: Prentice-Hall, Inc., 1963.

Hoffman, O. and G. Sachs. Introduction to the Theory of Plasticity for Engineers. New York: McGraw-Hill, 1953.

Lee, E. H. and P. S. Symonds (eds.). Plasticity: Proceedings of the Second Symposium on Naval Structural Mechanics. New York: Pergamon Press, 1960.

Nadai, A. Theory of Flow and Fracture of Solids. 2 vols. New York: McGraw-Hill, Vol. 1, 1950; Vol. 2, 1963.

Neal, B. G. The Plastic Method of Structural Analysis. New York: Wiley, 1956.

Phillips, A. Introduction to Plasticity. New York: Ronald, 1956.

Prager, W. and P. G. Hodge, Jr. Theory of Perfectly Plastic Solids. New York; Wiley, 1951.

Prager, W. An Introduction to Plasticity. Reading, Mass.: Addison-Wesley, 1959.

Sokolovskij, V. V. Theorie der Plastizität (German translation of Russian 2nd ed.). Berlin: Verlag Technik, 1955.

van Iterson. Plasticity in Engineering. New York: Hafner Publishing Co.

## 6.3. Plasticity—constitutive equations.

Besseling, J. F. A theory of elastic, plastic, and creep deformations of an initially isotropic material. Stanford University, SUDAER No. 78, April 1958.

Boyce, W. E. The bending of a work-hardening circular plate by a uniform transverse load. Quart. Appl. Math., *14* (1957) 277–88.

Bridgman, P. W. The compressibility of thirty metals as a function of pressure and temperature. Proc. Am. Acad. Art. Sci., *58* (1923) 163–242.

Bridgman, P. W. The thermodynamics of plastic deformation and generalized entropy. Rev. Mod. Phys., *22* (1950) 56–63.

Budiansky, B. A reassessment of deformation theories of plasticity. J. Appl. Mech., *26* (1959) 259–64.

Crossland, B. The effect of fluid pressure on the shear properties of metals. Proc. Inst. Mech. Engr., *169* (1954) 935–44.

Drucker, D. C. Stress-strain relations in the plastic range—A survey of theory and experiment. Report to Office of Naval Research, under contract N7-onr-358, Div. Appl. Math. Brown University, December 1950.

Drucker, D. C. A more fundamental approach to plastic stress strain relations. Proc. 1st. U.S. Natl. Congress Appl. Mech. (Chicago, 1951), 487–91.

Drucker, D. C. Stress-strain relations in the plastic range of metals-experiments and basic concepts. Rheology, *1* (1956) 97–119.

Drucker, D. C. A definition of stable inelastic material. J. Appl. Mech., *26* (1959) 101–6.

Handelman, G. H., C. C. Lin and W. Prager. On the mechanical behavior of metals in the strain-hardening range. Quart. Appl. Math., *4* (1947) 397–407.

Hencky, H. Zur Theorie plastischer Deformationen und die hierdurch im Material hervorgerufenen Nach-Spannungen, Z. ang. Math. Mech. *4* (1924) 323–34.

Hill, R. The mechanics of quasi-static plastic deformation in metals. Article in Surveys in Mechanics, G. K. Batchelor and R. M. Davies (eds.). Cambridge: University Press, 1956.

Hodge, P. G., Jr. Piecewise linear plasticity. Proc. 9th Intern. Congress Appl. Mech. (Brussels, 1956). Vol. 8 (1957) 65–72.

Ishlinskii, I. U. General theory of plasticity with linear strain hardening. In Russian, Ukr. Mat. Zh., *6* (1954) 314–24.

Kadashevich, I., and V. V. Novozhilov. The theory of plasticity which takes into account residual microstresses. J. Appl. Math. Mech. (Translation of Prikl. Mat. i. Mekh.), *22* (1959) 104–118.

Kliushnikov, V. D. On plasticity laws for work hardening materials. J. Appl. Math. Mech. (Translation of Prikl. Mat. i. Mekh.), *22* (1958) 129–60.

Kliusnikov, V. D. New concepts in plasticity and deformation theory. J. Appl. Math. Mech. (Translation of Prikl. Mat. i. Mekh.), *23* (1959) 1030–42.

Koiter, W. T. Stress-strain relations, uniqueness and variational theorems for elastic-plastic materials with a singular yield surface. Quart. Appl. Math., *11* (1953) 350–54.

Lévy, M. Memoire sur les equations generales des mouvements interieurs des corps solides ductiles au delà des limites ou l'élasticite pourrait les ramener a leur premier etat. C.R. Acad. Sci (Paris), *70* (1870) 1323–25.

Lode, W. Versuche über den Einfluss der mittleren Hauptspannung auf das Fliessen der Metalle Eisen, Kupfer, und Nickel. Z. Physik, *36* (1926) 913–39.

McComb, H. G. Some experiments concerning subsequent yield surfaces in plasticity. NASA TN D-396, June 1960.

Naghdi, P. M., F. Essenburg, and W. Koff. An experimental study of initial and subsequent yield surfaces in plasticity. J. Appl. Mech., *25* (1958) 201–9.

Naghdi, P. M. Stress-strain relations in plasticity and thermoplasticity. Article in Lee and Symonds (eds.), Plasticity (Bib. 6.2) 121–69.

Osgood, W. R. Combined-stress tests on 24 ST aluminum alloy tubes. J. Appl. Mech, *14*, Trans. ASME, *69* (1947) 147–53.

Prager, W. The stress-strain laws of the mathematical theory of plasticity—A survey of recent progress. J. Appl. Mech., *15* (1948) 226–33. Discussions on *16* (1949) 215–18.

Prager, W. The theory of plasticity: A survey of recent achievements (James Clayton Lecture). Proc. Instn. Mech. Engrs., *169* (1955) 41–57.

Prager, W. A new method of analyzing stress and strains in work-hardening plastic solids. J. Appl. Mech., *23* (1956) 493–6. Discussions by Budiansky, and Hodge, *24* (1957) 481–82, 482–84.

Prager, W. Non-isothermal plastic deformation. Nederl, K. Ak. Wetensch., Proc. *61* (1958) 176–92.

Prandtl, L. Ein Gedankenmodell zur kinetischen Theorie der festen Körper. Z. angew Math. Mech., *8* (1928) 85–106.

Ramberg, W. and W. R. Osgood. Description of stress-strain curves by three parameters. NACA TN 902 (1943).

de Saint Venant, B.: Mémoire sur l'établissement des équations différentielles des mouvements intérieurs opérés daus les corps solides ductiles au delà des limites où l'élasticité pourrait les ramener à leur premier état. C.R. Acad. Sci. (Paris), 70 (1870), 473–80.

Shield, R. T. and H. Ziegler. On Prager's hardening rule. Z. ang. Math. und Phys. (1958) 260–76.

Taylor, G. I. and H. Quinney. The plastic distortion of metals. Phil. Trans. Roy. Soc. (London) A, *230* (1931) 323–62.

Tresca, H. Mémoire sur l'écoulement des corps solides, Mém. prés. par div. Savants *18* (1868) 733–99.

von Mises, R. Mechanik der plastischen Formanderung von Kristallen. Z. angew. Math. Mech. *8* (1928) 161–85.

von Mises, R. Mechanik der festen Körper im plastisch deformablen Zustand. Göttinger Nachrichten, math.-phys. Kl. *1913* (1913) 582–92.

Ziegler, H. A modification of Prager's hardening rule. Quart. Appl. Math., *17* (1959) 55–65.

## **6.4.** Creep. See also Bib. 14.8, on p. 503.

Andrade, E. N. da C. On the viscous flow in metals and allied phenomena. Proc. Roy. Soc. (London), *84* (1910) 1.

Andrade, E. N. da C. The flow of metals under large constant stress. Proc. Roy. Soc. (London), *90* (1914) 329.

Andrade, E. N. da C. Creep of metals and recrystallization. Nature, *162* (1948) 410.

Andrade, E. N. da C. The flow of metals. J. Iron Steel Inst., *171* (1952) 217–28.

Finnie, I. and W. R. Heller. Creep of Engineering Materials. New York: McGraw-Hill, 1959.

Hoff, N. J. A survey of the theories of creep buckling. Stanford Univ., SUDAER No. *80* 1958.

**6.5.** Plasticity under rapid loading. See also Simmons[1.4].

Bodner, S. R. and P. S. Symonds. Plastic deformations in impact and impulsive loading of beams. Brown Univ. Div. Appl. Math. TR 61, 1960.

Fowler, R. G. and J. D. Hood. Very fast dynamical wave phenomenon. Phys. Review, *128*, 3 (Nov. 1962) 991–92.

Green, D. J. The effect of acceleration time on plastic deformation of beams under transverse impact loading. Brown Univ. Rept. A 11-112, ONR N7-onr-35801 (1956).

Perzyna, P. The constitutive equations for rate sensitive plastic materials. Quart. Appl. Math., *20* (1963) 321–32.

Reiner, M. and D. Abir. Research on the dependence of the strength of metals on the rate of strain. Israeli Technion Research and Development Foundation, TN 1 (Jan. 1961).

Rinehart, J. S. and J. Pearson. Behavior of metals under impulsive loads. Cleveland: American Soc. of Metals (1954).

Shewmon, P. G. and V. F. Zackay (eds.). Response of Metals to High Velocity Deformation. New York: Interscience Publishers, 1961.

Thomas, T. Y. On the velocity of formation of Lüders bands. J. Math. Mech., *7*, 2 (1958) 141–48.

von Kármán, T. and P. Duwez. The propagation of plastic deformation in solids. J. Appl. Physics., *21* (1950) 987–94.

**6.6.** Some selected papers on plasticity. See Lee and Symonds (Bib. 6.2) for a recent symposium in which many authors contributed.

Drucker, D. C. Plasticity. Art. in Goodier and Hoff (eds.), Structural Mechanics (Bib. 1.4) (1960), 407–55.

Hill, R. A variational principle of maximum plastic work in classical plasticity. Quart. J. Mech. Appl. Math., *1* (1948) 18–28.

Koiter, W. T. General theorems for elastic-plastic solids. Progress in Solid Mechanics, Vol. 1 (Sneddon and Hill, eds.), Ch. 4. Amsterdam: North Holland Pub. Co., 1960.

Murch, S. A. An extension of Duhamel's analogy to plasticity. J. Appl. Mech. *28* (1961) 421–6.

Prager, W. Stress analysis in the plastic range. Brown Univ., Div. of Appl. Math., TR 51, 1959.

**7.1.** Linear theory of elasticity. Basic references are listed on pp. 481–485. See especially Sternberg's article in Goodier and Hoff, Structural Mechanics (Bib. 1.4).

**7.2.** Torsion and bending. See Basic references, Bib. 1.2; especially, Biezeno and Grammel, Love, Sechler, Sokolnikoff, Timoshenko, Weber. See Goodier's article in Flügge, Handbook of Engineering Mechanics, Bib. 1.4. For the important subject of shear center, see the author's book, An Introduction to the Theory of Aeroelasticity (Bib. 10.5), p. 471–75, where references to work of Goodier, Trefftz, Weinstein, etc. are given.

**7.3.** Stress concentration due to notches or holes. See also Durelli,[1.4] Sternberg.[8.2]

Durelli, A. J., J. W. Dally, and W. F. Riley. Stress concentration factors under dynamic loading conditions. Illinois Institute of Technology, Dec. 1958.

Hardrath, H. F. and L. Ohman. A study of elastic and plastic stress concentration factors due to notches and fillets. NACA TR 1117, 1953.

Holgate, S. The effect of a hole on certain stress distribution in aeolotropic and isotropic plates. Proc. Cambridge Phil. Soc., *40* (1944) 172–88.

Isida, Makoto. On some plane problems of an infinite plate containing an infinite row of circular holes. Japan Soc. Mech. Eng. Bull., May 1960, 259–65.

Koiter, W. T. An infinite row of collinear cracks in an infinite elastic sheet. Ingenieur-Archiv, *28* (1959), 168–72.

Ling, Chih-Bing. Collected papers. Institute of Mathematics, Academia Sinica, Taipei, Taiwan, China, 1963. Contains many articles on notches and holes.

Neuber, H. Kerbspannungslehre, 1st ed. Berlin: Springer, 1937. (Translated as Theory of Notch Stresses. Ann Arbor: Edwards, 1946); 2nd ed., 1958.

Okubo, H. Stress concentration factors for a circumferential notch in a cylindrical shaft. Mem. Fac. Eng., Univ. Nagoya, *6* (1954) 23–29.

Peterson, R. E. Stress Concentration Design Factors. New York: Wiley, 1953.

Shea, R. Dynamic stress concentration factors. Watertown Arsenal Lab., TR 811.611 (1963).

Williams, M. L. Stress singularities resulting from various boundary conditions in angular corners of plates in extension. J. Appl. Mech., *19* (1952) 526–28.

**7.4.** Elastic waves. Kolsky[7.4] gives a lucid and attractive introduction. Ewing, Jardetzky, and Press[7.4] is comprehensive and contains a large bibliography. Davies,[7.4] and Miklowitz[7.4] also give extensive bibliography. See also Bib. 9.2, 11.1, 11.2.

Brekhovskikh, L. M. Waves in Layered Media. New York: Academic Press, 1960.

Cagniard, L. Reflexion et Refraction des Ondes Seismiques Progressives. Paris: Gauthiers-Villars, 1935.

Davies, R. M. Stress waves in solids, in "Surveys in Mechanics" (Batchelor and Davies, eds.). Cambridge: University Press, 1956, 64–138.

De Hoop, A. T. Representation Theorems for the Displacement in an Elastic Solid and Their Application to Elastodynamic Diffraction Theory. Thesis. Technische Hogeschool Te Delft. 1958.

Ewing, W. M., W. S. Jardetzky, and F. Press. Elastic Waves in Layered Media. New York: McGraw-Hill, 1957.

Garvin, W. W. Exact transient solution of the buried line source problem. Proc. Roy Soc. (London), *234* (1956) 528–41.

Hayes, M. and R. S. Rivlin. Surface waves in deformed elastic materials. Archive Rat'l Mech. and Analysis, *8* (1961) 358–80.

Knopoff, L. On Rayleigh wave velocities. Bull. Seism. Soc. America, *42* (1952) 307–8.

Kolsky, H. Stress Waves in Solids. Oxford: University Press, 1953.

Lamb, H. On the propagation of tremors over the surface of an elastic solid. Phil. Trans. Roy. Soc. (London). Ser. *A*, *203* (1904) 1–42.

Lamb, H. On waves due to a travelling disturbance, with an application to waves in superposed fluids. Phil. Mag. Ser. 6, *13* (1916) 386–99.

Maue, A. W. Die Entspannungswells bei plotzlichem Einschnitt eines gespannten elastischen Körpers. Z. f. angew. Math. Mech., *34* (1954) 1–12.

Miklowitz, J. Recent developments in elastic wave propagation. Appl. Mech. Rev., *13* (1960) 865–78.

Pekeris, C. C. The seismic surface pulse. Proc. Nat. Acad. Sci., *41* (1955) 469–80.

Pekeris, C. C. and H. Lifson. Motion of the surface of a uniform elastic half-space produced by a buried pulse. J. Acoustical Soc. America, *29* (1957) 1233–38.

**8.1.** Basic references to Chapters 8 and 9 are those of Bib. 1.2. Original references to Lamé, Galerkin, Popkovich, Neuber, Cerrute, Lord Kelvin, and Mindlin can be found in Westergaard[1.2]. References to recent works by Iacovache, Eubanks, Naghdi etc. can be found in Sternberg[8.1] (1960). See also extensive bibliography in Truesdell and Toupin.[1.2]

Almansi, E. Sull'integrazione dell'equazione differenziale, $\Delta^{2n}u = 0$. Annali di matimatica (III), *2* 1899. (See Frank and von Mises. Die Differential- und Integralgleichungen der Mechanik und Physik, Vol. 1, 845–62. New York: Dover Publications.

Boussingesq. J. Applications des potentiels à l'étude de l'équilibre et du mouvement des solides elastique. Paris: Gauthier-Villars, 1885.

Biot, M. A. General solutions of the equations of elasticity and consolidation for a porous material. J. Appl. Mech., *23* (1956) 91–96.

Cosserat, E. and F. Cosserat. Sur la théorie de l'elasticité. Annales de la Faculté des Sciences de l'Université de Toulouse, *10*, 1896.

Dorn, W. S. and A. Schild. A converse to the virtual work theorem for deformable solids. Quart. Appl. Math., *14* (1956) 209–13.

Finzi, B. Integrazione delle equazioni indefinite della meccanica dei sistemi continui. Rend. Lincei (6) *19* (1934) 578–84, 620–23.

Gurtin, M. E. On Helmholtz's theorem and the completeness of the Papkovich-Neuber stress functions for infinite domains. Archive Rational Mech. and Analysis, *9* (1962), 225–33.

Marguerre, K. Ansätze zur Lösung der Grundgleichungen der Elastizitätstheorie, Z. angew. Math. Mech., *35* (1955) 242–63.

Massonnett, C. General solution of the stress problem of the three-dimensional elasticity (in French). Brussels: Proc. 9th Intern. Congr. Appl. Mech., Vol. 5, 168–80, 1956.

Pendse, C. G. On the analysis of a small arbitrary disturbance in a homogeneous isotropic elastic solid. Phil. Mag. Ser. 7, *39* (1948) 862–67.

Sternberg, E. On some recent developments in the linear theory of elasticity, in Goodier and Hoff (eds.). Structural Mechanics, New York: Pergamon Press, 1960.

Sternberg, E. and R. A. Eubanks. On stress functions for elastokinetics and the integration of the repeated wave equation. Quart. Appl. Math., *15* (1957) 149–53.

Truesdell, C. Invariant and complete stress functions for general continua. Archive Rat'l Mech. & Analysis, *4* (1959/60), 1–29.

Weber, C. Spannungstunktionen des dreidimensionalen Kontinuums. Z. angew. Math u. Mech., *28* (1948) 193. Rediscovered Finzi's result.

## **8.2.** Bodies of revolution. See Yu[8.2] and Sternberg's reviews.[8.2]

Mindlin, R. D. Force at a point in the interior of a semi-infinite solid. Physics, *7* (1936), 195–202; Urbana, Ill.: Proc. 1st Midwestern Conf. Solid Mech., 56–59, 1953.

Papkovitch, P. F. Solution générale des équations differentielles fondamentales d'élasticité, exprimée par trois fonctions harmoniques. Compt. Rend. Acad. Sci. Paris, *195* (1932) 513–15, 754–56.

Sternberg, E. Three-dimensional stress concentrations in the theory of elasticity. Appl. Mech. Rev., *11* (1958) 1.

Yu, Y. Y. Bodies of revolution, ch. 41 of Handbook of Engineering Mechanics, W. Flügge (ed.). New York: McGraw-Hill, 1962.

## **9.1.** Complex variable technique. See list of Basic References, Bib. 1.2, especially Muskhelishvili, Green and Zerna, Sokolnikoff, Timoshenko, and Goodier.

Lekhnitsky, S. G. Theory of Elasticity of an Anisotropic Body (in Russian). Moscow, 1950.

Muskhelishvili, N. Investigation of biharmonic boundary value problems and two-dimensional elasticity equations. Math. Ann., *107* (1932) 282–312.

Savin, G. N. Concentration of Stresses around Openings (in Russian). Moscow, 1951.

Shtaerman, I. Y. The Contact Problem of Elasticity Theory (in Russian). Moscow, 1949.

**9.2.** Ground shock induced by moving pressure pulse. See also Bib. 7.4.

Ang, D. D. Transient motion of a line load on the surface of an elastic half-space. Quart. Appl. Math., *18* (1960) 251–56.

Chao, C. C. Dynamical response of an elastic half-space to tangential surface loadings. J. Appl. Mech., *27* (1960) 559–67.

Cole, J. and J. Huth. Stresses produced in a half-plane by moving loads. J. Appl. Mech., *25* (1958) 433–36.

Eringen, A. C. and J. C. Samuels. Impact and moving loads on a slightly curved elastic half-space. J. Appl. Mech., *26* (1959) 491–98.

Morley, L. S. D. Stresses produced in an infinite elastic plate by the application of loads travelling with uniform velocity along the bounding surfaces. British Aero. Research Comm. R and M 3266, 1962.

Sneddon, I. N. Fourier Transforms. New York: McGraw-Hill, 1951, 445–49.

**10.1.** There are many books on the calculus of variations. Bliss' introductory book is highly recommended for beginners.

Bliss, G. A. Calculus of Variations. Mathematical Association of America, 1925; reprinted, 1944.

Courant, R. and D. Hilbert. Methods of Mathematical Physics, Vol. I. New York: Interscience Publishers, 1953.

Gelfand, I. M. and S. V. Fomin. Calculus of Variations. Englewood Cliffs, N.J.: Prentice-Hall, Inc., 1963.

Gould, S. H. Variational Methods for Eigenvalue Problems; an Introduction to the Methods of Rayleigh, Ritz, Weinstein, and Aronszajn. Toronto: University Press, 1957.

Lanczos, C. The Variational Principles of Mechanics. Toronto: University Press, 1949.

Pars, L. An Introduction to the Calculus of Variations. New York: Wiley, 1962.

**10.2.** Variational principles in elasticity were known to Lagrange, and have been developed steadily. We give some more recent references. In examples of application, Lord Rayleigh's Theory of Sound (Bib. 1.2) is perhaps the most remarkable contribution. D. Williams' paper[10.2] illustrates the many ways a variational principle may be interpreted and stated.

Argyris, J. H. and S. Kelsey. Energy Theorems and Structural Analysis. London: Butterworth, 1960.

Gurtin, M. E. A note on the principle of minimum potential energy for linear anisotropic elastic solids. Quart. Appl. Math., *20* (1963) 379–82.

Hellinger, E. Die allegemeinen Ansatze der Mechanik der Kontinua. Art. 30, in F. Klein, C. Müller (eds.). Encyklopädie Mathematischen Wissenschaften, mit Einschluss ihrer Anwendungen, Vol. IV/4, Mechanik, 601–94. Teubner: Leipzig, 1914. Hellinger discusses oriented bodies (Cosserat), relativistic thermodynamics, electrodynamics, internal frictions, elastic after-effect, and finite deformation.

Langhaar, H. L. The principles of complementary energy in non-linear elasticity theory. J. Franklin Inst., *256* (1953) 255.

Levinson, M. and P. J. Blatz. Variational principles for the finite deformation of a perfectly elastic solid. GALCIT Report SM-5. Also, Levinson, Ph. D. Dissertation, California Institute of Technology, 1964.

Prager, W. The general variational principle of the theory of structural stability. Q. Appl. Math., *4* (1947) 378–84.

Reissner, E. On a variational theorem in elasticity. J. Math. and Physics, *29* (1950) 90–95.

Reissner, E. On a variational theorem for finite elastic deformations. J. Math. and Physics, *32* (1953) 129–35.

Reissner, E. On variational principles in elasticity. Symp. Calculus of Variations and Applications, Amer. Math. Soc., Apr. 1956, 1–7.

Reissner, E. Variational considerations for elastic beams and shells. J. Eng. Mech., ASCE, No. EM 1 (1962) 23–57.

Washizu, K. On the variational principles of elasticity and plasticity. Aeroelastic Research Lab., MIT, Tech. Rept. 25–18 (1955).

Washizu, K. Variational principles in continuum mechanics. University of Washington, Dept. of Aeronautical Engineering, Rept. 62–2 (1962).

Williams, D. The relations between the energy theorems applicable in structural theory. Phil. Mag. and J. of Sci., 7th Ser. *26* (1938) 617–35.

## **10.3.** Saint Venant's principle.

Boley, B. A. Some observations on Saint-Venant's principle. Providence, R.I.: Proc. 3rd U.S. Nat. Congress Appl. Mech., 1958, 259–64.

Boley, B. A. On a dynamical Saint Venant principle. ASME Trans., Ser. E-AM (1960) 74–78.

Friedrichs, K. O. The edge effect in the bending of plates. Reissner Anniversary Volume Ann Arbor: J. W. Edwards, 1949.

Frocht, M. M. and P. D. Flynn. On Saint Venant's principle under dynamic conditions. Illinois Inst. of Tech., Tech. Rept. 8 (1959).

Goodier, J. N. A general proof of Saint-Venant's principle. Phil. Mag. Ser. 7, *23* (1937) 607. Supplementary note *24* (1937), 325.

Hoff, N. J. The applicability of Saint-Venant's principle to airplane structures. J. Aeronautical Sci., *12* (1945) 455–60.

Horvay, G. Some aspects of Saint Venant's principle. J. Mech. and Phys. Solids, *5* (1957) 77.

Lichtenstein, L. Ueber die erste Randwertaufgabe der Elastizitätstheorie. Math. Zeitschrift, *20* (1924) 21–8.

Locatelli, P. Estensione del principio di St. Venant a corpi non perfettamente elastici. Atti. Accad. Sci. Torino, *75* (1940) 502, and *76* (1941) 125.

Naghdi, P. M. On Saint Venant's principle: elastic shells and plates. Univ. of Calif., Inst. of Eng. Research, TR 1 (1959).

de Saint-Venant, B. Mémoire sur la Torsion des Primes. Mém. des Savants étrangers, Paris, 1855.

Southwell, R. On Castighano's theorem of least work and the principle of Saint Venant. Phil. Mag. Ser. 6, *45* (1923) 193.

Supino, G. Sopra alcune limitazioni per la sollecitazione elastica e sopra la dimostrazione del principio del de Saint-Venant. Annali di mathematica pura ed applicata (1931) 91. Ser. IV, Tome 9.

Sternberg, E. On Saint-Venant's principle. Quart. Appl. Math., *11* (1954) 393–402.

Timoshenko, S. On the torsion of a prism, one of the cross sections of which remains plane. Proc. London Math. Soc. (2), *20* (1921) 389.

von Mises, R. On Saint-Venant's principle. Bull. Amer. Math. Soc., *51* (1945) 555.

Zanaboni, O. Dimostrazione generale del principio del de Saint-Venant. Atti. Accad. Lincei, Roma, *25* (1937) 117.

Zanaboni, O. Valutazione dell errore massimo cui da luogo l'applicazione de principio del de Saint-Venant in un solids isotropo. Atti. Accad. Lincei, Roma, *25* (1937) 595.

## 10.4. Variational principles for plasticity, viscoelasticity, and heat conduction bear close resemblance to those for elasticity.

Biot, M. A. Variational and Lagrangian methods in viscoelasticity. Deformation and Flow of Solids, Colloquium Madrid, Sept. 26–30, 1955, Grammel (ed.). Berlin: Springer-Verlag.

Biot, M. A. Variational principles for acoustic-gravity waves. Physics of Fluids, *6* (1963) 772–80.

Gurtin, M. E. Variational principles in the linear theory of viscoelasticity. Archive Rat. Mech. and Analysis, *13* (1963) 179–91.

Herrmann, G. On variational principles in thermoelasticity and heat conduction. Quart. Appl. Math., *21* (1963) 151–55.

Hodge, P. A new interpretation of the plastic minimum principles. Quart. Appl. Math., *19* (1962) 143–44.

Kachanov, L. M. Variational Methods of Solution of Plasticity Problems (Prikl. Mat. i Mekh., May–June 1959, 616). PMM—J. Appl. Math. Mech., *3* (1959) 880–83.

Onat, E. T. On a variational principle in linear viscoelasticity. Brown Univ. TR-74 (1962).

Pian, T. H. H. On the variational theorem for creep. J. Aero. Sci., *24* (1957) 846–47.

Pian, T. H. H. Creep buckling of curved beams under lateral loading. Proc. 3rd U.S. Nat. Cong. Appl. Mech. (1958) 649–54.

Sanders, J. L., Jr., H. G. McComb, Jr., and F. R. Schlechte. A variational theorem for creep with applications to plates and columns. NACA Report 1342 (1958).

**10.5.** Variational principles applied to nonconservative systems are discussed in the following references.

Bolotin, V. V. Nonconservative Problems of the Theory of Elastic Stability. New York: Pergamon Press, 1963.

Bisplinghoff, R. L., H. Ashley, and R. L. Halfman. Aeroelasticity. Cambridge, Mass.: Addison-Wesley, 1955.

Bisplinghoff, R. L. and H. Ashley. Principles of Aeroelasticity. New York: Wiley, 1962.

Fung, Y. C. An Introduction to the Theory of Aeroelasticity. New York: Wiley, 1956.

Fung, Y. C. Flutter. Chap. 63, Handbook of Engineering Mechanics, W. Flügge (ed.). New York: McGraw-Hill, 1962.

**10.6.** An unconventional application of variational principle.

Pian, T. H. H. and T. F. O'Brien. Transient responses of continuous structures using assumed time functions. Ninth Int. Cong. of Appl. Mech., Brussels, 1956. Also MIT Aeroelastic and Str. Res. Lab. TR 25–21 (1956).

**11.1.** Waves in beams and plates. See Abramson, Plass, and Ripperger's review.[11.1]

Abramson, H. N. Flexural waves in elastic beams of circular cross section. J. Acoustical Soc. of Am., *29* (1957) 42–46.

Abramson, H. N., H. J. Plass, and E. A. Ripperger. Stress propagation in rods and beams. Advances in Appl. Mech., *5* (1958) 111–94. New York: Academic Press.

Lamb, H. Hydrodynamics. New York: Dover Publications, 1879.

Mindlin, R. D. Influence of rotatory inertia and shear on flexural motions of isotropic, elastic plates. J. Appl. Mech., *18* (1951) 31–38.

Mindlin, R. D. and G. Herrmann. A one-dimensional theory of compressional waves in an elastic rod. Proc. 1st. U.S. Nat. Congress Appl. Mech., Chicago. (1951) 187–91.

Robinson, A. Shock transmission in beams. British Aeronautical Research Committee, R and M 2265 (Oct. 1945).

Timoshenko, S. P. On the correction for shear of the differential equation for transverse vibrations of prismatic bars. Phil. Mag. Ser. 6, *41* (1921) 744–46.

Timoshenko, S. P. On the transverse vibrations of bars of uniform cross-section. Phil. Mag. Ser. 6, *43* (1922) 125–31.

**11.2.** The theory of the Hopkinson's experiments, and the problem of spallation—i.e., a complete or partial separation of a material resulting from tension waves. See also Davies.[7.4]

Butcher, B. M., L. M. Barker, D. E. Munson, and C. D. Lundergan. Influence of stress history on time-dependent spall in metals. AIAA Journal, *2* (1964) 977–90.

Davies, R. M. A critical study of the Hopkinson pressure bar. Phil. Trans. Roy. Soc. (London), *240* (1946–1948) 375–457.

Hopkinson, B. A method of measuring the pressure produced in the detonation of high explosives or by the impact of bullets. Proc. Roy. Soc. (London), A, *213* (1914) 437–56.

Taylor, G. I. The testing of materials at high rates of loading. (James Forrest Lecture.) J. Institution of Civil Engineers. *26* (1946) 486–519.

**11.3.** On generalized coordinates and direct method of solution of variational problems, the following references are recommended.

Collatz, L. Eigenwertaufgaben mit technischen Anwendungen. Leipzig: Akad. Verlag, 1949.

Collatz, L. The Numerical Treatment of Differential Equations, 3rd ed. Berlin: Springer, 1959.

Collatz, L. Numerische und graphische Methoden. Vol. 11, 349–470, Encyclopedia of Physics, S. Flügge (ed.). Berlin: Springer, 1955.

Duncan, W. J. Galerkin's method in mechanics and differential equations. British Aero. Res. Comm. R and M 1798 (1937).

Duncan, W. J. The principles of the Galerkin method. British Aero. Res. Comm. R and M 1848 (1938).

Kantorovich, L. V. and V. I. Krylov. Approximate Methods of Higher Analysis (transl. from Russian by C. D. Benster). New York: Interscience Pub., 1958.

Ritz, W. Über eine neue Methode zur Lösung gewisser Variationsprobleme der mathematischen Physik. Z. reine u. angew. Math., *135* (1909) 1–61.

Sneddon, I. N. Functional Analysis. Encyclopedia of Physics, S. Flügge (ed.). Vol. 2. Berlin: Springer, 1955.

Temple, G. and W. G. Bickley. Rayleigh's Principle and Its Applications to Engineering. New York: Dover Publications, 1956.

Temple, G. The accuracy of Rayleigh's method of calculating the natural frequencies of vibrating systems. Proc. Roy. Soc. (London), A, *211* (1952), 204–24.

Trefftz, E. Ein Gegenstück zum Ritzschen Verfahren. Zurich: Proc. 2d Intern. Cong. Appl. Mech., 131–37, 1926.

**12.1.** Classical thermodynamics, with special reference to the mechanical properties of solids. There is much written on this subject; we list only a few references.

Born, M. Kritische Betrachtungen zur traditionellen Darstellung der Thermodynamik. Physik. Zeitschr., *22* (1921) 218–24, 249–54, 282–86.

Caratheodory, C. Untersuchungen über die Grundlagen der Thermodynamik. Math. Ann., *67* (1909) 355–86.

Epstein, P. S. Textbook of Thermodynamics. New York: Wiley, 1937.

Gibbs, J. W. On the equilibrium of heterogeneous systems. Trans. Connecticut Acad., *3*, 108–248 (1875), 343–524 (1877); The Collected Works of J. Willard Gibbs, Vol. 1, 55–371. New Haven: Yale University Press, 1906 (reprinted 1948).

Poynting, J. H. and J. J. Thomson. A Text-Book of Physics (Properties of Matter), 4th ed. Ch. 13, Reversible thermal effects accompanying alteration in strains. London: Ch. Griffin, 1907.

Thomson, Sir William (Lord Kelvin). Mathematical and Physical Papers. Cambridge: University Press, Vol. 1, 1882.

**13.1.** Thermodynamics of irreversible processes.

Biot, M. A. Theory of stress-strain relations in anisotropic viscoelasticity and relaxation phenomena. J. Appl. Phys., *25*, 11 (1954) 1385–91.

Biot, M. A. Variational principles in irreversible thermodynamics with application to viscoelasticity. Phys. Rev., *97*, 6 (1955) 1463–69.

Biot, M. A. Dynamics of viscoelastic anisotropic media. Proc. 2nd Midwest. Conf. on Solid Mechanics, Res. Ser. No. 129, Eng. Exp. Stn., Purdue Univ. (1955) 94–108.

Biot, M. A. Thermoelasticity and irreversible thermodynamics. Appl. Phys., *27*, 3 (1956) 240–53.

Biot, M. A. Linear thermodynamics and the mechanics of solids. Third U.S. Natl. Congress of Appl. Mech. (1958) 1–18.

Biot, M. A. New thermomechanical reciprocity relations with application to thermal stress analysis. J. Aero/Space Sciences, *26*, 7 (1959) 401–8.

Casimir, H. B. G. On Onsager's principle of microscopic reversibility. Rev. Mod. Phys., *17* (1945) 343–50.

Chandrasekhar, S. Stochastic problems in physics and astronomy. Rev. Mod. Phy., *15* (1943) 1–89, in particular 54–56.

Curie, P. Oeuvres de Pierre Curie. Paris: Gautheir-Villars, 1908.

De Groot, S. R. Thermodynamics of Irreversible Processes. Amsterdam: North-Holland Pub. Co., 1951.

De Groot, S. R. and P. Mazur. Non-Equilibrium Thermodynamics. Amsterdam: North Holland Pub. Co., 1962.

Meixner, J. Die thermodynamische Theorie der Relaxationserscheinumgen und ihr Zusammenhang mit der Nachwirkungstheorie. Kolloid Zeit., *134*, 1 (1953) 2–16.

Onsager, L. Reciprocal relations in irreversible processes. I. Phys. Rev., *37*, 4 (1931) 405–26. II. Phys. Rev., *38*, 12 (1931) 2265–79.

Onsager, L. and S. Machlup. Fluctuations and irreversible processes. Phys. Rev., *91*, 6 (1953) 1505–12. Part II, 1512–15.

Prigogine, I. Etude Thermodynamique des Phénomènes Irréversibles. Paris: Dunod., Liège: Ed. Desoer, 1947.

Prigogine, I. Le domaine de validité de la thermodynamique des phenomenes irréversibles. Physica, *15* (1949) 272–89 (Netherlands).

Prigogine, I. Non-linear Problems in Thermodynamics of Irreversible Processes. Stanford University: 1956. Heat Transfer and Fluid Mechanics Institute.

Prigogine, I. Introduction to Thermodynamics of Irreversible Processes, 2nd ed. New York: Interscience Pub., Wiley, 1961.

Prigogine, I. Non-Equilibrium Statistical Mechanics. New York: Interscience Pub., Wiley, 1962. Extensive use of Feynman diagram.

Voigt, W. Fragen der Krystallphysik. I. Über die rotatorischen Constanten der Wärmeleitung von Apatit und Dolomit. Gött. Nach. (1903) 87.

Ziegler, H. Thermodynamik und rheologische Problems. Ing. Arch., *25* (1957) 58.

Ziegler, H. An attempt to generalize Onsager's principle, and its significance for rheological problems. Zeit. f. ang. Math. u. Phys., *9* (1958) 748–63.

## **14.1.** Thermoelasticity—basic reference books.

Boley, Bruno A. and J. H. Weiner. Theory of Thermal Stresses. New York: Wiley, 1960.

Carslaw, H. S. and J. C. Jaeger. Conduction of Heat in Solids, 2nd ed. London: Oxford Univ. Press, 1959.

Gatewood, B. E. Thermal Stresses. New York: McGraw-Hill, 1957.

Hoff, N. J. (ed.). High Temperature Effects in Aircraft Structures. London: Pergamon, 1958.

Melan, E. and H. Parkus. Wärmespannungen. Vienna: Springer, 1953.

Nowacki, W. Thermoelasticity. Reading, Mass: Addison-Wesley, 1962.

Parkus, H. Instationäre Wärmespannungen. Vienna: Springer, 1959.

Parkus, H. Thermal Stresses, Ch. 43 (20 pp) of Handbook of Engineering Mechanics, W. Flügge (ed.). New York: McGraw-Hill, 1962.

Schuh, H. Heat Transfer in Structures. New York: Pergamon Press, 1964.

Timoshenko, S. and J. N. Goodier. Theory of Elasticity, 2nd ed. New York: McGraw-Hill, 1951.

**14.2.** Bibliography on Thermoelasticity.

Brahtz, J. F. and A. Dean. An Account of Research Information Pertaining to Aerodynamic Heating of Airframe. U.S. Air Force WADC TR 55–99 (1955), which includes a five-volume Part II on Bibliography.

**14.3.** Additional references. Goodier's papers[14.3] are recommended for readers who wish to obtain an intuitive physical feeling about thermal stress distributions in common problems.

Boley, B. A. Thermal Stresses, Chapter in Goodier and Hoff (eds.) Structural Mechanics. New York: Pergamon Press, 1960.

Goodier, J. N. On the integration of the thermo-elastic equations. Phil. Mag. Ser. 7, *23* (1937) 1017.

Goodier, J. N. Thermal stresses and deformations. J. Appl. Mech., *24* (1957) 467–74.

Hemp, W. S. Fundamental principles and methods of thermo-elasticity. Aircraft Engineering, *26*, 302 (April 1954) 126–7.

Jaeger, J. C. Conduction of heat in a solid with a power law of heat transfer at its surface. Proc. Camb. Phil. Soc., *46* (1950) 634–41.

Morgan, A. J. A. A proof of Duhamel's analogy for thermal stresses. J. Aero. Sci., *25* (1958) 466–67.

**14.4.** Aerodynamic heating is a subject of great concern to aerospace engineers. See Hoff (Bib. 14.1) and the following.

Bisplinghoff, R. L. Some structural and aeroelastic considerations of high-speed flight (19th Wright Brothers Lecture). J. Aerospace Sci., *23* (1956) 289–329.

Budiansky, B. and J. Mayers. Influence of aerodynamic heating on the effective torsional stiffness of thin wings. J. Aero. Sci., *23* (1956) 1081–93.

Eckert, E. R. G. Survey on heat transfer at high speeds. U.S. Air Force WADC TR 54–70 (1954). Very extensive bibliography.

Griffith, G. E. and G. H. Miltonberger. Some effects of joint conductivity on the temperatures and thermal stresses in aerodynamically heated skin-stiffner combinations. NACA TN 3699 (1956).

Hoff, N. J., M. Bloom, F. V. Pohle, et al. Theory and experiment in the solution of structural problems of supersonic aircraft. U.S. Air Force WADC TR 55–291 (1955).

Mansfield, E. H. The influence of aerodynamic heating on the flexural rigidity of a thin wing. British, Royal Aeronautical Establishment, Report Struc. 229 (1957).

Stainback, P. C. A visual technique for determining qualitative aerodynamic heating rates on complex configurations. NASA TN D-385 (1960). A temperature sensitive paint was used.

Thomann, G. E. A. and R. B. Erb. Some effects of internal heat sources on the design of flight structures. AGARD Report 208 (1958).

Yoshikawa, K. K. and B. H. Wick. Radiative heat transfer during atmosphere entry at parabolic velocity. NASA TN D-1074 (1961).

## 14.5. Biot's Lagrangian approach to thermomechanical analysis, and other approximate methods of analysis.

Biot, M. A. Variational and Lagrangian methods in viscoelasticity. Deformation and Flow of Solids, R. Grammel (ed.), IUTAM Colloquium, Madrid, 1955, 251–63. Berlin: Springer, 1956.

Biot, M. A. New methods in heat flow analysis with application to flight structures. J. Aero. Sci., *24* (1957) 857–73.

Biot, M. A. Variational and Lagrangian thermodynamics of thermal convection-fundamental shortcomings of the heat-transfer coefficient. J. Aero. Sci., *29*, 1 (1962) 105–6.

Biot, M. A. Lagrangian thermodynamics of heat transfer in systems including fluid motion. J. Aero. Sci., *29*, 5 (1962) 568–77.

Citron, Stephen J. A note on the relation of Biot's method in heat conduction to a least square procedure. J. Aero. Sci., *27*, 4 (1960) 317–8.

Goodman, T. R. The heat balance integral and its application to problems involving a change of phase. Trans. ASME, *80*, 2 (1958) 335–42.

Goodman, T. R. The heating of slabs with arbitrary heat inputs. J. Aero. Sci., *26*, 3 (1959) 187–88.

Goodman, T. R. and J. J. Shea. The melting of finite slabs. J. Appl. Mech., *27* (1960) 16–24.

Lardner, T. J. Biot's variational principle in heat conduction. AIAA Journal, *1*, 1 (1963) 196–206.

Washizu, K. Application of the variational method to transient heat conduction problems. MIT Aeroelastic Lab. Rept. No. 25–17 (1955).

## 14.6. Physical properties of materials vary with temperature. See various standard handbooks, and the following references.

Barzelay, M. E., K. N. Tong, and G. F. Holloway. Effect of pressure on thermal conductance of contact joints. NACA TN 3295 (1955).

Dorn, J. E. (ed.). Mechanical Behavior of Materials at Elevated Temperatures. New York: McGraw-Hill, 1961.

Goldsmigh, A., T. E. Waterman, and H. J. Hirschhorn. Handbook of Thermophysical Properties of Solid Materials. Vol. I, Elements. New York: Pergamon Press, 1961. (Reprint of WADC TR 58–476 (1959).)

Hoyt, S. L. Metal Data. New York: Reinhold, 1952.

The Reactor Handbook, Vol. I, Physics, Vol. II, Engineering. Washington, D. C.: United States Atomic Energy Commission, AECD-3646, U.S. Government Printing Office, May, 1955.

Tietz, T. E. Refractory metal alloys in sheet form: availability, properties, and fabrication. J. of Spacecraft and Rockets, *1* (May 1964) 225–33.

**14.7.** Thermal stress analysis of bodies with temperature dependent elastic constants usually require extensive calculations.

Hilton, H. H. Thermal stresses in bodies exhibiting temperature dependent properties. J. Appl. Mech., *74* (1952) 350.

Nowinski, J. Thermoelastic problem of an isotropic sphere with temperature dependent properties. Z. ang. Math. Phy., *10* (1959) 565–75.

Sokolowski, M. Thermal stresses in a spherical and cylindrical shell in the case of material properties depending on the temperature. Rozprawy Inzynierski, 169, *8*, 4 (1960) 641–67. USAF translation, MCL-1104/1 + 2, 1961:(Ad 267704).

Trostel, R. Wärmespannungen an Hohlzylindern mit temperaturabhangigen Stoffwerten, Ing. Archiv., *26* (1958) 134–42. Also, Ing. Arch., *26* (1958) 416–34.

**14.8.** Thermal stresses at higher temperature and higher stresses usually involve creep problems.

Bailey, R. W. The utilization of creep test data in engineering design. Proc. Inst. Mech. Engrs. (1935) 131–349.

Hoff, N. J. Approximate analysis of structures in the presence of moderately large creep deformations. Quart. Appl. Math., *12* (1954) 49–55.

Hoff, N. J. Buckling and stability. 41st Wright Mem. Lecture, J. Roy. Aero. Soc., *58*, 1 (1954) 1–52.

Johnson, A. E. The creep of a nominally isotropic aluminium alloy under combined stress systems at elevated temperature. Metallurgie, *40* (1949) 125–139. Article on Magnesium alloy, *42* (1951) 249–262. See also Inst. Mech. Engrs. Proc., *164*, 4 (1951) 432–47.

Kempner, J. Creep bending and buckling of non-linear viscoelastic columns. NACA TN 3137 (1954).

Marin, J. A survey of recent research on creep of engineering materials. Appl. Mech. Rev. *4*, 12 (1951) 633–34.

Mordfin, L. and A. C. Legate. Creep behavior of structural joints of aircraft materials under constant loads and temperatures. NACA TN 3842 (1957).

Odqvist, F. K. G. Influence of Primary Creep on Stresses in Structural Parts. Acta Polytechnica *125* (1953) Mech. Eng. Ser., *2*, 9 (also Kungl. Tekniska Hogskolans Handlingar Nr 66) 18 pp.

Wah, T. and R. K. Gregory. Creep collapse of long cylindrical shells under high temperature and external pressure. U.S. Air Force WADD TR 60-230 (1960).

A recent compendium is the following.

Hoff, N. J. (ed.). Colloquium on Creep in Structures, Stanford, Calif., 1960. New York: Academic Press, 1962.

**15.1.** Viscoelasticity. Sternberg's papers[15.2] and Gurtin and Sternberg[15.1] are particularly recommended.

Alfrey, T. Mechanical Behavior of High Polymers. New York: Interscience Publishers, 1948.

Bergen, J. T. (ed.). Viscoelasticity; Phenomenological Aspects. A Symposium. New York: Academic Press, 1960.

Biot, M. A. Stability problems of inhomogeneous viscoelastic media. Non-Homogeneity in Elasticity and Plasticity, proceedings of IUTAM. New York: Pergamon Press 1959.

Biot, M. A. and H. Odé. On the folding of a viscoelastic medium with adhering layer under compressive initial stress. Q. Appl. Math., *19* (1962) 351–55.

Bland, D. R. The Theory of Linear Viscoelasticity. New York: Pergamon Press, 1960.

Bodner, S. R. On anomalies in the measurement of the complex modulus. Trans. of Soc. of Rheology, *4* (1960), 141–57.

Chu, B. T. Deformation and thermal behavior of linear visco-elastic materials. Brown Univ. Div. Appl. Math. Tech. Rept. 1 and 2 (1958).

Coleman, B. D. and W. Noll. On the foundation of linear viscoelasticity. Mellon Inst. AFOSR TN 60–1367 (1960).

Ferry, J. D. Viscoelastic Properties of Polymers. New York: Wiley, 1961.

Flügge, W. (ed.). Plasticity and Viscoelasticity. Part 5, Chap. 46–54, Handbook of Engineering Mechanics. New York: McGraw-Hill, 1962. Articles by D. C. Drucker, A. Phillips, W. F. Freiberger, P. S. Symonds, A. P. Green, P. G. Hodge, Jr., J. E. Duberg, E. H. Lee, and J. Kempner.

Gross, B. Mathematical Structures of the Theories of Viscoelasticity. Paris: Hermann, 1953.

Gurtin, M. E. and E. Sternberg. On the linear theory of viscoelasticity. Archive Rat. Mech. Analysis, *11* (1962) 291–356.

Hsu, C. T., C. W. Chu, and C. C. Chang. Effect of Viscoelastic foundation on forced vibration of loaded rectangular plates. U.S. Air Force WADD TR 60–360 (1960).

Kolsky, H. and S. S. Lee. The propagation and reflection of stress pulses in linear viscoelastic media. Brown Univ. TR-5 (1962).

Lee, E. H. Viscoelastic stress analysis. Structural Mechanics, Goodier and Hoff (eds.) 456–82. New York: Pergamon Press, 1960.

Lianis, G. Constitutive equations of viscoelastic solids under finite deformation. Purdue Univ. Aero. and Engr. Sci. 63–11 (Dec. 1963).

Naghdi, P. M. and W. C. Orthwein. Response of shallow viscoelastic spherical shells to time-dependent axisymmetric loads. Q. Appl. Math. (1960) 107–21.

Naghdi, P. M. and S. A. Murch. On the mechanical behavior of viscoelastic-plastic solids. J. Appl. Mech., *30* (1963) 321–28.

Onat, E. T. and S. Breuer. On uniqueness in linear viscoelasticity. Brown Univ. Div. Appl. Math. TR—84 (1962).

Perzyna, P. The study of the dynamic behavior of rate sensitive plastic materials. Brown Univ. Div. Appl. Math. TR—77 (1962).

Prager, W. Linearization in visco-plasticity. Brown Univ., Div. Appl. Math. TR—64 (1960). Österreich. Ing. Archiv, *15* (1961) 152–57.

Prager, W. On higher rates of stress and deformation. J. Mech. Phys. Solids, *10* (1962) 133–38.

Schapery, R. A. Approximate methods of transform inversion for viscoelastic stress analysis. Proc. 4th. U.S. Nat. Congress Appl. Mech., 1962.

Williams, M. L. and R. A. Schapery. Studies of viscoelastic media. USAF Aero. Res. Lab., Dayton, Ohio ARL 62–366 (1962). Also, GALCIT SM 62–11, Calif. Institute of Tech.

## **15.2.** Influence of temperature variations on viscoelasticity.

Hilton, H. H. An introduction to viscoelastic analysis. Tech. Rept. AAE 62–1, Aero. Dept., Univ. Illinois (1962).

Koh, S. L. and A. C. Eringen. On the foundations of nonlinear thermo-viscoelasticity. Int. J. Eng. Sci., *1* (1963) 199–229.

Landau, G. H., J. H. Weiner, and E. E. Zwicky, Jr. Thermal stress in a viscoelastic-plastic plate with temperature-dependent yield stress. J. Appl. Mech., *27* (1960) 297–302.

Morland, L. W. and E. H. Lee. Stress analysis for linear viscoelastic materials with temperature variation. Trans. Soc. Rheology, *4* (1960) 233–63.

Muki, R. and E. Sternberg. On transient thermal stresses in viscoelastic materials with temperature-dependent properties. J. Appl. Mech., *28* (1961) 193–207.

Prager, W. Thermal stresses in viscoelastic structure. Z. ang. Math. Phy. *7* (1956) 230–238.

Sternberg, E. On the analysis of thermal stresses in viscoelastic solids. Brown Univ. Div. Appl. Math. TR—19 (1963).

## **15.3.** Reciprocal theorems and their applications.

DiMaggio, F. L. and H. H. Bleich. An application of a dynamic reciprocal theorem. J. Appl. Mech., *26* (1959) 678–79.

Graffi, D. Sul Teoremi di Reciprocità nei Fenomeni Dipendenti dal Temp. Ann. d Mathematica, 4, *18* (1939) 173–200.

Graffi, D. Sul Teoremi di Reciprocità nella Dinamica dei Corpi Elastici. Memoria della Academia della Scienze, Bologna, Series 10, *4* (1946–47) 103–111.

Goodier, J. N. Applications of a Reciprocal Theorem of Linear Thermoelasticity. Stanford Univ. Div. Appl. Mech. TR—128 (AD 260476) June 1961.

Greif, R. A. Dynamic Reciprocal Theorem for Thin Shells. AVCO. Res. and Adv. Dev. Div. RAD TM 63–80 (AD 426 712) Dec. 1963.

Knopoff, L. and A. F. Gangi. Seismic Reciprocity. Geophysics, *24* (1959) 681–91.

Lamb, H. On reciprocal theorems in dynamics. Proc. London Math. Soc., *19* (1888) 144–51.

Strutt, J. W. (Lord Rayleigh). Some general theorems relating to vibrations. Proc. London Math. Soc., *4* (1873) 357–68.

**16.1.** Finite deformations. See Eringen, Green and Zerna, Green and Adkins, Noll and Truesdell, Prager, and Truesdell and Toupin, in the Basic Reference list, Bib. 1.2, p. 482. See also Michal (Bib. 2.1) and Murnaghan (Bib. 1.4).

Doyle, T. C. and J. L. Erickson. Nonlinear elasticity. Advances in Applied Mechanics, *4* (1956) 53–115.

Kappus, R. Zur Elastizitat Theorie endlicher Verschiebungen. Z. f. ang. Math. u. Mech., *19* (Oct. and Dec. 1939), 271–85. 344–61.

Kirchhoff, G. Über die Gleichungen des Gleichgewichtes eines elastischen Körpers bei nicht unendlich kleinen Verschiebungen seiner Theile. Sitzungsberichte der mathematischnaturwissenschaftlichen Klasse der Akademie der Wissenschaften, Vienna, *9* (1852) 762–73.

Noll, W. On the continuity of the solid and fluid states. J. Rat. Mech. Analysis, *4* (1955) 3–81.

Rivlin, R. S. Some topics in finite elasticity, in Structural Mechanics (Goodier and Hoff, eds.), pp. 169–198. New York: Pergamon Press, 1960.

Rivlin, R. S. Mathematics and rheology, the 1958 Bingham medal address. Physics Today, *12*, 5 (1959) 32–36.

Truesdell, C. The mechanical foundations of elasticity and fluid dynamics. J. Rat. Mech. and Analysis, *1* (1952) 125–300, and *2* (1953) 593–616.

Truesdell, C. The Principles of Continuum Mechanics. Field Research Laboratory, Socony Mobil Oil Company, Inc. (Dallas, Texas), Colloquium Lectures in Pure and Applied Science, No. 5, 1960. These pungent and informal lecture notes are recommended for beginners and for those who want a taste of the lectures of this gifted speaker.

**16.2.** Theory of plates. See Timoshenko, Green and Zerna, Sechler, and Love in Bib. 1.2, p. 482.

Mansfield, E. H. The Bending and Stretching of Plates. New York: Pergamon Press, 1964.

**16.3.** The famous Kármán equation is given in this reference.

von Kármán, T. Festigkeitsprobleme im Maschinenbau. Encyklopädie der mathematischen Wissenschaften, Vol. IV, 4 (1910), Chap. 27, p. 349.

**16.4.** The theory of plates and shells has an extensive literature. One of the most important problems of plates and shells is stability or buckling. The following references contain extensive modern bibliography on this subject.

Flügge, W. Stresses in Shells. Berlin: Springer, 1960.

Fung, Y. C. and E. E. Sechler. Instability of thin elastic shells. Structural Mechanics, Goodier and Hoff (eds.). New York: Pergamon Press (1960) 115–68.

Koiter, W. T. (ed.). The Theory of Thin Elastic Shells. Symp. IUTAM, Delft, 1959. Amsterdam: North-Holland Pub. Co., 1960.

Mushtari, K. M. and K. Z. Galimov: Non-linear Theory of Thin Elastic Shells (transl. from Russian by J. Morgenstern, J. J. Schorr-Kon.) Washington, D.C.: U.S. Dept. of Commerce, 1961.

Novoshilov, V. V. The Theory of Thin Shells. Holland: Noordhoff, 1959.

von Kármán, T. The engineer grapples with nonlinear problems. Bull. Am. Math. Soc. *46* (1940) 615–83. Collected works of Theodore von Kármán, Vol. IV. London: Butterworths Sci. Publications, 1956.

Ziegler, H. On the concepts of elastic stability. Advances in Applied Mechanics, Vol. IV, 351–403. New York: Academic Press, 1956.

# AUTHOR INDEX

## I



## J



## K



## L

**Y**



**Z**

# SUBJECT INDEX

## T



## U



## V

## W



## Y